PNEUMATIC HANDBOOK

5th Edition

TRADE & TECHNICAL PRESS LTD.,
MORDEN, SURREY,
ENGLAND

PREFACE

Much water has flown under the bridge since air compression became a source of power in industry and much air has been compressed into service since the first edition of Pneumatic Handbook was published in 1959. The Pneumatic Handbook has become accepted as the most comprehensive, authoritative and unbiased reference work on pneumatic engineering readily available. The fact that thousands of copies of each edition have been sold throughout the world, is proof of its acceptance. In addition it has been translated into French, Italian and Spanish and has sold equally well in these languages. If the proof of the (compressed) pudding is in the pipeline, then the proof of the Pneumatic Handbook is in its sales.

For ease of reference this edition is divided into nine sections: **Section 1:** Principles, Properties, Performance, Gas Flow, Utilisation, Economics and Safety. **Section 2:** Compressors, Lubrication, Controls, Pressure Vessels, Heat Exchangers, Selection, Installation, Mains, Lines, Fittings, Conditioning Treatment. **Section 3:** Actuators, Cylinders, Motors. **Section 4:** Valves & Circuitry. **Section 5:** Hardware, Tools, Starters, Springs, Blowers & Fans, Lifts & Hoists, Timers, Vacuum Pumps. **Section 6:** Techniques, Systems Maintenance, Lower Temperature, High Pressure, Low Cost, Noise Control, Fluidics, Low Vacuum. **Section 7:** Gauges, Seals, Bellows, Diaphragms. **Section 8:** Surveys of: Compressors, Cylinders, Valves, Air Motors, Lines & Fittings. **Section 9:** Data, Formulae, Tables, Symbols. **Section 10:** Buyers Guide.

The contents of this 5th Edition have been compiled and edited over a period of twelve months, or more, from information submitted by many authors, companies and organisations to whom due acknowledgement is given, in particular to the British Compressed Air Society. The very latest information and data on trends and traits in the pneumatic industry of Europe is to be found in the monthly technical journal "POWER", produced by the publishers of this handbook.

THE PUBLISHERS

85461 068 5

Printed and published by TRADE & TECHNICAL PRESS LTD., CROWN HOUSE, MORDEN, SURREY, SM4 5EW. ENGLAND.

CONTENTS

SECTION 1

Principles and Practice

The Role of Compressed Air

Pneumatics Provides Efficient Power in a Convenient, Safe and Controllable Form

COMPRESSED air has made a tremendous contribution to the building of new industry, to its development and in ensuring it is adequately supplied with raw materials. It is in the almost unique position of being able to combine operating efficiency with safety and convenience and in these terms represents one of the most economic sources of industrial power today.

In terms of convenience pneumatic power is easily stored and most readily conveyed to the point of use. It is flexible in application as well as being compact and powerful. It needs no insulation against heat loss during transmission and produces no industrial hazards from fire or shock.

Consequently, compressed air has penetrated every branch of industrial activity from mining to food processing, from materials handling to machine tools.

It would be hard to imagine today's civil engineering, mining and construction industry without pneumatics, when one considers the isolated environments of Australian orefields or the rigorous conditions which impose strict self-dependency on a North Sea oil rig. One has only to consider the thousands of small portable air compressors in use in the road repair industry alone to gauge the extent of just one branch of the pneumatics industry.

Perhaps the biggest impact made on the manufacturing scene by pneumatics is in the field of the portable hand tool. In this form pneumatic power has few rivals in terms of power-to-weight ratio, design simplicity and low maintenance demands. Moreover, pneumatic tooling is very suitable to processes demanding short or intermittent full-power bursts and in such applications is unbeatable.

Mechanization and automation techniques powered by pneumatics have seen tremendous growth in recent years, particularly in parts handling, drilling, nut running and surface finishing. Pneumatics meets conditions demanding quick response, reliability, cleanliness and ease of control.

It therefore comes as no surprise to learn that pneumatics has a firm, expanding future. Indeed the United States Department of Commerce has forecast an annual increase of 13% in compressors alone.

Versatility provides a firm basis for applications growth

Compressed air as an energy source is barely a century old, but during its development is has had to compete with other forms of energy conversion such as hydraulics and direct electrical drives. But neither of these can compete in the hand tool field for instance. Pneumatic hand tools have an outstanding power-to-weight ratio and consequently beat other forms of portable tool on a minimum size and weight basis. A modern vane motor with an equivalent piston displacement of 1 cm^2 delivers about 0.27 hp of output power. This corresponds to 200 hp per litre/piston displacement — a remarkable level of compactness.

Durability and reliability are two of the more important factors in favour of the pneumatic handtool, which can be built simply and be robust enough for the toughest handling. Unlike electrical tools a pneumatic tool can be overloaded to stand still — a factor which is widely used in nut-running or setting screws — particularly if torque limits are preset. This applies also to rock drills, where numerous instances of overloading occur when varying hardness in rock structures are met.

Add onto these factors the pneumatic handtool's ability to work in explosive atmospheres and even under water where other power forms cannot. There it becomes easy to see why so much reliance is placed on these tools in the construction industry, process plant repair and in public works engineering, to name but a few applications.

Dramatic Development

Recently, the mechanization and automation of assembly processes in industry has been an area of dramatic development for the pneumatics handtool. Over the last decade the assembly of threaded fasteners has been influenced by a number of changing requirements, including far greater accuracy in the tightening of screw joints. 10% of these are termed 'critical' — joints which include wheel bolts, steering assemblies, suspension assemblies, brakes and so on. Then there are the 'economical' joints which do not directly affect the safety of the vehicle or its passengers but, if incorrectly assembled, would prove very expensive in terms of warranty or liability costs. Such joints include connecting rods, flywheels, bearing caps, cylinder heads and so on.

To comply with requirements, a number of automobile manufacturers request electronic monitoring equipment. On multi-nut runner spindles for instance, the spindles can be provided with transducers to sense torque and angle. Transducer signals are fed to an electronic control and monitoring cabinet. In this way screws can be tightened to a preset torque or to a preset angle and a permanent record of each joint is there to satisfy warranty demands.

In the engineering industry, drilling and tapping of holes represent a substantial and relatively costly part of machining operations with conventional drill presses. Many hours' drilling time is now being saved through the application of 'low cost automation' principles, in which pneumatics plays a central and vital role. Repetition hand drilling of work can be semi-mechanized by equipping standard jigs with pneumatic bench drill type or unit head type pneumatic drilling heads which, through a simple electro-pneumatic or pneumatic logic circuit, can perform drilling and tapping functions while the operator is loading and unloading other work from similar equipment or other machines.

It is possible for a mechanized drill rig to cut machining time by a hefty 70% while the cost of supervision is only 15% of that required with a manual tool or drill press.

Although pneumatic actuators and motors play the chief role in such automation activities, electronic controls will be replacing pneumatic valves and controls over the next few years.

Mechanizing Handling

Pneumatics also plays an important role in other branches of the engineering industry. In iron and steel mills, pneumatics plays a big part in handling and semi-automation. While hydraulics takes care of the heavier activities in the steel mill such as handling billets and blooms, in the smaller mills, where weight of billets may not exceed five tons, many small multi-pass rolling stands are being mechanized through pneumatic lifting tables. These tables, placed on both sides of a multi-level rolling stand, direct and catch the bloom as it is rolled and reduced through successive passes between each pair of rollers. Further down the line in the sorting and cooling banks, pneumatics again comes into play to lift the rolled products off the rollers, sort them and direct them into the banks.

Sophisticated systems for the materials handling industry combine the brute or gentle force of pneumatics with transducers, electronic level indicators and flowmeters to direct liquids and powders from one storage area to a processing area. The same goes for handling non-reactive materials in the transport industries such as in loading or unloading bulk grain railway wagons or transferring wet, ready made mixes in pre-cast concrete component plants.

Finishing Components

On the finishing side, the wider use of shot blast grit systems for cleaning metallic articles and the more sophisticated use of non-metallic grits for imparting varying degrees of finish to metallic and non-metallic articles are providing yet more scope for expanding the use of compressed air. In some applications a reliable supply of 'super-clean' air, free from moisture and lubrication oil carry-over is called for, for instance to treat non-ferrous components intended for use in the food industry. The same quality of clean air is continually demanded for use in food and packaging plants and in hygienic handling systems.

As an alternative to conventional painting,powder coating plants are expanding in Britain. This is shown by the number of companies starting their own powder coating sections. Also long-established paint sprayers and electroplaters are adding powder coating facilities using pneumatic applicators and, in some cases, pneumatic drives and control systems to apply the powder in pre-programmed deposition procedures.

Such growth in the use of pneumatic equipment has seen a revised view of the pneumatic power system come into play over the last ten to fifteen years. No longer is it common to see pneumatic supply systems grow haphazardly on a piecemeal basis. It is no longer a question of ensuring the efficiency of the existing air supply, but planning for future growth, and air consumption has a direct bearing on future economy. Added to this is the growing concern over energy conservation; only in a planned system is it possible to service the supply efficiently to prevent power losses. The big factor is the arrival of the factory or site 'ring main'. This is essentially the same concept as an electrical ring main, which can be tapped at any point for power take-off. Obviously, planning and costing such an installation is a much more scientific process when compared with the 'add-it-on-as-required' philosophy practised before.

Careful Planning Keeps Service Costs at Bay

It is estimated that 10% of all electrical power used in industry is employed in compressing air, so in the quest for higher efficiency in energy utilization, careful planning of the compressed air system is essential.

Today, a fully integrated compressed air installation can be planned and installed in a very short time, and while the larger companies always appear to be the 'leading lights' in this matter, the bulk of European industry still focusses on the medium-sized factory installation. It is the demands of these users that set the pace of the design and production of compressed air equipment.

Usually , the compressor and tool installations of the medium-sized user are undertaken piece-meal over a number of years. Under the circumstances this is understandable, but it can pose problems. It is not easy to match the growth of power to the needs of the factory and ensure maximum continuing economy. Sometimes a surplus of power is available whilst at others critical shortages may occur.

Consult the Supplier

The supplier is the possessor of the best all-round experience and is in the best position to advise the potential user of a new compressed air supply. However, mutual understanding between user and supplier is essential. It is useful to try and project requirements over several years and the long-term costs savings can be substantial.

Take the case of an installation with an immediate need for a compressor having an output of, say, 5 m^3/min (177 ft^3/min). The current cost of this unit will be around £3 500. But should demand rise within a year by the same amount so that demand stands at 12 m^3/min (425 ft^3/min) another compressor will be required. In that time, taking current inflation rates into account, the cost of a second unit will have risen by some 20%.

Purchasing a 12 m^3/min (425 ft^3/min) compressor now, together with a facility for full, half-load or no-load, and automatic control equipment, would cost about £6 500. So for a substantially lower investment over a long term coupled with similarly proportioned running costs, the user will have saved money, time and problems.

For a complete installation there are four main categories of equipment the user will have to consider; they are:

(i) Main power source including compressor, air dryers, preparation units, filters, couplings and air receivers and distribution. New factories tend to specify ring main systems allowing for a forecast growth in air demand.

(ii) Specially-designed production tools such as multi-spindle units covering key production operations. This is an area of increasing interest and importance.

(iii) Hand tools covering drills, hammers, wrenches, screwdrivers, chippers and so on.

(iv) Material preparation and mechanical handling including shot blasting and paint spraying in particular.

Selecting the Compressor

At the root of any pneumatic system is the compressor, and selection of the right type is of paramount importance. In this respect the user must give the maintenance engineer a say in compressor choice as it will be he who will be married to that compressor for the next ten years or more.

Buying a cheap inadequate machine will result in factory down-time, while over-specifying will escalate running costs.

Compressor technology has advances such that three low-power units can sometimes do a more efficient job than a large unit. And lately, the arrival of the compressor 'pack' allows air supply to be located at the source of demand without upsetting the factory environment.

The buyer has a choice of three types of compressor: the rotary vane, rotary screw and reciprocating piston. All types can be considered for constant load and fluctuating load conditions.

Many factories and sites call for air delivery in the medium range of 1.5 – 8.5 m^3/min (53 – 300 ft^3/min) at 7 bar (100 lb/in^2). All three forms of compressor are offered as a viable proposition by many makers and competition is often based on pricing. But price is not the best guide. Buying the cheapest can let the maintenance department in for trouble later on.

Running Costs is the Key

Total running costs are paramount, and the best long-term deal can only be got through the buyer supplying comprehensive information to the supplier. The following problem areas ought to be considered first:

* Running costs
* Reliability
* Quality of air
* Installation
* Maintenance
* Noise

Running costs depend on air line demand, quality of air, reliability of plant and maintenance. If loading fluctuates then off load power consumption is important.

Reliability means maintenance. While the reciprocating machine features high off-load power efficiency when compared with vane and screw types, it does lean more heavily on the maintenance bill. For example, valves have to be changed frequently. Oil carry-over from a lubricated piston compressor increases with piston ring wear, so piston rings have to be examined and renewed regularly.

If above-average air quality is demanded in terms of ultra-low moisture content and oil-free delivery, such as needed in shot blasting of nickel alloys or in food handling then the maintenance engineer will have to carry out regular preventative maintenance. Both reciprocating pistons and rotary screwtype compressors are used to provide oil-free air.

Rotary screw and vane compressor types can deliver air with low oil content without additional equipment or extra maintenance.

Dry compressed air can only be achieved with all three types working through air dryers. From a maintenance point of view, for average needs, the rotary screw machine is attractive as the newest models call for little other attention than a filter and oil change three times a year.

It is well worth while too to compare the cost of providing top quality air with consequent savings in tool life and tool part replacement costs. But often one will cancel out the other.

All told, there is little difference between the rotary and reciprocating types when some fluctuation has to be allowed for. This is based on comparative 2 000 hour operation with

Five Atlas Copco ZR3A rotary screw compressors in use at John Player & Sons, the cigarette maker in Nottingham. The Z Packs are self-contained, ready-to-run compressed air units. Easy to install and maintain, their oil-free compression chamber produces completely clean air. At Players the Z Packs have enabled the firm to dispense with the frequent oil filter changes needed to maintain the quality of air from previous compressors.

5.7 m^3/min (200 ft^3/min) machines. But the cost difference widens in favour of the screw machine when maintenance cost factors are brought into play. The replacement of three filters per year costs around £40 and is the only requirement for a screw compressor apart from the necessary labour to carry out the service.

Other cost considerations include the effect of the compressor on the airline. A reciprocating machine sets up pulsations which can produce pipe joint leakage and fatigue in gauge mechanisms. In this respect the rotary screw machine can offer some advantage.

Specifying the Compressor

Once factory demand characteristics have been identified there remains the decision of whether to buy one large compressor or two small ones. Off-load consumption is the main guide; if line demand fluctuates down to 70%–50% of full load it may be wiser to buy two or three small compressors fitted with automatic zero cut-out. One or two machines then automatically cut out when demand drops, leaving the remainder to operate at peak output and efficiency.

In one case, a manufacturer of steel storage systems had been served by one 14 m^3/min (500 ft^3/min) reciprocating compressor, often working at maximum output but at other times below 50% capacity. The harassed maintenance crew were faced with constant problems, with line demand fluctuating from 14 – 5.7 m^3/min (500 – 200 ft^3/min). Eventually, the company was faced with further expansion, and so selected three 7 m^3/min (250 ft^3/min) rotary screw compressors which took care of load fluctuation and growth with minimum maintenance requirements.

Finally, in compressor selection, when replacing an old compressor of some ten years more in age, it must be remembered that old compressors often work at lower efficiencies than modern compressors. For instance, a ten year old 17 m^3/min (600 ft^3/min) machine can often be replaced by a modern one of 15.5 m^3/min (550 ft^3/min). Two ten year old 5.7 m^3/min (200 ft^3/min) machines can be replaced by two modern 4.8 m^3/min (170 ft^3/min) compressors and so on. By monitoring compressor technology progress, a buyer can often save the company a substantial amount of money.

Pneumatic Developments Cater for Industry's Needs

Three major developments have occurred during the last decade or so which have influenced the design of compressed air installations. These are the increasing legislation governing health and safety at work, the growing need to conserve energy and increasing demand for high quality air.

The recent laws on noise pollution restrict noise to below 75–80 dB(A) in factories in certain European countries, and similar legislation is on the way in Britain, through the Health and Safety at Work acts. Also noise at the factory perimeter is limited by law to 45 dB(A) or lower in many industrialized countries. The chief outcome of this has been the re-appraisal of compressor design and the result is the 'pack' compressor.

Compressor installations used to be a rather noisy and often untidy assembly of various components linked by piping and wiring. Even in the best thought-out installations the result was still unprotected and space-consuming.

Installation was also costly as units required concrete foundations and a network of pipes to filters, air dryers, receivers, pulsation dampers, valves, aftercoolers etc.

Working drawings were necessary to ensure correct placing of such equipment and all installation work had to be carried out on site. Consequently, civil engineering costs were high, labour content considerable and space taken up often out of proportion to the required compressed air delivery in terms of a percentage of factory power needs.

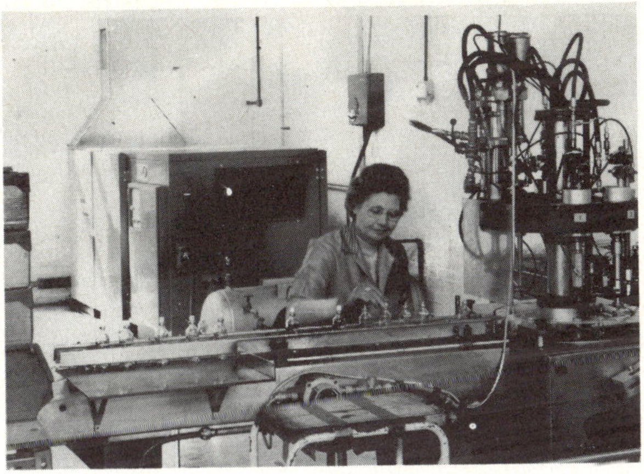

Compressor Pack boosts perfume packing — compressed air provided by an Atlas Copco GA310E high pressure compressor gives the power for bottling perfumes at the Eyelure Factory in Cwmbran, South Wales. Two 'Coster' Aerosol filling machines are used for this purpose.
Because of its low noise output — below 75 dB(A) at one metre — the Atlas Copco GA machine can be located very close to the production area — in this case only 12 ft from the aerosol filling machine.

Packaged Power Arrives

The first 'pack' to arrive on the scene was the portable compressor familiar in the streets of our cities. It contained a prime mover, a compressor, start-up equipment and sometimes an air receiver and fuel tank — all within an enclosed canopy.

The first real attempt to adapt this approach to industrial compressors came in the form of a baseframe or skid-mounted compressor, which considerably simplified installation. Usually the compressor was mounted on a steel, vibration-damped frame together with the prime mover, control panel, starter and aftercooler, piping and circuitry.

The modern pack unit — either reciprocating or rotary — is combined with accessories and control gear inside a steel, sound proofed canopy complying with CACI-PNEUROP specifications. Sound levels are generally at 75 dB(A) measured at 1 m from the machine.

There are now few restrictions on the type of compressor that can be packaged and packs can contain an aftercooler, air intake filter, moisture trap, starter, control panel, all interconnecting pipework and wiring, a flanged pipe connection to the main, air receiver, air drier and one overall dimensional drawing.

A cost-saving bonus is that only one installation drawing is required. If installation does not take place at once, there is no risk of parts being lost during storage and transport. The unit has merely to be placed and connected up when required. Low noise levels can also do away with the need for a compressor house. Although suppliers do recommend placing compressor packs on the factory floor, compressor environment should still be considered in terms of air temperature and humidity, relative to the compressor's design characteristics.

The arrival of the pack compressor has transformed the hiring of compressors as a pack can now be placed on site by forklift truck and plugged in where required. One company had specified a $14 m^3/min$ (500 ft^3/min) installation because for four days of the year it needed an extra 5.7 m^3/min (200 ft^3/min) to perform a particular operation. Now a 8.5 m^3/min (300 ft^3/min) plant working at a better efficiency level is adequate and the extra 5.7 m^3/min (200 ft^3/min) is hired and plugged in when required.

Conserving Pneumatic Energy

Since the oil crisis hit the industrial world like a bolt from the blue, responsible authorities have been engaged in looking for alternative energy sources, while making a great issue of conserving what conventional energy we have.

Electricy is consequently gaining in importance, and as compressed air is one of the most useful forms of energy, derived from electrical prime movers in industry and fossil fuel in construction and civil engineering, it is obviously valuable for energy conservation.

It is often forgotten that it is relatively simple to obtain very good overall efficiency — near 100% — with a compressed air installation. This can be done by recovery of the compression heat generated in the production of compressed air.

When compressed air is used in a factory to perform work we count on having at the pneumatic machine an absolute pressure of 7 bar; that is, the overpressure above atmospheric pressure is 6 bar. The compressed air temperature is the same as in the factory, say 20°C. We now consider this compressed air as it is expanded back to the atmospheric pressure of 1 bar, having performed work in an air motor with no loss. In theory, one would obtain work corresponding to 29 kWh for each 1 000 m^3 of free air (at 1 bar).

During expansion, the temperature of the air will drop from 20°C to −105°C. The greatly chilled exhaust air must now take from the factory air an amount of heat corresponding to work performed in the air motor, 41 kWh. In layman's terms, the compressed air, as it performs work under expansion, gives off internal energy and takes it back from the surrounding atmosphere after expansion. But in reality, a good air motor can only exploit about 15 kWh in the form of mechanical work from 1 000 m^3 of free air, that is, less than half of the theoretical energy content of the compressed air.

Taking into account all losses in the airline delivery, compressor action and so on, the production of 1 000 m^3 of free air needs a compressor working to compress air to 8 bar absolute pressure. During compression, all the work developed by the prime mover is converted into heat, which must be led off by cooling in the compressor cooling system. Since losses in the prime mover — say an electric motor — and the compressor are also in the form of heat, we can simply say that the electrical energy consumed is converted into heat in the compression chamber and must be removed from it.

Small and medium sized compressors are air-cooled, and heat is usually taken off with the ventilation air. On larger types water cooling performs the task. Now if the waste heat can be utilized for heating nearby facilities very good overall economy is possible. Say in one year 500 000 m^3 of free air are produced so that the total energy represented would be 500 x 125 = 62 500 kWh (125 kWh are required to produce 1 000 m^3).

If we can use this thermal energy for, say, eight months of the year, then the amount of heat would be 42 000 kWh which is roughly what two normal sized houses need for heating and warm water during a year. Or, it corresponds to the heat given out from 6 m^3 of fuel oil burned in a domestic furnace in a central heating system.

In the course of the year we obtain about 7 500 kWh worth of work from the 500 000 m^3. In all total utilization would be 42 000 + 7 500 = 49 500 kWh out of the 62 500 kWh consumed in the production of the compressed air. So overall efficiency could be 80%.

The efficiency of water-cooled installations can be higher, as heated water can be used for such things as laundries or shower rooms during summer too. About 100 kWh may be sacrificed in producing 1 000 m^3 in the large compressor, so some 80 kWh can be recovered from the cooling water. Add to this the work performed by the compressed air and overall efficiency tops 95%.

It should be profitable in many cases to utilize waste heat and show some profit in times of inflationary oil/electricity prices.

The cost of compressed air taken in isolation is somewhat higher than that of electricity, hence the greater need for conservation. It may not be possible to perform the same type of work with electricity for reasons already examined, but it is possible to reduce consumption.

One of Europe's largest motor car factories in Italy found that its consumption of compressed air increased from 370 m^3 to 850 m^3 per car over the period 1965 to 1972. By 1975, the factory had brought consumption down to 700 m^3 of free air per car through a carefully executed study and remodelling of the air supply.

Restrictory Leakage

Leakage is one of the greatest air consumers and can account for 5—10% of installed compressor capacity. Take a process industry, for instance, with an installed air capacity of 2 m^3/sec and leakage estimated at 5%. The consumption including leakage amounts to 1.5 m^3/sec during two day shifts and 0.5 m^3/sec during the night shift. A year's consumption amounts to

$$(1.5 \times 3\,600 \times 4\,000) + (0.5 \times 3\,600 \times 4\,000) = 29\ m^3 \times 10^6.$$

Total leakage during the same period amounts to

$$0.1 \times 3\,600 \times 8\,000 = 2.9\ m^3 \times 10^6$$ or 10% of total consumption.

In other words, twice the estimated leakage. For this 'moderate' leakage, energy costs will be 290 MWh on the basis that 1 kWh produces on average 10 m^3 of free air at 7 bar.

Regular leakage measurements will keep leakage within reasonable limits. By measuring when production is at a standstill and when compressor with known air flow output runs loaded in relation to the total running time needed to keep operating pressure at a normal value, one can calculate the true leakage. It is possible to localize leaks by closing off different sections during measurement. Regular checks should be made on local connections to air tools and equipment.

Another important factor governing leakage is maintenance of air tools, actuating cylinders and valves. Air consumption increases in worn equipment at the same time as output falls off. Correct operating pressure is also important. Increasing air pressure for temporary increases in output only accelerates wear rates, which ultimately increase air consumption. Reduced pressure means great pressure drops in airlines, and output of power from an air tool drops quicker than the rate of pressure fall. So again, for work out at lower pressure, consumption increases.

It is not merely a question of taking conservation steps at the compressor source; discipline in maintenance must apply throughout the system and air tooling if full conservation is to be realized.

Supplying Cleaner Air

The development of more sophisticated machines and processes which demand cleaner, drier air is proceeding steadily. High speed valving, actuators and monitoring equipment functions often deteriorate in the presence of even small quantities of oil and moisture. These substances collect sludge and rust particles and will block orifices in spools and other pneumatic components.

Compressed air is used in the food processing industries, and in breweries, for example, to move beer from vat to vat by pressurizing the vats. Cigarette manufacturers blow compressed air through every cigarette to test the degree of packing of the tobacco strands. In the iron and steel industry, non-ferrous metal and chemical industries, compressed air is used for changing and ventilating furnaces.

Oil bearing air would contaminate products in the first two examples and cause risk of explosion in the last case.

In most countries, the hygienic threshold limit values (TLV) for lubricating oil mist is 5 mg/m^3 (or 5 ppm). When the demands for oil-free air are absolute, or large quantities are needed, then the employment of non-lubricated compressors is imperative. In the standard reciprocating compressor about one third of the lubrication oil remains in suspension after passing through an aftercooler. The oil content of air leaving a well designed and maintained oil-flooded screw compressor is relatively low, a typical carryover being 16 ppm by weight. In some cases, the air leaving an additional built-in terminal filter can be as low as 1 ppm by weight.

If reciprocating compressors are employed, piston rings of carbon or polytetrafluoroethylene (PTFE) are usually a good solution, but microscopic particles resulting from wear have to be guarded against, while maintenance requirements are very high.

A recommended solution is the dry screw compressor which is now available in pack form, delivering up to 4.7 m^3/sec.

Soaking Up Moisture

Once the oil problem has been solved there remains humidity. There are several ways of drying compressed air: refrigeration drying, adsorption drying and so on.

Cooler driers cool the air to a pressure dewpoint of +2°C, at which point condensation occurs in the same way as in a conventional dryer. The air is then reheated and returned dry to the air network at room temperature.

Adsorption dryers work on the principle of adsorption of humidity and pressure dewpoints of –20°C to –40°C can be achieved.

The sorption dryer is one of the latest types for fitting to screw compressors. These employ heat generated by compression to regenerate the dryer agent. It is a type of adsorption dryer but works by combining physical adsorption and chemical reaction to remove water content from compressed air. The lowest pressure dewpoint for this system is –40°C.

The new Atlas Copco MD Sorption Dryer (right of picture) at work with the Atlas Copco Z Pack oil-free compressor. This combined unit provides air of the highest quality in an extremely compact installation. The unique feature of the MD Dryer is the slowly revolving sorption drum. Air is dried in one section of the drum, while regeneration is simultaneously taking place in the remainder – an operation which is accomplished without regeneration air losses or any external source of heat.

It is not always a black-and-white case of specifying either high or low quality air for a complete factory system. In cases where a mixture is required, say in an engineering components plant where 'standard' air is used for air tools but 'high quality' air is needed in finishing, then close consultation with the equipment supplier becomes of paramount importance.

It is in these three areas of compressor design, energy conservation and air quality that developments are continually taking place. Only close contact with the equipment supplier will result in getting the best, most economic deal for the job in hand.

Pneumatics' Advantages Consolidate Its Future Industrial Role

To gauge the future of compressed air, one has only to dwell on the current advantages of this power source as outlined at the beginning of this chapter: convenience, efficiency and safety. It is true that over the years the basic principles of pneumatic power have not altered dramatically, but the process of streamlining design and making genuine advances in efficiency of the pneumatic power system will continue.

Today, a fully-integrated compressed air installation can be planned and put into operation in a very short time. Its mechanics and physics are directly accessible and readily understood. On this basis, until something as safe, more convenient, more efficient and cheaper comes along, there can be no doubt over the future of pneumatic power.

Technical Trends

Technical trends will see an increasing demand for oil-free air in many functions, and its increasing use in the food, pharmaceutical and processing industries. There will be a corresponding increase in the demand for moisture-free air.

Compressor drive by direct-flanged electric motors is growing in importance. These save floor space, avoid alignment troubles and eliminate the 4–6% power loss due to drive belting. Resilient mountings are now available for reciprocating compressors up to 300 kW and screw compressors up to 650 kW. Packs are becoming more common and are growing in capacity. The range of accessories, including dryers and filters which can be built-in, will grow.

Finally, working pressures are increasing and the design of future compressors will be influenced by the available prime movers. The demands for high efficiency, low noise and purity of compressed air coupled with decreasing maintenance requirements will become more pronounced and will guide future compressor development.

The future may well belong to the rotary, oil-free, high speed, labyrinth-sealed, modular and fully automated compressors.

Properties of Air

DRY air is composed of a mechanical mixture of nitrogen and oxygen together with minor quantities of other gases – see Table I. Its composition remains substantially unaltered between sea level and an extreme altitude of about 20 kilometres, but its density decreases with increasing altitude, as well as varying with pressure and temperature. Under conditions of normal temperature and pressure the mass of dry air is 1.21 kg/m^3 (0.0765 lb/ft^3). It follows that at STP,

1 kg of air has a value of 0.818m^3

1 lb of air has a value of 13.07 ft^3

TABLE I – COMPOSITION OF DRY AIR

Component	Percent by Volume	Percent by Mass
Nitrogen	78.09	75.51
Oxygen	20.95	23.15
Argon	0.93	1.28
Carbon dioxide*	0.03	0.046
Neon	0.0018	0.00125
Helium	0.00052	0.000072
Methane	0.00015	0.000094
Krypton	0.0001	0.00029
Carbon monoxide*	0.00001	0.00002
Nitrous oxide*	0.00005	0.00008
Hydrogen*	0.00005	0.0000035
Ozone*	0.00004	0.000007
Xenon	0.000008	0.000036
Nitrogen dioxide	0.0000001	0.0000002
Iodine	2×10^{-11}	1×10^{-10}
Radon	6×10^{-18}	5×10^{-17}

* Varies slightly

At a specific temperature and pressure, the number of molecules in a unit volume is constant for any gas or mixture of gases. For a temperature of 0^oC and a pressure of one atmosphere, this number is 2.705×10^{19} molecules per cubic centimetre, or 7.84×10^{23} molecules per cubic foot.

At standard temperature and pressure, the mean velocity of gas molecules is of the order of 500 metre/second, with a mean free path between intermolecular collision, for air, of the order of 3×10^{-6} millimetres. The rate of collision under such conditions is of the order of 5×10^9 per second. The dissipation of energy resulting from such collisions is responsible for the pressure exerted by air (or any gas) on a surface immersed in it, or on walls containing the gas. Specifically, therefore, the pressure of any gas depends on its mass density, the number of molecules present and their mean velocity.

The effect of a change in temperature is to modify the value of the mean velocity. Hence, in the absence of any other change, the resultant pressure will vary with the temperature. Similarly, any change in volume will effectively modify the mass present and again affect the pressure. Thus pressure, temperature and volume are interrelated. This relationship can be expressed in the following form for a perfect gas (which assumes that the molecules are perfectly elastic, are negligible in size compared with the length of their mean free path, and exert no force on each other):

$$\text{Pressure (P)} = \frac{2}{3} \cdot \frac{n}{V} \cdot \frac{mv^2}{2}$$

when n = number of molecules in volume V
m = mean molecular mass
v = root mean square molecular velocity
V = volume

The quantity $\frac{n \cdot m}{V}$ is equal to the density of the gas. Thus

$$\frac{n \cdot m}{V} = \text{density} = \frac{1}{V}$$

where V is the specific volume

Hence

$$P = \frac{v^2}{3V}$$

or $PV = 0.3 \times v^2$

This relationship can be simplified by the fact that the mean molecular velocity, and thus the root mean square value (v), is directly proportional to temperature. Thus the fundamental equation can be written in the familiar form

$$PV = RT$$

where T is the absolute temperature
and R is the gas constant

TABLE II – GAS CONSTANTS AT 20°C (68°F)

Gas	GAS CONSTANT R			Ratio of Specific Heats
	kg/kmol	ft lb/lb	ft lb/slug	
Air	286.6	53.3	1715	1.40
Carbon dioxide	188.8	35.1	1130	1.30
Hydrogen	4124.5	766.0	24660	1.40
Oxygen	260.0	48.3	1555	1.40
Nitrogen	296.8	55.2	1776	1.40
Helium	2076.0	386.0	12420	1.67
Xenon	63.5	11.8	378	1.67

TABLE III – PHYSICAL CONSTANTS OF AIR

Molecular weight	28.96	kg/kmol
Density of gas at 15°C and 1 bar	1.21	kg/m^3
Boiling point at 1 bar	– 191 – 194	°C
Freezing point at 1 bar	–212 – 216	°C
Gas constant	286.9	J/K • kg
Critical pressure	37.8	bar
Critical temperature	-140.7	°C

TABLE IV – VAPOUR PRESSURE OF MOIST AIR
(pressure in millibars)

Temp. °C	Relative Humidity in Percent									
	10	20	30	40	50	60	70	80	90	100
–10	0	1	1	1	1	2	2	2	2	3
–5	0	1	1	2	2	2	3	3	4	4
0	1	1	2	2	3	4	4	5	5	6
5	1	2	2	3	4	5	6	7	7	8
10	1	2	4	5	6	7	9	10	11	12
15	2	3	5	7	9	10	12	14	15	17
20	2	5	7	9	12	14	16	19	21	23
25	3	6	10	13	16	19	22	25	29	32
30	4	8	13	17	21	25	30	34	38	42
35	6	11	17	22	28	34	39	45	51	56
40	7	15	22	30	37	44	52	59	66	74
45	10	19	29	38	48	57	67	77	86	96
50	12	25	37	49	62	74	86	99	111	123

The value of the gas constant for air and some other gases is given in Table II. Some other physical constants are given in Table III.

Atmospheric air normally contains water vapour, and the total pressure of the air is the sum of the partial pressures of the dry air and water vapour. The air is saturated when the partial pressure of the water vapour is equal to the saturation pressure of the water vapour at that temperature. The saturation pressure is dependent only on the temperature (see also Table IV).

When air is cooled at constant pressure, the *dew point* is reached, when the partial pressure is equal to the saturation pressure. Any further cooling will then result in water separating by condensation.

The presence of water vapour in atmospheric air leading to condensation implies practical problems both in compression and compressed air usage.

Basic Processes and the Gas Law

The three basic processes to which a gas may be subject are

(i) *Isotrope* — or change in volume under constant pressure, involving a corresponding change in temperature.

(ii) *Isotherm* — or change in volume under constant temperature, involving a corresponding change in pressure.

(iii) *Isobar* — or change in pressure under constant volume, involving a corresponding change in temperature.

The general relationship or perfect gas law for an ideal gas which applies in each case is

$$PV = RT$$

when P = pressure
V = volume
R = gas constant
T = temperature

For many engineering purposes the assumption that real gases conform to the perfect gas equation is satisfactory. The basic equation may, however, readily be extended to express the deviation of a real gas from a perfect gas when subject to compression, by introducing a compressibility factor (Z) when

$$PV = ZRT$$

The factor Z is derived directly as the 'departure' from the perfect gas relationship (ie $Z = PV/RT$), the extent of this being shown diagrammatically in Fig 1. It will be seen that there is a considerable area bounded by $Z = 0.99$ and $Z = 1.01$ where the pressure/temperature relationship for air follows the perfect gas equation within an accuracy of 1%.

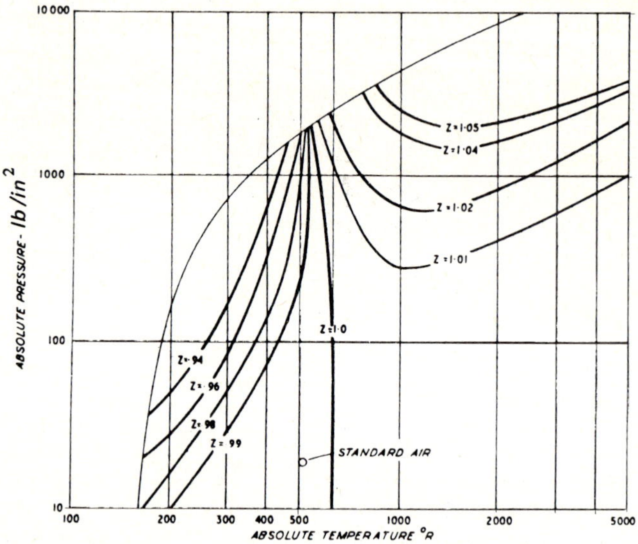

Fig 1

For exact solutions compressibility charts should be used, giving values of the compressibility factor (Z) for each specific gas or gas mixture. Generalized charts are also used, based on the assumption that the behaviour of all gases is virtually similar when their reduced pressure (P_R) and temperature (T_R) are the same, ie,

$$P_R = \frac{P}{Pc}$$

and

$$T_R = \frac{T}{Tc}$$

where Pc = critical pressure
Tc = critical temperature } see Table III

A generalized compressibility chart is shown in Fig 2.

Within the range Z = 0.99 – 1.01, it follows that for all practical purposes pressure and volume are inversely proportional at constant temperature, ie,

$$PV = \text{a constant}$$

This is consistent with *isothermal* conditions, where no heat is generated, or given up, by the compression or expansion process. The compressibility, or average bulk modulus in this case, is equal to the absolute pressure during compression.

Specifically, however, the change in volume of a gas, or a change in pressure, depends on the compression process, which can be expressed in the general form

$$PV^u = \text{a constant}$$

where u is an exponential determined by the compression process

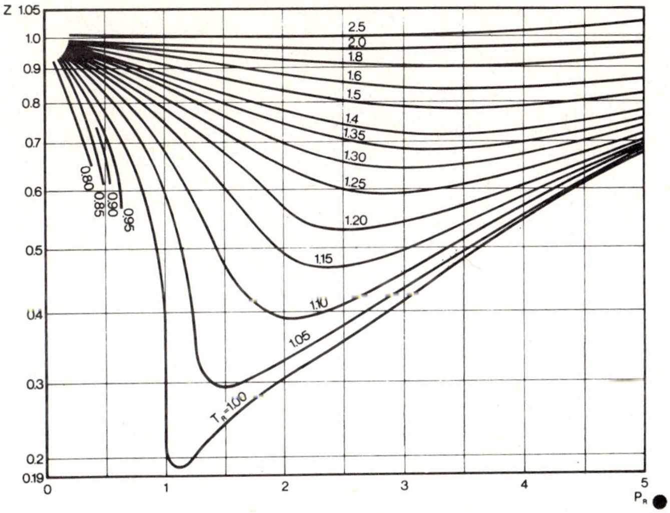

Fig 2 Generalized compressibility factor.

In the case of true isothermal compression the value of u is unity. With true *adiabatic compression,* where the compression is accompanied by a rise in temperature but with no heat exchange with the surroundings, the value of u is equal to the ratio of the specific heat of the gas at constant pressure, to the specific heat at constant volume. For air this gives u a value of 1.405.

In practice, neither exact condition of compression is likely to apply. Adiabatic compression is more desirable, since the heat generated by compression is tending to cause it to expand even whilst it is being compressed, hence the same reduction in volume will yield a higher absolute pressure. Thus, the value of u with practical compression can vary between 1 and 1.405, the condition appertaining generally being called polytropic compression. This implies that the compression process lies between true isothermal and true adiabatic compression – see Fig 2. With reciprocating compressors, the typical polytropic value of u will lie between 1.3 and 1.4

If the compression is adiabatic but is carried out slowly enough for equilibrium conditions to exist, the compression may be considered reversible and adiabatic, when u will assume the full adiabatic value. This case is normally referred to as isentropic compression.

The compressibility, or bulk modulus, in all cases where u is greater than unity, is equal to u times the absolute pressure.

Compression Efficiency

Compression efficiency can be defined as the ratio of the theoretical work required to compress a given amount of air to a specific volume, to the actual amount of work done by the compressor.

Since the units of pressure are F/L^2, and those of volume are L^3, work done can be expressed by the product of pressure and volume (units F x L).

The mean effective pressure (mep) with isothermal compression is

$$\text{mep} = P_1 \log_e \frac{P_2}{P_1}$$

$$= P_1 \log_e r$$

where r is the compression ratio

The isothermal indicated horsepower or *Isothermal Factor* is given by

$$K \cdot P_1 \log_e r = K \cdot P_1 \log_e \frac{P_2}{P_1}$$

K being a constant factor, depending on the units employed. Thus for Isothermal Factor defined as indicated horsepower for 100 ft^3 of piston displacement per minute in a theoretical cylinder with 100% efficiency, working from atmospheric air

$$\text{Isothermal factor} = 6.43 \log_e \left(\frac{P_2 + 14.7}{14.7}\right)$$

This formula, which represents the theoretical horsepower required to compress 100 ft^3 of free air isothermally per minute was originally widely used as a standard reference for compressor performance. It can also be used to correct for altitude. In this case the actual value of atmospheric pressure at altitude is substituted for P_1 in the basic formula.

More generally (or in metric units), the work required for isothermal compression is given as

$$\text{Work (joules)} = 10^5 \; P_1 \cdot V_1 \log_e \frac{OP_2}{P_1}$$

where P is the absolute pressure in bars
V is volume in m^3

For isothermal compression from atmospheric pressure

$$\text{Work (joules)} = 10^5 \; V_1 \log_e P_2$$

$$= 230.3 \, V_1 \log P_2$$

The power for isothermal compression follows by replacing volume in m^3 by volume flow in m^3/sec, viz

$$\text{power (kW)} = 230.3 \, V_F \log^1 /P_2$$

where V_F = volume flow m^3/sec
P_2 = joule pressure in bars

Power for *adiabatic compression* can be determined in a similar manner only in this case the mean effective pressure is

$$\text{mep} = \frac{u}{u-1} \times P_1 \left(r^{\frac{u-1}{u}} - 1\right)$$

$$= 3.463 P_1 \; (r^{0.29} - 1)$$

Thus power required for adiabatic compression is given by:

$$\text{power (kW)} = 230.3\ V_1 \log \left[\left(\frac{P_1}{P_2} \right)^{0.29} - 1 \right]$$

Indicated Power Factors

Indicated power factors for adiabatic compression can be derived for two or more stages of adiabatic compression, based on cooling back to the isothermal line between stages (ie assuming perfect intercooling).

For two-stage compression:

$$\text{Adiabatic indicated power} = K_1\ P_1 \left[\left(\frac{P_2}{P_1} \right)^{0.147} - 1 \right]$$

where K_1 = 3.02 for indicated *horsepower* per volume flow of 100 ft^3/min

= 143 for indicated power in *kilowatts* per volume flow of 100 m^3/min

For three-stage compression:

$$\text{Adiabatic indicated power} = K_2\ P_1 \left[\left(\frac{P_2}{P_1} \right)^{0.0956} - 1 \right]$$

where K_2 = 4.53 for indicated *horsepower* per volume flow of 100 m^3/min

For four-stage compression:

$$\text{Adiabatic indicated power} = K_3\ P_1 \left[\left(\frac{P_2}{P_1} \right)^{0.0725} - 1 \right]$$

where K_3 = 6.09 for indicated *horsepower* per volume flow of 100 ft^3/min.

= 286 for indicated power in *kilowatts* per volume flow of 100 m^3/min.

Adiabatic Head

The adiabatic head is given by the following:-

$$H_a = RT \frac{\left(r^{\frac{u-1}{u}} - 1\right)}{\frac{u-1}{u}}$$

$$= RT \cdot \frac{r^{0.286} - 1}{0.286}$$

Polytropic Ratio

This is defined as $\frac{k-1}{k}$ and is related to the adiabatic exponential u by the polytropic efficiency ηp

$$\frac{k-1}{k} = \frac{1}{\eta p} \times \frac{u-1}{u}$$

Polytropic Head

This follows in a similar manner to the adiabatic head, substituting k for u.

$$H_p = \frac{RT\left(r^{\frac{k-1}{k}} - 1\right)}{\frac{k-1}{k}}$$

Discharge Temperature

The temperature of air discharged from a compressor is given by

$$T_2 = T_1 + \frac{T_1\left(r^{\frac{u-1}{u}} - 1\right)}{\eta}$$

$$= T_1 + \frac{T_1\left(r^{0.286} - 1\right)}{\eta} \quad \text{(for adiabatic compression)}$$

where η = volumetric efficiency

T_1 and T_2 are absolute temperatures

A simplified version of this formula is often used for engineering calculations

$$T_2 = T_1\, r^{\frac{u-1}{u}}$$

$$= T_1\, r^{0.286} \quad \text{(for adiabatic compression)}$$

Volumetric Efficiency

The theoretical value of volumetric efficiency is given by

$$n = 1 - \left(r^{\frac{1}{u}} - 1\right) \times C$$

where r = compression ratio

C = clearance in cylinder, expressed as a decimal

Equivalent Free Air Volume

The relationship between free air volume and compressed air volume follows directly from the compression ratio, viz

$$\frac{\text{free air volume}}{\text{compressed air volume}} = \text{compression ratio}$$

or

$$\frac{V_{FA}}{V_C} = r$$

whence

$$V_{FA} = \frac{P_2 + P_1}{P_1} \times V_C$$

where P_2 = final pressure, absolute
P_1 = atmospheric pressure

For P_2 in lb/in^2

$$V_{FA} = \frac{P_2 + 14.7}{14.7} \times V_C$$

See also chapters on *Compressor Performance* and *Compressible Air Flow.*

Compressor Performance

THE theoretical compression cycle for both isothermal compression and adiabatic compression is shown in Fig 1. Isothermal compression is consistent with a long stroke, small bore compressor (cylinder) with complete cooling (so that there is no heat rise). Adiabatic compression is constant with a higher speed compressor with perfect insulation so that some of the energy supplied to the air is converted to heat and thus the temperature of the air will rise. The shaded area represents the saving in work of isothermal over adiabatic compression due to heat removal during compression.

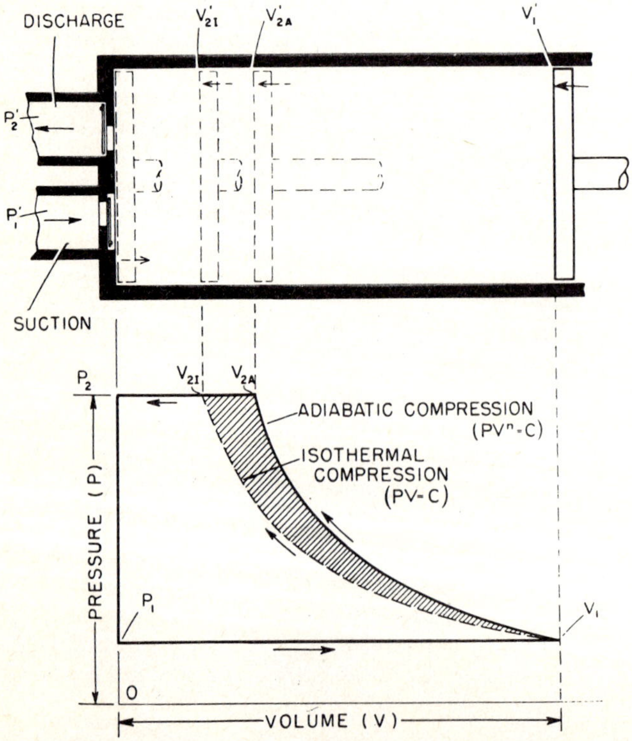

Fig 1 Pressure-Volume Diagram of Theoretical Compression Cycle. The figure illustrates pressure and volume relations and corresponding piston positions for adiabatic and isothermal compression in a theoretical single-stage single-acting compressor having no clearance. Areas on the pressure-volume diagram represent work input and the shaded area represents the saving in work of isothermal over adiabatic compression due to heat removal during compression. (Mobil Oil Co Ltd).

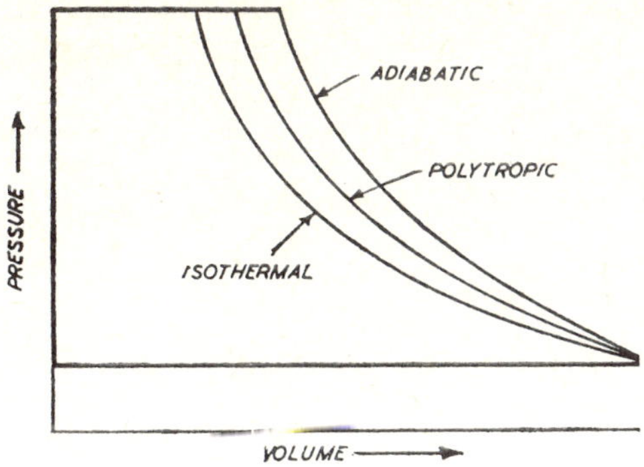

Fig 2

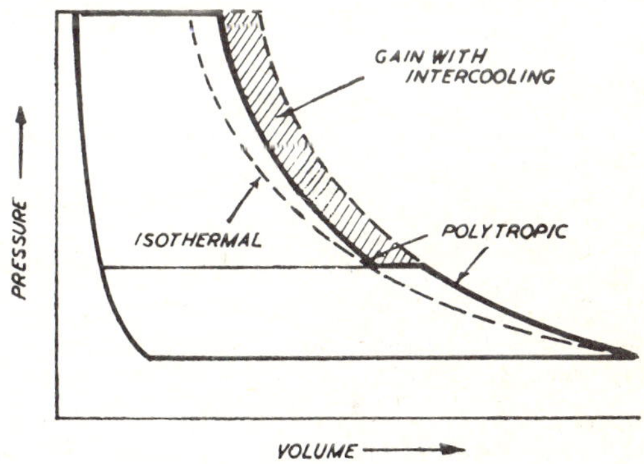

Fig 3

Practical compressors normally work by a polytropic process, ie intermediate between true isothermal and true adiabatic compression – Fig 2. There are definite advantages to be gained from achieving the final degree of compression required in two or more stages, with intercooling between stages. Multi-stage compression increases the volumetric efficiency and power requirements – power input being a minimum if the total work is divided equally over the stages. Cooling is particularly useful in decreasing the final air temperature and it also reduces the work input required to achieve the same degree of compression. This is exemplified in the generalized compression diagram of Fig 3 for two-stage compression and intercooling. The first stage polytropic curve approaches closely the adiabatic curve but at the conclusion of this stage air temperature is reduced to its original value by intercooling and the volume is also reduced. The second stage polytropic curve then follows more closely the isothermal line, the shaded area representing the input power saved by cooling. In practice the temperature reduction is more significant than the work-saving and the desirability of multi-stage operation is greater with increasing final pressure and capacity required.

To summarize, multi-stage compression

(i) limits temperature rise and thus improves efficiency

(ii) increases volumetric efficiency as the pressure ratio over the first stage is decreased.

Power requirements for multi-stage compression are minimal if the total work is divided evenly over the stages, although this can also be affected in practice by the degree of cooling achieved. With perfect intercooling minimum power is required when the pressure ratio of all the stages is the same.

As a general rule, the greater the number of stages the greater the efficiency possible because of the close approach that can be realized to isothermal compression. In practice, however, the number of stages employed may have to be limited on grounds of complication and cost.

Volumetric efficiency is a function of capacity normally specified in terms of free air delivered. This refers to the volume of free air at the intake pressure and temperature which the compressor will take in, compress and deliver at a stated delivery pressure. Output of a compressor refers to the free air delivered when the machine is working at the specified intake pressure and temperature and specified delivery pressure.

The volumetric efficiency is expressed by

$$M_V = K \left\{ 1 - v_r \left[\left(\frac{P_2}{P_1} \right)^{\frac{1}{n}} - 1 \right] - 1 \right\}$$

where K is a factor depending on heating of the air and valve losses (also varying with pressure ratio). A typical value is 0.96.

ϵ = relative clearance volume (typical values for average compressor designs 0.06 –0.12)

P_1 = intake pressure of the final stage

P_2 = discharge pressure of the first stage

n = polytropic exponent (typical value 1.3)

The significance of this parameter is shown in Fig 4.

Fig 4 shows the factor $\left[1 - \epsilon \left(\frac{P_2}{P_1} \right)^{\frac{1}{n}} - 1 \right]$ as a function of the first stage pressure ratio for various values of the relative clearance volume (ϵ).

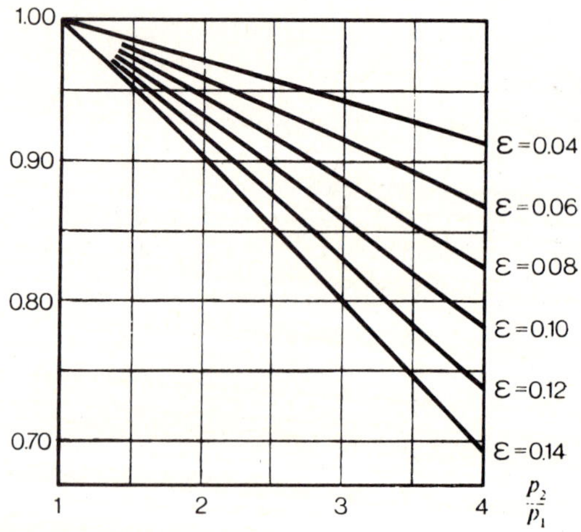

Fig 4

Compressor efficiency is defined by the ratio of the theoretical work required to compress a given amount of air to the actual amount of work developed on the air. The theoretical calculation may be based on isothermal compression (n = 1 in the basic formula) or adiabatic compression (n = 1.405), in which case the ratio would be referred to as the isothermal compression efficiency or adiabatic compression efficiency respectively.

The mechanical efficiency is the ratio of work done on the air to the actual work input to the machine and thus is a measure of mechanical losses. The overall efficiency is the ratio of the total power output to total power input, which can also be expressed as the product of the compression efficiency, mechanical efficiency and driver efficiency.

Thus volumetric efficiency largely governs the size of compressor required for a given duty, and overall efficiency the cost of the compressed air. Since annual power costs may approach one quarter to one third the initial cost of the compressor of a typical industrial system, overall efficiency can be an important factor in deciding on the most economical plant to install.

Displacement compressors induct and displace successive volumes of air (or gas) and work on the principle of increasing the pressure of a given volume of air (or gas) by reducing that volume. Dynamic compressors produce a static pressure rise by converting kinetic energy into pressure energy during the passage of air (or gas) through the machine. Mechanically, the difference is that displacement compressors incorporate a moving member or members, such as a piston, to produce displacement. The working element in a dynamic compressor is an impeller.

Reciprocating Compressor Performance

The displacement of a reciprocating compressor is expressed as the net volume actually displaced by the piston at rated machine speed or unit of time, or that of the first stage in the case of multi-stage machines. The capacity is defined as the actual volume rate of flow of gas compressed and delivered at the standard discharge point, referred to conditions of total temperature, total pressure and composition (eg humidity) prevailing at the standard inlet point.

The practical form of the volumetric efficiency formula is:-

$$M_V = K_r\left(r^{\frac{1}{n}} - 1\right) \times (C - A)$$

where r = compression ratio

n = polytropic exponent

C = cylinder clearance expressed as a decimal

A = correction factor for leakage, etc, typical values being 0.025 – 0.035 for lubricated machines to 0.07 for unlubricated machines.

The factor K is an empirical value of less than unity, allowing for other practical losses such as incomplete filling of the cylinder on each stroke. A typical value is 0.95.

Theoretical power requirements are given by

$$\text{power required} \propto PV\left(\frac{n}{1-n}\right)\left(r^{\frac{n-1}{n}} - 1\right)$$

where P = cylinder inlet pressure

V = volume flow at cylinder intake conditions

r_2 = compression ratio

In practical units:

$$\text{HP required} = \frac{144PV}{33000}\left(\frac{n}{1-n}\right)\left(r^{\frac{n-1}{n}} - 1\right)$$

where P is in lb/in^2 (gauge)
and V is in ft^3/min

$$\text{KW required} = 0.0024\,PV\left(\frac{n}{1-n}\right)\left(r^{\frac{n-1}{n}} - 1\right)$$

where P is in bar
and V is in m^3/min

In practice the actual power required is normally determined from an indicator diagram which will indicate the thermodynamic reversible adiabatic process, which can be generally classified as cylinder losses. Further losses can result from fluid friction through inlet and discharge valves, together with mechanical (friction) losses. In general, with good design and construction, compression efficiency can be expected to be of the order of 85 to 95%, and mechanical efficiency from 83 to 93% is typical. This means that overall efficiency is likely to be between 75 and 85%, compared with ideal isentropic compression.

Power saving is possible with multi-staging. Increasing the number of stages normally increases the overall efficiency and thus the economy of operation of the machine. It must be appreciated in such cases, however, that cylinder diameters are usually selected for specific intake and discharge pressures and operational pressures other than design values can have a marked effect on the performance of these machines.

Dynamic Compressors

Any turbo-machine operated at constant speed is capable of delivering a wide range of volumes with only a small variation in pressure. If volume is decreased by throttling on the discharge side, pressure will also decrease. Pressure developed when the discharge is completely throttled will depend on the machine design and operating speed, but will always be lower than the maximum pressure. It is thus a characteristic of aerodynamic machines that dangerously high pressures cannot be built up under any operating conditions and so relief valves are not necessary in the discharge lines. Regulation can be achieved either by throttling on the discharge side or variation in the impeller speed (or both).

The use of a valve offers the simplest form of control over pressure or capacity but any degree of throttling represents an irrecoverable power loss. Speed variation is, therefore, a more economic form of control when the compressor characteristics can be expressed in the form of a series of curves which may be superimposed on the demand curve to establish the required operating point Fig 5. In other words, by adjusting the speed, the compressor characteristics may be matched to any required values of pressure and capacity within the operating limits of the compressor and its driver. Efficiency will not be at a maximum, however, unless this operating point corresponds to the design value for maximum efficiency.

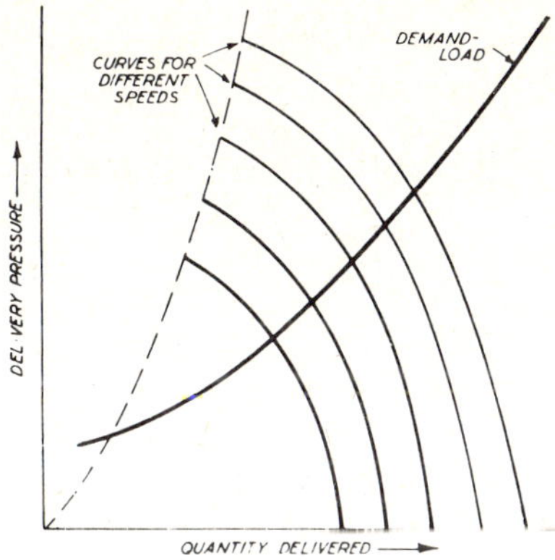

Fig 5

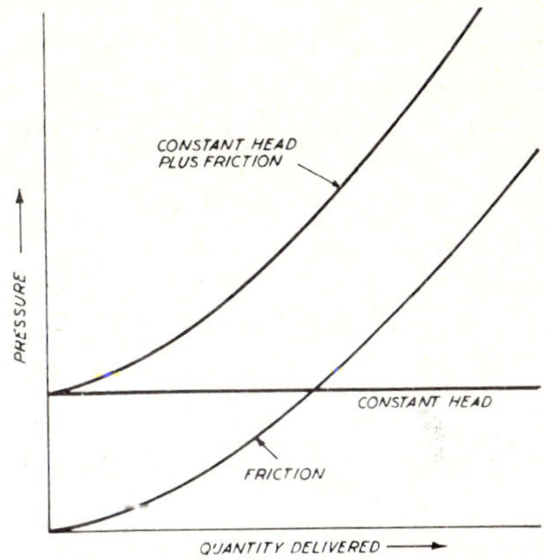

Fig 6

The variable speed, capacity and adiabatic head can be related in terms of specific speed, viz

$$\text{Specific speed } (n_s) = \frac{Q \times \text{rev/min}}{H_a\ 0.75}$$

where Q = inlet capacity

H_a = adiabatic head per stage

This is a relative quantity only and has little significance other than as a design specific speed or value at which a given impeller in a given type of casing shows its maximum efficiency. It then serves as an index of classification for the form of the impeller (in that type of casing). In fact, the compressor may be operated over a range of speeds, capacities and adiabatic heads, and thus a wide range of specific speeds. These actual or operating specific speeds are useful only as a basis for comparing performance relative to the performance obtained at the design specific speed.

The demand load itself will depend on the operating conditions and may involve working against a constant head or pressure, against frictional resistance or a combination of both. In the former case the demand load will be constant, and in the other two cases variable with capacity – Fig 6. The resultant demand curve may therefore range from a constant pressure value (operating against a constant head only – with negligible frictional resistance) to one rising steeply with increasing capacity. The demand curve may also vary under actual working conditions due to a change in head or frictional resistance. Equally the working point of the compressor can be adjusted by varying the demand curve, eg by means of a butterfly valve installed preferably near the compressor inlet or alternatively, near the discharge. Other methods of adjusting the compressor characteristics, and thus the pressure and capacity, include the use of adjustable diffuser vanes and adjustable inlet guide vanes.

Effect of Intake and Discharge Temperatures

Any change in ambient temperature will affect the efficiency and thus the rated performance of a compressor. A change in temperature will affect the intake volume required to produce a given volume of compressed air at a particular pressure which, from the gas law, implies that the relative intake volume will be directly proportional to the ratio of the ambient to design rated temperature.

Williams and James high-pressure compressor, compact model 175 is a V4 set, water cooled, delivering 30 s.c.f.m. at 3800 lb/in^2.

Northey type 242 CWD air or gas booster for maximum delivery pressure of 300 lb/in^2 (differential boost 40-50 lb/in^2 maximum).

HPC Spray 8 mobile compressor with twin horizontal air receivers

BroomWade V-Compact oil-free compressor

Since power requirement is directly proportional to the air volume indicated it will thus also be directly proportional to the temperature ratio.

In practice the final temperature of compressed air in a typical factory installation is usually substantially the same as the ambient temperature. If the compressor draws in air at this ambient temperature it will thus deliver the same quantity of free air at the point of application. If the intake temperature is lower than ambient, a proportionately smaller quantity of free air needs to be inducted to deliver a specified volume of free air at the higher end temperature and vice versa. Thus a definite advantage is realized if the intake air temperature is lower than the ambient, ie some power saving – (see Table I).

TABLE I – EFFECT OF INTAKE AIR TEMPERATURE
on compressor performance (relative to initial ambient air temperature of 15°C (60°F)

Initial Temperature °C	°F	Rated Delivery (%)
–29	–20	118.0
–23	–10	115.5
–18	0	113.0
–12	10	110.4
– 7	20	108.3
– 1	30	106.1
0	32	105.8
5	40	104.0
10	50	102.0
15	60	100.0
21	70	98.0
27	80	96.0
32	90	94.5
38	100	93.0
43	110	91.0
50	120	90.0
55	130	88.0
60	140	87.0
65	150	85.0
70	160	84.0

The theoretical discharge temperature from a compression is given by

$$T_d = T_i \, r^{\frac{n-1}{n}}$$

where T_d = discharge temperature (absolute)
T_i = intake temperature (absolute)

For adiabatic compression (n = 1.4)

$$T_d = T_i \, r^{0.286}$$

where r is the compression ratio.

A more accurate formula is

$$T_d = T_i + \frac{T_i\ (r^{0.286} - 1)}{\eta}$$

where η = volumetric efficiency.

In practice the discharge temperature will normally be reduced by cooling applied to the compressor and thus in practical systems may be of a similar order to the ambient temperature. Typical discharge temperatures (without cooling) relative to an initial intake temperature of 15°C (60°F) are given in Table II.

TABLE II – DISCHARGE TEMPERATURES
(Intake Temperature 15°C (60°F) – Intercooling to 15°C (60°F)

Pressure		Single Stage		Two-Stage		Three-Stage	
bar	lb/in²	Temperature °C	Temperature °F	Temperature °C	Temperature °F	Temperature °C	Temperature °F
0.70	10	62	143	–	–	–	–
1.38	20	96	205	–	–	–	–
2.00	30	124	255	–	–	–	–
2.75	40	145	293	–	–	–	–
3.51	50	168	334	84	183	–	–
4.1	60	187	368	91	196	–	–
4.8	70	203	398	98	208	–	–
5.5	80	219	426	103	218	–	–
6.2	90	233	452	109	228	–	–
7.0	100	247	476	114	237	–	–
7.5	110	260	499	119	246	–	–
8.0	120	271	520	123	254	83	182
9.0	130	280	538	127	261	85	185
9.5	140	298	560	131	268	88	191
10.0	150	302	578	135	275	91	196
14.0	200	349	659	151	303	99	211
17.0	250	384	726	163	326	107	225
34.0	500	–	–	207	404	132	269

Effect of Altitude

Ambient pressure and temperature both decrease with altitude, as given in Table III. These changes affect the pressure ratio of compressors (and air tools), which can appreciably modify performance. The extent to which this is realized will depend to some extent on the type and design of compressor, but the following are typical percentage reductions in capacity and power consumption for two-stage compressors working at 8 bars.

Compressor type	% reduction for each 1 000 metres increase in altitude	
piston-type	capacity	power consumption
medium size, air cooled	2.1	7.0
large piston-type, water cooled	1.5	6.2
screw, oil injected	0.6	5.0
large screw, water cooled	0.3	7.0

It should be noted that increasing altitude will also have an effect on the performance of internal combustion engines, driving compressors, usually roughly equivalent to a 1% loss of power for each 100 m of altitude (which can be halved by turbo-supercharging). The actual effect of high ambient temperature is rather more marked – typically a 4% power loss for every $10^{\circ}C$ rise in ambient temperature. This cannot be offset by turbo-supercharging. The performance of electric motors is not affected by altitude, but the lower density of air will result in poorer cooling. To offset this it may be necessary to reduce the shaft power at altitudes above 1500m. Electric motor shaft horsepower should also be reduced by 5% for every $5^{\circ}C$ ambient temperature rise above $40^{\circ}C$, again from cooling considerations.

TABLE III – STANDARD ATMOSPHERE ACCORDING TO NASA

Altitude m	Pressure bar	Temperature °C	Density kg/m³
0	1.013	15.0	1.225
100	1.001	14.4	1.213
200	0.989	13.7	1.202
300	0.978	13.1	1.190
400	0.966	12.4	1.179
500	0.955	11.8	1.167
600	0.943	11.1	1.156
800	0.921	9.8	1.134
1000	0.899	8.5	1.112
1200	0.877	7.2	1.090
1400	0.856	5.9	1.069
1600	0.835	4.6	1.048
1800	0.815	3.3	1.027
2000	0.795	2.0	1.007
2200	0.775	0.7	0.986
2400	0.755	−0.6	0.966
2600	0.737	−1.9	0.947
2800	0.719	−3.2	0.928
3000	0.701	−4.5	0.909
3200	0.683	−5.8	0.891
3400	0.666	−7.1	0.872
3600	0.649	−8.4	0.854
3800	0.633	−9.7	0.837
4000	0.616	−11.0	0.819
5000	0.540	−17.5	0.736
6000	0.472	−24.0	0.660
7000	0.411	−30.5	0.590
8000	0.356	−37.0	0.525

Compressible Gas Flow

AT MODERATE pressures, Mach numbers and temperatures, gases can be considered as behaving like perfect fluids (ie density variations are negligible), when the general gas equation applies

$$PV = RT$$

where P = pressure
V = volume
T = absolute temperature
R = gas constant

The numerical value of the gas constant (R) is given by

$$R = \frac{\bar{R}}{M}$$

where $\bar{R}$ = universal gas constant 8315 $\frac{J}{K \cdot kmol}$

M = molecular weight of the gas in $\frac{kg}{kmol}$

For air, $M = 28.96 \, \frac{kg}{kmol}$

Thus $R_{air} = 286.9 \, \frac{J}{K \cdot kg}$

(For absolute temperature in ^{o}K and pressure in lb/in^2, the value of R for air is 11.9).

For other gas mixtures, the gas constant of the mixture can be determined as

$$R = x_1 \cdot R_1 + x_2 \cdot R_2 + \ldots\ldots x_n \cdot R_n$$

R_1, R_2, etc = individual gas constants

where x_1, x_2, etc = concentration of constituents by *weight*

$$\text{or} \quad R\frac{1}{R} = \frac{y_1}{R_1} + \frac{y_2}{R_2} + \ldots \frac{y_n}{R_n}$$

where y_1, y_2, etc = concentration of constituents by *percentage volume.*

General Flow Solutions

The general flow equation is

$$\rho_1 \cdot V_1 A_1 = \rho_2 \cdot V_2 A_2$$

where ρ = mass density of gas
V = flow velocity
A = cross sectional area of flow

This implies that the mass flow is constant, or

$$\frac{P_1}{RT_1} \cdot V_1 \cdot A_1 = \frac{P_2}{RT_2} \cdot V_2 \cdot A_2$$

This provides specific solutions for one unknown, the other values being known. Note that the product VA also defines the flow rate. With flow through a parallel pipe or duct, working is simplified by the fact that $A_1 = A_2$ — Fig 1. Analysis can, however, be extended to cover convergent or divergent flow, although the case of flow nozzles or orifices needs separate analysis.

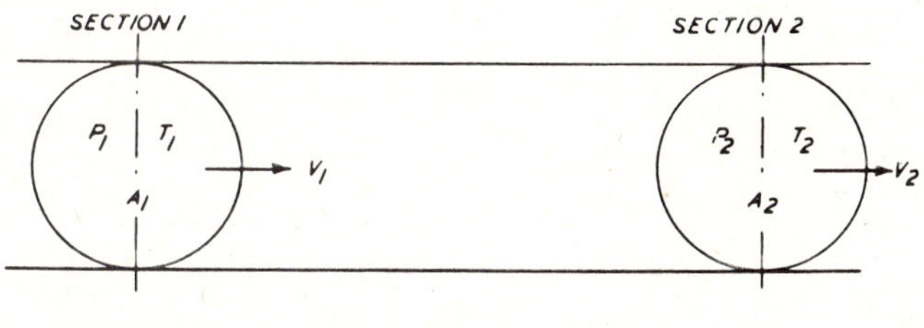

Fig 1

For steady adiabatic flow the energy equation is:-

$$\frac{V^2}{2} + h = \text{a constant}$$

where $\frac{V^2}{2}$ is the kinetic energy and h is the enthalpy

Flow is possible between two extremes. At one extreme, velocity is zero and temperature is a maximum, since all the kinetic energy is converted to enthalpy. The speed of sound is also a maximum (stagnation point or stagnation state). At the other extreme, the velocity is a maximum and the temperature falls to absolute zero, all the enthalpy being converted into kinetic energy. The speed of sound is then zero (zero temperature state). Between these extremes the practical flow may be subsonic, transonic, or supersonic — see Fig 2 — although the zero temperature can never be reached (ie a hypothetical condition).

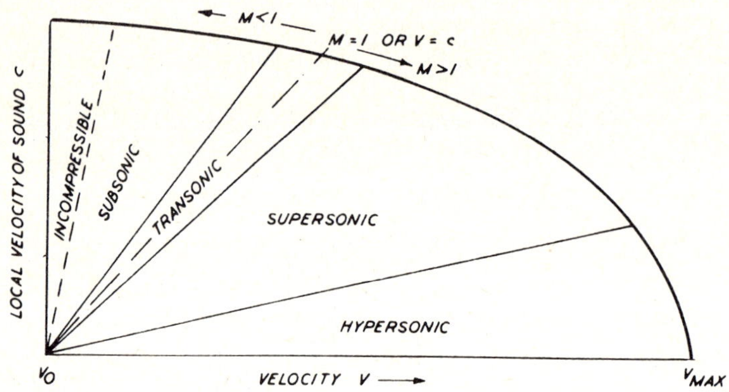

Fig 2

At the stagnation state

$$\frac{O^2}{2} + h_o = \frac{V^2}{2} + h$$

or

$$h_o = \frac{V}{2} + h$$

From the general gas relationships it follows that the stagnation temperature (T_o) is given by

$$T_o = T + \frac{V^2}{2sp}$$

where sp = specific heat at constant pressure

Alternatively,

$$\frac{T_o}{T} = 1 + \frac{k-1}{2} \times M^2$$

where M = Mach number = $\frac{V}{c}$

c is the reference sonic velocity

The stagnation pressure can be derived as

$$\frac{P_o}{P} = \left(\frac{T_o}{T}\right)^{\frac{k}{k-1}}$$

$$= \left(1 + \frac{k-1}{2} M^2\right)^{\frac{k}{k-1}}$$

This may be expanded in the form

$$P_o = P + \frac{\rho V^2}{2}\left(1 + \frac{M^2}{4} + \frac{2-k}{24} M^4 + \ldots\ldots\right)$$

The quantity $\left(1 + \frac{M^2}{4} + \frac{2-k}{24} M^4 + \ldots\ldots\right)$

represents the effect of increased gas density due to compressibility, generally termed the *compressibility factor.* Values of the compressibility factor range from unity at very low Mach numbers (where there are no compressibility effects), up to 1.276 as the Mach number approaches 1 (velocity approaches the speed of sound in the gas). Besides increasing the dynamic pressure of compressible flow, compared with incompressible flow, the rising value of the compressibility factor can also affect the flow velocity through ducts with varying area. The relationship between area and velocity changes is, in fact, a function of the local Mach number, and can be rendered in the form

$$\frac{dA}{dV} = \frac{A}{V} \quad (M^2 - 1)$$

With subsonic flow, a decrease in area produces an increase in flow velocity and vice versa (similar to incompressible flow) ie area and Mach number changes are opposite. The flow velocity may be sonic only at a constant section. With supersonic flow, a decrease in flow area produces a decrease in flow velocity, and vice versa, ie area and Mach number changes are the same.

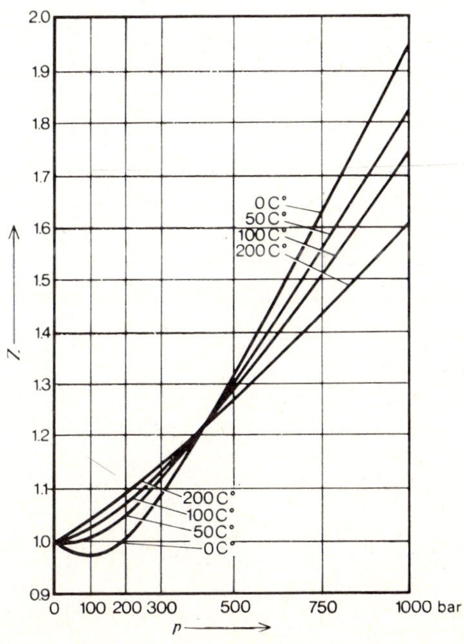

Fig 3 Compressibility factor Z.

All real gases deviate from the general gas law approaching the critical point (ie flow velocity approaching M = 1). The gas equation under those conditions is expressed as

$$PV = ZRT$$

where Z is the compressibility factor

The compressibility factor for dry air is shown in Fig 3.

Flow from Stagnation Conditions

Gas compressed and stored in a reservoir is essentially under stagnation conditions, where velocity is zero and the pressure and temperature are known (or can be determined). Where the reservoir is used as a supply, the velocity, temperature and pressure at any other section of flow are determined basically from the following relationships:-

Velocity at any arbitrary section

$$V = \sqrt{2sp\,T_o\left(1 - \frac{P}{P_o}\right)^{\frac{k-1}{k}}}$$

where sp = specific heat at constant pressure

Alternatively, for adiabatic flow, the velocity at any section can be determined from the temperature at that section

$$V = \sqrt{2sp\,T_o\left(1 - \frac{T}{T_o}\right)}$$

Flow rate

Flow rate can be determined as the mass flow, ie, mass flow = $VA\rho$ or directly as the product of V and A in numerically consistent units

$$\text{Dimensions are } \frac{L}{T} \times L^2 = \text{flow rate} = \frac{L^3}{T}$$

Pressure at any arbitrary section

$$P = \frac{P_o}{\left(1 + \frac{k-1}{2} M^2\right)^{\frac{k}{k-1}}}$$

Temperature at any section

$$T = \frac{T_o}{1 + \frac{k-1}{2} M^2}$$

At any (constant) section where the flow is sonic the flow conditions are described as critical, yielding a critical temperature (T^*) and a critical pressure (P^*), where

$$\frac{T^*}{T_o} = \frac{2}{k+1} \qquad \text{(Adiabatic or isentropic flow)}$$

$$\frac{P^*}{P_o} = \left(\frac{2}{k+1}\right)^{\frac{k}{k-1}} \qquad \text{(Isentropic flow only)}$$

For air, where $k = 1.4$, the value of critical pressure is $\left(\frac{2}{2.4}\right)^{\frac{1.4}{0.4}} = 0.528$

That is, the critical pressure is 52.8% of P_O. Similarly the critical temperature can be calculated as 83.3% of T_O.

Critical area:

The relationship between the critical area (A^*) or throat area where $M = 1$ and the area of any other section (A) is given by

$$\frac{A}{A^*} = \frac{1}{M} \; \frac{\left(1 + \frac{k-1}{2} \cdot M^2\right)^{\frac{k+1}{2(k-1)}}}{\frac{k+1}{2}}$$

$$= \frac{1}{M}\left(\frac{1 + 0.2\,M^2}{1.2}\right)^3 \qquad \text{for air}$$

Nozzle Flow

Flow through a nozzle is limited by the pressure and temperature, as well as the force of the nozzle. The relationship between the pressure before the nozzle (P_1) and the actual pressure (P^*) is

$$\frac{P^*}{P_1} = \psi$$

where ψ is a flow factor which varies with the isentropic exponent – see Fig 4.

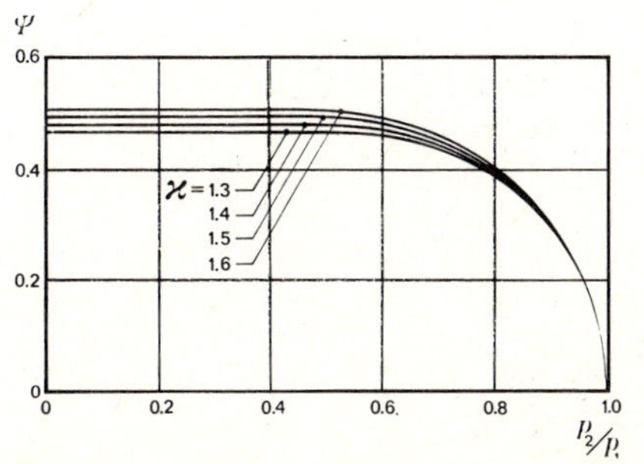

Fig 4 Flow factor ψ

Flow is super critical when the pressure downstream of the nozzle (P_2) is less than P^*. The mass flow through the nozzle is given by

$$\text{mass flow (kg/s)} = \alpha \cdot A\ P_1\ \psi \sqrt{\frac{Z}{RT_1}} \quad \times 10^5$$

where α = nozzle coefficient
A = flow area in m^2

Flow at the throat of a nozzle, supplied by a reservoir or similar source under stagnation conditions, will be sonic if the critical pressure is greater than the receiver pressure – see Fig 5a. This means that the flow will be critical. The flow velocity follows from calculating the critical temperature, from which

$$\text{flow velocity } c = 49\sqrt{T^*} \text{ ft/sec}$$

(The velocity of sound in air = c = $49\sqrt{T}$ feet per second where T is the absolute temperature in degrees Rankine).

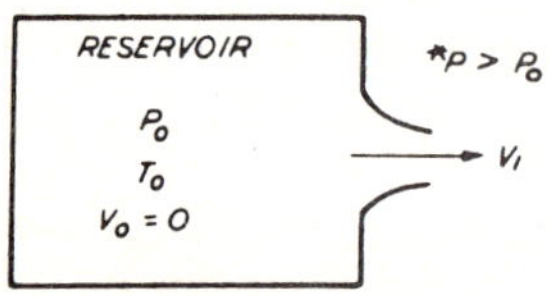

Fig 5a

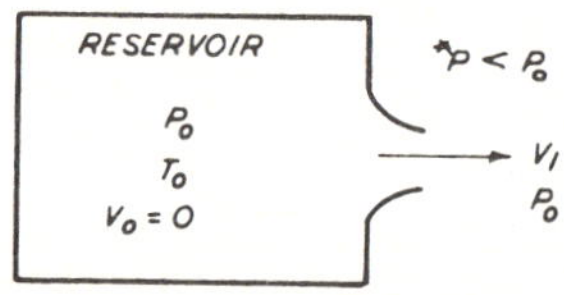

Fig 5b

If the critical pressure is less than the receiver pressure, then the flow cannot be critical – Fig 5b. In this case the flow will be subsonic and the exit pressure will equal the receiver pressure. The temperature can be calculated from the general formula, or from

$$T_1 = T_o \left(\frac{P_1}{P_o}\right)^{\frac{k-1}{k}}$$

The velocity is likewise calculated from the general formula.

Similar analysis applies where the nozzle is of convergent-divergent form – Fig 6. In this case it is first necessary to establish whether the flow is critical or not (at the throat). The throat velocity can be determined accordingly; and from this the final exit velocity from the divergent section.

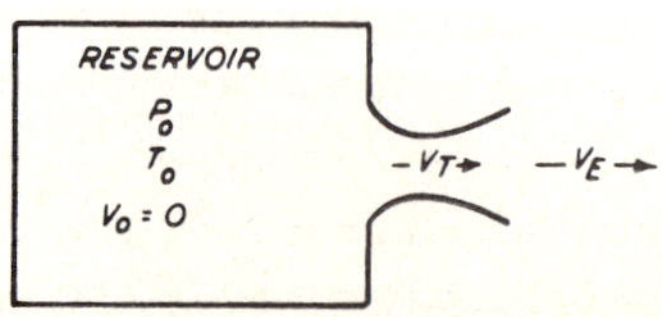

Fig 6

Flow through a nozzle can be rendered directly in terms of flow rate and a discharge coefficient, this being a convention for engineering calculations. The complete nozzle formula is:-

$$\text{mass flow} = A \cdot E\, c \cdot \delta \cdot d^2 \sqrt{h}\ \sqrt{P_2/T}$$

where A is a constant depending on the units employed

E = coefficient for the velocity of approach

$$= \frac{1}{\sqrt{1 - m^2}}$$

$$\text{where } m = \frac{\text{cross sectional area of nozzle}}{\text{cross sectional area upstream}} = \frac{A_2}{A_1}$$

c = nozzle coefficient

δ = expansibility factor allowing for the change in air density which occurs during acceleration through the nozzle

$$= \frac{1 - 0.07h}{13.6\,P_2} \quad \text{for values of m circa 0.16}$$

(and h in inches wg and P_2 in inches of mercury)

d = diameter of nozzle

h = pressure drop across nozzle

P_2 = absolute pressure on downstream side of nozzle

T = absolute temperature on downstream side of nozzle

If T is in °R, P_2 in inches of mercury, h is in inches wg and d is in inches, a value of A = 0.1148 gives the mass flow in units of lb/sec.

For a specific nozzle profile, the formula can be simplified by the use of a nozzle constant appropriate to that particular geometry and nozzle size. Rendered as a solution for conventional flow rate (Q)

$$Q = K\,(T_1/P_1)\sqrt{h}\ \sqrt{P_2/T_2}$$

where K = nozzle constant

T_1 = absolute temperature at specified inlet point

T_2 = absolute temperature at nozzle or specified point downstream

P_1 = absolute pressure at specified inlet point

P_2 = absolute pressure at nozzle or specified point downstream

h = pressure drop across nozzle

Simplified Orifice Formulas

An orifice is a simple form of nozzle, formed by a circular hole cut in a thin flat plate. Flow can again be determined with reference to an empirical discharge coefficient or orifice coefficient. This will be much lower than for nozzles because of the less streamlined flow but, due to the simpler form of the nozzle, will be less subject to variation. Thus nozzle coefficients may vary between 0.90 (or less) to 0.995, depending on size and geometry, whereas an orifice coefficient can be expected to be of the order of 0.61, regardless of size, and to differ only if the orifice has a well rounded, as opposed to a sharp, entry.

Very much simplified formulas can therefore be applied to assess the discharge of air through orifices, and the following are generally satisfactory for straightforward engineering calculations:-

(1) For upstream pressures above 1 bar

$$Q\ (\text{ft}^3/\text{min})\ \text{for sharp-edged orifice} = \frac{218 \times A \times P_u}{\sqrt{460 + T}} \qquad \text{(a)}$$

$$\text{or} \quad = \frac{172 \times d^2 \times P_u}{\sqrt{460 + T}}$$

$$Q\ (\text{ft}^3/\text{min})\ \text{for rounded entrance orifice} \mathrel{\hat{=}} \frac{417 \times A \times P_u}{\sqrt{460 + T}} \qquad \text{(b)}$$

$$\text{or} \quad \mathrel{\hat{=}} \frac{263\ 5 \times d^2 \times P_u}{\sqrt{460 + T}}$$

(2) For upstream pressures below 1 bar

$$Q\ (\text{ft}^3/\text{min})\ \text{for sharp-edged orifice} = \frac{210 \times A \times P_u}{\sqrt{460 + T}} \qquad \text{(c)}$$

$$\text{or} \quad = \frac{166 \times d^2 \times P_u}{\sqrt{460 + T}}$$

$$Q\ (\text{ft}^3/\text{min})\ \text{for rounded entrance orifice} \mathrel{\hat{=}} \frac{324 \times A \times P_u}{\sqrt{460 + T}} \qquad \text{(d)}$$

$$\text{or} \quad \mathrel{\hat{=}} \frac{255 \times d^2 \times P_u}{\sqrt{460 + T}}$$

where A = orifice area, in^2
P_u = upstream pressure, lb/in^2
d = orifice diameter, inches
T = upstream air temperature ^oF

Note that for an upstream air temperature of 15°C (60°F) these formulas further simplify to:-

1 (a) $Q\ (ft^3/min) = 11.9\ A \cdot P_u$
or $= 9.4\ d^2\ P_u$

(b) $Q\ (ft^3/min) \triangleq 18.3A \cdot P_u$
or $\triangleq 14.4\ d^2\ P_u$

2 (c) $Q\ (ft^3/min) = 11.5\ A \cdot P_u$
or $= 9.05\ d^2\ P_u$

(d) $Q\ (ft^3/min) \triangleq 17.7\ A \cdot P_u$
or $\triangleq 13.92\ d^2\ P_u$

Compressible Flow Through Pipes (Isothermal)

The real flow of pressurized air (or gas) through pipes differs appreciably in a number of important characteristics from the flow of liquids in pipes. Pressure, for example, drops at an increasing rate along the pipe, rather than with a constant pressure gradient. At the same time velocity tends to increase up to a maximum defined by $M = 1/\sqrt{k} = 0.845$ for air, but subject to a limiting or maximum length, which must correspond to the end of the pipe. At this point the pressure gradient is infinite, ie, the pipe is effectively closed.

The limiting or minimum pressure can be determined from the initial Mach number (M_1) and initial pressure (P_1), viz

$$P^* = P_1 M_1 \sqrt{k}$$

The limiting length is given by:-

$$L^* = \frac{D}{f}\left(\frac{1}{kM_1{}^2} - 1 - \log_e \frac{1}{kM_1{}^2}\right)$$

where D = pipe diameter
f = pipe friction factor

At any section 2 distance L from section 1 the following relationship applies:

$$f\frac{L}{D} = \frac{1}{kM_1^2}\left(1 - \frac{P_2}{P_1}\right)^2 - 2\log_e\frac{P_1}{P_2}$$

This equation normally requires to be solved by successive approximations.

The velocity at this section can be determined from the requirements of continuity:-

$$V_2 = \frac{V_1\rho_1}{\rho_2} = V_1\frac{P_1}{P_2}$$

These equations enable solutions to be worked out for any length of pipe, or at any station(s) along the length of pipe, within the limiting length L^*. If L^* is less than the actual pipe length, then initial conditions must be adjusted to bring the real length equal to, or less than, the limiting length.

Compressible Flow Through Pipes (Adiabatic)

Similar treatment applies, although the corresponding formulas are more complicated and may require working as a series of approximations in order to reach real values in particular cases. In general, the pressure, temperature and velocity will always be slightly less than those for isothermal flow, but the limiting length will be similar. The differences are usually small enough to be negligible, except at higher Mach numbers, and thus, for simplification of calculation, isothermal formulas can be used for subsonic adiabatic flow.

Limiting Velocity

This is given directly by:

$$\frac{V^*}{V_1} = \frac{1}{M_1}\sqrt{\frac{2\left(1 + \frac{k-1}{2}M_1^2\right)}{k+1}}$$

Limiting Temperature

$$\frac{T^*}{T_1} = \frac{2\left(1 + \frac{k-1}{2}M_1^2\right)}{k+1}$$

Limiting Length

$$L^* = \frac{D}{f}\left(\frac{1 - M_1^2}{kM_1^2}\right) + \frac{k+1}{2k}\log_e\left(\frac{(k+1)M_1^2}{2\left(1 + \frac{k-1}{2}M_1^2\right)}\right)$$

Limiting Pressure

$$\frac{P^*}{P_1} = M_1\sqrt{\frac{2\left(1 + \frac{k-1}{2}M_1^2\right)}{k+1}}$$

FLUID FLOW THROUGH PIPES AND TUBES

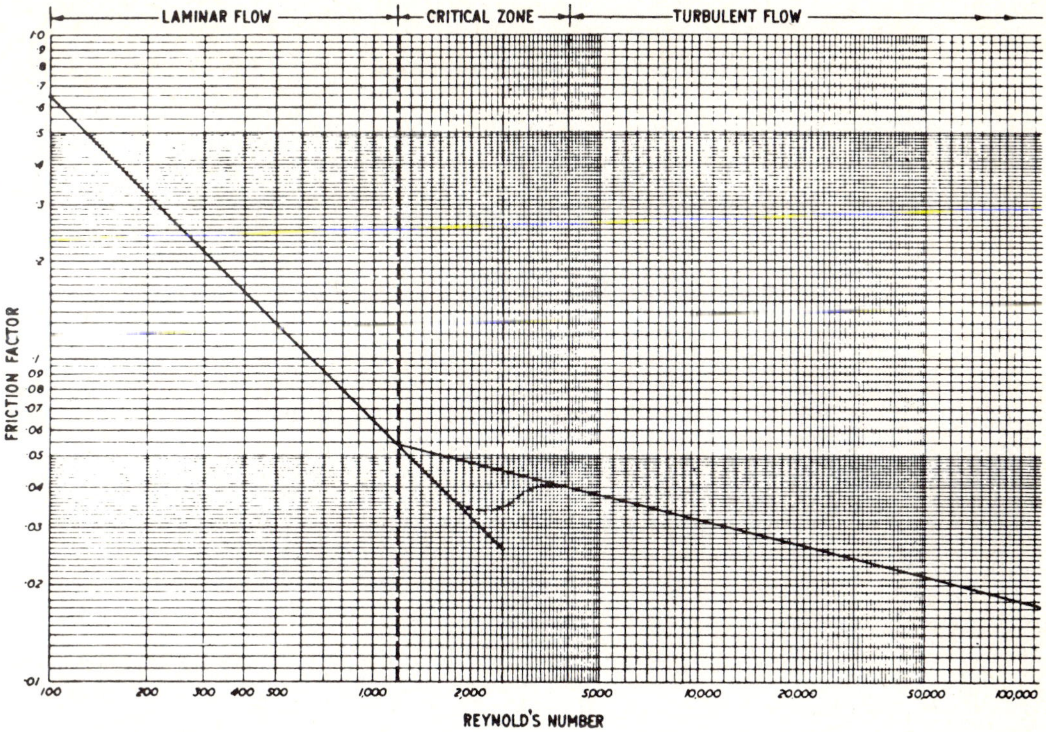

Fig 7

Friction Factor (f)

The friction factor is dependent on the Reynold's Number of flow and pipe roughness. It can be assumed independent of Mach number.

The Reynold's Number will be constant for isothermal flow, but may vary with adiabatic or isentropic flow (and certainly vary with diabatic flow), in which case a mean value can be assumed.

The Reynold's Number is a dimensionless quantity, given arithmetically by

$$RN = \frac{\rho VD}{\mu}$$

or

$$\frac{VD}{\nu}$$

where ρ is the mass density, μ = viscosity, ν = kinematic viscosity — in units consistent with velocity (V) and pipe diameter (D)

For air at standard temperature $\nu = 1.568 \times 10^{-4}\ ft^2/sec$

Thus $RN = 531.5\ VD$

$= 530\ VD$ (with sufficient practical accuracy)

where V is in ft/sec

and D is in inches

The friction factor (f) is common for all fluids (ie gases and liquids) and is normally determined from empirical charts — see Fig 7. For laminar flow, the friction factor is dependent only on Reynold's Number and is numerically equal to 64/RN, laminar flow being defined by the Reynold's Number not exceeding 2 000. For turbulent flow and smooth bore pipes, the Reynold's Number can be calculated from the following empirical formula —

$$f = \frac{0.3164}{RN^{0.25}}$$

Utilization of Compressed Air

THE GENERAL characteristics of compressed air systems offer specific advantages over the performance characteristics or requirements of alternative systems capable of combining the functions of speed, thrust and control and, perhaps to a lesser extent, motor duties. Circuits may range from simple control valve and cylinder combinations to complex interlocking groups, combining the function of logic with all sequencing, interlocking and outputs automatically controlled in a predetermined manner. Control functions can be performed by standard low cost components included in a suitable circuit design, or in separate control circuits operating at very low power levels. The latter is a characteristic of fluidic circuits where pneumatics are in direct competition to electronics as a logic control medium. Either method can offer low cost automation.

Compressed air generated by a compressor is normally stored in a receiver, the volume of which is selected to sustain a given demand within predetermined pressure levels. Various methods of compressor control may be adopted to ensure that the receiver remains charged within prescribed limits. With such a system no work is done when there is no motion initiating demand. Thus, unlike hydraulics, output forces can be held 'open' to the reservoir by stationary actuators without the expenditure of work. This is essentially a feature of the 'elastic' nature of the working fluid (compressed air).

The basic operating principle of a pneumatic system is simple. Air is allowed to expand in the actuator or pneumatic device, giving up its energy content and exhausting to the atmosphere. Thus, compressed air is used only once on a single-pass basis.

The utilization of air as a working fluid demands work input to compress it. For general services air is used at pressures of the order of 5.5 to 7 bar (80 to 100 lb/in^2), the power input required by typical compressors being of the order of 20 hp per 100 ft^3 of free air delivered per minute (6 hp per $metre^3$ FAD per minute), although this can vary somewhat depending on the efficiency of the compressor and ambient conditions (eg temperature and altitude).

The relatively low fluid pressures employed means that output forces obtainable from air cylinders are lower than those of oil-hydraulic cylinders of similar size. Thus larger cylinders have to be employed for comparable work output — eg a 225 mm diameter air cylinder would be needed to provide the same output force as a 50mm oil-hydraulic cylinder operating in a 140 bar system. Some comparative data are drawn in Table I.

TABLE I – OUTPUT FORCES FROM LINEAR ACTUATORS

Cylinder Bore		Pneumatic Cylinder* at 100 lb/in² (gauge) (7 bar)		Hydraulic Cylinder† at 1 000 lb/in² (70 bar)		at 2 000 lb/in² (140 bar)	
in	mm (nominal)	lb	kg	lb	kg	lb	kg
1	25	79	36	790	360	1580	720
2	50	314	142	3140	1420	6280	2840
3	75	707	320	7070	3200	14140	6400
4	100	1257	570	12570	5700	25140	11400
5	125	1964	890	19640	8900	39280	17800
6	150	2827	1280	28270	12800	56540	25600
7	175	3849	1745	38490	17450	76980	34900
8	200	5027	2280	50270	22800	100540	45600
9	225	6362	2885	Not normally used in larger sizes (except for heavy duty presses)			
10	250	7854	3560				
12	300	11310	5125				
14	350	15394	6985				
16	400	20106	9150				
18	450	25447	11550				
20	500	31416	14200				

(Pneumatic columns, from 8 in bore downward: Hydraulic cylinders generally favourable above this point)

*Maximum static thrust. Maximum working thrust up to 90% of this value
†Maximum static thrust. Maximum working thrust up to 98% of this value

Four Atlas Copco ET6E Compressors provide the compressed air used in GKN's Tremorfa Works Cardiff. The factory produces 600,000 tonnes of coil steel per annum and uses compressed air around the clock for its production processes.

Atlas Copco PNS 1200 Compressor in use on Britain's largest jetty. The higher pressure air is used to remove spoil from the drilling heads which are cutting into the sea bed at the Hunterston Ore Terminal Jetty.

The speed of response of a pneumatic system can be very high, the full theoretical value, neglecting friction, being that of the speed of sound. This can give both rapid control and rapid actuator movements if required. Control speeds are still very much lower than those provided by electronics, with a theoretical potential for working at the speed of light; but in fact, where electronic control circuits involve induction or magnetic operating principles,pneumatics can be many times faster, when arranged in a comparable circuit. From the practical point of view a wide speed range is available, ie pneumatic actuator outputs can range from low to high speeds, with a reasonable degree of control achieved by simple means. Hydraulics cannot compete with pneumatics in terms of speed response.

The weight of pneumatic components is generally low, since thin-wall construction can be used as the internal pressures handled are not very high. The same applies to lines. Nylon tubing, for example, can readily replace metallic lines where weight saving is important.

Lack of lubricity offered by the working fluid (compressed air) is an inherent disadvantage compared with oil-hydraulics, where the working fluid is also a lubricant. However, compressor lubrication is not so critical as hydraulic pump lubrication and can easily be applied at the machine itself. Effective lubrication for pneumatic actuators and devices can readily be provided by injecting lubricant into the compressed air supply after compression (preferably at a point in the line

Special rig for North Sea operations BroomWade V-Major oil free compressor receivers, controls etc. in a special rig for North Sea oil operations.

Five Atlas Copco ZR3A rotary screw oil-free compressors are being used by John Player & Sons, the cigarette manufacturer in Nottingham.

shortly in front of the device), or from lubricators attached to the device. The only disadvantages of this are that lubricant is continually lost from the system when it is working (being exhausted to the atmosphere with the air), and exhaust air is laden with lubricant rather than being perfectly 'clean'. In practice, neither is usually a real disadvantage.

Pneumatic systems can be used at extremes of temperature which preclude the use of oil-cylinders (except with special fluids).

As far as low temperature working is concerned, however, pneumatics can have distinct disadvantages. Unless the air is specially conditioned by dryers capable of removing the majority of the water vapour present, internal icing up and blockages can be a problem in pneumatic systems exposed to ambient temperatures falling below freezing point.

Pneumatics lends itself probably better than any other available medium to the ready mechanization of special machines and plant. If a certain operation or series of operations has to be done regularly and by one or two people, then it is a serious candidate for automation by pneumatics. Broadly speaking, hydraulics is used to give man unlimited strength whilst pneumatics is probably at its best when it takes over direct from the human operator. It can then, depending on the application, reduce fatigue, speed up or provide full or semi-automatic working.

See also chapter on *Low Cost Automation.*

Economics

AIR is a zero-cost gas, thus the total energy cost is represented by the energy spent in compressing it (and, in some systems, by the additional energy cost in treating it).

Specific energy requirements can be expressed directly in terms of kilowatt hours or HP hours per unit volume of compressed air delivered, which can be determined directly as theoretical values from the compression ratio involved (see Fig 1). The energy unit of kwh/m^3 or kwh/ft^3 is generally applicable to stationary compressors (and other types driven by electric motors). In the case of compressors driven by internal combustion engines, the specific energy can be expressed in units of litres or gallons of fuel per hour/unit volume, which now seems to be preferred to HP input. The true cost of energy must, however, take into account the efficiency of both compressor and driver.

Equally, the time cost of supplying compressed air must take into account a number of other factors, viz:

(i) Power costs
(ii) Labour
(iii) Maintenance

} Operating costs

(iv) Depreciation
(v) Interest on capital investment.

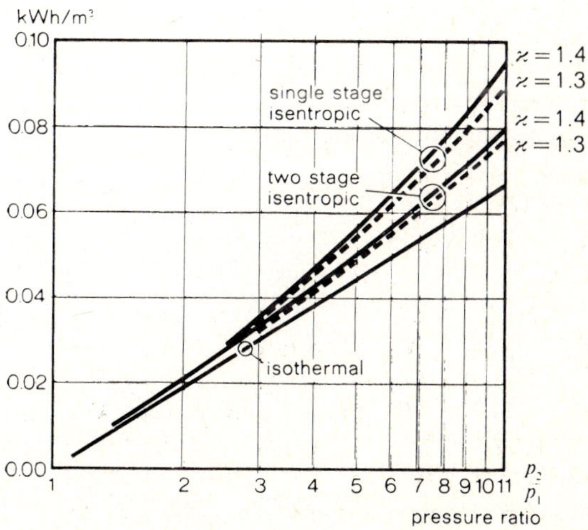

Fig 1 Specific energy requirements for compression of air from atmospheric pressure. (Atlas Copco).

Of these, power costs are normally by far the largest item, embracing the importance of reducing power consumption wherever possible – eg by optimized unloading of the compressor; and also suitable choice of compressor size. The latter can be a significant parameter in the case of an installation where demand is likely to increase, since efficiency tends to increase, and thus specific power requirements decrease, with increasing compressor size.

As an example, consider an installation using three 7 m^3/min (250ft^3/m) compressors where the demand has increased beyond the supply capacity available. This is met by the addition of a joint 7 m^3/min (250 ft^3/m) compressor. Table 1 compares power requirements at this level with those for a single 28 m^3/min (1000 ft^3/m). The latter would show a substantial saving in power costs. The picture is not complete, however, without taking the other cost factors into consideration, e.g. (ii) and (iii) would also be favourable in the case of the 28 m^3/min (1000 ft^3/m) compressors, (iv) and (v) would be unfavourable. There is also the question of amortization of the three existing compressors, although these could probably be used advantageously for back-up duties or to meet further expansion; or as an alternative to the large compressor at periods when demand is light.

TABLE I: POWER COSTS COMPARED FOR 28m^3/min (1000ft^3/min) FAD DELIVERED

Compressor Capacity	Compressor Input Power (average)	Total Power Requirement	Energy Requirement	40 hr week Power Requirements	(48 weeks) cost*
4 x 250 ft^3/m	60 hp each	240	180 kW	7200 kW/hr	£144
1 x 1000 ft^3/m	200 hp	200	150 kW	6000 kW/hr	£120 †

*Electric meter driver at 2p per unit

†Annual saving (based on 48 weeks) = £1152

Compressor Power Input Required

The input power required to generate compressed air at the normal air line pressure of 7 bar (100 lb/in^2) is of the order of 20 hp per 100 ft^3/min. This will vary somewhat with the type of compressor and its overall efficiency, and also with the actual pressure generated – see Table II.

The true cost of generating this input horsepower depends on the efficiency of the driver. Electric motors are usually the most economical type of driver, since they are capable of achieving efficiencies in excess of 90%, offsetting what is apparently a high 'fuel' cost.

A further allowance may have to be made for drive losses, where the driver cannot be directly coupled to the compressor. Thus i/c engines are relatively low speed machines and are suitable for direct coupling to lower speed compressors (eg reciprocating compressors). They would need to be geared up to drive high speed rotary compressors, or such compressors could suffer a loss of efficiency with direct drive. In the case of electric motors, driver speeds are again limited to a maximum of 3000 rev/min with synchronous motors on 50 cycle supply.

An important point is that where demand is high and fairly continuous, efficiency figures can be very significant. A relatively small difference in compressor, driver or drive efficiency can represent a substantial difference in compressed air costs over a period of a year or more.

TABLE II: APPROXIMATE HP REQUIRED TO GENERATE COMPRESSED AIR

FAD ft^3/m	PRESSURE lb/in^2 and bar 60 (4.2)	80 (5.6)	100 (7)	125 (8.75)	FAD m^3/min
10	1.5(1.6)	1.7(2.0)	1.9(2.2)	2.1	0.28
20	2.9(3.3)	3.4(3.9)	3.8(4.4)	4.3	0.57
30	4.4(4.9)	5.1(5.9)	5.7(6.6)	6.4	0.85
40	5.9(6.5)	6.8(7.8)	7.6(8.8)	8.5	1.13
50	7.9(8.2)	8.6(9.5)	9.6(11.1)	10.7	1.42
60	8.8(9.8)	10.3(11.7)	11.5(13.3)	12.8	1.70
70	10.3 (11.4)	12.0(13.7)	13.4(15.5)	14.9	1.98
80	11.8(13.0)	13.7(15.6)	15.3(17.7)	17.0	2.27
90	13.2(14.7)	15.4(17.6)	17.2(19.9)	19.2	2.55
100	14.7(16.3)	17.1(19.5)	19.1(22.1)	21.3	2.83
200	29.4(32.6)	34.2(39.0)	38.2(44.2)	42.6	5.66
300	44.1(48.9)	51.3(58.5)	57.3(66.3)	63.9	8.50
400	58.8	68.4	76.4	85.2	11.33
500	73.5	85.5	95.5	106.5	14.16
600	88.2	102.6	114.6	127.8	16.99
700	102.9	119.7	133.7	149.1	19.82
800	117.6	136.8	153.8	170.4	22.65
900	132.3	153.9	171.9	191.7	25.49
1000	147	171	191	213	28.32

Compressor Regulation

Either constant or variable speed control may be used to advantage, where demand is fairly continuous, for this will usually produce a minimum overall cost for the compressed air consumed. Variable speed control is rather better than constant speed control in this respect, although demanding a more complex control arrangement.

Stop and start control is simpler and generally more economical on large systems subject to intermittent demand. Here the compressor charges the receiver in the normal way, switching off when a predetermined pressure is reached and remaining idle until the receiver pressure has fallen to a predetermined low level. The compressor is then automatically restarted and continues running until maximum charge has been reached once more.

The economies of stop and start control depend largely on the number of stops and starts per hour, which is related to the receiver size and demand. The greater the number of stops and starts per hour, the higher the wear and tear on the compressor and driver, and the control gear. For any receiver size, the maximum number of stops and starts per hour will be realized when the air consumption figure is one half of the FAD (Free Air Delivered) of the compressor. To hold the number of stops and starts to a reasonable level, a generous receiver size is usually adopted, eg typically about twice that required with constant speed regulation. See also chapters on *Compressor Controls* and *Receivers and Pressure Vessels.*

Air Mains

Distribution of the compressed air is by pipe, or air main, from a receiver following the compressor which may be either a single line terminating in a dead end, or a ring main. In both cases the main is normally pitched in the direction of air flow so that condensed water will be carried along away from the receiver end, to collect in traps or drains installed at suitable points. This is an essential feature of economical operation as maximum efficiency and maximum appliance life is achieved only by the use of clean and preferably dry compressed air.

One of the most important control parameters in the overall economics of a compressed air system is optimum line sizing. Delivery of compressed air through the air mains, branch lines, and to industrial tools and appliances, immediately results in a pressure drop, and thus energy loss. These losses are minimized by optimum pipe sizing based either on a specific design pressure drop, or an 'acceptable' pressure drop (usually of the order of 5 to 10%). It is important to remember in this connection that pressure drop (and thus line sizing parameters) are dependent on air velocity and thus also governed by demand.

The subject of line sizing is detailed in the chapters on *Compressed Air Mains* and *Air Lines.*

Air Leaks

Compressed air systems are, in general, very reliable and instantaneous; failures are rare. Because of the relatively low operating pressure, developing faults tend to produce a proportionately high pressure drop, giving an early indication of progressive failure by sluggish operation, loss of force, etc, although the system itself will usually keep on working. This can be a 'disadvantage', as it is surprisingly common for compressed air systems to be continuously worked, even when obviously in need of major attention. Minor faults or leakages continue to be ignored and work is continued almost indefinitely with diminishing overall efficiency – routine maintenance being ignored and preventative maintenance not even considered.

Air leaks are not hazardous, but they can be extremely wasteful of power. In a typical system as much as 10% of the power requirements at any period can be made up of air leaking to waste. Table III gives some idea of the leakage possible through various size holes, with a rough indication of costs.

TABLE III : AIR LOSS AND POWER WASTAGE BY AIR LEAKS*

Line Pressure		Leakage in ft^3/min				Approximate Power Wastage in kW/hr			
hour	lb/in^2	1/16in	1/8in	¼in	½in				
		3.5	1.45	59	235	0.65	2.6	7.8	25.4 *
7.0	100	6.5	26	104	415	1.25	5.2	18.5 *	59 *
10.0	150	9.5	37	150	600	1.6	6.4	26 *	104 *

* = *two stage compression*

*B. C. Regan (Spirax Sarco Ltd)

It is a simple exercise to determine the magnitude of the leaks in a compressed air system at some time during the day, perhaps during the lunch break when the plant has no programmed air requirements, by checking what percentage of time the compressor is on-load. This on-load time and power is being used exclusively to maintain leaks. A little extra thought and money on ensuring a good initial installation will result in a leak-free, less expensive compressed air plant. Apart from savings on running costs, a leak-free system will defer the need to purchase additional expensive and probably unnecessary compressor plant.

Actuator Efficiencies

Conversion efficiency achieved by an actuator powered by compressed air, ranges from about 90% down to 65% or lower in the case of linear actuators, depending primarily on the speed of operation. Conversion efficiency is also affected by the compression ratio, or actual air pressure at the actuator, normally reaching a maximum at pressures of 7—10 bar (100—150 lb/in^2) with conventional designs of actuators. Moderate pressures of this order reduce pumping costs and also the stresses on lines and components subject to internal pressure.

Actuators

In general a pneumatic system is the least expensive of any type where linear motions are required, provided the output force required is within the practical range of air cylinders. Compressed air is also unique in its suitability for powering certain types of tools such as percussion tools for road-breaking and rock drilling, etc, and is directly competitive with electric motors for powering small workshop tools.

Linear actuators powered by compressed air are inherently fast acting, but unrestricted high speeds can bring about high pressure drop at the inlet port, with consequent loss of effective pressure applied to the piston. Rarely, however, is a linear actuator called upon to work at its fastest speed, and efficiency can be improved by restricting the speed attained via simple throttling controls. Thus an 80% conversion efficiency is readily achieved in the majority of applications, with static thrusts or break-out forces equal to the full theoretical performance. Whilst this working efficiency is lower than that obtained with hydraulic systems, it is still high, and the overall economics of compressed air as a working or 'power' fluid is often better than that of hydraulics, or indeed any alternative system. Further advantages are often apparent with the convenience and simplicity of installation offered by compressed air systems.

In the case of positive displacement actuators, air demand can be calculated from first principles. Thus, in the case of air cylinders, air consumption is derived directly from the cylinder size and piston speed, viz:-

$$Q\ (FAD) = kD^2Vr$$

where D is the piston diameter
V is the piston speed
r is the compression ratio of the compressed air
k is a constant determined by the units employed

This formula is often more usefully written in terms of time of operation (t) in seconds, based on mean piston speed over the stroke, (S) viz

$$Q\ (FAD) = \frac{k^1 D^2 S \times P_a}{t}$$

Alternatively, air consumption may be calculated directly from the geometry of the cylinder and the number of strokes per minute –

$$Q = 0.7854\ D^2 \text{ x stroke x strokes per minute}$$

In the case of a double-acting cylinder, the air consumption on the return stroke is reduced by virtue of the entering rod volume, when the corresponding value of Q is obtained by using the quantity $(D^2 - d^2)$ instead of D^2 in the formula, where d is the diameter of the rod. The difference may, however, be ignored, and the total air consumption per cycle taken as twice that for the outgoing stroke. With a through-rod cylinder, consumption is calculated on the basis of the true displacement.

The displacement demand calculated directly from cylinder geometry represents the volume of compressed air consumed per stroke or cycle. The equivalent consumption in terms of Free Air Delivered (FAD) is given by multiplying the displacement by the compression ratio (r), ie

$$r = \frac{P_2 + P_1}{P_1}$$

Where P_2 = compressed air gauge pressure
P_1 = atmospheric pressure

Compressor capacity is normally given in m^3/minute or ft^3/m FAD (free air delivered). If compressor capacity is rated in terms of free air *displaced,* this displacement figure must be multiplied by the compressor efficiency to obtain the true FAD figure.

Air Tools

In the case of pneumatic tools and other appliances, air consumption figures are normally based on empirical data – see *Data Section.* These are normally maximum demand figures, consistent with continuous operation. Actual demand will depend on the *time factor* and *work factor* involved.

The time factor is concerned directly with the time the appliance is in use. Thus, tools and appliances are normally used intermittently rather than continuously, when a work factor can be based on observation over a suitable period of use. In the case of linear actuators or devices allied to a work cycle, the time factor can be established more accurately by time and motion studies, or calculated directly from the cycle time, if appropriate.

The work factor is related to the actual period of time during use when the appliance is consuming air. This can vary from 1.0 where the appliance is continuously driven by air when being used, to appreciably less where the demand is other than constant. For example, actual demand may vary according to the work involved, particularly if the load changes, or if the speed of operation is varied by a throttle control.

The product of the time factor and work factor represents the true load factor, or true demand from the compressor, from a single tool or device. The total demand is thus the sum of all the individual air consumption figures multiplied by their respective load factors. This will normally be considerably less than the sum of all the maximum air consumption figures for all the appliances connected to a system. Load factors can be expressed as a percentage or decimal fraction – see Table IV. The true demand is then represented by the sum of the individual maximum demands of all the appliances connected to the system, when each is multiplied by its respective load factor.

TABLE IV : LOAD FACTORS

Work Factor	TIME FACTOR 0.1	0.2	0.3	0.4	0.5	0.6	0.7	0.8	0.9	1.0
0.1	0.01	0.02	0.03	0.04	0.05	0.06	0.07	0.08	0.09	0.10
0.2	0.02	0.04	0.06	0.08	0.10	0.12	0.14	0.16	0.18	0.20
0.3	0.03	0.06	0.09	0.12	0.15	0.18	0.21	0.24	0.27	0.30
0.4	0.04	0.08	0.12	0.16	0.20	0.24	0.28	0.32	0.36	0.40
0.5	0.05	0.10	0.15	0.20	0.25	0.30	0.35	0.40	0.45	0.50
0.6	0.06	0.12	0.18	0.24	0.30	0.36	0.42	0.48	0.54	0.60
0.7	0.07	0.14	0.21	0.28	0.35	0.42	0.49	0.56	0.63	0.70
0.8	0.08	0.16	0.24	0.32	0.40	0.48	0.56	0.64	0.72	0.80
0.9	0.09	0.18	0.27	0.36	0.45	0.54	0.63	0.72	0.81	0.90
1.0	0.10	0.20	0.30	0.40	0.50	0.60	0.70	0.80	0.90	1.00

The reduced air consumption in each case is derived directly by multiplying the maximum air consumption figure (column 2) by the load factor. The final total arrived at is 35.4% of the demand, as given by summing all the individual maximum air consumption figures. Provided the time and work factors have been correctly established, this lower figure would apparently meet all the requirements.

In practice this is not necessarily so, since the study does not analyse the instantaneous demand at any particular time. It merely assumes that the time factors are employed simultaneously, whereas they are not. For any time factor less than unity, there will be periods when the actual demand is equivalent to a time factor of 1.0, ie when the appliance is consuming air at maximum rate. If two or more of these periods coincide, as they are very likely to, then the true total momentary air consumption figure will be higher than the reduced figure, although the *total* demand over a period will equal that of the reduced figure.

In general, it is usually satisfactory to estimate on the basis of total reduced air consumption and allow a generous size of receiver to take care of momentary peak demands. In specific cases it may be necessary to analyse the momentary peak demands in more detail and estimate mains sizes (the latter particularly) rather than design to marginal values.

Actual demand will also vary with pressure and ambient air temperature (and compressor delivery will also vary with air intake temperature), but these factors can generally be ignored if the compressor capacity and receiver volume required is generously estimated in the first case. Some readjustment of capacity requirements (and loss of compressor capacity) may have to be made for working at altitude, however.

Corrosion

Corrosion is an economic factor in that it affects depreciation and possibly working efficiency of actuators, etc.

The main corrosion problem is that compressed air is normally saturated with water vapour, which is inevitably deposited in the system when the delivered air cools. This favours the use of non-ferrous metals for lines and components, or corrosion protection treatment to ferrous metal surfaces exposed to the 'wet' air. Corrosion protection is also given by lubricants resistant to water-washing, which will remain adhered to internal surfaces. The maintenance of the cylinder bore and rod surface in good condition is important in linear actuators, in order to give good seal life.

Safety

COMPRESSED air is inherently a 'safe' form of energy, free from the fire hazard associated with fossil fuels and oils and the potential hazard of electrical supplies or high pressure steam. Pneumatic systems can be used at temperatures, or under conditions, which would preclude the use of hydraulics or electrical equipment — eg at high temperatures which would be beyond the working range of conventional hydraulic fluids, or in hazardous conditions which would call for the use of special fluids or expensive isolation and protection of electrical gear. There is no fire hazard with pneumatics, except in very high pressure systems. Here 'dieselling' can be a problem.

The same virtue applies generally to pneumatically operated devices and machines. They are relatively free of hazard and do not normally call for special provision of 'emergency stop', unless specifically required by relevant safety standards or the mode of working. This does not mean that pneumatic devices are automatically safe, for hazards are inevitably present with any working machine, particularly should it malfunction. It is essential, therefore, that a machine operator should be aware of the abnormal motions a machine may perform under certain circumstances and how to deal with them. This can apply particularly in the case of air cylinders.

The main hazard associated with an air cylinder is that its speed of action is governed by the exhaust opening and/or external load. Abnormal functioning is usually caused by a change in one of these 'control' parameters — usually the load. This can render the cylinder potentially dangerous as long as it is under supply pressure.

If, for example, the load is abnormal, the normal cylinder movement will be slowed, or even stopped. With the supply still connected pressure will build up on the piston, so that if the load is suddenly removed the piston will shoot forward at uncontrolled speed. This could cause injury to anyone in its path — the 'classic' subject in this case being an operator removing the blocking load with hands or feet. The correct action in such a case is both to shut off the supply *and* depressurize the cylinder before attempting to remove the blocking load.

Emergency stop controls are commonly provided on all complex machines and on larger cylinders and devices rated for medium and heavy duties. On light duty machines, 'emergency stop' is usually via the manual control valve. Where an emergency stop is provided this should be in a readily accessible position, separate or clearly distinguished from the other controls. It is also desirable that this control should have a locking or 'hold' action. Thus the usual form of 'emergency control' is a pressbutton. Once pressed 'stop' should remain effective even when the button is released. Restarting can then be initiated by a separate control after the system or service has been cleared as safe, at the same time resetting the emergency stop button.

The mode of 'emergency stop' must also be considered in some detail, based on the characteristics and requirements of the device concerned. It may be desirable to stabilize the stop position, eg by locking the air in the cylinder on either side of the piston, or even more rigidly by means of a hydraulic check unit. In such cases the 'emergency stop' circuit would be designed accordingly. In other cases it may be necessary to avoid an abrupt stop as this could cause damage to the machine or introduce a further potential hazard. Here the desirable stop movement would be controlled, delayed or positioned (eg movement inched to a 'safe' position).

Reset characteristics can be equally important. The main requirement with air cylinders is that once reset has been initiated the possibility of excessive piston velocities must be avoided. One way in which this can be done is by using a three-way valve connected as a shuttle valve to a throttle valve in the supply line. The initial air supply on reset is thus throttled, enabling pressure to build up slowly in the cylinder. Once the required working pressure has been reached the three-way valve is automatically switched over to provide full flow – see Fig 1. An example of an emergency stop system with automatic resetting of main valves with slow replenishment of air is shown in Fig 2.

The simplest method of achieving an emergency stop is by using a three-position valve as the control, shutting off the supply and opening the system to the atmosphere. At the same time the associated electrical circuits can be tripped by the valve movement or broken by a pressure switch.

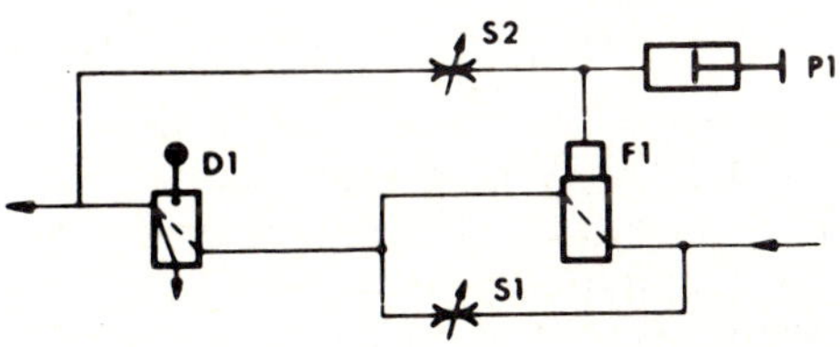

Fig 1

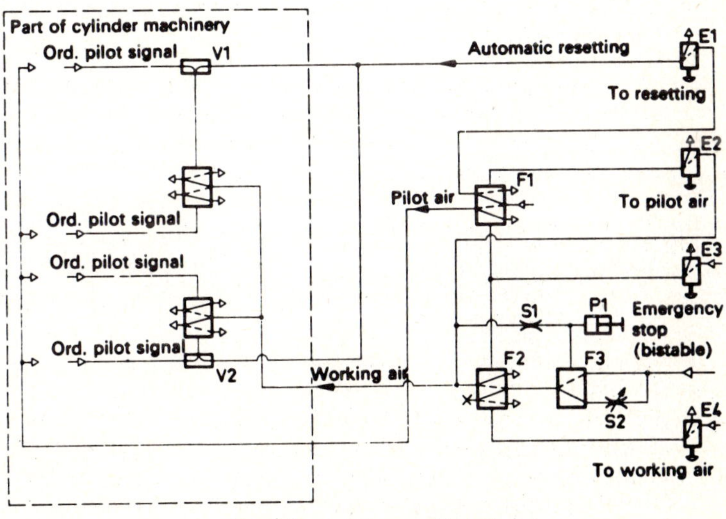

Fig 2 (from Mecman-Technique)

This basic arrangement can then be modified as necessary to incorporate other action, eg throttling, and the valve selected accordingly. Thus a four-position valve would normally be required to provide locking action or any *necessary* control. For example, it could be hazardous to evacuate a pneumatic clamp on emergency shutdown as this could cause the workpiece to be ejected violently. In other cases additional motions may be necessary to render the system safe on shutdown. Fig 3 shows three further safety circuits for cylinders.

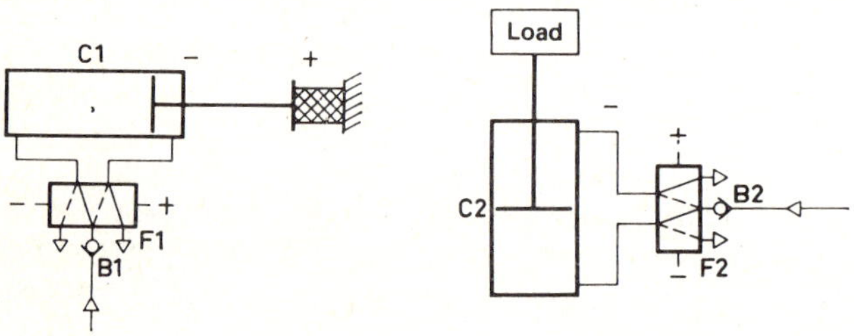

Fig 3 C1, C2 – cylinders.
B1,B2 – non-return valves.

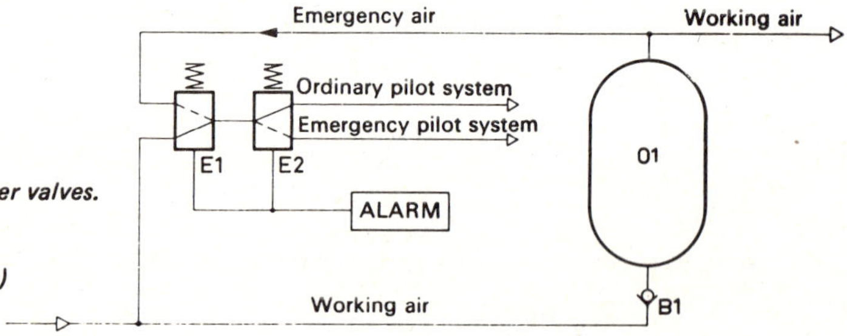

Fig 4 F1, F2 – changeover valves.
B1 – non-return valve.
01 – reservoir.
(from Mecman Technique)

Systems in which loss of mains pressure can lead to hazard (eg incorporating machines with clamping components) can readily be protected by automatic emergency stop sensors, such as pressure switches. Fig 4 is a simple example using a non-return valve as a sensor to protect the clamping cylinder from loss of pressure. Once clamping pressure has been built up by the supply, any loss of pressure in the supply line will have no effect since there can be no backflow (and thus loss of pressure) through the non-return valve.

Fig 4 shows another type of circuit providing automatic stop in the event of the mains pressure falling below a certain level. At a given lower pressure, valves E1 and E2 are switched over. The reservoir now supplies working air to the service line(s), to move them to a 'safety' position controlled by the emergency pilot signal.

Safety, in other words, can be built into most pneumatic systems, at very little additional expense. The more complex the system or the heavier the mechanical output forces involved, the more desirable the need for 'safety switching', either manually or automatically. The primary aims are to remove any hazard from the system; and at the same time to maintain all necessary control over the movements or cycle involved. The first provides direct operator protection, and the latter can both provide further protection for the operator and safeguard the machine elements involved. Built-in safety can thus also provide economic benefits.

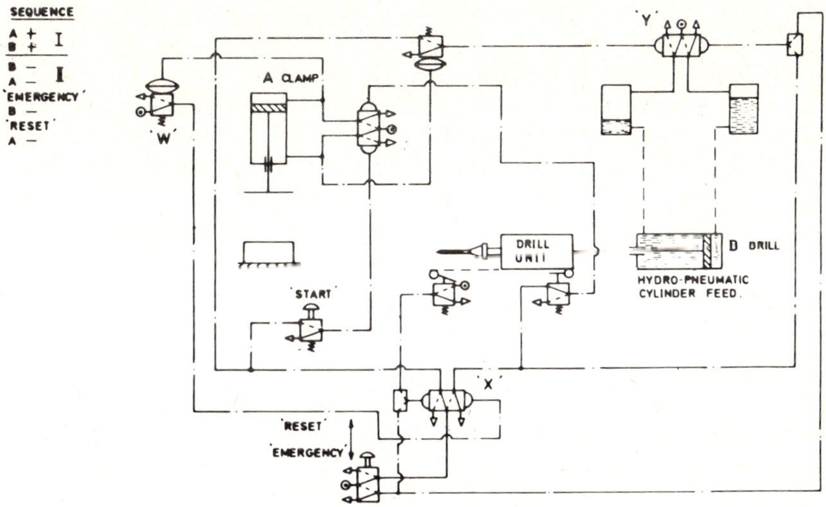

Fig 5 Example of 'safety' circuit with push-button emergency control for stopping and reversing an electro-hydraulic duct.

Examples of Fail-Safe Cylinder Circuits

The actual control may vary from simple open/closed condition to that of modulating control by a servo cylinder. Obviously, it must be decided at what level the pressure is classed as failed, and once the pressure has dropped to below the minimum which is essential to carry out the desired function, failure must be registered. Therefore it is very important to have an adjustable pressure switch, which can be set to switch at the desired failure pressure. Fig 6 shows a typical circuit.

Normally, in the healthy state the cylinder is controlled by the 5-port valve, via the two double pressure operated 3-port valves. In the event of the primary air failure, the pressure switch will sense and switch, reversing the two 3-port valves. The front end of the cylinder will now be vented

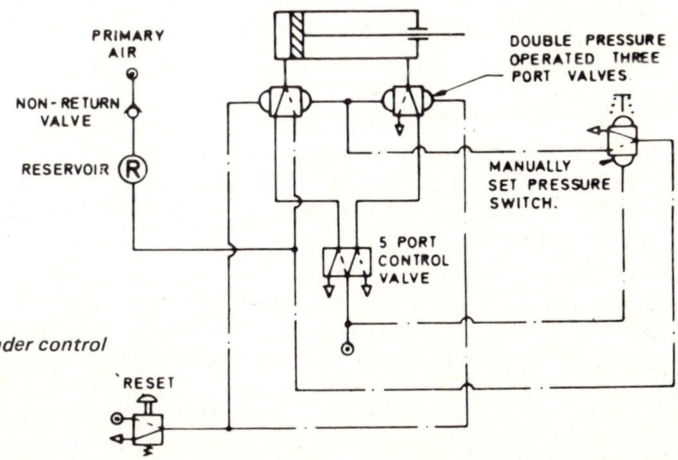

Fig 6 Double-acting cylinder control

and the reservoir, which has been charged during healthy conditions, charges the rear end of the cylinder, causing the cylinder to outstroke. The non-return valve is necessary to ensure that the reservoir pressure is not fed back into the air main in the event of primary air failure.

Obviously, the cylinder circuit can be designed to fail to the instroke position, if required, by merely changing over the valve connections. When the primary air supply is restored the reset valve is operated to set the circuit to normal function. Fig 7 shows a similar arrangement but in this circuit a servo or positional cylinder is being used.

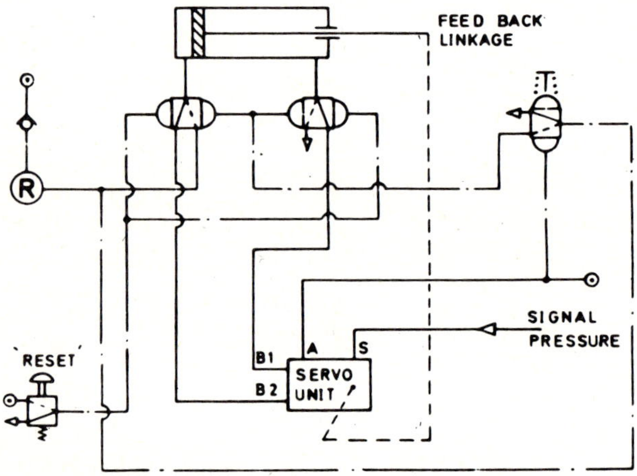

Fig 7 Servo cylinder control

In some applications a reset valve may not be required and it would be more advantageous for the system to reset automatically when the primary air supply is restored. This arrangement is obtained by using differential operated 3-port valves as shown in Fig 8 and of course this also can be applied to the servo system.

The reservoir size is very important as it is necessary to ensure that when the cylinder has made its fail safe movement by utilizing the reservoir air, the pressure in the reservoir is still above the safe working pressure for the system. It can be seen that if the size of reservoir is calculated incorrectly the pressure could drop during a fail safe movement to below the safe working pressure

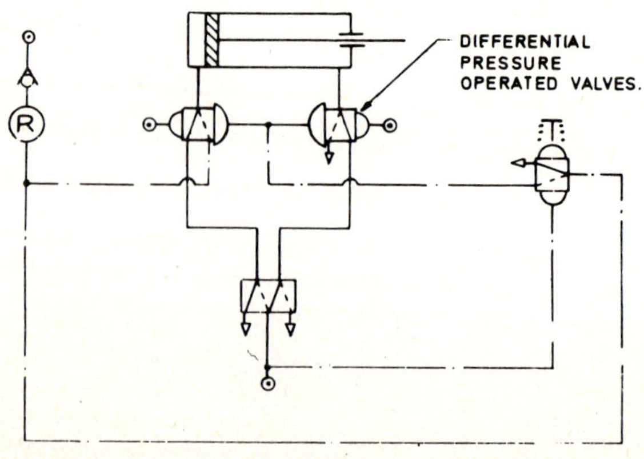

Fig 8

of the system, and therefore the cylinder could stall in an unsafe position. It would therefore be useful to investigate the calculations necessary to arrive at the correct size of reservoir.

It should be noted that the double pressure operated valves and the reservoir should be mounted as near to the cylinder as possible, so avoiding losses in the pipe line and reducing the distance/time lag in the emergency system. The pilot line from the pressure switch should also be of minimum length and 3/16 in o.d. tube size is recommended. Obviously, the location of the pressure switch should be such that it cannot be tampered with.

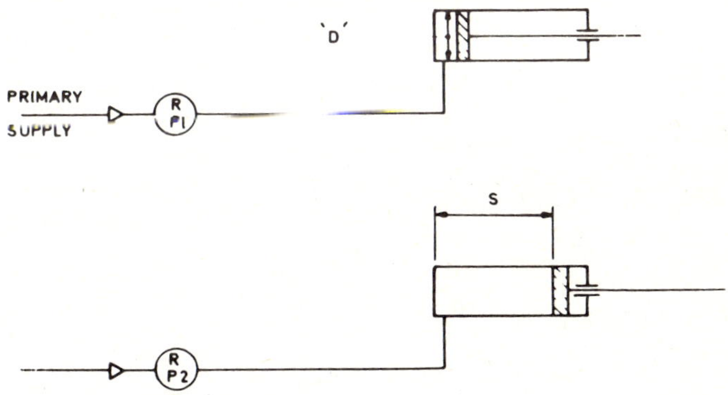

Fig 9

Cylinders Failing to the Hold Position

On occasions it is required to hold the cylinder in the position occupied at the time of air failure. This requirement generally applies to servo cylinders which normally operate between the stroke extremities.

The solution of this problem is achieved without the use of a reservoir, as shown in Figs 12 and 13. Normally the two poppet valves are held open by a signal from the pressure switch. In the event of air failure the signal from the pressure switch is removed and consequently the two poppet valves close under spring pressure. The air in the cylinder is now trapped, resulting in the cylinder being pneumatically locked. This locked or held position is only relative to the terms of reference already outlined.

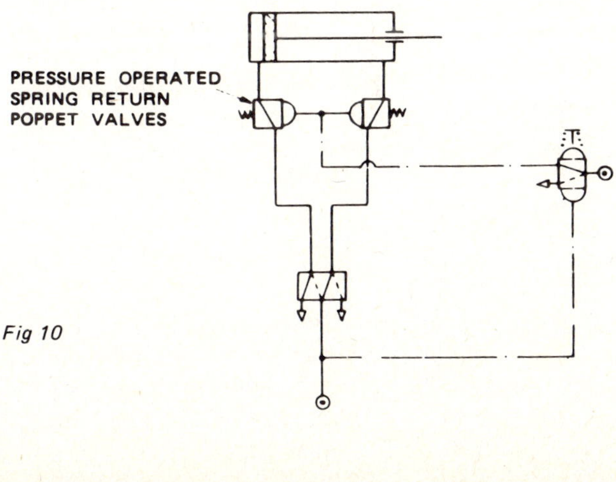

Fig 10

If it is found necessary to override this condition manually prior to resumption of primary supply, a stop valve can be added to the circuit as shown in Fig 11. This is normally closed, but by opening the stop valve during the lock condition the two ends of the cylinder are connected and therefore the pressures are balanced out and it is possible to move the cylinder manually.

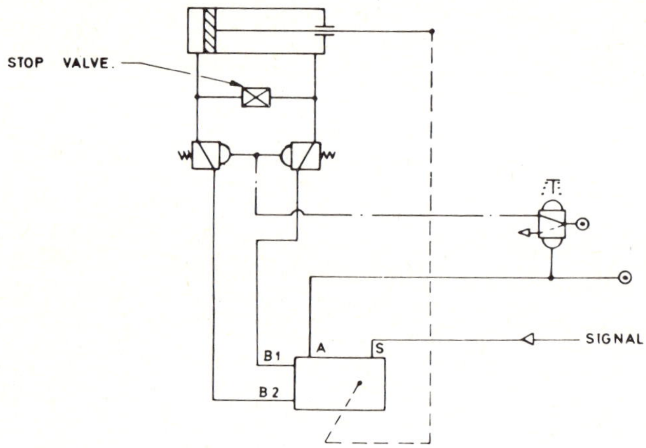

Fig 11

Checking Safety Arrangements

For safety reasons it is very important that the fail safe arrangements are checked at regular intervals to ensure that they function satisfactorily. This can be provided for by the inclusion of an override valve in the line between the primary supply connection and the pressure switch (Fig 12).

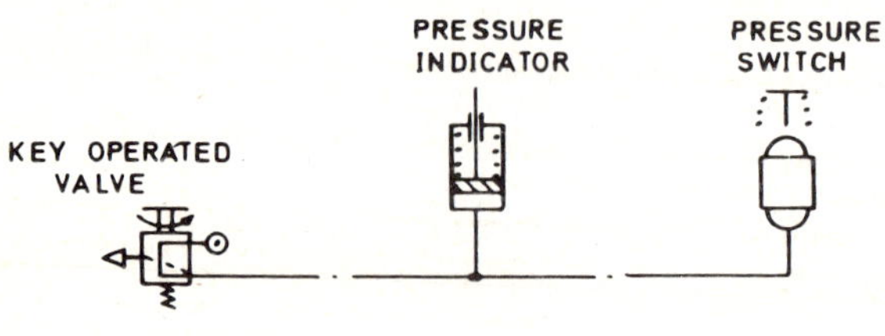

Fig 12

Inspection can now be carried out by suitable personnel, who, on turning the key in the valve, would simulate the loss of primary supply pressure by putting the pilot line to atmosphere. An indicator could be fitted to supply a visual means of ensuring that after the test the primary supply has been restored.

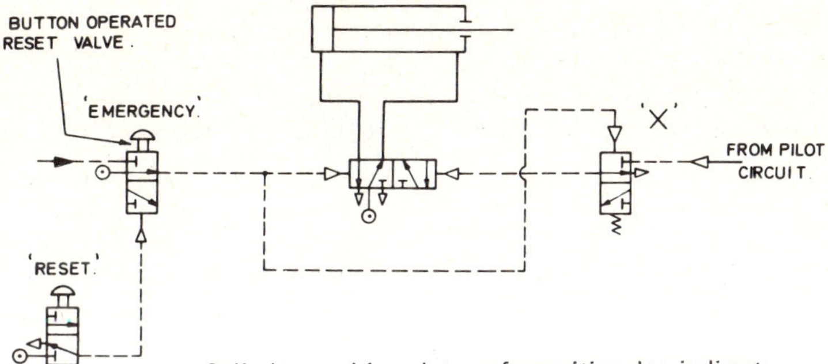

Cylinder positioned to safe position by indirect manual operation (using double pressure operated valve).

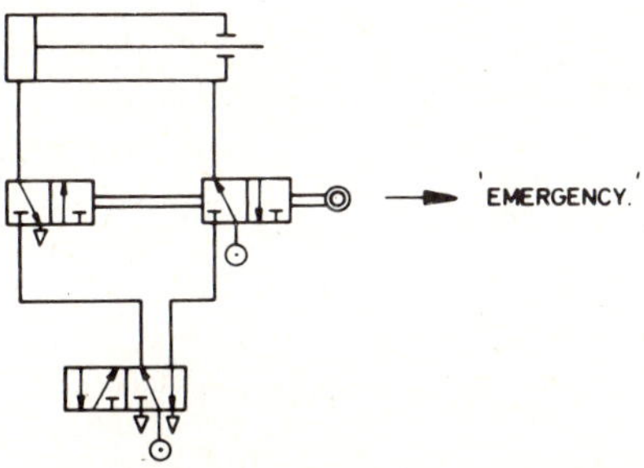

Cylinder positioned to safe position by direct manual operation.

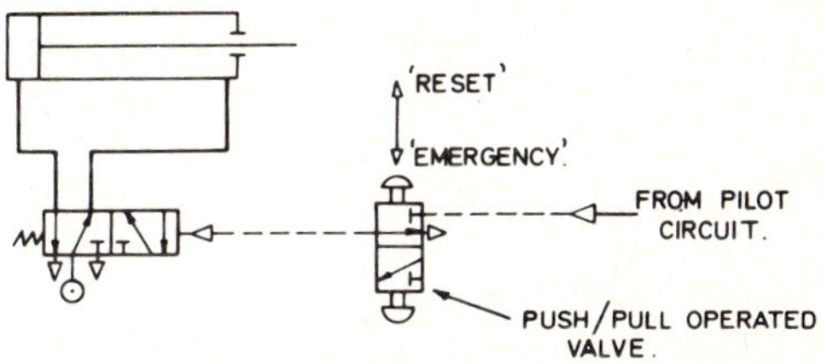

Cylinder positioned to safe position by indirect manual operation (using auto return valve).

Fig 13 Safety circuits for cylinders (Martonair). All circuits shown in emergency condition.

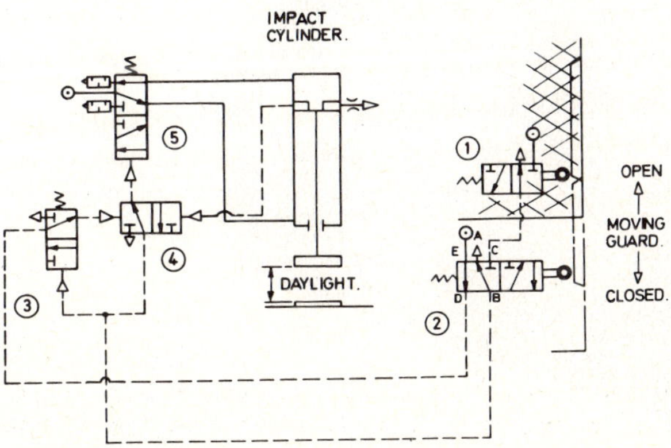

Fig 14 Impact cylinder circuit (Martonair).

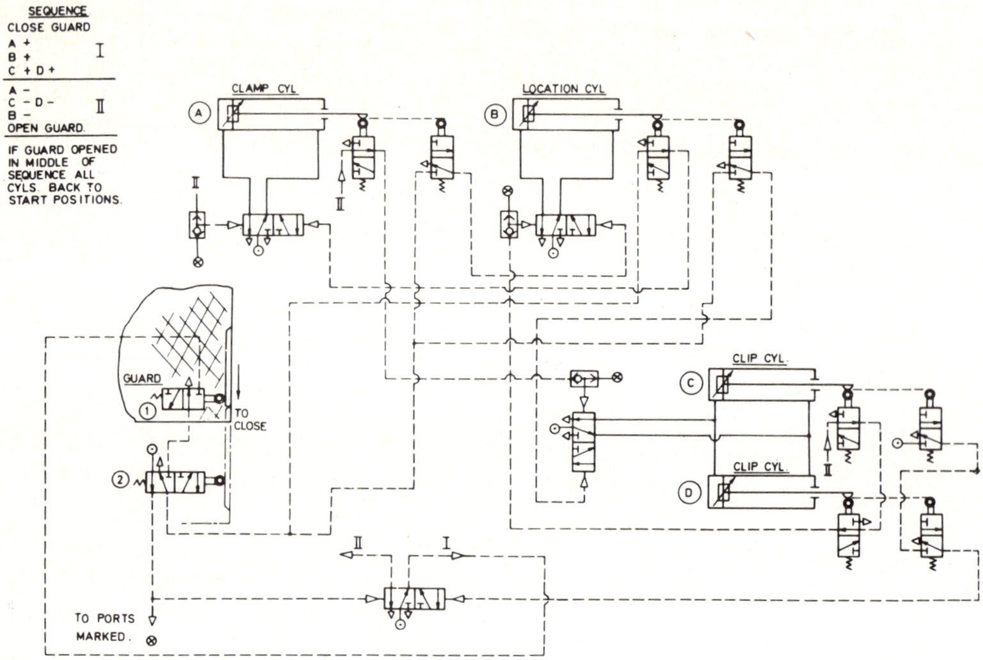

Fig 15 2-group cascade circuit (Martonair)

'Bi-Manual' two handed safety control system provides a signal suitable to operate single-acting cylinders or pilot control valves to operate double-acting cylinders.

SECTION 2

Air Supply

Compressors

COMPRESSORS can be classified under two main types depending on their manner of working. These two categories, in turn, lead to sub-types, eg

i) Displacement compressors
   - a) reciprocating
   - b) rotary
   - c) diaphragm and others

ii) Dynamic compressors
   - a) radial flow
   - b) mixed flow
   - c) axial flow

See also Table 1.

Displacement compressors induct and displace successive volumes of air (or gas) and work on the principle of increasing the pressure of a definite volume of air (or gas) by reducing that volume. Dynamic compressors produce a static pressure rise by converting kinetic energy into pressure energy during the passage of air (or gas) through the machine. Mechanically, the difference is that displacement compressors incorporate a moving member or members, such as a piston, to produce displacement. The working element in a dynamic compressor is an impeller.

Bristol Pneumatics BP 2101 compressor

Other types of compressor may be referred to by duty. A booster compressor is one which inducts air (or gas) which is already compressed and delivers it at a higher pressure. They may be single- or multi-stage machines. A blower is essentially a low pressure compressor delivering air at a pressure not greater than 3.5 bar (and often less). A vacuum pump is also essentially a compressor working in reverse mode, ie to evacuate rather than compress a given volume of air or gas.

TABLE I – COMPRESSOR CHARACTERISTICS

	Type	Maximum Pressure (gauge)		rev/min	Compression Ratio	Maximum Capacity Range		Remarks
		bar	lb/in^2			m^3/min	ft^3/min	
RECIPROCATING	Single-stage	10	150		2.5 to 3.5			Single or double acting
	2-stage	70	1 000		per stage			Multi-cylinder in various configurations
	3-stage	175	2 500					
	4-stage	350	5 000	up to 1·500				
	5-stage	500	7 000		Up to 6 per stage for intermittent operation			
	6-stage	550 to 700	8 000 to 10 000					
VANE	Single-stage	4	60	250 (large)		1.4–5.7	50–200	Water cooled above 30 lb/in^2
	2-stage	up to 10	up to 150	up to 3 500 (small)		1 400	50 000	
	Oil-cooled	up to 8.5	up to 120	up to 3 000				Higher efficiency than 'dry' types
ROTOR	Roots, single-stage	up to 1	up to 15	250 (large)		1 400	50 000	Used mainly as blowers
	Roots, 2-stage	up to 2	up to 30	up to 6 000 (small)		1 400	50 000	
	Screw, single-stage	up to 4	up to 60	3 000 to 25 000		3–140	100–5 000	Compression ratio 3.5:1 to 5:1
	Screw, 2-stage	up to 10	up to 150			3–420	100–15 000	Compression ratio up to 12:1
LIQUID RING	Single	up to 1.5	up to 20	up to 3 600		0.5–30	10–1 000	Low volumetric efficiency
	Series (2-stage)	up to 2–3	up to 30–45					
DIAPHRAGM	Mechanical, single-stage	4	60			0.7	25	
	2-stage	7	100			0.2	6	
	Hydraulic, single-stage	19	220			1.7	60	
	2-stage	1 400	20 000			0.6	20	
	3-stage	2 100	30 000			0.4	15	

Reciprocating Compressors

The reciprocating compressor is the oldest type, and one which has been subject to continual development and improvement. It is suitable for operating over a very wide range of speeds, with a practical maximum of about 10 bar (150 lb/in^2) delivery pressure from a single-stage unit up to 70 bar (1 000 lb/in^2) from a two-stage machine. Multi-stage reciprocating compressors may be built for special purposes capable of supply delivery pressures up to or in excess of 700 bar (10 000 lb/in^2). By carefully selecting the number of stages the designer can also produce a machine which approaches the ideal or isothermal compression curve more closely than any other type, with the possible exception of the very large volume axial flow compressors.

In the case of larger compressors operating with closed cycle cooling cooling is accomplished by passing the water through a mechanical cooler or a cooling tower. Again temperature control can prove advantageous, particularly to avoid too low a temperature being developed in the second stage cylinder.

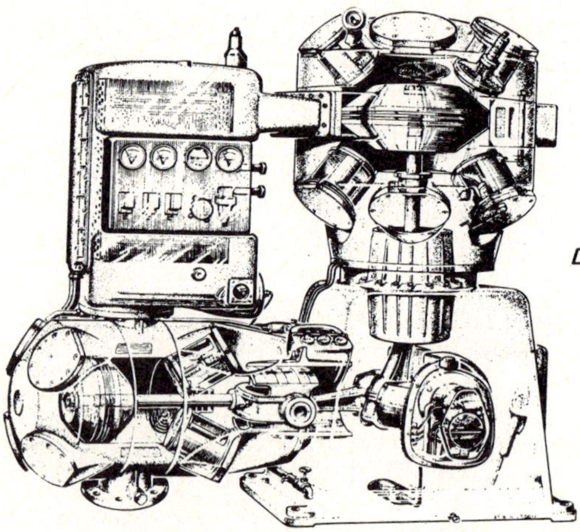

Double-acting, two-stage, water cooled air compressor.

In the double-acting compressor the space on the other side of the piston is also enclosed and both sides are used for compression, giving two compression strokes for each revolution of the crankshaft. Alternatively, one side of the piston may be used for one stage and the other side for another stage in a two-stage compressor. Individual cylinders may also be used for multi-stage compression, disposed in a number of arrangements.

The greatest number of reciprocating compressors are single-stage or two-stage, ranging from fractional horsepower units up to very large machines with input power requirements of the order of 3000 horsepower. The smaller compressors are usually single-stage of the single or vee-twin layout with air cooled cylinders and are powered by electric motors.

Intermediate sizes comprise a whole variety of different configurations. Thus vertical compressors may comprise one or more cylinders in-line, vee, 'W' and 'H' arrangements for multi-cylinder units, and also 'L' configuration which has both vertical and horizontal cylinders disposed around a common crankshaft.

The 'angle' arrangement of cylinders offers certain specific advantages, notably reduced bulk and weight and superior machine balance (since all the primary forces can be accurately balanced). The 'L' configuration is advantageous for larger machines with the low pressure cylinder vertical and the high pressure cylinder horizontal, facilitating installation, assembly, maintenance and dismantling.

Atlas Copco today – one of the largest organisations in the world specialising in compressed air equipment – manufactures and distributes portable compressors and contractors' tools for the contracting, civil engineering and local authority markets; rock drilling and loading equipment for the mining, quarrying and civil engineering markets; and stationary compressors, industrial tools, paint spraying equipment and pneumatic components for all forms of industry.

With sales companies in 40 countries, representatives in a further 85, 16 manufacturing plants in Sweden, 12 more worldwide and 18,000 employees across 5 continents it's obvious that, for Atlas Copco, things just ain't what they used to be!

Great British Atlas Copco...

Atlas Copco (Great Britain) Ltd., in extensive headquarters at Hemel Hempstead, Herts., distributes the complete range of compressed air products through its network of 10 wholly-owned Compressed Air Centres to the entire British Isles.

Thanks to this decentralised system, we can get much closer to our customers.

By understanding their problems and solving them with the assistance of ample local stocking, spares and servicing facilities.

At Hemel Hempstead, Atlas Copco also make a range of paving breakers and down-the-hole drilling equipment for the world market.

And, again thanks to our close customer-contact, we've kept ahead of market trends – using the talents of our Central Laboratories to develop a wide range of environmentally-orientated compressed air equipment...

Including dust-control equipment for industrial and quarrying applications; the Bubble Barrier to contain oil-spillages in harbours; the Limno system for regenerating stagnant lakes; and silenced paving breakers, rock drills and compressors.

And in case you hadn't guessed, it was Atlas Copco who introduced ergonomics into the design of industrial hand tools – typical!

Stationary Compressors

A comprehensive series of piston and screw industrial compressors in a wide range of pressures and capacities from 2-12,000 cfm (1-5,700 l/s). The range includes normal, high and low pressure versions, together with oil-free models for specialised applications.

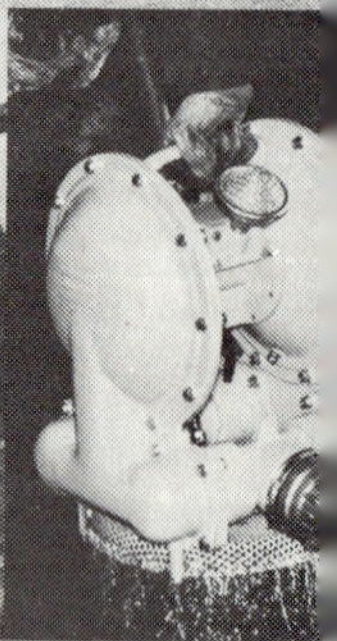

...but baby, look at us now!

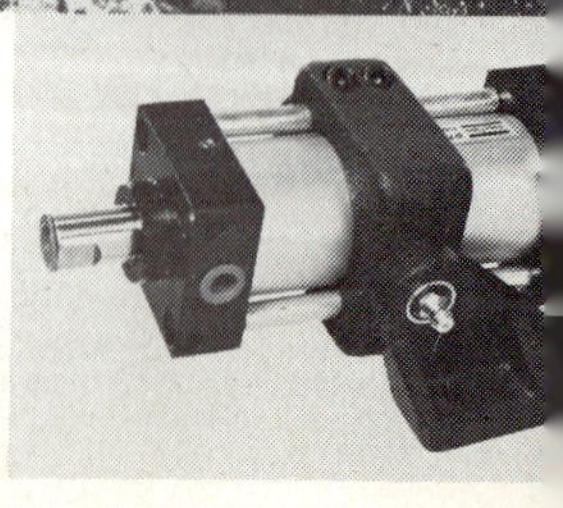

Compressed Air Plants
Self-contained air plants are delivered to site ready to 'plug in' and begin work. The range includes piston, screw and oil-free models with outputs from 90-12,000 cfm (40-5,700 l/s).

Gas Compressors
Reciprocating and turbo models for all technical gases in the chemical and hydro-carbon industries.

Dryers
Desiccant and refrigerant types with capacities to 4,600 cfm (2,100 l/s).

Portable Compressors
A complete range of portable compressors includes piston and rotary-screw machines. All models are available in silenced form, as are oil-free and high-pressure versions. Capacities range from 85-1,500 cfm (40-700 l/s).

Paving Breakers
Light, medium and heavy SUPER TEX paving breakers, standard and with integral robust silencers.

Cobra and Pionjar
Petrol-engine driven motor drill/breakers.

Pumps
The range of pumps includes centrifugal and diaphragm models for contracting, mining and many industrial purposes.

Airline Accessories
Hoses, couplings, pressure gauges, filters, regulators, lubricators, tool balancers etc., etc., for optimum output and economy from the compressed air system.

Pneumatic Components
An extensive range of air motors, cylinders, directional control valves and logic systems.

Air Tools and Equipment for Industry
A comprehensive range of tools, including drills, screwdrivers, impact wrenches, grinders, sanders, saws, chipping hammers, rammers, engraving pens. Multi-spindle nut runners with precision torque-control.

Shot Blast, Paint Spraying, Air Hoists
Shot-blast units for rapid preparation of large areas. Spray guns, automatic, electrostatic and hot spraying systems, spray booths. A wide range of air-operated hoists.

Equipment for Mining, Quarrying and Civil Engineering
Hand-held, mechanised and tracked rock drills; rock dust control equipment; hydraulic booms; tunnelling machines; drill steels and bits; loaders and dumpers; winches; lamps.

Stores and Workshops
Atlas Copco's headquarters at Hemel Hempstead boasts vast warehouse facilities to feed the network of Compressed Air Centres country-wide with local stocks of equipment and spares. Our workshops are engaged in assembling and modifying compressors and drill rigs to customers' specifications.

"...the Bristol compressor system achieves a higher volume of useable air for a given horse power...And that quite simply is what compressed air is all about"

Bristol Duplex industrial compressors are of double acting two stage design, airflow cooled. Moving parts are fully balanced to reduce vibration and provide smoother running characteristics. The Bristol Duplex system compressing air **twice** in the same cylinder, achieves improved efficiencies giving a high volume of usable air for a given horse power. There is a wide range of models with capacities from 3½ c.f.m. to 400 c.f.m. at 100 p.s.i. working pressure. Speed and pressure variable within certain limits to match customers' requirements.

Illustrated Model 308/ATS/60. Free air delivery 48 c.f.m. at 100 p.s.i. guaranteed to B S.726.

BRISTOL INDUSTRIAL COMPRESSORS

BRISTOL PNEUMATIC LIMITED
Fishponds Bristol England
Telephone: Bristol 657511
Telex: 44641

MIDLAND PNEUMATIC

Range of equipment

Your pneumatic control problem won't worry us. We've probably solved something similar before and with the most up to date equipment available. Our range of valves, cylinders and air line fittings is extensive.

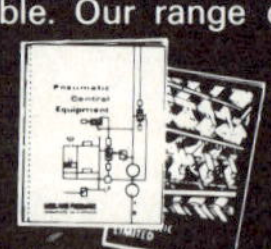

Send for our comprehensive catalogue.

Midland Pneumatic Ltd
Wednesfield, Wolverhampton, WV11 3DZ, England
Telephone: 0902 732501/2.

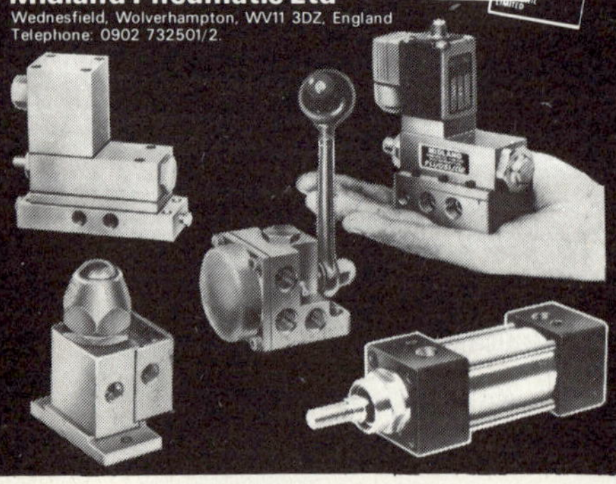

COMPTON OIL-FREE COMPRESSORS

A range of units designed to meet a wide variety of applications for compression, circulation and vacuum.
Compton compressors have the unique diaphragm principle which completely isolates the moving parts and ensures absolutely oil-free operation.
For complete details of the Compton range send for our fully illustrated brochure or consult us with your particular requirements.

DAWSON, McDONALD & DAWSON LIMITED
Compton Works, Ashbourne, Derbyshire, DE6 1DB
Tel: Ashbourne 3184 (STD code 03355)
Telex: 377072

Larger reciprocating compressors are commonly horizontal double-acting, tandem, duplex, etc. Basic two-stage layouts are shown in Fig 1, and two specific layouts in Fig 2. Variations include transferring the sidethrust at the piston end of the connecting rod to the cylinder itself ('trunk' type), or relieving the piston of the sidethrust by means of a cross head. When the space below the piston of a double-acting cylinder is used for pumping air, it is virtually essential to use a crosshead to convert the rotary action of the crankshaft into reciprocating motion in order to obtain a satisfactory seal of the piston rod where it passes through the compression chamber underneath the piston. The crosshead ensures that all sidethrust from the crankshaft is taken up in the crosshead guide and not transferred to the piston and cylinder. Single-acting reciprocating compressors are normally of the trunk-piston type, whilst differential-piston double-acting compressors can be of either type.

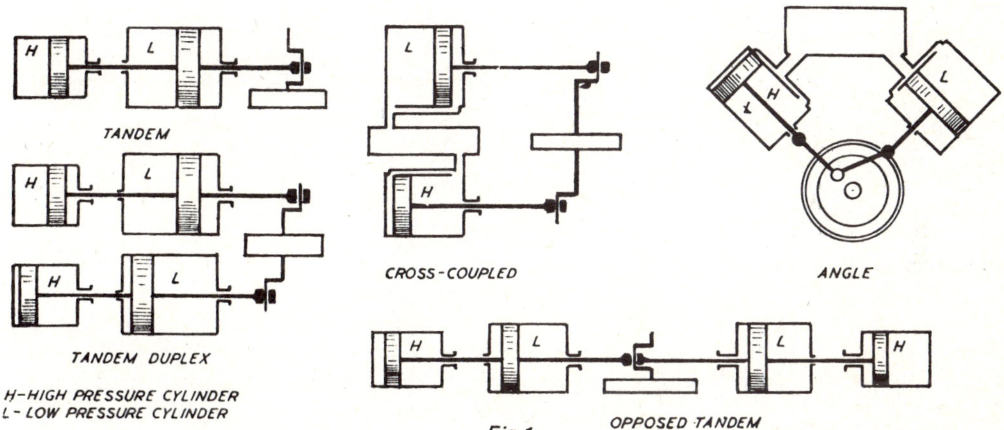

Fig 1

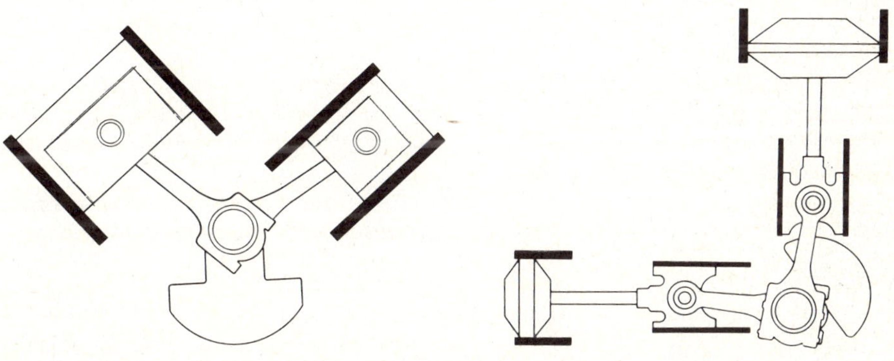

V-type Fig 2 L-type

Differential pistons may also be used as shown in simplified form in Fig 3. Here one stage of compression takes place in the annular space between the shoulder of the piston and the corresponding shoulder in the cylinder.

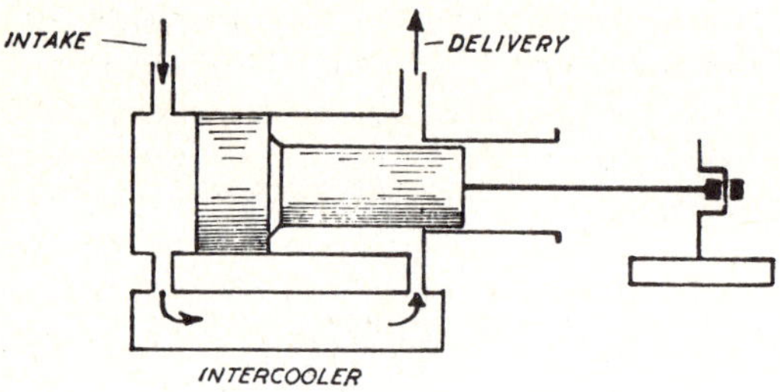

Fig 3

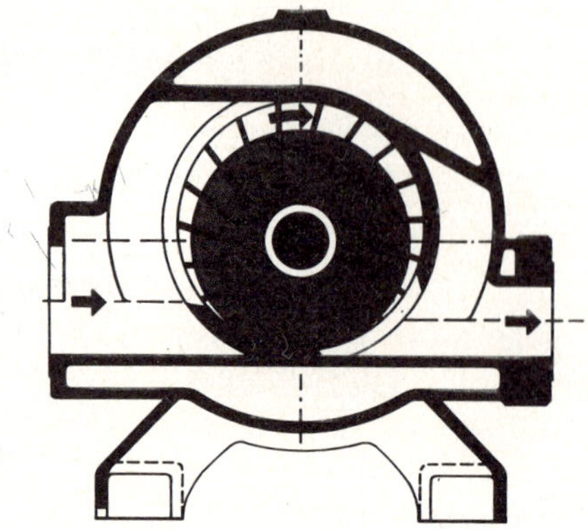

Fig 4

The use of multi-stage reciprocating compressors is normally confined to higher pressure requirements. Three-stage units for continuous duty are produced capable of discharge pressures up to 175 bar, whilst output pressure can be further increased, if necessary, by additional staging. The design tends to be much more complex than in the case of single- and two-stage reciprocating compressors, but again may embrace either single- or double-acting cylinders and trunk pistons or crosshead configuration.

For continuous duty compression ratio per stage is usually of the order of 2.5 to 3.5 Compression ratios of 6 or more per stage may, however, be used on multi-stage machines intended primarily for intermittent duties.

Rotary Compressors

This classification is normally applied to rotary machines with positive displacement characteristics, as distinct from centrifugal and axial flow machines. These include sliding vane compressors and liquid ring pumps where positive displacement characteristics are calculated by trapping the air (or gas) being handled in a series of individual cells which decrease in volume as rotation proceeds, thus providing compression. Under the same general classification rotary machines which do not have true positive displacement may be included — eg intermeshing rotor and screw pumps.

Vane Compressors

There are numerous variations on the design of some compressors, but all follow the general form of a cylindrical rotor with a series of radial or near radial slots cut into it. The rotor is mounted eccentrically in a cylinder. Each slot in the rotor carries a blade, the outer edge of which is kept in contact with the cylinder by centrifugal force. As the rotor turns the space between each pair of adjacent vanes alternatively grows and contracts. The inlet port is positioned to let in the air where the space is opening up, and the delivery port opens where the space closes. In the smaller rotary vane machines, the blades are usually made of hardened steel. Due to the large centrifugal force acting on the cylinder, the larger machines usually have a floating cylinder liner or rings, which are slightly less in diameter than the cylinder bore. Consequently the blades are in contact with this floating member, which is free to rotate in the cylinder, with the result that in theory all wear between the blades and cylinder is eliminated.

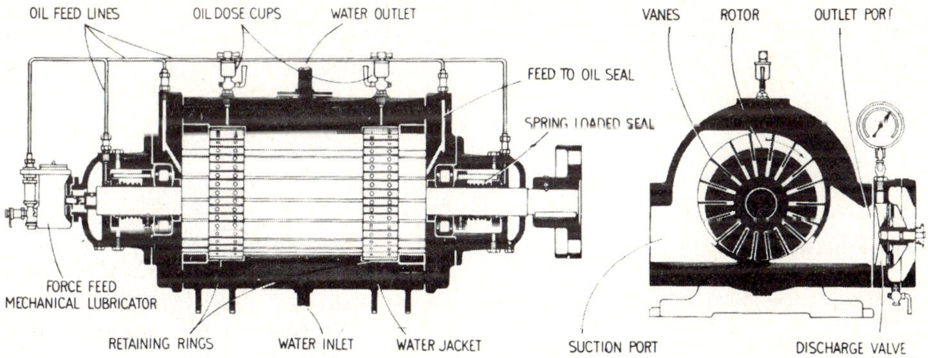

Fig 5 Single-stage sliding vane compressor

Fig 5 is a diagrammatic representation of a (Wittig) vane compressor, also shown in the form of a practical machine in Fig 6. Vane compressors do not have valves, but a check valve may be installed after the compressor. Capacity may be controlled by a throttle valve in the air supply line, although automatic stop-start regulation is usually more economical.

Vane-type compressors may be single- or two-stage, cooled by water, air or oil. Single-stage vane pumps are generally used for pressures up to 4 bar (60 lb/in^2) and two-stage machines for pressures up to 10 bar (150 lb/in^2) whilst operating speeds may range from as low as 250 rev/min up to 3500 rev/min, speed normally being inversely proportional to size. Being a well balanced machine, a vane compressor is well suited for direct connection to a relatively high speed driver (eg an electric motor). Its efficiency, however, is generally less than that of a reciprocating compressor and can decrease substantially with age through blade wear.

The vane compressor is generally most efficient at lower working pressures. Capacity range covered by vane compressors is normally 1.5–6 m^3/min (50–200 ft^3/min).

Normal methods of cooling are either by air or water. In the former the casing is usually finned to increase the external surface area and thus the rate of heat dissipation. With water cooling, the water jacket is normally cast integral with the casing and a circulatory flow of clean cold water established so that the temperature of the discharged air does not exceed 150°C (300°F). Cooling requirements are not as severe as with reciprocating compressors since there are no restrictions on the discharge side and incoming air does not come into contact with a heated surface to raise its initial temperature. Water cooling is normal for vane-type compressors where the discharge pressure exceeds 2 bar (30 lb/in^2). In the case of two-stage machines, intercooling is normally provided, either by a separate intercooler, as in the case of reciprocating compressors, or by an intercooler built into the base of the machine.

Oil-cooled rotary vane-type compressors are a comparatively recent development, the principle being that oil is injected in relatively large quantities into the compressor and recovered on the discharge side and recirculated. Besides providing the most effective form of cooling by acting as a heat transfer medium in direct contact with the air, flooding the machine with oil considerably reduces internal leakage and improves volumetric efficiency. Volumetric efficiencies of oil-cooled vane compressors, in fact, are usually directly comparable with those of reciprocating machines, whilst the former are considerably more compact and lighter for a similar duty.

The quantity of oil circulated and injected is normally designed to provide the maximum acceptable degree of cooling and in some cases may be of the same order as the volume of air delivered at the discharge end. To remove the oil at the discharge end it is usual first to interrupt the flow with baffles on which the oil can impinge, then to pass the discharge through a separator or filter to remove as much as possible of the remaining oil. Oil separation is assisted by the fact that the discharge temperature is low so the oil is in the form of droplets rather than aerosol size particles and vapour, and oil removal to the order of 99.0–99.9% is usually possible by simple mechanical separation

Medium or large capacity vane compressors running at high speeds normally incorporate a method of avoiding contact between the vanes and casing wall in order to reduce wear and minimize frictional losses and overheating. A widely used method is to fit restraining or floating rings over the vanes, the internal diameter of these rings being slightly less than the cylinder bore. The geometric difference is calculated to preserve a minimum clearance between the vane tips and the casing walls under operating conditions – see Fig 6. The rings actually rotate with the vanes but the peripheral speed of each vane tip varies with the degree of extension. The rings rotate at constant speed. There is thus some relative motion between vane and rings, but the contact between rotor and rings is essentially a rolling motion.

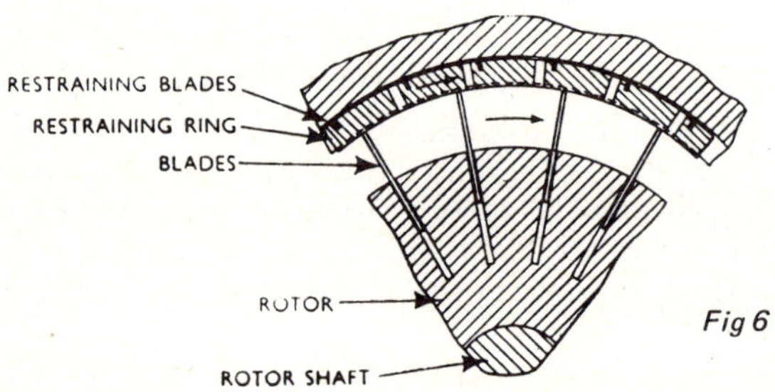

Fig 6 Vane compressor with sliding rings (Hick Hargreaves).

The number of blades is usually limited to a maximum of 10 or 12 with high speed designs providing discharge pressures of the order of 7 bar (100 lb/in^2). A lesser number – eg 8 blades – may be necessary where relatively thick blades are used. To increase volumetric efficiency it is also desirable to spread the pressure difference between inlet and outlet over two rotors, leading to a two-stage rotor. It should be noted that unlike other compressor types intercooling is not required between stages (since continuous cooling is provided by the oil in both stages) and the only significance of staging is in improving volumetric efficiency.

Liquid Ring Compressors

The liquid ring compressor is based on a rotor with a series of fixed blades, mounted in a cylinder and so arranged that clearance between the blade tip and the cylinder varies cyclically on each revolution of the rotor – Fig 7. The cylinder is part-filled with a liquid, usually water when the machine is handling air. During operation, the water is carried around the cylinder by the action of the rotor blades. Due to the centrifugal force, the water forms a solid ring around the cylinder, the inner wall of which varies in its distance from the rotor, as in fact the cylinder wall does. Thus, the space between the blades varies cyclically, as with the sliding vane compressor.

Compressors from 17 c.f.m. to 680 c.f.m. up to 200 p.s.i.

We started the quiet revolution in Air Compressors!

With over a quarter of a century of pioneering experience behind us we offer our "FLUIDAIR" range for maximum reliability with minimum maintenance. Silently and efficiently "FLUIDAIR" compressors can be relied upon to handle your air demand day in – day out. Fully acoustic enclosures available. If you are a time, cost and space conscious engineer yourself, talk to one of our air engineers about your particular needs. He is backed by a nationwide technical Sales and Service organisation.

WHITTAKER HALL LIMITED,
RADCLIFFE, MANCHESTER.
TEL: 061-723-2421 TELEX: 667071
Regional Offices, at Reading, Birkenhead and Glasgow.

ALUP KOMPRESSOREN

Super noise-suppressed industrial compressors

The guarantee for:

1. High degree of energy efficiency
2. Valve life of up to 5000 working hours
3. Minimal, troublefree maintenance
4. Easily surveyable operation
5. High operating safety and prompt service
6. Environmentally friendly

We also supply:

Compressed air refrigeration dryers and super line oil separators

Please send for detailed leaflets or personal consultancy service

ALUP KOMPRESSOREN
D-7316 Köngen/Neckar, FR Germany
P.O.Box 281, Telephone (070 24) 89 01
Telex 7 267 215

Hull
Leicester
Nottingham
Sheffield

ATC Airtools and Compressors LIMITED

Compressed Air & Painting Systems

We were formed in 1965 to give a comprehensive service to the users of compressed air power and paint finishing. Through local depots situated in major industrial areas, the company designs, installs and maintains compressed air and painting systems.

Customers may use the free consultancy service in on–site discussions with a sales engineer. Urgent requirements are dealt with at the depots trade counters.

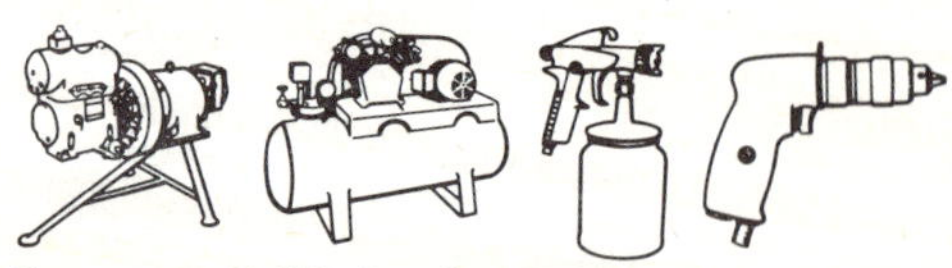

Compressors 0-600 cfm Compressed Air Equipment
Pipework Installations Spraybooths and Ovens

Easy parking at all depots
Head office: 89-91 Nottingham Road Loughborough LE11 1ES
Tel: (0509) 66011

WILLIAMS & JAMES (ENGINEERS) LIMITED

Head Office and Main Factory

No 2 Factory

No 3 Factory

For information on how W & J can help meet your requirements for air compressors contact Sales Engineering at the address opposite.

WILLIAMS & JAMES (ENGINEERS) LTD, have been engaged in high reliability engineering for over half a century. For much of that time their major effort has been in the development and production of air and gas compressors.

Many types have been manufactured but there has been a planned concentration upon machines having higher than average reliability and delivering 'good quality' air.

From modest beginnings as a family concern there has been steady, unspectacular growth up to the present day, when as a public company production is centred in a spacious modern factory adjacent to the administrative offices. Two additional factories in the same area have been established to help meet the heavy, growing demand for their products.

An early association with the then infant control and automation industry led to the introduction of the 'oil-free', dry cylinder compressor which is now in extensive use. Many types of high pressure compressors are produced and cater for a wide variety of applications.

Nowadays, it is true to say that W & J compressors and equipment have achieved a very high reputation for reliability and efficiency and may be found located in virtually every corner of the world on offshore drilling platforms, in factories, chemical plants, refineries, ships, airports, laboratories — in fact, anywhere that there is a need for such compressors and the vital efficient service to support them.

Williams & James believe that they have an obligation to their customers to advise them upon plant selection, to ensure that the right compressors and ancilliary equipment are used for a particular application; to advise them upon its installation and use and to furnish a first class spares and service facility for its future support.

WILLIAMS & JAMES (ENGINEERS) LIMITED

TYPE 695

OUTPUTS up to 200 c.f.m.

Pressures up to 150 p.s.i.

The Williams & James Type 695 compressor is designed to provide top quality air for instrumentation and control, breathing air and other demanding applications. Being air cooled, it has found many applications in arid, cold and marine environments because it does not involve the extra installation and maintenance costs of water-cooled machines.

The 695 is a two-stage, two cylinder compressor incorporating carbon rings running in dry cylinder bores. Integral with the compressor are air-blast cooled inter and after-coolers. Crankcase motion work is force lubricated, and dynamic balancing with vee construction gives smooth, vibration-free running and high reliability.

190 Series

Outputs up to 120 c.f.m. Pressures up to 125 p.s.i.

The Williams & James 190 series are oil-free compressors developed to provide high-quality air for instrumentation and control, surface demand diving, breathing air for hazardous environments, in desert and Arctic areas and marine applications. This series can be supplied in one, two or four cylinder versions with 5-inch bores and stroke. Cylinders are all fitted with carbon rings. The one and two cylinder units are vertical, while four cylinder versions are in vee form. Crankcase motion work is force lubricated in all cases and is fitted with integral air-blast after-coolers. A large proportion of spares are interchangeable with those for the Williams & James 140/150 series of normally lubricated compressors, which reduces stock-holding requirements where both types are in use.

TYPE 175

Outputs up to 40/35 c.f.m.

Pressures up to 2,500/4000 p.s.i.

The Williams & James Type 175 is a rugged, high output compressor suitable for both air and gas duties. It is used for pneumatic gun operation for offshore seismic surveys. With associated filtration and drying equipment it is also widely used for breathing air bottle charging and for handling oxyhelium mixtures for deep diving.

Type 175 is a vee-form, four cylinder, three-stage compressor with water or radiator cooling, incorporating inter-and-after coolers and moisture separation. There is a high degree of spares interchangeability with 140/150 and 190 series models.

WILLIAMS & JAMES (ENGINEERS) LIMITED,
Chequers Bridge, Gloucester GL1 4LL, England.
Telephone : Gloucester (0452) 28431. Telex : 43233

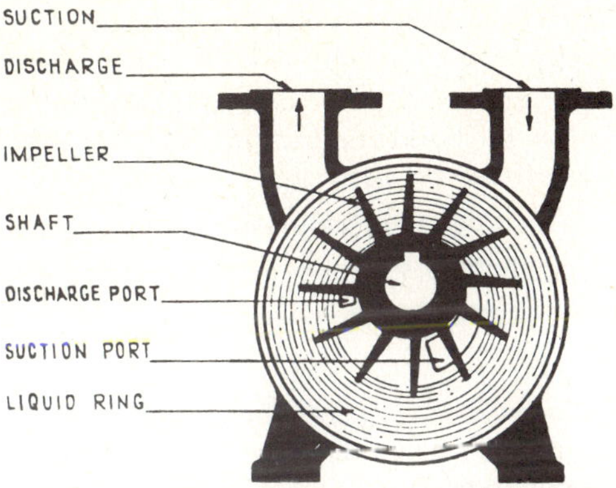

Fig 7

Unlike the sliding vane compressor it is possible, by suitably designing the cylinder wall, to obtain several cycles of expansion and compression per revolution. In practice this is usually limited to one or two. During operation, water may be either supplied from an outside source, or recirculated through a heat exchanger. This type of machine is relatively inefficient, but due to its oil-free operation, the liquid ring compressor has applications in chemical engineering or where the air or gas being handled has corrosive elements in it. It is also used as a vacuum pump.

Liquid ring compressors are made in sizes from 1–80 m^3/min (12–940 ft^3/min). Operating speeds range from 300–3000 rev/min depending on the size of the machine.

Model 695 low pressure oil-free compressor manufactured by Williams and James.

Diaphragm Compressors

Diaphragm compressors are generally classified as either mechanically driven or hydraulically actuated, depending on the method of deriving diaphragm movement. Mechanical diaphragm compressors are manufactured in smaller capacities than hydraulically actuated units and consequently overlapping of the two types only occurs over a restricted range.

Mechanical machines are less expensive to produce, are very compact and can be used on applications involving sub-atmospheric pressures.

Hydraulically actuated units are able more easily to generate high pressures than are mechanical machines as the latter are usually limited by considerations such as bearing loading.

Apart from the fact that the compact form of a diaphragm unit enables it to be built directly onto an electric motor or driver (or hydraulic pump unit in the case of a hydraulically-operated machine), the main advantages offered by diaphragm compressors are:-

(i) Complete gas-tightness with only static sealing involved.

(ii) Isolation of the gas being handled and thus the possibility of 100% oil-free compressed gas supply.

The main limitations of the type are low delivery rates and a limited compression ratio. The latter applies particularly to mechanically-operated diaphragms where the diaphragm is normally of synthetic rubber. Hydraulically-operated diaphragms may employ metallic diaphragms and be capable of generating much higher pressures. Multi-staging is possible with both types.

A further method of obtaining high pressure working with a mechanically-operated diaphragm compressor is to enclose the compressor and driver within a pressure vessel. This vessel is pressurized to a level equal to that present at the inlet. The same pressure is thus present on the underside of the diaphragm, reducing the pressure differential on the diaphragm when working.

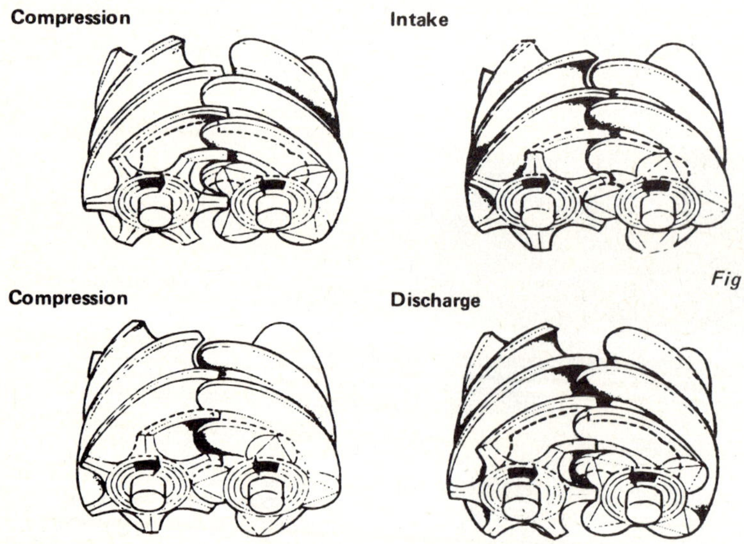

Fig 8 Sequence of working of a screw compressor

Screw-Type Displacement Compressors

Machines of this type are based on the Lysholm Screw comprising two intermeshing rotors with helical lobes. The inlet is located at one end of the rotors. As the rotors revolve, the space between the unmeshing lobes increases, allowing inlet air to fill up the intervening space, until the male lobe is disengaged from the female lobe along its whole length. The helices of the male and female rotors are designed to permit complete charging of the inter-lobe space before they remesh. On completion of the filling operation, the inlet ends of the male and female rotors pass the inlet port and become sealed in the casing. With continuing rotation, the male and female lobes begin to re-engage each other, the volume of this space is reduced and compression begins. This compression increases with continuing rotation until the built-in pressure ratio is reached and the air is discharged through the outlet port, see Fig 8.

Clearances between the male and female lobes, between lobes and casing, and between the ends of the rotors, are kept to the minimum, so that the minimum of slip or leak-back from the high pressure side to the inlet side is obtained. This machine has the advantage over the reciprocating compressor of mechanical simplicity, as there is no contact between the male and female rotors and/or the casing. Therefore, no internal lubrication is required and oil-free air is obtained. In addition, as inlet and discharge valves are not required, maintenance is greatly simplified.

Capacity regulation can be provided by a valve that controls the air supply to the compressor, or more simply by varying the speed. The delivery pipe is normally fitted with a check valve so that the system is automatically closed when the compressor is stopped or idling. In the latter case air is conducted from the high-pressure side to the intake side via a suitable channel to prevent the compressor working as a motor.

Screw compressors are suitable for high speed operation and thus deliver high capacities from a small machine. They can be used to generate higher pressures than a Roots blower. Pressure ratios of 3.5:1 to 5:1 can be obtained in a single stage and 12:1 in two-stage machines. Thus single-stage

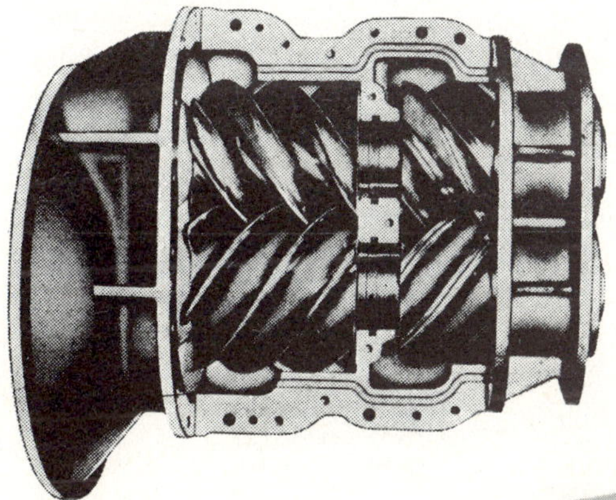

Two-stage screw compressor (Atlas Copco)

Cut-away view of BroomWade RA75 screw compressor.

machines are suitable for delivery pressures up to 4.2 bar (60 lb/in^2) and two-stage machines for deliveries up to 10 bar (150 lb/in^2). Operating speeds are normally high, since this enables the size of the machine to be reduced, and may range from 3000 to 25000 rev/min. Typical capacity range is from 2.8m^3/min (100 ft^3/min) FAD up to 425 m^3/min (15000 ft^3/min) FAD.

The volumetric and isentropic efficiency of a rotary screw compressor are shown in Fig 9.

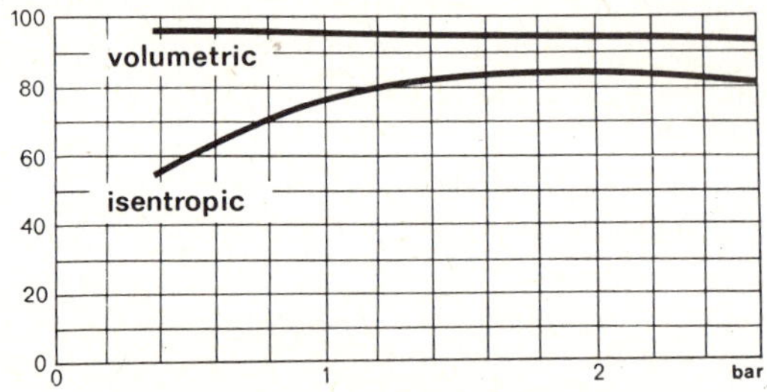

Fig 9 The isentropic and volumetric efficiences of a rotary screw. (Atlas Copco).

Lobe-Rotor Compressors

The typical rotor type compressor employs a pair of two lobed rotors running on separate parallel axes and driven by external gearing so that small but positive clearance is maintained at all times between the intermeshing rotors and casing. The best known type is the Roots blower – Fig 10 – although the principle is applied to other rotor shapes and, to a much more limited extent, three or more intermeshing rotors on parallel axes.

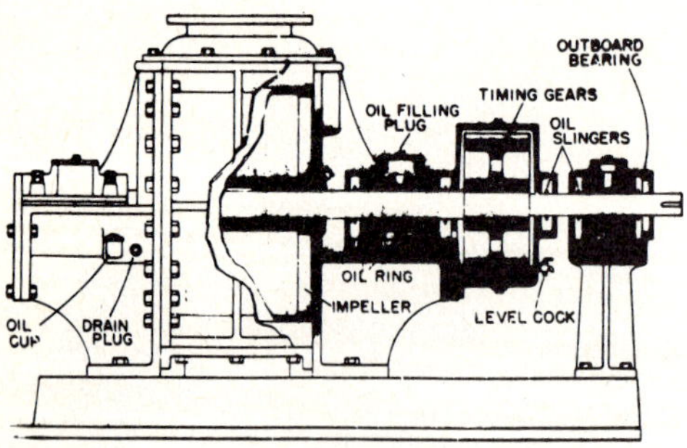

Fig 10 Roots blower.

Volumetric efficiency depends on maintaining the smallest possible clearances between the rotors and between rotors and casing. Delivery pressure depends on the back pressure in the system since no actual compression takes place in the compressor. This normally results in poor efficiencies at pressures above 0.6 to 0.7 bar (8–10 lb/in^2) although pressures up to 2 bar (30 lb/in^2) can be obtained with reasonable efficiency on two-stage machines. Special versions of Roots blowers may be capable of providing pressure of up to 3 bar (45 lb/in^2).

The principal advantages of this form of blower are simple construction and high operational reliability – particularly for the low pressure delivery of air or gases.

Speed of operation of Roots blowers ranges from about 250 rev/min for large machines to 6000 rev/min for small i/c engine superchargers. High speeds are usually desirable in order to reduce the size of the machine and increase volumetric efficiency, leakage rate being largely independent of the actual size of the machine, but increasing substantially with increasing pressure. Back pressure or backflow is also responsible for a large rise in temperature relative to the rise in pressure.

At delivery pressures up to about 0.6 bar (8 lb/in^2) efficiencies compare favourably with most other types of compressor, but due to increasing losses the maximum pressure obtainable with a single stage is of the order of 1 bar (15 lb/in^2). Two-stage machines, used for pressures up to 2 bar (30 lb/in^2), are normally provided with intercooling to maintain a reasonable discharge temperature.

Dynamic Compressors

Dynamic compressors are essentially turbo-machines in which compression is achieved partially by compressing the air by a reduction in relative volume through the rotor and partly by increasing the kinetic energy of the air in respect of its absolute velocity. This kinetic energy is subsequently converted into pressure energy by diffusion in the stator. Types involved include both radial flow (centrifugal) and axial flow machines. They are, in fact, dynamic machines rather than positive-displacement devices and thus differ in characteristics from both reciprocating compressors and other types of rotary machine, working on positive displacement principles. They also differ from the other types in that no attempt is made to seal the rotor or compressing element.

In the centrifugal compressor a multi-bladed impeller is rotated at high speed in a casing. Air trapped between the blades is accelerated and thrown outwards, leaving the blade tips with increased pressure and high velocity to enter a diffuser ring. Here there is a region of increasing area in the direction of flow, resulting in a reduction in velocity and a substantial increase in pressure. From the diffuser the air enters the volute casing with further reduction in velocity and increase in pressure. At the same time flow through the impeller is maintained by a reduction in pressure at the inlet, drawing in a continual supply of air as long as the impeller is rotating. Inlet air may enter from one side of the impeller only, or from both sides simultaneously, when the compressor is known as a single-flow or double-flow type respectively. The advantage of double-flow is that thrust forces on the impeller are balanced.

The pressure which can be obtained in a single stage is limited by the maximum permissible peripheral speed of the impeller, which is typically of the order of 300 metre/sec (1000 ft/sec). This largely limits single-stage capabilities to a maximum delivery pressure of about 6 lb/in^2, although higher pressures can readily be achieved by adding more stages. Thus, typically, 2.5 bar (35 lb/in^2) may be produced in five stages and very much higher pressures can be achieved if required with further staging. Equally, multi-staging is not always associated with higher delivery pressures as multi-stage machines may be specifically designed for low pressure applications.

A distinction is also drawn between 'low' and 'high' pressure aerodynamic machines. Where the delivery pressures does not exceed about 2 bar (30 lb/in^2) such machines are generally termed 'blowers', and where delivery pressure is 2 bar (30 lb/in^2) or more they are termed compressors.

Discharge pressure and volume of air delivered are co-related. Thus if at a given speed the back pressure is reduced, the volume delivered will increase, and vice-versa – see Fig 11. If the back pressure increases to a value above that generated by the compressor then a condition known as 'stall' or 'surge' occurs and gas flows back into the machine. If the pressure then drops the machine starts to deliver again and if this sequence is allowed to continue serious damage to the machine could result. Practical operating limits as regards characteristic curves, are, therefore, terminated in a surging limit – see Fig 12. To avoid surging it is desirable that the controls of the machine should incorporate a provision for isolating it from the mains when operating conditions approach the surge point, and for restoring it to operation when the surge pressure has fallen.

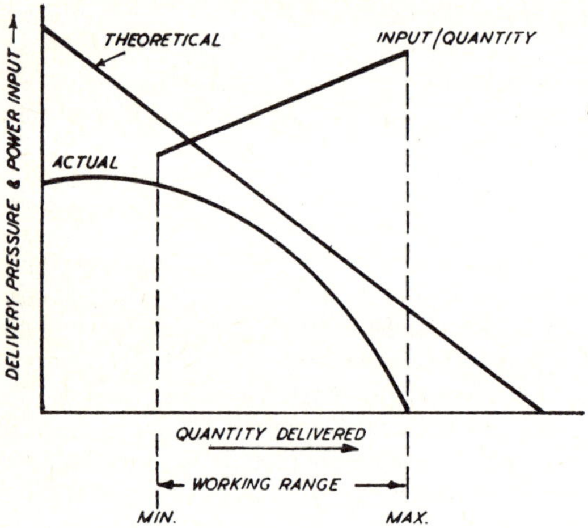

Fig 11

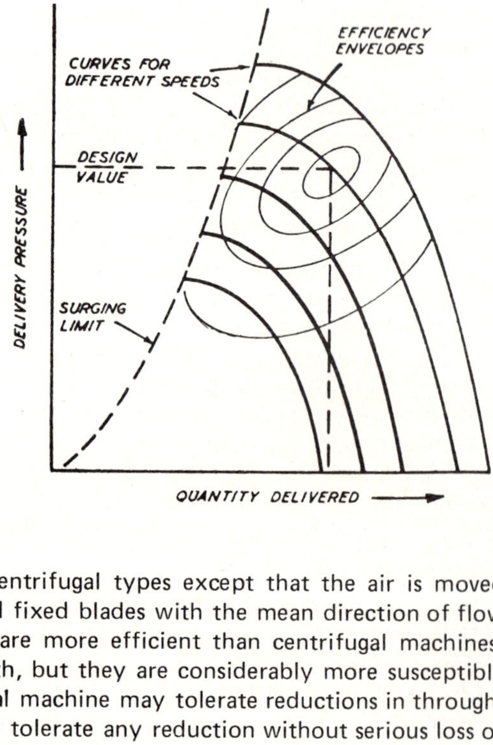

Fig 12

Axial Flow Compressors

Axial flow machines are similar in principle to centrifugal types except that the air is moved through a series of rows of alternating rotating and fixed blades with the mean direction of flow parallel to the axis of the rotor. In general they are more efficient than centrifugal machines, largely because of the simpler and shorter flow path, but they are considerably more susceptible to changes in throughput. Thus, whilst a centrifugal machine may tolerate reductions in through-flow of as much as 30%, axial machines can barely tolerate any reduction without serious loss of

Fig 13

compressors--exhausters

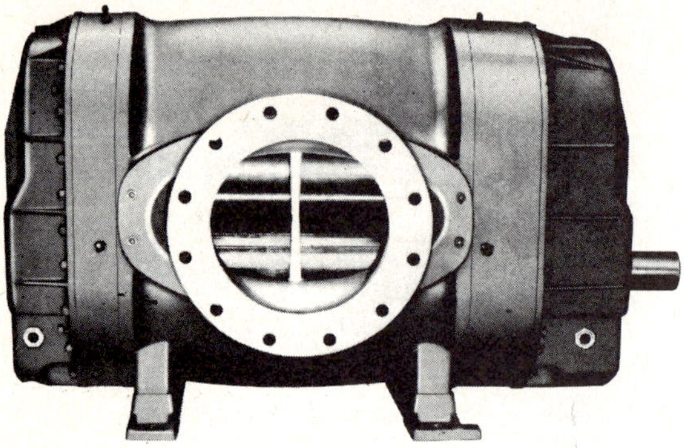

The Donkin MkV Roots type Compressor is an improved range of positive displacement machines. It is designed to give a pressure rise of 1.03 bar (15 lbf/m^2) from atmosphere on all sizes with a capacity range of 16.8 m^3/m (600 cfm) to 510 m^3/m (18,200 cfm). As an exhauster, vacuums up to 380 mm Hg (15" Hg) can be achieved under normal dry operation and 560 mm Hg (22" Hg) with water injection to the inlet side.

An SD90C Centrifugal Blower as supplied for a pulp and paper mill in Canada. It is driven by an electric motor through epicyclic speed increasing gear.

Typical duty:

Volume flow 820 m^3/m of air (29,000 s.c.f.m.)

Working pressure 0.55 bar (8.0 lbf/in^2 g)

Blower speed 7090 r.p.m.

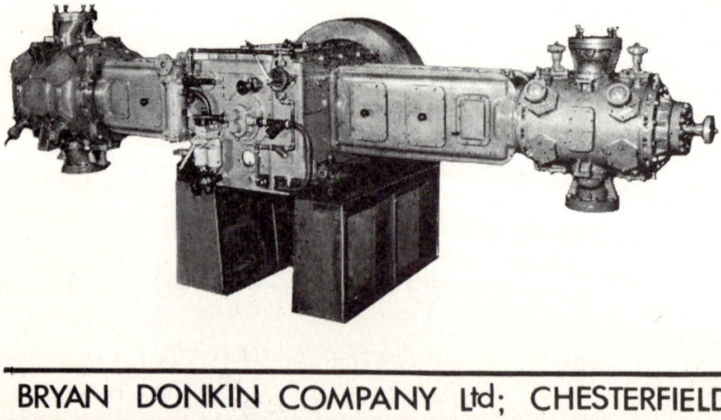

A 2HB2GT two stage, oil free, dual duty argon/nitrogen compressor. Typical duty:-

Pe 0.02 Kgf/cm^2 (0.3 lbf/in^2)

Pa 3.18 Kgf/cm^2 (45.3 lbf/in^2)

Capacity 4.4x1162 Nm^3/hr (10^4 s.c.f.h.)

BRYAN DONKIN COMPANY Ltd; CHESTERFIELD. Phone (0246) 73153 & 70151

valves and regulators..

Non-Return Valve.
Size range 2" to 48" diam.
Working pressure 100 lbf/in^2
(7.03 Kgf/cm^2)

Donkin 680 Regulator.
An extremely versatile, sensitive and stable regulator giving 40% extra capacity.
Sizes 2" to 8" bore.
Maximum pressure —
Inlet 30 lbf/in^2
Outlet 4" to 56" w.g.

Fig. 50 Ball Valve.
A quick operating valve with wide field of application where 100% cut-off is required.
Especially suitable for the isolation of gases.
Sizes ½" to 6".

Fig.994 Hot Gas Valve.
This valve is one of the very few valaves available for continuous duty at elevated temperatures, especially in the size ranges offered.
2" to 42" bore diameter.
Working pressure 20 lbf/in^2.
Temperature 750°C.

donkin

valves, regulators and compression plant for gas and process installations

BRYAN DONKIN COMPANY Ltd; CHESTERFIELD. Phone(0246) 73153 & 70151

A.P.E-BELLISS Ltd

The total service in service air

A.P.E-Belliss offer a total service in medium to heavy-duty service air.

Air or water cooled compressors are available, lubricated or oil-free in capacities from 200 to 3000 cfm (5.7 to 84 Nm^3/min). They can be skid-mounted, free-standing or supplied as completely packaged units. The outstanding qualities of the range are massive dependability, easy installation and maintenance and a rationalised build-up which facilitates servicing with speed and economy.

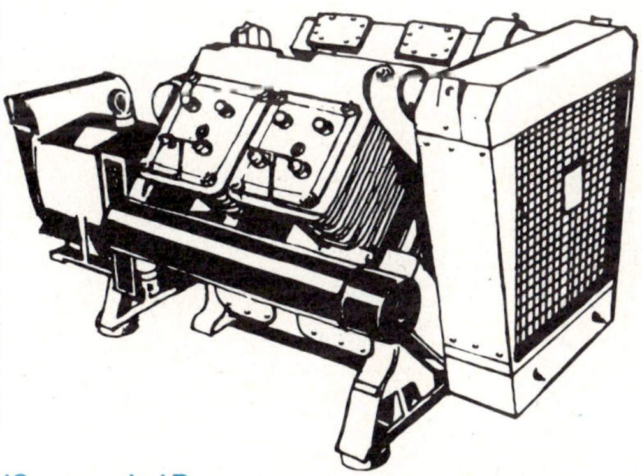

'Commander' Range

Air-cooled, medium duty packaged compressor for easy instant installation. 5 models available: 3 V-twins (200, 300 and 420 cfm) and 2 V-four cylinder models (400 and 600 cfm). All models are designed for continuous duty – inside, outside or even underground in a flameproof version.

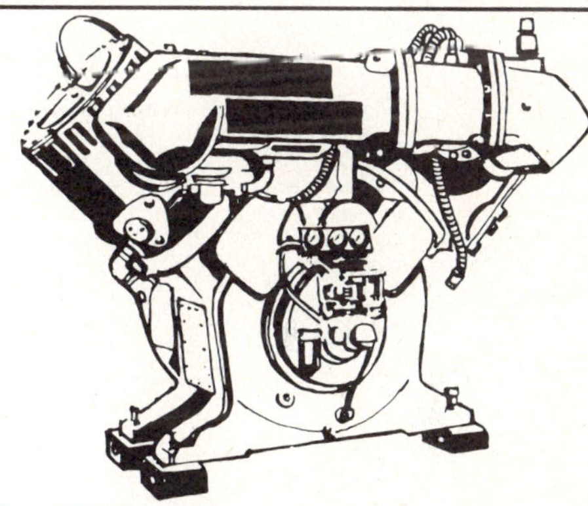

'V' and 'W' Compressors

Conforming to metric standards these compressors incorporate every advance made possible by modern technology.

They are renowned for long-term dependability and low-vibration, eliminating the need for special foundations. Available lubricated or oil-free, in single-stage, two-stage and three-stage form.

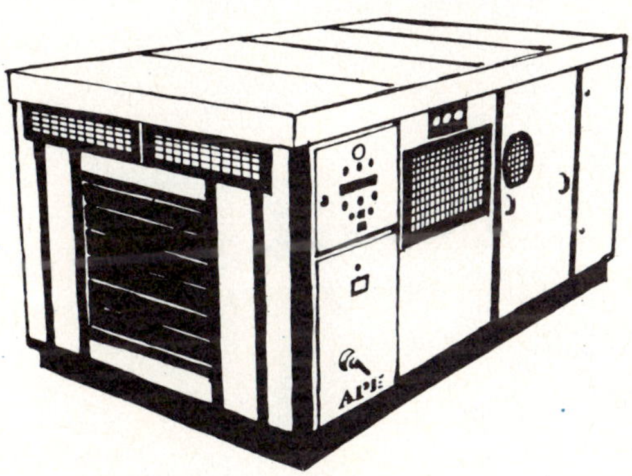

New PBM Series Medium Duty Packaged Compressors

Models provide outputs of 300, 420 and 600 cfm respectively under designations PBM 30/42/60. These are acoustically packaged compressors, air-cooled and self-contained for simple installation. The silencing is extremely effective making these units compatible with most working environments.

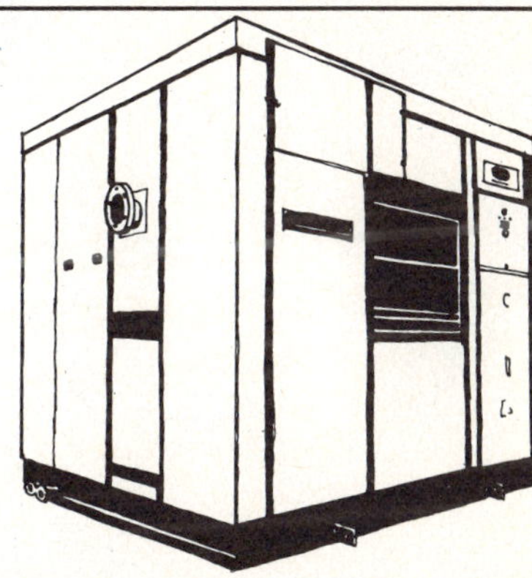

New PVH Series Heavy-Duty Packaged Compressors

Acoustically packaged with all ancillaries these units represent a major advance for users of heavy-duty service air. No special foundations are needed. Silencing is most effective. Capacities from 750 to 1000 cfm.

A.P.E-BELLISS Ltd

Icknield Square, Birmingham B16 0QL. Tel: 021 454 3531
Telex: 337507

A subsidiary company of AMALGAMATED POWER ENGINEERING LTD. APE

efficiency. They are particularly suited for handling large volumes of air but centrifugal machines are better able to handle wide ranges in quantity where the variation in delivery pressure or compression ratio is relatively small.

Basic elements involved are a row of moving blades attached to a rotor followed by a row of fixed blades attached to the casing, the pitch of the two rows of blades being in opposition – Fig 13. Curvature applied to the blades has the effect of increasing the sectional area through which the gas is flowing with resulting pressure increase as in the diffuser of a centrifugal compressor.

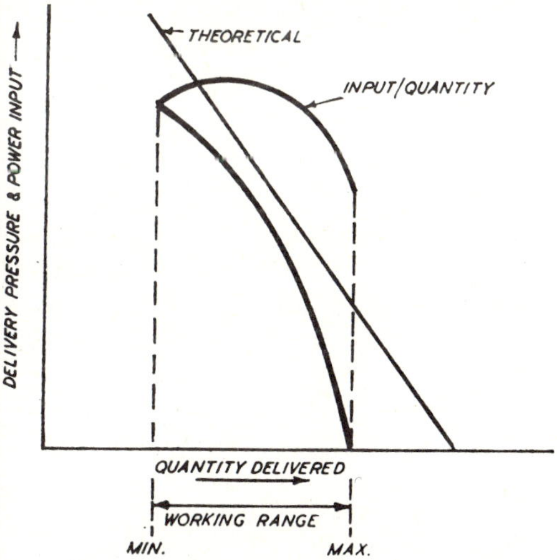

Fig 14

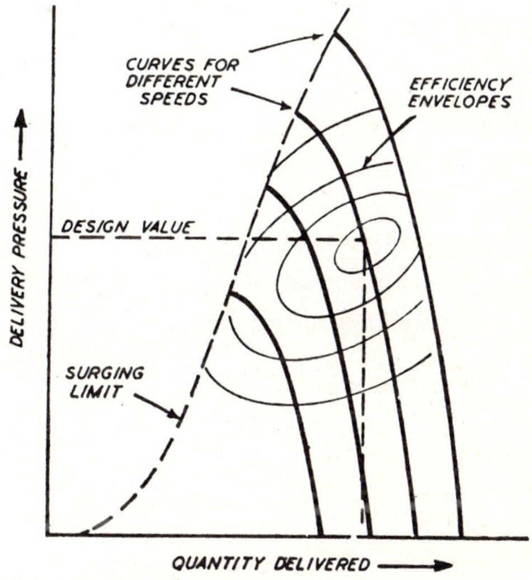

Fig 15

At the same time work is being done on the gas by the rotating blades, increasing its absolute velocity. The process is repeated throughout the length of a succession of blade sets with velocity of the gas relative to the moving blades decreasing and the absolute velocity of the gas increasing. The fixed blades also act as diffusers providing a further increase in pressure to the gas passing through them. The quantity of air delivered is more influenced by delivery pressure than is the case in centrifugal machines – see Fig 14 – and the characteristic curves are somewhat flatter – see Fig 15.

Axial flow machines offer a considerable economy in overall size over centrifugal machines for handling large volumes of air at high efficiency. A centrifugal machine is, however, normally employed where the quantities to be handled are at the lower end of the range covered by aerodynamic machines.

Compressor Lubrication

LUBRICATION is an extremely important parameter in compressor performance, its function being to dissipate frictional heat, reduce wear on sliding and stationary surfaces, reduce internal leakage and protect parts from corrosion — in addition the lubricant can serve the duty of flushing away wear products, defects and moisture.

Both the choice of lubricating system and the choice of lubricant are important, particularly as lubrication requirements differ with different types of compressor and are in different sections of the machine in the case of reciprocating compressors.

The significance of correct lubrication is that it ensures

(i) more continuous and more efficient working of the compressor

(ii) reduced maintenance costs

(iii) reduced power consumption

(iv) reduced lubricant costs.

Splash lubrication with controlled feed. (Atlas Copco)

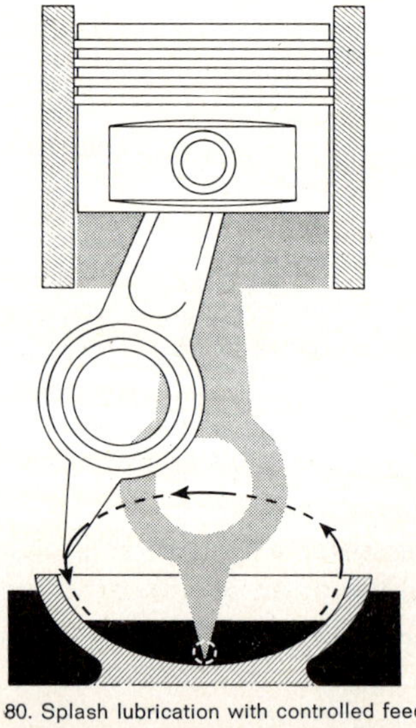

80. Splash lubrication with controlled feed

In the case of reciprocating compressors the simplest arrangement is *splash lubrication,* whereby a projection on the connecting rod big end dips into oil held in the crankcase. The crankcase thus acts as an oil reservoir and must be closed with the exception of a small breather hole. Lubricant is dispersed mainly in the form of an oil mist reaching the bearings, surfaces and cylinders. For best control the splash oil should be taken from a trough. Failing this, some provision should be made to maintain the crankcase oil at a constant level. Water which collects in the bottom of the crankcase should be drawn off at regular intervals.

Splash lubrication is largely limited to small single-acting sectional compressors. It has a number of inherent disadvantages, notably:-

(i) the lubricant cannot be filtered and lubricant life is limited

(ii) lubrication during starting up may be marginal

(iii) the bearing film of lubricant is thin and is not supplied under pressure

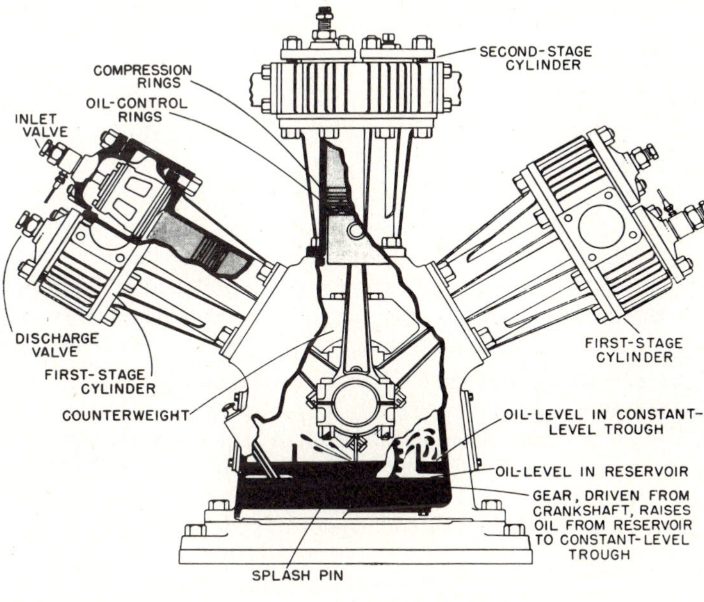

W-type compressor with splash lubrication of cylinders and crankcase.

Flood lubrication is widely used on horizontal compressors for lubricating bearings and cross-head. In this case the oil flow is under gravity after first being elevated by a disc, oil ring, chain or collar on the crankshaft which lifts it to a trough with distribution tubes or to the upper part of the journal(s). Basically, this provides better control of the amount of oil distributed than splash lubrication but the pressure developed is very low and not generally sufficient for a high speed compressor.

Force-feed lubrication utilizing a circulatory system is normally to be preferred. Feed is taken from an oil reservoir by a plunger or gear pump. Actual circulation may be by gravity, when the oil is pumped to an overhead reservoir and the circulation system is slightly more pressurized.

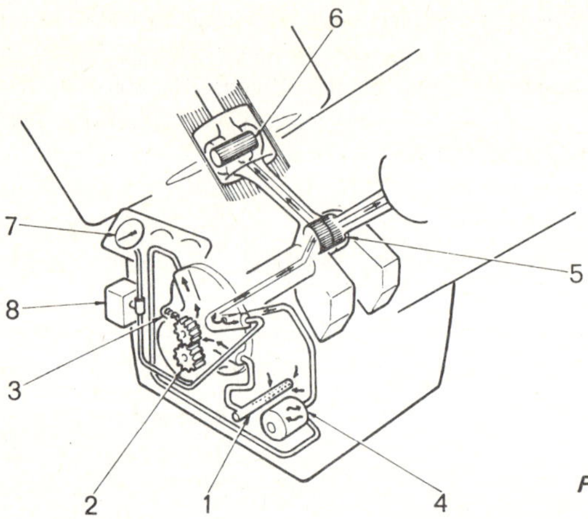

1 – oil strainer
2 – gear-type oil pump
3 – overflow valve
4 – oil filter
5 – crank bearing
6 – crosshead pin
7 – oil pressure gauge
8 – oil pressure switch

Force-feed lubrication (Atlas Copco)

Alternatively, the oil may be fed directly from the pump under pressure through pipes to the various lubricating points. Oil pressure is kept constant by a spring-loaded overflow valve. For cylinder lubrication of crosshead compressors special piston-type lubricators are normally used.

Sight-feed drip oilers are occasionally used to lubricate small parts. These are similar to cylinder lubricators and can be regulated within close limits provided they are kept in good condition. Siphon wick oilers may also be used when it is important that there should be no undue variation in oil level in the container. Grease lubrication may be provided at certain points (eg to lubricate rolling bearings). In this case the lubricating points are either fitted with grease nipples or screw-down grease cups.

In the case of cylinders the ideal oil feed is that which will provide efficient and effective lubrication with the minimum amount of oil, at the same time maintaining a perfect piston seal – an excess of oil is undesirable. A practical check is to service the discharge valves periodically and examine their appearance. They should not be dry or show signs of rust (insufficient oil) or be excessively wet (excess of oil). They should have a wet appearance and be oily to the touch. Excess of oil can also be noted by the appearance of pools of oil lying in the cylinders, or surplus oil on the connecting rods.

Rotary Compressors

Adequate lubrication is necessary in the case of vane-type compressors to minimize friction and wear on the vanes and/or wear ring and to assist in sealing. Quite high bearing pressures may be exerted by the leading edges of the vanes at the trailing edges of the slots, calling for a lubricant which forms a tenacious oil film. Bearing lubrication is normally separate, preferably by a force-feed lubricator although this is not allied to a circulatory system.

Oil for lubricating the vanes may be fed to the casing via the air intake, direct to the casing or to the rotor through its shaft. The first method is more usual, loading the intake air with oil by an atomizer. (Discharge air may be used to inject the oil into the intake). The other two methods involve force-feeding, either direct to the ends of the rotor or through a hole along the centre of the shaft. In the latter case interconnecting radial holes drilled along the length of the shaft or at the centre of the shaft throw out oil under centrifugal force.

Similar methods of lubrication may be employed in screw-type compressors although direct oil-injection is the usual method employed. Other types of rotary compressor (eg Roots blowers) may only require lubrication of shaft bearings and timing gears. Plain bearings can be lubricated by ring oilers and timing gears by splash lubrication. Where rolling bearings are used, bearings at the timing gear end may be lubricated by oil and those at the other end by grease.

Ring lubrication (Atlas Copco)

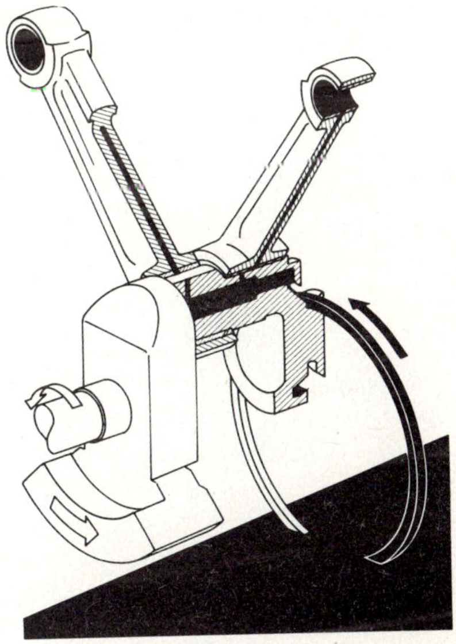

Dynamic Compressors

Dynamic compressors work with running clearance between the rotor and casing; thus only the bearings need lubrication. Bearings are normally accommodated in sealed housings outside the main compressor casing and may be lubricated by ring oilers, force-feed (particularly for larger high-speed compressors), or grease.

Lubricants

The primary requirements of a compressor lubricant are that it should

(i) reduce wear on sliding or rotating surfaces to a minimum
(ii) have good oxidation stability and not form gums or lacquers
(iii) prevent corrosion
(iv) be resistant to water-washing
(v) not emulsify with water

These requirements can be met by a carefully refined mineral oil suitable formulated, although the use of synthetic lubricants of phosphate ether have increased. Oil viscosity is also a significant parameter because of the high operating temperatures which may be involved and the consequent 'thinning' effect on the lubricant. Thus moderate to high viscosities are normally called for.

To meet all of these requirements a high grade oil is essential, formulated to match the type and characteristics of the compressor. In the case of reciprocating compressors, strictly speaking a higher grade of oil is required for cylinder lubrication than for bearing lubrication, although a single oil would normally be used for convenience of purchase and stock (and to avoid possible errors in use). There are exceptions, however – eg two different grades of oil are often used for the lubrication of multi-stage water cooled high pressure reciprocating compressors. There may also be a specific difference in the requirements for different lubricating systems. For example, lubricants used in systems applicable to reciprocating compressors need to be soluble, whereas lubricants used with oil-injection systems (a crew compressor) should not be volatile.

As far as the user is concerned, suitable oil grades are normally specified by the compressor manufacturer or by the oil companies. This, however, represents only the first part of the requirement. The others are that the lubricant should be stored under suitable conditions, be kept clean during handling and dispensing and applied in correct quantities.

Bristol Pneumatic BP 2113 compressor

Compressor Controls

IT IS A desirable economic feature that the output of a compressor should be subject to some form of automatic control so that it adjusts to varying demand levels. This is normally achieved by incorporating a sensing unit in the receiver or delivery line which governs the response of the control system. The control system used may be continuous or work on an on-off or *discontinuous* basis.

Continuous regulation is widely applied to i/c engine driven compressors, particularly portable compressors. The principle employed is of speed regulation of the driver to obtain continuous control of the compressor pressure – eg by means of a slide valve sensing the final pressure and controlling the position of the throttle valve – see Fig 1.

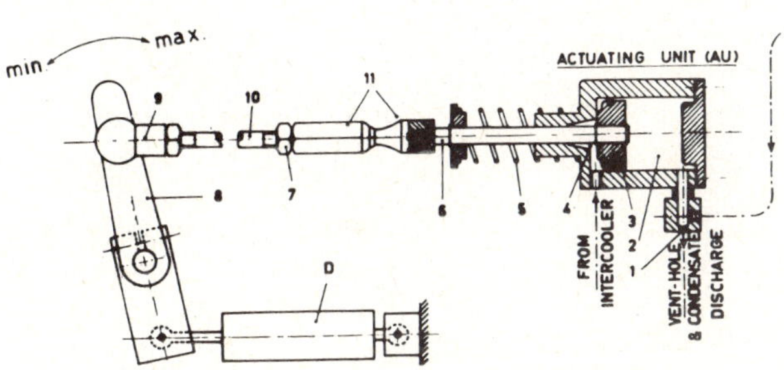

Fig 1 ***Speed regulating system for i/c engine compressor driver.*** *(Atlas Copco).*

D – damper, engine speed control lever
1 – air vent-hole and condensate discharge
2 – chamber, converted air receiver pressure
3 – piston
4 – chamber, constant intercooler pressure
5 – return spring, piston
6 – piston rod
7 – check nuts, ball joints
8 – engine speed control lever
9 – ball joint
10 – operating rod, actuating unit
11 – ball-and-socket joint

The first practical speed regulating control system was introduced in 1939, coupling running speed to air demand, and allowing the engine to idle when the demand for air was zero, thus reducing wear and tear on both engine and compressor and also reducing overall fuel consumption. This was followed by stepped controls enabling the engine speed and compressor speed to vary exactly, according to demand. This has subsequently become a more or less standard feature on all larger sizes of portable reciprocating compressors, although there are difficulties in adapting a similar form of control to rotary machines.

In the case of stationary compressors, the most common arrangement is discontinuous control. The simplest form of *discontinuous regulation* device is a blow-off valve on the delivery side. The compressor works continuously, but when a specified maximum pressure is reached, the valve opens to allow compressed air to be blown off to the atmosphere, instead of being delivered to the receiver. Such a control system is applicable to all types of positive displacement compressors.

Single and two-stage reciprocating compressors are commonly regulated by *unloading,* using a device of the type shown in Fig 2. Pressure from the receiver is transmitted to the underside of the piston of the device, which acts against a spring set at a pre-determined pressure. When the pressure in the receiver exceeds this figure, the spring compresses and allows air to flow to a piston on top of the suction valves, which in turn depress a claw which holds the suction valves off their seat, so that compression cannot take place in the cylinder. This is termed *suction valve unloading.*

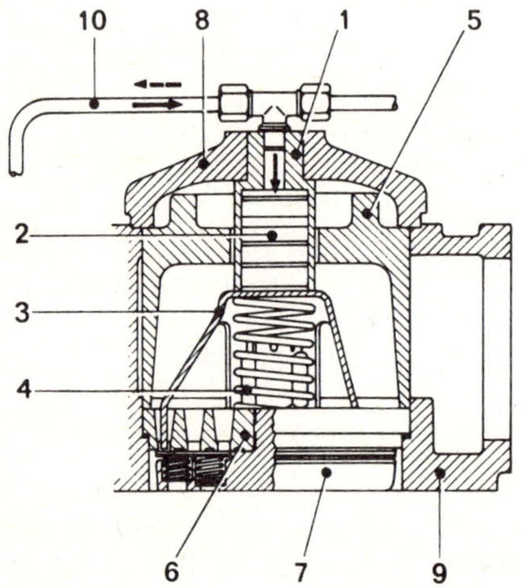

1 – unloading cylinder
2 – unloading piston
3 – unloading claw (depress the valve discs and keeps the suction valve open during the unloading period)
4 – unloading spring (returns unloading claw at start of loading)
5 – valve retainer
6 – seat of suction valve
7 – suction valve guard
8 – suction valve cover
9 – cylinder head
10 – compressed air to and from unloading device

Fig 2 Unloading device for piston compressors.

When double-acting compressors are used, by having two air relays set at different pressures, it is possible to unload each side of the piston in turn, giving full load, half load, and zero load. This is termed *three step unloading.*

In addition to three-step unloading there is a variation called *clearance pocket unloading.* The pockets are brought into communication with the cylinder by valves which enlarge the clearance volume and so reduce the output of the compressor. A normal method on double-acting compressors is to have a combination of clearance pocket and suction valve unloaders to give 100%–75%–50%–0% unloading. This is called *suction valve unloading.*

A further method of controlling the output is by throttling the intake, but this method requires more hp at part load conditions than the methods described above.

A different form of unloading is adopted in the case of screw compressors. The capacity is regulated by throttling the air inlet and, at the same time, opening a blow-off valve. The pressure side is shut off with a non-return valve. Under load the blow-off valve is closed and the throttle valve fully open. The unloading device is operated by a spring loaded automatic air relay of the same type as that used on reciprocating compressor controls. A complete installation is shown in Fig 3.

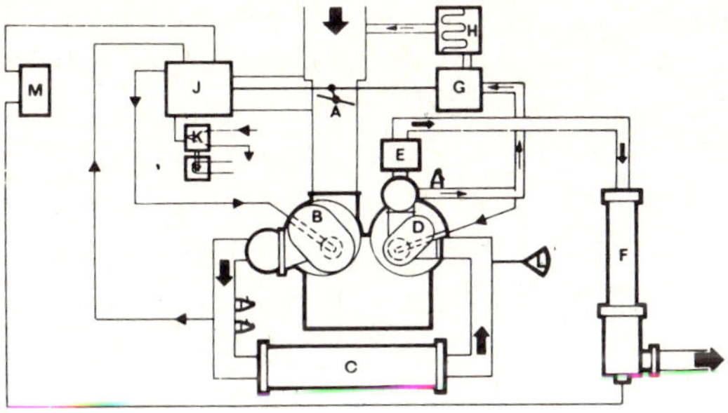

A – throttle valve
B – low-pressure stage
C – intercooler
D – high-pressure stage
E – non-return valve
F – aftercooler
G – blow-off valve
H – blow-off cooler
J – servo valve
K – solenoid valve
L – temperature switch
M – regulator.

Fig 3 Capacity unloading of screw compressors. (Atlas Copco).

Compressor controls may also be allied to monitoring devices, or monitoring systems, capable of providing automatic shut-down in the event of failure of the system approaching a dangerous condition. The number of monitoring devices required varies with the type and complexity of the system, and the exact number can be decided according to need.

Additionally, compressors driven by electric motors may be fitted with automatic *stop-start controls* which may also incorporate delay circuits – to time electronically the necessary interval after unloading the compressor and to stop the driver automatically after a preset idling time of the compressor.

All three main functions – stop-start, unloading and built-in delay circuits – may be built into a single regulator unit.

Temperature Controls

Where compressors employ closed-circuit water cooling (either thermo-syphon or forced circulation), additional cooling can be provided from the mains or similar cold water supply via a temperature regulator. This allows cold 'make-up' water to be supplied to the cooling tank at a controlled rate, hot water being displaced through an overflow. A typical installation of this type is shown in Fig 4.

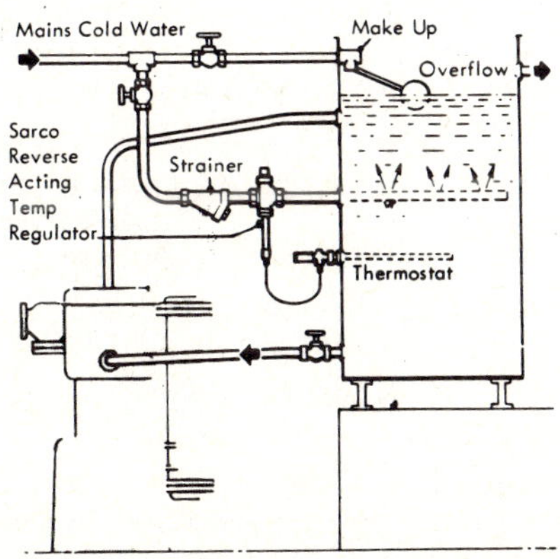

Fig 4 Temperature regulator controlling introduction of water to holding tank. (Sarco).

Automatic temperature control can also be applied to mains water cooling systems, the purpose in this case being both to eliminate wastage of water and to prevent excessive cooling which could cause condensation within the cylinder. A typical installation applied to a two-stage compressor is shown in Fig 5. It is important that the temperature regulator cannot shut off the supply of cooling water completely. This can be guarded against by incorporating an adjustable bleed in the valve itself (ensuring a small flow when the valve is shut); or the use of a bypass, the flow through which is controlled by a needle valve.

In the case of larger compressors operating with closed cycle, cooling is accomplished by passing the water through a mechanical cooler or a cooling tower. Again temperature control can prove advantageous, particularly to avoid too low a temperature being developed in the second stage cylinder.

An important consideration in the case of cooling towers is the possibility of freezing, with consequent cessation of water circulation calling for the compressor to be shut down. This possibility can be avoided by using a temperature regulator valve commanding a supply of steam to

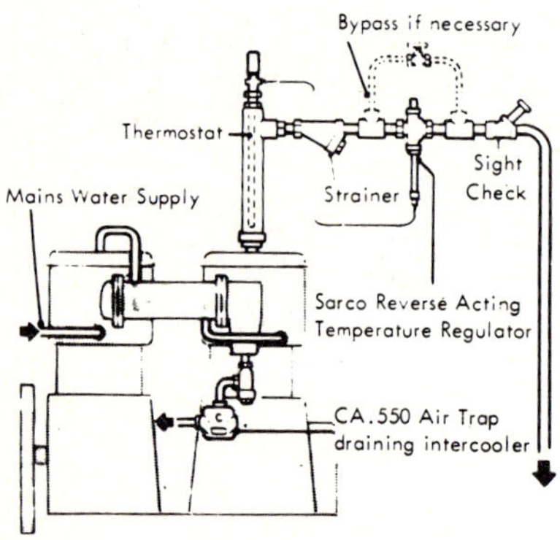

Fig 5 Temperature regulator controlling mains water cooling of two stage compressor. (Sarco).

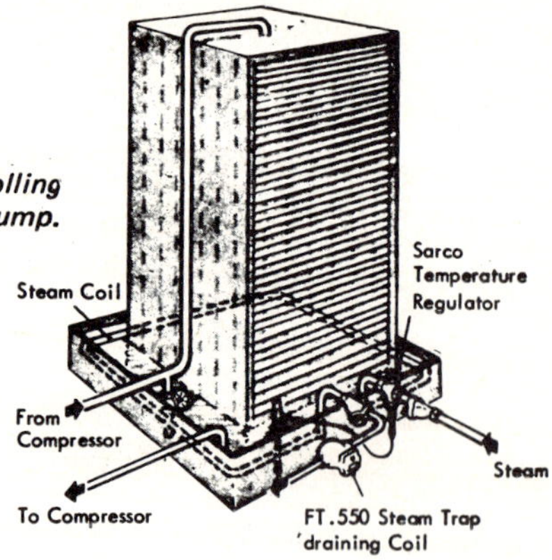

Fig 6 Temperature regulator controlling temperature of cooling tower sump. (Sarco).

a heating coil in the sump. The regulator can be set to open when the water temperature falls below, say, 1.7°C (35°F), so that no steam is used until near-freezing conditions are approached — Fig 6. To guard against the possibility of the cooling tower freezing solid, the line from the diversion control valve can be valved to return water from the compressor direct to the sump instead of the top of the tower under low temperature conditions.

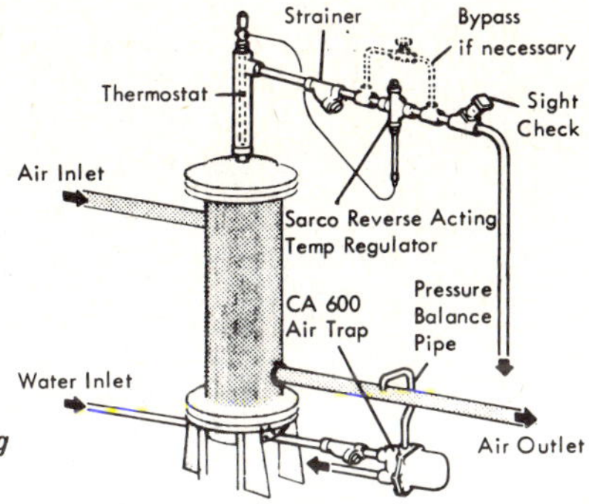

Fig 7 Temperature regulator controlling aftercooler. (Sarco).

Temperature regulators can also be used to advantage on water cooled aftercoolers. The use of a temperature regulator enables the water consumption to be set to the point of maximum efficiency, and maximum water economy. An example of the use of a temperature regulator controlling main water supply to an aftercooler is shown in Fig 7.

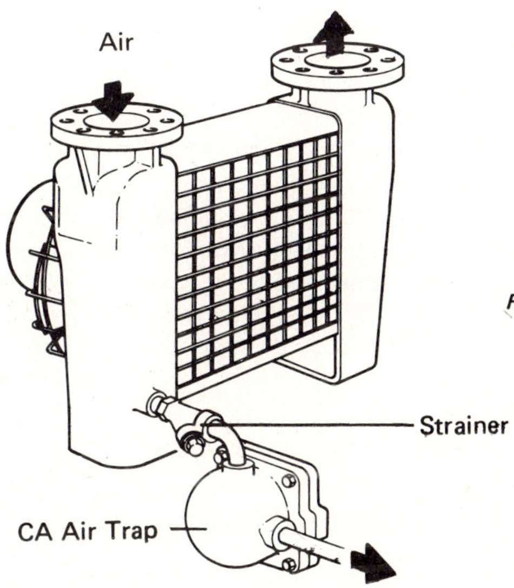

Fig 8 Air blast aftercooler (Sarco)

Water cooled aftercoolers are usually the most efficient and the lower the air temperature they can produce the better. Air cooling has also come more to the fore in recent years, particularly in areas where cooling water is expensive or not readily available. An air-blast aftercooler is then usually the first choice. The principle on which these work is that an electric motor driven fan directs a stream of air over a bank of finned tubes through which the compressed air flows — e.g. see Fig. 8.

Receivers and Pressure Vessels

AN air receiver following a compressor not only provides storage capacity for the compressed air delivered but also serves to equalize pressure variations in the supply lines. It may also provide a useful source of additional cooling and collection of residual condensate in the compressed air delivered by the compressor.

A compressor is normally followed by one (or more) receivers designed to provide storage capacity for the compressed air delivered. This capacity is also used to equalize pressure variations in the supply lines and reduce pressure pulsations, hence the capacity required is related not only to the compressor capacity, but also to the demand and the regulating system. A receiver may also provide the supplementary duty of increased cooling and collection of residual condensate.

The useful storage capacity of a receiver is the volume of free air which can be drawn from the receiver over the required, or permissible, working pressure range. The size of the receiver must, therefore, be proportioned to provide the storage capacity required. For constant demand the volume of the receiver need not be greater than the output of the compressor, in terms of output at working pressure, since the receiver need only operate as a pressure damper to maintain uniform pressure. Where the consumption is only intermittent, or variable, the receiver must have a margin of storage capacity relative to the compressor output, to adjust to demand and damp out pressure fluctuations.

For constant demand the theoretical size of receiver required is:-

$$\text{Volume of receiver } (V_R) = \frac{C_o \times \text{atmospheric pressure}}{P_a}$$

$$= \frac{C_o \times \text{atmospheric pressure}}{P_g + \text{atmospheric pressure}}$$

where P_a = working pressure, absolute

= gauge pressure P_g + atmospheric pressure

If the demand is constant and steady, the size need not be proportioned to the output of the compressor, **since** under such conditions the function of the receiver is merely to damp out pulsation pressures, so as to maintain a constant pressure. Normally, however, a receiver volume matched to the compressor output is regarded as the minimum size required (calculated on the basis of the compressor output in one minute). If the demand is variable, then a larger receiver volume is necessary to prevent rapid fluctuations of pressure. The assessment of a suitable size is largely arbitrary for general applications, and is commonly taken as three times the minimum size (ie three times V_R calculated on the basis of compressor output).

Thus, for variable demand:-

$$V_R = \frac{3 \times P_a \times C_o}{P_g + P_o}$$

where P_o = atmospheric pressure

P_g = working pressure, gauge

The optimum size of receiver will also vary with working pressure, tending to be larger for low pressures and smaller for high pressures. Special calculations for receiver size are required for cyclic working.

The useful storage capacity of a receiver is the volume of free air which can be drawn from it over a permissible range of pressure drop. This can be determined arithmetically as

$$\text{Useful storage capacity} = \frac{V_R \times \Delta P}{P_o}$$

where ΔP = permissible pressure drop

P_o = atmospheric pressure

Storage capacity and the time over which this capacity can be delivered are given by the relationship

$$V_R = \frac{P_o \times V_d \times t}{\Delta P}$$

$$\text{or} \quad t = \frac{V_R \;\; P}{P_o \times V_d}$$

where V_d = air volume requirement (volume of air drawn)

t = time in minutes over which this air is to be drawn

This avoids any possible misinterpretation of 'pressure drop' since specific figures are used, ie P_2 is the minimum supply pressure required. On the other hand the original formula applies where permissible pressure drop is specified. The two formulas are, of course, identical since

$$P = P_1 - P_2$$

With the *bar* as the standard unit of pressure

$$V_R = \frac{C_o}{P_g + 1} \quad \text{(constant demand)}$$

$$= \frac{3 \times C_o}{P_g + 1}$$

where P_g = gauge pressure in bars

For a pressure drop of 1 bar

$$t = \frac{V_R}{V_d}$$

For a pressure drop of 15 lbf/in²

$$t = \frac{V_R}{V_d} \quad \text{(approx)}$$

For a pressure drop of 10 lbf/in²

$$t = 0.7\frac{V_R}{V_d} \quad \text{(approx)}$$

These formulas assume that no air is supplied to the receiver during the time interval at which air is being drawn off. If the receiver is being continuously supplied with air from the compressor at a constant rate during this interval, the appropriate formula is

$$\text{time (minutes)} = \frac{V_R \times \Delta P}{V_a \times (V_d - S)}$$

where S is the delivery of the compressor, in the same units as V_d

In English units:

$$t \text{ (minutes)} = \frac{V_R \times \Delta P}{14.7(V_d - S)}$$

For a permissible pressure drop of 15 lb/in² this simplifies to:-

$$t \text{ (minutes)} = \frac{V_R}{(V_d - S)} \quad \text{(approx)}$$

And for a permissible pressure drop of 10 lb/in^2 :-

$$t \text{ (minutes)} = \frac{0.7V_R}{(V_d - S)} \quad \text{(approx)}$$

In metric units:

$$t \text{ (seconds)} = \frac{0.07V_R \times P}{V_d}$$

where V_R = receiver volume in metres^3
V_d = delivery required in metres^3
ΔP = pressure drop in bar

Thus, for a permissible pressure drop of 0.7 bar

$$t = \frac{0.05V_R}{V_d} \quad \text{(approx)}$$

For a permissible pressure drop of 1 bar

$$t = \frac{0.07V_R}{V_d}$$

In practice it may be necessary to restrict the receiver inlet so that an excessive amount of air may not be withdrawn under heavy demand. Thus, even with continuous charge, the charging rate may be relatively low in such cases, ie when operating under heavy demand, and it may be safer to size the receiver on the basis that it does not receive any charge during the time of heavy demand.

Cycling Times with Stop-Start Control

For a given compressor and a given receiver the maximum number of starts, with the compressor operating on automatic stop-start control, will be obtained when the demand (V_d) is half the FAD of the compressor. Basically, the compressor running time will be 30 minutes in every hour, and during the charging period half of the FAD will be absorbed by the demand and the other half will go to charge the receiver. The actual number of stops and starts will depend on the size of the receiver.

$$\text{Free air available from receiver} = \frac{V_R \, \Delta P}{P_a}$$

$$V_d = S/2$$

where S = FAD of compressor

Time taken to use up receiver air

$$= \frac{V_R \, \Delta P}{P_a} \times \frac{1}{S/2}$$

$$= \frac{2V_R \, \Delta P}{P_a \cdot S}$$

This represents one half the total time

$$\text{Hence total time} = \frac{4\,V_R\,\Delta P}{P_a \cdot S} = \frac{60}{n}$$

$$\text{where } V_R = \frac{15 P_a\,S}{n\,\Delta P}$$

where n = number of stops and starts

For a pressure drop of 1 bar (14.5 lb/in^2) $V_R = \dfrac{15S}{n}$

For a pressure drop of 0.7 bar (10 lb/in^2) $V_R = \dfrac{22S}{n}$

From this the receiver size can be determined consistent with a given maximum number of stops and starts. General recommendation is that the maximum number of stops and starts should not exceed 12 to 15 starts per hour, using conventional automatic stop-start switches. If necessary, however, much higher values of n are possible using modern and switchgear.

Useful data on receiver sizes are summarized in Tables I, IIA and IIB.

The receiver is normally placed as close to the compressor as possible in order to keep the discharge pipe short and eliminate surge effects. In practice the receiver follows the intercooler and separator and may, in fact, be located in a separate block in order to be sited in a cool location. This does not nullify the requirement above.

TABLE I – TYPICAL RECEIVER SIZES

Receiver Volume		Matching Compressor Capacity (FAD)			
		Constant Speed Control*		Stop-Start Control†	
ft^3	m^3	ft^3/min	m^3/min	ft^3/min	m^3/min
4.5	0.125	45	1.25	25	0.71
11	0.31	110	3.1	55	1.6
20	0.57	190	5.4	100	2.8
34	0.96	340	9.6	170	4.8
55	1.6	550	15.6	275	7.8
100	2.8	1 000	28	500	14
150	4.2	2 000	57	1 000	28
225	6.4	3 000	85	1 500	42.5
325	9.2	4 500	127	2 250	63
430	12.2	6 000	170	3 000	85

Note:

* This is consistent with the general rule that a receiver capacity of one-tenth the compressor FAD should be adequate for average systems.

† This is consistent with the general rule that with automatic stop-start control the receiver size can be one half that necessary with constant speed control.

Note also that with increasing compressor capacity the relative size of receiver can usually be reduced from the above proportions.

TABLE IIA – USEFUL CAPACITY OF AIR RECEIVERS (cubic metres of free air)

Receiver Volume m^3	Permissible Pressure Drop – bar							
	0.25	0.5	0.75	1	1.25	1.5	2	2.5
1	0.25	0.5	0.75	1	1.25	1.5	2	2.5
2	0.5	1	1.5	2	2.5	3	4	5
3	0.75	1.5	2.25	3	3.75	4.5	6	7.5
4	1	2	3	4	5	6	8	10
5	1.25	2.5	3.75	5	6.25	7.5	10	12.5
6	1.5	3	4.5	6	7.5	9	12	15
7	1.75	3.5	5.25	7	8.75	10.5	14	17.5
8	2	4	6	8	10	12	16	20
9	2.25	4.5	6.75	9	11.25	13.5	18	22.5
10	2.5	5	7.5	10	12.5	15	20	25

TABLE IIB – USEFUL CAPACITY OF AIR RECEIVERS (cubic feet of free air)

Receiver Volume ft^3	Permissible Pressure Drop – lb/in^2							
	5	10	15	20	25	30	35	40
1	0.34	0.7	1	1.3	1.7	2	2.3	2.7
2	0.68	1.4	2	2.7	3.3	4	4.7	5.4
3	1.0	2.0	3	4	5	6	7	8
4	1.4	2.8	4	5.2	6.6	8	9.4	10.8
5	1.7	3.4	5	6.7	8.4	10	11.7	13.4
6	2.0	4.0	6	8	10	12	14	16
7	2.4	4.8	7	9.4	12	14	16	19
8	2.7	5.4	8	11	13	16	19	21
9	3.0	6.0	9	12	15	18	21	24
10	3.4	6.8	10.2	13.5	17	20	23.5	27
20	6.8	13.6	20.4	27	34	40	47	54
30	10.2	20.4	30.6	40.5	51	61	71	81
40	13.6	27.2	40.8	54	68	82	94	108
50	17.0	34.0	51	68	85	102	118	136
60	20.4	40.8	61.2	81.5	102	122	143	163
70	23.8	47.6	71.4	95	119	143	167	190
80	27.2	54.4	81.6	109	136	163	190	217
90	30.6	61.2	91.8	122	153	183	216	244
100	34.0	68.0	102	136	170	204	238	272

A stop valve should not be fitted between the compressor and receiver unless a safety valve is also incorporated between this valve and the compressor, to guard against the possibility of the compressor being started up with the stop valve closed. When a compressor is connected to an air mains in common with other compressors, a safety valve must be fitted between the compressor and the first stop valve, for similar reasons.

Examples of vertical receivers (Drury Eng. Co. Ltd)

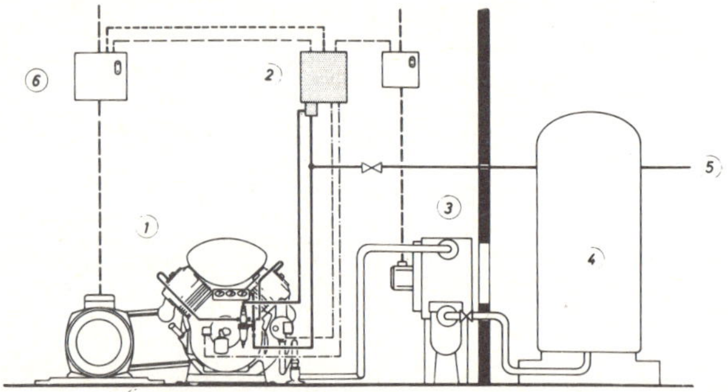

1. *Compressor/motor unit*
2. *RD regulator*
3. *Aftercooler*
4. *Receiver*
5. *Delivery line*
6. *Starting equipment*

Compressor installation showing recommended position of receiver (Atlas Copco)

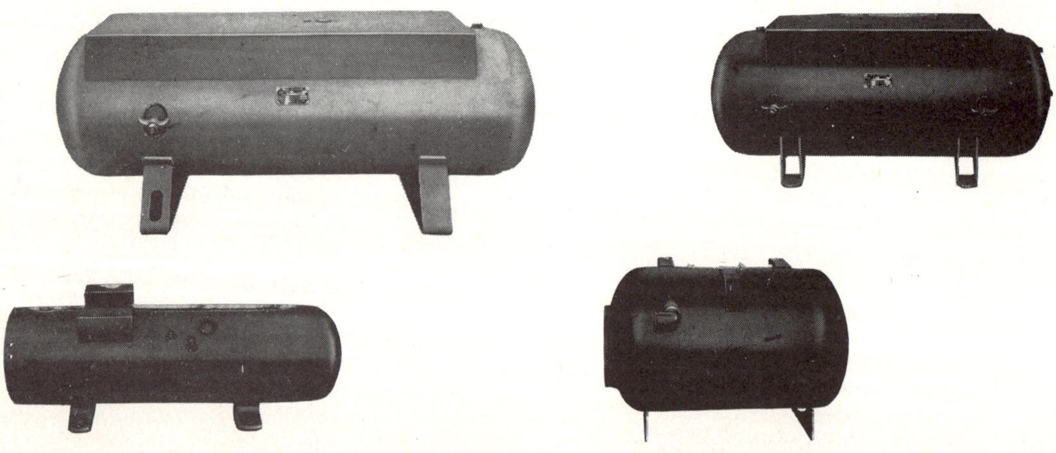

Examples of horizontal receivers (Drury Eng. Co. Ltd)

Pressure Vessel Design

The usual form of pressure vessel is a cylindrical shell with dished ends. Material stress in the cylinder walls can be determined as

$$S = \frac{PD}{2t}$$

where S = material stress
P = internal pressure
D = overall diameter of shell
t = wall thickness

Expressed in terms of wall thickness required

$$t = \frac{PD}{2S} = \frac{PR}{S}$$

when R = radius of shell

This formula is valid where P is the maximum permissible internal pressure consistent with a maximum permissible material stress, provided:-

(i) The material stress figure allowed is below the limit of proportionality of the material.
(ii) The material stress figure is determined for the maximum service temperature of the pressure vessel.

The formula does not make any allowance for stress raisers, effects of corrosion, and method of construction. These can be allowed for separately, or taken into account in determining the maximum permissible material stress.

Recommended practice is to base maximum permissible material stress on the upper yield point of the material, with a factor of safety applied to this value, viz

$$\text{Maximum permissible stress} = \frac{\text{upper yield point}}{f}$$

For ISO/TC semi-killed and killed structural steels, f = 1.92; ie maximum permissible material stress = 0.521 x upper yield point. For pressure vessel steels and ISO/TC17 structural steels, f = 1.6; ie maximum permissible material stress = 0.625 x upper yield point.

Typical upper yield point values are shown in Tables IIIA, IIIB and IIIC.

With *welded* construction an additional safety factor is normally called for — typically 0.8 for conventional manual and machine welding with normal inspection. Where the welding is subject to strict control and inspection, a higher value of 0.9 may be used.

An additional allowance may also be made for corrosion reducing the effective thickness with age. This takes the form of a simple addendum to calculated minimum thickness, a typical value being 0.75 mm (0.03 in).

TABLE IIIA – PRESSURE VESSEL STEELS

Specification	Upper Yield Point at Temperature of 20°C		100°C		150°C		200°C	
	lb/in²	bar	lb/in²	bar	lb/in²	bar	lb/in²	bar
UK								
BS 1501 –								
151 Grade 23B	29 300	2 060	25 000	1 760	24 500	1 720	23 000	1 640
161 Grade 23B	29 300	2 060	25 000	1 760	24 500	1 720	23 300	1 640
211 Grade 26B	34 300	2 410	29 700	2 090	29 300	2 060	28 700	2 020
211 Grade 28B	37 000	2 600	32 000	2 250	31 600	2 220	30 800	2 170
Germany								
DIN 17155 –								
HI	32 700	2 300	–	–	–	–	25 600	1 800
HII	37 000	2 600	–	–	–	–	29 900	2 100
HIII	39 800	2 800	–	–	–	–	34 100	2 400
17Mn4	41 250	2 900	–	–	–	–	35 500	2 500
Sweden								
INSTA –								
1B, C, D (SIS 1330)	34 100	2 400	32 000	2 250	30 600	2 150	29 900	2 000
2B, C, D (SIS 1330)	37 000	2 600	34 800	2 450	33 400	2 350	32 000	2 200
3B, C, D (SIS 1430)	39 800	2 800	37 700	2 650	35 600	2 550	34 800	2 400
4B, C, D (SIS 2101)	48 400	3 400	46 200	3 250	44 800	3 150	44 100	3 000
USA								
ASTM A 285B	27 000	1 900	25 400	1 800	24 200	1 700	20 000	1 400
ASTM A 201A	30 000	2 100	27 700	1 950	27 000	1 900	25 000	1 750

TABLE IIIB – ISO/TC 17 SEMI-KILLED AND KILLED STEELS

Designation	Upper Yield Point at Temperature of 20°C		125°C		200°C	
	lb/in²	bar	lb/in²	bar	lb/in²	bar
Fe 37B	34 000	2 400	29 500	2 070	–	–
Fe 42B	37 000	2 600	33 000	2 320	–	–
Fe 44B	40 000	2 800	36 000	2 520	–	–

TABLE IIIC – ISO/TC 17 STRUCTURAL STEELS

Designation	Upper Yield Point at Temperature of 20°C		125°C		200°C	
	lb/in²	bar	lb/in²	bar	lb/in²	bar
Fe 37C–D	34 000	2 400	–	–	25 600	1 800
Fe 42C–D	37 000	2 600	–	–	30 000	2 100
Fe 44C–D	40 000	2 800	–	–	33 000	2 300
Fe 50D	51 000	3 600	–	–	42 500	3 000

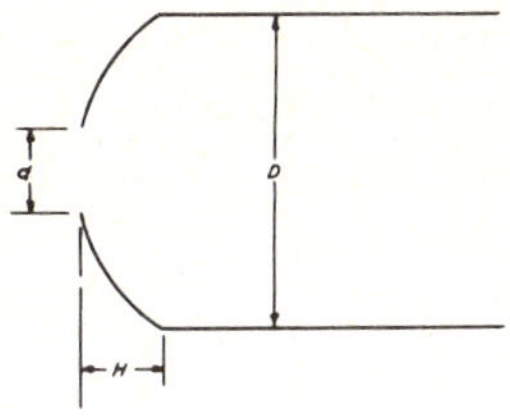

Fig 1

End Shapes

The form of the ends of pressure vessels is normally dished, when the stress formula is modified by a shape factor, ie

$$\text{Minimum thickness of shell } (t) = F \frac{RP}{S} + e$$

where F = shape factor

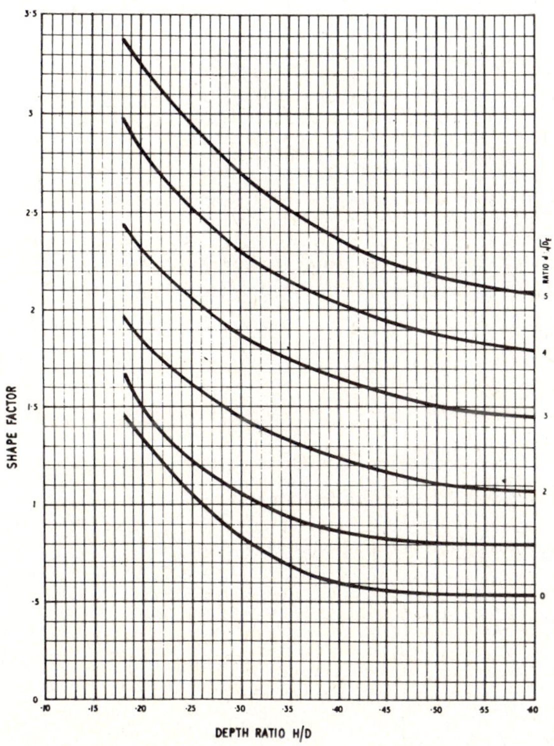

Fig 2

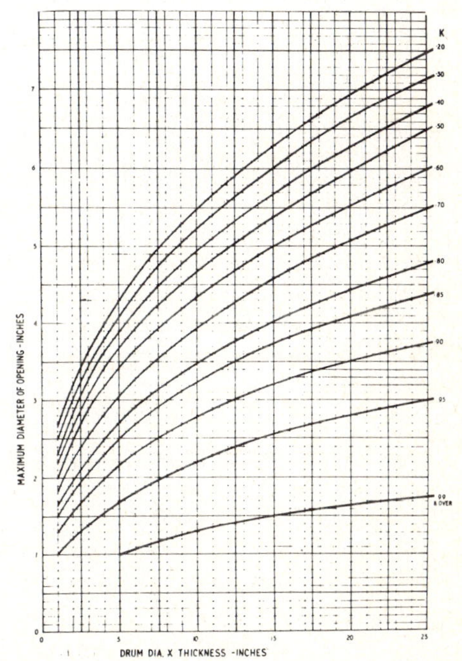

Fig 3

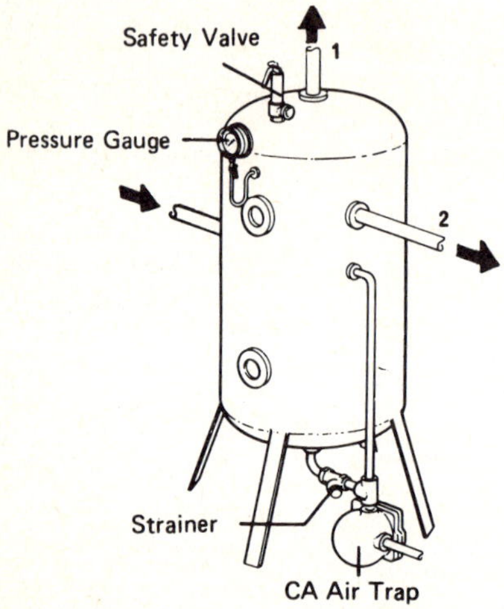

Typical vertical receiver with air trap (Sarco)

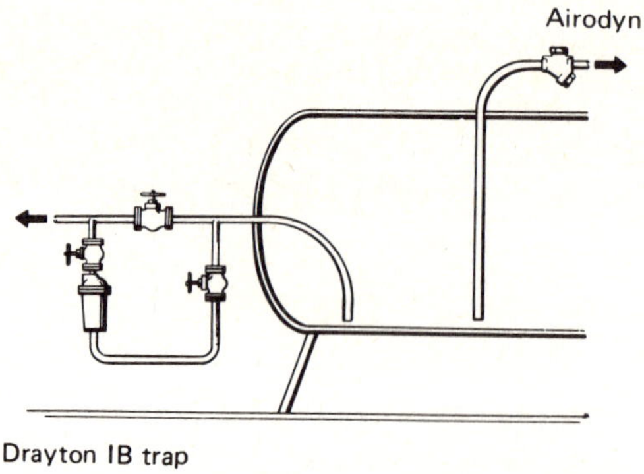

Small horizontal receiver with alternative trapping arrangements (Sarco)

The shape factor is related to the depth ratio of the dished (drumhead) end, expressed as the height of the drumhead (H) divided by the diameter (D) – see Fig 1. It also takes into account the weakening effects of any opening which may be necessary, the opening being of diameter d. In the absence of an opening, the shape factor is substantially less than unity, with increasing depth ratio. If there is an opening the effect is most readily expressed in terms of the ratio.

$$d : \sqrt{D} \text{ (as in Fig 2)}$$

Solutions in such terms are based on the opening being unreinforced.

It is necessary to establish a maximum permissible diameter for unreinforced openings because the shape factor increases rapidly with increasing diameter of opening for a given shell diameter. This is specified in the PNEUROP code in terms of a factor K related to the product of drumhead diameter and thickness. The value of K is calculated from the formula

$$K = \frac{DP}{1.82St}$$

where S is the maximum permissible material stress.

Maximum permissible diameters for unreinforced openings are then given by Fig. 3.

Heat Exchangers

COOLING may be advantageously, or necessarily, applied to compressors to achieve the following ends:—

(i) Maintain working clearances on machine components.

(ii) Limit the amount of heating applied to the inducted charge, thus improving compression efficiency.

(iii) Cool the compressed air between stages to improve compression efficiency.

(iv) Cool the delivered air to remove water (after cooling).

All practical compressors work under adiabatic conditions with some heat loss. The amount of air heating is, however, considerable. This may not be an immediate disadvantage at low compression ratios, except that the air will be saturated with water and aftercooling could remove much of this. Unless the machine itself specifically demands cooling, therefore, a low compression ratio compressor may work satisfactorily uncooled (eg a single-stage centrifugal blower). If the machine does demand cooling (eg reciprocating compressors), the necessary heat exchange may be provided by finned cylinders with or without forced circulation. The chief limitation which applies then is that if the machine is not adequately cooled it will transfer heat to the inducted charge, which can seriously reduce the efficiency of the compressor.

Interstage Cooling

Where more than one stage of compression is involved, it is desirable from the point of view of maximum efficiency that each stage does an equal amount of work. This can only be realized if the temperature of the air fed into the second stage is similar to that of the air inducted into the first stage. Machine cooling will remove only a proportion of the heat generated by compression, because only a proportion of the air passing through the compressor will come into contact with the cooled surfaces of the cylinder walls. Thus, with an efficient water cooling system applied to a reciprocating compressor cylinder, only about 10—15% of the heat generated by compression will actually be removed. Cooling of the air itself, or intercooling, is thus necessary between stages, to bring the air temperature down to more acceptable levels. This represents an independent problem, involving the use of a separate heat exchanger.

Air Cooling

In the case of very small single-stage compressors, natural air cooling may be sufficient, using finned cylinders. The performance of such a compressor can be expected to vary somewhat with changes in ambient temperature, as the cooling available is only relative and not specific. Forced draught cooling is generally preferred, eg generated by an engine driven fan or from a power take-off on the compressor spindle, as this can provide both more effective heat dissipation and more consistent heat transfer performance. If the compressor is located in a warm atmosphere it may be advantageous to draw the cooling air from a lower temperature source, eg through ducting led to the outside of the building. The degree of cooling required by a compressor is appreciably higher than that considered the optimum with an internal combustion engine. The aim is to lower the compressor temperature as far as possible, rather than to maintain it as a steady, but fairly high 'running' temperature.

Forced draught air cooling may be adequate for both small and large compressors, particularly if the latter are low capacity or multi-stage machines. With multi-stage machines, the increase in temperature per stage is lower than that of a single-stage machine. Air cooling may, however, have distinct limitations where the compressor has a moderate to high capacity and is run continuously.

TABLE I – SINGLE-STAGE COMPRESSION

(Inlet Temperature 21°C (70°F)

Volume	Isothermal Compression Absolute Pressure	Isothermal Compression Air Temperature °C	Isothermal Compression Air Temperature °F	Adiabatic Compression Absolute Pressure	Adiabatic Compression Air Temperature °C	Adiabatic Compression Air Temperature °F
V_1	P_1	21	70	P_1	21	70
$V_2 = V_{½}$	$P_2 = 2\,P_1$	21	70	$P_2 = 2.64\,P_1$	115	239
$V_2 = V_{¼}$	$P_2 = 4\,P_1$	21	70	$P_2 = 6.97\,P_1$	240	463

Water Cooling

Most larger single-stage compressors, and the bulk of multi-stage compressors, are water cooled. In the case of reciprocating compressors, the principle adopted is basically similar to that of internal combustion engine cooling, using water jackets for the cylinder barrels and cylinder heads. These can either be cored into the cylinder and head castings, or added as fabricated jackets. An alternative method sometimes used is to immerse the whole cylinder in a bath.

The temperature of the cooling water should be as low as possible and the rate of water change high to reduce the compressor temperature as much as possible. In a closed system this will demand the use of a further heat exchanger to cool the circulating water and maintain lower water inlet temperature, which again can follow automobile practice – eg a radiator exposed to a forced draught provided by a fan. In larger installations other types of water cooler, such as cooling towers, may be used to ensure a supply of cold water to the cooler inlet.

Centrifugal compressors are generally cooled by one of three methods:–

(a) *Diaphragm cooling* – which implies that the return and diffuser passages are water jacketed with a supply of coolant water through these jackets.

(b) *Injection cooling* – which is direct cooling produced by spraying cold water into the air passing through the machine.

(c) *Intercooling* – applied between stages.

Diaphragm cooling has the advantage that the cooling of each stage can be separate and regulated or adjusted accordingly, to give close control over temperature.

Injection cooling has the advantage that it reduces the temperature rise of the gas during compression without removing heat from the system.

Intercooling has little advantage unless the stage compression exceeds about 3:1, when it can reduce power consumption. It can be applied between successive stages, or at intervals of two or three stages, as necessary.

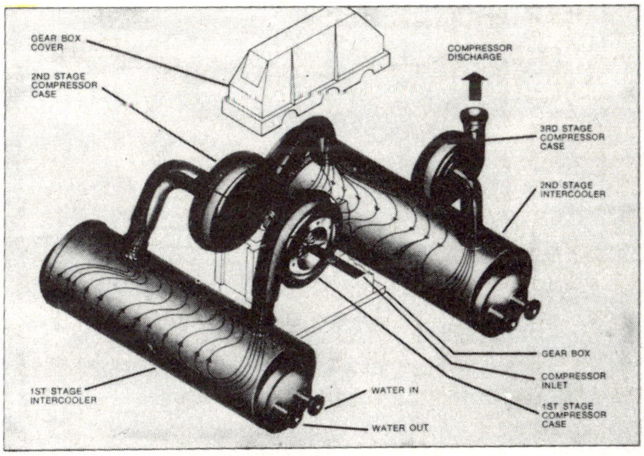

Example of compressor intercooling

Where the compressor itself is water cooled, then it is usual for the intercooler also to be water cooled. In this case the intercooler normally takes the form of a nest of straight tubes in a cylindrical shell. Cooling water is then fed through these tubes, passing several times up and down the length of the intercooler. Compressed air from the preceding stage flows through the shell and over the tubes, and is further directed across the bank of tubes by baffles as necessary.

Two-shell intercooler construction is often preferred as providing two stages of intercooling and higher heat transfer for the same amount of cooling water. Each shell is a complete heat exchange unit, with its own nest of tubes and supply of cooling water. The compressed air flow is through each unit in turn, ie series flow. For most efficient cooling, with the lowest water flow, the coldest water is supplied to the final unit.

Other types of cooler include co-axial pipe designs and submerged coils. In the latter case a continuous coil of tubing carrying the air is submerged in a tank through which cooling water is circulated.

Water cooled intercoolers may be used with air cooled compressors, although the more usual arrangement here would be an air cooled intercooler. Air cooled intercoolers may also be used with water cooled compressors.

An air cooled intercooler normally comprises one or more lengths of finned tubing between headers. These lengths may be straight or coiled, depending on the heat exchange area required. Cooling is then assisted by forced draught, eg generated by a fan.

All intercoolers will condense out water and so must be provided with a drain. Treatment varies from the fitting of a simple drain tap at the lowest point of the intercooler, or the fitting of a drain leg, to the incorporation of a specific separator or separator chamber in the design of the intercooler.

Heat Exchanger Calculations

The general heat transfer equation is

$$AW = A R K V^{0.6} S \Delta t$$

where A = a constant*
R = exchange rate value
K = average heat transfer coefficient*
V = ventilating airspeed*
S = total heat exchange area*
Δt = temperature difference* between the final temperature of fluid and surroundings.

*numerical values of these parameters are dependent on the units employed — see Table II.

TABLE II

Parameter	Practical Units Metric	Practical Units English
Power loss (Δ W)	kilowatts	horsepower
Constant (A)	0.0048	0.00068
Exchange rate value (R)	$10V^{0.6}$ *	$10V^{0.6}$ *
Ventilating air speed (V)	metres/sec	feet/sec
Average transfer coefficient (K)	kcal/hour/m^2/°C	BTU/hour/ft^2/°C
Total heat exchange area (S)	$metres^2$	$feet^2$
Temperature difference (Δ t)	°C	°C

*Specific value expressed in terms of ventilating air speed

The exchange rate (R) is a function of the ventilation present. In a fully enclosed system R = 1. In such a case, if the ventilating air speed (V) was zero (ie static air conditions inside the enclosure), there would be no heat transfer and the enclosure would simply assume the temperature of the fluid. In practice, the enclosure surface radiates some heat and V would assume a positive but small value, due to ascending air currents inside the enclosure, but the amount of cooling produced would be very small.

In a fully open or exposed system, surrounded by static air, the value of R approximates to 2. Any natural draught will increase this value until, with slightly forced ventilation, a value of 8 will be reached. With strong forced draught the value of R may rise to as high as 12 — eg this is typical of the exchange rate value achieved with a car-type radiator and fan cooling. A specific value of R can be calculated as $10 \times V^{0.6}$.

The average transfer coefficient is less readily determined as, besides the design of the heat exchanger and the materials used, this is also dependent, to some extent, on the fluid velocity and the temperature difference. Typical values normally lie between 3 and 8 kcal/hour/m^2/per degree C.

TABLE III – TYPICAL COOLING WATER REQUIREMENTS (Gallons per 100 cu ft free air)

Component	Pressure		Water Temperature			
	lb/in^2 (gauge)	bar(gauge)	16°C (60°F)	21°C (70°F)	27°C (80°F)	32°C (90°F)
Compressors:						
Cylinder Jacket:						
Single-stage	40	2.8	0.5–0.6	0.7	0.8	0.9
	60	4.2	0.7–0.8	1.0	1.1	1.2
	80	5.6	0.9–1.2	1.3	1.4	1.5
	100	7.0	1.2–1.5	1.6	1.7	1.8
Two-stage	80–120	5.5–8.5	1.0	1 1	1.2	1.4
Intercoolers:						
In series with cylinder jackets	80–120	5.5–8.5	2.8	3.4	4.0	4.5
Separate	80–120	5.5–8.5	2.5	3.0	3.5	4.0
Aftercoolers:						
Single-stage	80–120	5.5–8.5	4.0	4.5	5.2	6.0
Two-stage	80–120	5.5–8.5	2.5	3.0	3.5	4.0

TABLE IIIA – TYPICAL COOLING WATER REQUIREMENTS (Litres per cubic metre free air)

Component	Pressure		Water Temperature			
	bar(gauge)	lb/in^2 gauge	16°C (60°F)	21°C (70°F)	27°C (80°F)	32°C (90°F)
Compressors:						
Cylinder Jacket:						
Single-stage	2.8	40	0.8–1.0	1.15	1.3	1.45
	4.2	60	12.–1.3	1.6	1.8	1.95
	5.6	80	1.45–1.95	2.1	2.25	2.4
	7.0	100	2.0–2.4	2.6	2.75	2.9
Two-stage	5.5–8.5	80–120	1.6	1.8	1.95	2.25
Intercoolers:						
In series with cylinder jacket	5.5–8.5	80–120	4.5	5.4	6.5	7.3
Separate	5.5–8.5	80–120	4.0	4.85	5.7	6.5
Aftercoolers:						
Single-stage	5.5–8.5	80–120	6.5	7.3	8.5	9.7
Two-stage	5.5–8.5	80–120	4.0	4.85	5.7	6.5

The ventilating air speed is not readily determined quantitively. A typical value is commonly assumed for static installations, based on empirical fan data.

The total heat exchange area can readily be manipulated. Since the power loss is directly proportional to the surface area S, it follows that heat exchange and cooling will be directly increased in porportion to the surface area available. For convenience, the heat exchange surface area required can be related to the temperature difference required and the exchange rate value in the form of a graph – Fig 1. This assumes typical or average values for the other parameters involved. Such a graph can be used directly to assess the area of the radiating surfaces required to produce the temperature difference desired, or to estimate the temperature difference likely to be achieved by a particular design.

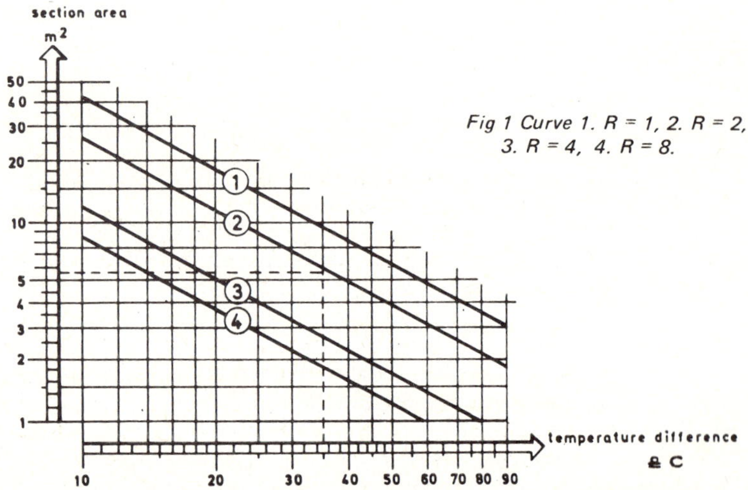

Fig 1 Curve 1. R = 1, 2. R = 2, 3. R = 4, 4. R = 8.

Similar analysis can, of course, be applied to water cooling. In this case the same general equation applies, except that the average transfer coefficient is much higher. The ventilating speed then applies to the flow velocity of the circulating water. However, the heat gained by the water must itself be dissipated in a closed circuit, so that it can be returned to the inlet at the original (low) inlet temperature, if consistent cooling is to be maintained. Thus, although water cooling may dramatically reduce the size of heat exchanger required, ie total heat exchange area for direct cooling, a further heat exchanger is then required to cool the water.

Water cooling requirements are more generally solved on empirical or semi-empirical lines. The amount of heat exchange area available may be dictated by constructional considerations, or based on experience. Any final adjustment is then available through the flow rate. The resulting outlet temperature of the water must then be dealt with as necessary, by a second heat exchanger if the water is to be recirculated.

See also chapter on *Compressor Controls.*

Selection of Compressors

THE AIR delivery and pressure required will normally dictate the most suitable compressor for a particular duty. Thus, *reciprocating compressors* are the general choice for deliveries up to 8.5 m^3/min (300 ft^3/min) at pressures up to 7 bar (100 lb/in^2) or slightly above. Larger deliveries demand the use of two-stage reciprocating compressors in the range 8.5–100 m^3/min (300–3 500 ft^3/min). Higher pressures can be provided by three-stage machines.

Higher capacity demands are readily met by *centrifugal compressors,* although here, however, efficiencies may not be attractive at capacities below about 170 m^3/min (6 000 ft^3/min). Capacities offered by centrifugal compressors extend up to about 2 800–3 500 m^3/min (100 000 –125 000 ft^3/min), or well into the range also covered by *axial flow compressors.* Thus the two overlap in a strictly competitive range, although the axial flow type becomes increasingly attractive as the capacity required increases. Thus a very high capacity compressor is the normal choice although the pressure range available may be more limited than with a centrifugal type. The lower limit of capacity at which the axial flow compressor usually ceases to be competitive is about 280 m^3/min (10 000 ft^3/min).

Other types of compressor – eg vane, Roots, screw and liquid ring – may have specific attractions. *Vane compressors,* for example, are best suited to low volume medium pressure duties, but may be directly competitive with reciprocating compressors up to, and well beyond, delivery of 100 m^3/min (3 500 ft^3/min). Modern designs of screw compressors can also be directly competitive with both reciprocating compressors and turbo machines with deliveries up to 600 m^3/min. Some general comparisons are summarized in Tables I and II.

Cost Parameters

As a general rule, initial cost may be a primary parameter where the demand is relatively small and/or intermittent. This would generally favour a reciprocating machine. Compressor efficiency may be insignificant, particularly if use is intermittent and off-loading is incorporated. Where the demand is high and continuous, overall efficiency may be more significant than initial cost.

Efficiency figures quoted arbitrarily can be misleading. With good design, most compressor types should prove capable of achieving overall efficiencies of between 70% and 85%, relative to the theoretic power required for ideal isentropic compression with no friction, fluid or other

TABLE I – COMPRESSOR CHARACTERISTICS – PRESSURE, CAPACITY AND SPEED

Type	Sub-Type	Pressure bar	Pressure lb/in^2	Delivery m^3/min	Delivery ft^3/min (FAD)	Speed rev/min	Remarks
Reciprocating	Single-stage	up to 7	up to 100	below 8.5	below 300		Generally most economic type of compressor for capacities up to 3 500 ft^3/min
	Two-stage	up to 7	up to 100	over 8.5	over 300		
	Multi-stage	up to 10	up to 150 or higher	over 8.5	over 300		
Diaphragm	Single-stage	4	60	up to 0.7	up to 25		
	Two-stage	7	100	up to 0.2	up to 6		
	Three-stage		Higher pressures when hydraulically actuated				
Vane	Single-stage	up to 3.5	up to 50–60			250–3 500 depending on size	Water cooled above 30 lb/in^2
	Two-stage	up to 10	up to 120–150	1.5–1 500	50–50 000		
	Oil-cooled	up to 8	up to 120			up to 3 000	
Roots	Single-stage	up to 1	up to 15	up to 1 500	up to 50 000	250–6 000 depending on size	Blowers
	Two-stage	up to 2	up to 30				
Lysholm screw	Single-stage	up to 4	up to 60	17–550	600–20 000	300–25 000	Essentially high speed machines
	Two-stage	up to 10 or higher up to 30	up to 150 or higher up to 450				
Liquid ring	Single-stage	up to 1.5	up to 20	up to 150	up to 5 000	up to 3 500	Large volumes at low pressure or low volumes at higher pressures
	Two-stage	up to 3	up to 30–45				
Centrifugal	Single-stage	0.5	6	6.5 upwards	230 upwards	up to 25 000	Blowers
	Four-stage	2.0	30	up to 3 500	up to 125 000		
	Five-stage	2.0	35	170–3 000	6 000–100 000		Water cooled
	Multi-stage	7.0	100				
	Multi-stage in series	above 2.0	above 35	up to 3 000	up to 100 000		Compression ratio 1.5 to 1.6 per stage
Axial flow		As required	As required	As required	As required		Efficiency tends to increase with increasing capacity

TABLE II —COMPRESSOR CHARACTERISTICS INFLUENCING SELECTION

	First Cost	Efficiency	Installation	Vibration	Noise Level	Remarks
Reciprocating	Low to moderate	High		Reduced with L & W layouts	Very high	Usual choice for capacities up to 100 m^3/min (3 500 ft^3/min)
Diaphragm	Low	High		High	High	Limited capacity but particularly suited to handling toxic or corrosive gases
Vane	Low to moderate	Slightly inferior to reciprocators	Simple	None	Low (especially water cooled type)	Best suited to medium pressure and low volume duties
Roots	Low	Low but improves with increasing size	Simple	Low if rotors properly balanced	High	Most suited to low pressure (blower) high volume requirements
Screw	High	High (directly comparable with reciprocators)			Very high	Modern designs can be competitive with reciprocating compressors and turbo machines
Centrifugal	High	Poor in small sizes; high in larger sizes	Requires suitable bed	Balance very important due to high speeds	Can be low with filtering	Essentially high speed machines for efficient performance. Not generally competitive below 170–200 m^3/min (6 000 ft^3/min)
Axial	High	High	Requires suitable bed		Can be low with filtering	Unrivalled performance for high outputs. Not generally competitive below 280 m^3/min (10 000 ft^3/min)

mechanical losses. Some types of compressor are inherently less efficient than others, however, and all are influenced by machine size, speed and other operating parameters. Thus, where overall efficiency is a major parameter and therefore exact determination of efficiency is required, this is normally given as a guaranteed performance. To derive any realistic comparison between large capacity compressors of the centrifugal and axial flow type (or for comparing vane compressors with other types), it is virtually essential to use empirical efficiency figures.

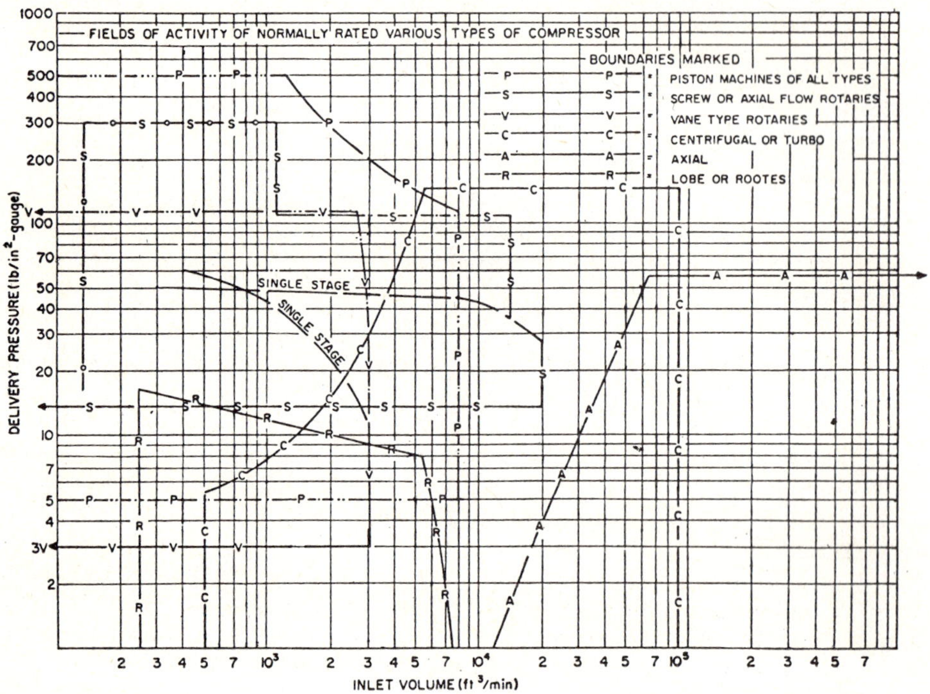

Fig 1

Fig 1 is a general guide only as to the likely efficiencies available from different types of compressor over a range of sizes (inlet volumes).

Efficiency also depends on cooling and intercooling. Good cooling system design reduces water cooling requirements and cost. Fan cooling should be cheaper, but tends to be rather less efficient and more restricted in application.

Other cost parameters which may or may not be significant are:—

(i) *Mounting requirements* — which can vary with the type and size of compressor. The cost of suitable foundations is also substantially higher with some types than others.

(ii) *Driver cost* — again variable with type and size of machine. Optimum compressor speed is also significant here, for this governs the suitability for direct drive, or the necessity for gearing up or down.

(iii) *Depreciation and maintenance* — which can be major overall cost factors after the first year.

Comparative costs are summarized in Table III.

TABLE III – GENERAL GUIDE TO COMPARATIVE COSTS

	Reciprocating							
	Vertical	Horizontal	V or W	Vane	Roots	Screw	Centrifugal	Axial Flow
Initial cost	Moderate	Moderate	Low	Lowest	Very low	Low	Moderate	Slightly lower than centrifugal
Mounting cost	Can be very high*	Can be very high	Low to moderate	Very low	Lowest	Lowest	Low	Low
Driver cost	High	High	Moderate	Very low	Moderate	Very low	Lowest	Lowest
Main-tenance	Moderate to low	Moderate to low	Moderate to low	Blade and retaining ring wear can be high	Little required	Low	Low	Low

*in larger sizes (over 20 hp)

The general rule commonly adopted for assessing compressor life is ten years or 100 000 hours, although this may be reduced substantially where the compressor is working under particularly adverse conditions. This can be taken as a typical figure from which to calculate a realistic estimate for depreciation. The figure of 100 000 hours can be used for a compressor which is run continuously, and ten years for a compressor run intermittently; the lower hours in this case are offset by the deterioration likely to occur during periods of idleness.

Working time between overhauls or replacement can be considerably shorter, although logically all critical parts of a compressor should be designed for the same 10 year/100 000 hour life. Certain types, however, may have parts with an inherently shorter life – eg the life of the blades on a vane compressor may be of the order of one year only, or even less. Also, of course, reliability, as expressed in terms of continued troublefree operation, can be a characteristic of the individual design and quality of construction.

Vibration

Rotary machines are inherently smoother running than reciprocating machines. The latter may require particular attention to prevent excessive vibrations (and noise) being generated, eg by the application of anti-vibration mounts so that the machine is suitably isolated. In the case of rotary machines, the vibration level depends primarily on the degree of dynamic imbalance present in the rotating elements. Effective vibration damping can add appreciably to the total cost of installation.

Noise

Some types of compressor tend to be inherently noisier than others although a general description of this particular parameter is impossible, particularly as the noise spectra generated by different machines may vary appreciably.

With some types air noise can be reduced substantially by simple treatment, eg with a filter type silencer applied at a suitable point. With others this may not be possible, except by isolating the machine in a complete enclosure. 'Machine' noise transmitted through the casing may also vary appreciably in level and spectra with different types.

The subject of noise (and vibration) control is a highly specialized one, requiring knowledgeable technical treatment to achieve any realistic improvements.

Lubrication

Lubrication is seldom a problem with modern compressor designs. Pressure lubrication is now generally employed on modern reciprocating compressors. Some types can run dry and would be a normal choice where 'oil free' compressed air is required.

Valve Design

Good valve design ensures low-loss performance and reliability. This can be a particularly important factor in the case of reciprocating compressors.

Control

The most common form of control is *discontinuous regulation.* The degree of control possible depends on the type of compressor and may influence the choice of compressor type in some applications. The ability to match a compressor with associated plant may also be a consideration.

See also chapter on *Compressor Controls.*

BroomWade V-Major oil-free compressor

Installation of Compressors

IDEALLY a compressor should be located where it can induct clean, dry and cool air. Another primary consideration is that a compressor is inherently a noisy machine and, unless specifically designed or treated for noise control, should be located in a 'noisy' area of a factory or workshop. In the case of a permanent installation it is generally desirable (and virtually necessary in the case of larger machines) to have the compressor in a separate well ventilated room or chamber, inducting air from the outside of the building.

Basic treatment which can be applied is the fitting of an air intake filter of a suitable type and size to ensure that air reaching the compressor is clean; and a separate air intake silencer to reduce 'intake noise', which is usually a predominant noise-generator. Further noise treatment may, or

Fig 1 Principles for the construction of a compressed air pipe system. (Atlas Copco).

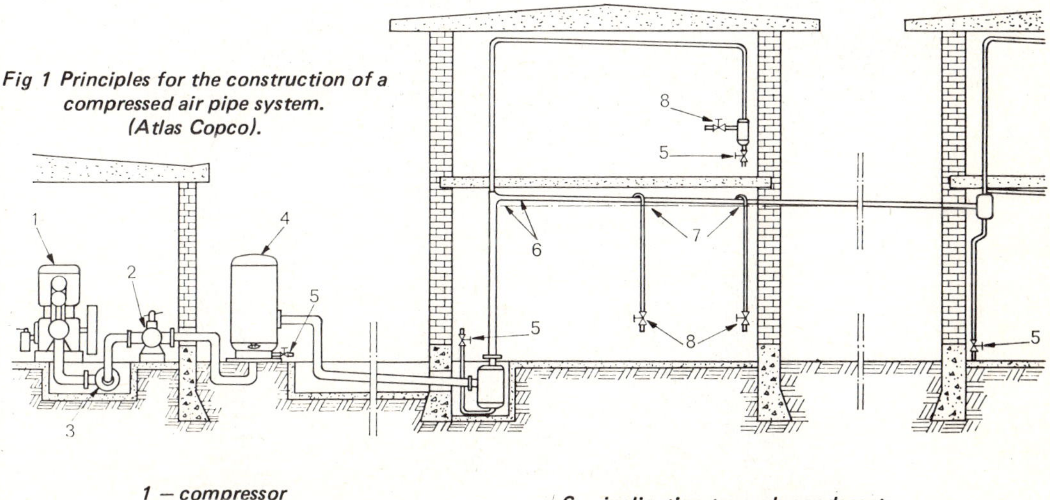

1 – compressor
2 – aftercooler
3 – discharge pulsation damper
4 – air receiver
5 – condensate draining valve
6 – inclination towards condensate drainer
7 – branch pipe connections
8 – outlet valve

may not, have to be applied to the machine itself and the room (eg the provision of adequate ventilation can establish noise-escape paths which may have to be dealt with). Specific treatment for noise reduction is described in the chapter on *Noise Control.*

A complete compressor installation normally involves compressor and drive motor, aftercooler, moisture separator and receiver. The latter needs to be located in a cool area and so is frequently located outside the compressor chamber, as in the typical arrangements shown in Figs 1 and 2. However, an important point to be considered is that the length of the discharge pipe between compressor and receiver should be kept as short as possible.

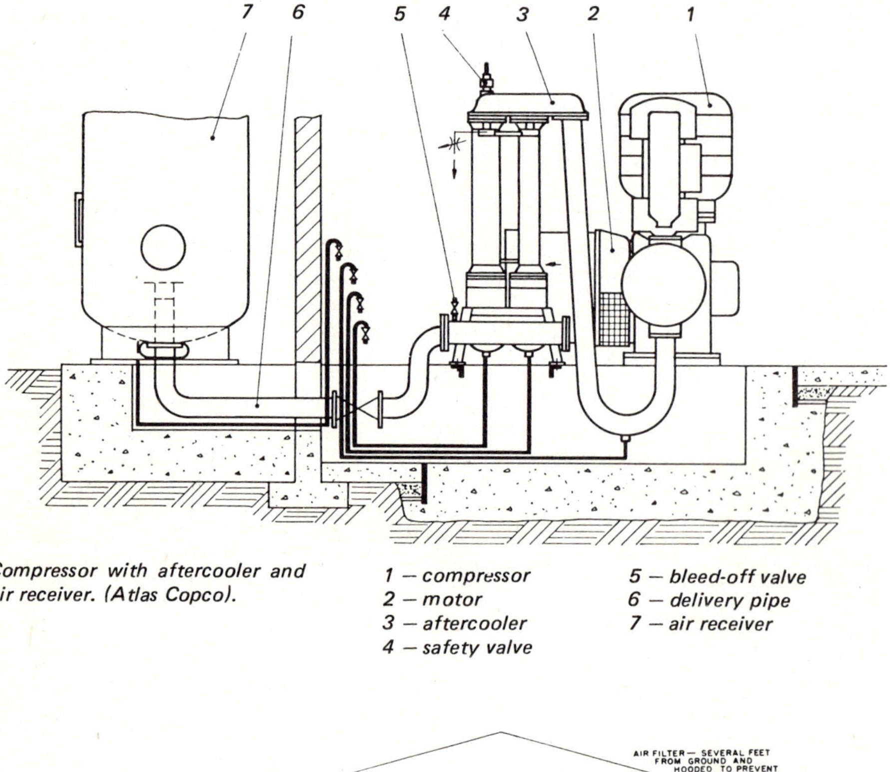

Fig 2 Compressor with aftercooler and air receiver. (Atlas Copco).

1 – compressor
2 – motor
3 – aftercooler
4 – safety valve
5 – bleed-off valve
6 – delivery pipe
7 – air receiver

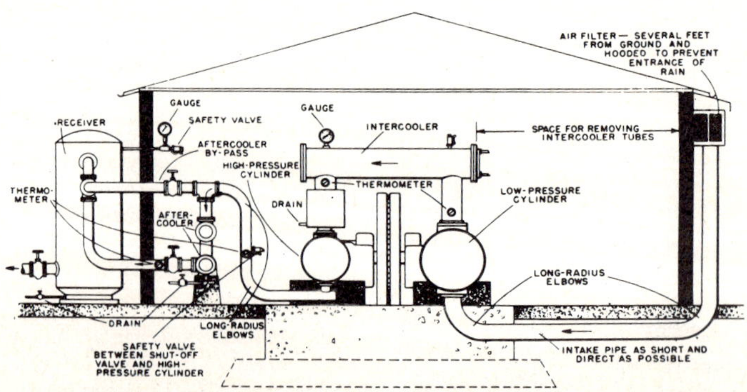

Fig 3 Typical two-stage compressor installation.

Intake System

The intake system must be designed with low pressure drop in mind, implying a limit on intake air velocities. At the same time intake flow should be smooth rather than pulsating as this will appreciably reduce intake noise and resonance effects.

Low pressure drop can be ensured by suitable pipe sizing restricting intake flow velocity to:—

(i) 5–6 metres/sec (16–20 ft/sec) in the case of single-acting compressors

(ii) 6–7 metres/sec (20–22 ft/sec) in the case of double-acting compressors.

Combined with a venturi-type intake silencer these intake velocities can be increased by up to 10%, this form of treatment becoming increasingly used. A typical arrangement of intake piping with venturi silencer is shown in Fig 4. A venturi silencer with double paper filters is shown in Fig 5.

Air intakes for large compressors should be located as high as possible, with particular attention paid to avoiding sound reflection on adjacent walls or roofs. Where more than one compressor is installed in the compressor house, each compressor should have its own separate intake pipe.

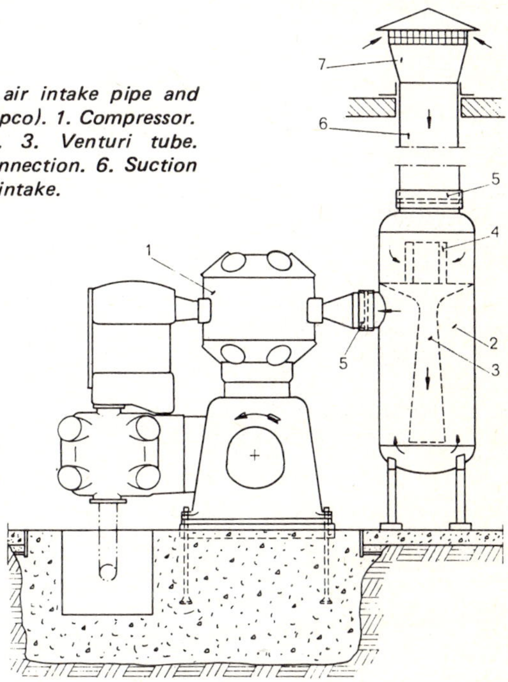

Fig 4 Arrangement of air intake pipe and venturi silencer (Atlas Copco). 1. Compressor. 2. Silencer with filter. 3. Venturi tube. 4. Filter. 5. Flexible connection. 6. Suction pipe. 7. Air intake.

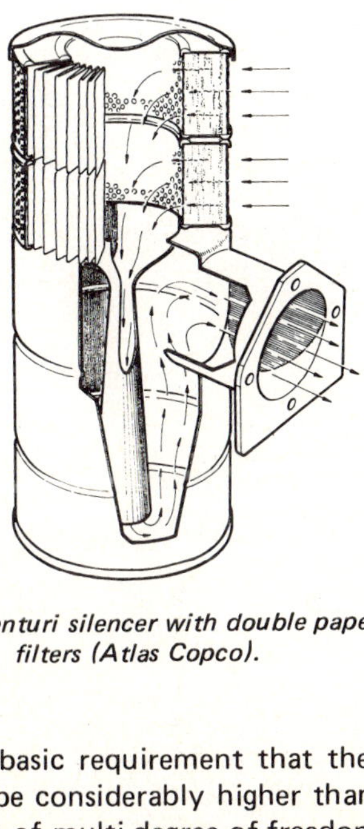

Fig 5 Venturi silencer with double paper filters (Atlas Copco).

The basic requirement that the lowest component present in the forcing frequency spectrum should be considerably higher than the resonant frequency of the spring-mass systems, applies. In the case of multi-degree-of-freedom systems, however, several resonant frequencies may apply to the spring-mass systems. Also a limit to the softness of the springs which may be used is set by the necessary lateral stability of the system.

Crank arrangement	Forces			Couples		
	primary		secondary	primary		secondary
	without counterwts.	with counterwts.		without counterwts.	with counterwts.	
Single crank 	F_1	$0.5 \cdot F_1$	F_2	—	—	—
Two cranks at 180° cylinders in line 	—	—	$2 \cdot F_2$	$F_1 \cdot d$	$0.5 \cdot F_1 \cdot d$	—
opposed cylinders 	—	—	—	—	—	—
Two opposed cylinders on one crank 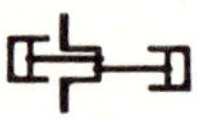	$2 \cdot F_1$	F_1	—	—	—	—
Two cranks at 90° 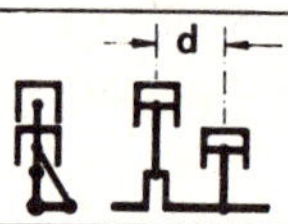	$1.41 \cdot F_1$	$0.71 \cdot F_1$	—	$0.71 \cdot F_1 \cdot d$	$0.35 \cdot F_1 \cdot d$	$F_2 \cdot d$
Two cylinders on one crank cylinders at 90 	F_1	—	$1.41 \cdot F_2$	—	—	—
Three cranks at 120° 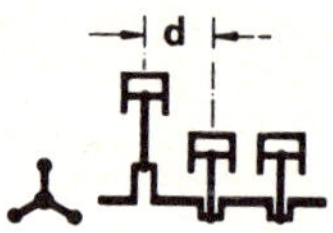	—	—	—	$3.46 \cdot F_1 \cdot d$	$1.73 \cdot F_1 \cdot d$	$3.46 \cdot F_2 \cdot d$
Four cylinders cranks at 180 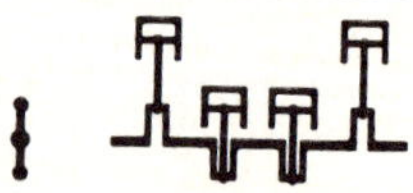	—	—	$4 \cdot F_2$	—	—	—
cranks at 90 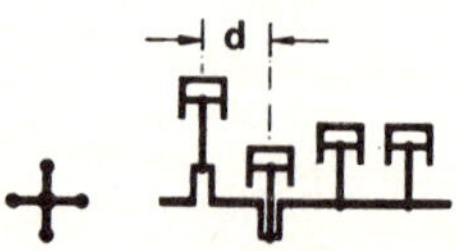	—	—	—	$1.41\ {}_2 F_1 \cdot d$	$0.71 \cdot F_1 \cdot d$	$4 \cdot F_2 \cdot d$

In this table:

F_1 is the primary inertia force in N

F_2 is the secondary inertia force in N

r is the crank radius in mm

n is the shaft speed in rpm

w is the weight of the reciprocating parts in one cylinder in kg

l is the length of the connecting rod in mm and

d is the cylinder center distance in mm

$$F_1 = \frac{\pi^2}{1000 \cdot 30^2} \cdot w \cdot r \cdot n^2 = \frac{w \cdot r \cdot n^2}{91188} \qquad F_2 = \frac{r}{l} \cdot F_1$$

(The couples are expressed in N · mm)

Fig 6 Inertia forces associated with different reciprocating compressor configurations (Atlas Copco Ltd).

In the case of medium and large size compressors delivery pipes can be laid on the floor, or preferably in trenches covered by floor plates. In both cases pipes should be properly supported with sufficient freedom for thermal expansion. Long pipes can also be suspended flexibly from wall brackets. They should never be attached solidly to walls or ceilings.

Where a tuned delivery pipe is specified, its dimensions are specified by the compressor manufacturer and must be strictly adhered to. If this is not practicable, then a pulsation damper must be installed with the shortest possible delivery pipe between the damper and aftercooler.

Ideally delivery pipes for medium-sized compressors and large stationary compressors should be welded and provided with flanged connections for fittings. With smaller compressors threaded pipe connections are more commonly used.

Due consideration must be given to ventilating the compressor house to dissipate heat generated by the compressor and driver. This is particularly necessary in the case of air-cooled units. With water cooled compressors, or air-cooled compressors with water-cooled intercooler, the amount of radiant heat dissipated is usually negligible since the majority of heat is carried away by the cooling water.

Stationary compressors should always be installed on a firm base where ground conditions are good.

Installation is seldom a problem with compressors below a horsepower rating of 20. Larger machines, however, may need special foundations, the size and cost of such foundations increasing with compressor size, particularly in the case of reciprocating compressors with large out-of-balance forces. Very considerable savings are possible in such cases with 'L', 'V' or 'W' configurations due to the considerable reduction of such forces with these layouts. Large centrifugal and axial flow type compressors are far less troublesome since thay are essentially balanced machines and the foundations are required to take little more than the weight of the machine and driver.

Larger compressors will, however, require the use of lifting or hoisting equipment to erect them, and also to dismantle or remove sections for inspection and maintenance.

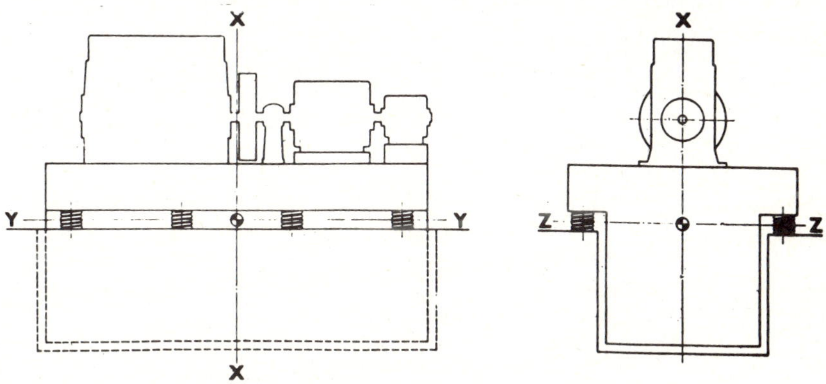

Fig 7 System with three planes of symmetry.

An analysis of the inertia forces and couples generated by different configurations of reciprocating compressors is given in Fig 6. These may need to be considered in detail in deciding on a suitable method. If the machine is to be mounted on a spring-suspended inertia block (concrete block), this will be equivalent to several-degrees-of-freedom system with three places of symmetry and six possible modes of vibration — Fig 7. Three of these modes will be translational with vibratory movement along the principal axes, and rotational movements about the principal axes of inertia. Due to the symmetry of the system all six modes will be uncoupled, provided the mountings are in the same horizontal plane as the centre of gravity.

The stiffness required from the isolator can be determined directly from the formula:—

$$K = a P f_n^2$$

where a = 0.0315 for load P in pounds, giving K in lb/in
or a = 0.004 for load P in kilograms, giving K in kg/cm.

This is an approximate formula but accurate enough for general solutions with symmetrical installations (ie with isolators placed symmetrically with respect to the centre of gravity of the mass to be isolated). The load on each isolator is then equal to the total load (mass) of the machine divided by the number of isolators.

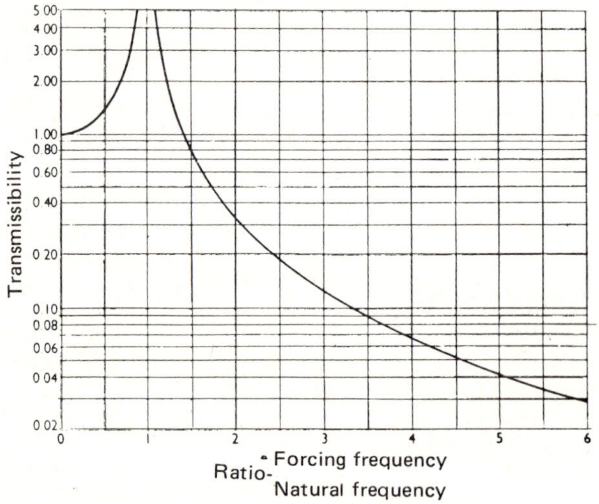

Fig 8

From this formula it is possible to calculate the isolator stiffness required to achieve a specific resonant frequency for the spring-mass system. This latter figure is estimated on the basis of sufficient separation (below) from the lowest exciting frequency. The degree of isolation achieved can then be estimated directly from Fig 8 (in the case of uncoupled springs), or Fig 9 once a suitable value for resonant frequency has been established.

$$\text{Isolation} = (1 - \text{transmissibility}) \times 100\%$$

Isolators or isolator materials are not generally specified in terms of spring stiffness (K), but rather in terms of static deflection under static load, viz

$$K = \frac{P}{d}$$

where P = load
d = static deflection.

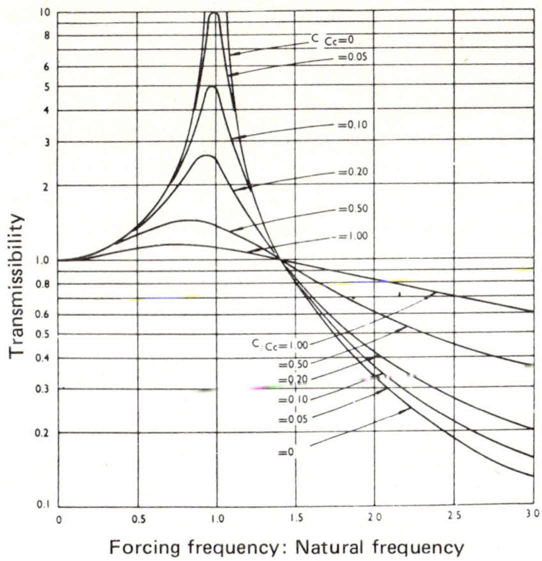

Fig 9. Transmissibility in terms of the ratio of forcing frequency to natural frequency for a viscously damped system.

The required stiffness can be selected on this basis. Alternatively, the natural frequency can be derived directly from the static deflection:—

$$f_n = 3.13\sqrt{\frac{1}{d}}\,\text{Hz}$$

where the static deflection of the isolator is in inches.

$$85\sqrt{\frac{1}{d}}\,\text{Hz}$$

where the static deflection of the isolator is in millimetres.

This formula applies only where the dynamic stiffness is the same as the static stiffness, and damping is present. Damping requirements may have to be considered separately. The damping present will directly affect the degree of isolation achieved at particular frequencies, thus:—

$$f_n \text{ (with damping)} = \sqrt{\frac{K}{M}} \times \sqrt{1 - (C/C_c)^2}$$

$$= \frac{C}{\sqrt{2KM}}$$

Two further effects may be present:—

(i) *Wave Effects* — originating from the generation of longitudinal standing waves in the springs. The effect is to produce peaks in the transmissibility curve at higher forcing frequencies. Such effects are most marked in springs with little internal damping. Wave effects are seldom significant in materials with relatively high internal damping.

(ii) *Foundation Reacting* — where the foundation is not perfectly rigid (or its mass is such that it does not behave as a practical rigid body). This will have a modifying effect on the resonant frequency of the spring-mass system. Mathematical analysis in this case is extremely complex and a practical approach is generally simplest, eg

(a) ensuring that the foundation is sufficiently massive to act as a rigid body in practice.

(b) direct measurement of the resonant frequency of the system where the foundation is not rigid.

Compressed Air Mains

THE choice for a main air distribution system lies between a dead-end main or a ring main. A dead-end main has the advantage of distributing the compressed air over the shortest possible run, reducing pressure drop and economizing on pipe cost. Its main disadvantage is its lack of flexibility, although the service supplied by the compressor can be extended by the use of a split flow main – Fig 1.

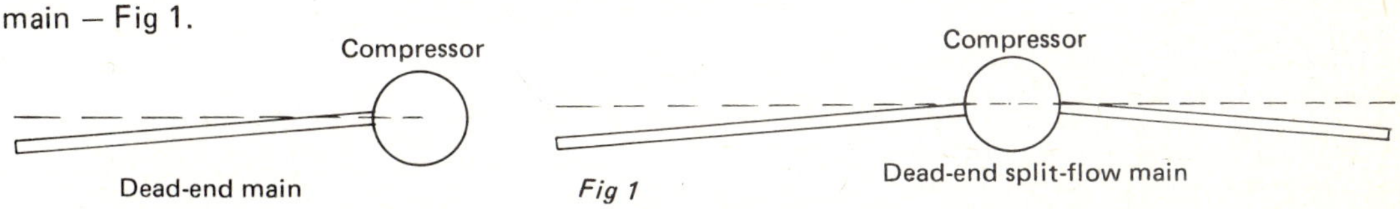

Fig 1

A ring main has the advantage that being in the form of a 'closed circuit', the velocity of air in the main will be reduced and the pressure drop should be less. The ring main itself may be of appreciable volume and capable of acting as a form of 'reservoir' to assist in maintaining pressure at any particular point of use where there is a sudden draw-off. Also, if the system is extended by the addition of another branch, this branch can draw its supply from both ends of the ring main, reducing air velocity and minimizing pressure drop. A ring main system thus offers maximum flexibility with functional advantages and is generally the preferred choice (Fig 2).

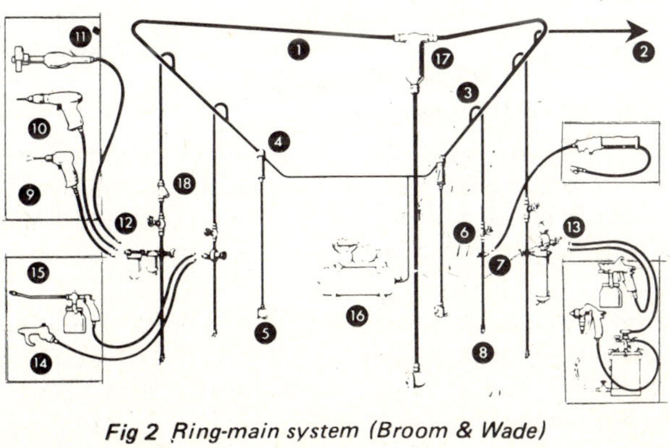

Fig 2 Ring-main system (Broom & Wade)

1 – ring main
2 – branch main
3 – take-off points
4 – filter/separator in main
5 – automatic drain
6 – isolating valves
7 – connecting point
8 – drain cock
9,10,11 – tools
12 – separator/lubricator
13 – separator/reducing valve
14 – blow gun
15 – oil/water jet
16 – compressor
17 – moisture trap
18 – solids separator

Regardless of the type of main, the most important consideration is that the pipeline should be of suitable size to pass the maximum volume of air likely to be required by the system without any significant pressure drop. Pressure drop represents a power loss, not just in terms of output power at the take-off points, but also input power required (and consequently higher operating costs) to maintain the desired working level. A further advantage of a low pressure drop is that the operating pressure range of the compressor can be significantly reduced.

Pressure drop in a straight run of pipe is given by the following approximate formula

$$\text{pressure drop} = \frac{L\ V^2}{K\ R\ D^5}$$

where L = length of pipe

V = volume of free air passing through the pipe

K = a constant, depending on the units employed for L,V and D.

R = compression ratio at entrance of pipe

D = bore of pipe

In terms of the significant working parameters pressure drop is:

directly proportional to pipe length

directly proportional to the *square* of the free air delivered

inversely proportional to the *fifth* power of the pipe diameter (more accurately to $D^{5.3}$).

It follows that pipe diameter is the most important control parameter; doubling the pipe size, for example, reduces the pressure drop by a factor of nearly 40. This favours the choice of a very generous 'over size' pipe diameter, particularly as the cost of piping only increases in approximately simple proportion to pipe diameter.

Using a generously proportioned air main will bring other benefits as well. Although, ideally, the compressor house with its aftercooler and receiver should produce air at ambient temperature from which all surplus water and oil have been removed, this very seldom happens in practice. The compressed air distribution pipework, therefore, acts in very nearly all cases as an additional cooling surface and as a result of this, water and oil will separate out throughout its length. The higher the velocity of the air through the main, the more likelihood there is that any moisture which has formed in it will be whipped up by the air stream, atomized and carried on. Here then is another reason for keeping the size up and the velocity down.

Theoretically, at least, pressure drop would be reduced to a minimal figure by the selection of a very large diameter pipe. This, however, is not necessary or even practical. Excessively oversize pipes can cause air (and time) to be wasted in filling the extra volume, although actual losses in this respect can be relatively insignificant. Saving unnecessary extra cost of pipelines is more important.

As a general design rule a pressure drop of the order of 5% per 30 m (100 ft) of pipe is satisfactory, bearing in mind that actual pressure drop is also dependent on air velocity through the pipe which in turn is governed by demand. If further extension of the main is likely, then 'oversizing' by about 50% on the original figure should be considered.

TABLE I – MAXIMUM PERMISSIBLE FLOW THROUGH STANDARD MEDIUM WEIGHT PIPE
(C. A. Norgren Limited)

lb/in^2	PIPE SIZE (inches)								
	1/8	1/4	3/8	1/2	3/4	1	1¼	1½	2
5	0.58	1.3	2.9	6	9	18	39	57	112
10	0.95	2.1	4.8	10	15	29	62	93	180
20	1.67	3.7	8.2	16	27	50	107	160	310
30	2.32	5.2	11.7	22	38	70	150	225	435
40	3.00	6.8	15.0	28	48	91	190	285	550
50	3.75	8.2	18.5	35	60	112	235	350	670
60	4.40	9.8	22.0	42	71	132	280	408	800
70	5.10	11.3	25.5	48	881	152	322	475	920
80	5.80	13.0	28.0	54	91	171	365	520	1 050
90	6.50	14.5	32.3	61	102	190	405	600	1 180
100	7.20	16.2	36.0	67	113	215	455	660	1 300
125	8.90	20.0	44.3	83	140	270	560	820	1 600
150	10.70	24.0	53.0	100	168	320	665	970	1 900
175	12.20	28.0	62.0	115	196	375	770	1 150	2 200
200	14.00	34.0	72.0	130	225	420	880	1 300	2 500
250	18.00	40.0	90.0	155	280	525	1 100	1 600	3 200

This Table can be used as a general guide in determining sizes of equipment in compressed air circuits. Flow rate figures are given in ft^3/min and show the permissible maximum air flows consistent with the following pressure losses in 100 ft of pipe:-

Sizes up to and including ½ in — 15% of applied pressure
Sizes ¾ in to 2 in inclusive — 10% of applied pressure

Note: Continuous flow rates should not exceed 75% of the maximum permissible figures.

TABLE II – VARIATION OF PRESSURE DROP WITH FLOW RATE

Design rating for continuous flow ... ... ↓

Flow Rate %	50	60	70	75	80	90	100	110	120	150
Pressure Drop lb/in^2	2.5	3.6	4.9	5.6	6.4	8.1	10	12.1	14.4	22.5
Pressure Drop bar	0.2	0.25	0.35	0.40	0.45	0.57	0.7	0.85	1.0	1.6

Design maximum permissible flow rating ... ... ↑

Mains sizing can be determined directly from theoretical or empirical data, or based on accepted recommendations. 'Maximum' flow rates can be quoted for specific pipe sizes, established on the basis that this maximum rate corresponds to a pressure drop of 10% – see Table I. For continuous flow it would then be advisable to base the pipe size on a 75% flow rating. Further, the continuous flow figure should be based on the maximum continuous flow demand and not on the average, to allow for the possibility of all services fed by the main being used simultaneously. Table II shows how departures from design flow rate can affect the pressure drop, calculated for a 7 bar (100 lb/in^2) supply pressure.

The maximum flow rate required can be based on 1.5 to 3.0 times the average flow, depending on the number of outlets in use at any one time and the particular geometry of the system.

A specific maximum pressure drop figure may be adopted to cover the requirements of the system. The choice of this figure is somewhat arbitrary. A compressor is normally capable of supplying a higher pressure than that needed for efficient operation of the various devices it feeds, but against this is the fact that pressure drop represents lost work or wasted power input. Logically, the system designer should aim at achieving a pressure drop of no more than 3–8%, as representing efficient working. In practice a higher figure is perfectly acceptable, and more realistic requirements are:-

(i) The lowest practical pressure drop in the main distribution system (eg in a ring main).

(ii) A maximum of 10% pressure drop at the farthest point in the system in the device itself, although in particular circumstances a higher figure of up to 15% pressure drop may well be tolerated at such a point.

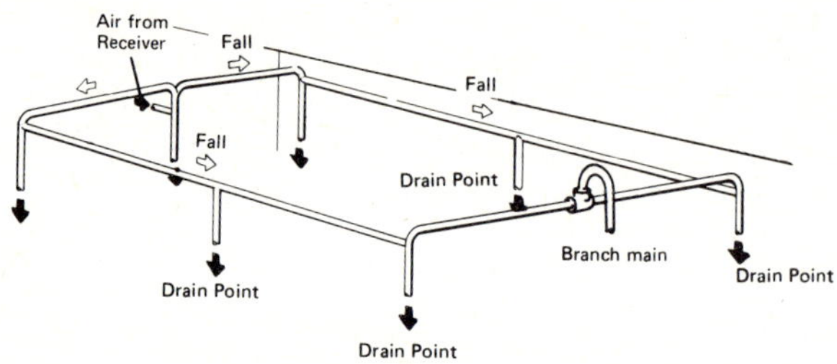

Fig 3

It is also important that a ring main be given a fall in the direction of the airflow (Fig 3). A fall of 1 in 100 or 1 in 120 is generally enough to ensure that water separating out is carried along to a low point by both gravity and airflow. Moisture traps and/or drains can then be installed at suitable points along the run and at each individual low point. Automatic traps or drains are preferred as these cannot become overfilled and thus inoperative.

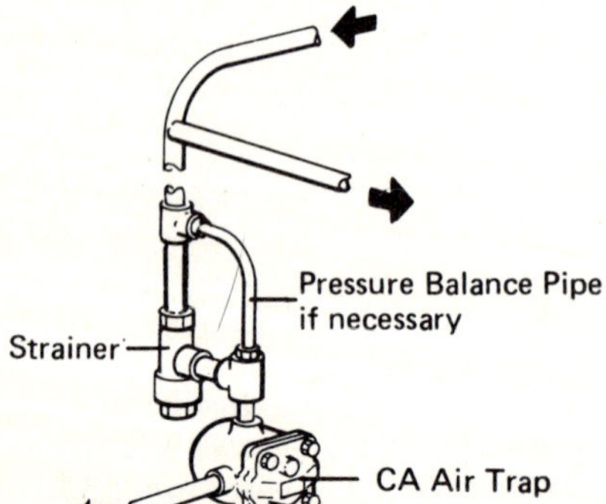

'Spirax' Air Trap draining a relay point

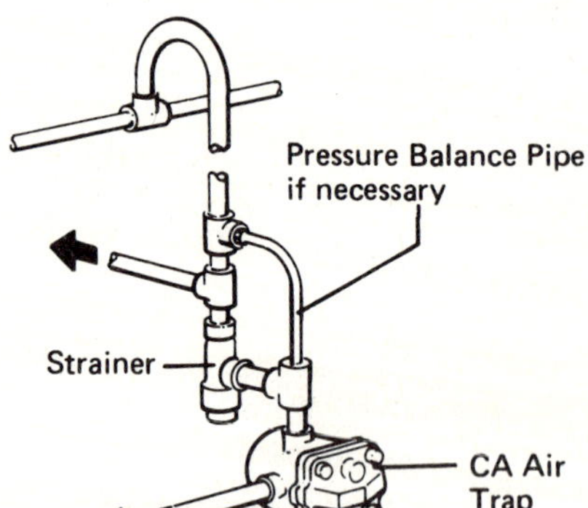

'Spirax' Air Trap draining a branch main

Fig 4

Branch mains or individual take-off points should emerge in the form of an upward loop – Fig 4. This will help contain any condensate present in the ring main and minimize the carry over of moisture to tool points, etc. A filter/separator should be fitted in the ring main at a high point between the compressor and the first take-off point, to reduce further the moisture entering the main and branch lines. This should be of the automatic type.

Each branch and take-off point should have its own isolating stop valve and drain or drain cock at the blind end. Filters, separators and lubricators should be in the individual lines feeding tools or appliances requiring protection and/or lubrication. The choice of these devices will depend on the type of tool or appliance concerned. Tools normally require a moisture separator/lubricator in the feed line. For paint spraying a moisture separator/reducing valve unit is required, the latter enabling the working pressure to be controlled at the take-off point. Blow guns should also be used via a pressure-reducing valve to control the force of the air jet.

Where the main involves any appreciable length of run, flange connections and expansion joints should be used throughout the system as necessary. The run should also be above ground and accessible so that any leaks can readily be detected and rectified.

Pressure drop is adversely affected by restrictions in the flow. Thus the interior of all pipes should be smooth and clean, and bends should be gradual with generous radii. Sharp elbows, tees and similar fittings should be avoided and all fittings used should be full-bore. Similar considerations apply to valve and hose connections.

Adequate pipe support is necessary for ring mains. As a general guide, the following recommendations for support spacing apply:-

Main diameter		**Pipe run between supports**	
mm	**in**	**metres**	**feet**
25	1	2.4	8
32	1¼	2.4	8
38	1½	2.7	9
51	2	3.0	10
76	3	3.4	11
102	4	3.7	12
127	5	4.0	13
152	6	4.3	14

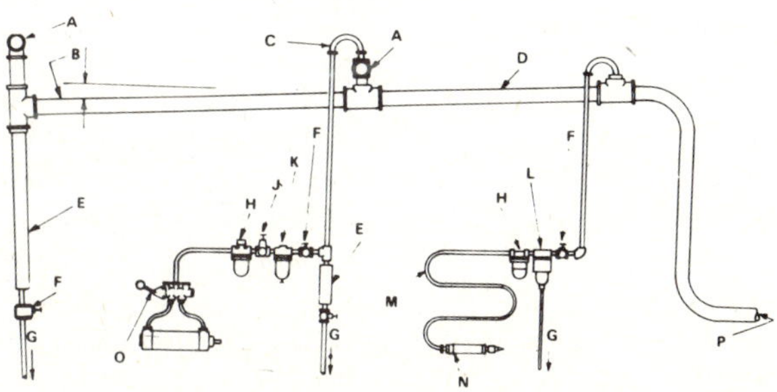

A – branch main
B – pitch with flow
C – wide pattern return bend
D – main
E – trap or water leg
F – stop valve
G – drain
H – lubricator
J – regulator
K – filter
L – automatic drain air filter
M – flexible hose
N – air drill
O – hand-operated valve and air cylinder
P – from receiver

Fig 5 Dead-end main (Maxam).

A typical example of a dead-end mains system is shown in Fig 5. The *low level* dead-end main, shown in Fig 6, can offer particular advantages for factory installations where each main is supplied by its own compressor. The main is run at a height of 0.45–0.60 m (18–24 in) between back-to-back tool stations or under a bench run, and carried on floor-mounted stanchions as necessary. A spacing of 3–4 m (10–15 ft) apart is generally satisfactory.

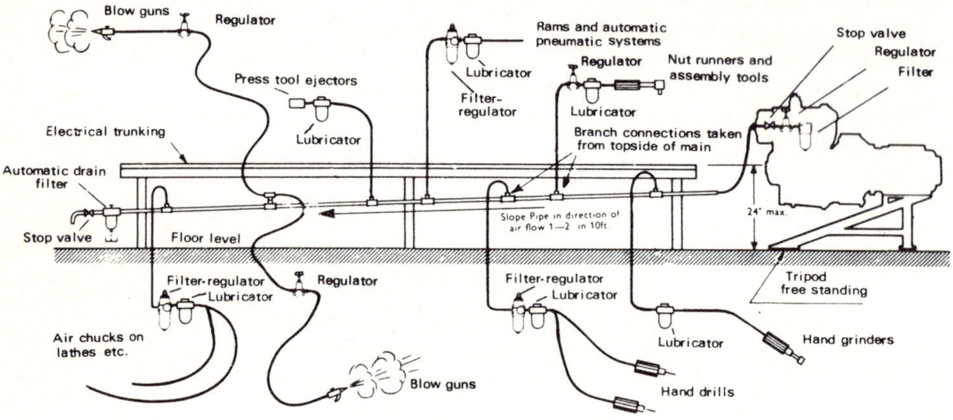

Fig 6 Low level main (Norgren).

Outdoor Air Lines

Outdoor air lines present special problems – notably the need to protect the outside of pipes with an anti-corrosive paint or similar treatment, and the desirability of burying the lines in the ground at frost free depth to prevent freezing of condensate and pipe obstruction or blockage developing at low ambient temperatures. Underground installation can be costly, particularly as inspection wells are also necessary to gain access to condensate traps and valves.

Overground installation, with the pipes suspended on posts or supported on plinths is generally cheaper and simpler. Quick-coupling pipes are recommended in this case. This form of joint has an expansion-contraction capability to counteract both thermal movement and some misalignment. With overground lines, however, particular attention must be given to condensate separation where this is a risk of freezing.

It is also recommended that all outdoor pipes should be of welded steel construction, hot-galvanized and fitted with flanged joints for water traps and valves.

Pipe Materials

Steel pipes are normally used for mains. Steam quality steel piping is used for sizes from 20–75 mm (¾ in to 3 in) diameter, commonly with screwed joints. For larger sizes, 75–150 mm (3–16 in) diameter, the traditional choice lies between cast iron piping with flanged joints and welded steel piping with flanges welded on.

For smaller lines, eg up to about 28 mm (1.1/8 in) bore, copper piping or nylon tubing is normally preferred. Diameters employed are normally selected from the following nominal bore sizes:-

1/8 in, 3/16 in, 3/8 in, 1/4 in, 1/2 in and 3/4 in matching BSP threads

5 mm, 8 mm, 12 mm, 15 mm and 20 mm

Copper pipes for threading are o.d. sizes and in British practice consistent with BSP pipe thread standards. Pipes for use with compression fittings may be o.d. sizes or nominal bore sizes, fittings being readily available for each type.

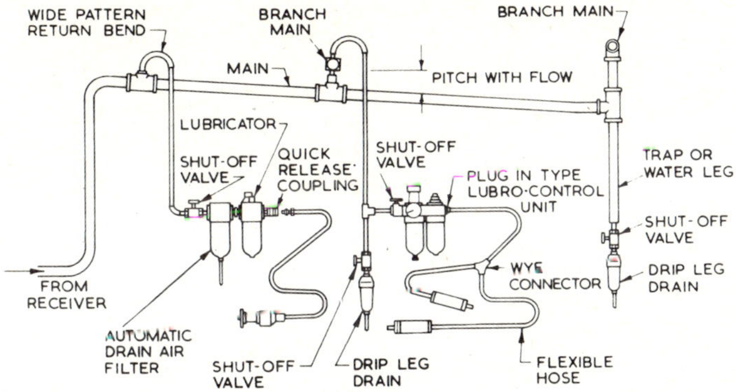

Correct airline installation. (Norgren).

Summary

The following factors are important in estimating the line sizes required, taking into account the losses incurred by fittings, etc. These can be rendered in equivalent straight pipe lengths, and the estimate of the total pressure drop can be based on total effective pipe length.

(i) Undersizing of the main supply lines lowers efficiency of all points in the system and restricts the use of tools and appliances at all points in the system.

(ii) Undersizing of branch mains only affects the performance of local systems connected to it.

(iii) Bad design of tool point take-offs will affect only the operation of that individual tool or appliance.

See also chapter on *Air Lines.*

Air Lines

THE EFFICIENCY of any pneumatic system fed through pipe lines depends very largely on the pipe size adopted. Pipes which are too small will choke the flow, resulting in excessive pressure drop.

Pipe *length* is also significant, since pressure drop is directly proportional to length. Pressure drop figures, in fact, are commonly quoted in terms of pressure drop per unit length (eg lb/in^2 per 100 ft). Pipe length, however, is relatively insiginficant as a design control parameter compared with pipe bore size, since pressure drop is inversely proportional to $(bore)^5$, approximately. In other words, a small change in bore size can have a marked effect on pressure drop, whereas even doubling the pipe length will only result in doubling the pressure drop.

Flow and Pressure Drop

The relationship between flow rate and pressure drop ($P_1 - P_2$) is given by the following formula:

$$Q = 18.062\frac{T_O}{P_O}\sqrt{\frac{(P_1^2 - P_2^2)\ (d)^{5.33}}{\frac{TL}{5\ 280}}}$$

where Q = ft^3/hr at pressure base P_O and temp

T_O = temp, ^OF absolute defining the ft^3 of air (eg 460 + 60^OF)

P_O = lb/in^2 absolute defining the ft^3 of air

P_1 = inlet line pressure (lb/in^2 db)

P_2 = outlet line pressure (lb/in^2 ab)

d = internal diameter of line (inches)

T = absolute flowing temp, ^OF + 460

L = length of line (ft)

This formula is complex amd tedious to use, but can give accurate results when analysing critical circuits. For general work, simpler formulas are used, or pressure drop figures are obtained from empirical or semi-empirical data.

YOU'VE GOT TO KNOW THE SIGNS IF YOU WANT TO PUT SAFETY FIRST

When you use a component with the Dyna-Quip stamp of approval you'll know it's a product that combines safety with the efficiency industry expects.

Here are just a few examples of products carrying the Dyna-Quip stamp of safety.

DYNA-QUIP SAFETY BLOW GUNS

Air Saver Safety Blow Guns supply low pressure air with high volume to safely and efficiently remove all 'chips'. They use 30% compressed air and 70% entrained room air effecting considerable economy. Air spills out of exhaust ports in event of blockage.

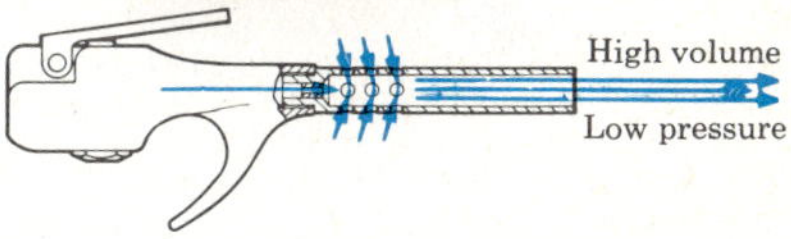

DYNA-CON QUICK RELEASE HOSE COUPLINGS

Plugs can fly at up to 75 mph when uncoupled under normal pressures. Because Dyna-Con is a valve as well as a coupling, downstream air is automatically bled off before uncoupling to eliminate the hazard of whipping hoses and plugs.

DYNA-QUIP VALVES

A range of ball valves in different materials to suit fluids of varied pressures and temperatures. Highly effective and safe. Exhaust type models protect downstream equipment in the 'off' position by bleeding down residual pressure. Locking handle available on all models.

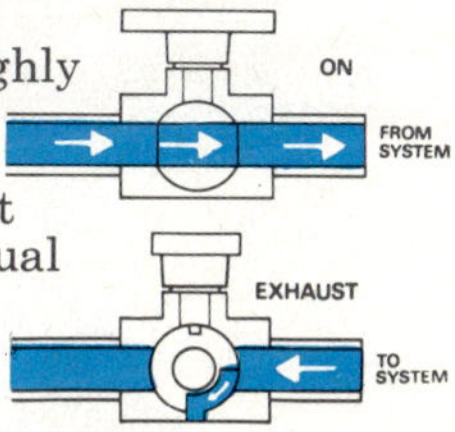

For full details contact:

dyna-quip ltd.

Shipston-on-Stour, Warwickshire, England.
Tel: 0608 61676 Telex: 83208

Norgren Shipston products

An IMI Company

A generally acceptable simplified formula, expressed as a direct solution for pressure drop is:

$$\Delta P = \frac{C.L.Q^2}{r.D^x}$$

where ΔP = pressure drop
C = a coefficient of friction
L = pipe length
Q = flow rate
r = compression ratio at pipe entrance
D = pipe bore
x = an exponential, the theoretical value of which is 5.33, although 5.0 is more conveniently used for non-critical calculations.

A specific value of C for smooth bore pipes is:

C = 0.31 for ΔP in lb/in^2, L in feet, Q in ft^3/sec and D in inches

Thus for English units and Q in ft^3/min:

$$\Delta P = \frac{0.00284\,Q^2}{rD^{5.33}}$$

The main limitation of this formula is that it does not take into account the fact that the friction factor varies with Reynold's Number of flow and thus the diameter of the pipe. A summary of friction factors (F) determined for different Reynold's Numbers and expressed in the parameters free air flow and pipe diameter is given in Table I. The appropriate factor can be used in the following simple formula for estimating pressure drop

$$\Delta P/100 \text{ ft pipe length} = \frac{14.7F}{\text{pipe inlet pressure } (lb/in^2 \text{ ab}) + 14.7}$$

This will generally give satisfactorily accurate figures for average design requirements, assuming that the pipes are smooth bore, straight and in good condition.

Where comparatively large flow rates are involved with low initial pressures (eg blower applications), the following simplified formula can be used:—

$$\Delta P = \frac{F}{P_m} \times \frac{L}{1\,000}$$

where P_m = mean pressure (absolute)
F = a friction factor

$$= \frac{7V^{1.85}}{D^5}$$

where V = flow velocity in ft/sec
D is in inches
ΔP and P_m are in lb/in^2

TABLE I – COEFFICIENT OF FRICTION (F)

Free Air Flow (CFM)	PIPE BORE – INCHES					
	½	¾	1	1¼	1½	2
5	1.27	0.12	0.05	–	–	–
10	5.07	0.78	0.22	0.05	–	–
15	11.40	1.76	0.49	0.11	–	–
20	20.20	3.04	0.87	0.20	0.09	–
25	–	5.00	1.36	0.32	0.14	–
30	–	7.04	1.96	0.45	0.20	–
35	–	9.59	2.66	0.62	0.27	–
40	–	12.50	3.48	0.81	0.36	–
45	–	15.90	4.40	1.02	0.45	0.12
50	–	–	5.45	1.26	0.56	0.16
60	–	–	7.83	1.82	0.80	0.22
70	–	–	10.70	2.47	1.09	0.29
80	–	–	13.90	3.23	1.43	0.38
90	–	–	17.60	4.09	1.81	0.48
100	–	–	–	5.05	2.23	0.60
110	–	–	–	6.11	2.70	0.72
120	–	–	–	7.27	3.22	0.86
130	–	–	–	8.53	3.78	1.01
140	–	–	–	9.89	4.38	1.17
150	–	–	–	11.40	5.03	1.34

Pressure Drop Through Bends and Fittings

Pressure drop through bends and fittings can only be determined accurately by empirical tests since it is specific to the internal geometry involved.

TABLE II – RESISTANCE COEFFICIENTS OF FITTINGS

Fitting	Order of Resistance Coefficient
Large elbow	0.33
Small elbow	0.42
90° sharp elbow	0.50–1.00
Large tee, straight through	0.33
Large tee, branch flow	1.00–1.33
Small tee, straight through	0.40–0.50
Small tee, branch flow	1.30–1.50
Cross, straight through	0.35–0.50
Cross, branch flow	1.00–1.40
Close return bend	0.60–0.90

Fitting resistance can be expressed in the form of a friction factor or *fitting resistance coefficient*, which can then be used for calculating pressure drop in the same formula as for pipes. Typical resistance coefficients are given in Table II. In the case of proprietary fittings, this will only give accurate results if the specific value of the resistance coefficient has been determined (at the anticipated Reynold's Number of flow, ie representative flow velocity for a given fitting size).

It is generally more convenient and simpler to calculate fitting resistance in terms of equivalent length of straight pipe – ie the length of straight pipe, of the same bore size as the fitting, which will have the same resistance (and thus pressure drop) as the fitting. Again such values can only be determined on empirical lines, but the following formula will be found to be in close agreement with the results obtained in practice:–

$$L_f = KfD^{1.2}$$

where L_f = equivalent length of straight pipe

K is a constant = 43.7 for L in feet and D in inches
= 4.34 for L in metres and D in centimetres

f = fitting resistance coefficient

D = pipe or fitting bore size

The accuracy of this formula depends on the accuracy of the frictional coefficient used. For non-critical applications tabular data are usually referred to (see Table III).

TABLE III – RESISTANCE OF PIPE FITTINGS IN LENS OF EQUIVALENT PIPE LENGTH

Type of Fitting	Nominal Pipe Size									
	inch									
	½	¾	1	1¼	1½	2	2½	3	4	5
	mm									
	12.5	19	25	32.5	37.5	50	62.5	75	100	125
Elbow	0.85	1.20	1.6	2.2	2.5	3.5	4.5	6.0	8.0	10.5
90° Bend (long)	0.50	0.60	0.8	1.1	1.5	2.0	2.5	3.0	4.0	5.0
Return bend	1.50	2.00	2.5	3.5	4.0	5.5	6.5	8.5	12.0	16.0
Globe valve	2.50	3.50	4.5	6.5	8.0	11.0	13.0	17.0	24.0	31.0
Gate valve	0.35	0.45	0.6	0.9	1.0	1.3	1.6	2.1	3.0	4.0
Run of standard tee	0.40	0.60	0.8	1.1	1.3	1.7	2.2	2.8	4.0	5.0
Through side outlet of tee	1.70	2.30	3.0	4.5	5.2	7.0	9.0	12.0	16.0	21.0

Pressure Drop Through Flexible Pipes

Theoretically, at least, if a flexible pipe has a smooth bore and a straight run, the pressure drop will be the same as that of a rigid smooth bore pipe of the same bore size. Flexible pipes are, however, liable to manipulation through various forms of bends, with consequent increased and variable resistance to flow; and the flow itself may be of a fluctuating or pulsating nature. Line sizing in this case is thus normally based on practical recommendations – see Table IV. Likely losses can be estimated from tabular data. Only in critical systems is it likely to be necessary to determine the actual pressure drop produced by specific tests on a sample length of hose manipulated under design working conditions.

TABLE IV – TYPICAL PRESSURE DROP IN FLEXIBLE AIR HOSE

(Pressure drop in lbf/in² per 100 ft of hose)

Hose Bore Diameter	Gauge Pressure lbf/in²	FREE AIR FLOW (ft³/min)													
		20	30	40	50	60	70	80	90	100	110	120	130	140	150
½″ with couplings at each end	50	4	10	20	36	–	–	–	–	–	–	–	–	–	–
	60	3	8	17	30	47	–	–	–	–	–	–	–	–	–
	70	2	7	14	25	40	–	–	–	–	–	–	–	–	–
	80	2	6	12	22	35	50	69	–	–	–	–	–	–	–
	90	2	5	11	19	30	44	61	82	–	–	–	–	–	–
	100	1	5	10	17	27	39	54	73	–	–	–	–	–	–
	110	1	4	9	15	24	35	49	67	89	–	–	–	–	–
¾″ with couplings at each end	50	1	2	3	5	7	9	13	17	23	28	–	–	–	–
	60	1	1	2	4	6	8	10	14	17	22	–	–	–	–
	70	0.5	1	2	3	5	6	8	11	14	18	22	–	–	–
	80	0.5	1	2	3	4	6	7	9	12	14	18	21	–	–
	90	0.5	1	1	2	3	5	6	8	10	12	15	18	–	–
	100	0.5	1	1	2	3	4	5	7	9	11	13	16	19	22
	110	0.25	0.5	1	2	3	4	5	6	8	10	12	14	17	20
1″ with couplings at each end	50	0.25	0.5	1	1	2	2	3	4	5	7	10	14	–	–
	60	0.25	0.5	1	1	1	2	2	3	4	5	7	8	11	14
	70	–	0.25	0.5	1	1	1	2	3	3	4	5	6	8	9
	80	–	0.25	0.5	0.5	1	1	2	2	3	3	4	5	5	7
	90	–	0.25	0.5	0.5	1	1	1	2	2	3	3	4	5	6
	100	–	0.25	0.5	0.5	1	1	1	2	2	2	3	4	4	5
	110	–	0.25	0.5	0.5	0.5	1	1	1	2	2	3	3	4	4
1¼″ with couplings at each end	50	–	–	0.25	0.5	0.5	1	1	1	1	2	–	–	–	–
	60	–	–	–	0.25	0.5	1	1	1	1	2	2	2	3	–
	70	–	–	–	0.25	0.5	0.5	1	1	1	1	1	2	2	3
	80	–	–	–	–	0.25	0.5	0.5	1	1	1	1	1	2	2
	90	–	–	–	–	0.25	0.5	0.5	1	1	1	1	1	1	2
	100	–	–	–	–	–	0.25	0.5	0.5	1	1	1	1	1	1
	110	–	–	–	–	–	0.25	0.5	0.5	1	1	1	1	1	1
1½″ with couplings at each end	50	–	–	–	–	–	0.25	0.5	0.5	0.5	1	1	1	1	1
	60	–	–	–	–	–	–	0.25	0.5	0.5	0.5	1	1	1	1
	70	–	–	–	–	–	–	–	0.25	0.5	0.5	0.5	1	1	1
	80	–	–	–	–	–	–	–	–	0.25	0.5	0.5	0.5	1	1
	90	–	–	–	–	–	–	–	–	–	0.25	0.5	0.5	0.5	1
	100	–	–	–	–	–	–	–	–	–	–	0.25	0.5	0.5	0.5
	110	–	–	–	–	–	–	–	–	–	–	0.25	0.5	0.5	0.5

Losses in End Fittings

Losses developed in end fittings can have a significant effect on the overall pressure drop, as in addition to the free flow area an *orifice loss* is also involved. This latter figure can be expressed in terms of an *orifice coefficient,* the value of which will always be less than unity (and will thus represent a relative loss or throttling effect).

Where the orifice coefficient for the end fitting is known, then the airflow through the fitting can be calculated from the following simple formula, which is sufficiently accurate for general use:

$$\text{air flow} = A.c.d\sqrt{P}$$

where c = orifice coefficient
d = orifice diameter
P = air pressure at the entry to the orifice (ie taking into account the total pressure drop to this point)

A is a constant which has the following specific values

A = 115 for c and d in inches and pressure in lb/in^2
= 2.55 for c and d in millimetres and pressure in bars.

Pipe Materials

Steel pipes are normally only used for air mains. For branch pipes or smaller lines up to about 25 mm bore copper piping or nylon tubing is commonly employed, with flexible lines at tool take off points. Typical sizes of metallic pipes are summarized in Tables VA, VB and VC.

TABLE VA – NOMINAL BORE SIZE STEEL TUBES TO BS1387

Nominal Size in	Actual Bore (in) Medium	Actual Bore (in) Heavy	Approximate o.d. in
1/8	0.226	0.178	13/32
1/4	0.338	0.290	17/32
3/8	0.476	0.428	11/16
1/2	0.623	0.575	27/32
3/4	0.839	0.791	1 1/16
1	1.060	0.996	1 11/32
1 1/4	1.401	1.337	1 11/16
1 1/2	1.633	1.569	1 29/32
2	2.066	2.002	2 3/8

TABLE VB – COPPER PIPE SIZES FOR AIR LINES

o.d. in BS2017	Bore in	o.d. mm	Equivalent in
1/8	0.075	–	–
3/16	0.131	5	13/64
1/4	0.178	8	5/16
3/8	0.279	–	–
1/2	0.410	12	15/32
3/4	0.655	16	5/8
1	0.872	22	7/8
1 1/4	1.122	28	1 1/8

o.d. in BS659	Bore in	o.d. mm BS2871	Bore mm
3/16	1/8	–	–
–	–	5	3.4
11/32	1/4	8	6.4
15/32	3/8	–	–
19/32	1/2	12	10.4
27/32	3/4	15	13
–	–	–	–
–	–	–	–

TABLE VC – EQUIVALENT PIPE SIZES

Nominal Bore Steel (in)	Equivalent Copper o.d.
1/8	1/4
1/4	3/8
3/8	1/2
1/2	3/4
3/4	1
1	1 1/4

TABLE VII – PROPERTIES OF NYLON TUBING

o.d. in	i.d. in	Burst Pressure at 20°C lb/in^2	bar	Min Bend Rad	Weight (lbs) per ft
Light Gauge					
1/8	0.095	1 000	67	3/4	0.0023
3/16	0.137	1 000	67	1.1/4	0.0060
1/4	0.190	1 000	67	2	0.0100
5/16	0.242	1 000	67	2.1/4	0.0140
3/8	0.295	1 000	67	3.1/2	0.0200
1/2	0.406	900	60	4.1/2	0.0370
5/8	0.500	800	54	6	0.0500
Heavy Gauge					
1/8	0.075	2 250	150	3/4	0.0140
3/16	0.107	2 250	150	1.1/4	0.0033
1/4	0.150	2 250	150	1.3/4	0.0090
5/16	0.182	2 250	150	2.1/4	0.0230
3/8	0.225	2 250	150	3	0.0310
1/2	0.375	2 000	133	4	0.0600

Preferred sizes of nylon tubes are: (see also Table VI)

o.d. (ins)	i.d. (ins)	o.d. (mm)	i.d. (mm)	BSP Fitting Size
1/8	0.058	5	3.3	1/8
3/16	0.122	8	5.5	1/4
1/4	0.165	12	8.5	3/8
5/16	0.213	15	12.0	1/2
3/8	0.245			
1/2	0.369			
3/4	0.587			

Flexible nylon tubes may be used directly for small diameter hose or reinforced with braid for larger hose sizes. Coiled nylon tubes are also used for flexible lines, the tubing in this case being of small bore (eg up to 5/8 in) formed in the shape of a close-wound coil. This results in an extremely flexible line which can also be extended considerably in length, retracting and re-stacking to its original coil form when released.

Another flexible form of air line hose is PVC reinforced with terylene. Working pressure rating is similar to that of plain nylon tubing, eg 17.5–12 bar (250–170 lb/in^2), depending on size. Bore sizes up to about 25 mm (1 in) are readily obtainable. Working temperature range is, however, limited to about -10^oC to $+40^oC$. Reinforcement may be mesh type (ie yielding a transparent tube) or an opaque braid. Many of the earlier limitations of plastic hoses, eg particularly lack of flexibility at lower temperatures, have been overcome with later formulations. Spiral wire reinforced plastic tubes may be used for hose duties, the latest types offering a smooth rather than a corrugated bore. Spiral wire reinforced plastic tubes resist deformation under negative (suction) as well as positive internal pressures.

Rubber hose is generally used for flexible lines where a wider working temperature is required or larger hose sizes are needed. Air line hoses are normally manufactured to BS924 Type A or B requirements (or American or Continental equivalents). The basic construction comprises a seamless inner tube of oil-resistant rubber, wrapped with plies of cotton and a smooth outer cover. Typical sizes and ratings are:–

Type A – sizes 4.8–9.5 mm (3/16 in to 3/8 in) bore – pressure rating 12 bar (170 lb/in^2) maximum.

Type B – size 12.5–63.5 mm (½ in to 2½ in) bore.

Typical pressure ratings:–

mm	in	bar	lb/in^2
4.8	3/16	32	450
6.3	1/4	31	440
8.0	5/16	26	375
9.5	3/8	22	310
12.5	1/2	19	275
19.0	3/4	16	225
25.0	1	14	200

Rubber hoses and flexible lines for suction services normally feature wrapped ply construction. The fabric reinforcement is laid in the form of several close layers or plies, rather than braid, producing a more rigid construction. Where greater flexibility is required, the plies may be separated or cushioned. This also improves the kink-resistance of the hose. Metallic wire reinforcement may also be used in the form of a wire spiral, embedded or semi-embedded in the liner. Only fully embedded spirals produce a smooth bore tube.

Plain (unreinforced) polythene tubing is more flexible than nylon tubing. Typical maximum pressure rating is 7 bar (100 lb/in^2) for 15 mm (5/8 in) o.d. tubing and up to 14 bar (200 lb/in^2) for 5 mm (3/16 in) o.d. tubing. Flexible polythene tubing is not suitable for pressure use at temperatures much above 75^oC, nor for rigid installations except where adequate support by pipe clips can be arranged.

Flexible PVC tubing is only suitable for low pressures and has a maximum service temperature rating of 70^oC. Its flexibility can compare with, or is better than, that of polythene tubing.

Tube and Pipe Bending

Small diameter metallic pipes and tubes can be manipulated by hand for bending, but care is needed to ensure that the tube is not kinked or otherwise deformed, which will restrict the air flow. The absolute minimum for the radius of bend is three times the tube o.d., and a higher figure should always be adopted for hand bending. Larger radius bends will also have lower frictional losses. Fillers may be used to support the inside of pipes and tubes for hand bending. These are effective only if the tube is filled 'solid'. Low melting point metallic fillers are generally to be preferred, and are readily removed by gentle heat after bending. The material is re-usable. Sand is not recommended as a filler due to the difficulty of ensuring its complete removal unless pressure cleaning equipment is available.

Proprietary pipe benders are available for manipulating larger sizes of pipes and tubes and also for handling all thin walled metallic tubing, which is difficult to bend without local deformation. These work on one of the following principles and can produce satisfactory cold bends.

Compression bending — where the tube is clamped and effectively put in compression during bending. It is most suitable for thicker walled tubing and larger bend radii. Its main advantage is that the construction of the bender is simple and the bending operation can be completed rapidly.

Draw bending — where the tubing is clamped to a grooved form and literally wound to the bend by the action of a moving shoe. It is suitable for the tightest bends, but there may be some deformation of the section through flattening.

Press bending — a simple and rapid technique, where the tube is manipulated by rollers working against a reaction block. It is mostly suited to thicker walled tubing and generous bend radii.

Roll bending — particularly suited to the manipulation of thin walled tubes, involving the passing of the tube through a series of offset grooved rollers. Some of the rollers are power driven, the others are idlers. The driven rollers are brought progressively nearer to each other, until the required radius of bend is obtained. Such machines can also produce complete 360° bends and coils.

Pipe Supports

Permanent installations in rigid (metallic) tubing must always be adequately supported so that no strain is put on the compressor or other devices to which they are attached. Supporting clips should, preferably, be of the type with elastomeric liners or similar feature, which clamp the pipe or tube rigidly, but at the same time allow for axial movement to accommodate contraction or

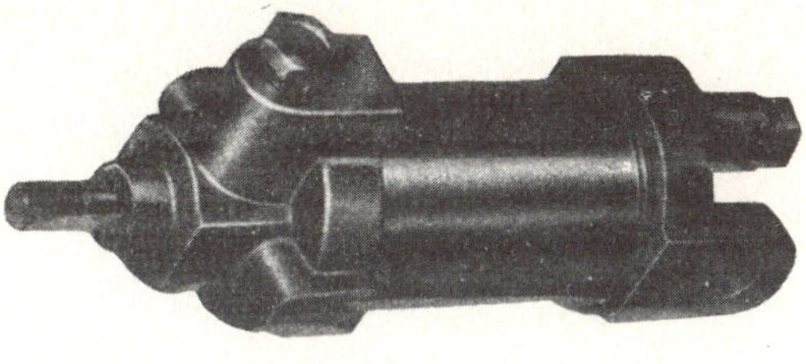

Lang CA cylinder.

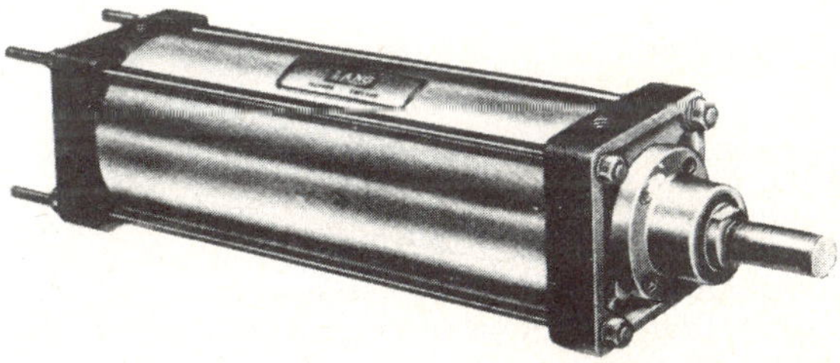

Lang heavy duty 'S' cylinder.

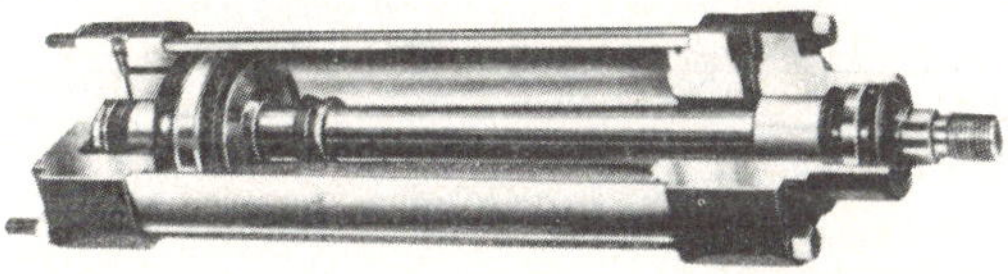

Cut away section of tie rod cylinder. Available 1½, 2¼, 3, 4, 5, 6, 8, 10,12 and 14 inch bore.

expansion. Alternatively, for shorter runs, thermal expansion and contraction can be accommodated by bends or, in the case of large pipes, the fitting of expansion joints.

Clip spacing is generally less critical in the case of nylon tubing, although again the question of avoiding snagging is important. It may also be desirable to anchor the tubing securely on either side of bends to prevent whipping or excessive displacement which could follow an interruption of flow.

Clips should be used on pipes close up to fittings, etc, which are not themselves mounted to save the pipe from carrying the weight of the fitting. Logically, too, the complete line system should be laid out as neatly and tidily as possible, although this consideration is often modified by the respective lengths of lines involved. However, the shortest distance between two points is not necessarily the best run for a line. Conversely, designing a 'tidy' system by using a multiplicity of elbows and fabricated bends to group runs of lines, may represent unnecessary cost and unnecessary additional pressure drop.

In certain systems it may be desirable that all piping be electrically bonded. This can be done by using pipe clips which incorporate a small tongue or similar projection contacting the surface of the pipe, when the pipe may be earthed through the clip. With plain clips, a bond can be formed by inserting a layer of metallic gauze between the pipe and clip and connecting the gauze to a suitable point.

Simple pipe clips are commonly manufactured in brass or copper. These may be shaped to hold individual pipes of a specified overall diameter, or up to six pipes in a single clip. Unless insulated from the pipe, eg with a plastic liner, pipe clips should always be of similar metal to the pipes to avoid any possibility of electrolytic corrosion.

See also chapter on *Compressed Air Mains.*

Air Line Fittings

STEAM-QUALITY steel piping is used for compressed air mains up to 78 mm (3 in) diameter, usually with screwed joints. For larger sizes of air mains cast iron pipes with integral flanges or steel piping with welded-on flanges are the usual choice. Copper piping or nylon tubing is used for smaller lines. Copper pipes for threading are o.d. sizes, consistent with BSP pipe threads. Both copper fittings and nylon tubing are used with compression fittings, when tube sizes may be o.d. or nominal bore.

A selection of brass fittings from Drallim Tube Couplings.

Solderless connections are widely favoured for pipe sizes up to 19 mm (¾ in) diameter. The simple olive joint is effective (Fig 1) provided the tube projects well through the olive. Tightening the nut then produces a tight compressive lock without deforming the tube wall, this being particularly suited to rigid tubing. Another type of proprietary coupling is shown in Fig 2 which employs a resilient seal.

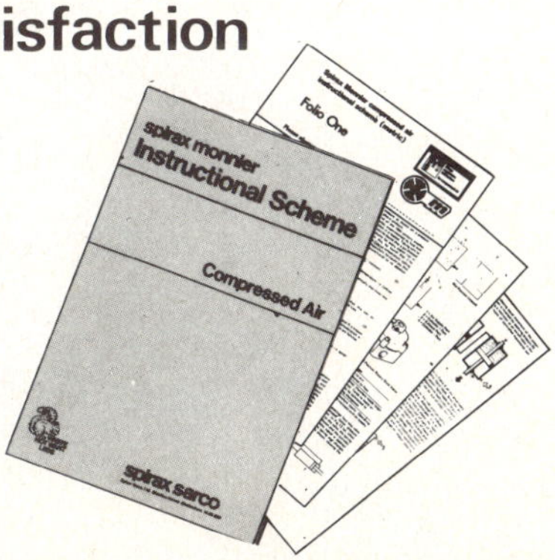
spirax monnier
Instructional Scheme
Compressed Air
Folio One
spirax sarco

great inventions

The Science of Biology, the study of living things and how they function (as in "Private Eye") began with the microscope, which enabled scientists to observe what was hitherto unobservable (as in "Penthouse"). Galileo constructed a simple one, but it was not until the 17th Century that the microscope came into regular use. Anthony Van Leeuwenhoek (1632–1723) devoted much of his long life (a generous beer-man) to biological research (not for washing machines) and the making of microscopes. It was he who first observed and described the red corpuscles in the blood, the lens of the eye and the structure of hair (which ran long in the West End).

TTP has also devoted itself to the study of moving things, like gears and motors, pumps and valves, Lulu and Fifi, noise and vibration, instruments and controls, Lulu and Fifi, industrial fasteners and fabrication, hydraulics and pneumatics, Lulu and Fifi, fluids and seals, fire protection and security, Lulu and Fifi – in fact, they have been observing industrial power and engineering (and Lulu and Fifi) for 44 years and what they don't know about the subjects aforementioned, ain't worth knowing. It is all written down, edited and published in the forty or more technical books and journals of TTP. Why not ask for our books and journals list – they are all there (so are Lulu and Fifi).

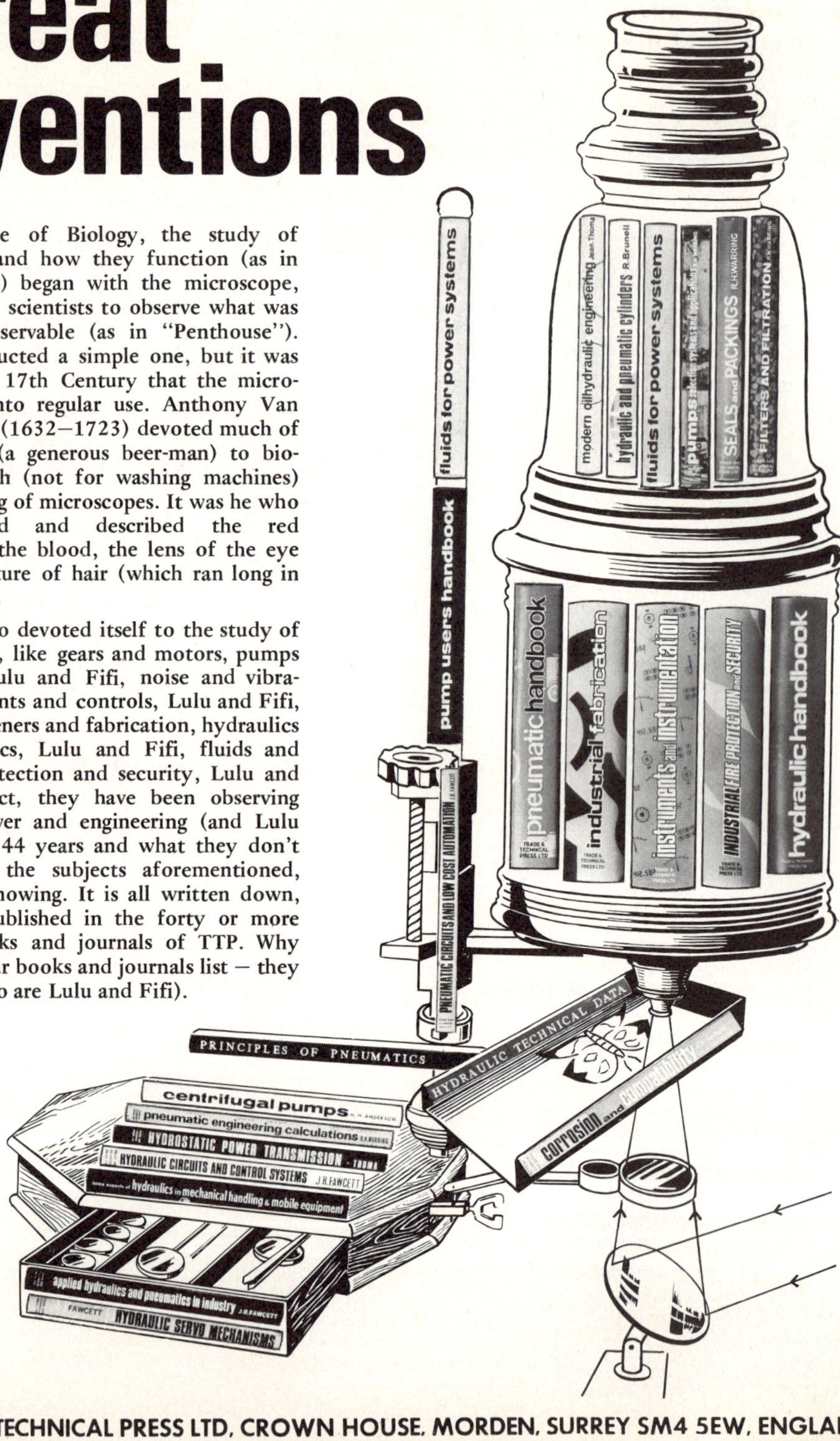

TRADE & TECHNICAL PRESS LTD, CROWN HOUSE, MORDEN, SURREY SM4 5EW, ENGLAND.

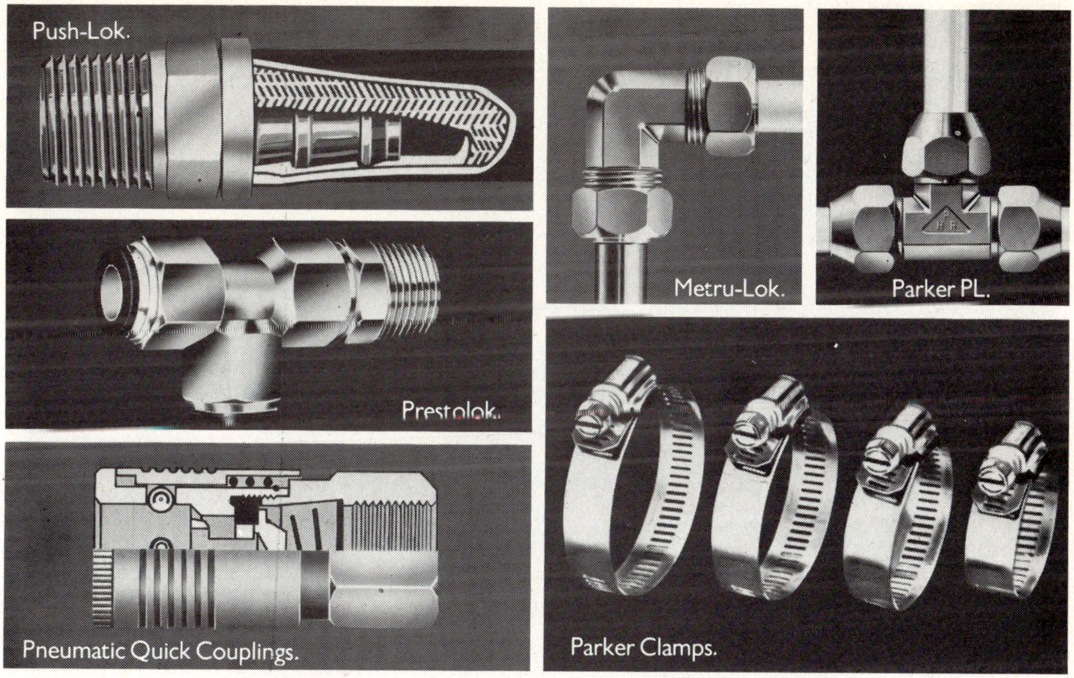

You can't afford to miss our tube connections

Nobody puts more than Parker into tube and hose fittings. Into designing them. Testing them. Refining them. And into providing a range that's wide enough to cope with every job.

That's why Parker fittings are helping industries maintain maximum productivity and profitability everywhere in the world.

Post us the coupon for more details. Unless you're specifying Parker already, you can't afford not to.

Parker PL. Brass fittings for all plastic tubes in most metric dimensions. Leakproof until tube burst pressure. Can be assembled and disassembled time and again.

Metru-Lok. For both brass and plastic tubes, Metru-Lok fittings give really fast connections. Simply insert the tube and tighten a nut. The connection is leakproof, vibration resistant and can be remade up to 15 times.

Push-Lok. Inexpensive hoses and fittings that let you assemble low pressure lines in seconds, without clamps, clips, bands or wires.

Pneumatic Quick Couplings. With an automatic fluid control valve they make fast connection and disconnection by hand absolutely foolproof.

Parker Clamps. In a wide variety of sizes including miniature. No spot welds to rupture. High torque gives high sealing pressure. The range meets stringent military and marine specifications.

Prestolok. Brass fittings for nylon tube. The tube only has to be pushed into the fitting for a positive seal and can be released simply by pressing the black outer ferrule.

Parker
Fluid Connectors

TO: Parker-Hannifin (U.K.) Ltd.,
Haydock Park Road, Derby DE2 8JA
England.
Telephone Derby (0332) 365631.
Telex 37427.
Please send me full details of all Parker tube fittings.

Name______________________

Address______________________

PH 77

Fig 1 Metal compression couplings with conventional olive (left) and bi-cone olive (right). (Drallim).

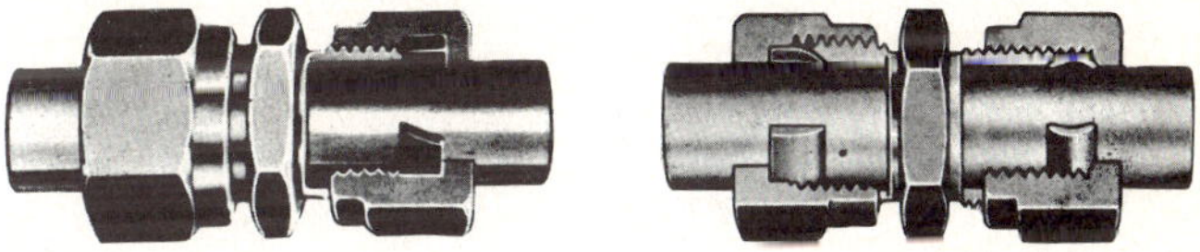

Fig 2 Tube coupling with resilient seal in union nut. (Drallim).

Various other types of compression fitting are used. Where the tube material is ductile, a collet may be used, tightened up by a nut, to produce an annular deformation of the tube wall to lock the tube within the coupling. The same type of fitting can be used with nylon and plastic tubes, only in this case, support for the inner wall may be provided by an additional inner sleeve – Fig 3.

Fig 3 Stages of assembly of metal compression coupling used with plastic tube. (Drallim).

Various designs of 'instant' couplers have also been produced for use with nylon tubes. The simplest is a barbed fitting over which the tube is pushed, the barbs restraining the tube against outward movement. Other types employ various forms of locking to hold the tube secure once pushed in place. Nylon tube can only be removed from a barbed coupling by cutting. With other 'instant' fittings, however, it may be possible to remove and replace the tube many times over, if required.

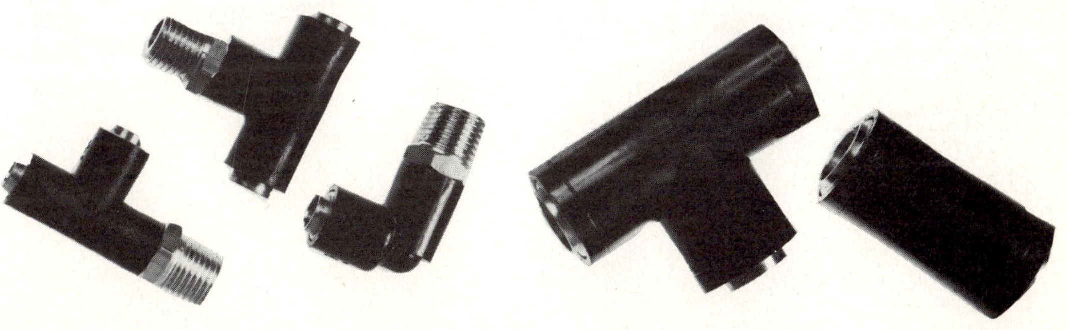

Examples of Martonair Instant Couplers.

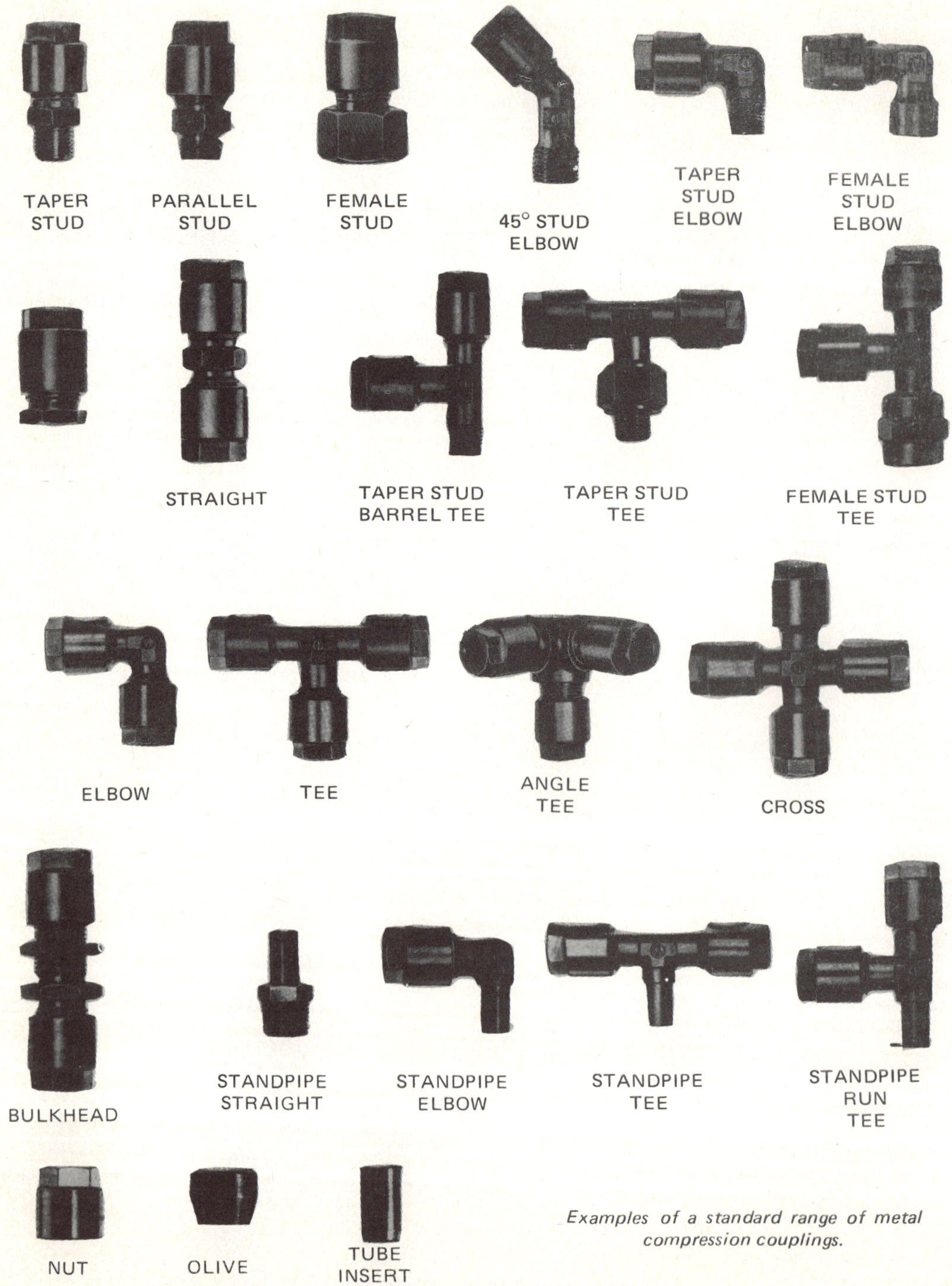

Examples of a standard range of metal compression couplings.

Screwed connections are generally restricted to metallic tubes of larger diameter and thicker walled sections. Taper threads (eg BSP taper or API taper) are usually preferred. The best seal is given by having the taper on the male rather than the female part of the joint. These can assemble with either taper or parallel internal (female) threads. Parallel external (male) threads are less suitable for tight joints, except in 'longscrew' form. Longscrew threads, used for joints, incorporate a soft material which is compressed onto the surface of the external thread by tightening a backnut against a socket.

All the solderless joints described — compression type and screw thread — produce joints which can be disassembled and reassembled repeatedly, without replacement of components (although olives and rings may be replaced on remaking a broken down joint).

Nylon tubing is finding increasing application for branch and service lines up to about 12.5 mm (½ in) diameter, and less frequently in larger sizes (up to 1 in diameter). In addition to permanent installations, traditionally piped in metallic tubing, nylon tubing is particularly suitable for connecting lines where slight movement may be experienced.

Apart from its lower cost and ease of manipulation, nylon tubing has other advantages over metallic tubing, particularly the absence of fatigue. It is, however, more limited in its service temperature range, and is not suitable for pressure rating at temperatures in excess of 90—100°C. The actual softening point of nylon is of the order of 140—150°C with an internal pressure of 60—70 lb/in^2 (4.2—5 bar), reducing to circa 50°C with an internal pressure of the order of 250 lb/in^2 (17.5 bar).

See also *Survey of Air Line Fittings.*

Separators

Separators may be fitted in distribution mains for removal of suspended water droplets, preferably associated with an automatic drain. A ball float trap is commonly preferred because of its positive shut-off in the absence of water and positive lift when the trap is filled with water.

Separators may also incorporate filter elements designed to remove solid contaminants and/or oil droplets. Units of this type are more properly called filter/separators.

Filters

Air line filters are normally designed to provide both filtration and separation, ie function as filter/separators. The size, degree of filtration provided, and to a lesser extent the form, vary depending on whether the filter/separator is to be used in the main, branch lines, or individual air lines.

The geometry of the filter body is so shaped that air entering is directed into a swirling motion, the centrifugal force thus imparted to the incoming air tending to throw larger solids and the bulk of the water or oil droplets out of the stream to impinge on the walls. From here the separated liquid falls under gravity to the bottom of the bowl or settling chamber. The airstream meanwhile is directed onwards through a suitable filter element which functions purely as a mechanical screen and traps and retains all solid particles larger in diameter than the pore or screen size of the filter element.

The satisfactory performance of an air line filter thus depends on dynamic considerations governing the efficiency of the centrifugal vortex in removing water and oil vapours, and the mechanical or purely static action of a conventional filtering element to remove the remaining solid particles not washed down with condensate. A further refinement is the provision of a quiescent zone or baffle (or both) which ensure that condensate collecting in the bottom of the bowl remains isolated from the main airstream and thus cannot accidentally be entrained again and carried on to the filter element. *Separation* is thus largely a matter of filter bowl shape and internal geometry. *Filtration* depends on the use of a conventional element.

Satisfactory performance from a filter/separator demands that the filter bowl must be emptied before the contaminants settling out have reached such a level that they are exposed to the vortex flow, with the possibility of remixing with the air. Also the filter element must be cleaned, or replaced, periodically before it has become clogged.

The emptying of the bowl can be done manually by opening a drain cock on the bottom, when the contaminants are blown out under air pressure. Contaminant level is usually visible (through a transparent bowl), or a service indicator or tell-tale device is fitted to indicate when the settling chamber needs emptying. The use of a transparent bowl normally limits the maximum service temperature to about $50^{o}C$ ($120^{o}F$) for high service temperatures, and for pressures above 10 bar (150 lb/in^2), metal bowls are normally used. Metal bowls may also be specified where rough usage is anticipated.

Air line filters are now commonly available with automatic drain mechanisms, operated by a float or similar device. Once the liquid level in the bowl has reached a predetermined height, the float opens the drain mechanism allowing the liquid contents to be discharged under air pressure. Filters with automatic drains are particularly advisable for use in remote or inaccessible locations.

Filter Elements

Filter element materials commonly employed include wire mesh or metal cloth, sintered metal and porous ceramics. More limited use is made of resin-impregnated papers, synthetic fabrics, and porous plastics (eg sintered high density polythene and mechanically reinforced polyurethane foam).

Simple metal cloth elements have the advantage of providing a relatively open flow path, with consequent low pressure drop, although they are generally limited in performance to coarse filtering (eg 60 microns upwards). Used in disc assembly, or in conjunction with paper elements, fine filtering down to 10 and 5 microns respectively can be provided at the expense of increased pressure drop. Sintered elements with controlled pore size, which filter in depth, can provide filtering down to 5 microns or better, if required. Filtering in depth also has the advantage that the element life between cleaning or replacement is generally prolonged.

The majority of air line filters are designed with 'tee' configuration. In-line filters are, however, also produced, with the advantage of being more compact and lighter in weight, although their performance is generally restricted to that of filtering with no provision for working as a separator.

Particular applications for in-line filters are on portable equipment and in flexible lines.

Filter Performance

For average use in air lines supplying mechanical equipment a cut-off of 40 microns is usually adequate, and even higher values may be tolerated where a virtually unrestricted flow is required. A 40 micron cut-off can be given by comparatively simple wire cloth elements, although for protecting tools and appliances with close working clearances, control valves, etc, a very much lower cut-off may be required. Sintered metal elements are particularly favoured for low cut-off filters because they can be produced with controlled and consistent porosity so that the cut-off is positive rather than somewhat arbitrary. Their only real limitation is that such elements are virtually impossible to clean by conventional methods and so usually have to be returned for special cleaning, and in the meantime they must be replaced by a clean element. Under normal

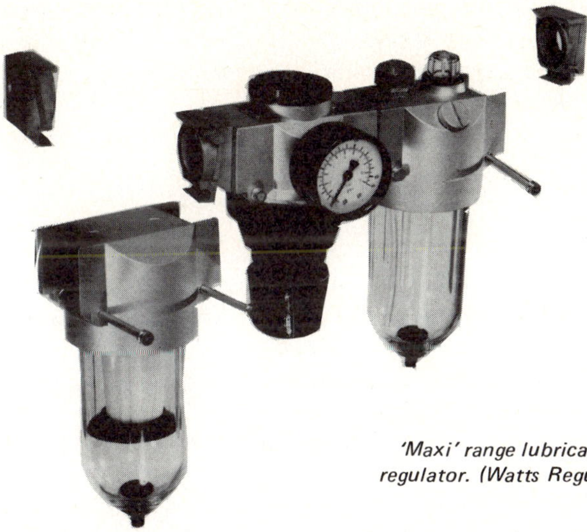

'Maxi' range lubricator, filter and regulator. (Watts Regulator (UK) Ltd).

service conditions, and with regular 'servicing' in respect of cleaning, such elements are, however, capable of virtually indefinite life. Alternative low-cut-off filters may employ felt, paper, cloth or wire mesh elements or combinations of various materials in specially arranged forms, according to the performance required.

Where absolutely clean and dry air is required, a special filter may be called for with a cut-off as low as 1 micron. In addition a further element may be incorporated made of a suitable absorbent cotton, to 'wring out' remaining traces of moisture and vapours. An alternative, when dry air is specifically called for is the use of an adsorption dryer in the system.

Filter Sizing

The size of air line filter required is normally dictated by the element design and construction, as it affects the pressure drop. Element areas must also be adequate to maintain a low velocity at maximum capacity so that the remaining contaminants will tend to be separated on the surface of the element and drop into the settling chamber, rather than impinge on the element. High velocity flows through the element can carry contaminants through with the air, whereas the same type of contaminant may be stopped by the element at lower flow velocities.

Typically the element open area is usually made 20 to 30 times greater than the cross sectional area of the pipeline so that at maximum capacity rating the pressure drop does not exceed 0.35 bar ($5 lb/in^2$) on flow rates of up to $2.8 m^3/min$ ($100 ft^3/min$) and only slightly higher (eg 0.42 bar ($6 lb/in^2$) at higher flow rates (eg up to $14 m^3/min$) ($500 ft^3/min$). Actual element size is then related to the effective transmission of the element.

In practice, capacity rating is dependent on the permissible or desirable pressure drop. For a given size of filter, capacity rating may range between a ratio of 5:1 or more, depending on the pressure drop. The figure of 0.2 bar ($3 lb/in^2$) maximum pressure drop is typical of good practice, but for various reasons a much lower figure may be preferred, in which case, for the same flow rate, the size of the filter will have to be increased. Also for any given size of filter (element) the capacity rating will tend to decrease with lowered cut-off point. Thus, basically, the cut-off required determines the type of element preferred or necessary, and the pressure drop which is acceptable or necessary as a design figure determines the actual size of the filter (element). Such data will be specific to a particular design of filter. Some units may be rated for capacity on a specific pressure drop basis, whilst others may be rated for capacity on a nominal pressure drop basis – eg probably of the order of 0.2 bar (3 lb/in^2) or possibly even 0.35 bar (5 lb/in^2). In the latter case the actual pressure drop in service will be lowered if a filter with a larger capacity is used on the particular service.

Proprietary air line filters are normally rated for a given flow range at a stated (maximum) pressure. Such sizing is normally consistent with a pressure drop of not more than 0.35 bar (5 lb/in^2) on flow rates up to 2.8 m^3/min (100 ft^3/min) and only slightly higher at higher flow rates (eg 6% pressure drop up to 11 m^3/min (400 ft^3/min). Element size is determined accordingly, relative to the effective transmission of the element. Thus for a given size of element the capacity rating will tend to decrease with lowered cut-off point.

Capacity rating can also be taken as dependent on pressure drop. Thus for a given filter the actual pressure drop achieved can be reduced by reducing the capacity rating. Some units are, in fact, rated for capacity on a specific pressure drop basis rather than on capacity at a nominal maximum pressure drop.

Typical regulator/filter. (Norgren).

Martonair exhaust filters.

Norgren's Air Aces

reach Olympian heights

Drip Leg Drain

Miniature Filter

'Designer' Regulator

11–918 Precision Controller

OLYMPIAN 'PLUG-IN' SYSTEM

'Designer' Lubricator

'Designer' Filter-Regulator

The Norgren Olympian system of plug-in compressed air processing equipment–filters, pressure regulators, filter-regulators and lubricators–is today's simplest method of installing an efficient compressed air system for your particular operation. In fact, installation, maintenance and interchange is as near to instant as you can find, anywhere.

And the Norgren Air Aces offer you even more than that. The Designer range, for example, developed to allow system designers to save space and maintain a good, neat appearance without sacrificing efficiency. Plus a complete range of 'miniatures'.

But we can't go into all the diverse Norgren compressed air processing products here. Why not get in touch? You'll find that service and delivery live up to the high standards set by our products.

C. A. NORGREN LTD.

SHIPSTON-ON-STOUR, WARWICKSHIRE, ENGLAND.
Telephone: 0608 61676 Telex: 83208

A Norgren Shipston product

An IMI Company

Air travels better with Norgren

How tight can you be with your compression fittings?

Nitex fittings ensure controlled compression — eliminating the danger of damage caused to nylon tube by over-tightening. No ferrule misalignment is possible, since the liner — which supports the bore of the tube preventing its collapse — is manufactured as an integral part of the ferrule. A permenent union is formed — ideal for lasting and temporary installations. Nitex assemblies can be dismantled and re-made time and again without detriment to the seal. Nitex fittings are also suitable for use with copper, bundy and steel tubes when tubing nuts and olives of conventional design are supplied.
Fully illustrated literature available on request.

a subsidiary company of Imperial Metal Industries Limited

COMPRESSION JOINTS LTD
Oldmixon, Weston-super-Mare, Somerset. Phone: W-s-M 24171

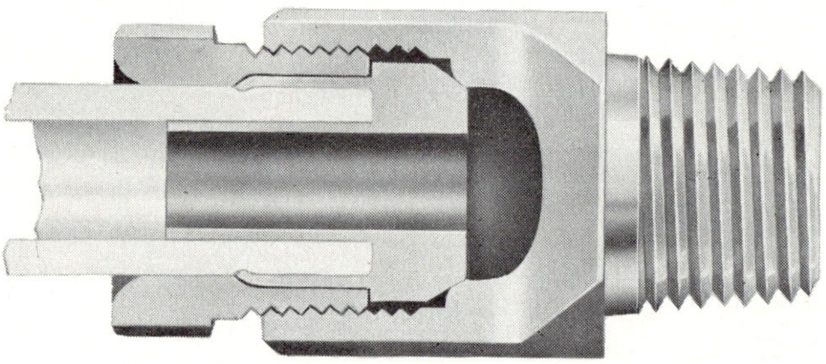

Nitex Compression Fittings

178B

AIR LINE SERVICE UNITS

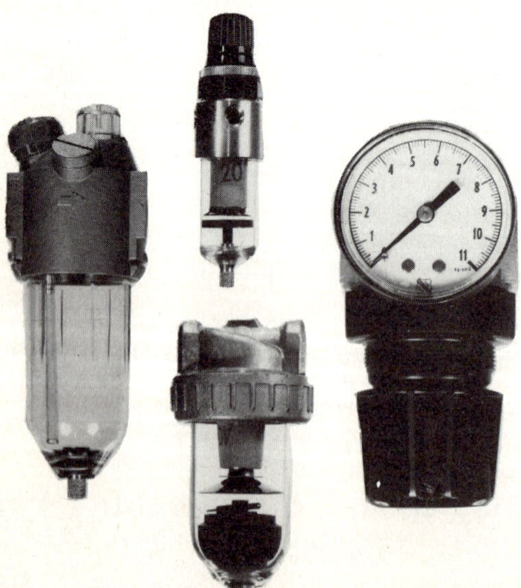

A RANGE OF AIR LINE FILTERS/ REGULATORS/LUBRICATORS, AND INTEGRAL FILTER/REGULATOR

Sizes — 1/8″ to 2½″ B.S.P.

High Efficiency and performance characteristics, flow rates up to 1000 cfm and pressures of 200 p.s.i. ensure that our range is imminently suitable for OEM or general plant installations.

Filters and Lubricators fitted with polycarbonate bowls, as standard. Metal bowl guards or metal bowls available, if required.

WATTS REGULATOR (UK) LIMITED
London Road, Thrupp, Stroud,
Gloucestershire, England.

This must be the largest pneumatic 'push in' fittings system and it's ex stock.

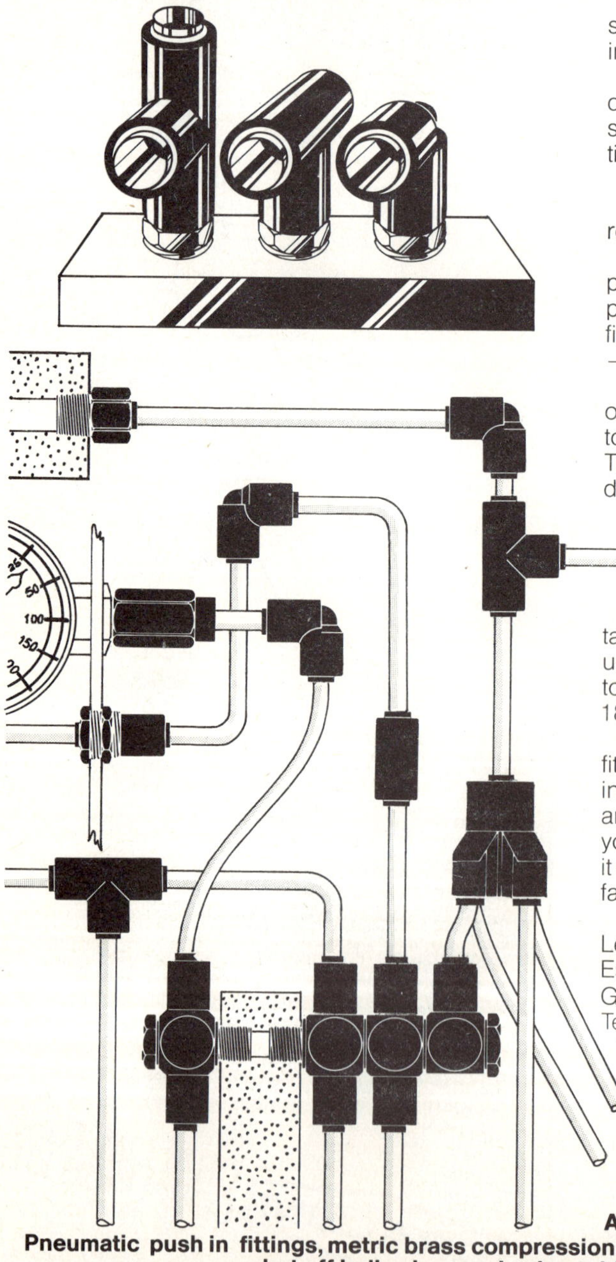

The Legris LF3000 is a complete system of 'push in' fittings for every installation.

But more important in most applications we can show considerable cost savings, by drastically cutting assembly time and increasing productivity.

If you don't believe it, we'll prove it.

There's no tightening or inspection required and every fitting has been tested.

With the LF3000 series you can position units with closer centres compared to conventional compression fittings reducing panel or cabinet sizes —another cost saving.

All stud elbows and tees swivel on an 'O' ring enabling assembly close together with no need for swing space. Tubes can be orientated without disturbing the thread seal.

The LF3000 is available with BSP tapered thread on stud fittings and for use with metric 4 to 14 mm or inch ⅛″ to ½″ OD tubes. Working pressure is 18 BAR.

We can't show all the 30 types of fittings, in numerous sizes—metric or inch, plus modular or DIN rail connectors and Multi-Way Couplers. But we can tell you it's the largest range available, it will cut assembly costs and there is a fast delivery.

For all the facts write or telephone Legris (UK) Limited, Ashville Trading Estate, The Runnings, Cheltenham, Glos GL51 9PT.
Tel: 0242 43103
Telex: 43335.

LEGRIS

Available worldwide in 29 countries.

Pneumatic push in fittings, metric brass compression fittings, quick release couplings, shut off ball valves and nylon tubing & cutters.

Absorbent Filters

Whilst filter/separators are capable of removing oil particles from compressed air, additional treatment is necessary if oil-free air is required for a specific application. The best solution is to employ an oil-free compressor, but where the cost of this is not considered justified, oil filters may be used in the lines instead. Because oil can exist in a variety of particle sizes and generally includes a proportion of sub-micron sizes (ie oil haze), filters with absorbent elements are generally necessary, supplementing mechanical separation or filtering of the larger oil particles.

Careful selection of the physical properties of this absorbent material will enable a filter to be designed which will have an extremely low pressure drop yet show high efficiency of removal of sub-micron particles.

Filters of this type may use conventional mechanical methods of separation, such as direction change, centrifugal action, high velocity impingement, velocity drop and wire mesh demisters, together with an organic absorber element designed specifically to absorb the hydrocarbon oils normally used in compressors.

The absorber material has also to be chosen so that it is unaffected by the presence of water vapour — ie it will go on absorbing even if saturated with water, with the oil displacing the water. It is made in the form of a replaceable cartridge element and can be designed to change colour as it becomes saturated with oil, thus providing a visual indication of the state of the cartridge and when it needs replacement.

Filters of this type are capable of over 99% efficiency as regards oil removal and have proved satisfactory for obtaining nominal oil-free air from conventional compressors for many processes which would otherwise demand the use of an oil-free compressor.

Oil removal is, of course, only required in specific cases. Normally the presence of oil in a compressed air supply is desirable for tool or machine lubrication and oil may, in fact, be injected into the supply for this purpose.

Special miniature filters are made for use with circuits incorporating fluidic devices to provide the degree of protection required. These may provide for oil mist removal as well as mechanical filtering to a fine degree.

See also chapter on *Air Conditioning and Treatment*

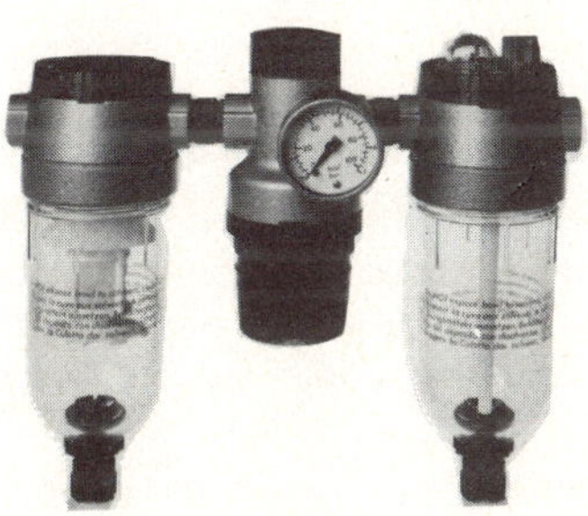

Spirax Monnier International range of air service units.

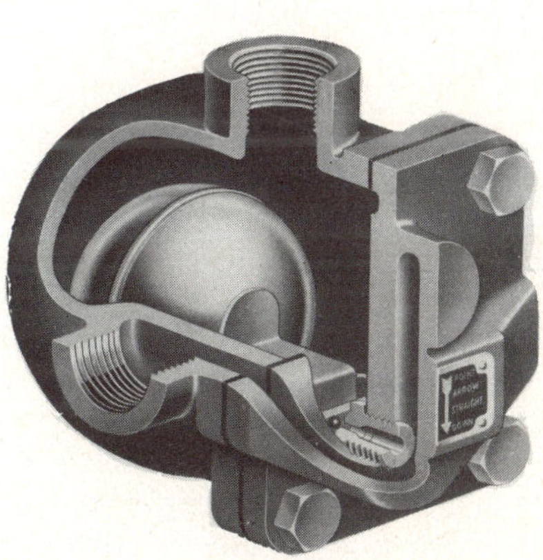

CA 550 Compressed air drain trap.

Air Conditioning and Treatment

THE following treatments are essential to ensure efficient, trouble-free working with compressed air:-

(i) Prefiltering of air prior to the compressor.

(ii) Separation of water from the compressed air, after the compressor.

(iii) Water separation in ring mains and individual lines.

The following further treatments may, or may not, be required:-

(iv) Filtering of individual lines — generally highly desirable but may not be considered essential in all systems.

(v) Injection of lubricant into lines — normally an essential requirement in the case of air tools and similar working devices to ensure adequate lubrication. In less critical cases the normal oil content of compressed air carried from the compressor may provide suitable lubrication for the system.

(vi) Separation of oil from the compressed air — only called for when an oil-free delivery is required and the compressor is not an oil-free type. In the latter case, oil separation may still be called for to remove any traces of oil or oil mist present.

(viii) Drying — normally restricted to specialized applications. However, effective drying of the compressed air before delivery can reduce or eliminate the need for water separators in the system.

Initial filtration of aspired air is provided by a conventional intake filter preceding the compressor. This reduces the accumulation of solid deposits which could otherwise build up inside the compressor, or the compressor lubricant, with a proportion carried through the compressor into the delivered air. Without intake filtration the quantity of solids inducted can be considerable — eg a pound or more a week in a medium sized compressor working in an ordinary town atmosphere; or as much as 14 to 22 kgs (30 to 50 lbs) a week with a large compressor working in a heavily contaminated industrial atmosphere. These dust particles can range in size from a few microns to up to 100 microns in diameter — or more, and commonly include a high proportion of abrasive matter — see also Fig 1.

Intake filters may be of the 'dry' or 'wet' type. Compressor manufacturers will normally specify which type and size of filter to use for specific applications, or supply suitable general purpose

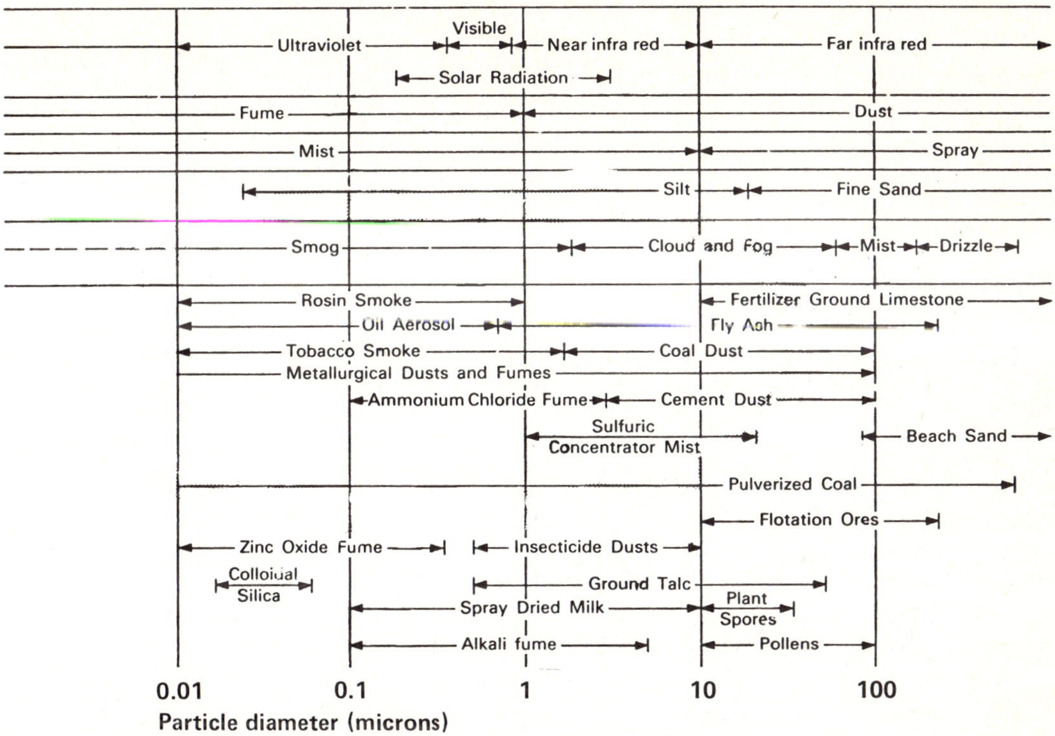

Fig 1 Sizes of typical contaminants in air.

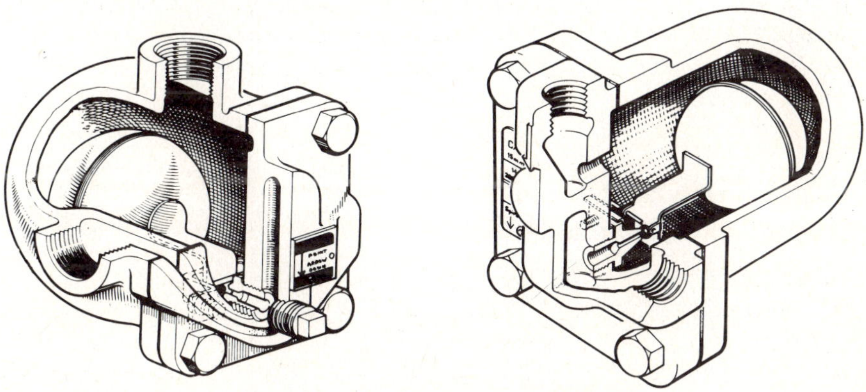

Ball float type compressed air drain traps (Spirax Sarco)

filters as standard fittings. The intake filter may also be combined with a silencing chamber (filter silencer), or a separate air silencer fitted after the filter to suppress a compressor noise. Where air is inducted from an outside source the filter would be fitted at the point of air entry to the building rather than on the compressor itself. In this case it is important that the pipework between the filter and compressor be clean and free from corrosion. If any doubts exist on this score, an additional filter or simple filter screen can be fitted at the compressor inlet.

A moisture separator normally follows the aftercooler to complete the compressor installation, before the air is delivered to the receiver. Even with intercooling, air leaving the compressor is inevitably saturated with water vapour and condensed water carried over. (Fig 2).

Separators are usually simple mechanical devices designed to impose a sudden change in velocity on the airstream, causing water and oil droplets to be thrown out of the main airstream and collect in the vessel, free from the influence of the airstream. Once 'dumped' in this manner the liquid cannot be re-entrained in the airstream unless the vessel overfills. Thus, provision must be made to drain off the condensate periodically. This can be accomplished manually, or automatically, the latter being the preferred method.

The drain should either be at the lowermost point of the vessel, or near the top of the 'high' level. Oil and water will separate out in the condensate, with the oil floating on top. If the drain point is towards the bottom, but not actually at the bottom, water can be drawn off leaving a residue of oil. This could, in time, degenerate into a viscous mass, or even emulsify, to cause trouble. Thus 'bottom draining' is only fully reliable if all the contents are withdrawn through the drain. With a 'high' level drain the oil is drawn off first, but a residue of water is left. This is the preferred method with continuous draining as, in this case, other tapping points could lead to the vessel eventually tending to fill entirely with oil.

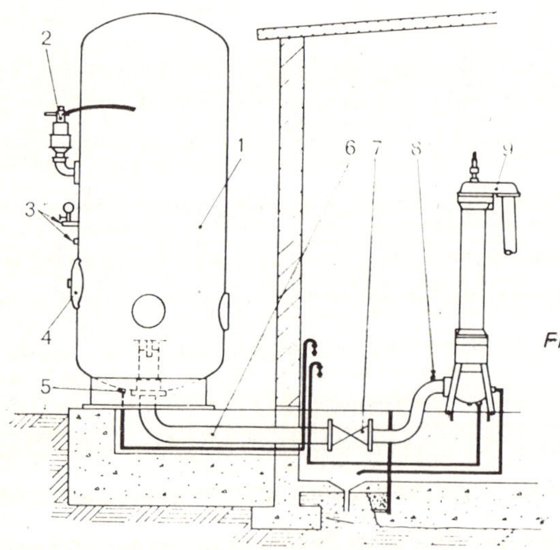

1 – air receiver
2 – safety valve
3 – connection for regulating pipe
4 – inspection cover
5 – connection for draining
6 – delivery pipe
7 – shut-off valve
8 – bleed valve
9 – aftercooler

Fig 2 Typical air receiver and aftercooler. (Atlas Copco).

Various types of automatic drain valves have been produced which automatically releases water collected from the air line when filled. A ball float type valve is commonly used since this maintains a positive shut-off, opening only when enough water has collected to lift the float, and closing again immediately the water has been ejected. The other main type is of 'aerodynamic'

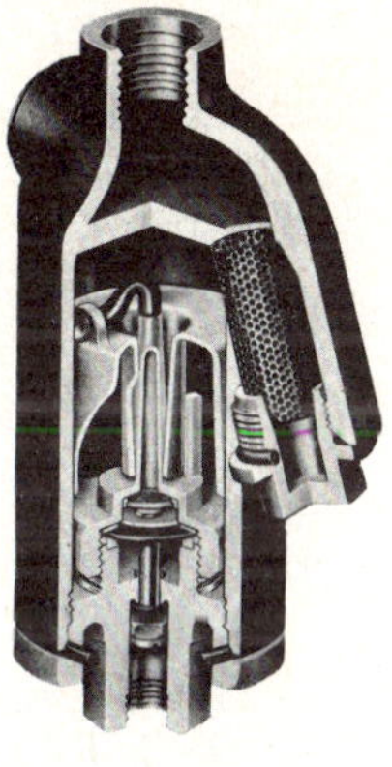

Spirax Dri-line air trap

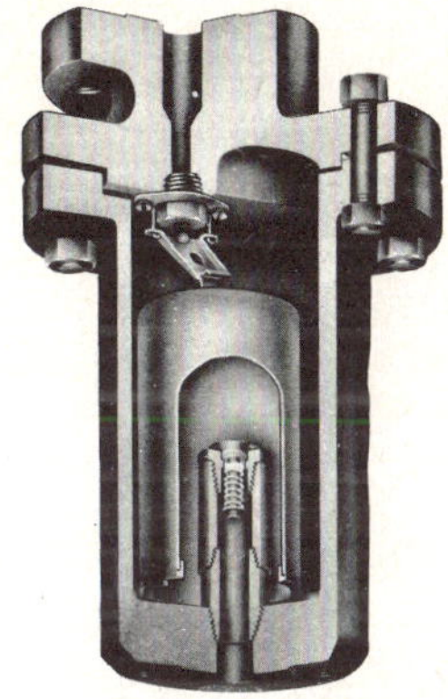

Spirax Drayton inverted bucket trap

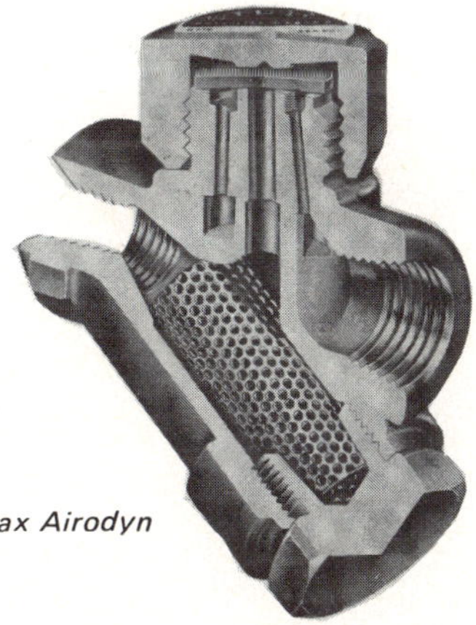

Spirax Airodyn

design which blasts open to eject the contents of the trap when a specific quantity of water has collected. The main difference between the working of the two types is that the float valve type produces a continual discharge, and the blast type a heavier discharge at intervals.

Two limitations may apply to mechanical drains (or traps) – the possibility of becoming 'air locked', and limited ability to handle water contaminated with oil. The former condition may be countered by fitting a pressure balance pipe. Where oil contamination results in a build up of emulsion which makes the automatic drain sluggish in operation (or even inhibits satisfactory operation), more frequent cleaning is necessary, or a modified form of valve used capable of handling heavy contamination. Some types of mechanical traps may also be susceptable to blocking or jamming open in the presence of scale or larger solid contaminants, although many are designed to have self-cleaning valve actions.

The fact that the aspired air is filtered, and that a proportion at least of any solids carried through the compressor will be withdrawn from the airstream by the separator, does not eliminate the necessity for filtering the delivered air via line filters. These can, in any case, provide a higher degree of protection than an intake filter without excessive pressure drop, as well as taking care of any contaminants originating in the compressor. Thus, a large system may require additional high flow rate filters in the main and branch main lines,as well as individual line filters. Alternatively, filters may be used in individual lines for specific protection of individual appliances or devices. The latter has the advantage that pressure is dropped only through a single filter to each take-off point rather than subjecting the main line flow to pressure drop through filters. On the other hand, main line and branch line filters can reduce the number of individual filters required in the system.

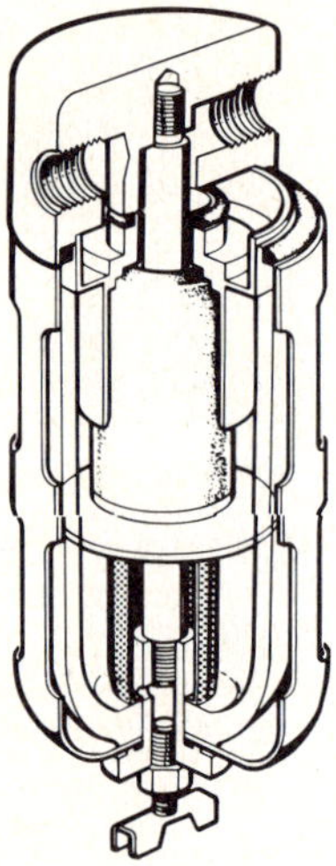

Spirax – Monnier compressed air filter

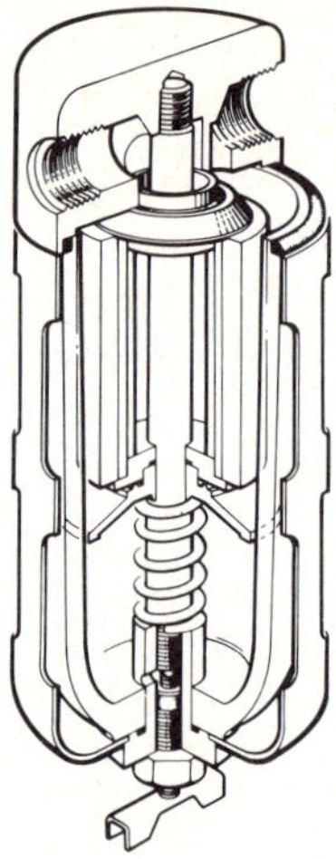

Spirax – Monnier Coalescina filter

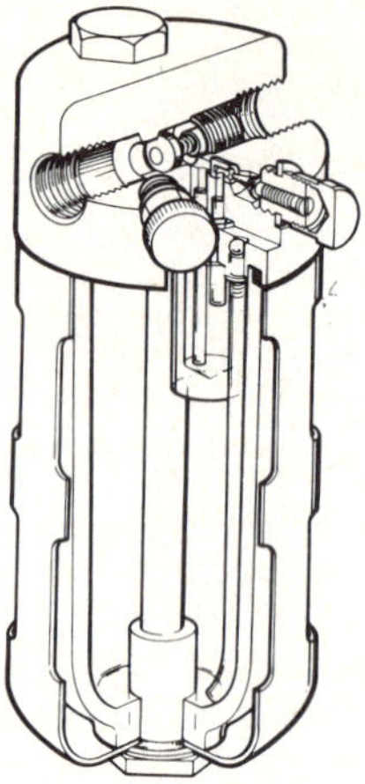

Spirax – Monnier lubricator

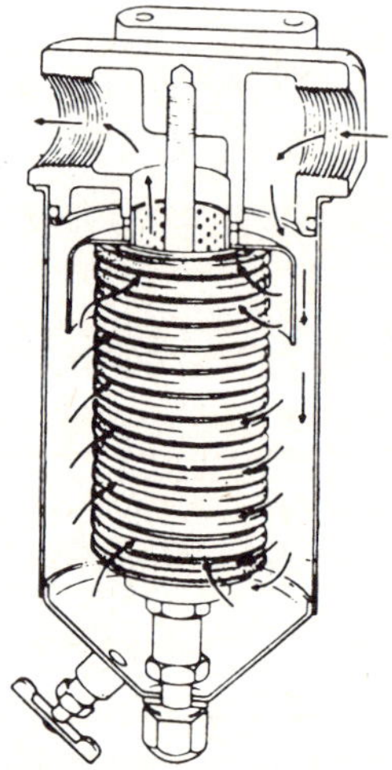

Felt element air line filters (Air Maze Ltd).

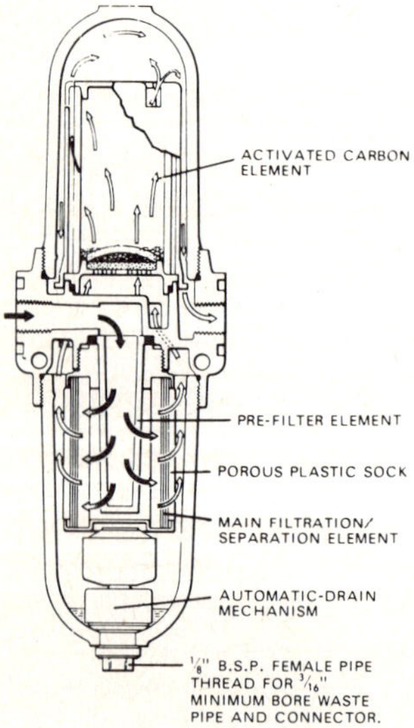

Norgren oil-removing filter.

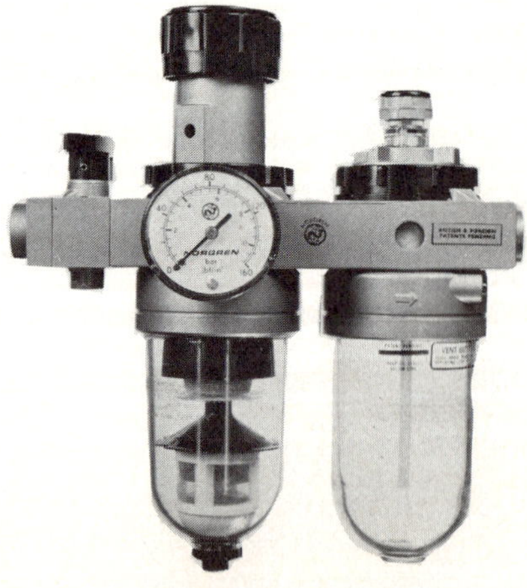

Norgren filter, regulator, lubricator.

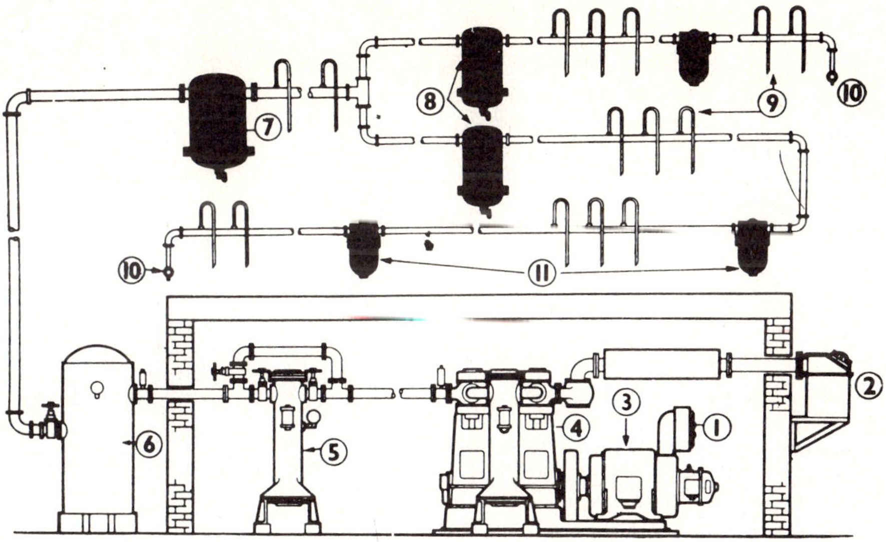

1.Dry fabric filter protecting electric motor cooling air supply. 2.Air filter protecting compressor intake. 3.Electric motor. 4.Compressor. 5.Aftercooler. 6.Receiver. 7.Main pipeline filter/separator. 8.Smaller main filters in branch lines. 9.Supply points. 10.Drain cock. 11.Individual air line filters. (Vokes Limited).

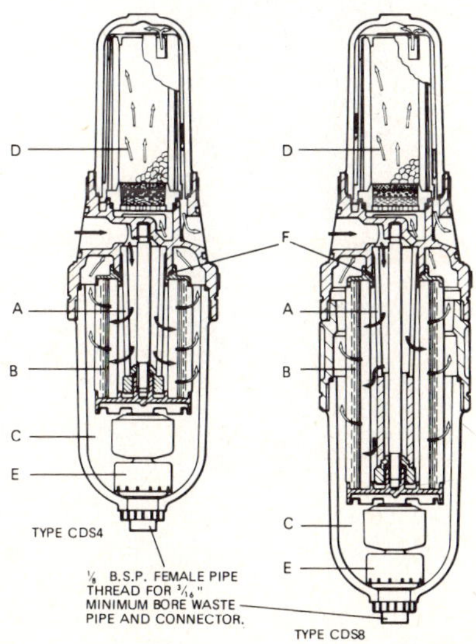

A – pre-filter element
B – main filter/separator
C – quiet zone
D – activated carbon element
E – sealed drain unit

Norgren series F50 filters.

Lubrication

Lubrication of actuators, tools, etc, can usually be most conveniently provided for by injecting lubricant into the air line, after filtering and separating at the take-off point. Three basic types of lubricators are employed, viz –

(i) wick type

(ii) oil-fog

(iii) oil-mist

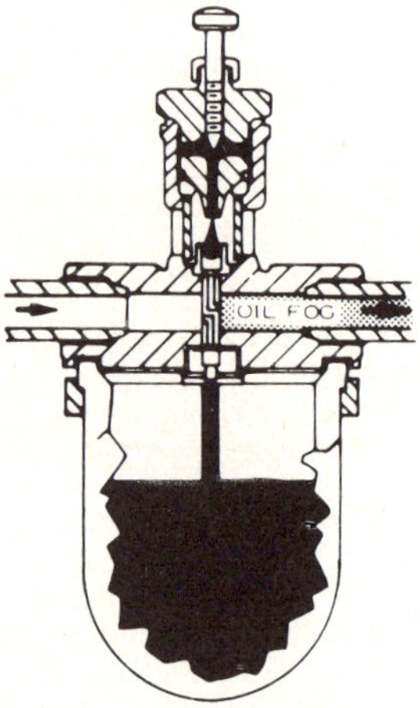

Oil fog lubricator (Norgren)

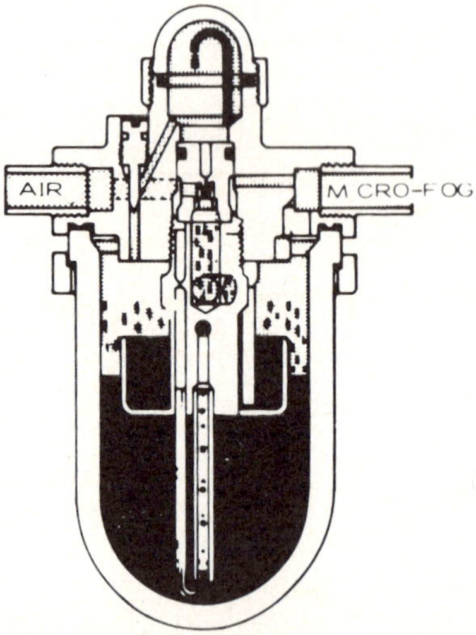

Micro-fog lubricator (Norgren).

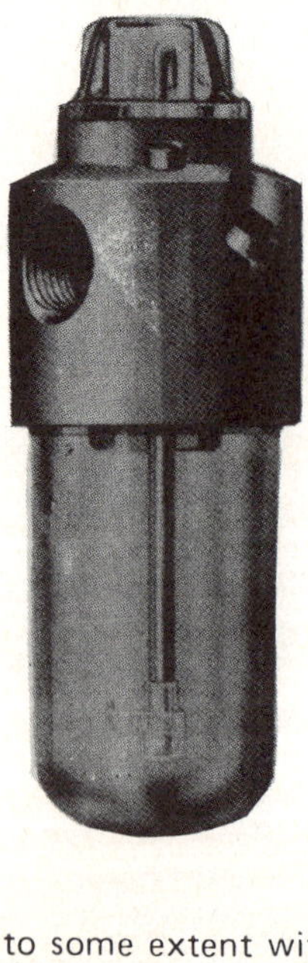

'Perma-fog' miniature size lubricator (Watts Regulator).

The type of lubricating oil varies to some extent with the particular application involved, eg heavy duty pneumatic tools may require a different type of oil to light duty tools. In general, however, a good quality lubricating oil within the viscosity range of 80–350 seconds Redwood No 1 is suitable for use in most pneumatic circuits, provided it has good resistance to water washing and will continue to flow freely at low temperatures.

Good resistance to water washing can be obtained by compounding a straight mineral oil with emulsifying and adhesive agents. These agents enable the oil to form emulsions with any water present, capable of good adhesion to all the lubricated surfaces. The presence of these emulsions also acts as a barrier to prevent condensate settling on the internal surface of tools, to cause rusting. It also inhibits corrosive attack by acid condensates. Straight mineral oils, in general, lack this property and would need to be introduced in far larger quantities to combat water washing.

A low pour point is essential, since for adequate air tool lubrication it is essential that the lubricant should be able to flow freely under its own weight at very low temperatures. It is also important that the oil remains fluid enough at such temperatures to feed easily through air line lubricators and integral oiling devices.

Drying Air

The standard method in general use for drying air is the use of an aftercooler, although this has the disadvantage of reducing both the volume of air and its energy content. Greater efficiency is achieved if the air is pre-cooled in a refrigerating machine, when it is possible that the power saving in the compressor can be greater than the power consumption of the refrigerator, thus comparing favourably with the conventional use of a system with an aftercooler. The working temperature of such a refrigerator is usually of the order of 1 to 3 degrees C, ie just above freezing point, to avoid freezing of the coils.

An alternative, and simpler, method of drying is by supercompression. Here the compressor delivers air at a substantially higher pressure than the system pressure required, fed through a water separator. After the separator the air is allowed to expand. It carries only that amount of water equivalent to saturation at supercompression, hence the actual water content of the delivered air is determined by the initial compression. The practical disadvantage of such a system is that it can be extremely wasteful of power, ie in compressing the air initially to a higher pressure than that finally required. Aftercooling or pre-cooling is, therefore, a more general choice; or absorption or adsorption drying where very dry air is required.

Dryers

Where water-free air is required, a dryer can be used to treat the delivered air. Apart from specialized applications where the use of dry air is essential, drying can also be advantageous for outdoor installations which may be susceptible to the freezing of supply lines in winter, causing blockage due to internal ice formation; or to enable any system to be washed at temperatures below freezing point. Dry air may also be essential for the efficient operation of sensitive control devices.

Commercial dryers are normally heatless; they apply dessicants for *adsorption drying*, or are based on *refrigeration units.* Adsorptive dryers are more commonly used since they require no power, (except for types which employ heat regeneration). Dessicant dryers are readily capable of removing 99.9% or more of water vapour present, together with oil and other fumes and vapours, and thus do not necessarily require the use of a prefilter, although this is often used. Refrigerant dryers have a more moderate drying performance and are suitable for indoor use only.

Dessicants retain their solid form up to 99.9% or more water removal although it is seldom necessary to work dessicants to their limit. Moisture removal of the order of 75 to 90% is usually adequate for most purposes — ie to reduce the humidity of the air sufficiently to avoid vapour condensing to form water in the system. De-rating the drying requirement also enables a simpler — and cheaper — form of dryer to be employed.

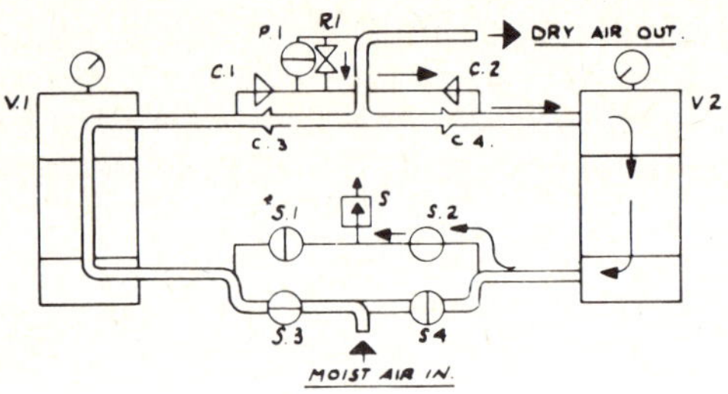

Fig 13 Operation of adsorption dryer (Rimer-Birlec)

V1	–	*Vessel drying*
V2	–	*Vessel reactivating*
S1, S2 *S3, S4*	–	*Solenoid valves for controlling air entry and purging*
C1, C2 *C3, C4*	–	*Check valves controlling dry air exit and purge air supply*
R1	–	*Regulator for air purge*
P1	–	*Pressurisation solenoid valve*
S	–	*Purge air outlet silencer*

Pressure operated valves are used on larger sizes

For continuous use, an adsorption dryer normally consists of two separate cylinders holding dessicant, interconnected in such a way that whilst one of them is being used to dry the air the other is being reactivated. A typical circuit is shown in Fig 3. Wet compressed air enters one vessel and the dessicant adsorbs the moisture vapour in the air which finally goes through a super dry zone before passing into the discharge line. A small proportion of this super dry air is back flushed at atmospheric pressure through the other wet vessel still retaining the heat of adsorption from the previous drying cycle. This creates an environment in which the moisture is removed from the dessicant which reaches equilibrium with the bone dry air before the end of the cycle. The fully automatic change-over cycle is arranged so that bed disturbance is eliminated by reverse pressurization before the change. An electrical sequence timer controls the change-over. A typical cycle would be ten minutes.

An alternative arrangement is to extend the operating cycle, with each cylinder fitted with a heater for force drying during the reactivating part of the cycle. In a typical cycle of four hours each cylinder would in turn provide drying for two hours. The balance of two hours would then be equally divided between reactivation and cooling.

Normally, when a dryer is used, it follows the receiver, where it may be preceded by an oil pre-filter to remove further oil passed by the separator. This arrangement is generally suitable where the receiver is sized to maintain pressure fluctuations within about 10% of the mean value. If larger fluctuations are likely to be present the dryer would have to be designed for the lowest pressure, with some loss of performance. Alternatively, the dryer could be placed in front of the receiver, followed by a pressure balancing valve to sustain the pressure against receiver pressure fluctuations. For the fullest efficiency from the dryer, drying should take place at the highest possible pressure.

Also the temperature of air flowing into the dryer should not exceed 38°C (100°F); the lower the temperature the longer is the dessicant life.

Dessicant dryers are produced in a wide range of sizes, with capacities up to about 280 $metres^3$/min (10 000 ft^3/min).

Ultra-Clean Air

Special types of filters may be employed where ultra-clean air is required (eg for critical processing, control or instrumentation work). These are based on specially constructed filter elements with a very fine cut-off, generally preceded by a coarser prefilter, and with particular attention given to eliminate re-entrainment of contaminants. A separate prefilter may also be installed in the delivery line.

TABLE I – MOISTURE CONTENT OF SATURATED COMPRESSED AIR

Air Temp °F	Air Pressure – Pounds per Square Inch (lb/in^2) gauge												
	0	30	40	50	60	70	80	90	100	110	120	130	150
32	4.34	1.43	1.16	0.98	0.85	0.75	0.67	0.61	0.55	0.51	0.47	0.44	0.38
40	6.00	2.09	1.60	1.36	1.16	1.03	0.91	0.84	0.77	0.70	0.65	0.60	0.52
50	8.69	2.87	2.35	1.93	1.70	1.52	1.43	1.23	1.12	1.02	0.95	0.89	0.79
60	12.41	4.10	3.32	2.80	2.35	2.13	1.90	1.77	1.58	1.46	1.35	1.26	1.10
70	17.85	5.90	5.05	4.08	3.52	3.10	2.66	2.52	2.28	2.10	1.95	1.82	1.59
80	24.97	8.25	6.80	5.70	4.87	4.35	3.88	3.52	3.20	2.93	2.78	2.53	2.23
90	30.81	10.20	9.33	7.90	6.78	6.00	5.33	4.90	4.42	4.06	3.76	3.50	3.13
100	43.40	14.30	12.70	10.80	9.28	8.25	7.32	6.93	6.03	5.57	5.15	4.80	4.20
110	64.50	21.30	18.80	14.60	12.60	11.10	9.85	8.98	8.12	7.46	6.93	6.45	5.66
120	87.00	28.70	23.20	19.70	16.80	14.80	13.30	11.60	10.90	10.10	9.30	8.67	7.56

Example: 1 000 cu ft of saturated compressed air at 100 lb/in^2 and 120°F contains 10.9 fluid ounces of water in the form of water vapour. This same quantity of air when cooled to 60°F at 100 lb/in^2 contains 1.58 fluid ounces of water vapour. The difference between the two figures, or 9.32 fluid ounces, is released from the air in the form of liquid water and must be removed from the system with a suitable filter to prevent its passage into air tools, bearings, cylinders or other pneumatically serviced devices. The quantity of water which must be removed from the compressed air system is directly proportional to the amount of compressed air passing through the system. Therefore, in a system supplying 500 ft^3/m the quantity of liquid water to be removed would be 500/1 000 x 9.32, or 4.66 fluid ounces per minute. If the system handled 500 ft^3/m the quantity of moisture to be removed would be 280 fluid ounces per hour.

TABLE II – GENERAL LUBRICATION REQUIREMENTS

Device or Tool	Part	Method of Lubrication	Lubricant
Air cylinders	(i) piston assembly	air line lubricator	oil
	(ii) linkages	hand oiling	oil
Hoists	(i) cylinders	air line lubricator	oil
	(ii) movements	manual	oil or grease
Vane motors	(i) vane tips	built-in oilers and/or air line lubricators	oil
	(ii) bearings	built-in or manual	oil
Piston motors	(i) cylinders	built-in lubricators or air line lubricators	oil
	(ii) bearings	splash	crankcase oil
Turbine motors	bearings	built-in lubricators	oil
Small tools (miscellaneous)		air line lubricators	oil
Percussive tools	(i) pistons	air line lubricator	oil
	(ii) valves	built-in oiler manual	

Filtration by Micro Fibre Filters

Specific advantages offered by micro fibre coalescing filter elements are low costs and high efficiency in removing not only oil but also water and sub-micron particles from compressed air.

When used in conjunction with oil lubricated compressors they can be acceptable alternatives to non-lubricated machines whenever oil free air is needed in existing or new installations. Recent advances in filtration/separation technology have removed the original problem area asssociated with the use of micro fibre coalescers – chemical attack.

Micro fibre coalescing filter elements are bombarded by oil and water aerosol. The liquid aerosol found in compressed air can become acidic due to inadequate maintenance of compressors or atmospheric pollution. Research has proved that acidic mineral oils and synthetic lubricants such as phosphate ester and phosphate diester can attack the polyurethane foam normally fitted to the outside of micro fibre elements to ensure drainage of coalesced liquids. This outer sheath can more adequately be described by the term – anti re-entrainment barrier. Further, compressor oil becomes more and more acidic with each day of use.

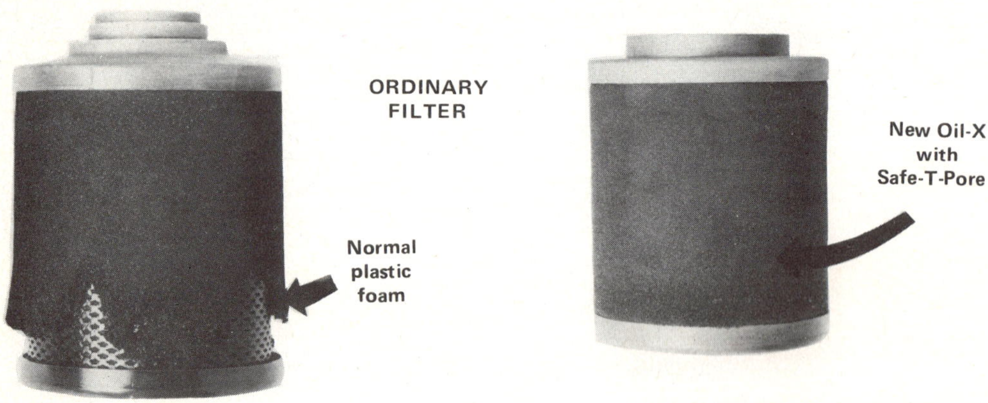

Fig 4

Other sources of chemical attack are phosphate ester and phosphate diester synthetic lubricants. They normally have a longer service life and are a lower fire risk than mineral oils. Unfortunately some of the component parts of ordinary coalescing oil removal filters such as sealants, seals and the polyurethane foam plastic anti re-entrainment barrier are not compatible with synthetic lubricants.

Recently micro fibre filter coalescers with exceptional chemical resistance have been developed. Comparative performance with an ordinary polyurethane foam filter is shown in Fig 4.

Service Life

Service life of a micro fibre coalescer relates directly to (a) the total mass of particles in the air stream; (b) the size of those particles and (c) the degree of degradation of the oil being removed.

The prime cause of particulate contamination of compressed air is atmospheric pollution. In addition to solid particles, the compressor will suck in tar and acid mists, solvent vapours plus hydrocarbon vapours from unburnt aviation fuel and exhausts from trains, trucks and cars. A normal compressor intake filter cannot deal with these small particles and mists, also oxidation of lubricating oils is inevitable and occurs in all types of oil lubricated compressors. Two additional problems exist with old, badly maintained reciprocating compressors where

(i) lubricating oil breaks down to produce a sticky varnish-like substance which can block fine micro fibre coalescers quite quickly and

(ii) extremely fine carbon particles are generated within the compressor as oil is converted into carbon dust on 'hot spots' in the cylinder head and valves.

To prevent short service life it is necessary to consider each installation separately and, where necessary, make use of high dirt holding prefilters. Operating conditions vary considerably from industry to industry and it is generally accepted that two grades of micro fibre coalescer are required to cope adequately with the broad spectrum of applications. To simplify this description these two grades have been called A and AA.

Grade A preferably has a deep bed graded density structure capable of removing all particles down to 1 micron and should reduce oil content of the air to less than 0.5 ppm. By grading the density and size of the fibres, dirt holding capacity is dramatically increased and large particles are trapped at the surface. This prevents premature blocking of the fine fibrous inner structure which is needed to trap small particles and coalesce oil aerosol. They are intended for use as main line filters for general ring main protection and also act as prefilters to grade AA in installations where a high dirt burden or badly degraded oil is found. A typical main line filter is shown in Fig 5.

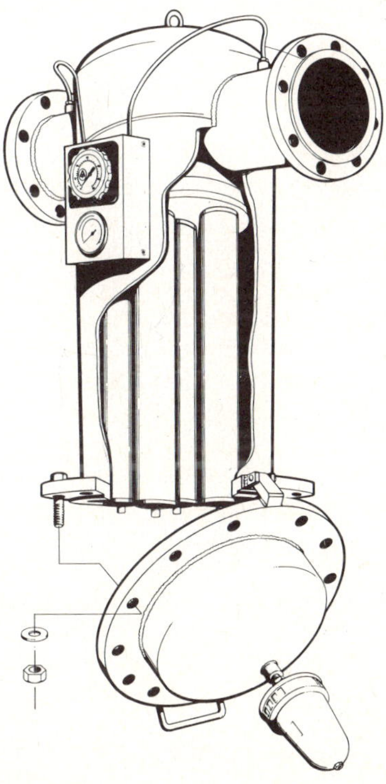

Fig 5

Grade AA is usually manufactured from extremely fine micro fibre capable of (a) removing all particles down to 0.01 micron and (b) ensuring that the maximum remaining oil content is less than 0.01 ppm. They can be used without prefilters in screw or rotary vane oil flooded compressor installations where the compressor has a built-in oil separator and the condensed water vapour burden is low. Conditions can,however, vary from month to month and normal practice would be to install a duplex or packaged filtration system incorporating grade A and AA coalescers.

The Ideal Oil Removal Filter

Taking all operational and contamination problems into account it is easy to appreciate that the ideal oil removal filter must have the following characteristics:

- (i) high dirt holding capacity to ensure long service life
- (ii) resistance to acidic mineral oils
- (iii) compatibility with synthetic lubricants such as phosphate ester and phosphate diester
- (iv) capacity to withstand high surge pressure differentials
- (v) ability to withstand reverse air flow without filter element collapse
- (vi) capability to deal with extremely fine atmospheric pollution, as small as 0.01 micron
- (vii) be constructed so as to prevent premature plugging by carbon and degraded oil carried over from badly maintained compressors
- (viii) low operating pressure differential
- (ix) production of air with less than 0.01 ppm oil content
- (x) low capital, running and maintenance costs

The type of compressor oil, temperature of the air entering the micro fibre coalescer and line pressure must be taken into consideration when determining the type and location of micro fibre coalescing filters within a compressed air system.

Temperature

Removing oil from compressed air at temperatures above 45–50°C may not appear to have been successful if the air temperature falls after the air has passed through the micro fibre coalescer. Oil vapour, a gas, cannot be stopped by coalescing filters and as the air temperature falls it will condense to form liquid downstream of the filters. Compressed air at 7 bar produced by a mineral oil lubricated compressor contains ten times more vapour at 45°C than at 21°C. It is therefore important to locate oil removal filters at a point where a subsequent reduction in compressed air temperature is not possible; *or* vapour adsorption filters, normally designated AC, should be fitted to remove condensable hydrocarbons. General protection from vapour can be obtained by installation of AC filters in the ring main immediately downstream of grade AA micro fibre coalescers. Wherever possible, however, vapour adsorption filters should be installed at the point of use rather than in the ring main. To obtain maximum service life from vapour adsorption cartridges air temperature should be kept as low as possible, 26°C or less.

Cost Comparison

Atmospheric pollution, condensed water vapour, rust particles and particles due to wear of the compressor itself are normally present in compressed air produced by non-lubricated or oil lubricated compressors. A line filter capable of eliminating this contamination is therefore essential regardless of the type of compressor chosen. Free-standing, packaged filtration systems are now available which in use require no special skills or high level of filtration knowledge. The 'Filterpac' shown in Fig 6 comprises a duplex micro fibre coalescing unit incorporating grade A and AA elements. The unit is fitted with automatic drains to discharge coalesced liquids and has a built in pressure differential gauge to indicate the condition of the micro fibre coalescers. These packaged units reduce the cost of installation and a change of micro fibre coalescing elements can be achieved in minutes.

Fig 7 shows a graph of potential savings available to those who select a packaged filtration system plus oil lubricated screw compressors rather than the equivalent single stage water cooled machines. The graph is based on a range of American manufactured compressors.

Fig 6

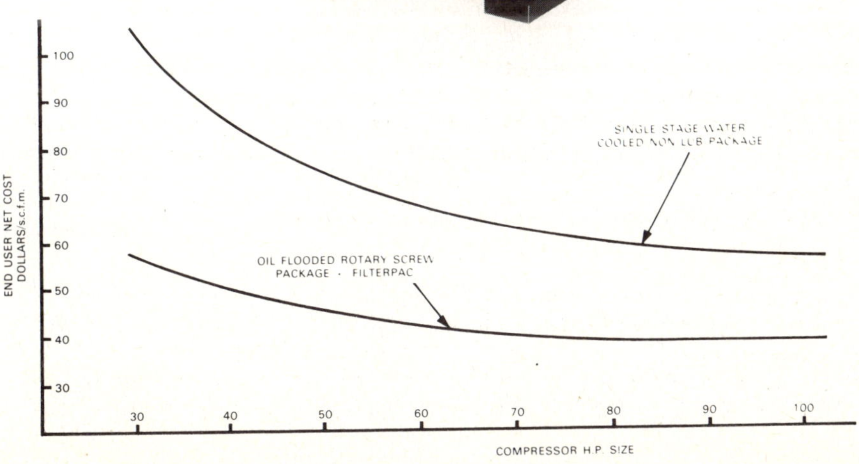

Fig 7

Installation Recommendations

The diagrams and information shown are intended as a general guide. Readers are recommended to obtain specific advice on each installation being contemplated.

1. NORMAL RING MAIN PROTECTION

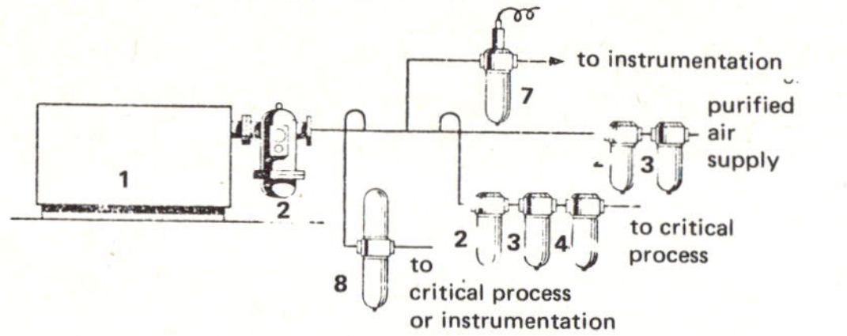

1. Oil lubricated screw compressor with built-in aftercooler.
2. Grade A or PF line filter.
3. Grade AA line filter.
4. Grade AC adsorption filter for extremely critical process applications.
5. Refrigeration dryer.
6. Twin tower desiccant dryer.
7. Combined filter/heater. Combats condensation in instrumentation lines.

2. STRAIGHT FORWARD ALTERNATIVE TO NON-LUBRICATED COMPRESSOR

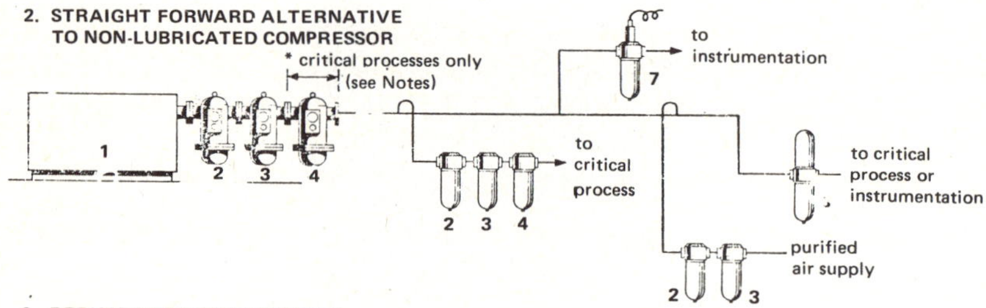

3. REDUCED DEW POINT SYSTEM

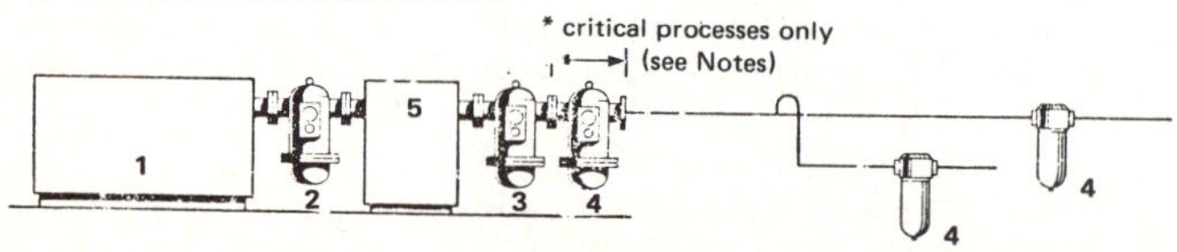

4. EXTREMELY LOW DEW POINT SYSTEM

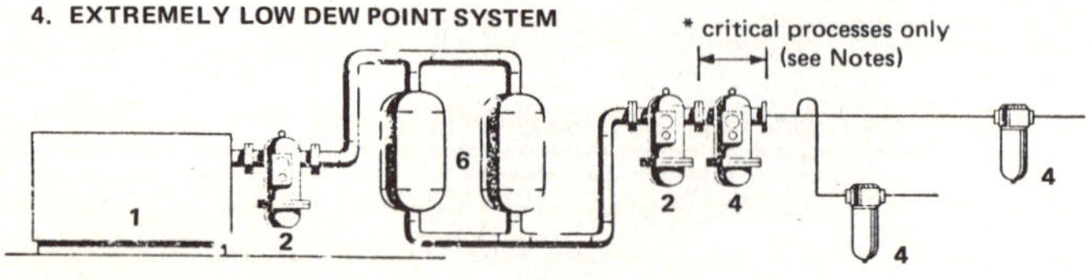

Notes

In piston compressor installations it is normal to fit a receiver immediately downstream of the aftercooler.

In rotary vane and piston compressor installations the use of an efficient aftercooler sited immediately downstream of the compressor is essential.

In systems 1 and 2 grades AA and AC filters are added as terminal filter/separators to remove water and oil-vapour which may have condensed due to a fall in temperature of the compressed air as it passes through the ring main. The filter/heater is shown as a less expensive, low maintenance alternative to a refrigeration or adsorption instrument air dryer.

In systems 2, 3 and 4, a grade AC main line filter should be installed only when large volumes of process air are required for a critical process. Wherever possible grade AC should be used as terminal filters.

In system 4 shown above a grade A filter is used to trap desiccant dust introduced into the air stream by the dryer.

It is important that all terminal filters are installed as close as possible to the process or plant using the purified air.

Untreated mild steel pipe work is not recommended for ring mains. Rust and pipe scale can break free from the inside of untreated pipes and reduce the service life of filter elements.

SECTION 3

Actuators

Air Cylinders

AN AIR cylinder consists essentially of a tube sealed at both ends by covers and fitted with a piston and piston rod. Compressed air admitted through a port at one end produces movement of the piston; displaced air is exhausted through a second port at the other end.

An air cylinder may be *single-acting,* capable of developing in one direction only, or *double-acting,* capable of being pressurized from each end alternately and thus of developing an output force in both directions – Fig 1. In this case the output force in one direction (rod moving inwards) will be less than that in the other direction due to the effective loss of piston area blanked off by the rod. Equal output forces can be provided by a *through-rod* cylinder (Fig 2). Through-rod geometry may also be chosen for greater stability of working (ie the rod is supported by two bearings).

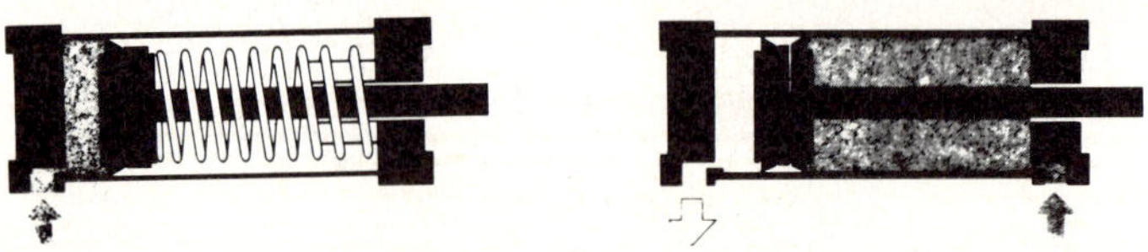

Fig 1 Single-acting (left) and double-acting cylinders.

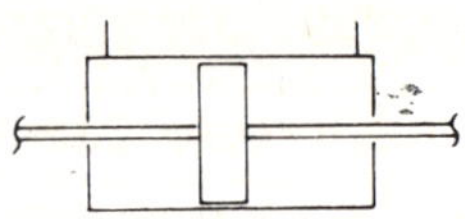

Fig 2

The terms positive (+) and negative (–) strokes may be used to describe the outward and return stroke motions respectively. That portion of the cylinder which is pressurized to produce the stroke is similarly described – ie the positive chamber producing a + stroke, and the negative chamber producing a – stroke – see Fig 3.

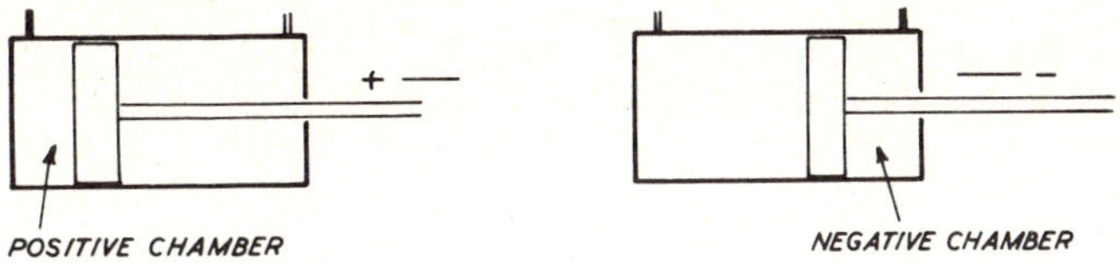

Fig 3

The theoretical force of thrust available from an air cylinder is directly proportional to the piston area and available air pressure. Thus in the case of a single-acting cylinder with pressure applied to the positive chamber.

$$+\text{Force} = 0.7854 \times D^2 \times P$$

where D is the cylinder bore
P is the pressure applied to the positive chamber

In the case of a spring return single-acting cylinder the output force is reduced by the spring force acting in opposition to piston movement.

In the case of a double-acting cylinder the force on the return stroke is reduced by the fact that part of the piston area is masked by the rod, viz

$$-\text{Force} = 0.7854 \times (D^2 - d^2) \times P$$

where d is the rod diameter

With a through-rod cylinder, this reduced force applies to both the positive and negative strokes.

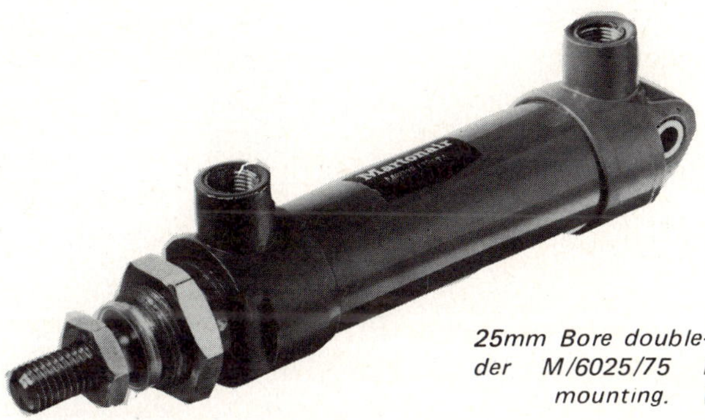

25mm Bore double-acting midget cylinder M/6025/75 integral rear eye-mounting. (Martonair)

Theoretical figures are normally quoted for cylinder performance, since these can be calculated directly from the cylinder geometry – see Tables 1A and 1B. They are often taken as the *static thrust rating* for the cylinder, although this will only be correct if the pressure figure used is that actually developed in the positive (or negative) chamber – not the nominal line pressure. A more realistic determination of static thrust is given by

$$+\text{Force} = 0.7 \times D^2 \times P$$
$$-\text{Force} = 0.7 \times (D^2 - d^2) \times P$$

where P is the line pressure

This provides a nominal allowance for line pressure drop.

TABLE IB – THEORETICAL THRUST OF INCH SIZE CYLINDERS
(Thrust in Pounds)

Cylinder Bore mm	Pressure bar 1	2	3	4	5	6	7	8	9	10
12	1.1	2.3	3.4	4.5	5.7	6.8	7.9	9.0	10.2	11.3
16	2.0	4.0	6.0	8.0	10.0	12.0	14.0	16.0	18.0	20.0
20	3.1	6.3	9.4	12.6	15.7	18.8	22.0	25.1	28.3	31.4
25	4.9	9.8	14.7	19.6	24.8	29.4	34.3	39.2	44.1	49.0
32	8.0	16.0	24.0	32.0	40.0	48.0	56.0	64.0	72.0	80.0
40	12.6	25.2	37.8	50.4	63.0	75.6	88.2	100.8	113.4	126.0
50	20	39	59	78	98	118	137	157	176	196
63	31	62	94	125	156	187	218	250	281	312
80	50	100	151	201	251	301	351	402	452	502
100	79	157	236	314	393	471	550	628	707	785
125	123	245	368	491	614	736	859	982	1104	1227
150	176	352	528	704	880	1056	1232	1408	1584	1760
200	314	628	942	1256	1570	1884	2218	2512	2826	3140
250	492	984	1476	1968	2460	2952	3434	3936	4428	4920
300	708	1416	2124	2832	3540	4248	4956	5664	6372	7080
350	962	1924	2886	3848	4810	5772	6734	7696	8658	9620
400	1258	2516	3774	5032	6290	7548	8806	10064	11322	12580
450	1590	3180	4770	6360	7950	9540	11130	12720	14310	15900
500	1965	3930	5895	7860	9825	11790	13755	15720	17775	19650

TABLE IA – THEORETICAL THRUST OF METRIC SIZE CYLINDERS
(Thrust in Kilograms)

Cylinder Bore ins	Piston Area in^2	Air Pressure – lb/in^2 40	60	80	100
$\frac{1}{2}$	0.196	7.85	11.75	15.7	19.6
$\frac{7}{8}$	0.601	24	36	48	60.1
$1\frac{1}{2}$	1.77	71	106	142	177
$2\frac{1}{4}$	3.98	159	239	318	398
3	7.07	283	424	565	707
4	12.57	503	754	1005	1257
5	19.64	785	1178	1571	1960
6	28.27	1131	1696	2262	2827
8	50.27	2011	3016	4021	5026
10	78.54	3142	4712	6283	7854
12	113.1	4524	6786	9048	11310
14	154	6158	9236	12315	15394
16	201	8042	12064	16085	20106
18	254	10160	15240	20320	25447
20	314	12560	18850	25120	31416

Metric Conversion – Pressure $lb/in^2 \times 0.07031$ = bar

The only real value of a static thrust figure is to indicate the likely maximum 'break out' force available from an air cylinder. As soon as stroke motion is induced the thrust force will be reduced by friction (mechanical loss) and compressed air pressure drop through the inlet port (reduction in effective pressure in the operating chamber). The overall thrust reduction is, typically, of the order of 10–15% at slow stroke speeds up to as much as 50% or even more for high stroke speeds.

This can present problems in cylinder size selection, since the thrust produced by a given size of cylinder can be modified by the use to which it is put (and particularly the speed of working). For non-critical applications, air cylinders are commonly sized on the basis of a theoretical thrust rating of 25% to 50% in excess of the working force required. It is better to select an oversized cylinder rather than the reverse, since the thrust provided by an 'undersized' cylinder can only be uprated by increasing the line pressure (which is seldom a practical solution), or by reducing the operating speed.

In cases of uncertainty, a solution is to select a suitable size of cylinder with the performance assessed on a lower figure than the pressure available. The cylinder is then fed through a pressure reducing valve. If necessary, the cylinder force can then be increased via the pressure control valve.

For more critical applications it is generally necessary to consider in detail:

(i) actual power requirements, taking into account losses due to positive and negative accelerations.

(ii) any additional power requirements which may be necessary to ensure constant motions under varying load conditions.

(iii) how to calculate the actual value of the effective pressure under working conditions.

Air cylinders are generally produced in three different grades or specifications, namely *light duty, medium duty* and *heavy duty.* Such a classification is based on the rigidity of cylinder construction and its suitability for accommodating working loads normally associated with these three grades. It does not refer to the output performance of the cylinder, which will be similar for cylinders of similar bore size operating at the same working pressure.

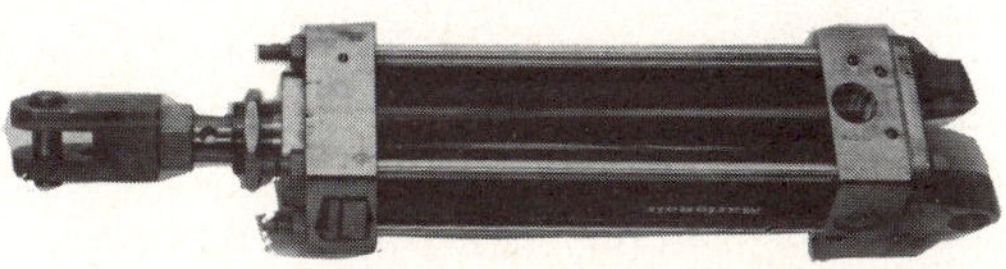

Heavy duty cylinder with 'F' and 'D' mountings.

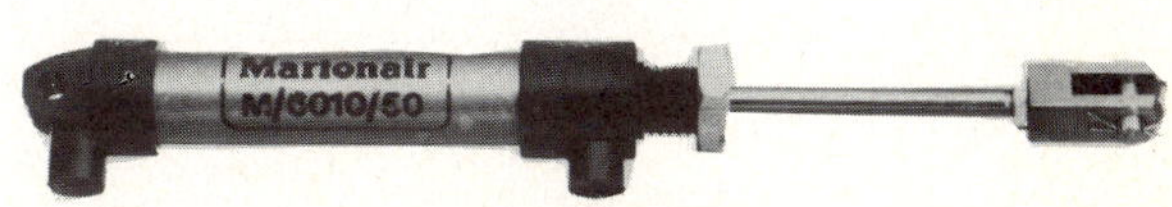

Midget cylinder with 'F' mounting (Martonair)

The majority of air cylinders are designed for working with maximum air pressures of the order of 10 bar (140 lb/in^2), although the usual operating pressure is of the order of 6–7 bar (80–100 lb/in^2) which is constant with the typical delivery air pressure available from the bulk of standard production compressors.

Air Consumption

The air consumption of a cylinder can be derived directly from the cylinder geometry. It is normally expressed in terms of free air delivered.

\+ stroke

$$\text{ft}^3/\text{min (FAD)} = \frac{0.7854\ D^2 \times S \times \text{SPM} \times (P + 14.7)}{1278 \times 14.7}$$

where D = cylinder bore in inches
S = stroke in inches
SPM = strokes per minute
P = pressure in lb/in^2 gauge

\+ stroke

$$\text{m}^3/\text{min (FAD)} = \frac{0.7854\ D^2 \times S \times \text{SPM} \times (P + 1)}{10^9}$$

where S and D are in millimetres
P = pressure in bar

– stroke

$$\text{ft}^3/\text{min (FAD)} = \frac{0.7854\ (D^2 - d^2) \times S \times \text{SPM} \times (P + 14.7)}{1728 \times 14.7}$$

where d = rod diameter in inches

– stroke

$$\text{m}^3/\text{min (FAD)} = \frac{0.7854\ (D^2 - d^2) \times S \times \text{SPM} \times (P + 1)}{10^9}$$

where d = rod diameter in millimetres

In the case of double-acting cylinders the reduction in air consumption on the – stroke is commonly ignored and taken as twice that of the + stroke.

Total consumption is calculated by multiplying the consumption per stroke by the number of strokes per minute. It is often more convenient to render consumption directly in terms of the operating time to complete a stroke, when the corresponding formulas are:

$$\text{ft}^3/\text{min (FAD)} = \frac{0.00186\ D^2 \times S \times (P + 14.7)}{t}$$

$$\text{m}^3/\text{min (FAD)} = \frac{D^2 \times S \times (P + 1)}{t \times 10^4}$$

where t = operating time in seconds

Operating Time

The operating time of an air cylinder is controlled partly by load inertia and partly by the size of the inlet port. The latter produces a 'throttling' effect, with pressure drop increasing as the piston speed (and hence air delivery through the port) increases. The basic relationship between consumption and piston speed is:

$$Q = D^2 \times V \times r$$

where V = piston speed

r = compression ratio

Q = consumption (FAD)

Without throttling, piston velocity is finally restricted by the pressure drop through the inlet port. With small inlet ports and light external loads, pressure drops of the order of 2–3 bar (30–40 lb/in^2) may readily be realized, resulting in a halving of the output force available from normal line pressures. Individual cylinders will, therefore, have a characteristic speed performance, determined by their bore size and port size.

The fitting of a silencer to the exhaust port of a cylinder control valve can also have a speed control effect. This will be dependent on the size and type of silencer, ie its effective permeability. In practice the effect of a suitable size of silencer capable of giving a significant reduction in exhaust noise on the piston can vary from zero to a maximum of about 20% loss of speed.

Normally the speed range likely to be available from a cylinder is 0.5–2.5 metres/sec (1.6–8 ft/sec). Attempting to operate at lower or higher speeds can produce difficulties.

Speed Control

The speed of operation of an air cylinder can be controlled by throttling either the inlet or outlet air or both. The degree of speed control available from such simple methods is suitable for a wide range of applications although the actual response time will still be variable if the load is variable, because of the varying compressibility of air against different loads. Actual control of speed as such, however, is often less important than reducing the speed of operation to a satisfactory level or avoiding excessive operating speeds which increase wear on the cylinder and waste air.

Speed control accomplished by metering the inlet air will only produce a constant speed if the resisting force is constant – Fig 4. Basically, pressure will first build up in the cylinder until the force available is greater than that of 'stiction' (static friction) and resistive load, when the piston will start to move. It will then continue to move forward at a more or less constant speed if the resistive load is constant, with speed governed by the rate of air admission. Any increase or decrease in resistive load will cause the speed of movement to slow down or increase respectively.

Speed control by metered exhaust – Fig 5 – will tend to give an initial rapid movement followed by a slowing down as air is compressed on the exhaust side of the piston. The movement

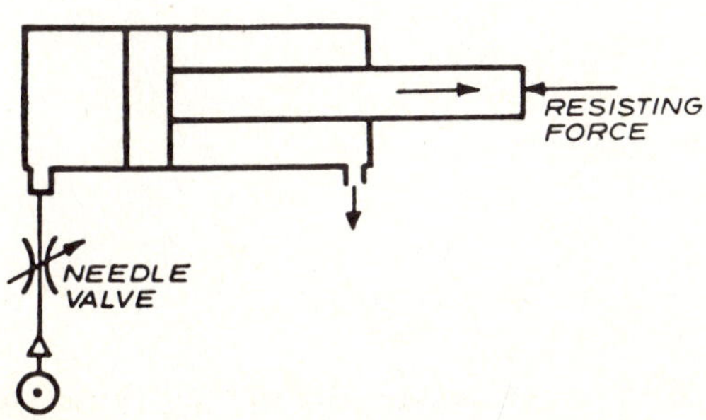

Fig 4 Metered inlet speed control

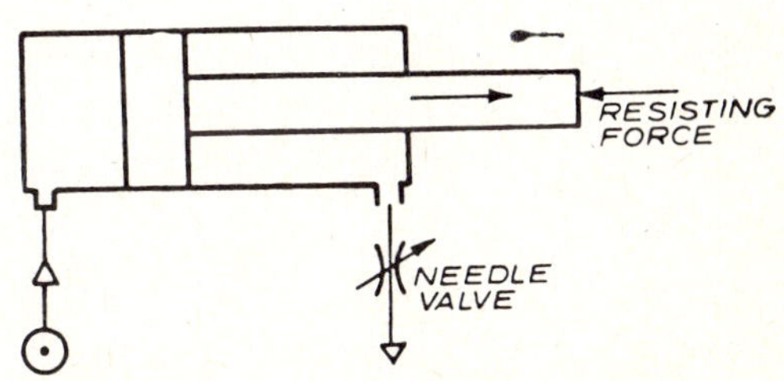

Fig 5 Metered exhaust speed control.

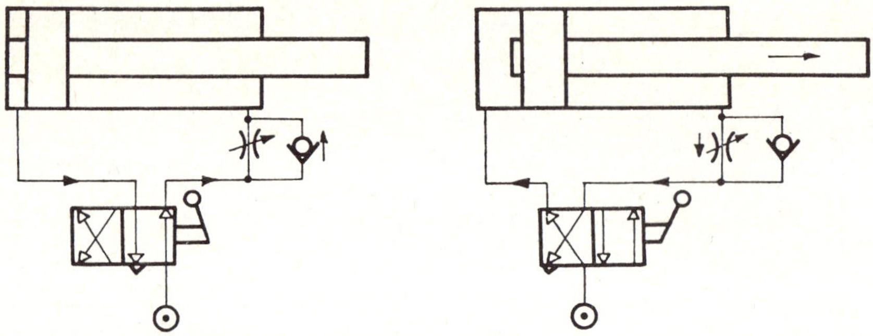

Fig 6 Meter-out speed control with rod end connected to air supply immediately prior to reversal. The view on the left shows conditions before starting the forward stroke and on the right when the piston is moving at controlled speed.

is thus likely to be jerky. Much better control is provided if the modified circuit of Fig 6 is used where the piston is held in the retracted position under pressure and, when the control valve is operated, the piston will start to move forward through pressure applied to the head before the air has had time to leak away from the rod end. The result is a freedom from initial jerk and smooth movement throughout. Also the tendency to jerk forward should the resisting force be removed is minimized.

The general recommendation is that speed control should preferably be applied by throttling the exhaust from a cylinder. A combination of both methods may, however, be used in specific applications.

For more precise speed control, and particularly to take care of fluctuations in load and/or line pressure, surge damping may be applied to the system. This would normally demand the use of an 'oversize' system with a surge tank in the circuit, forming the equivalent of an accumulator in a hydraulic circuit. For greater accuracy of control, however,a check unit would be employed.

A typical check unit consists of a small hydraulic cylinder with both ends connected to a suitable size of reservoir. The hydraulic cylinder is mechanically connected to the air cylinder so that flow in and out of the hydraulic cylinder (displaced by its piston movement) is a function of the air cylinder movement. By inserting regulating valves in the hydraulic cylinder circuit the movement is checked or regulated, thus regulating the speed of movement of the whole assembly. Control and rigidity of the system is then directly comparable with that of an all-hydraulic system.

Fig 7 Hydro check unit using an air-hydraulic cylinder.

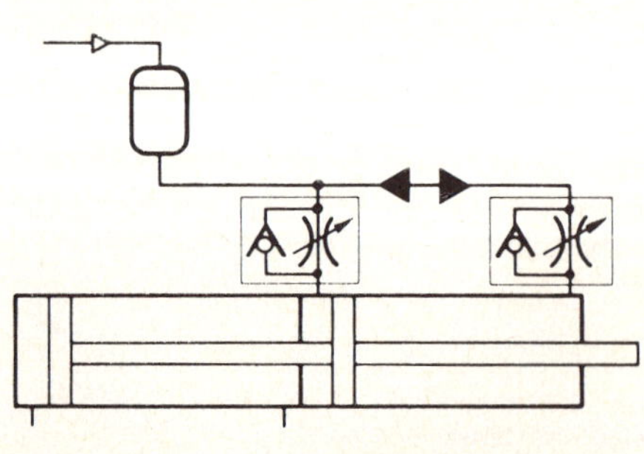

The hydraulic cylinder can be a separate entity (with mechanical linkage) or built integral with the air cylinder in a common unit, more usually known as an air-hydraulic cylinder. Speed of movement is then controlled by a metering valve in the hydraulic circuit whilst variations in oil volume through temperature changes or leakage are catered for by the reservoir – Fig 7.

Further variations are possible, such as bypassing the (hydraulic) control valve for a proportion of the movement to give a rapid initial movement followed by checking and positive control. It should be noted, however, that the controlled speed as established by a particular setting of the (hydraulic) metering valve will still be influenced by the actual value of the resisting load, although speed variations with a fluctuating load will be greatly smoothed. In a more sophisticated system it is possible to couple the metering adjustment to the actual resistive load to make the system fully self-compensating and so maintain constant speed. For most practical applications adequate precision in speed control can usually be obtained in such systems by employing an oversize cylinder and so rendering it less subject to load fluctuation effect, when the proportionately smaller variations can be dealt with adequately by the check unit.

With light resistive loads, and without check units, the capability of air cylinders to perform rapid stroking motions can also be a disadvantage in that considerable kinetic energy may be dissipated in the form of shock at the end of a stroke. This can set up severe stresses in the cylinder itself or its end covers, or on the frame of the machine to which the cylinder is attached. This can be overcome by providing cushioning to produce gradual deceleration of the piston as it approaches the end of its stroke. Thus, in general, cushioned cylinders are to be preferred to reduce shock and 'hammer' at the end of the operating stroke, unless there are obvious reasons for employing a non-cushioned type.

Cushioning

Since air cylinders are fast-acting and the piston will tend to accelerate over the whole of the stroke (unless restricted by throttling or load), considerable shock loads can be developed if the piston is allowed to impact on the end of the cylinder. To avoid this a decelerating device or *cushion* is commonly incorporated in the cylinder geometry. This normally takes the form of a cushion chamber incorporated in the end of the cylinder, into which a nose section on the piston enters – Fig 8.

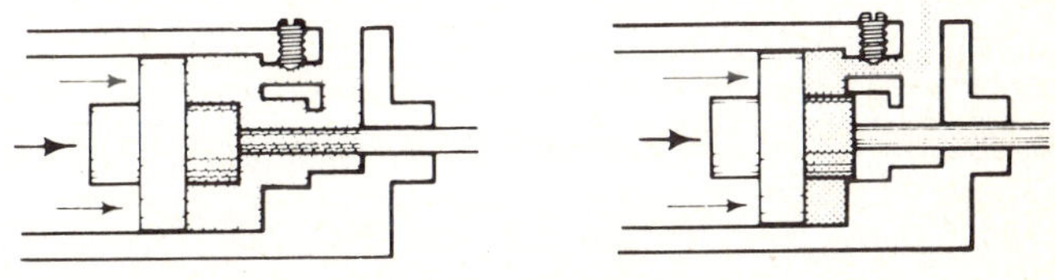

Fig 8A Principal of air cylinder cushion (CompAir Maxam Ltd).

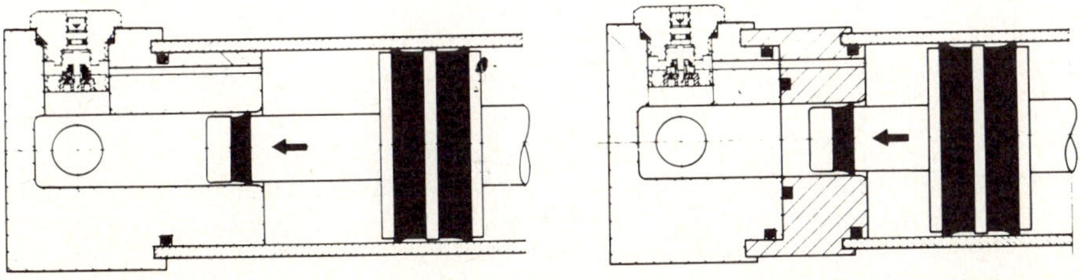

Fig 8B Extended cushions for cylinders (Mecman)

The effect is that as the piston approaches the end of its stroke and the nose section enters the chamber, air is trapped in this chamber, from which it can only escape through a controlled bleed. This provides adjustable damping or 'braking' over the length of the cushion travel, with the kinetic energy of the piston mass dissipated in the form of heat.

Provision must also be made for reverse air feed to act on the full area of the piston, otherwise initial acceleration and break-out force available will be very low. A common arrangement is to fit a non-return valve in the end of the cover so that reverse flow is directed straight to the face of the piston, the end cover or piston having some form of fluting to ensure that pressurized air is distributed over the whole face of the piston.

Detail design of cushions varies with different manufacturers. The cushion length is usually only a matter of 25 mm (1 inch) or so in most cylinders. The cushion boss or piston nose is usually tapered to provide a smooth, progressive entry, but provides a plunger-type seal once it has fully entered the cushion chamber. Alternatively, ring seals may be used in place of plunger-type seals.

A cushion can be fitted to one or both ends of a cylinder. The cushion chamber may be incorporated in the end cover(s) or be in the form of a cushion collar.

Variations in integral cushion design include the use of a spring-loaded face seal, which forms its own cushion automatically when brought into contact with a seating surface at the end of the cylinder, and flexible seal type cushions. In the latter case the seal is made flexible enough to collapse when return flow is applied, thus opening the full area of the piston to the flow. Other types, incorporating a sliding plate valve, may permit adjustment of the cushion length to meet different working conditions.

Integral cushions have certain distinct limitations, the main being that at speeds above about 4.5–5.5 metres/second (15–18 feet/second) they are virtually ineffective, unless the cushion length is made impractically long. For cushioning higher piston speeds external cushioning is normally required. This can take the form of flow control where restriction is applied to the exhaust flow over a selected portion of the stroke, starting at a point well before that practicable with integral cushions. This normally involves the use of a pilot valve, suitably positioned for operating by the cylinder movement, controlling a diverter valve which applies restriction to the exhaust flow. If necessary this set up can be duplicated or triplicated, etc, to provide a number of different stages of cushioning.

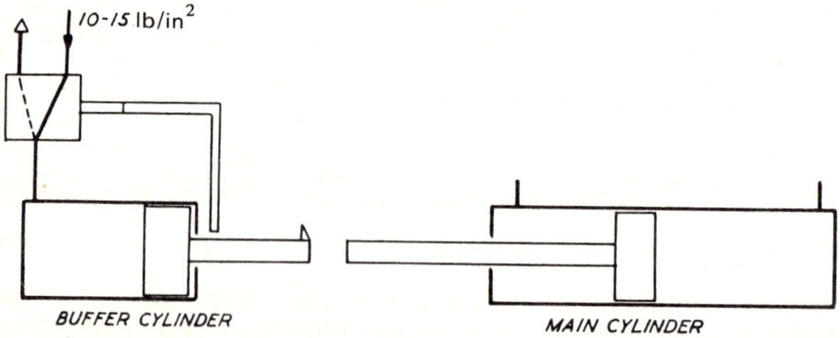

Fig 9

An alternative arrangement is the use of a buffer cylinder. This is a separate cylinder pre-pressurized and proportioned to absorb the excess momentum – Fig 9. The main requirement is that the buffer cylinder should have reserve capacity, and that the maximum pressure realized in the buffer cylinder should not exceed the design rating (which can be very much higher than that for conventional air cylinders). With suitable design, cushioning of high speeds and forces can be provided by quite short stroke buffer cylinders prepressurized to the order of 0.7 to 1 bar (10 to 15 lb/in^2).

The valving circuit associated with a buffer cylinder must be capable of reversing the main control valve at the end of the buffer stroke, otherwise the high pressure generated in the buffer cylinder will tend to reverse the travel of the load.

Other limitations of integral cushions include the fact that they are generally ineffective where a cylinder is fitted with quick exhaust valves or is pre-exhausted, as the amount of air trapped in the cushion is likely to be inadequate for proper cushioning; there is also the practical problem that the bleed may be subject to clogging by contaminants unless periodically attended to. Further of course, integral cushions are only effective if the full stroke of the cylinder is used.

Nevertheless cylinders with integral cushions have a wide application and are generally capable of giving excellent performance in the majority of systems. Piston speed limitations is less likely to be a factor than may appear at first sight, since air cylinders are seldom required to operate at their fastest speed, and the cylinder operating speed is frequently controlled.

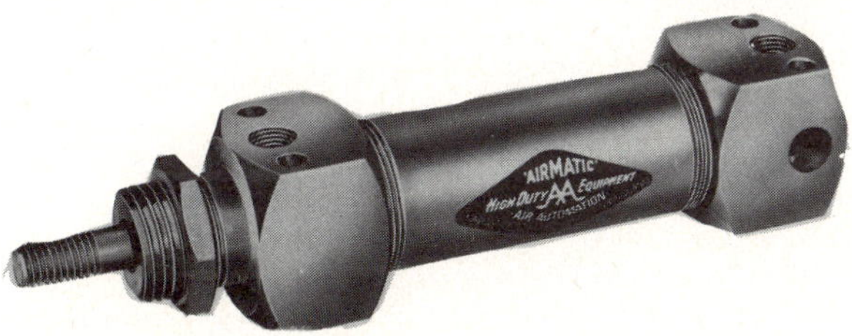

¾in and 1in bore universally mounted air cylinders with nitrile rubber mechanical cushions (Air Automation).

Multiple Cylinders

Tandem cylinders are produced in two different configurations – separate cylinders with the piston mounted on a common rod, and a common cylinder with separate pistons, back-to-back. These are known as *front connected* and *rear connected* respectively – Fig 10.

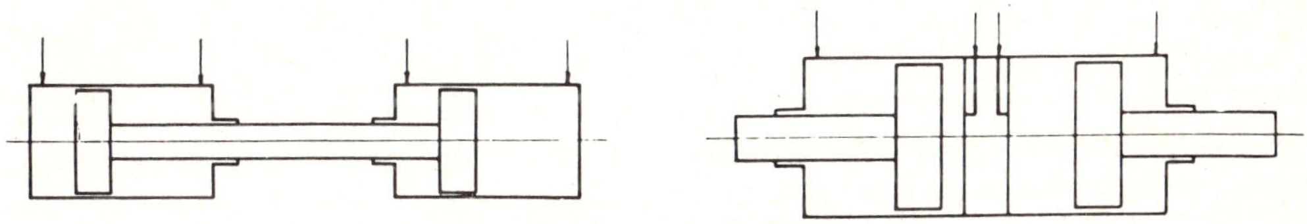

Fig 10 Twin cylinders-front connected (left) and rear connected (right).

Tandem cylinders may be used where there is insufficient space available to install standard cylinders of the appropriate diameter necessary for the force required. Control is generally similar to that of a conventional cylinder, except that compressed air is fed simultaneously to the two cylinder chambers.

Three-position cylinders can be used for operating mechanisms which require three fixed stop positions – ie one at an intermediate position as well as a stop position at each end. The basic configuration is shown in Fig 11. This is really another form of tandem cylinder, except that the two piston assemblies can move independently. The blind end cylinder (referred to as the 'stop' cylinder) provides the intermediate stop position on its + stroke, and again limits the – stroke movement of the main piston if held in this position. The two cylinders may be of equal, or different, diameters, depending on the power ratios required. See also Figs 12 and 13.

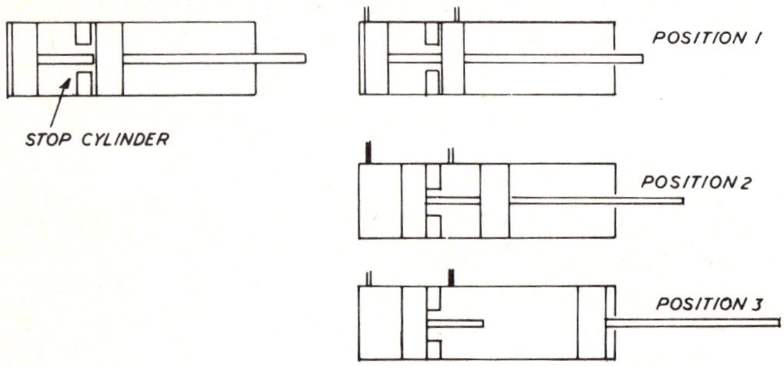

Fig 11

Fig 12 Three-position cylinder (Mecman)

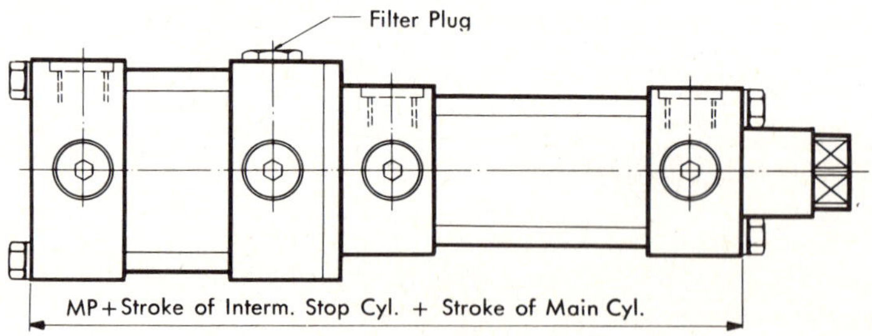

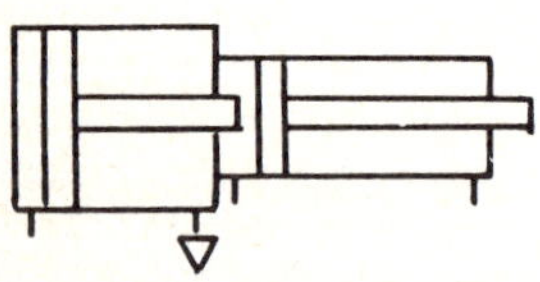

Fig 13 CETOP symbol for three-position cylinder.

Adjustable Stroke Cylinders

For certain applications, a cylinder with an adjustable stroke may be required. In its simplest form this can comprise a standard double-rod cylinder with one rod end threaded where it emerges from the cylinder. Check nuts are then run on to this thread to limit the available movement, and thus the stroke of the cylinder, to that required. In such designs it is important that the threads are not allowed to run into the end gland as otherwise the seal will be damaged resulting in air leakage. This can be avoided by extending the rod so that the threaded length never enters the gland, and fitting a sleeve over the rod extension (or an extended housing to the cylinder end) to act as a stop for the check nuts.

Various other methods of stroke adjustment can be used, according to the circumstances and requirements, ranging from adjustable external stops which are not part of the cylinder as such (and are thus applicable to standard cylinders) to piston travel limit via a large stud screwed into the end cover and locked with a check nut. In such cases the stud must be of sufficient size and strength to absorb the continual impact of the piston without damage to either component; in the case of double-acting cylinders it will be necessary to provide a seal for the stud thread.

Martonair cable cylinder (produced in 2in, 2½in, 3in and 4in diameter sizes) with maximum stroke lengths up to 3 metres.

Positioning Cylinders

A *positioning cylinder* is one designed to provide intermediate movements or positioning of the piston. It is normally produced by adding an external control and/or indicating device known as a *position controller* (Fig 14). A position controller provides a means of measuring or determining the position of the piston in the cylinder. Within limits, the position is largely unrelated to the load on the cylinder, with the position controller itself controlled by a variable pressure air supply. This can either be varied manually, or automatically by a suitable instrument or control gear, depending on the application involved. The most common application is where a quantity has to be varied continually, such as in air-operated valves,speed variators and similar devices.

Martonair fixed range positioning cylinder.

F — airline filter
R — pressure regulator
P — air line purifier
L — lubricator

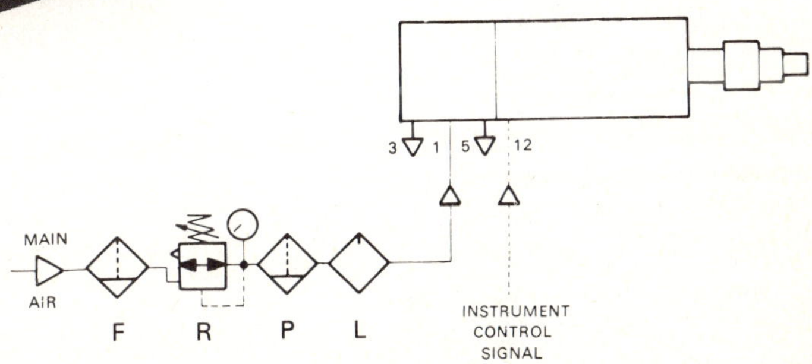

Martonair fixed-range positioning cylinder and circuit connectors

Individual cylinders may also be adapted for mechanical or pneumatic signalling and control, governed purely by piston movement. The simplest method is to employ limit switches or strokers fitted externally and operated by the rod movement. See also Fig 14.

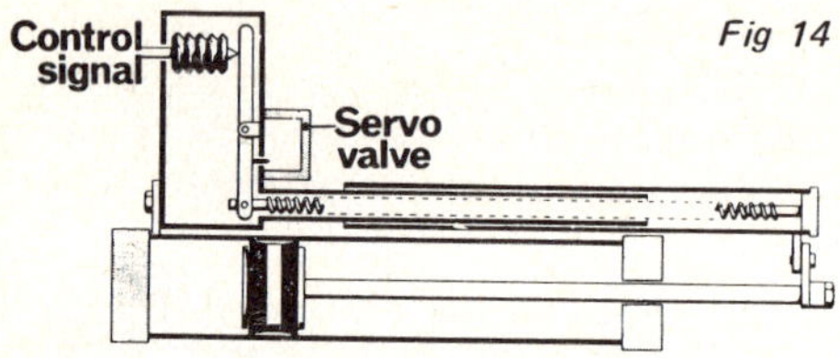

Fig 14 Principle of operation of a positioning cylinder (Atlas Copco)

Impact Cylinders

An 'impact' cylinder is one arranged so that the speed developed is high enough for impact work, such as forging, piercing, etc.

It is possible to achieve this mode of working with a conventional cylinder fitted with extra large inlet and outlet ports and suitable valving. Such a system is shown in Fig 15. This provides for the air on the underside of the piston to be quickly exhausted, while simultaneously supplying air to the other side. The piston speed obtained can be varied by adjusting the appropriate valves. It is generally advisable – or even necessary – to use a reservoir of at least equivalent capacity to that of the cylinder displacement and which is connected close to the cylinder in order to avoid too high a pressure drop.

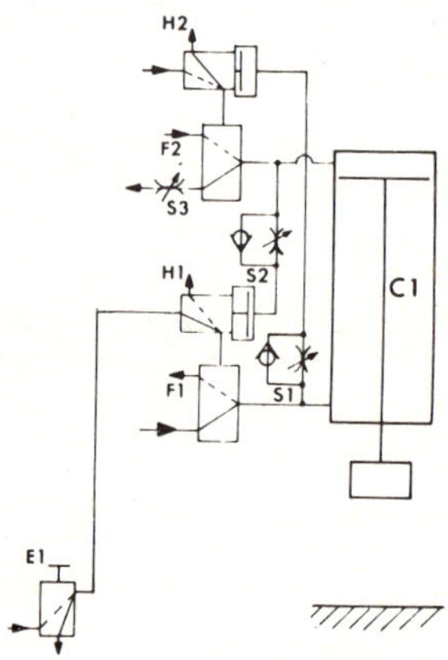

Fig 15 Circuit for operating an air cylinder as an impact cylinder (Mecman).

An alternative approach is to incorporate a reservoir section in the cylinder, connecting directly to a large inlet port valve. Such a design is shown in Fig 16A. Initially the piston rod end is charged with air at normal line pressure with the annulus between the piston seat and centre piece connected to exhaust. The control valve is then opened and pressure builds up in the reservoir. This acts on the piston seat, which is 1/n that of the piston area (ie A/n where n is the ratio of the piston to annulus area). The thrust force downwards is thus P x A/n.

The piston rod end of the cylinder is connected to exhaust when the pressure in this end falls to less than the supply pressure P. When it has fallen to just below P/n the piston moves away from the poppet seat. Immediately this happens the whole of the piston is exposed to the reservoir pressure causing the piston to move rapidly downwards. Maximum piston speed is reached within the first 25–35 mm of the stroke and is of the order of 8 metres/sec (25 ft/sec) with a supply pressure of 6 bar.

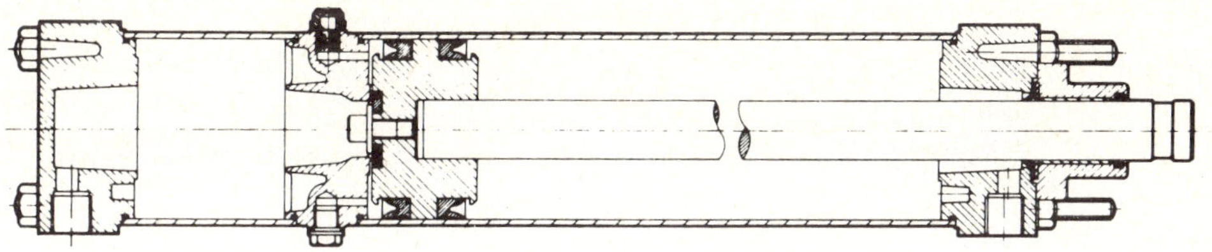

Fig 16A Martonair impact cylinder.

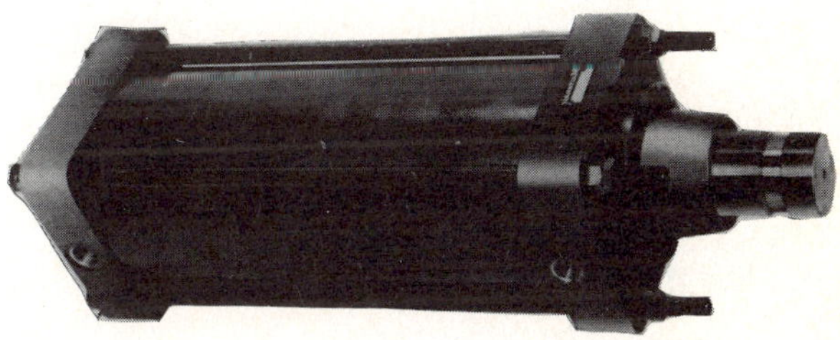

Martonair impact cylinder

Fig 16B Martonair 14-ton impact bench press incorporating a 4 in bore impact cylinder.

Because of the high shock loads developed, a necessary requirement for all impact cylinders is that they be mounted rigidly. For even higher energy outputs pairs of cylinders can be mounted in opposition and operated simultaneously. With proper alignment, shock loading can be negligible in such cases, all the energy being absorbed by the workpiece. Some impact cylinders are fitted with a cushion to prevent damage if they are used for unsuitable work.

Diaphragm Cylinders

A diaphragm cylinder is basically a large diameter, short stroke cylinder fitted with a diaphragm instead of a piston. Construction normally takes the form of two shallow steel pressings or castings clamped or bolted together with the diaphragm secured between the two halves. The output rod is attached to the centre of the diaphragm. Such cylinders may be single-acting or double-acting – see Fig 17.

Fig 17 Single and double-acting diaphragm cylinders (Atlas Copco)

Diaphragm cylinders are capable of generating very high forces with very short strokes and are commonly called *thrusters.* Theoretical thrust available is equal to the product of the applied air pressure and effective diaphragm area. Stroke is normally limited to a maximum of about one third the cylinder diameter. Normal mounting position is with the rod horizontal or pointing downwards.

Linear Motors

Air cylinders are essentially linear actuators. If adapted for continuous self-reciprocation they are known as *linear motors.* This form of motion can be provided by fitting a conventional double-acting cylinder with trip valves or similar devices which switch the control valve automatically at the end of each stroke and thus maintain continuous reciprocation – see Fig 18.

Fig 18 Circuit for air cylinder operating as a linear motor (Mecman).

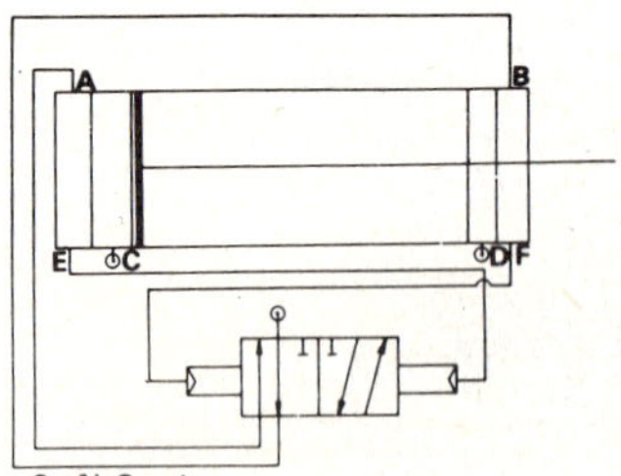

Air cylinders are most commonly assembled from hard drawn metallic cylinder tubes with fabricated end covers. The rod end cover is extended to incorporate a bearing for the piston rod and rod seals. The end covers also normally incorporate the cylinder ports and cushion chambers where fitted. Choice of constructional materials varies both with the duty and size of cylinder – see Table II.

Various methods are used for attaching the end covers, including screwing, welding and mechanically fastening (eg screwing into a square cylinder body). Tie-rod assembly is the most widely adopted method, enabling substantial end covers to be used, square in shape, with the assembly bolted up with the tie rods in tension. These rods are normally of high tensile steel. Prestressing of the tie rods by tightening up is limited by the ability of the cylinder tube to resist buckling.

The main requirement is that the end covers cannot move when the cylinder is under internal pressure. With tie rods, this can present problems on long cylinders, when one solution adopted is to add a support ring at the centre into which the tie rods are studded. This effectively halves any deflection or elongation of the rods. Another manufacturers' solution is to use a central support ring which bows the full length tie rods outwards slightly.

TABLE II – AIR CYLINDER MATERIALS

	CYLINDER CLASSIFICATION		
	Light Duty	Medium Duty	Heavy Duty
Cylinder tube	Plastic Hard drawn metal tube (i) Aluminium (ii) Brass	Hard drawn brass tube Cast aluminium	Hard drawn metal tube (i) Brass (ii) Steel Cast brass Cast bronze Cast iron Cast steel Welded steel tube (largest diameters only)
End covers*	Machined from: (i) Aluminium (ii) Brass Cast aluminium	Machined from: (i) Aluminium (ii) Brass (iii) Bronze Cast aluminium Cast brass Cast bronze Cast steel	High tensile castings
Tie rods	High tensile steel	High tensile steel	High tensile steel
Piston	Machined from: Aluminium Cast aluminium	Machined from brass Machined from bronze Cast aluminium Cast brass Cast bronze Cast steel	Cast aluminium Cast brass Cast bronze Cast iron Cast steel Machined from brass Machined from bronze Aluminium forging
Piston rod	Mild steel Mild steel (chrome plated)	Steel Steel (chrome plated) Hardened steel Stainless steel	Steel (chrome plated) Hardened steel Stainless steel
Piston seal*	O-ring Cup seals	O-rings Cup seals Proprietary seals	Cup seals U-rings C-rings Proprietary seals
Rod seal	O-rings Hat seal U-rings	U-rings	U-rings Chevron seals Proprietary seals
Mounts	Cast aluminium Machined from light alloy Reinforced plastic	Cast aluminium Cast brass Cast bronze	High tensile steel (cast or fabricated)

†Seal selection is influenced by piston diameters as well as duty – larger diameters require 'heavy duty' seals

* Preferably in same material as cylinder tube to minimize risk of electrolytic corrosion.

Pistons for light and medium duty cylinders are usually aluminium alloy castings. Aluminium castings or forgings are used for heavy duty cylinders with more robust sections. Alternatively, medium and heavy duty cylinder pistons may be fabricated from cast iron, brass or bronze.

The piston itself may be one piece, with machined grooves for ring seals, or two or three piece construction depending on the type of seals used (eg some types of seal require the piston to be disassembled for fitting). Cups, U-rings and C-rings are the favoured types of seal for medium and heavy duty pistons and for all larger diameter sizes. Wear rings may be incorporated in the piston assembly in addition to seals, and/or the piston length increased for better support in the case of long-stroke cylinders.

For satisfactory seal life a fine cylinder bore finish of the order of 15 microinches or better is required, particularly when using elastomeric seals. Finer finishing is desirable on cylinders intended for low operating speeds. Hard chrome plating followed by honing and polishing may be used to protect cylinder bores.

One problem which has arisen over the adoption of CETOP specifications is that metric rod threads are not the isometric standard pitch used throughout industry and thus it is not always readily possible to obtain such items as nuts or taps and dies to conform to cylinder threads on CETOP cylinders. Thus there is some preference for producing metric cylinders with metric cylinder rods standardized on the isometric coarse pitch thread which is in general use wherever metric threads are used.

Piston rods are normally En8 steel (or a similar mild steel specification) ground and polished and chrome plated. Chrome plating provides suitable hardness, but hardened steel rods may be employed in some designs. Stainless steel, ground and polished, is now more favoured as an alternative. Regardless of the material used, a high finish is essential in order to minimize wear on the rod seals. This also favours the use of a non-corrosive rod surface.

Because of the relatively low thrust forces involved, rod diameter:cylinder diameter ratio is considerably lower than that of hydraulic cylinders; this has the advantage of reducing the piston area loss on a double-acting cylinder, as well as reducing rod weight (and thus the tendency to sag). However, the rod diameter size is governed by end load and rod length, the diameter being chosen so that the critical length of the rod is not exceeded.

For simplicity of production, rod diameters are usually standardized in smaller numbers than cylinder sizes — ie one 'standard' rod size may serve two or more 'standard' cylinder sizes. Alternative (larger) rod sizes may be needed for specific duties, and particularly for longer strokes. Loading limits for standard rods are also specific, although not always directly given in the cylinder specification. All heavy duty cylinders require larger rod:piston diameter ratios than other types. Long stroke cylinders also present a particular problem since, as well as requiring a larger rod:piston diameter ratio to prevent buckling under load, the weight of a long rod will tend to make it sag in the case of horizontal cylinders. If this weight problem is likely to be critical, a hollow (tubular) rod can be used. It is usual in the case of longer stroke cylinders to find that the length of the rod bearing is increased to provide better support.

Rod bearings used range from simple plain bearings in a cast iron end cover to inserted metal sleeves in bronze or sintered bronze. The bearing assembly also invariably incorporates the rod seals. Simple seals are suitable for light duty cylinders. Medium and heavy duty cylinders normally employ multiple seals or seal sets. A separate wiper seal may also be incorporated at the outer end of the bearing assembly.

TABLE IIIA – TYPICAL STANDARD ROD DIAMETERS

INCH SIZE CYLINDERS

Cylinder Diameter in	Rod Diameter in
$\frac{3}{4}$	$\frac{5}{16}$
1	$\frac{3}{8}$
$1\frac{1}{2}$	$\frac{1}{2}$ or $\frac{5}{8}$
2	$\frac{5}{8}$ or $\frac{3}{4}$
$2\frac{1}{2}$	$\frac{3}{4}$
3	1
4	$1\frac{1}{4}$
5	$1\frac{1}{4}$ or $1\frac{1}{2}$
6	$1\frac{3}{8}$, $1\frac{1}{2}$ or $1\frac{3}{4}$
8	$1\frac{1}{2}$ or $1\frac{3}{4}$
10	$1\frac{3}{4}$, 2 or $2\frac{1}{4}$
12	$2\frac{1}{4}$, $2\frac{1}{2}$ or 3
14	$2\frac{1}{2}$ or 3

METRIC SIZE CYLINDERS

Cylinder Diameter mm	Rod Diameter mm
12	4
16	6
20	8
25	12
32	12
40	16
50	20
63	20
80	25
100	32
125	32
150	35
200	50
250	60
300	70
350	80
400	100
450	110
500	120

TABLE IIIB – TYPICAL CYLINDER PORT THREAD SIZES

CYLINDER SIZE RANGE		PORTS (in)
Metric (mm)	English (in)	
12 – 40	½ – 1½	BSP $\frac{1}{8}$ (Some M5)
50 – 63	2 – 2½	BSP ¼
80 – 125	3 – 5	BSP $\frac{3}{8}$
150	6	BSP ½
200	8	BSP ¾
250	10	BSP 1
300	12	BSP 1¼
350 – 400	14 – 16	BSP 1½
450 – 500	18 – 20	BSP 2

Plastic Cylinders

Plastics are used in the construction of some light duty cylinders. Construction may be all-plastic except for the piston rod, or mixed metal and plastic. For example, a plastic cylinder tube and end covers with a metal piston and stainless steel rod may have advantages when working in a corrosive ambience. Plastic materials used include nylon, polyacetal, polycarbonate, rigid PVC and reinforced epoxy resins.

Apart from corrosion resistance, plastic cylinders can also offer cost advantages. However the higher moduli associated with plastic materials can restrict the working temperature range of plastic cylinders and substantially reduce the maximum service temperature rating.

Cylinder Mounts

The method of mounting an air cylinder can affect the maximum rating or maximum thrust loading of the cylinder. Choice of mount may also be affected by the construction of the cylinder, eg the rod construction may favour tie-rod mounting as the most straightforward choice; or the construction may embody specific mounts – eg integral centreline mounts. In general, however, whatever the construction, standard ranges of cylinders are designed which take attachments to adapt any cylinder to a variety of different mounts (see also Fig 19).

Mounts can be classified under three basic types:-

(i) rigid mounts which absorb the force on the cylinder centreline.

(ii) rigid mounts which absorb the force on a plane parallel to, but removed from, the cylinder centreline.

(iii) pivot mounts.

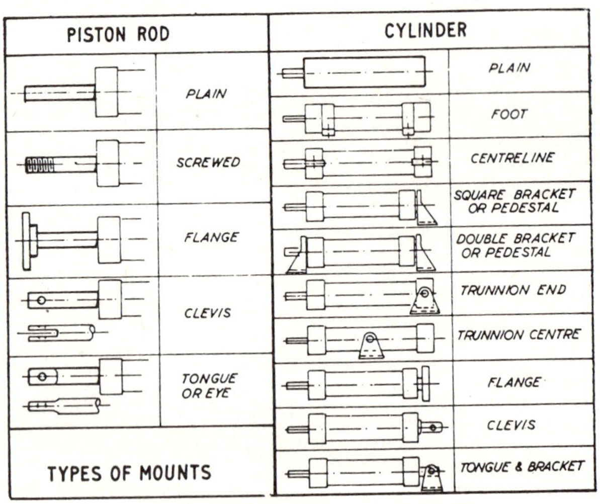

Fig 19

Note : Plain cylinders may have a screwed neck for nose fitting. Single foot mounts (centrally positioned) may be used on some light duty cylinders.

Category (i) mounts are generally to be preferred where the forces to be absorbed are heavy. They also usually provide the most rigid form of mounting with a minimum of complication. Provided there is no possibility of rod misalignment, the mounting bolts are only stressed in simple tension, with no compound stresses.

For cylinders of tie rod construction, mounting on the tie rods themselves provides a category (i) mount and is particularly attractive where mounting space is limited. An additional plate may, however, be added for flange mounting for ease of alignment. Further possibilities are summarized in Table IV.

TABLE IV – CYLINDER MOUNTS

Mount	Position	Remarks
Flange	(i) Rod end (ii) Blind end	Normally the most rigid and strongest method of mounting since mounting bolts are in simple tension, provided no misalignment of rod is present (or possible during operation).
Foot	(i) Single (ii) Double	In centre of cylinder – for light loads only. Rigid mounting but mounting bolts are under compound stress. A common and often the most convenient method of mounting a plain cylinder.
Pedestal or lug	(i) Single (ii) Double	Similar to flange mounting but heavy bending loads and compound stresses are transferred to base of pedestal and base fixing bolts. Similar to double foot mounting.
Centreline	(i) Single (ii) Double	Suitable for light duties only. Rigid mounting which absorbs force on cylinder centreline. Mounting bolts are stressed largely in simple shear.
Tie-rod	(i) Rod end (ii) Blind end	Identical characteristics to flange mounting
Trunnion	Single (centre or end)	Pivot mounting which absorbs force on cylinder centreline. Trunnion fixing bolts are in compound stress (tension and shear).
Screwed Nose	Rod end	Rigid mounting with simple stressing.
Clevis	Blind end	Pivot mounting allowing for some misalignment of the rod whilst absorbing force on cylinder centreline.
Tongue and bracket	Blind end	Similar to clevis but with bracket fixing bolts in compound stress (tension and shear).

Fixed category (i) mounts derived from cylinder end positions will normally exhibit a tolerance for some rod misalignment at the end of the cylinder stroke, but little or none at the other end of the stroke. With long stroke cylinders it may therefore be desirable, or necessary, to provide additional mounting support for the free end of the cylinder to prevent sagging and imparting compound stresses to the mounting bolts. Double mounts are always desirable where a long stroke is completely overhung, and very long cylinders may benefit from an additional centre support.

Category (ii) mounts are often the most convenient, but have distinct limitations where heavy forces are involved, and particularly heavy shock loads. Such loads can tend to distort the cylinder, imparting heavy compound stresses to the mounting bolts. The platform to which the mounts are bolted may also need stiffening or support to resist bending loads which may be transmitted to it through the mounts, caused primarily by distortion of the cylinder.

A possible method of overcoming this is to fit keys or pins to take the shear loads, leaving the mounting bolts in simple tension. Sometimes a thrust key can be incorporated on the cylinder itself, see Fig 20. which can be fitted directly into a keyway cut in the surface of the mounting member. When using such methods of accommodating shear loads only one end of the cylinder should be keyed in order to allow for expansion and contraction due to changes in temperature and pressure.

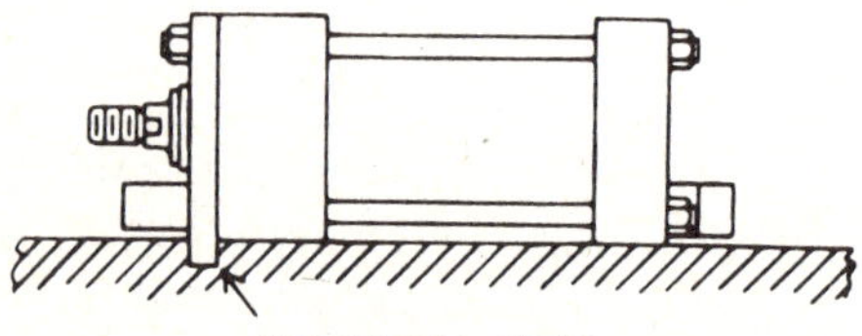

Fig 20

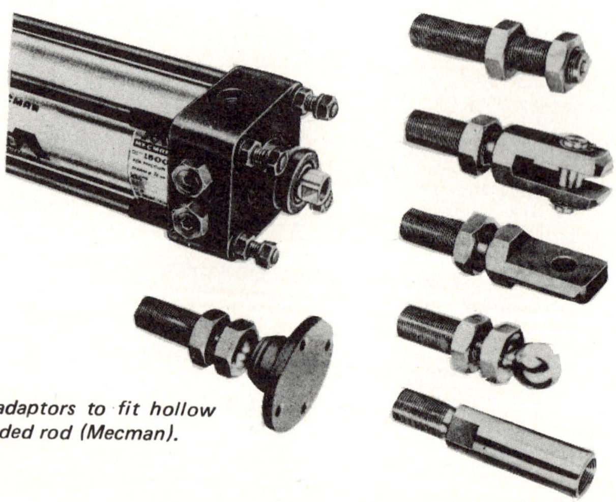

Fig 21 Screw-in adaptors to fit hollow drilled and threaded rod (Mecman).

A category (iii) mount is normally required where the piston rod is required to traverse a curved path rather than a straight line. It may also be used to accommodate a degree of rod misalignment where the motion is basically linear.

Where the curved motion or misalignment motion is confirmed to one plane, simple pin mounts such as provided by a trunnion or clevis are satisfactory, the pin being mounted on the centreline of the cylinder. A clevis or tongue and bracket mount is invariably fitted to the blind end of the cylinder. Trunnion mounts can be at either end or in the centre of the cylinder.

Centre or rod end trunnion mounting is generally preferred on long stroke cylinders as it has a greater resistance to rod buckling. Pins are normally designed to resist plain shear loads and thus trunnions or clevis ends should not be fitted with self-aligning bearings as these could introduce bending loads on the pins. Pivot mountings incorporating universal joints or swivel joints are produced, however, and may have to be used where the path described by the rod traverses more than one plane.

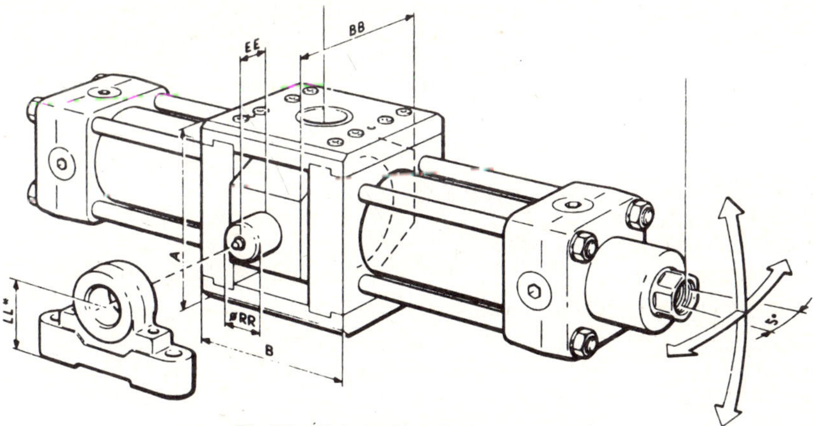

Fig 22 Universal cylinder mount for flexible and shear-free installation (Mecman).

Rod ends can be plain or screwed, clevis or tongued, or fitted with a flange – Fig 21. Alternatively, a universal ball joint fitting can be used to accommodate rod movement in more than one plane. Similar considerations apply as mentioned previously – ie self-aligning bearings should not be used to support pinned rod end joints where the pin is only designed to accommodate plain shear loads – see also Fig 22.

Air-Hydraulic Cylinders

HYDRO-PNEUMATIC working is achieved by mechanically coupling a pneumatic cylinder and a low pressure hydraulic cylinder so that they have a common movement, normally initiated by compressed air applied to the air cylinder, which is a double-acting type. The normal configuration is back-to-back mounting with a common piston rod – Fig 1. The hydraulic (oil) cylinder is also double-acting, but with a closed loop circuit. Control of the rate of flow of oil through this circuit governs the speed of operation of the tandem cylinder.

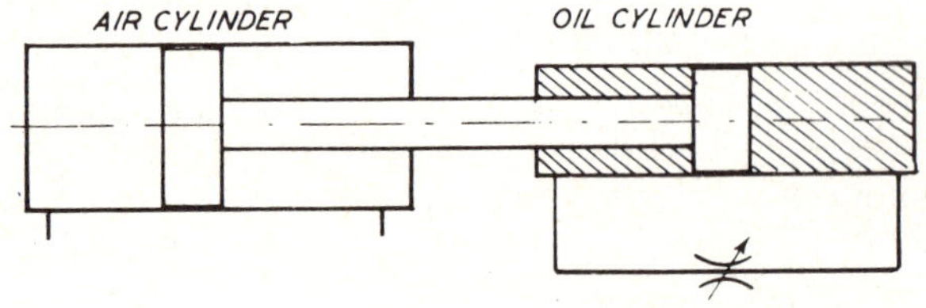

Fig 1

In practice it is essential that air cannot leak into the hydraulic circuit, otherwise this would cause 'spongy' operation. Thus two sets of rod seals are employed, one on the air side and one on the oil side. Provision also has to be made on the hydraulic side to accommodate any changes in working volume of the oil – eg by the entry and withdrawal of the rod in the simple system of Fig 1; and also oil volume changes due to ambient temperature changes.

A logical design configuration is to make the air cylinder the blind end of the combination so that the hydraulic cylinder then has a through-rod giving equal displacements with either direction of motion. It is then possible to reduce the volume of the compensating oil reservoir to minimize proportions, or even eliminate it entirely if the cylinder will be working in a substantially constant temperature.

Speed Control

Throttling may be provided on both sides of the hydraulic piston for adjustable speed control in both directions, or on one side only with a flap valve for differential speed control with fast movement on the opposite stroke, as in Fig 2.

The big M.
High quality components for industrial mechanisation with pneumatic and hydraulic systems
MECMAN
7000
TILL
FRÅN
MECMAN
MECMAN LTD.,
206 ELGAR ROAD, READING,
BERKSHIRE. Tel. 0734/84639
Austria · Belgium · Canada · Denmark · Finland · France · Germany · Hungary · Italy · Japan ·
The Netherlands · Norway · Portugal · Spain · Sweden · Switzerland · USA · Yugoslavia.

AIR HYDRAULIC CYLINDERS (B)

Reciprocating Compressor or Vacuum Pump

Typical Reciprocating and Rotary Compressors and Vacuum Pumps

COOPERGOLD

Reciprocating Oil-Less compressors
7 atu continuous and up to 100 L/M

Bell & Gossett
Reciprocating Oil-Less compressors
Pressures up to 12 atu and up to 330 L/M
Vacuum Pumps to 620 mm Hg.

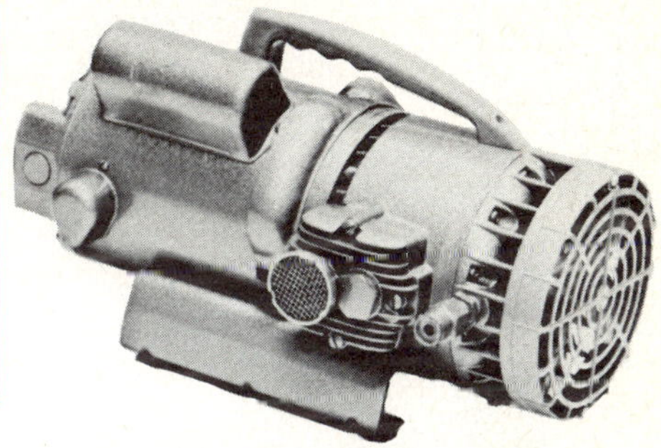

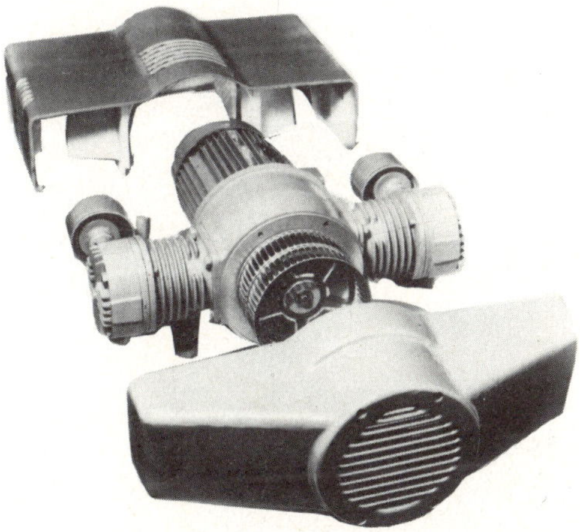

EUROPA E30

Reciprocating Oil-Less compressor
7 atu continuous and up to 220 L/M

Reed Valves
Teflon Rings and Skirts
Double acting fan
Lightweight – 65 Kg.
Compact – with streamlined cover.

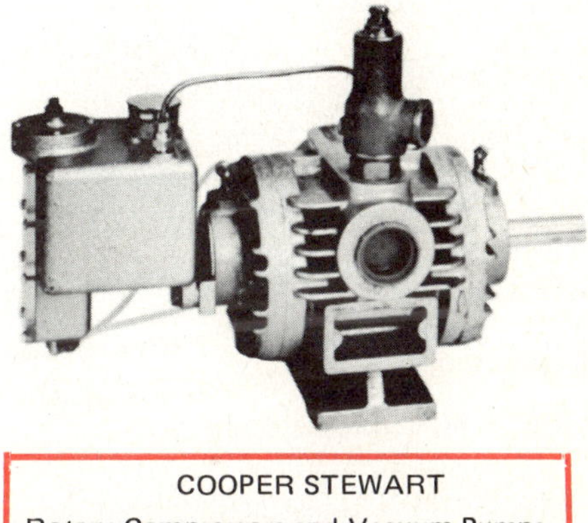

COOPER STEWART

Rotary Compressors and Vacuum Pumps
Pressures up to 2 atu
F.A.D. to 8 000 litres per min
F.A.I. to 7 500 litres per min
Vacuum to 620 mm Hg

Belcon Industrial Estate, Bingley Road, Hoddeston, Herts. Tel: 61-67191

ITT

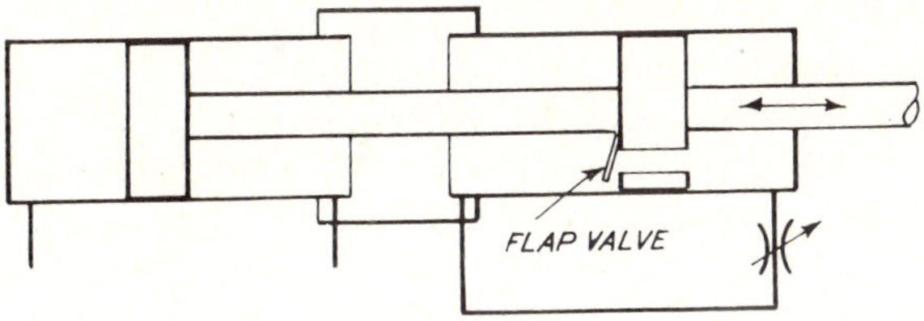

Fig 2

Basic movement controls possible with the control unit(s) located in the hydraulic circuit are:—

Speed, one direction — one throttle valve or restrictor.

Speed, both directions — two throttle valves or restrictors.

Fast Approach, with final controlled approach — (i) throttle and bypass valve in parallel (ii) mechanical (applicable only in case of separate cylinders).

Fast Return — flap valve in piston.

Fast Approach, controlled return — flap valve in piston plus throttle valve in return flow line.

Controlled Approach, fast return — throttle valve in one line; flap valve in piston.

Various other combinations of the above are possible, eg fast approach with controlled final approach, followed by fast return. In that case the control units required would be those for both types of movements, on appropriate sides in the hydraulic circuit — see Fig 3.

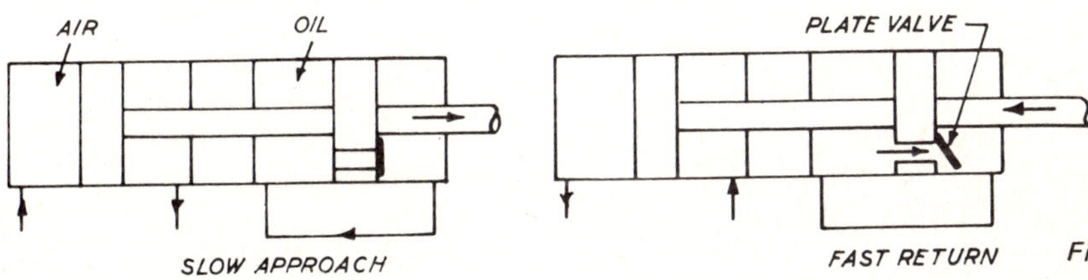

Fig 3

In general:—

Throttle valve (or restrictor) provides speed control (in appropriate line).

Throttle valve with bypass valve provides differential speed control (in appropriate line).

Flap valve in piston provides uncontrolled fast approach, or fast return (depending on which side of the piston it is fitted).

Further controls can be provided in the pneumatic circuit, which is the 'working' circuit. However, unloading or pressure relief valves cannot be used in the air circuit, since these could be operated by the damping produced by the hydraulic circuit, causing loss of working pressure. If unloading is required on the air side, this would have to be done via microswitches operated by mechanical movement, or suitable pilot valves.

All air-hydraulic cylinders are likely to have certain limitations as regards control of speed and feed, compared with pure hydraulic cylinders. Thus they have a more limited application and become increasingly unsuitable, the more sophisticated the movement control required.

Typical applications of air-hydraulic cylinders include feeds for drills, milling machines, circular saws, etc and similar operations where the power supply available is compressed air, output forces required are relatively light, but a rigid feed is required.

Intensifiers

Intensifiers are normally a special type of air-hydraulic cylinder whereby low pressure air can be used to generate a supply of high pressure fluid. The high pressure fluid may be air (or a gas), or a liquid (eg hydraulic oil or water). Intensifiers may be used directly as a source of low volume high pressure fluid, or as high pressure hold devices. In the latter case initial pressurization is achieved from a pump, the system then being switched over to the intensifier to produce and hold the final high pressure.

A diagrammatic representation of a typical air-hydraulic intensifier is given in Fig 4. Air pressure is used to drive the piston in a low pressure cylinder, the piston rod forming a plunger for a very much smaller diameter high pressure cylinder. Pressure intensification is directly related to the ratio of the cylinder bores, a typical intensifier ratio being of the order of 50:1. This would yield a hydraulic pressure of the order of 350 bar (5 000 lb/in^2) from a conventional air line supply at 7 bar (100 lb/in^2). If necessary, very much higher pressures can be obtained from air-hydraulic intensifiers. Normally a manufacturer would have a basic air cylinder and vary the diameter of the hydraulic cylinder to give the desired high pressure output. Output volume, of course, reduces with reducing diameter of the high pressure cylinder for a given stroke (available from the air cylinder), as well as the speed of reciprocation. Speed falls off as pressure builds up, until reciprocation finally ceases, when the loads on the low pressure piston and high pressure plunger are balanced. Reciprocation of the air cylinder is achieved by any suitable form of valve gear.

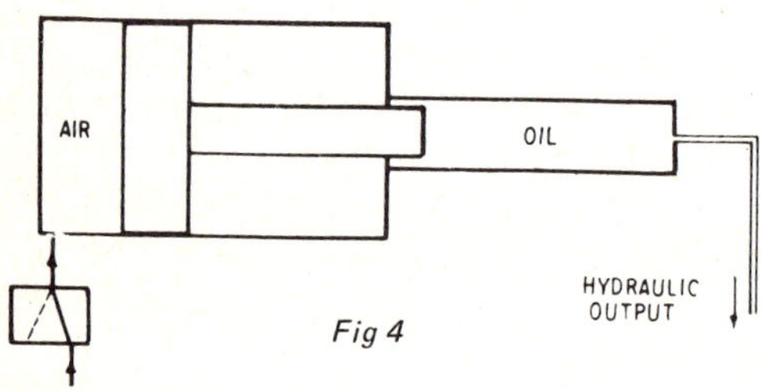

Fig 4

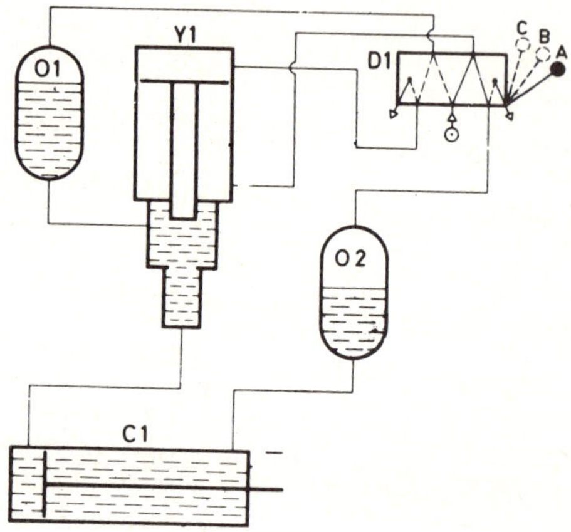

Fig 5

A typical circuit using a 6-way directional control valve is shown in Fig 5. In the starting position D1 is in the position A. C1 and Y1 are in the withdrawn position. When D1 is moved to position B, reservoir O1 is pressurized and simultaneously reservoir O2 is vented. C1 moves outwards under low pressure. When D1 is moved to position C, Y1 moves downwards under air pressure, and hydraulic pressure actuating the cylinder C1 is intensified. Return movement is accomplished by pressurizing O2 and venting O1.

Rotary Actuators

THE TWO basic forms of pneumatic rotary actuator are:-

(i) The torque cylinder or semi-motor.

(ii) Single or paired diaphragm cylinders.

The typical form of a torque cylinder is shown in Fig 1. It comprises two cylinders, back-to-back, with a common piston rod. The rod is cut with, or carries, a toothed rack engaging with a pinion mounted transversely in the centre section. Linear movement of the piston is thus translated into limited rotary output. Rotary movement is usually restricted to a maximum of 360°, or less. Nevertheless this is considerably more than that possible from a conventional cylinder connected to a lower output where the maximum angular displacement available is usually 120° or less.

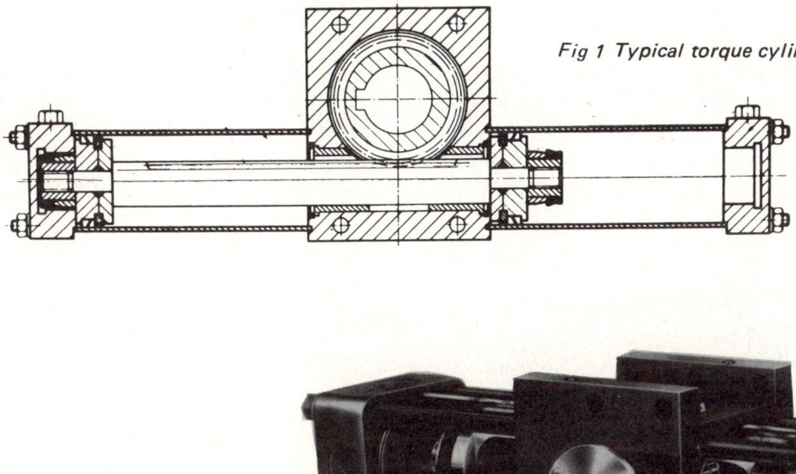

Fig 1 Typical torque cylinder in section.

Mecman torque cylinders.

A particular advantage with the torque cylinder configuration is that for constant applied pressure the torque output is constant over the whole range of rack travel. Cushioning may or may not be incorporated at each end of the cylinder, depending on requirements.

A paired cylinder diaphragm-type rotary actuator is shown in Fig 2. Here two conventional diaphragm cylinders are mounted on a common centrepiece carrying a toothed wheel. Each cylinder rod terminates in a ratchet engaging this wheel. The selected diaphragm cylinder is operated in an automatic oscillating mode, driving the toothed wheel continuously in a series of progressive rotary movements. Selecting the opposite cylinder reverses the direction of motion.

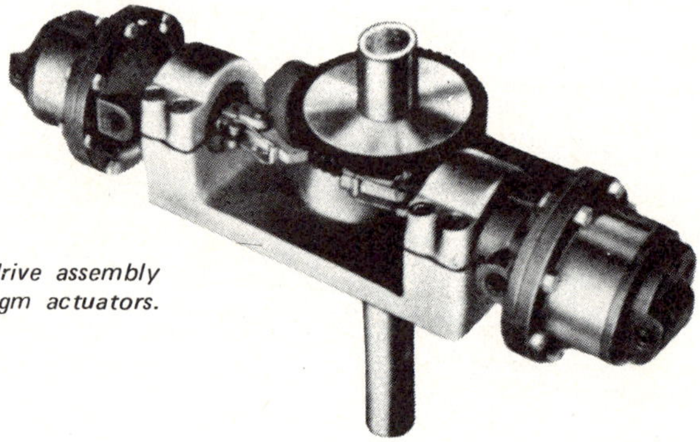

Fig 2 Telektron rotary drive assembly powered by two diaphragm actuators.

Since diaphragm cylinders are essentially high-thrust devices, high output torques can be realized. Also this type of rotary actuator is capable of continuous rotation in either direction.

Rotating Cylinders

Two basic types of cylinder can be described under this heading. The true rotating cylinder is one where cylinder and piston are locked in such a manner that no relative rotary motion can occur between them. The complete cylinder is then mounted on a distribution gland which allows the cylinder to rotate relative to a workpiece. The other type of 'rotating cylinder' is one where the design provides for the piston to rotate freely relative to the cylinder with no loss of working characteristics. This implies that the piston seals must be capable of accommodating rotary as well as sliding motions.

'Norbro' pneumatic valve actuator. Pistons move together, or apart, driving the output spindle with a rack and pinion.

Air Motors

AIR motors are an attractive form of driver, competitive with electric motors, i/c engines and hydraulic motors in many fields of application where power requirements do not exceed 30 hp. They have a particularly wide range of applications at low power levels down to and including fractional horsepower sizes. Specific advantages are:

(i) relatively low initial cost

(ii) low running costs

(iii) high reliability with minimal maintenance requirements

(iv) safety, particularly for operation in hazardous areas since there is no spark hazard

(v) inherent overload protection

(vi) low inertia and ease of control

(vii) high power/weight ratio usually associated with rugged construction

Air motors can be operated continuously for long periods,provided they have an adequate air supply and satisfactory lubrication. They are very unlikely to overheat because the expansion of exhaust air has a cooling effect. They can operate at the extremes of ambient temperature, although for sub-zero temperature working special seals may be required. They can be operated under water, down to depths of 12 m (40 ft) or more; in this case vents and exhaust must be led to the surface to avoid excessive back pressure being developed, and additional seals may be necessary to prevent water entering the motor.

Power is normally compressed air from a compressor or compressed air line. For short-term running the motor may be fitted with its own air supply in the form of an air receiver — this is only viable if high pressure air is available as otherwise the size of receiver required would be too large.

Air motors can also be operated by any other clear gas at a suitable pressure, provided lubrication requirements can be met — eg made gases or even process gases where a suitable differential pressure can be extracted from the system without interfering with the process requirements.

Control of Air Motors

The control of any air motor is usually simple and direct. For most applications an ordinary throttling valve is adequate for variable speed control in conjunction with a four-way valve if reversing is required. This will provide full control of both speed and direction. Special control valves are

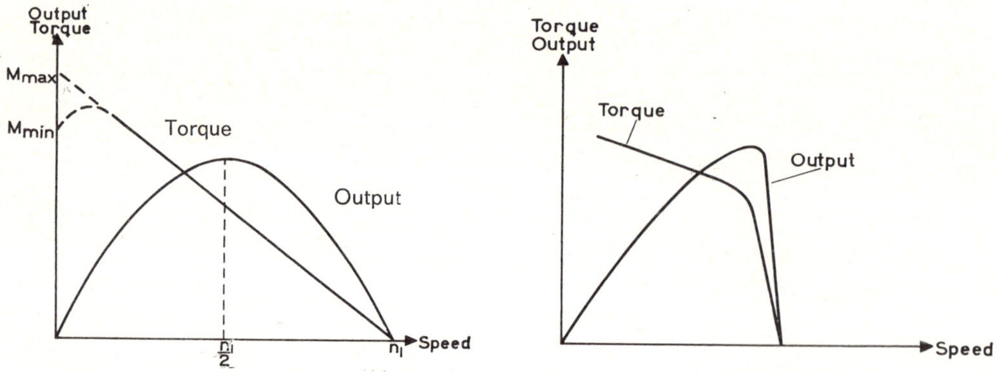

Fig 1 Characteristics of ungeared (left) and geared vane-type air motor. (Atlas Copco).

normally employed, however, rather than standard air line control valves, in order to realize the optimum characteristics of the motor (ie to achieve progressive inlet and exhaust opening). Idling speed can be controlled by throttling or by a centrifugal governor which keeps the speed constant — see Fig 1.

Whilst the speed/torque characteristics are largely determined by the type of motor, most manufacturers offer a range of gearboxes for direct (or integral) coupling to their motors, thus providing a range of speed/torque characteristics from a single unit. Another accessory commonly offered is an automatic brake for use in lifting, hoisting or similar duties.

There is no limit to the number of times an air motor can be started, stopped or reversed except that incurred by the speed of operation of the control valve. Air motors accelerate rapidly and will run as fast as the air supply and load resistance permit. Speed can be controlled by simple throttle or pressure regulating valves which adjust airflow rate or pressure respectively. The possible speed regulation is normally down to 10% or better of the full-throttle full-pressure speed against a given load. When controlled to any particular speed, however, actual running speed will vary if the load varies.

Exhaust

For normal industrial applications exhaust is vented directly to atmosphere via a silencer. In 'clean' areas it may be desirable or necessary to remove fine oil mist from the exhaust air by fitting a suitable filter. In ultra-clean areas the exhaust can be led outside the building. In this case a suitable size of exhaust line must be fitted in order to avoid excessive backpressure which would otherwise reduce the power of the motor.

General Performance Characteristics

Broadly speaking, the torque and power characteristics of an air motor can be regarded as similar to those of a permanent magnet or series wound d.c. electric motor, with maximum torque at stall, decreasing substantially linearly with speed. Maximum power is then achieved at around 75% of the free running (no load) speed. With the application of load, the running speed decreases (from the free running condition) until the torque balances the load. With a constant load, speed can be further adjusted by variation in air pressure supplied. On the other hand, with varying load the motor speed will continually adjust to the load. These basic characteristics are modified in detail by the type of air motor concerned and the specific design characteristics.

TABLE I – AIR MOTOR CHARACTERISTICS

Motor Type	No Load Speed rev/min	Typical Operating Speed Range rev/min †	Torque	Torque Curve	Initial Torque	Sizes (Typical Productions)
Vane	up to 20 000	7–15 000	Low	Flat	V.Low*	Fractional HP up to about 20 BHP
Piston	up to 2 000	600–1 500	Medium to High	Flat to Steepish	High	Fractional HP up to about 25 BHP
Diaphragm	up to 100	0–20	Extremely High	Steep	Extremely High	Fractional HP
Turbine	up to 120 000	50 000–80 000	Extremely Low	Flat	Negligible	Fractional HP
Gear	up to 3 000	1 500–2 000	Low	–	Low	1–20 HP

† Direct drive or spindle speed *Depends on number of vanes

TABLE II – PERFORMANCE PARAMETERS

Motor Type	Max.BHP developed at (%) No Load rev/min *	Speed at Max. BHP	Reduction Gearing Practical	Torque/ Speed Ratio	Speed Control	Torque with falling speed
Vane	45–55	High	Yes	Low	Air pressure	Rising
Piston	55–65	Low to Moderate	Yes	Medium to High	Air pressure	Rising
Diaphragm	20–25	Very Low	–	Extremely High	Air Flow rate	Rising steeply
Turbine	75–90	Extremely High	No	Extremely Low	Air pressure	Falling
Gear	50–60	Moderate	–	Low	Air flow rate and air pressure	Straight
Linear	–	High †	–	–	Exhaust restriction	–

* Approximate guide only † Without exhaust restriction

Where constant speeds have to be held against varying loads, speed control can set particular problems. Such applications may well influence the choice of motor, or the detail design of a particular type of motor – eg they may call for a motor with a steep torque curve so that any variation in load will be accompanied by only a minimum readjustment of speed. Specific speed control may have to be adopted in other cases.

Types of Air Motor

The majority of modern air motors are *vane* or *piston* configuration. Other types include diaphragm motors, turbine motors, gear motors and linear motors.

In addition to general purpose motors (vane and piston types particularly), air motors are commonly produced for specialized duties – eg tool drives, valve actuators, starters for internal combustion engines and gas turbines, winches, hoists. In addition, the field of application of air motors is partially overlapped by rotary and semi-rotary actuators, and air motors may also be used as drivers for rotary, semi-rotary and linear actuators.

Piston Motors

Piston motors are the most common type employed for industrial drives. The great majority are multi-cylinder units which may be in-line, vee, H, flat-four, radial or other variants. There are also further variations on conventional reciprocating engine design, such as crankless motors employing a rotating piston-cylinder assembly with slotted piston ends traversing an elliptic cam ring. In the main, however, the vee is the most popular configuration for four-cylinder motors, with the radial configuration equally popular for three or more cylinders. Radial configuration offers some possibility of weight reduction and is frequently adopted for larger motors. Either an odd or even number of cylinders can be used, eg, 3, 4, 5 and 6 cylinder radials are all common types.

The vee-four motor normally has the two cylinder banks at 90° to each other. Drop forged two-throw crankshafts are normally used with two connecting rods on each crank pin. Distribution valves may be oscillatory or rotary, there being little to choose between them. Valve timing can be arranged to provide equal power in each direction of rotation, or may be biased to provide asymmetrical timing and increased speed and power in one direction of rotation.

The typical radial motor has all connecting rods mounted on a common crank – eg see Fig 2. Power and exhaust strokes are controlled by suitable distribution valves and a two-stage exhaust is often adopted. In this case the inlet valve opens just after top dead centre and remains open for

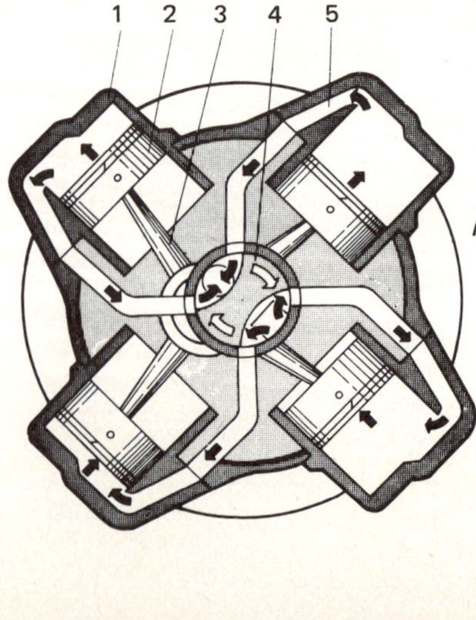

1 – cylinder
2 – piston
3 – connecting rod
4 – rotary valve
5 – air channel to cylinder

Fig 2 Typical construction of a piston-type air motor. (Atlas Copco).

approximately 160°. The inlet valve then closes and a large primary exhaust valve opens quickly to release the high pressure air to the atmosphere. After a further 40° or so of rotation, this valve closes and a smaller exhaust valve opens to continue evacuation of the compressed air. Reversal of direction of rotation is obtained by interchanging the inlet and secondary exhaust connections. It should be noted that the radial configuration particularly lends itself to porting (including a shaft port) instead of the use of mechanically operated valves.

Piston motors characteristically develop high starting and initial torque, falling with increasing speed. Both torque and bhp are directly dependent on supply air pressure. Starting, optimum running and stall torques are generally related to supply pressure as shown in Fig 3. Torque characteristics can be modified to some extent by detail design, eg valve timing and valve sizing. As they are positive displacement machines, air consumption is directly related to speed, ie the air consumption/speed curve is substantially linear for any given supply pressure. Power output is of the order of 5–10 kW/dm^3 cylinder volume, depending on the size and design.

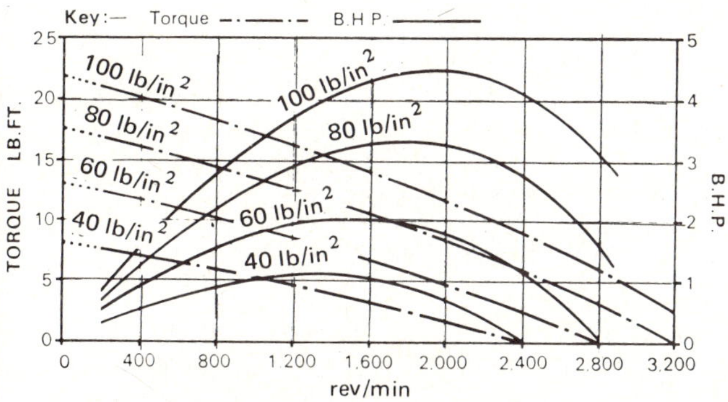

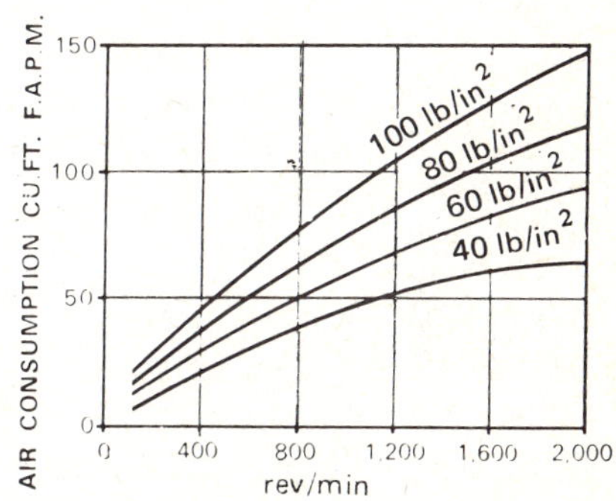

Fig 3 Performance curves for typical piston motor ***(Globe Model RM 200).***

Running speeds are generally low to moderate, although smaller piston motors may be designed to run at high speeds. In this case speed reduction with torque multiplication may be incorporated via reduction gearing to an output shaft.

The overall characteristics of piston motors make them suitable for an extremely wide range of applications. With modification, standard designs can often work satisfactorily down to supply pressures as low as 0.35 bar (5 lb/in^2). Continuous lubrication is essential and the method of lubrication will depend largely on the individual design and the need (if any) for separation of the various lubricants used. Thus a lubricant containing an emulsifying agent may be desirable in cylinders to prevent 'water washing', but would not be suitable for splash lubrication in a crankcase.

Variations on piston motor designs include swashplate motors, face-cam motors and other forms of axial piston motor. In general such designs have been largely superseded by vane motors.

Typical applications of piston motors are as follows:-

Fractional horsepower Motors – portable and bench tools;
light lifting, winching and hoisting duties;
mechanical handling.

Larger sizes – general driver duties, with a particular application where loads are heavy and/or fluctuating and where good low speed torque characteristics are important.

Vane Motors

Vane motors are usually lighter, more compact and cheaper than piston types, and cover a power range from fractional sizes up to about 30 hp. The fractional horsepower units are used extensively on portable tools and on numerous multiple tool applications. The compactness of such designs is emphasized by the fact that a modern vane motor may be capable of producing 200 kW (dm^3) cylinder volume. Their power to weight ratio is outstanding. Internal vane motors are also characterized by very high spindle speeds with low starting torque.

With 20:1 reduction gearing the vane motor will show substantially the same peak speed and peak speed torque as a piston motor, ignoring the gearing losses – eg see Fig 4. The latter will tend to offset the higher efficiency (ie lower air consumption per horsepower) of the vane motor. Thus, in practice, there is considerable overlapping in the application of these two main types due to the extensive use of reduction gearing to adjust both speed and torque levels to specific operating requirements.

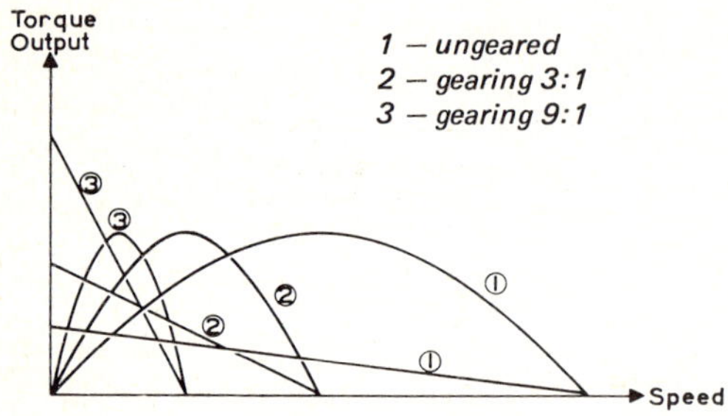

Fig 4 Diagram showing how the characteristics of an ungeared motor are changed with gearing. (Atlas Copco).

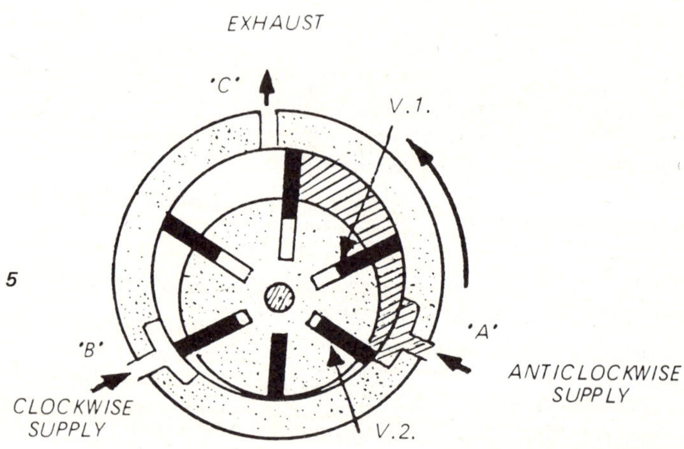

Fig 5

The operating principles of a reversible internal vane motor are shown in Fig 5. A rotor shaft carrying the vanes is mounted eccentrically in the power chamber with the vanes arranged to slide radially, to seal against the power chamber wall. The vanes are usually mechanically or pneumatically sprung against the casing to provide a sealing force for starting; this force is increased by centrifugal action when the motor speed rises. Air is supplied to the rotor chamber via either Port A or Port B, depending on the direction required.

If air is supplied to Port A, the rotor is driven in an anti-clockwise direction due to the effective thrust area of vane 1 being considerably greater than that of vane 2. This effective thrust ratio is maintained until vane 1 passes Port C, and the pressure is released to atmosphere. Any residual pressure that remains in the chamber between vane 1 and vane 2 after vane 2 closes exhaust Port C, is exhausted via supply Port B, which is fully vented when air is being supplied to Port A.

For clockwise rotation, air is supplied to Port B and Port A is vented.

The supply and exhaust ports can be in the cylinder casing as shown, or alternatively in the end caps. Most motors have some or all of the ports in the cylinder wall, and, as the vanes pass over them, particularly at the high speeds involved, uneven wear on the vanes is often experienced. The vanes, however, are manufactured from a range of materials that give a good working life, and are easily replaced.

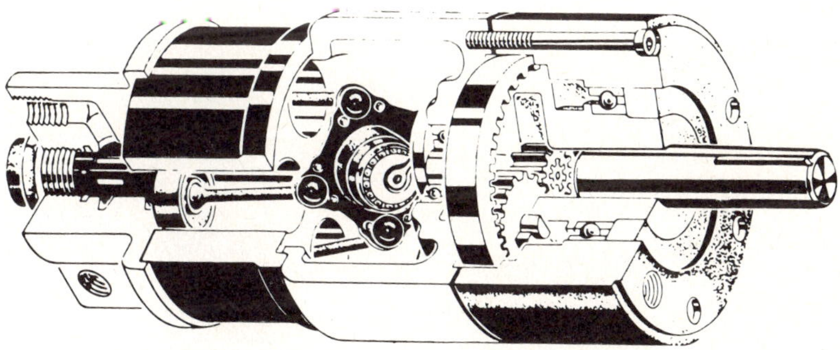

Swashplate or axial-piston motor. (Gardner-Denver).

The number of vanes employed is not a critical factor, but there are generally three or more. The greater the number of vanes, the less the internal leakage and the higher the starting torque, although friction and wear are increased. Good starting torque also depends on the vanes sealing at low rotational speeds. This is normally provided by 'spring loading' the vanes in a radial direction, either pneumatically or mechanically.

Typical torque and power output characteristics of an internal vane motor are shown in Fig 6. Torque is substantially linear, although the starting torque available is invariably lower than the theoretical maximum or stall torque. Also torque values are relatively low, good power output depending on high running speed. The peak of the bhp curve may occur at 50% of the no-load (maximum) speed or somewhat higher, depending on the detail design and, particularly, the porting. Motors of this type, in fractional horsepower sizes, commonly have free running speeds of the order of 20 000 rev/min, although some may incorporate governors to limit the no-load speed attained. Power-to-weight ratio is good although starting torque is low and low speed performance relatively poor. A vane motor would not, therefore, be used for applications requiring low and creep speeds.

Chief applications of vane motors are:-

Fractional horsepower sizes – portable air tools, multiple light duty tools, etc.

Larger sizes – larger air tools, pump drivers and general driver duties where light starting loads only are involved.

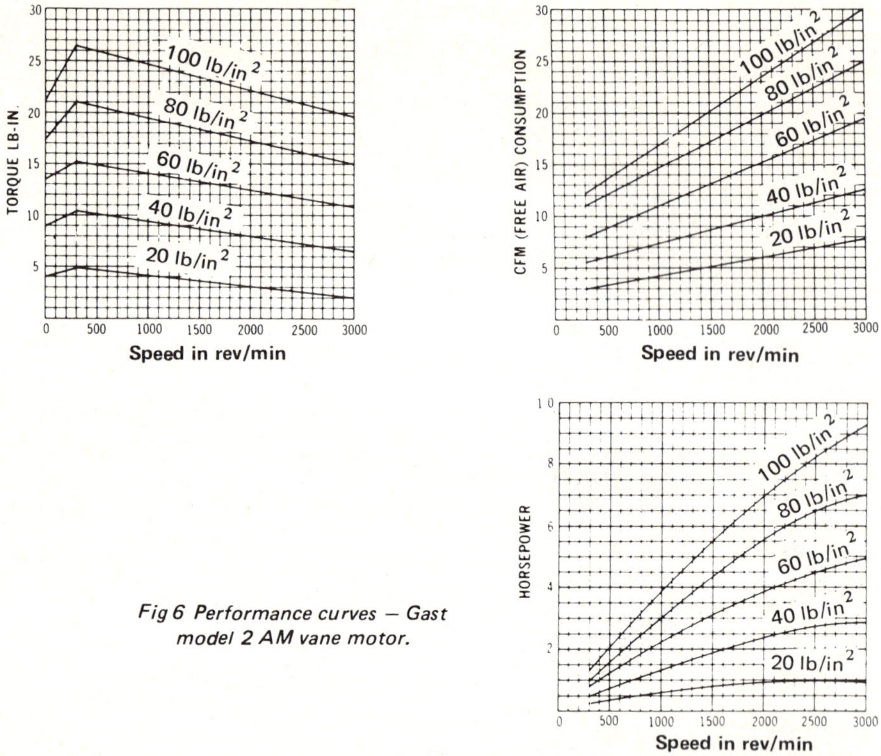

Fig 6 Performance curves — Gast model 2 AM vane motor.

Included under the general classification 'vane motors' are external vane motors, although these are less common. The cam-vane motor, shown diagramatically in Fig 7, is based on a single vane rubbing on a solid rotor, thus avoiding the uneven wear of vanes running over port openings. It is far more limited in performance, however, with an inherent 'dead-spot'. Thus to provide continuous rotation, a flywheel type load is required; two or more such motors may be connected in series to promote smoother and continuous running.

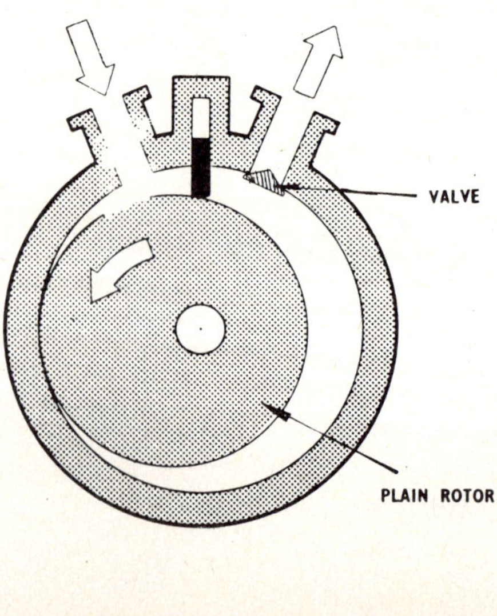

Fig 7

Other variations include the use of lobe-rotor designs, literally in the form of pumps or compressors working in reverse. These have been specifically developed for high temperature work, but have little general application.

Diaphragm Motors

A diaphragm motor is basically a reciprocating device with limited stroke. Rotary output motion is derived via a ratchet and pawl – Fig 8. Essentially speed of working is low, but very high torques may be developed. For example, for the same amount of air consumption a diaphragm motor is capable of developing up to 800 times the torque of an ungeared vane motor, or up to 40 times the torque of a piston motor. Drive output is normally fractional horsepower, however, because of the low speed of working.

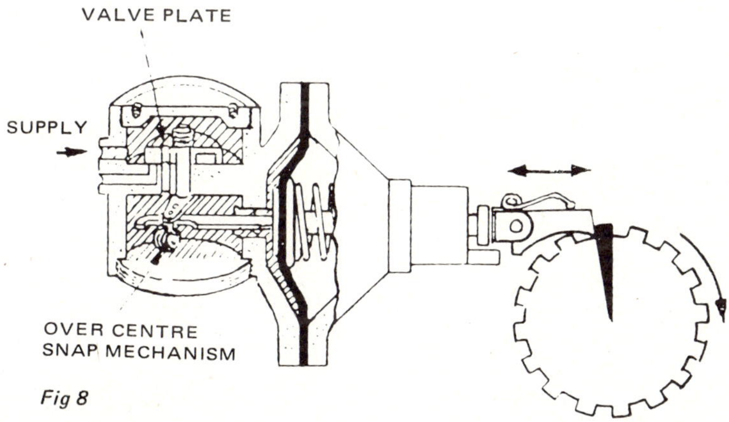

Fig 8

This diagram shows a typical valve chest associated with the diaphragm movement. The centre of the diaphragm is rigidly attached to a rod, one end of which forms the thrust shaft while the other admits high pressure air to the diaphragm chamber and then opens the chamber to exhaust as soon as the diaphragm has been displaced over its maximum travel. The return motion of the diaphragm is commonly produced by a spring. Alternatively, the diaphragm may be driven by compressed air in both directions, ie high pressure air being admitted alternately to each side of the diaphragm, with corresponding provision to exhaust each side of the chamber in opposite phase. The power chamber is completely sealed, enabling the motor to be operated on very restricted air flows so as to produce extremely low speeds with no loss of thrust or torque. Further, since the thrust is always acting tangentially to the ratchet wheel, there is a little variation in the torque generated throughout a cycle, which enables the motor to be accurately positioned at any point in the working stroke.

Typical performance characteristics are as shown in Fig 9. Very high initial torque is available, torque falling off rapidly with increasing speed, although the speed range is strictly limited. Peak hp is normally developed at circa 20 rev/min for motors of this type. In addition to the ratchet wheel diameter size variation (which affects speed as well as torque), motors of this type are also produced with two and three diaphragms operating on the same shaft to increase the thrust and final torque.

Diaphragm motors are particularly suitable for high torque, low speed applications – eg for valve actuators. Here, high initial torque is required to crack open the valve. The falling load will then enable the motor to build up to a good running speed when the pressure differential across the valve

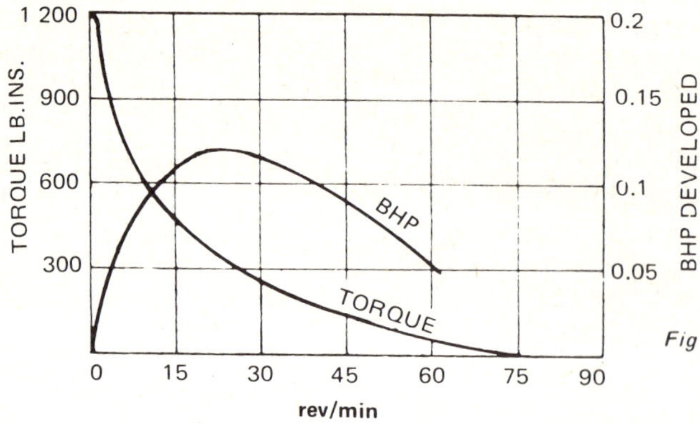

Fig 9

has dropped. Additionally, if used as a valve positioned on an automatic control system, the maintenance of output with restricted supply air flow maintains full power operation of the valve as the balanced condition is approached. Diaphragm motors can, of course, also be used directly as vibrators.

Diaphragm motors are particularly efficient (although less so if the diaphragm is driven in both directions by pressurized air), and because of the null-leakage characteristics,air consumption is directly related to bhp or speed, but not torque.

Turbine Motors

The turbine is essentially a very high speed motor, capable of generating only minimal torque. Torque, in fact, is virtually negligible, except at mid-speeds, with maximum speeds of the order of 50 000 to 80 000 rev/min. Applications are thus distinctly limited. Torque multiplication by reduction gearing is not a viable proposition, because the load involved would be of the same order as the torque available. Hence the main application of turbine motors is where very high speeds can be used with direct drive against comparatively light loads – eg high speed pencil-type grinders and dental burrs. Like other types of air motor a turbine can, of course, be stalled without damage. Some turbine motors may be fitted with mechanical governing devices to limit the maximum speed which can be attained.

Turbine motors operate with clearances throughout, with no sliding or rubbing parts other than the spindle bearings. Bearing friction is thus the limiting factor governing free running speed. Turbines can operate equally well on dry air (ie without lubricated air), provided the bearings are self-lubricating or have separate lubrication. This can be an advantage in specialized applications.

Gear Motors

Meshing gear motors are a further type, but are not directly competitive with piston or vane motors. The relatively low efficiency (high leakage loss) of straight spur gears can be largely overcome by employing helical gears.

Gear-type air motors are generally characterized by moderate running speeds, typical operating speeds being of the order of 1 500 to 3 000 rev/min with direct drive. Output speed can be reduced and torque multiplication obtained by the use of reduction gearing. The application of gear-type motors is, however, strictly limited and few commercial versions are produced.

Typical geared, rotary vane air motor.

Air motor driven priming pump – diesel engine (Marine).

Radial piston air motor.

Linear Motors

The linear motor is essentially a double-acting cylinder associated with suitable control gear to provide automatic reversal at the end of each stroke and thus continuous reciprocal motion. This control may be initiated by mechanical triggering, or pneumatic or electronic signalling (eg controlling a pilot-operated or solenoid-operated valve respectively) – Fig 10.

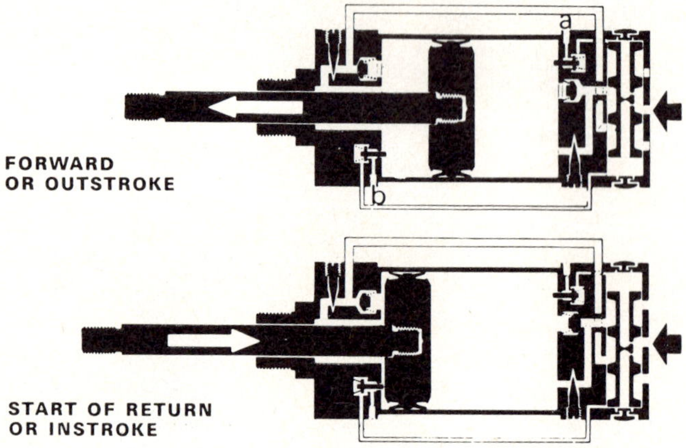

Fig 10 Reciprocating air motor (Martonair).

The reciprocating output of the cylinder (ie rod movement) may be used directly or to drive a crank for final rotary output. In the latter case, the crank throw must obviously be matched to the stroke available, or rather the crank throw will, in fact, govern the actual cylinder stroke achieved. The versatility of the linear motor may be further improved by providing variable speed of stroking. This is usually done by fitting adjustable restrictors on the exhaust side controlling the rate of escape of pressurized air from the non-pressurized side of the piston, and thus the backpressure, which effectively controls the speed.

Strictly speaking, a linear motor adapted to rotary final output is no longer a linear motor as such. Connected to a crank mechanism it is really a single cylinder piston motor and its characteristics will be similar. If a short stroke cylinder is used, coupled to a ratchet drive, then its performance characteristics will approximate to those of a diaphragm motor.

See also chapters on *Air Motors* and *Lifts and Hoists.*

SECTION 4

Valves and Circuitry

Pneumatic Valves

PNEUMATIC valves are generally divided into three main categories defined by function, viz

(i) Directional control valves.

(ii) Flow control valves.

(iii) Pressure control valves.

In addition there are a number of specialized valve types which do not fall into these general categories, including valves employed as 'logic' elements.

Directional Control Valves

A directional control valve function can be described by the number of port openings or 'ways' and the number of positions of the valve. The simplest type of valve is the two-way, ie pressure 'in' at P, and a connecting port for outlet (A) – Fig 1. Such a valve is either normally closed or normally open, ie it has two positions. It is then a two-way, two-position valve, also referred to as a straightway valve.

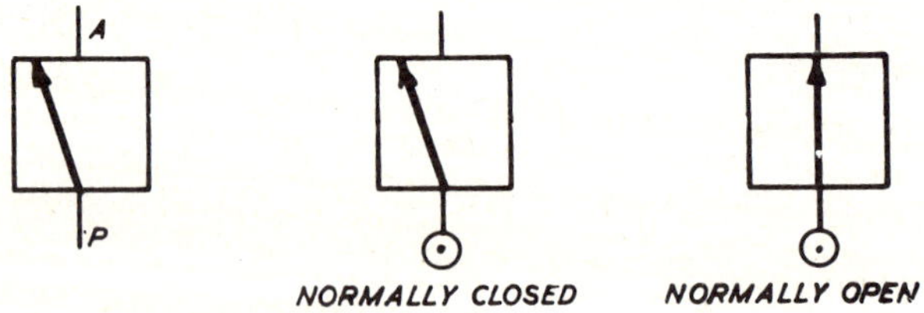

Fig 1

In the normally closed mode it is an 'off-on' switch for any device connected to A, actuating the valve supplying air to the device, and restoring the valve shutting off the supply again. Since air cannot exhaust downstream in the normal position, it is limited to switching devices which have a free exhaust, eg air motors.

In the normally open position it is an 'on-off' switch. In other words, actuation of the valve isolates the line or service connected to A. Again it should be noted that this can only be applied

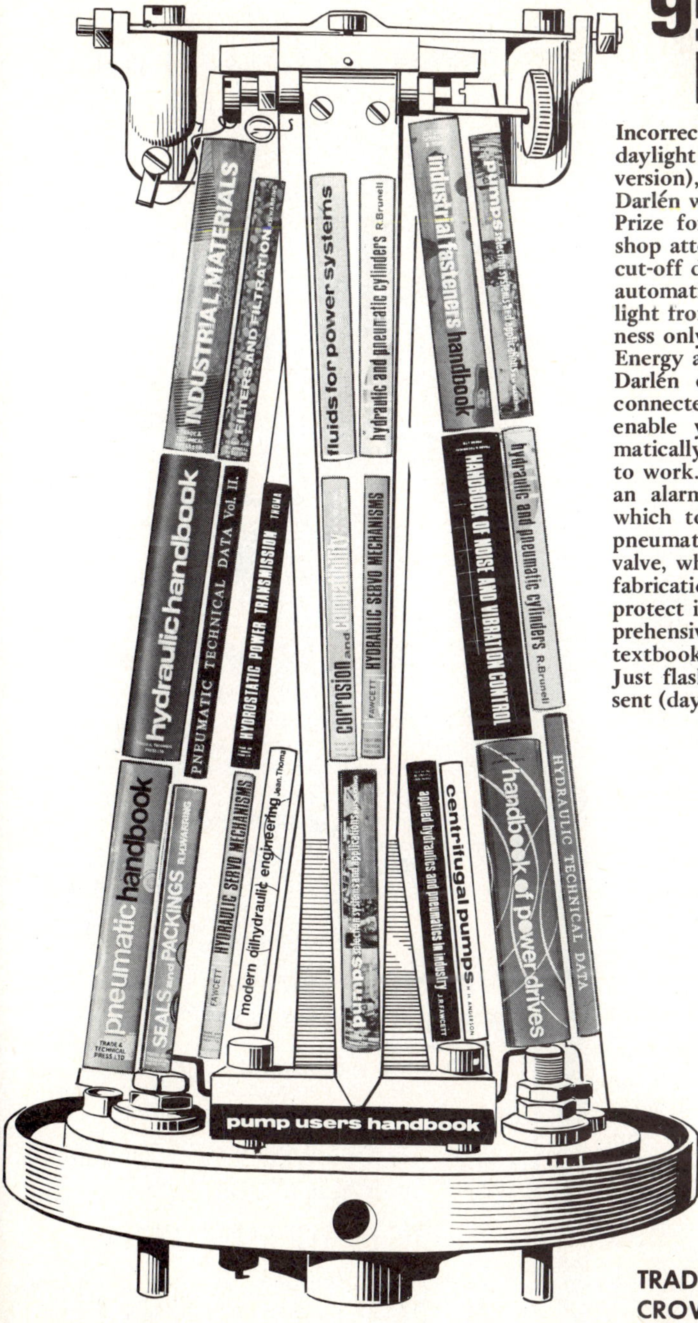

great inventions

Incorrectly called 'the sun valve', for it works in daylight (UK version) as well as sunlight (Sahara version), it was invented in 1907 by Gustaf Darlén who, 5 years later, was awarded the Nobel Prize for his work in physics (not as a Boot's shop attendant). This valve, connected through a cut-off device to the gas supply, provided the first automatic navigational beacon, ensuring that the light from the beacon burnt during hours of darkness only. Had there been (or Benn) a Minister of Energy at that time, he would no doubt have had Darlén dubbed a Noble fellow. Such a valve connected to your angle-poise reading light would enable you to burn 'the midnight oil', automatically switching off when it was time to go to work. This would serve a dual purpose; firstly, an alarm clock and secondly, illumination by which to read how to invent an oil hydraulic, pneumatic or mechanical, noise controlled sun valve, what pumps, industrial materials, fasteners, fabrication and instruments to use and how to protect it from fire and theft — all subjects comprehensively covered in the technical journals, textbooks and handbooks published by TTP. Just flash your beacon, and a books list will be sent (daytime only).

TRADE & TECHNICAL PRESS LTD,
CROWN HOUSE, MORDEN, SURREY
SM4 5EW, ENGLAND.

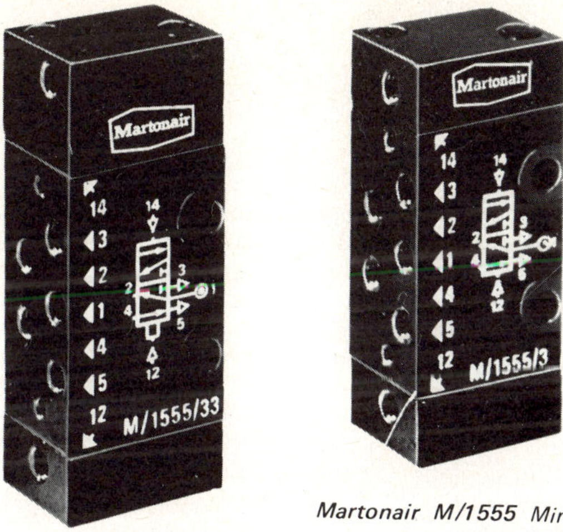

Martonair M/1555 Mimaker directional control valves.

to a service from which downstream air does not have to be exhausted. To perform such a function, eg to use a two-way, two-position valve for a single-acting cylinder control one valve would have to be used to control the incoming air and a second valve to control the exhaust.

This duty can be performed by a single three-way, two-position valve, shown in simple diagrammatic form in Fig 2. In the normally closed mode, the supply air is stopped but the outlet is connected through to the third port, to allow downstream air to exhaust through the valve. Actuation of the valve connects the pressure inlet to A, shutting off the exhaust port. Basically, such a valve can be used as an 'off-on' control for a device connected to A which requires pressurizing in one direction only and makes provision to exhaust downstream air when switched 'off' through the same connecting line – eg a single-acting cylinder with spring return.

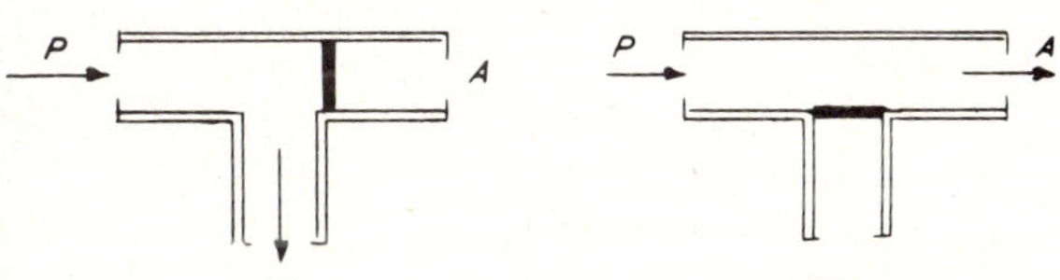

In the normally open mode, pressurized air flows directly through the valve, with the exhaust port closed. When the valve is actuated the pressure supply is cut off and downstream air can flow back through the valve to exhaust it. It is thus an 'on-off' switch, controlling a device which has to remain pressurized until the valve is operated, when the device can exhaust downstream air back through the valve.

The three-way, two-position valve is thus a logical choice for a single-acting cylinder circuit control or for any other simple circuit where downstream air has to be exhausted. It is not necessarily a complete answer to such circuits and there are cases where the use of two two-way, two-position (straightway) valves may be preferred.

¼in quick exhaust valve (CompAir Maxam)

Switching capability can be further extended by increasing the number of ways (ports) and, if necessary, the number of positions. Thus, the four-way, two-position valve of Fig 3 has two through-connections. In one position P is connected to A, and B can exhaust downstream air through E. In the second position P is connected to B, and A can exhaust downstream air through E. Thus a four-way, two-position valve can be used to operate a double-acting cylinder or any other device requiring alternate pressure and exhaust on two connecting lines.

Whilst this is a logical control valve choice for double-acting cylinder, again there are instances where the use of two three-way, two-position valves may be preferred, eg for close coupling or mounting the valves directly on the cylinder ports.

The five-way, two-position valve of Figs 4 and 5 is similar to the four-way valve, except for the provision of an additional exhaust port. Thus, when either A or B is switched to exhaust, it operates through a separate exhaust. This can have specific advantages in particular applications.

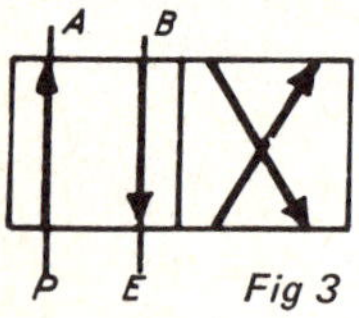

Fig 3

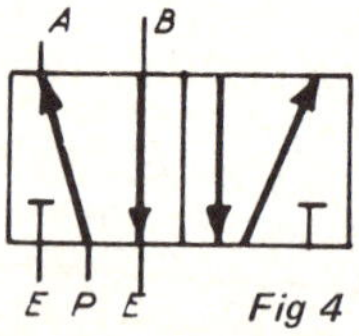

Fig 4

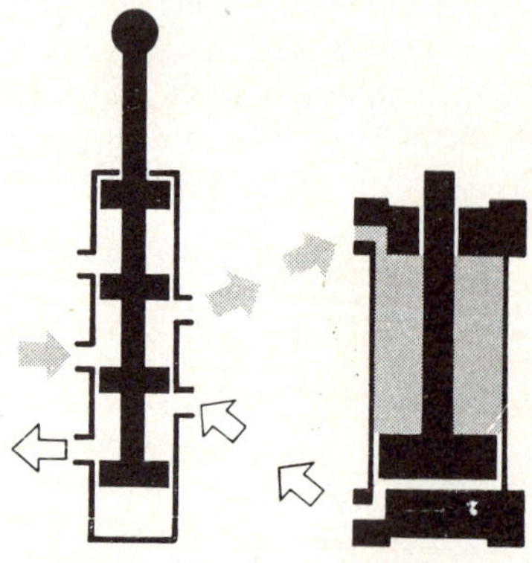

Fig 5

The other directional valve is the four-way, three-position valve, which can have three possible modes – Fig 6. In the first (left) the function is similar to a four-way, two-position valve except that an additional position is available with all ports blocked, ie no flow is possible through the valve in either direction. This is usually the normal position. A typical application is the control of a double-acting cylinder in which 'hold' facility is also available, ie pressurized air can be trapped on both sides of the piston.

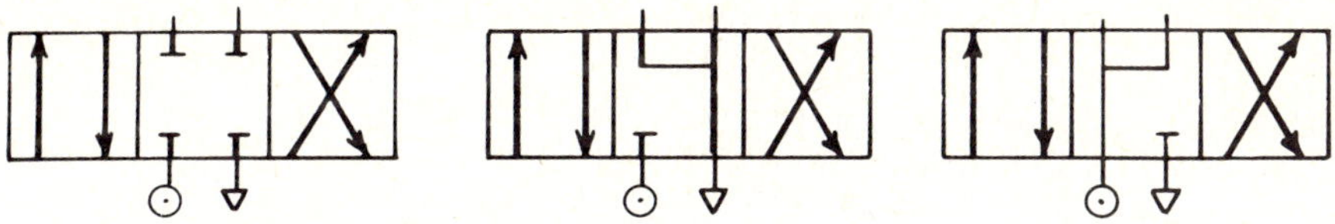

Fig 6

In the second mode (centre), conventional four-way switching is available from the two extreme positions, but the mid position shuts off the supply and connects both downstream lines to exhaust through the valve. Again this type of control is used with double-acting cylinders where it is required to 'free' the piston by exhausting air from both sides of the piston.

In the third mode (right), normal switching is available from the two extreme positions but the centre position provides pressurized air to both downstream lines, with the exhaust closed. This provides an alternative hold facility for double-acting cylinders with both sides of the cylinder continuously pressurized. This will not, of course, give true 'hold' on a single-rod cylinder, unless the pressure is dropped in the line to the blank end side of the piston to compensate for the loss of effective piston area on the rod side.

Standard Valve Symbols

Symbols used for designating valves on pneumatic circuits are conventionally in the form of adjacent squares, each square representing one position of the valve. Thus a two-position valve would have two adjoining squares and a three-position valve three adjoining squares. Interior connections between points in each position are then indicated by an arrow or arrows within the appropriate square, together with some indication of unconnected points.

Numerous variations on this basic method of diagrammatic representation have been used, and still continue to be used by various individual valve manufacturers. (The previous diagrams, Figs 1–6, are also non-standard). These comments also apply to the marking or ports. Standardization on ISO symbols is now generally agreed, as given in ISO R.1219 (based largely on the original CETOP code). In addition to standardizing diagrams, main ports are also identified by numbers rather than letters, viz (see also Fig. 7).

The main ports are identified as follows:–

1 = Normal inlet i.e. main supply port.

2 = Normal outlet port on 3/2 and 3/3 valves.

2 & 4 Normal outlet ports on 5/2 and 5/3 valves.

3 Normal exhaust port on 3/2 and 3/3 valves.

3 & 5 Normal exhaust ports on 5/2 and 5/3 valves.

Port 1 is the main supply port, other odd numbers denote exhaust ports, and even numbers outlet ports. Port 3 is always internally connected to port 2 when the valve is in one position. In the reverse position on a 5 port valve port 5 is always connected to port 4.

The pilot control ports are identified as follows:–

'10', '12', and '14'.

Fig 8 shows ISO symbols for two-position, three-position and five-position two-way valves. Additional symbols are also appended to the valve diagram to show the method of actuation (and also return, eg spring-return) – see Fig 9. See *Data Section* for a complete presentation of valve symbols.

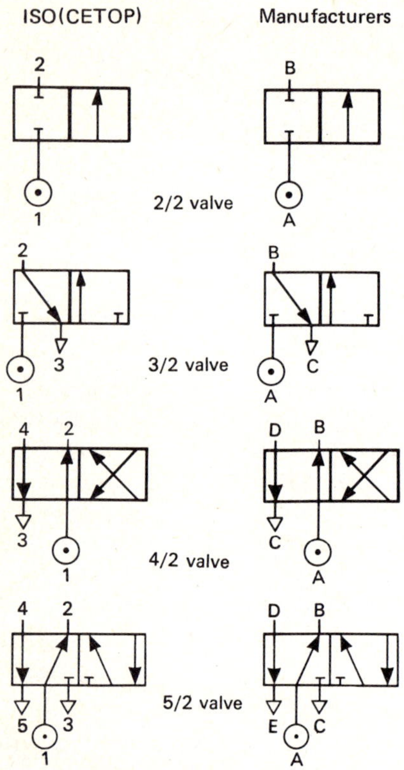

Fig 7 Comparison of ISO (CETOP) and one manufacturer's valve diagrams.

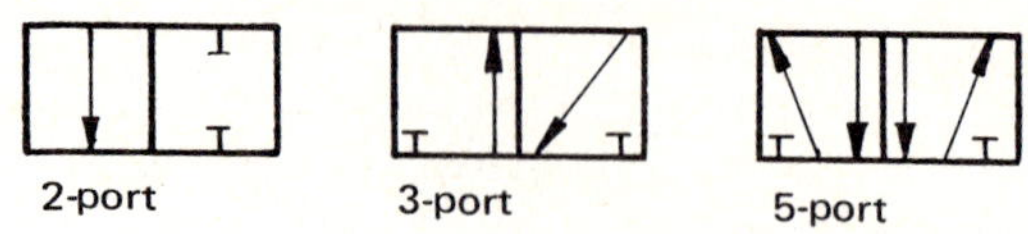

Fig 8

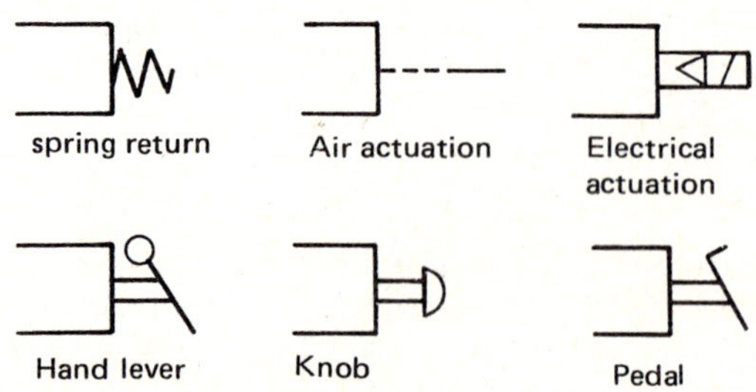

Fig 9

Valve Construction

Valve types are normally described by function (as above). Valve sizes are normally designated by the size of the port tappings. This offers a specific match between pipe size and valve – eg a ¼in valve would have ports tapped to take ¼in bore pipe tapped externally with BSP or NTP thread. The range of sizes most commonly covered is from 3 mm to 25 mm (1/8 in to 1 in) although main line valves are produced in larger sizes.

Valves can also be categorized by their form of construction, which can have an appreciable effect on the valve characteristics.

Poppet valves are a logical choice for switching elements, since they require minimum movements to achieve full opening, offer low resistance to flow when in the open position and have good sealing properties when closed. They can also be made to seal adequately for long periods. They do require relatively large operating forces, however.

A simple poppet valve, designed for a pressure of 10.5 bar (150 lb/in^2) may require an opening force as large as 10 kgf (22 lbf) because of its unbalance. On the other hand, the fact that full opening is achieved with only small travel gives rapid full flow response to signal, which can be of particular advantage in high speed applications and for pilot circuits.

Martonair poppet valves.

A limitation of poppet valves is restricted switching ability. The basic poppet valve is capable only of on-off switching function. To perform a three-way or four-way function, a combination of poppet valves is needed. Fig 10 shows a three-way poppet valve designed for push operation. The plunger is drilled for part of its length and is spring biased in the exhaust position. When actuated, the plunger first seals off the exhaust port on the exhaust valve disc, and then pushes the disc off its seat, admitting pressurized air. On release, the disc reseats and the exhaust port is again uncovered.

Alternative configurations for a three-way valve include the use of a plunger and ball instead of a mushroom valve. Four-way function is normally accomplished by combining two three-way valves in a common body or connected in parallel.

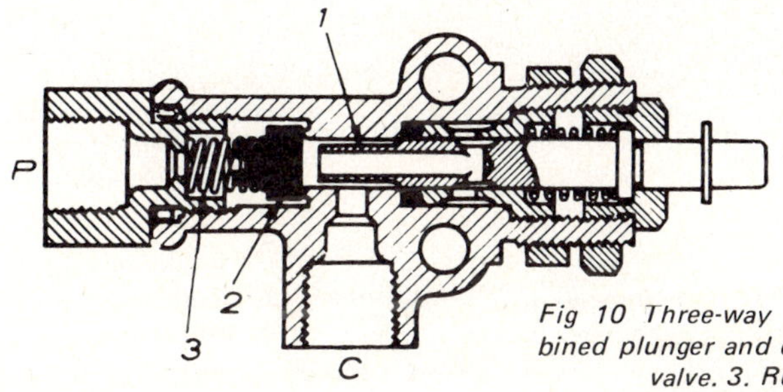

Fig 10 Three-way push valve. 1. Combined plunger and exhaust port. 2. Inlet valve. 3. Return spring. (Benton & Stone Ltd).

Spool Valves

Spool valves are attractive because they offer extreme flexibility in interconnection via a simple sliding mechanism. They are also relatively simple and economic to manufacture, although for adequate sealing fine surface finish is required on both the spool and barrel bore, with close tolerances to ensure practical minimum clearances. In practice, spool and valve are usually lapped to fit.

Spool valves embody two basic elements, a cylindrical barrel in which slides a plunger. The plunger provides a sliding seal with axial movement covering or uncovering interconnected port openings. Port blocking is provided by lands or full diameter sections on the spool with intervening waist sections which provide port interconnection through the barrel. This makes it easy to provide multi-way and multi-position switching.

A possible limitation of such a valve is that in certain applications and particularly with larger spool valves, the precision bore and fit may be distorted by externally applied loads.

Manufacturing problems may be simplified in the case of pneumatic valves by fitting the spool glands with seals, for example O-rings located in grooves. This eliminates the need for close tolerances between the bore and spools. A disadvantage in this case is that the sealing rings must slide over the ports where they may be subject to cutting action or distortion. Variations, primarily again, in the case of air valves, include the use of 'U' seals, bushed bodies and ports drilled with a number of fine holes where sealing rings pass, and a linered construction with a liner floated on simple seal rings. All are aimed at reducing the possibility of damage or distortion of sealing rings sliding over port openings.

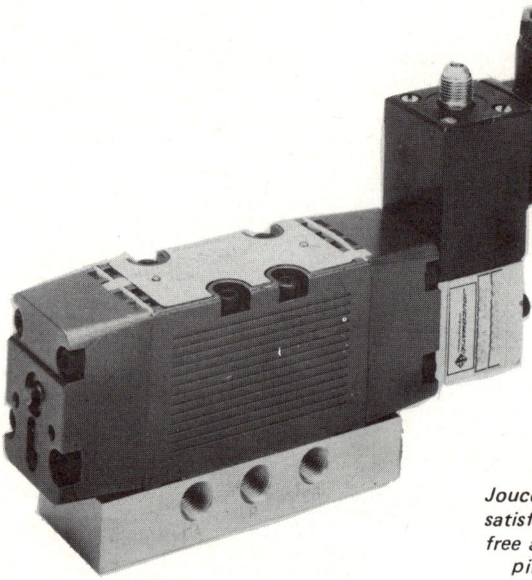

Joucomatic 541 spool valve, operates satisfactorily without 'stiction' on oil-free air. Version shown is single solenoid pilot operated with spring return.

Slide Valves

Slide valves offer an alternative approach equivalent in effect to an unwrapped spool. Ports are normally drilled in the fixed slide and matching recesses in the sliding member. The two members are then lapped to fit and held in contact by spring loading. Various methods may be used to reduce friction forces. Thus, in some valves, the sliding member is suspended on spring plates to prevent metallic contact. In other cases, friction forces are reduced by incorporating hydrostatic pressure balance within the valve. The principal limitation of slide valves is that they demand extreme precision of manufacture in order to realize leakage rates comparable with those of other valve types. Slide valves are also sometimes known as plate valves or sliding plate valves.

Rotary valves are another form of plate valve or sliding valve with the moving element rotating over a fixed plate to open or block ports drilled in that plate. Channels cut in the rotary plate can provide transverse flow. Reaction-forced compensation can be introduced by the use of deflector veins to reduce friction. However, all sliding plate valves, ie slide valves and rotary valves, are basically restricted to two-, three-, or four-way operation. Rotary valves have the particular advantage that they are readily adapted to multi-outlet working. For this reason they are often favoured as sequence valves in pneumatic circuits. Spool and plate valves are also commonly used in combination in two-stage valves. Two-stage valves are intended to overcome the practical limitations of power response imposed by single-stage operation. A two-stage valve merely employs one valve as a first stage and a second as a second stage, normally combined in one body or unit.

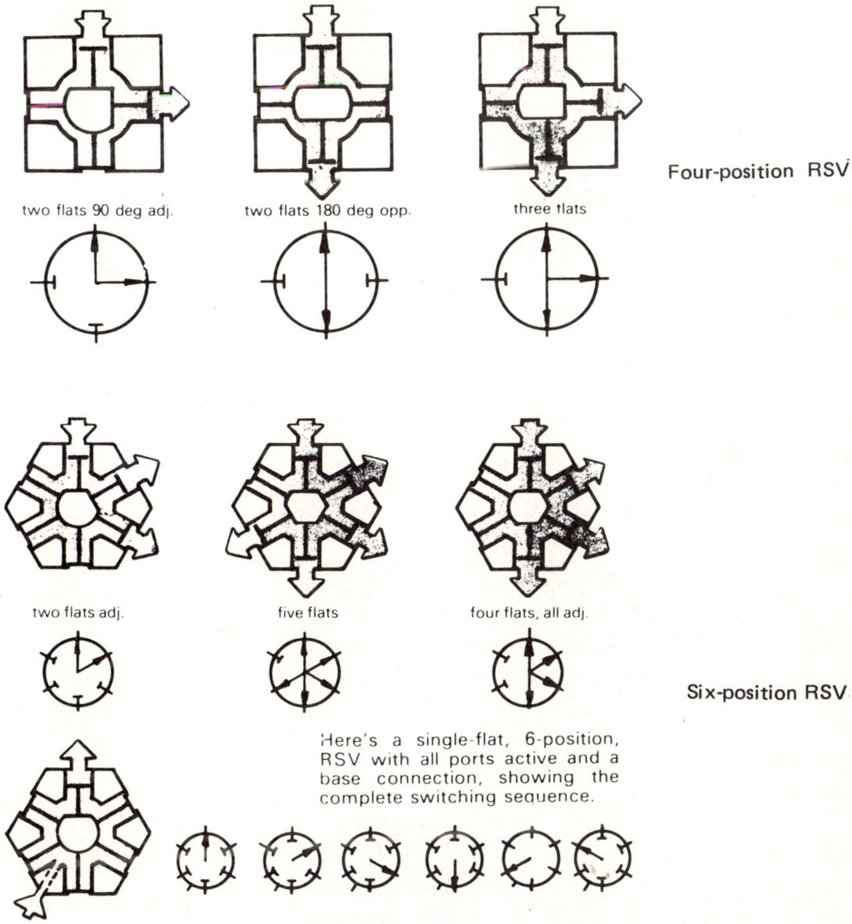

Fig 11 Drallim RSV rotary switch valve shown in simple diagrammatic form.

An extremely versatile and unique form of rotary valve is shown in Fig 11. Basically this consists of a square or hexagonal section body with four or six peripheral ports communicating with a central chamber. An operating spindle passes through this chamber, operating spring-loaded poppets in individual ports by cam action. A number of valves of this type may be ganged to form a multibank valve.

Some of the latest models from the extensive range of rotary switch valves (RSVs) from Drallim Controls Ltd. A multiport manual distributor, the valve is used for circuit selection in pneumatic systems. Rear centre is an RSV ganged with a rotary switch, allowing simultaneous making and/or breaking of pneumatic and electrical circuits.

Poppet valves, slide valves and spool valves are of three principal types. Each is subject to considerable variations in geometry, detail design, materials and methods of construction depending on the preferences of individual manufacturers. Spring-loaded ball valves are not excluded but these are normally confined to simple flow control devices used in pneumatic circuits. They have the same limitations as poppet valves in that they can only provide two-way action, with further mechanical disadvantages. Provided these can be overcome by suitable detail design, ball valves can be considered as a direct alternative to poppet valves. Both types may be combined in a single valve, eg a poppet (or plunger) valve combined with a ball valve in a three-way body.

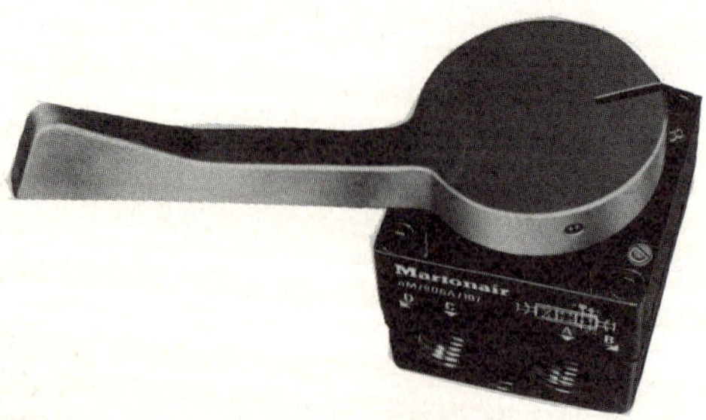

Martonair rotary lever operated plate valves.

Shuttle Valves

The shuttle valve is an 'automatic' type of directional control valve with three-way, two-position characteristics. This is based on a simple three-port body with a moving element in the form of the free spool or shuttle, or a ball – Fig 12. A shuttle valve allows either one of two pressurized inlets

COMPRAX Automatic Valves
Automatic Plate Valves for air and gas compressors, pumps and blowers. Special high-lift types for non-return valves in pipelines. A comprehensive range of valves is available to suit all operating conditions. Frictionless guided valves for non-lubricated machines a speciality. Extra strong high pressure valves and enclosed safety guards. Valves to customers' designs to suit particular requirements. Spare parts for all valve types. Streamline valves for 'Sulzer' Diesel Engines. Pilot Valves and Regulating Monitoring Equipment for Compressors.
Now also the complete range of COMPRAX HOERBIGER Pneumatic instruments and controls.
Sole Licensees of all 'Hoerbiger' Patents
COMPRESSOR ACCESSORIES LTD.
7c, Factory Centre Kings Norton
Birmingham B30 3JD England
Telephone: 021-458-1124
Telex: 338835 Compax Bham
Comprax

(whichever is the greater value) ie the higher pressure inlet moves the shuttle to a position where it blocks the other inlet. It is thus more of a special valve than a true directional control valve and is basically used for logic OR switching.

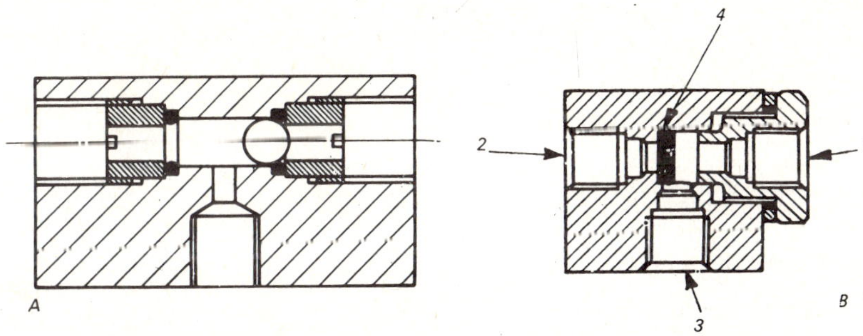

Fig 12 (A) Shuttle valve with ball element. (Air Automation). (B) Shuttle valve allowing either one or two pressures to be applied to a single outlet without mutual interference. (CompAir Maxam)

Flow Control Valves

Flow control valves are essentially throttling devices. Their geometry can range from that of a fixed orifice to adjustable needle valves or shaped grooves, etc, capable of providing progressive throttling. Their basic action may also be combined with a second function such as non-return.

Check valves, or non-return valves, are designed to provide unrestricted flow in one direction when pressurized, but close to ensure positive sealing in the absence of pressure. The common form is a suitably mounted spring loaded plunger housed in a cylindrical body seating on an internal annulus to provide a one-way seal, readily lifted off its seat by pressure on the face of the piston. The plunger end is usually fitted with an O-ring to provide the actual sealing surface. The main requirement (apart from positive sealing in the one-way direction) is that the annulus and piston should offer minimum resistance to airflow in the open position.

Pressure Control Valves

Pressure control valves may be designed to reduce the pressure to a desired level, or to maintain it at that level. A common form of construction incorporates a spring-loaded diaphragm pressurized in opposition to the spring. In the absence of pressure, the valve is held open by the spring – Fig 13. When pressure is admitted, it builds up on the outlet side until the force generated by the pressure on the diaphragm equals that of the control spring. The spring can thus be adjusted to reduce the effective outlet pressure from a maximum downwards, with reasonable accuracy of setting. In practice, the valve is simply opened enough to allow sufficient air to pass to meet the particular demand.

An alternative type of diaphragm pressure-reducing valve, which has the added facility of exhausting returned downstream air, is shown in Fig 14. This also incorporates a tappet control for pressure instead of a screw.

Fig 15 shows a more sophisticated type of pressure regulation valve which is pilot operated and has a continuous bleed off, ie is never completely closed. Regulation of the accuracy to ±0.5% is possible with such a device, for reasonable variations in flow and inlet pressure.

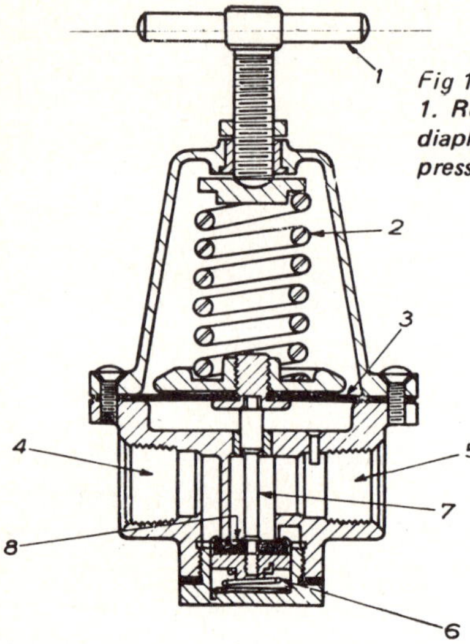

Fig 13 Reducing valve for pressures up to 100 lb/in^2. 1. Regulating handle. 2. Pressure spring. 3. Rubber diaphragm. 4. High pressure inlet. 5. Reduced pressure outlet. 6. Valve return spring. 7. Push rod. 8. Synthetic valve washer in metal holder.

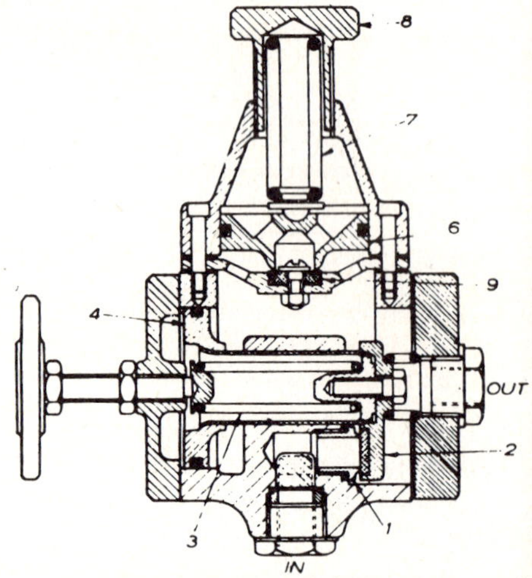

Fig 14 Balance valve for maintaining constant pressure in a loaded cylinder. 1. Inlet valve seat. 2. Inlet valve. 3. Inlet pressure spring. 4. Inlet valve piston. 5. Inlet pressure adjustment. 6. Relief valve piston. 7. Relief pressure spring. 8. Relief pressure adjustment. 9. Relief valve. (Martonair).

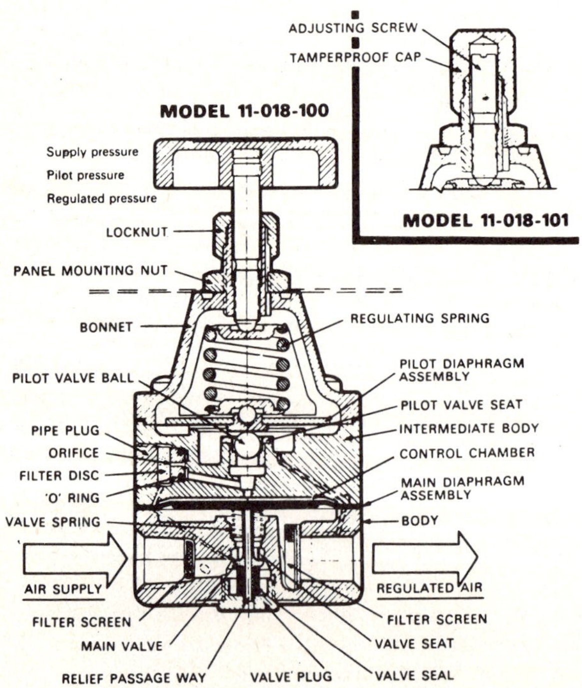

Fig 15 Precision regulating valve with pilot valve having continuous bleed controlling the pressure acting on the main valve control diaphragm. (Norgren)

Valve Actuation

Valve actuation, or movement of the valve element from one position to another, can be performed in the following ways:—

(i) Manual actuation.
(ii) Mechanical actuation.
(iii) Solenoid actuation
(iv) Air actuation
(v) Self actuation

Combinations of these methods may also be used.

From the practical point of view manual and mechanical operation can be considered basically similar in that a mechanical force is applied directly to the moving member of the valve. With lever movements the mechanical advantage can readily be adjusted. However, the particular application may demand careful consideration of the working conditions, and the possible necessity of incorporating spring detents to hold a valve in a particular position until reversed by the next movement on the part of the operator.

Solenoid operation is an electro-mechanical system. The solenoid can be mounted directly on the valve, or directly connected to it by a lever, so that when an electrical switch is closed the solenoid is energized to apply direct force to the valve movement. This is *direct* solenoid actuation.

Double-solenoid actuation may be necessary with certain valves, *eg* three-position valves. If the valve is spring controlled, one solenoid is energized to shift it to one position and the second to switch it to the other position.

Double solenoids may also be used for *momentary* working where a valve position has to be held for an appreciable time. To reduce the current drain and solenoid heating, the valve is shifted by momentary solenoid actuation, the movement being mechanically latched. The valve then remains in this position until unlatched by momentary actuation of a second solenoid releasing the catch, when it returns under spring action.

An alternative form is *solenoid pilot actuation*. Here the solenoid movement is utilized to open a small *pilot* valve, which then directs air pressure onto the main valve piston to initiate main valve movement. With this system a relatively small solenoid can be used to actuate quite a large valve. Another advantage is that the pilot (and its operating solenoid) does not necessarily have to be close to the main valve, but can be mounted at some remote and more convenient point to provide optimum location of both pilot and main valves. A solenoid and small pilot valve, when assembled as an integral unit, are generally referred to as *solenoid pilot*.

In the case of air operation this may be direct, indirect, or pilot operated. With a direct air-operated valve, signal pressure is applied directly to a piston formed by an end cap or full-bore section on the end(s) of the valve spindle. Air-operated movement can be in one direction (with spring return), or in both directions. Either method is suitable for establishing two-position movements. For three-position movements a central position can be given by applying signal pressure to both ends simultaneously, releasing pressure on alternative sides, as required.

The signal pressure required is generally substantially less than the line pressure carried by the valve. Thus, lower signalling pressures can be used in many cases. However, if the signal pressure available is too low, the valve may be operated indirectly by a low pressure air signal through a suitable actuator, such as a diaphragm device. Again, one-direction or two-direction actuation is possible, by feeding signal air pressure to one or both sides of the diaphragm, respectively, with a central position obtained by pressurizing both sides of the diaphragm simultaneously.

Maxam 1/8in BSP panel mounted pilot valves.

TABLE I – ALTERNATIVE CONFIGURATIONS OF SOLENOID PILOT VALVES

Configuration	Normal Condition	Spring Action	Spring Strength	Remarks
TWO-WAY	Closed	Holding plunger against pressure force	High	Requires relatively large coil and current
	Open	Return plunger only	Light	High pressure/flow for same size coil and current
THREE-WAY	Closed	Holding only, no pressure force present	Very Light	Highest pressure/flow for same coil size and current
	Open	Return plunger only	Light	Directly comparable with 2-way, open configuration

Basically, pilot operation, or air-pilot operation, as it is more correctly described, provides the main valve with a pneumatic actuator, capable of moving the valve when the low pressure pilot signal is applied. The Pilot valves used are normally three-way or four-way configuration, connecting the two chambers alternately to pressure and exhaust for pressurized operation with plug removed. Alternatively, exhaust operation would require the use of two two-way pilot valves which allow the chambers to exhaust when opened. In this configuration the stopper is removed and the chambers are permanently pressurized through the main inlet port.

Air-actuated valves can also be designed to provide a delay in shifting the valve after receipt of pilot air ('timed-in') or cessation of the pilot supply ('timed-out'). For timed-in operation, air is admitted slowly so that a few seconds are required for enough air to be accumulated in the head to shift the valve. For timed out operation the process is reversed, causing a delay in the exhausting of the "pilot" air from the head. The valve thus remains actuated for a few seconds after the "pilot" air supply is turned off. The delay mechanism can be adjusted to provide delays of varying duration. The possibility of having an automatic sequence of operations without the need for operator control is obvious with this type of valve. "Pilot" air can be supplied automatically by some other operation of the machine and the subsequent step can be delayed for a pre-set period of time.

Valves may be produced as integral units, or designed for mounting on a suitable common base-plate or on a manifold. In the first case they may be generally described as 'line mounted' or 'in line mounting'. The other two forms are referred to as subplate or baseplate mounting, and manifold mounting, respectively.

With subplate or manifold mounting the valve body normally has no tapped pipe connections. All internal ports are brought to the base of the valve and connection is completed by mounting on a matching subplate or manifold carrying corresponding ports. Joints in such cases are normally sealed with gaskets or O-rings.

Subplate mounts may be designed to bolt or stack together to form complete modules without the need for interconnecting piping. Similarly, manifolds may be designed to accommodate a given number of valves.

See also chapter on *Pneumatic Circuits* and *Data Section* for standard symbols.

Pneumatic Circuits

A PNEUMATIC circuit is normally distinguished by being open-ended, ie the compressed air is exhausted to the atmosphere after passing through the actuator. In practice exhaust facilities are usually provided by the control valve or valves employed in the circuit, requiring only a single line connection between valve and actuator. It is also a general character of such circuits that no power is consumed in a system under pressure with no motion, unless there is a deliberate bleed or accidental leakage. Here depressurization is not necessarily required to save power after a movement.

Actuators normally comprise air cylinders where linear motions are required, or air motors for rotary motions and drives. The variety of circuits which can be derived in the former case is virtually infinite, ranging from simple stroking motions with manual control to fully automatic working allied to sophisticated automatic or programmed control systems.

The basic circuit for the simple control of a single-acting cylinder is shown in Fig 1. Control is provided by a 3-way valve capable of directing the cylinder connection either to supply or exhaust. The action is obvious and straightforward, although in a complete circuit the control valve itself may be manipulated in such a manner as to provide automatically timed movements, sequencing, etc, as necessary.

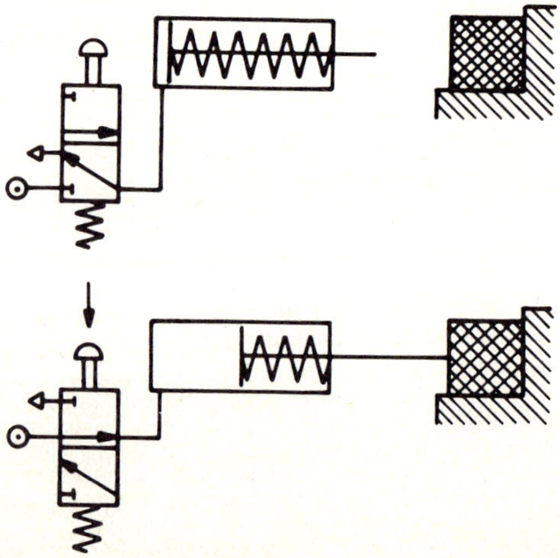

Fig 1

Note that circuits are always drawn so that they show the neutral or 'at rest' position of the various circuit elements ie valves, in their normal (unoperated) position and actuators in their un-actuated position. This particular circuit diagram shows a manually operated valve with two stable positions. If the valve was fitted with a spring for mechanical return, the cylinder would return to its original position as soon as normal actuation of the valve ceases.

The basic circuit shown in Fig 2 shows two 3-way valves used to operate a double-acting cylinder. If necessary, one-way restrictors can be mounted in either or both lines to provide speed control when extending or retracting, or in both directions of motion. With both valves returned to neutral the cylinder is depressurized.

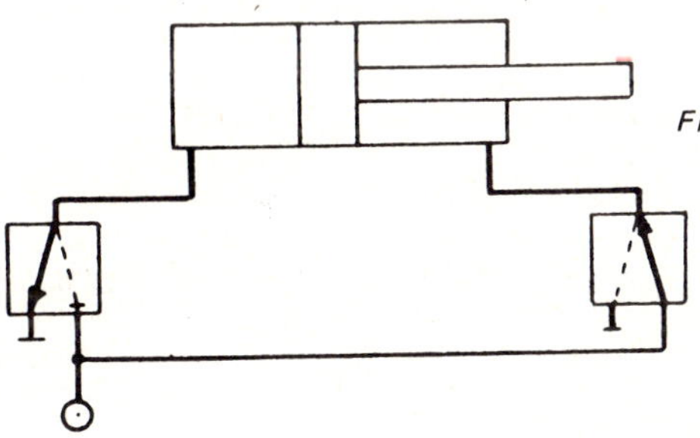

Fig 2

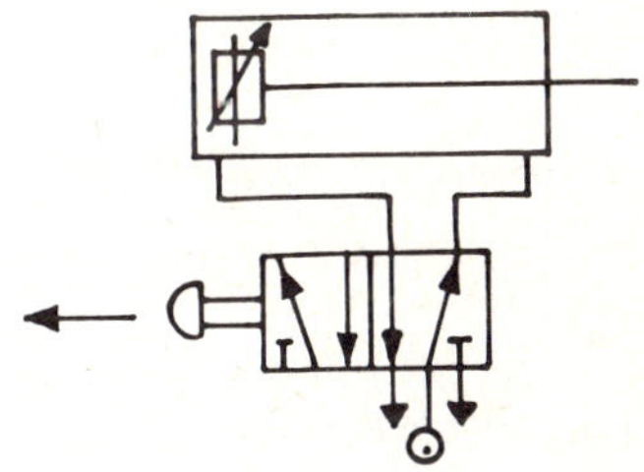

Fig 3

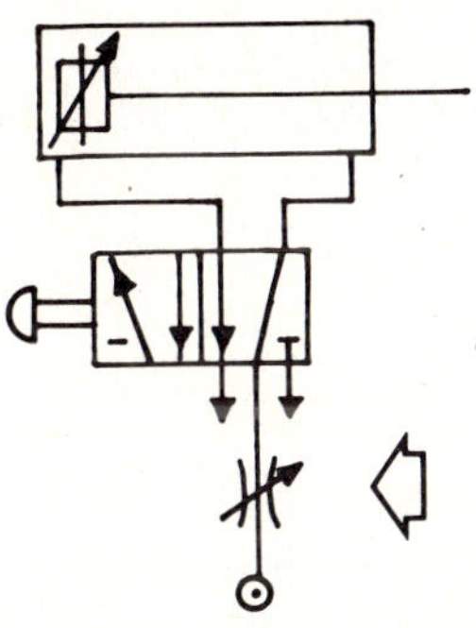

Fig 4

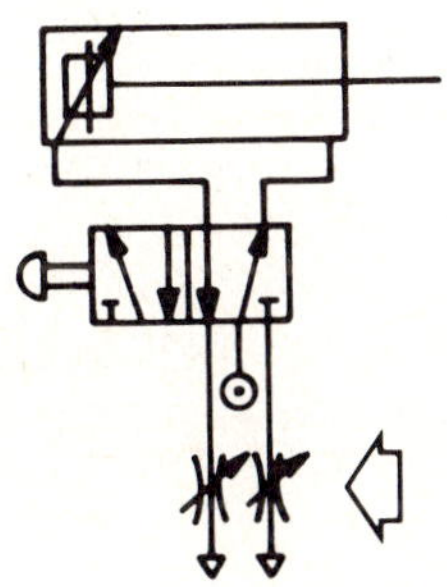

Fig 5

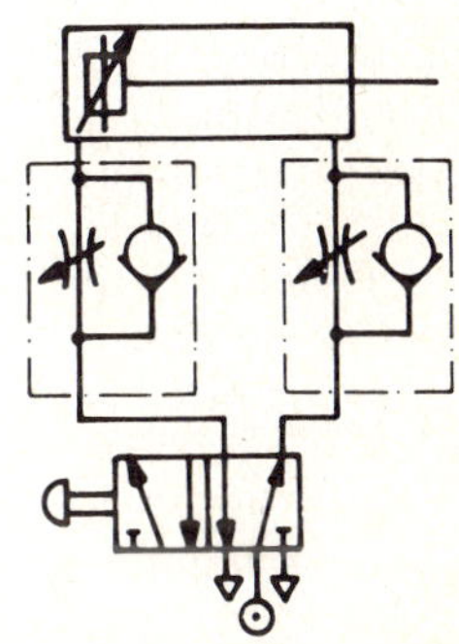

Fig 6

More usually a double-acting cylinder would be controlled by a two-position 5-way valve, as in Fig 3. Fig 4 shows this circuit extended to embrace speed control with a restrictor placed prior to the valve in the supply line, and Fig 5 with restrictors in the outlet. The latter has the advantage that speed can be regulated in either direction independently. Again, in practice, a better solution is usually to regulate the exhaust air from the cylinder with one-way restriction valves, as in Fig 6.

Automatic return in the case of a manually operated control valve can be initiated by spring return of the valve. If the control valve is pilot operated (ie from a separate pilot valve), automatic return can be provided by an additional mechanically operated return valve – as in Fig 7.

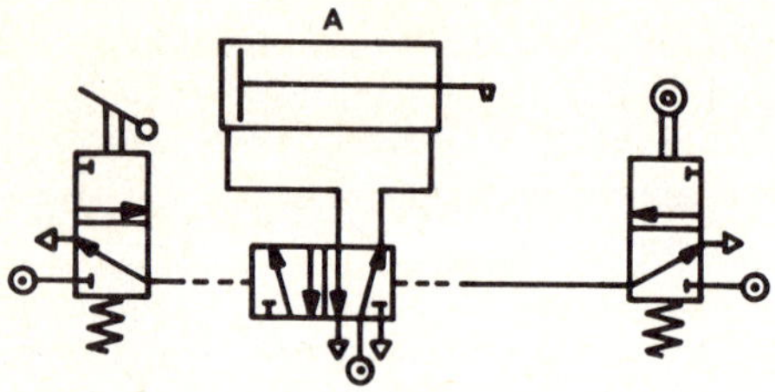

Fig 7

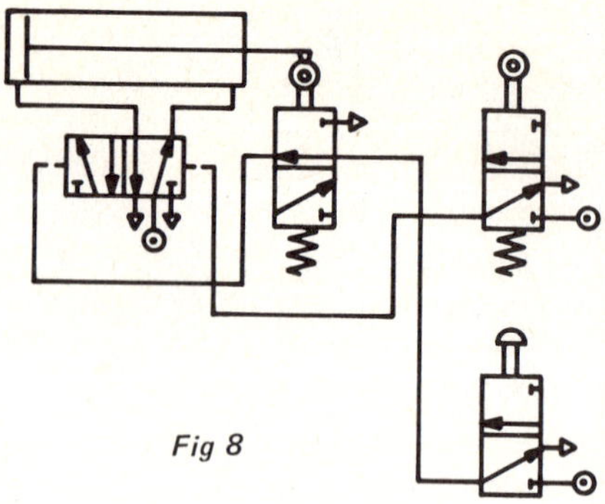

Fig 8

To extend cylinder coupling to automatic reciprocation, an additional pilot valve is needed, as shown in Fig 8. Using a lever-operated start-stop valve, this has a further provision that if this valve is returned to stop, the supply is cut off and the piston stops in the extended position. With the start-stop valve left in the 'on' position, the cylinder will continue to reciprocate.

Numerous other variations are possible on this type of circuitry. Mechanically tripped valves offer simplification of the circuit since they need no separate controlling air supply.

Two faults can readily arise in the design of more complicated pneumatic circuits. The first is 'over design' where the system is unnecessarily complicated, or incorporates an excessive number of components (some of which may actually be redundant when considered in paired action). If the problem is re-thought, rather than the original concept pursued, resulting in increasing complexity, a very much simpler solution will often emerge. The second is the possibility of the paper design producing a locked circuit, or definite circuit fault, by omitting to appreciate that a signal can remain uncancelled in a particular part of the circuit, with locking effect. Ironically, this often arises from a determination to avoid the fault of 'over design', and can be worse in the final result. An 'over designed' circuit will perform the functions required, but an 'over simplified' circuit may not work in practice.

Three-Position Valves

Three-position valves can be used either to provide 'locking' or 'relieving' of a cylinder. A basic 'locking' circuit is shown in Fig 9, and a basic 'relieving' circuit in Fig 10.

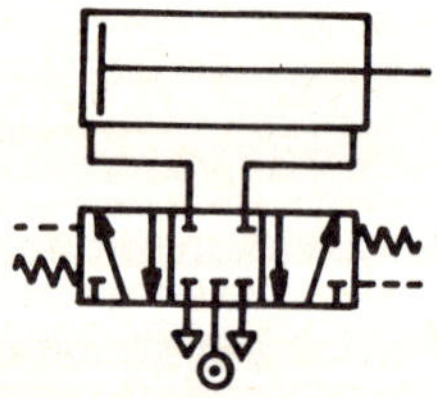

Fig 9

Fig 10

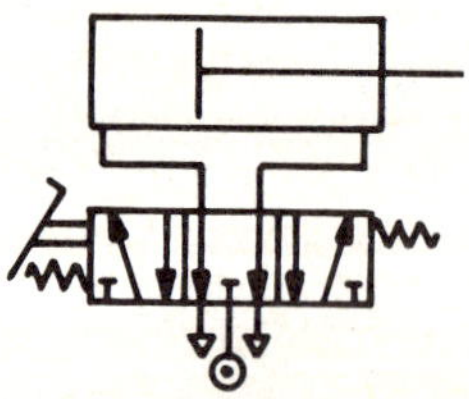

'Two-hand Release'

This description is given to the type of circuit used in presses and similar heavy-duty actuators to eliminate the risk of starting when the operator has his hands in a particularly dangerous position. To eliminate such risk, the circuit is designed so that actuation is only possible when both hands are placed on control valves.

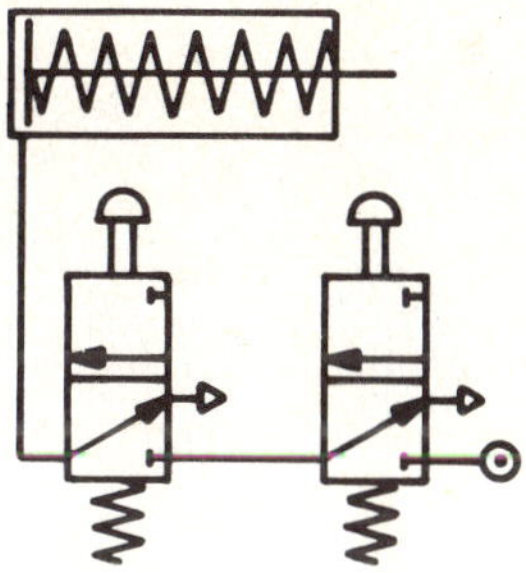

Fig 11

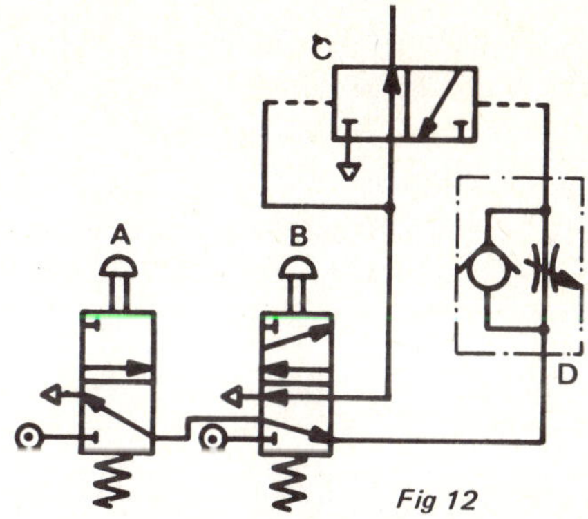

Fig 12

The basic circuit shown in Fig 11 is not a satisfactory solution since the circuit is capable of single handed actuation if one of the valves is pneumatically actuated. The actuator circuit shown in Fig 12 is 100 per cent safe, working as follows:-

If only A is actuated, the air will go to C via B and D at which D switches, which only causes the out-signal to be bled in C and not through B as before. If only B is actuated, the air goes to C via D at which C switches, and the out-signal is bled through C and not through B–A. If A and B are actuated simultaneously, the two pipes to C are aired, but due to the restriction D the signal from A–B will always be first and the out-signal is a fact.

The out-signal can be directly connected to a single-acting cylinder. If a double-acting cylinder is to be controlled, the out-signal has to be connected to a 5-port working valve with mechanical spring return (monostable operating valve) that controls the double-acting cylinder.

The above, and the following examples are by Atlas Copco.

Stop Devices

In Fig 8 the working unit will stop in its minus position when the control valve is closed since the position transmitting valve will then not receive any air. If the control valve were instead connected to the position transmitting valve in the plus position, the working cycle would stop in that position.

If the unit is only to finish the movement started when the stop signal is given, the air to both the position transmitting valves has to pass through the control valve as per the diagram in Fig 13a.

If the cylinder is to adopt its minus position regardless if it moves plus or minus, the connection is to be as shown in Fig 13b.

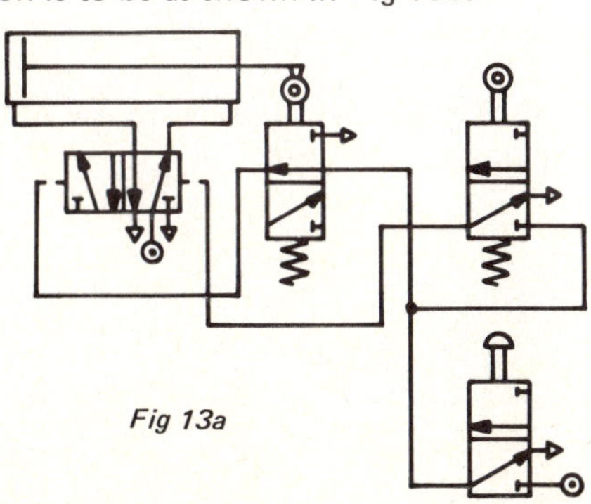

Fig 13a

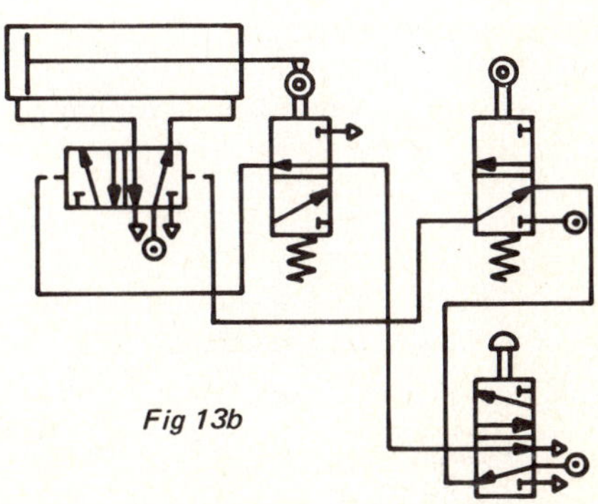

Fig 13b

If the cylinder is to stop instantly, the inlet air to the working valve can be cut off, but this brings the following disadvantages:

No control of the position in which the cylinder stops

No force on the cylinder

No control of the movement when the process is started again

These disadvantages can be eliminated if the connection is done according to Fig 14.

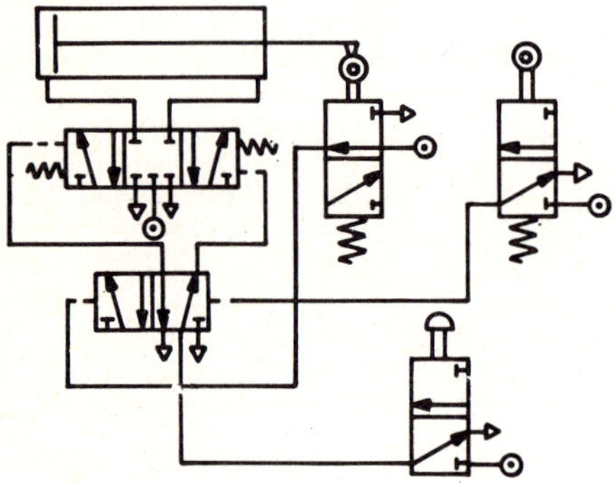

Fig 14

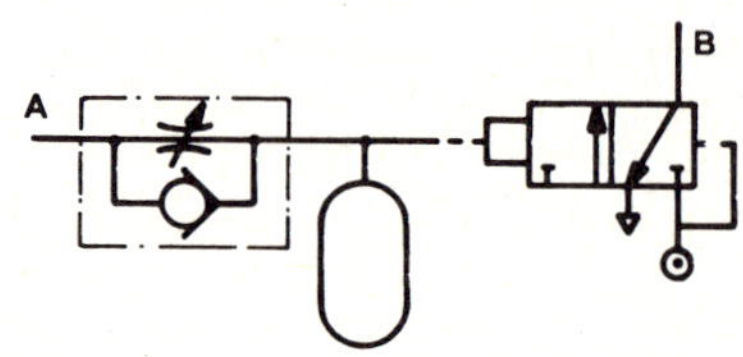

Fig 15

Time Delay

If a cylinder, for example, is to stop in its plus position for a certain period of time before it returns or if, for construction reasons, a position transmitting valve cannot be placed at the plus position, a time-delaying circuit can be used. Such a circuit is obtained by means of a one-way-restriction valve, an air reservoir and a differential valve Fig 15.

The circuit functions so that when air is connected (in-signal) owing to the restriction valve, a certain time will pass before the pressure in the reservoir becomes so high that the differential valve switches. The delayed out-signal is then received. When the in-signal ceases, the differential valve is reset, as the air in the reservoir is exhausted via the check valve. Time delays from 0 to 5 minutes can be obtained with this type of circuit by adjusting the restriction valve and the volume of the reservoir.

An alternative to the time circuit shown in Fig 15 is to regulate the pressure on the smallest area of the differential valve which gives one further setting possibilitv Fig 16.

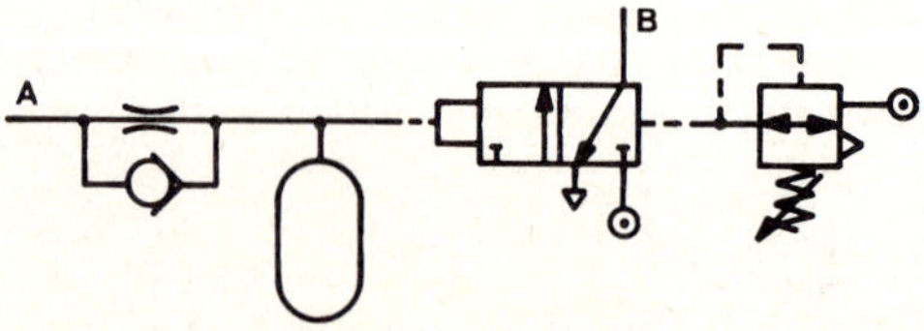

Fig 16

The circuit diagram Fig 17 shows a cylinder that remains in its plus position for a certain time because of the time circuit that delays the acknowledgement signal from the position transmitting valve in the plus position. And the connection diagram, Fig 18 shows a cylinder that returns automatically after a certain time, after having received a signal to go plus.

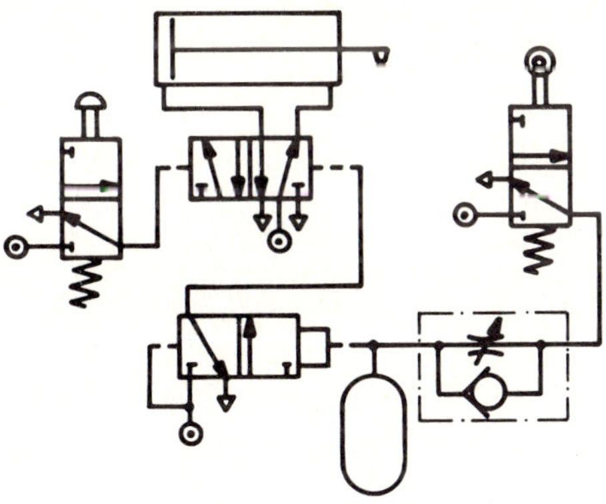

Fig 17

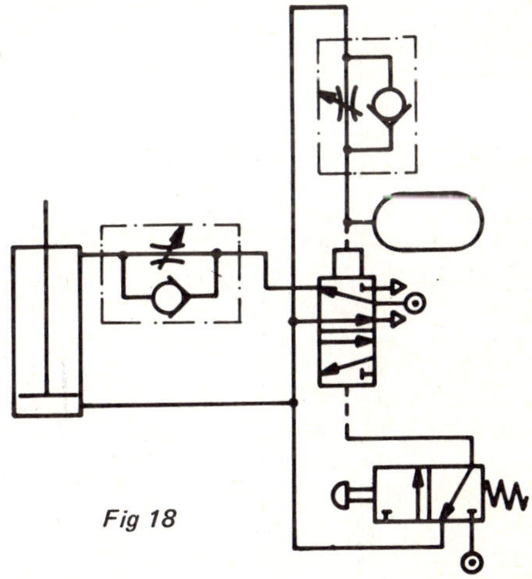

Fig 18

The connection system works as follows:

When the control valve is actuated, air runs to the working valve that switches and the air flows to the plus chamber of the cylinder. At the same time the air flows to the time circuit which, after a certain time, makes the working valve switch and the cylinder goes minus. In order to obtain a distinct switch of the working valve, the system is provided with a back feed of the air for the plus movement of the cylinder via the control valve to the working valve. This back feed makes it necessary for the working valve to be of differential type as can be seen from the Fig 18.

Pulse Circuit

There is often a demand for a signal to be cut out after a certain time, regardless of the duration of the in-signal, eg for cleaning. This is achieved in the way shown in Fig 19.

When the in-signal A comes, it goes directly through the valve which is of "normally open" type. At the same time, air flows via the restriction valve to the actuating part of the valve. Depending on the restriction and the volume after the restriction valve (normally no extra volume is needed, as the volume of the conduit is sufficient), the valve switches after a certain time and the out-signal B ceases, irrespectively of whether the in-signal A remains. When A disappears, the signal conduit is bled through the check valve.

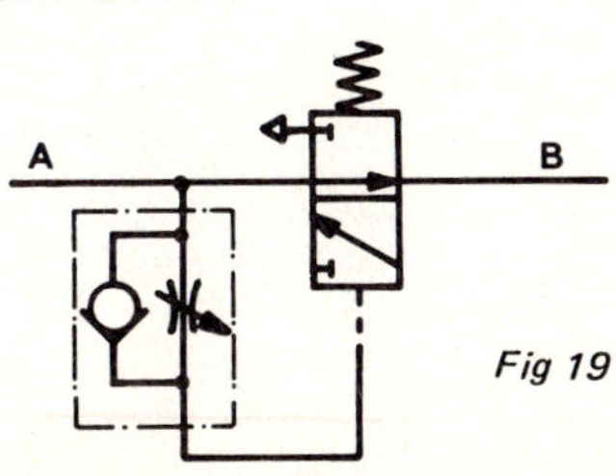

Fig 19

Control by Pressure Drop

In Fig 18 an example was shown of how to make a cylinder return without using a position tran mitting valve in the plus position. Another way of obtaining return without a position valve is t sense the pressure in the minus chamber of the cylinder. This is done by inserting a 3-port diffe ential valve, the larger area of which is connected to the de-airing minus chamber Fig 20.

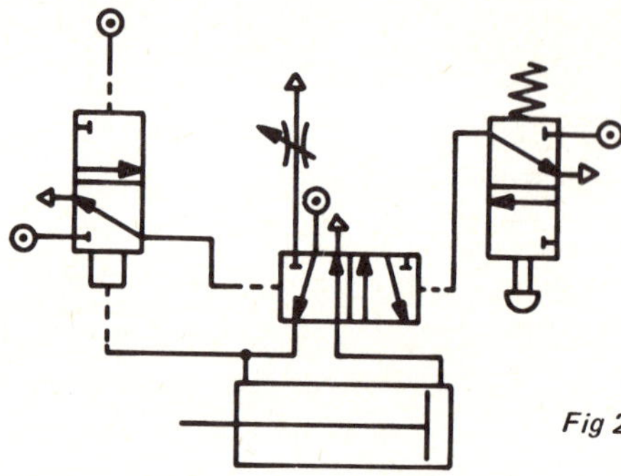

Fig 20

When the minus chamber is under pressure ie during the minus movement, and when the pisto is in its minus position, there is an overpressure in the minus chamber. But, also during the pl movement, there is a certain overpressure in the minus chamber, and consequently the differe tial valve will also be actuated during the plus stroke.

To avoid the pressure in the minus chamber decreasing too fast and thus returning the cylind before it has reached the plus position, a restriction valve has to be connected to the cylinder co nection where the pressure is indicated. When the piston reaches its plus position, or continue movement is otherwise obstructed, the pressure in the minus chamber decreases and the diffe ential valve switches, as does the 5-port valve which controls the cylinder that goes minus. Th pressure indication described can, of course, be used in both cylinder directions.

In this kind of control, the cylinder should not be loaded with more than 50% of th theoretical cylinder force.

Control of multi-position cylinders

The connection diagram Fig 21 is valid for control of 3-position cylinders and Fig 22 fo 4-position cylinders.

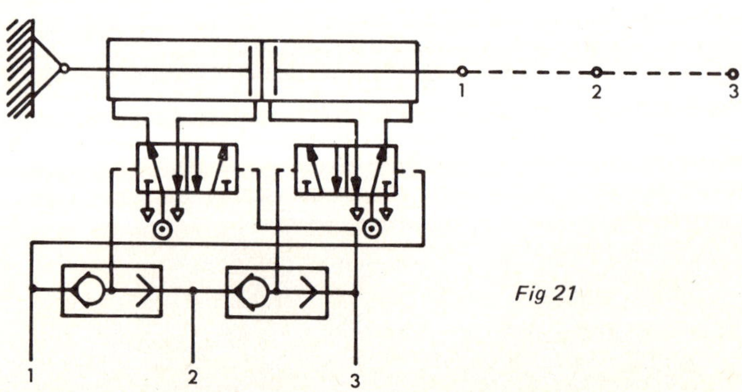

Fig 21

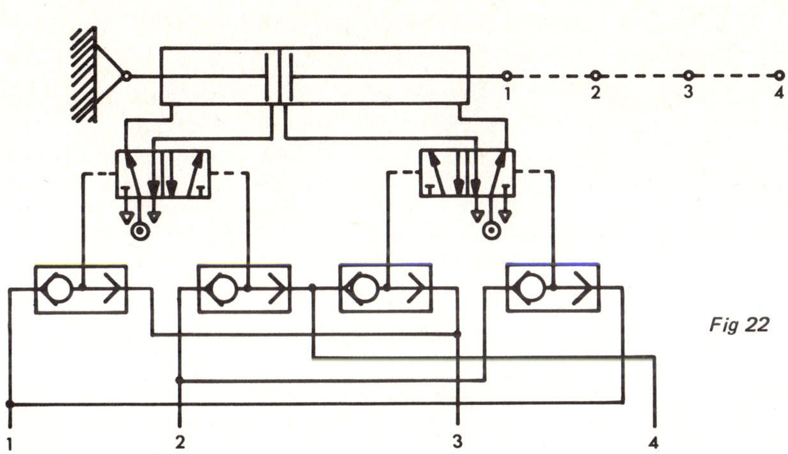

Fig 22

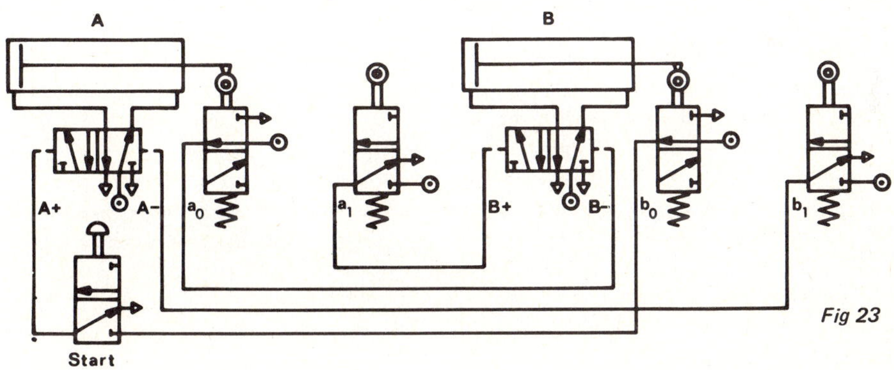

Fig 23

With the connections shown in the figures there is an in-signal for each position the piston can opt. Irrespective of the position of the piston, a signal to a certain inlet will result in the piston d adopting the position of the in-signal concerned.

Fig 23 shows a more complex circuit where two cylinders are to be operated in sequence +, B+, A–, B–. Circuit design becomes increasingly complex when an additional function is quired – eg to obtain the following sequence A+, B+, B–, A–. Such a circuit is shown in Fig 24, sed on the initial premise:

(i) a_0 is connected via "Start" to A+

(ii) a_1 is connected to B+

(iii) b_1 is connected to B–

(iv) b_0 is connected to A–

To achieve the desired sequence an auxiliary function is to take away signal b_0 from A– fore A+ demands a signal. Further, is to take away a_1 from B+ before B– demands a signal.

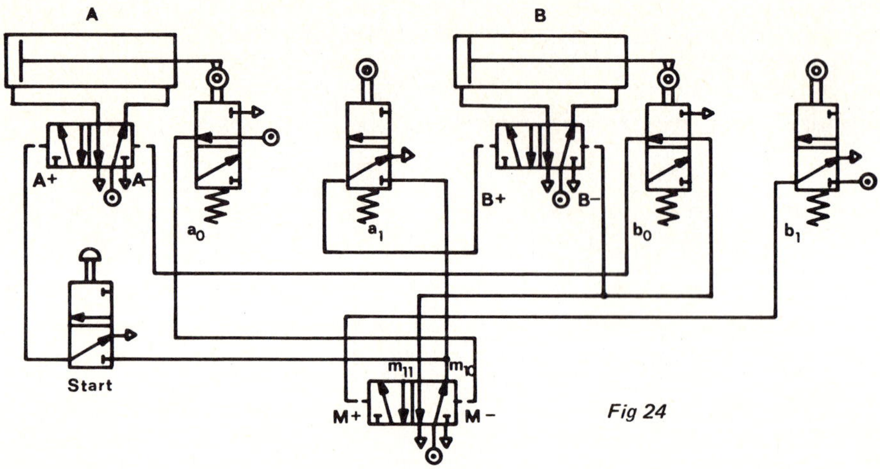

Fig 24

This auxiliary function can be performed by a 5-port 2-position double air-controlled valve as follows: air is supplied to b_0 while a_1 is bled in one position of the 5-port valve, in the other position, air is supplied to a_1 while b_0 is bled.

If the connections are made so that when cylinder A comes minus, a_0 gives a signal to the side of the auxiliary function that bleeds b_0, and when cylinder B comes plus, b_1 gives a signal to the side that bleeds a_1, the demand is met that b_0 shall be taken away from A– before A+ gets a signal, and that a_1 shall be taken away from B+ before B– gets a signal.

For more complicated sequences it may be necessary to insert further auxiliary valves. A major problem often involved is to appreciate all the blocking signals that have to be broken. A practical method of solving sequence problems is described in the chapter *Low Cost Automation.*

Hydro-Pneumatic Circuits

THE principal application of hydro-pneumatic circuits is for movement stabilization, a basic circuit being shown in Fig 1. Here the air is the directly working medium, with the hydraulic circuit providing clamping for uniform piston movement and speed control via the restrictor valve in the hydraulic circuit.

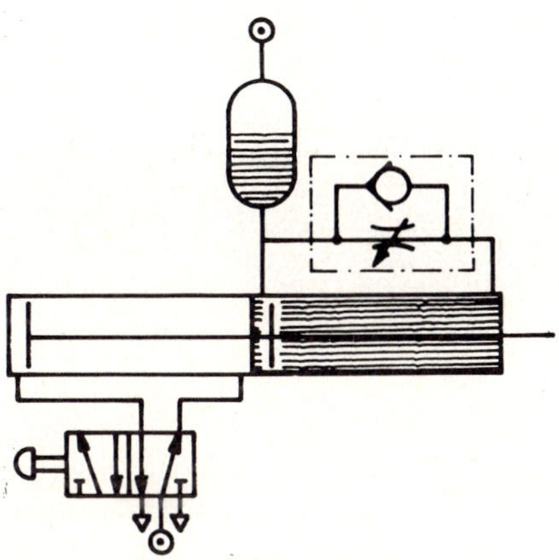

Fig 1

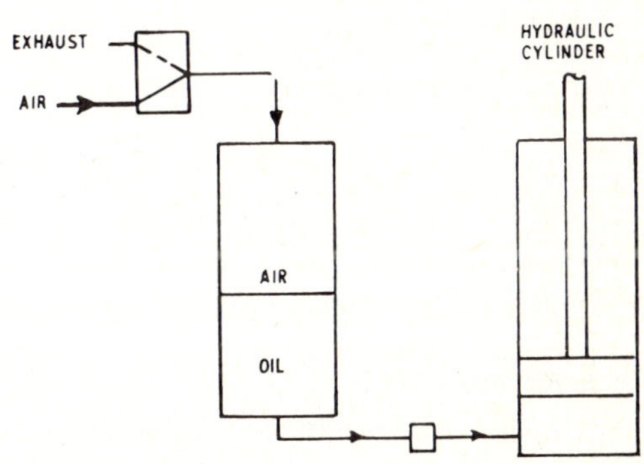

Fig 2

An alternative basic is shown in Fig 2 which differs in that oil is now the working medium, with air only working indirectly. Such a system can be extended to double-acting motion by using a second air/oil reservoir and control valve, as shown in Fig 3. This is further shown in the form of a continuous circuit diagram in Fig 4. Speed control is possible in air in both directions by inserting air-way restrictor valves between the air/oil reservoirs and the hydraulic cylinder.

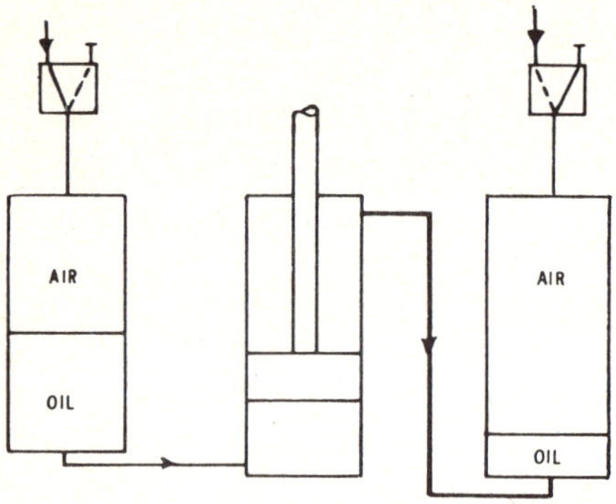

Fig 3

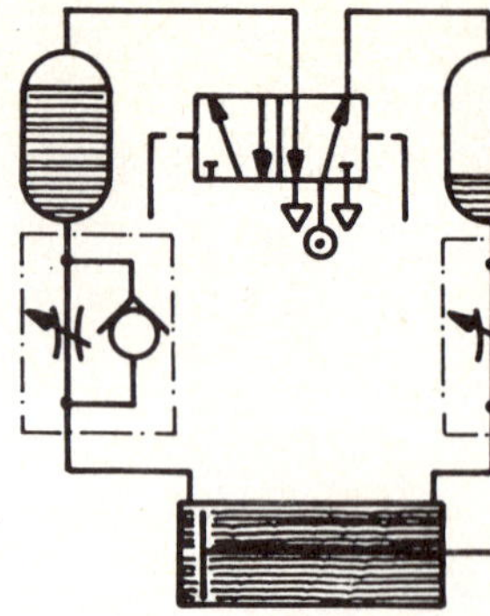

Fig 4

Hydro-check Cylinders

Two examples of basic hydro-pneumatic cylinder circuitry are shown in Fig 5. In the left hand diagram speed is controlled in the + stroke of the cylinder. If the – stroke is to be controlled, the non-return restrictor valve is inserted in the other hydraulic cylinder line.

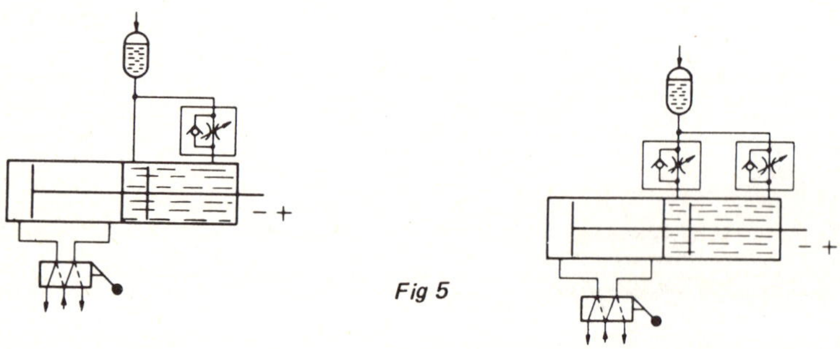

Fig 5

The second diagram shows the same circuit adapted to speed control in both directions. However, one throttle valve may substitute for both non-return restrictor valves where equal speed in either direction is called for, or can be accepted.

Air Hydraulic Conversion

In certain applications requiring rigidity of movement, a hydraulic cylinder can be employed connected indirectly to the compressor air supply thus providing air-hydraulic working without the necessity of using an air-hydraulic cylinder.

The basic arrangement is shown in Fig 6. For small cylinders it may be possible to use the filter bowls as oil reservoirs. If the working volume of the hydraulic cylinder exceeds two thirds of the capacity of the filter bowls, however, it will be necessary to add two oil reservoirs.

The following are important points to note to achieve satisfactory working:-

(i) The hydraulic cylinder should be a low-pressure type with low 'break-out' force and friction. Internal leakage must be minimal, which may call for the use of special seals.

(ii) Filters should be used, even if reservoirs are necessary for additional oil capacity, since these act as baffles to prevent emulsification of the oil.

(iii) If filters only are used, (ie if they provide adequate oil capacity), the oil level in either filter bowl should not be allowed to rise above the bottom of the filter element, otherwise oil may be lost through the exhaust port of the main valve.

(iv) Reservoirs should always be mounted at a higher level than the cylinder to provide self-bleeding action when filling.

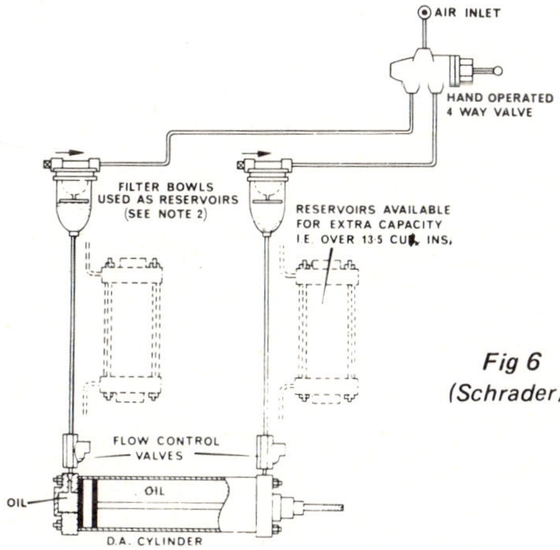

Fig 6
(Schrader)

Air Hydraulic Pumps

The same principle that operates the air-hydro pump can be used to pressurize a hydraulic system, either for general working or for pressure test use. Pressures up to 1000 bar (15000 lb/in^2) can be generated for the latter purpose – very much higher pressures can be generated by specially designed pumps.

Small air-hydraulic pumps of this type are widely favoured as direct replacements for normal hydraulic power, both in portable and fixed installations. Air supply can be from compressed air mains, or bottled air (or nitrogen). The fitting of an air pressure regulator enables the hydraulic pressure to be adjusted over a wide range, dependent on the physical size and geometry of the pump.

Air-Hydro Pump

The Madan 'Ghurka' air-hydro pump operated by compressed air (minimum 10 lb/in^2, maximum 100 lb/in^2). Constructionally it consists of a top double-acting air cylinder connected to a hydraulic piston in a bottom cylinder, the two cylinders forming an integral unit. The double-acting air cylinder is worked as a linear motor by a changeover mechanism, thus producing reciprocation of the hydraulic cylinder which forms the pump.

Fluid outputs of up to 55500 l/hr (12000 gal/hr) are obtainable, depending on the size of the unit. The air side of the pump can be provided with a pressure regulator enabling the inlet air pressure to be raised. This controls the speed and output of the pump, as well as the hydraulic pressure generated. Performance characteristics are given in Fig 7.

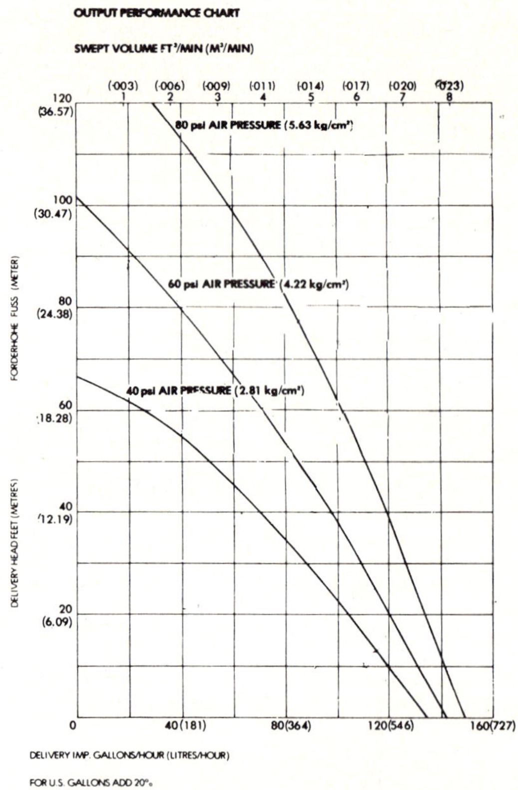

Fig 7 Madan 'Ghurka' standard air-hydro pump and performance chart.

The particular advantage of an air-hydro pump is that it is entirely flameproof and can be used to pump flammable fluids. It is also generally cheaper than a centrifugal pump of equivalent performance.

Typical applications include:-

Pollution disposal, Pressurizing engine bearings, Garage petrol tank cleaning, Transfer of bulk liquids to header tanks, Sludge dispersal from engine sumps, Tank and bilge cleaning, Low pressure testing, Civil engineering projects.

SECTION 5

Hardware

Air Tools

THE majority of roadbreaking, contractors' and mining tools powered by compressed air are of the percussive or reciprocating type. They work on the principle of impingement of compressed air on alternate sides of a piston confined within a cylinder, the reciprocating or oscillating motion of the piston being used to actuate the tool. The tool itself may either be an extension of the piston or a rigid attachment to the piston, as in the case of hammer-type tools; or the work head may form a separate part of the tool detached from the piston and free to move independently of the head, as in the case of rock drills, riveters and the bulk of the heavier percussive tools. The majority of modern rock drills, etc, also incorporate a self-rotating action, and thus differ in this respect from concrete breakers.

Percussive Tools

The theoretical energy (e) per blow delivered by a percussive tool is

$$e = \frac{m \cdot v^2}{2} \text{ Nm}$$

where m is the mass of the piston
v is the velocitv with which the piston strikes the tool shank

The theoretical power delivered, at an impact rate of n/s, is thus

$$\text{power} = \frac{m \cdot v^2 n}{2s} \text{ Nm}$$

In practice, energy per blow is usually the most important parameter, since this needs to be selected to suit the work requirements, but the impact rate may also be significant because of its effect on material characteristics (eg in riveting). The stroke of the piston may also be a significant factor in the design of the tool in order to obtain the required output. The impact energy achieved may also be limited by the amount of recoil which can be tolerated without making the tool difficult to handle.

In order to produce oscillation of a piston in a percussion type of tool suitable porting or valving is necessary to deliver air under pressure to alternate sides of the piston. In the case of light tools, and where a high rate of oscillation is desirable (as in scaling hammers and dressing hammers), compressed air intake and exhaust may be achieved via ports in the cylinder wall, the openings of which are controlled by movements of the piston. Such types are sometimes referred to as valveless tools. For heavier tools or where better control is required, mechanical valving is normally employed, of which there is a considerable variety in design. In general, however, the three main types are:-

(i) Thimble or tubular valves

(ii) Plate valves

(iii) Flapper valves

Tubular valves are extremely straightforward in construction and work on the 'compression throw' principle with a high flow efficiency. Their chief limitation is that their relatively high mass limits the speed at which satisfactory valve action can be maintained, hence they are less suited for fast hitting hammers. Plate type (circular disc) and flat type valves are generally quicker acting, but tend to have a somewhat lower flow efficiency. The type and design of the valve chosen is, therefore, usually representative of the manufacturer's own solution to the necessary compromise between flow efficiency and operating speed, consistent with the requirements of the particular tool involved.

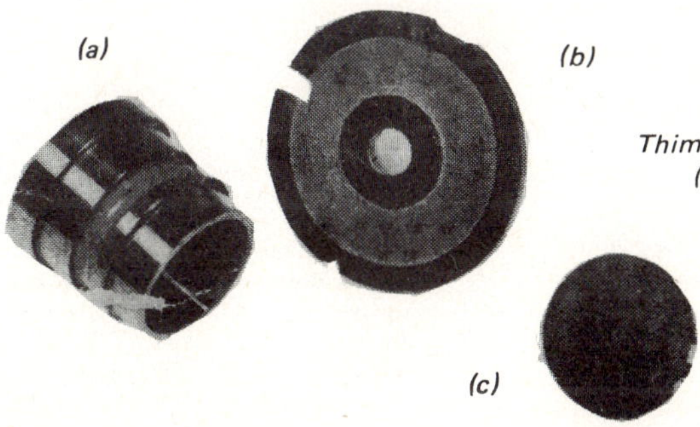

Thimble valve (left), plate valve (centre), flap valve (right).

Rotary Tools

Rotary pneumatic tools form a further important group. These again may be heavy duty tools for contractors' or mining applications, or specific workshop tools such as screwdrivers, nut runners, impact wrenches and grinders.

The drive in the case of rotary tools is normally an air motor, driving a work spindle either directly or through reduction gearing. As with percussive tools, the work spindle may be attached permanently to the motor unit, or may be in the form of a detachable fitting. Thus the same basic air tool may be adapted, through various fittings, to power a variety of rotary work heads. It is more usual, however, to find such tools designed as specific rather than as multi-purpose tools.

In the majority of cases, and particularly with smaller rotary tools, vane motors are normally preferred because of their favourable torque/speed characteristics and simple, compact design. Piston motors may be employed for larger tools where higher powers and torque are required at lower operating speeds. Turbine motors are limited in application by their very high running speeds and low torque and are thus used mainly for lightweight ultra high speed grinders and burrs.

Since there may be considerable differences in size, scope and performance of the various tools produced under the main classifications, a logical breakdown is to group such tools under specific duties or applications in the three main fields:-

(a) Contractors' tools

(b) Mining and rockworking tools

(c) Workshop tools

To complete such a classification further tools and devices can be included which are powered specifically by air motors (as opposed to rotary tools which are built around an air motor as an integral unit). A comprehensive classification on this basis is given in Table I.

TABLE I – CLASSIFICATION OF AIR TOOLS BY PURPOSE

Type	SUB-GROUPING		
	Contractors' Tools	**Mining Tools**	**Workshop Tools**
Percussive	Breakers Diggers Tampers Hammers Spaders Shuttering vibrators Stone working tools Road rippers Pile drivers Picks Steel extractors	Rock drills Coal picks Drifters Stopers Breakers Diggers Spaders Sheeting drivers Spike drivers Steel extractors	Scaling hammers Caulking hammers Chipping hammers Riveting hammers Weld flux scalers Paint scrapers Chisels Vibrators Engraving tools Rivet cutters Rivet busters
Rotary	Shank grinders Concrete vibrators Sump pumps Sludge pumps Conveyors Rock drills	Coal drills Stone drills Vibrators Sludge pumps Sump pumps Air motors	Grinders Pencil grinders Vibrators Screwdrivers Woodborers Deburrers Shears Nibblers Tappers Reamers Nut runners Stud runners Rotary files Burrs Impact wrenches Tube expanders
Machines	Compressors Portable compressors	Compressors Portable compressors	Compressors Portable compressors
Air motor driven	Concrete vibrators Hoists Winches Pumps	Vibrators Riddles Conveyors Loaders Coal tables Pumps Hoists	Vibrators Sifters Hammers Hoists Winches Pumps Tube expanders
Cylinders	Tool feeds Jacks Hoists Air legs	Tool feeds Jacks Hoists	Clamps Vices Forging hammers Fettling grips Chucks Jacks

Contractors' Tools

The majority of tools in this category come under the general heading of breakers. They are basically hammers used in combination with chisels or steels for demolition or cutting on various materials. The shape of the steel will depend on the material to be worked, a pointed steel being used on concrete and similar hard materials, and a wide chisel-edge for softer materials such as asphalt. For digging or trenching work a spade-shaped steel can be fitted, whilst further attachments may adapt the breaker for ramming, pile-driving, etc.

Tools of this type normally range in weight from about 14 to 40 kg (30 to 90 lb), depending on the specific duty intended which also influences the degree of robustness of construction. Arbitrary classifications are 'light', 'medium' and 'heavy' duty.

Heavy breakers are used for demolishing concrete foundations and walls, breaking concrete pavement, trenching in very hard ground, etc. Medium breakers are used for breaking light concrete pavements and floors, tarmac and similar surfaces. Light breakers are used for light demolition work on floors, paving and masonry walls, and as general utility 'trimming' tools.

Air consumption in the case of breakers will vary considerably with size and duty, and may range from 0.93 m^3/min (30 ft^3/min) at 6 bar (80 lb/in^2) for light breakers to 2m^3/min (70 ft^3/min) at 6 bar (80 lb/in^2) for heavier tools.

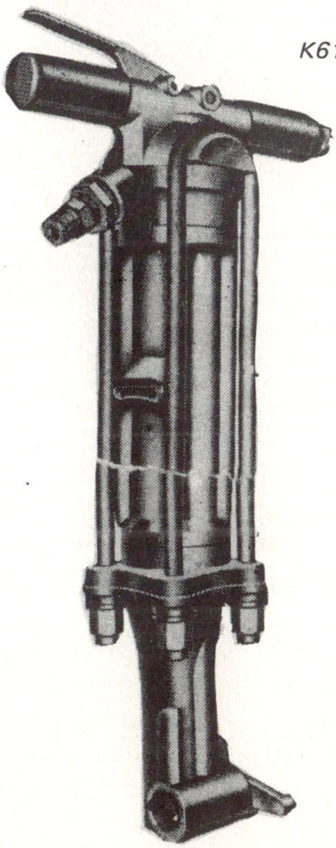

K61 Speedline paving breaker for medium work. (Joy-Sullivan Ltd).

'Universal Drilrig' for deep hole rock drilling. (Holman Bros. Ltd).

All breakers employ solid steels which are driven without rotation and thus provide, in effect, a straightforward chiseling action for breaking, driving or tamping. A wide variety of different tools may be used, depending on the work involved. Fig 1 shows a selection of steels used for road ripping. Further attachments can include pipe or sheeting drivers, spike drivers, light pile drivers, etc, depending on the capability of the tool — eg sheeting and light pile driving would normally be limited to heavy breakers.

Fig 1 Road ripper accessories. Top row (left to right) chisel edge steel (asphalt cutting), wide curved scoop, wide flat spade, digger. Bottom row (left to right) plug and feathers, tamper, chisel pointed steel, frost wedge, pointed steel.

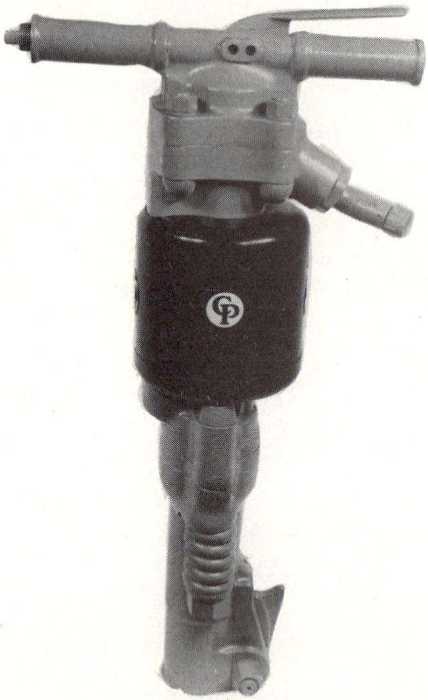

Heavy duty breaker from Consolidated Pneumatic, designed for low noise output.

Various types of tool retainers are used such as a latch or spring as well as link locks and patented retainers. The size of the tool shank varies according to the class of the breaker, eg from 25 mm by 100 mm (1 in by 4¼ in) in the case of light breakers to 30 mm by 150 mm (1¼ in by 6 in) for heavy breakers — see Fig 3.

In the case of breakers, a T-shaped handle is normally employed with the throttle valve conveniently located above one side of the handle. The throttle action normally also operates a built-in lubricator as adequate internal lubrication is an essential feature with such tools. This does not necessarily preclude the use of an air line lubricator. In fact, the additional use of an air line lubricator is generally recommended.

Spaders (Diggers)

When designed specifically as a spader or clay digger the tool is normally smaller than a concrete breaker and generally has a D-shaped handle. Provision may also be made to fit an extension handle so that the tool may be manipulated and controlled when digging narrow trenches. All types of spaders consist essentially of a light to medium duty size hammer body capable of taking a variety of different spade, scoop and chisel shaped steels. Air consumption of a typical light duty spader is usually of the order of 0.7–0.85 m^3/min (25–30 ft^3/min) at 6 bar (80 lb/in^2) and that of a medium duty spader 0.85–1.13 m^3/min (30–40 ft^3/min) at 6 bar (80 lb/in^2).

A spader may also be used for light demolition work, or as a pick, by fitting a moil or chisel point instead of a spade – see Fig 2. It may also be produced in specific forms, viz:-

Clay digger – for digging or trenchwork, fitted with a D-shaped handle and inside or outside throttle control trigger.

Trench digger (Trencher) – with an extended T-shaped handle for open trench evacuation, the length of the handle allowing the operator to stand upright.

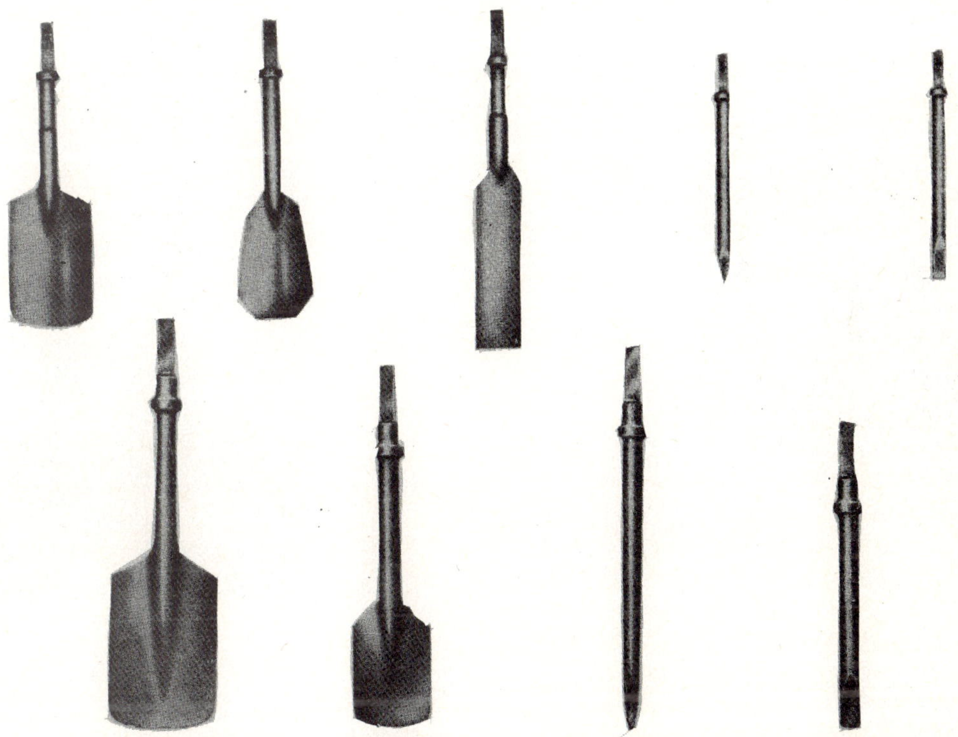

Fig 2 Range of accessories. Top row (left to right) curved scoop, flat spade, digger blade, moil point, chisel point. Bottom row (left to right) curved scoops, moil point, chisel point. (Holman Bros.Ltd).

Tampers

A tamper is a pneumatic hammer fitted with a tamping bar. Size and weight can vary considerably, but a typical tool of this class would weigh some 13.8–18 kg (30–40 lb) with an air consumption of the order of 0.57 m^3/min (20 ft^3/min) at 6 bar (80 lb/in^2).

Special designs of tamper may be used for work in confined areas, also multiple tampers for working over large unrestricted areas. Thus a duplex tamper would comprise two separate tampers, parallel mounted with a common handle, and a triplex tamper three separate tampers similarly mounted. The triplex arrangement is much to be preferred for multiple operation since the tamper blows can be desynchronized to produce denser compaction.

Rammers

A rammer is essentially the same tool used for such work as consolidating back-fill in trenches and excavations and general compacting work which may come under a contractor's specifications. The typical contractors' tool of this type has an air consumption of the order of 1 m^3/min (35 ft^3/min) at 6 bar (80 lb/in^2). Smaller rammers, with correspondingly reduced air consumption, are used for foundry work (see chapter on *Workshop Tools*). Large rammers may be trolley mounted for ease of manipulation.

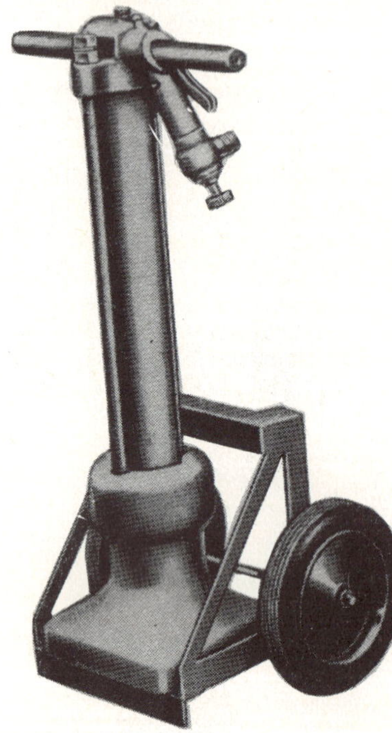

TR2 'Jumping Jack' pneumatic rammer (Broom & Wade Ltd).

C.P.5 pneumatic vibrator (Consolidated Pneumatic Tool Co. Ltd).

Concrete Vibrators

Light, fast-acting pneumatic hammers are used for concrete vibration by being bolted or clamped to the shuttering. They are normally relatively small in size (eg about 4.5 kg (10 lb) weight) and any number of hammers may be attached to the shuttering, depending on the thickness and surface area of concrete involved. Rammers also are sometimes used for concrete vibration although they are less effective than vibrators for this class of work since the latter may have speeds of up to 5 000 blows per minute.

Stone-Working Tools

Special stone-working tools for light or medium dressing range in weight up to about 2.7 kg (6 lb) with air consumption figures from 0.7–0.85 m^3/min (5–10 ft^3/min) at 6 bar (80 lb/in^2). They are normally high-speed hammers (often of valveless construction) and the tool tip may be permanently attached and hardfaced, or detachable. Slightly larger tools of this type may be used for roughing and bushing, with air consumption figures of up to 0.42 m^3/min (15 ft^3/min) at 6 bar (80 lb/in^2).

In addition, contractors' work may involve the use of tools more properly classified under mining tools, or rock drills.

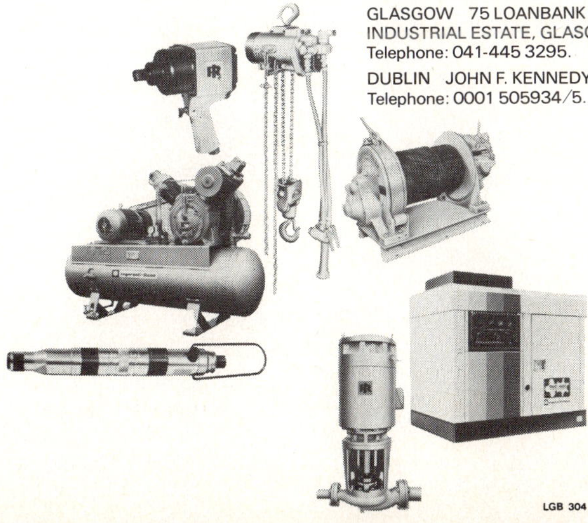

Thors

TOOLS DIVISION

STEWART WARNER LTD.,
TYNEMOUTH, TYNE & WEAR, NE24 07E.
Tel: North Shields (08945) 73181 Telex; 53602
Telegrams: Thortools Tynemouth Tyne & Wear

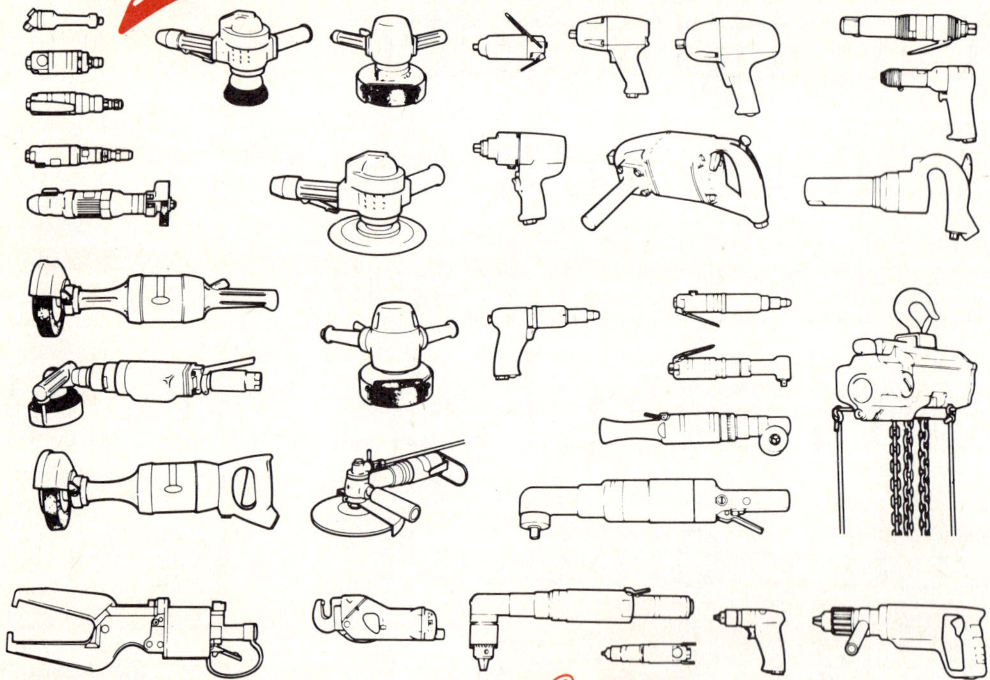

AIR OPERATED TOOLS
. . . for Industrial requirements . . .

Thors *. . . known the world over . . .* Grinders, Screwdrivers, Nutsetters, Tappers, Impact Wrenches, Drills, Hammers, Hoists, Suspension Balancers . . .

LES OUTILS PNEUMATIQUES
. . . adaptes aux besoins de l'Industrie . . .

Thors *. . . connus dans le monde entier . . .* Meuleuses, Visseuses, Boulonneuses, Taraudeuses, Cles a Chocs, Perceuses, Marteaux, Palans — Equilibreurs de Suspension . . .

DRUCKLUFT WERKZEUGE
. . . in allen Produktionsbereichen der Industrie . . .

Thors *. . . weltweit bekannt . . .* Schleifmaschinen, Drehschrauber, Winkelschrauber, Gewindeschneider, Schlagschrauber, Bohrmaschinen, Meisselhammer, Hebezeuge, Federzuge . . .

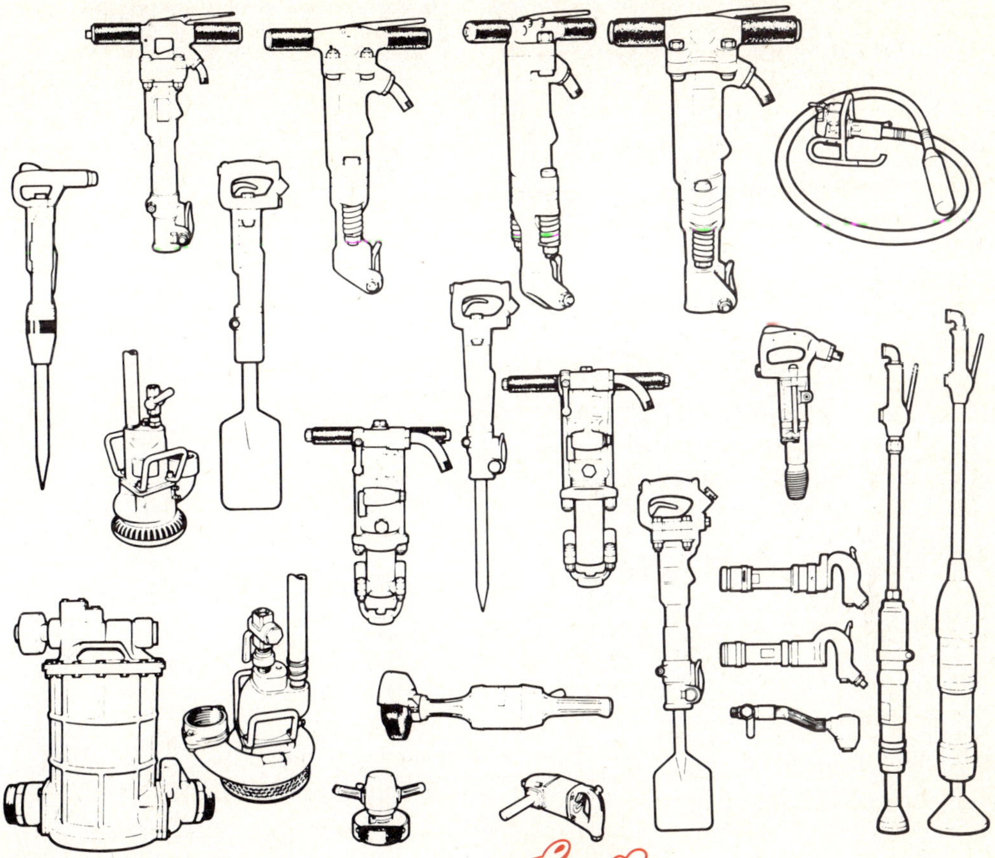

Mining Tools

The general heading 'mining' embraces underground work in mines (coal and ore mining, etc) and also quarrying and similar work involving the cutting of stone, rock, etc. The principal tools involved are rock drills and picks, although various types of contractors' tools may also be employed, notably concrete breakers and spaders or diggers, and possibly rammers.

As a rough classification, portable or handheld rock drills can again be grouped as 'light', 'medium' and 'heavy', typical weights being 16 kg (35 lb), 20 kg (45 lb) and 25 kg (55 lb) respectively. Heavier rock drills of 34 kg (75 lb) weight or more are sometimes referred to as 'sinkers', particularly when employed for shaft sinking.

Rock drills may also be classified as 'wet' or 'dry' types depending on the method employed for cleaning the drill hole. 'Wet' drills are normally used for underground mining where the drilling dust is damped down by a water spray applied under pressure from an outside source led down to the bottom of the hole through an internal water tube. Dry drills utilize compressed air for hole cleaning, by either continuous blowing or a pulsating flow of compressed air supplied through the steel during drilling.

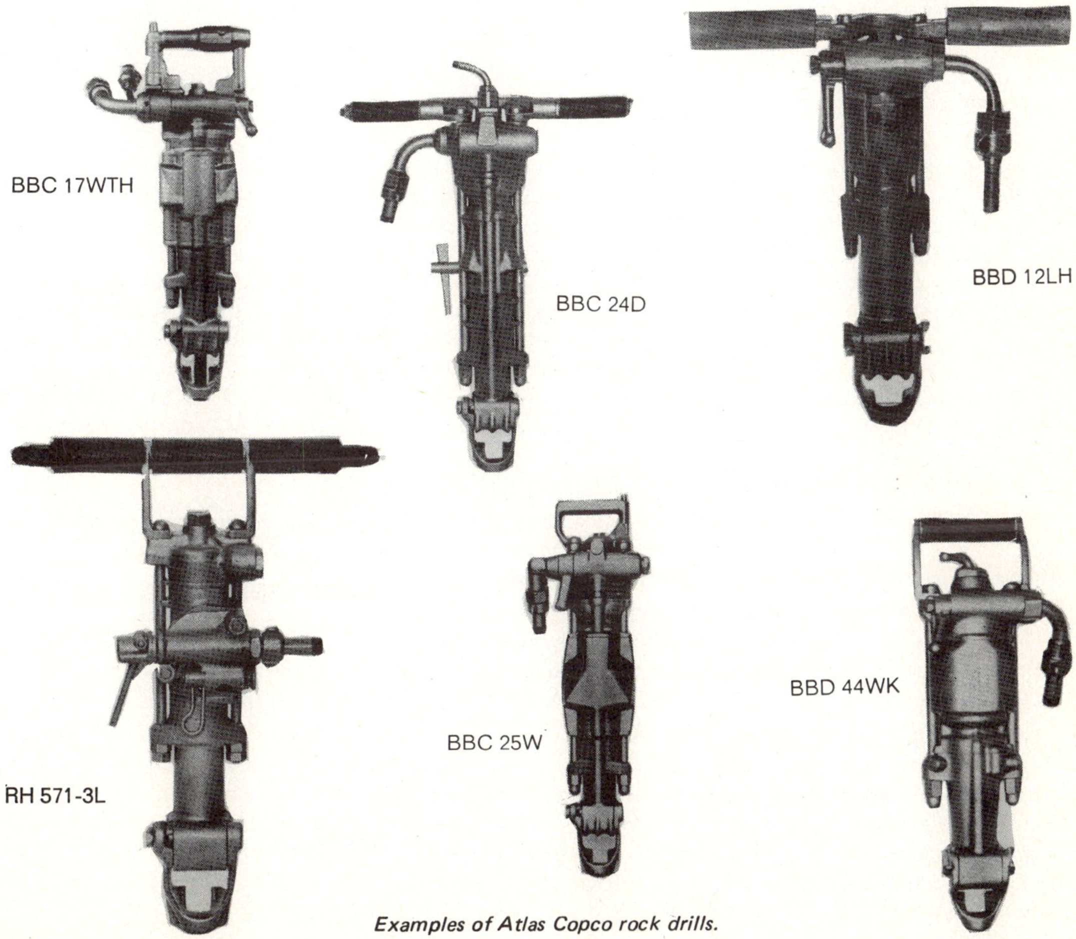

Examples of Atlas Copco rock drills.

BBD 91W

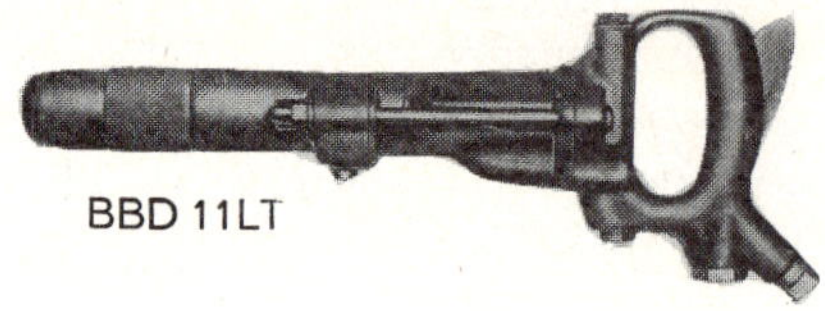

BBD 11LT

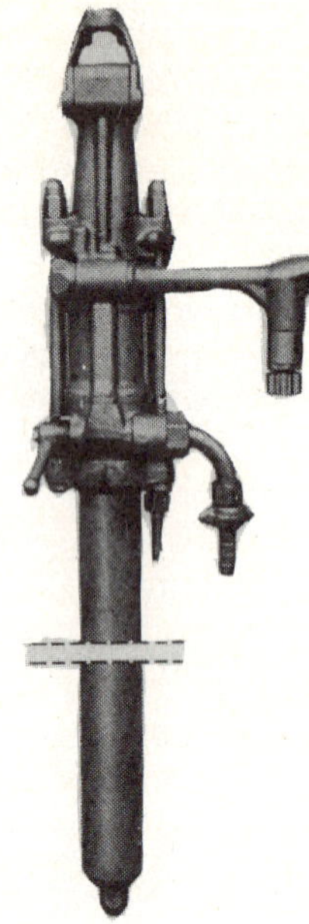

BBD 46WS

Examples of Atlas Copco rock drills.

Rock Drills

A rock drill is a percussive hammer driving a drill bit. Although the form and construction are similar to those of a concrete breaker one significant difference is that on the upstroke spiral flutes on a twist bar or rifle bar engage with a nut on the top of the piston causing the drill bit to be rotated through part of a turn. The other end of the rifle bar accommodates pawls operating in a ratchet ring, disengaging on the downstroke to allow the drill to be driven straight without rotating. Each return stroke (only) is thus accompanied by partial rotation of the drill steel so that the resulting hole is circular. The rate of rotation of the drill is arranged to suit the conditions under which it will be used and some designs may incorporate provision for adjusting the rate of revolution. Other designs may incorporate a control releasing the ratchet so that starting and stopping of drill rotation is under the control of the operator. In the case of very heavy duty drills – eg large capacity drills of large diameter – rotation of the drill may be achieved by a separate air motor (usually piston type) because of the high torque required to rotate the drill.

It will be appreciated that although it is capable of cutting a circular hole the actual action of a rock drill is that of continual hammer blows crushing the rock. To keep the hole free and reduce the risk of the drill jamming, compressed air or water can be forced into the hole through a hollow drill, blowing or washing away the crushed rock parts as they are produced.

In the case of compressed air blowing the operator has various methods of directing compressed air to the bit end, such as opening a transfer valve to limit piston travel and/or direct air to the bit end, or holding the piston at mid stroke to allow live air to travel down to the drill shank, etc.

With wet drilling – also known as dustless drilling since the water fed into the hole damps down the dust formed by the drilling action – the water stream may be led down through an internal water tube. A separate valve controls the water flow and this may be used with compressed air blowing – ie cleaning of the air is done by air blowing, and damping of the dust by the water stream. An alternative method still used is an attachment fitted to the side of the drill which directs an atomized spray of water onto the drill and into the hole. All have certain limitations as regards true 'dustless' drilling.

TABLE II – ROCK DRILL CATEGORIES

Category Weight	Plug 9 kg (20 lb)	Light 16 kg (35 lb)	Medium 20 kg (45 lb)	Heavy 25 kg (55 lb)	Sinkers 34 kg or more (75 lb or more)
Duty	Plug and feather work light hole drilling	Coal mining, small hole drilling	General drilling work	Deep rock drilling	Shaft sinking
Depth		1.8–2.4 m (6–8 ft)	5–6 m (up to 16–20 ft)	7.6 m (up to 25 ft)	7.6 m (up to 25 ft)

Some improvement in this respect may be achieved by venting the drill, ie providing a passage for compressed air which has leaked past the piston to the bit end to vent to atmosphere rather than escape under pressure along the shank of the drill.

The actual feeding of the drill can be done by the operator or by means of compressed air cylinders known as pusher legs (Fig 3). In either case, to withdraw the drill after it has cut to any

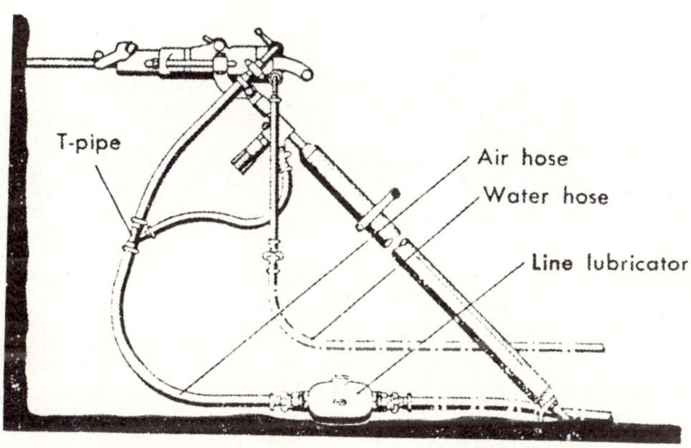

Fig 3 Diagram of connections – an Atlas Copco rock drill mounted on a BMK pusher.

Speedline rock drills at work. (Joy-Sullivan Ltd).

appreciable depth it may be necessary to withdraw it with the drill operating. In this situation the drill steel may be pulled downwards to a position where it is no longer struck by the piston, in which event the piston blow is 'cushioned' by the air trapped between the piston and front end of the cylinder. Without such a cushioning effect the drill could be damaged.

Drill steel and special equipment. (Atlas Copco).

Coal Picks

Coal picks are lighter percussive type tools fitted with removable pick steels and, as well as breaking down coal, may be used for light demolition work, etc. Like rock drills they may be provided with water jets to reduce the amount of dust raised by the steel. The most common form of water feed is by an attachment of equi-spaced nozzles around the bottom of the tool directing a fine spray of water onto and around the pick steel. Tools of this type are known as 'wet' picks. Those which have no water jet attachment are normally termed 'dry' picks.

Sizes are arbitrarily classified as 'light', 'medium' and 'heavy', with typical air consumption figures 0.6–0.7 m^3 (20–25 ft^3/min), 0.85 m^3 (30 ft^3/min) and 1.0–1.13 m^3 (35–40 ft^3/min) respectively at 6 bar (80 lb/in^2).

(For breakers, spaders and rammers also used in mining and quarrying, see Contractors' Tools).

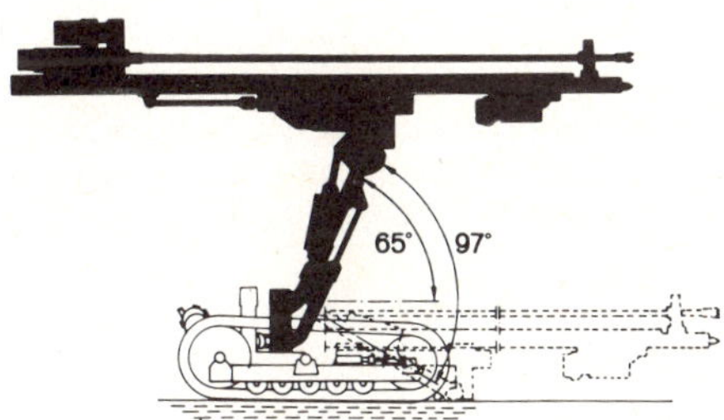

The boom can be raised 65° above the horizontal and lowered 32° below it

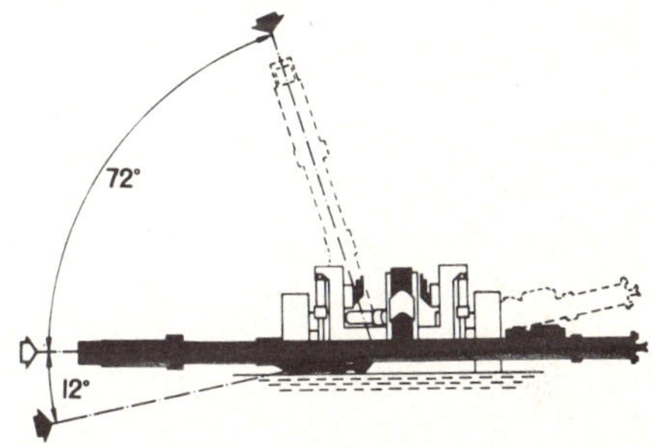

Horizontal holes can be drilled ahead of the tracks

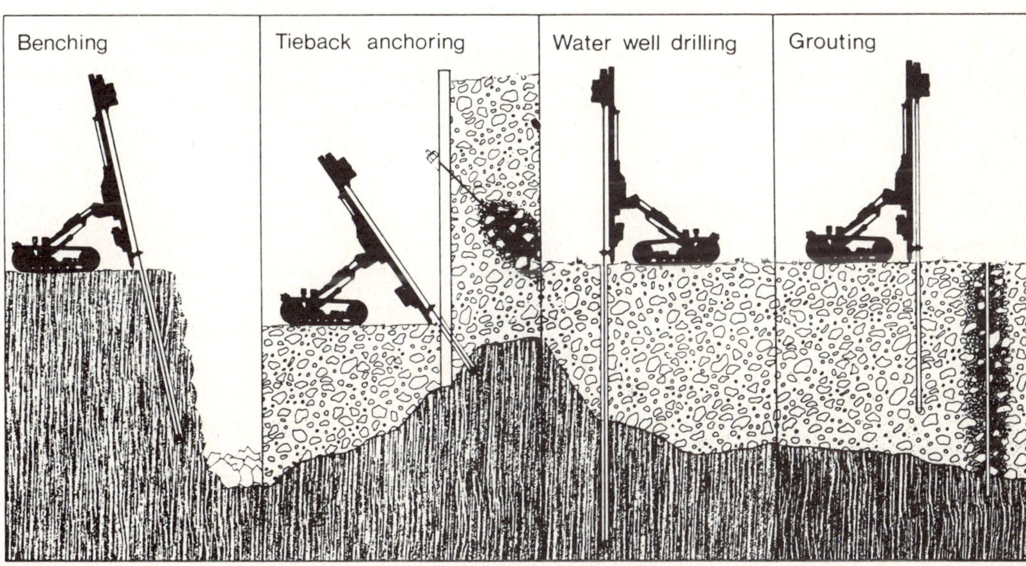

Atlas Copco boom system rig.

Drifters

A drifter is a rock drill specially adapted for horizontal boring, mining or tunnel driving, etc, although it may also be suitable for angle drilling. It may be supported on a column or crossarm, or fitted to jibs or booms for enhanced mobility, in which case a number of drills may be mounted on the same chassis or trolley. Multiple drilling rigs of this type have particular application for quarry work as well as tunnelling and underground working. Standard drifters a꜀e made for wet or dry drilling with controlled blowing, or in vented form.

Stopers

Stopers are used mainly for overhead vertical or near-vertical drilling in underground workings and comprise a percussive-type air drill attached to a pneumatic cylinder which acts as a telescopic leg both to support the drill at the working height required and to provide forward feed. Modern types of stoper may also be provided with automatic self-rotation in a similar manner to that standardized for rock drills (ie by means of a rifle bar and ratchet mechanism) although the design differs in other details (eg stopers may employ a collared chuck for retaining the steel and omit the usual anvil action for increased speed of drilling). In other respects they are similar to rock drills and are made for both wet and dry drilling.

Self-propelled Drills

In mining and quarrying, considerable use is made of self-propelled drifters and similar drills, and such tools can also be mounted on a wheeled cradle or trailer which can be pushed or towed to the position required. The drill itself is mounted on a boom and control over drill position, angle and rate of feed may be manual or provided by pneumatic or hydraulic power.

Fully mobile, self-propelled units may or may not incorporate a compressors as part of the rig and may be based on a wheeled or crawler tractor configuration. In such larger units hydraulic power is often employed for raising and lowering the boom, the hydraulic pump being driven by the engine; or all motions may be operated entirely from a compressed air supply. Drilling rigs of this type may also be designed in the form of attachments for mounting on standard tractors, with provision made for both percussive and rotary drilling by fitting an adaptor and rotary bit in place of the hammer unit.

Rotary Tools

In the case of a rotary tool the driver is in the form of an air motor built into the tool, the high power/weight ratio of such motors making them particularly adaptable to maintaining speed under load. Free running speed as well as load speed can be controlled through the use of a governor or similar device, thus enabling constant speed working to be maintained, if necessary under varying load. This is particularly applicable in the case of such tools as grinders and rotary drills. Gear trains and clutches may be added to all types of motor, with the torque multiplication obtained by reduction gearing especially significant in the case of torque wrenches, nut setters, etc — tools subject to high torque loading.

Rotary tools may be grouped in the same way as percussive tools, and individual tools are discussed under their respective headings. See also chapter on *Workshop Tools*.

Rotary Contractors' Tools

There are considerably fewer rotary tools in this category than percussive ones, although a variety of types have been developed for specific applications. These embrace, mainly, concrete vibrators, shank grinders, sump pumps, hoists and winches (although the last two are more properly classified as miscellaneous pneumatic devices).

Concrete Vibrators

Rotary type concrete vibrators are normally of the 'internal' type for placing within the mass of concrete – ie the concrete is dumped over the machine to completely immerse it or the vibrator so designed that it can readily be immersed in a mass of concrete. The concrete mass is then vibrated by operating the machine for a predetermined period of time (usually about half a minute or slightly less).

The basis of a concrete vibrator is usually an eccentric weight mounted on low friction bearings and capable of being rotated at high speed by an air motor. The unit is fully enclosed within a steel casing which is completely sealed, thus preventing entry of either water or concrete. Both delivery and exhaust (air return) lines are attached to the casing so that exhausted air is carried away from the machine and not vented into the concrete mass. Sizes vary from about 62.5 mm to 160 mm (2½ in to 6 in) diameter, with weights between about 11.5 and 45.4 kg (25 and 100 lb).

A vane type motor is normally employed for this type of tool with the speed limited by a centrifugal governor linked to the air control valve.

Shank Grinders

The shank grinder is a form of rotary grinder with a specially shaped recessed cup grinding wheel for trueing, squaring up and deburring the ends of steels used on percussive type tools. The wheel is normally driven by a vane motor fitted with a centrifugal governor and provision is made to oscillate the wheel across the workpiece, eg by employing a pivoted grinding head manipulated by a hand lever. Other controls may be provided for feeding and clamping the steel or workpiece in position. The centre of the wheel may also be fitted with a reaming bit.

Air-operated sump pump type 38 for industrial duties. (Broom & Wade Ltd).

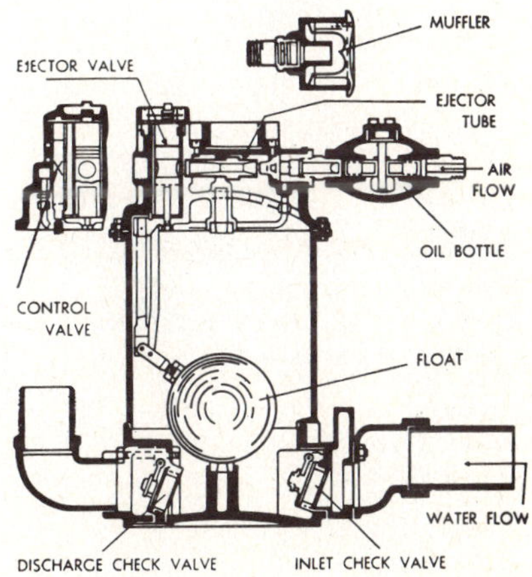

Ejector type sludge pump. (Consolidated Pneumatic Tool Co.Ltd).

Sump Pumps

Sump pumps are designed expressly for dewatering excavations, etc, where they may be called upon to handle sludge and mud as well as clear water. It is also desirable that the geometry of the unit be arranged to pump down to the lowest possible level.

A sump pump thus consists of a suitable pump (almost invariably a simple self-priming centrifugal type) powered by a suitable driver (an air motor in the case of a pneumatic unit). Both piston and vane type motors may be used, although the latter are usually preferred. The motor spindle is then usually connected directly to the pump at motor speed.

Speed control is accomplished by a centrifugal governor or similar device, control being necessary to prevent the motor (and pump) overspeeding should it be left running after pumping dry. Equally, the pump must be suitable for running 'on snore' without damage, unless provision is made for automatic cut-off and restart.

The compressed air sump pump has the additional advantage that where high heads are involved delivery can be assisted by air lift by injecting the air stream into the pump outlet. In this manner greater heads can be achieved than those delivered by the pump alone. This is not necessarily a feature of sump pumps as such, but may be incorporated on units designed for high heads.

Atlas Copco 'Cobra Super' motor drill/breaker being used to drive a submersible pump.

Workshop Tools

AIR-OPERATED workshop tools can be divided handily into rotary tools generally powered by air motors and percussive tools operating on a similar principle to rock drills, etc.

Rotary tools can be divided under the following headings:-

(i) grinders and abrasive tools;

(ii) drills;

(iii) screwdrivers, nut runners, impact wrenches, etc.

There are also a considerable number of specialist tools developed for particular industries and standard types of air-operated tools adapted to specific duties.

Percussive tools such as chipping hammers, rammers, core busters, riveting hammers, etc, are all basically the same type of tool, but with differences in stroke and diameter of piston and speed, according to the duty specified. Scaling hammers are designed for fast speeds and low impact energy blows. Chipping hammers have a high speed with a medium or short stroke. Riveting hammers normally have a long stroke and a slower speed of working.

All such hammers normally have a throttle valve regulating the air supply so that the hammer may be started slowly if desired, and subsequently operated at any speed from slow to maximum. The return stroke of the piston is cushioned to make the tool easier to hold and manipulate.

Grinders and Abrasive Tools

These, basically, comprise an air motor carrying a grinding wheel or abrasive wheel, etc, normally on the shaft extension, or in some cases driving the wheel through reduction gearing. The fitting of buffing wheels, wire brushes, etc, is usually a simple alternative, and thus the basic grinder becomes a multipurpose abrasive tool in this respect. Grinders may be made as portable or fixed units, depending on size and duty requirements. Portable units predominate numerically.

Most modern grinders are fitted with governors to ensure an optimum operating wheel speed without exceeding the maximum.

Vitrified bonded wheels are limited to a maximum peripheral speed of 33 m/s (6 500 ft/min) permitted speed when idling. Conventional resin-bonded grinding wheels permit maximum peripheral speeds up to 48 m/s (9 500 ft/min). Glass-reinforced grinding wheels and specially sectioned wheels may be capable of safe operation at peripheral speeds up to 80 m/s (16 000 ft/min).

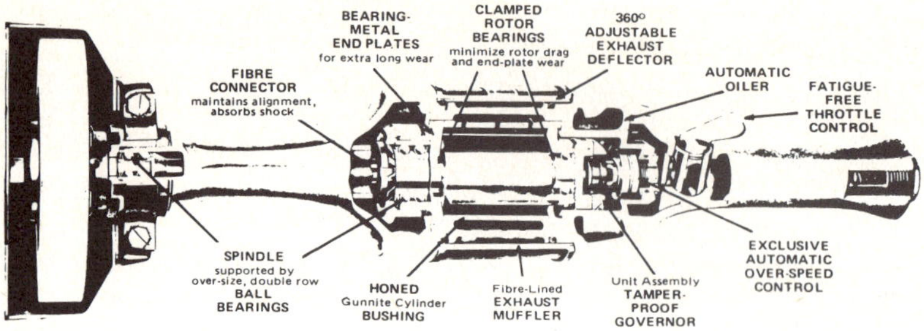

Straight grinder shown in section (Thor Tools).

Grinders may be generally classified as —

(a) straight or horizontal grinders

(b) vertical or surface grinders

(c) angled or angle-headed grinders, referring to the attitude of the grinding wheel relative to the body in portable tools, and

(d) pedestal grinders where the complete tool is mounted as a fixed unit, normally with foot control of the motor to permit the workpiece to be manipulated with both hands.

Other types of 'fixed' grinder for bench mounting are basically portable grinders mounted in suitable stands.

Sizes of portable grinders range from very small, light, high-speed units running up to 100 000 rev/min (in which case turbine motors may be used) and with a maximum wheel capacity of about 2 in (50 mm) diameter, to large, heavy-duty models running at up to 4 000 rev/min with a wheel capacity of up to 8 in (200 mm) diameter. There is a very wide variety in individual detail design, and also in the production of 'specific type' grinders — eg for particular types of dressing work.

Right angled hand grinder. (Desoutter Bros. Ltd.).

Die-grinders for small wheel sizes may employ vane type motors for speeds up to 40 000 rev/min although 20 000 rev/min is the more usual maximum when a wheel is fitted. With the high operating speeds burrs, rotary files and abrasive points are the normal attachments. Typical duties for this class of tool include deburring, touching up welds, finishing castings, etc. The very light point grinders with speeds of up to 70 000 rev/min employ turbine motors.

Conventional grinders, or 'straight' grinders are normally classified by their wheel capacity (ie grinding wheel size) related to the maximum permissible or 'safe' speed for the wheel. Speed control is provided by a governor to restrict the air supply as load decreases, and vice versa. By changing the tension of the governor spring or the size of the governor weights the speed of the motor can be altered to suit various types or sizes of wheel. Speed governors also restrict the no-load air consumption of the motor and safeguard the vanes and cylinder against undue wear.

Sanders

Vertical sanders are essentially the same as vertical grinders with the exception that a sanding disc and pad is substituted for the grinding wheel. Standard disc sizes are 5 in (125 mm), 7 in (175 mm) and 9 in (225 mm) diameter. Standard speeds normally range from 1 000 to 10 000 rev/min, depending on the size of the tool and the duty involved.

For lighter work the right angle sander is particularly attractive since the tool body and handle are at right angles to the disc. This may take the form of a right angled attachment for fitting to a normal grinder body or a right angled drive incorporated as an integral feature of the body, with or without reduction gearing. Speeds normally range from 2 000 to 10 000 rev/min, depending on the size and duty.

Horizontal grinders may be adapted as belt sanders (or belt grinders) by the fitting of an attachment comprising, basically, a large diameter driving pulley separated from a jockey wheel with an abrasive belt looped between the two wheels. The driver wheel is also used as the contact wheel for working and is commonly made of rubber.

Orbital sander (Sundstrand)

In addition to disc and belt sanders, portable air-operated orbital sanders are produced for both wet and dry sanding operations. A typical tool of this type weighs 4–5 lb (1.8–2.25 kg) with an air consumption of 6–10 ft^3/min (0.17–0.28 m^3/min) and an oscillating rate of 5 500 cycles per minute, and is often designed for moderate air pressures (eg 50–75 lb/in^2) (3.5 to 5.25 bar). Larger machines with opposed pads reciprocating in opposite directions are also available for production line work, etc.

Rotary Files

Rotary files and burrs are used for light finishing work, weld finishing, deburring, etc. Spindle speeds are somewhat lower than that common with grinders, from 1 000 to 3 000 rev/min being usual when dealing with hard metals. Higher speeds can, however, be used on softer materials and also when using carbide burrs.

For small and intricate work the burr or rotary file may be driven via a flexible extension shaft for maximum ease of manipulation, and also to permit access of the tool into restricted areas. Speeds are usually higher with small diameter burrs, and in specialized applications, such as dental drills, the tool may be powered by a turbine motor.

Dust Extractors

Dust extractors may be fitted as separate devices to grinders, sanders, wire brush machines, fettling chisels, etc, to extract at source dust created by the tool. They are normally designed specifically to suit a particular type of tool — eg grinding wheel, sanding disc, chisel, and comprise, basically, a suction nozzle or extractor head connected by a flexible line to the extractor.

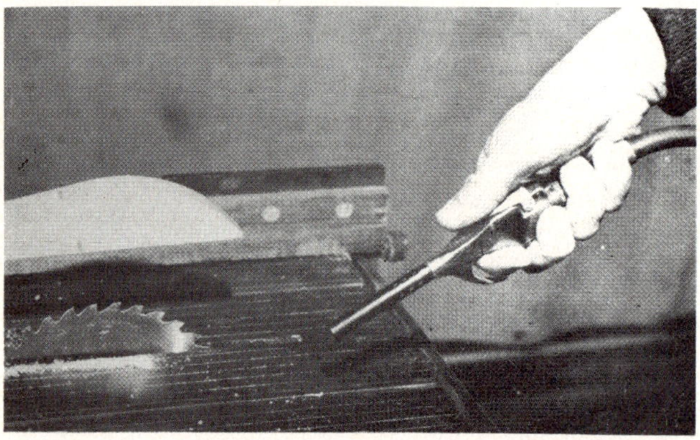

Blow gun (Dyna-Quip Ltd)

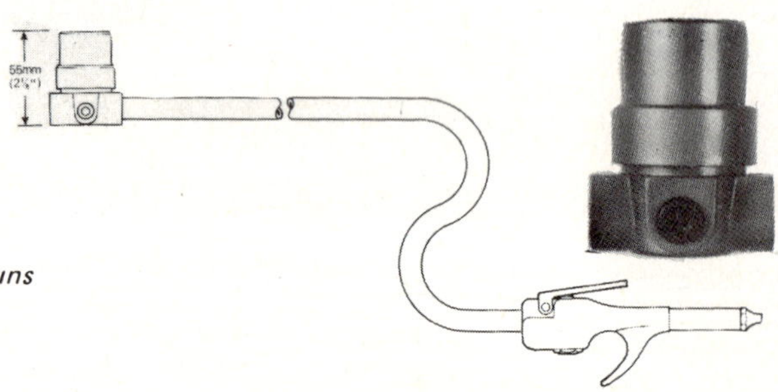

Norgren safety regulator for blow guns

Drills

Sizes range from the smallest lightweight drills weighing less than 1 lb (0.45 kg) with a drill capacity of up to 1/8 in (3 mm) diameter to large, heavy duty tools which may weigh up to 150 lb (68 kg) and have a drill capacity of 3 in (75 mm) diameter. Free-running speeds may vary from 15 000 rev/min in the case of the small light duty drills down to 75–125 rev/min in the case of heavy-duty tools. Speed control is usually provided via a throttle valve and most types are available either with uni-directional or reversible rota ion, the latter being a necessary feature for tapping and other duties. Non-reversible drills may be preferred for specific operations, eg reaming.

Although described as drills, tools of this type can take a variety of bits other than drills, and thus may be equally suited for reaming, tapping, stud setting, stud driving, tube rolling, etc. Many tools so described are, in fact, air drills with the appropriate attachment. Similarly, other attachments may be available for drilling, etc, with fixed angle heads, 30°, 45° and 90°, and adjustable angle heads. The shape may vary from purely cylindrical form to pistol grip tools, either or both capable of taking a variety of attachments.

Performance of air drills, particularly in the lightweight sizes, is generally superior to any other type for virtually all drilling jobs, with the lightweight and high power/weight ratio making for faster working and easier handling in the case of portable tools. Air drills can also be readily adapted to self-feed or mechanical feed, either as portable units or mounted in fixtures, as well as being particularly suited to multi-drilling rigs.

Larger heavy-duty air drills are normally fitted with a Morse taper socket chuck, sizes ranging from No 1 or No 2 in the case of 3/16 in (14 mm) and 7/8 in (22 mm) drills up to No 5 in the case of 3 in (75 mm) capacity drills. The chuck may also be fitted with an ejector pin to enable the drill to be removed readily. Large capacity drills also normally incorporate some form of self-feed.

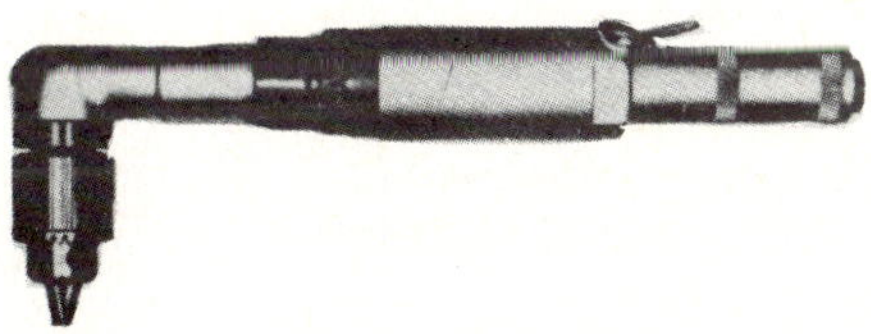

Angle drill (Thor Tools).

Desoutter model RB5–P–2100 multi-purpose air drill with new pistol grip body.

Wood Borers

Wood borers, again, are basically air drills designed for slower running speeds and fitted with reversible motors to assist in backing off and withdrawal of the auger bit. Typical size ranges and speeds are:

up to 1 in (25 mm) auger capacity, 750 – 1 000 rev/min

up to 2 in (50 mm) auger capacity, 600 – 750 rev/min

up to 3 in (75 mm) auger capacity, 250–350 rev/min

Tools in this category are widely used for heavy timber construction. For lighter wood boring duties standard air drills may be used with auger bits, driven through a reduction gearbox if necessary.

Desoutter Super Atom series motor which is quiet, easy to set up and operate. Externally adjustable clutch. Choice of speeds : 1 000, 1 500 and 2 000 rev/min. This tool is reversible. Once set, the torque output remains constant.

Woodworking

The range of pneumatic powered tools for timber and woodworking duplicates that of electric portable tools, with certain advantages. For example, it is readily possible to provide reversible motors in air-powered tools (eg woodborers). In addition, portable pneumatic tools are light in weight for their work capacity, generally reliable in operation, safe to use under conditions where an electric motor to electric supply may represent a hazard, and have low operating costs. Also an air-powered tool can be stalled without damage, unlike an electric powered tool.

Woodborers – these are designed specifically for accommodating a range of auger bits with lower rotational speeds than drills. The usual capacity range is from ¼ in (6.4 mm) up to 4 in (100 mm) diameter (auger size) and in various sizes of shank. Various auger shapes are used, according to the timber to be bored – eg a Scotch auger for rough timbers, a round nose auger for angular boring, special augers for green-heart, etc. The larger woodborers are usually reversible as this is a highly desirable feature where large holes are being drilled.

Drills – the pneumatic drill is the counterpart of the electric drill and for woodworking may be fitted with a wood bit chuck. Working capacity usually ranges up to about 7/8 in (22 mm) auger diameter, according to the size of the tool. The drill may or may not be reversible. In the latter case it is generally recommended that the auger diameter be restricted to a somewhat lower figure, especially when long augers are being used.

Saws – tools available in this class range from pneumatic-powered chain saws for contractors' work to small portable or bench type circular saws. Reciprocating saws are also produced, usually in smaller sizes for fitting with standard hacksaw blades. A particular feature of a pneumatic powered chain saw is that it can be used under water. In this case a safety throttle is advisable so that the saw motor stops immediately should the diver (operator) accidentally lose his grip on the tool.

Adze – for use as an adze a timber slicer can be fitted to certain types of pneumatic hand drills. The light overall weight of the resulting tool makes it particularly easy to manipulate compared with any other form of powered adze.

Screwdriving, nail driving, sanding and similar duties can be also be performed by specific and standard pneumatic tools, the range of such tools and attachments being very wide.

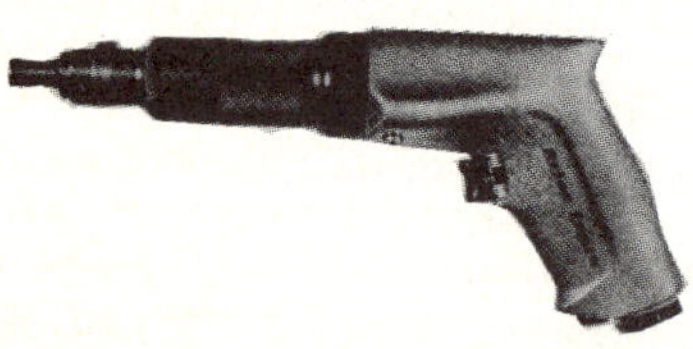

Pneumatic screwdrivers (Atlas Copco)

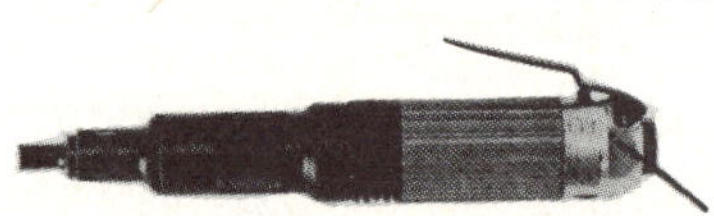

Screwdrivers

Screwdrivers are normally powered by vane motors driving through planetary reduction gearing to reduce the spindle speed to a suitable value. The motor may or may not be reversible. There is considerable variety in detail design and shape – eg straight or offset handles. Although they are light and compact, provision may also be made for suspension of a tool by lightweight balancers to reduce operator fatigue on a production run, whilst other devices such as pneumatic or magnetic pick up of screws may be incorporated. Numerous clutch and torque control methods are also employed.

With geared drives speed and torque are directly related, ie a low speed gives higher torque, and vice versa. Choice of torque and speed depends on the nature of the assembly, the materials involved and the speed and accuracy required. In some designs of screwdriver there is provision to adjust torque. Ideally the choice should be a screwdriver which is fast, easy to handle, and adaptable to different jobs (unless employed at a specific point on a production line where the duty requirement does not change).

With direct drive the screwdriver bit is connected directly to the motor via the reduction gearing, hence maximum torque is equal to the stall torque of the motor multiplied by the reduction gearing, the air pressure being regulated so that the tool will stall when the screw is driven home. A simple clutch may, however, be fitted in order to make it possible for the operator to present the tool to the screw with the motor running, although the driver is still classified as a direct drive type since there is no limitation on applied torque below the stall torque.

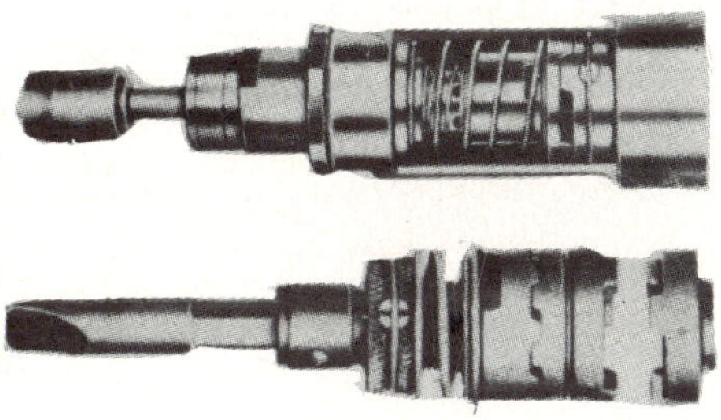

Uni-tork (top) and adjustable clutches for screwdrivers/nut-setters (Thor Tools).

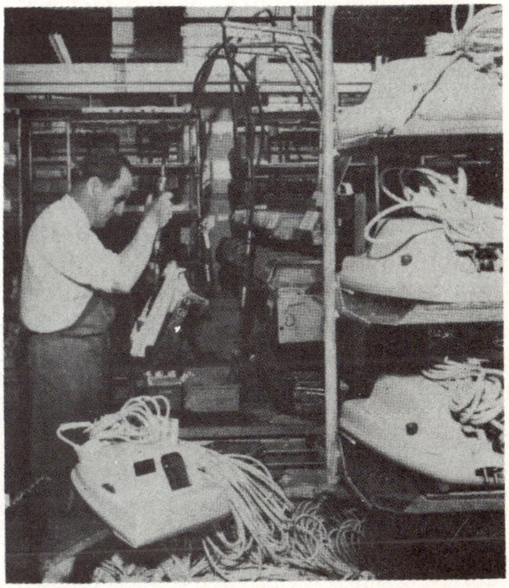

Desoutter screwdrivers in use on a vacuum cleaner assembly line.

Where torque limiting is required, either a positive clutch or a torque-limiting clutch can be incorporated. In the former case the clutch is, usually, of the dog type held apart by a spring, the clutch being engaged by forward pressure on the tool. Releasing pressure on the work allows the clutch to disengage, thus the motor keeps running although the screw is not being driven home. In practice this means that the operator can regulate the amount of torque transmitted by varying the amount of downward pressure applied to the tool.

With a torque-limiting device the clutch is a little more complicated but can be preset to slip at a given torque level. The tool will then transmit torque only up to that level, after which the drive will slip. There are numerous designs of torque-limiting clutch which provide this facility, the most common usually employing two sets of opposing clutch faces, one set providing the engaging clutch and the other the slipping clutch disengaging at a predetermined torque level.

Screwdrivers with torque-limiting clutches are generally to be preferred, particularly for light tools used for work on finished surfaces or where precise fastening torques are required, as they are far less liable to slip or damage the fastener and do not place a premium on operator skill.

Nut Setters

Basically these are identical to screwdrivers with a socket bit replacing the screwdriver bit. They are produced in straight or angled forms and with a handle or grip shape and size consistent with the working loads and reactions likely to be involved.

Torque control may be even more important in order to avoid over-tightening, in which case a torque-limiting clutch is to be preferred. Motors are invariably made reversible to enable the runner to be used for assembling and removing nuts. Larger sizes of nut runners need not be fitted with a clutch since in such cases the operator can more readily handle the stall torque of the tool without losing control.

The inherent characteristics of both screwdrivers and nut setters make them particularly suited for multi-point working assembled on a suitable jig or fixture for production line work. Close spacing can be achieved by stacking the individual tools at different levels or positions, as necessary, with extension bits to bring all the sockets into the same plane as the work. In multiple units it is also possible to arrange for a pre-set sequence of tightening individual nuts if required.

Reamers

Special tools are not normally produced for reaming since this duty can be performed by a standard drill reducing the capacity by approximately 25%, ie the reaming capacity can be taken as 75% of the drill capacity.

Virtually any size of drill can be used for reaming with such a reduction in capacity. Smaller drill sizes, however, may exclude tapping as a permissible duty.

Heavy duty tappers (Consolidated Pneumatic Tool Co).

Tappers

Standard air drills, nut setters, screwdrivers, etc, may be used as tappers within the capacity of the motor, which will be lower than the capacity for drilling. The reduction in capacity is usually of the order of 40–20% – eg typically a drill with a drilling capacity of 9/16 in (14 mm) will have a tapping capacity of 5/16 in (8 mm) to 7/16 in (11 mm), a size with a 1 in (25 mm) drilling capacity would have a tapping capacity of 3/4 in (20 mm), and so on.

In addition, air tools may be specifically designed as tappers, when it is usual to provide a higher reverse speed for withdrawal of the tool.

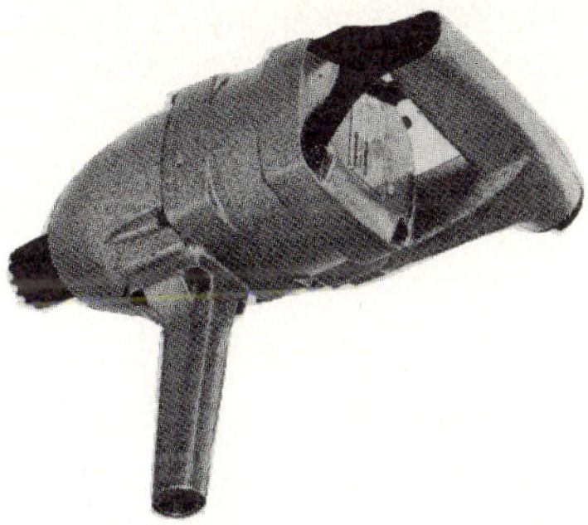
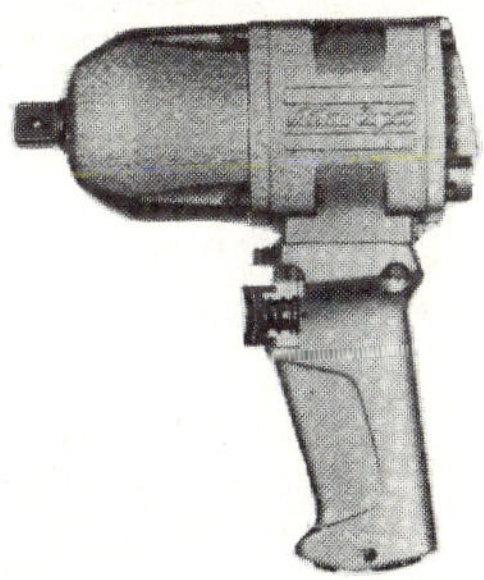
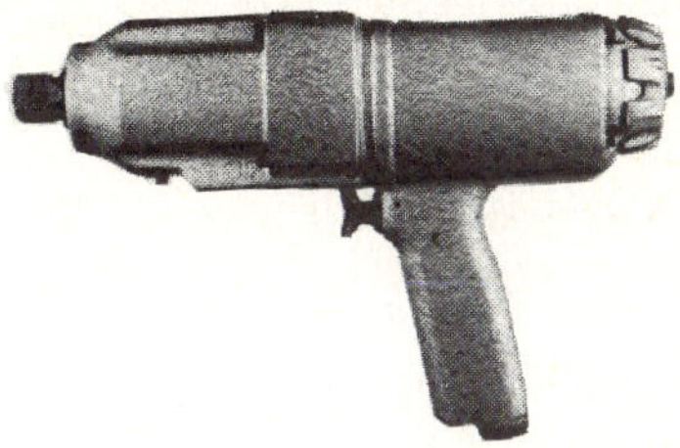

Impact wrenches (Atlas Copco)

Typical impact wrench in use on car production line (BroomWade).

Impact Wrenches

The impact wrench is merely another form of the screwdriver or nut setter except that with increasing torque load the driving or running action is automatically converted from a smooth torque into a series of rapidly repeated impacts. Besides making it possible to exert very high tightening (or loosening) torques (very much higher than those provided by a continuous drive), torque reaction on the operator is practically eliminated since the turning effort at this stage is developed by impact blows instead of sustained torque.

Typical sizes of impact wrenches range from about 1½ lb (0.7 kg) in weight with a bolt capacity of 7/32 in (5.5 mm) diameter, up to 600 lb (270 kg) with a bolt capacity of 12 in (300 mm) diameter.

The correct choice of socket is particularly important in the case of an impact wrench for this is subject to considerable stressing and needs to be stronger than conventional sockets of similar size. Sockets for impact wrenches are normally made of heat treated alloy steel.

Gardner-Denver torque control right angle wrench giving torques from 29 ft lbs – 98 ft lbs. with better than ± 5% repeatable accuracy.

For convenience of handling, impact wrenches commonly have a D-shaped handle with, often, a side handle. However, since the torque is imparted as a series of very rapid impact blows torque reaction is comparatively low, as far as the operator handling the tool is concerned.

In smaller sizes, impact wrenches may have D-shaped lever or pistol grip handles.

Impact wrenches may also be used for stud driving, screwdriving in the case of larger wood-screws, machine screws and self-tapping screws, and various other duties, using suitable bits. Stud runners are usually standard forms of nut runners fitted with stud setting attachments.

A torque control ratchet wrench from Gardner-Denver combining high torque repeatability with access to awkward bolt heads.

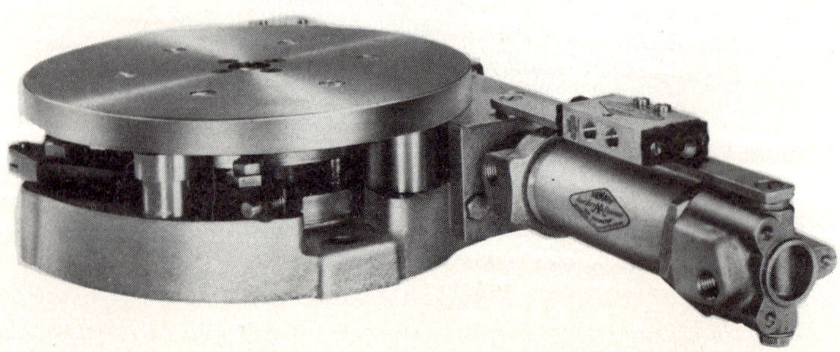

Rotary indexing table (Air Automation)

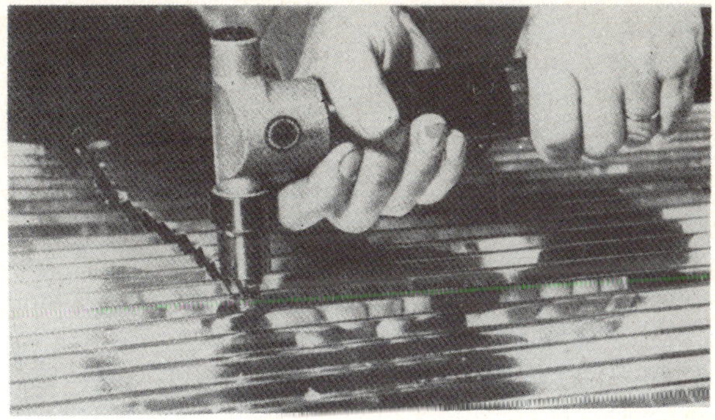

Nibbler (Atlas Copco)

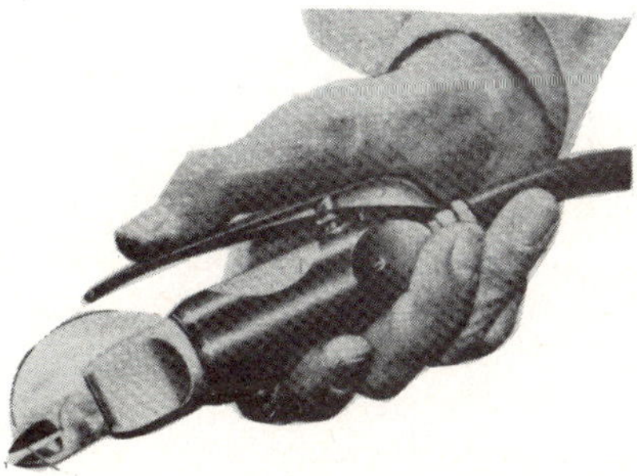

CP 42 Air Nip pneumatic wire cutter (Consolidated Pneumatic Tool Co).

Shears and Nibblers

Two basic types of action are employed in metal-cutting tools of this type –

(i) a definite shear action provided by two blades working in the manner of a pair of scissors; and

(ii) a nibbling action provided by a cutter punching out small portions of the sheet material as it progresses along the cut line.

Power in either case is normally provided by a vane type motor driving through reduction gearing (usually through a right angled head).

In the case of shears the cutting blades are visible during operation, thus a marked line can readily be followed. With a nibbler the cutting elements are normally obscured by the head, demanding the use of a template for accurate cutting. Both types of tool, however, have the same application and in addition to sheet metal working can be used for cutting other materials such as non-metallic sheets too hard to be sawn.

Typical sizes range from about 3–15 lb (1.4–2.25 kg) in weight, individual tools normally being rated according to their capacity for cutting various gauges of metal. The larger models have a capacity of up to 12 swg (2.6 mm) in sheet steel and may achieve cutting speeds of up to 10 ft/min (3 m/min) in this thickness of mild steel sheet. Nibblers may have a slightly higher capacity, eg up to 8 swg (4 mm) in mild steel sheet.

Tube Expanders

Air drills of 1 in (25 mm) and upwards may also be used as tube expanders and are able to expand up to about 150% of the drill capacity – eg a 1 in (25 mm) capacity air drill may also be rated for tube expanding up to 1½ in (40 mm). Impact wrenches also may be used for tube expanding operations. For heavy duty tube expanding work, however, special tube expanders are normally used which may employ either a vane type or piston motor. Free speed is usually very low with extra heavy duty tube expanders (eg 20 to 40 rev/min) with stall torques ranging up to 3 000 ft/lb (415 m/kg).

Circular Saws

This type of tool is purely a conventional circular saw driven by an air motor via a spur train or worm reduction gearing (in the case of portable units). It can be considered the equivalent in size and scope of a portable electric saw, although generally lighter and more compact for a given capacity because of the higher power/weight ratio of an air motor.

Motors used are normally of the vane type. Various additional attachments may be fitted for expanding the scope of the saw for grooving, rebating, etc, similar to electric motor powered saw sets. Reciprocating blade saws may also be powered by air motors as jigsaws, power hacksaws, etc, as well as for use as power files. In such cases exhaust air is usually direct to blow over the workpiece and so clear the cutting area of dust and swarf.

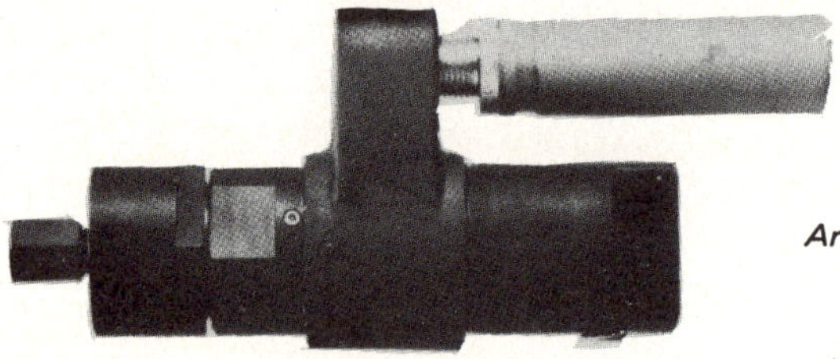

Aro power and milling motor.

Tool Chucks

Air-operated chucks offer many advantages over normal manual chucking arrangements, particularly as regards the speed with which workpieces can be mounted on or removed from the machine.

In a typical air-chuck, operation of the chuck is controlled by movement of a piston in an air cylinder, this movement being transferred by toggle linkage to the chuck jaws. Air pressure to the cylinder is normally under the control of a reducing valve enabling the chuck jaw pressure to be regulated as necessary, to avoid distortion or crushing of the workpiece. Further pressure controls may be provided to enable the operator to switch from the high jaw pressure necessary for roughing to a lower jaw pressure for fine finishing, thus relieving the chuck of continual high pressure clamping and consequent strain.

Similar principles of operation are embodied in the pneumatic vice, made in sizes for both workshop and foundry use, pneumatic clamps, etc. Again speed of operation is the main advantage offered by such tools. Line pressure is multiplied by the cylinder diameter (ie the pressure multiplication factor is equal to the cross sectional area of the piston in^2, relative to a line pressure in lb/in^2). Additional pressure multiplication can be obtained by inserting a compound toggle mechanism between the cylinder and the vice or clamp. Clamping pressures of up to 15 000 lb/in^2 (1 000 bar) can therefore be developed without resorting to excessive cylinder bore sizes — ie overall multiplication factors of up to 150 are readily achievable. With high pressure multiplication however, jaw movement is usually strictly limited and may amount to only a few thousandths of an inch over the initial setting.

Forging Tools

Pneumatic tools and machines under this heading comprise drop forging hammers, power hammers and forging presses. Cylinder sizes for drop forging hammers may range up to 48 in (1.22 m) and strokes from 36–72 in (0.91–1.83 m) with the maximum rate of working ranging from about 200 blows per minute for the smallest sizes down to 60 blows per minute for the largest sizes (although older designs may be much slower working). Forging presses are cylinder-actuated machines and may be all air-operated for small capacities, or powered by air-hydraulic cylinders for higher ratings.

Wide use is also made of a variety of other air tools for forge work, such as scaling hammers, chippers, air-hoists, etc, which are dealt with under the relevant sub-headings.

Foundry Work

A wide variety of pneumatic tools is used for foundry work, embracing both portable tools and static installations such as forging hammers, fettling grips, sand sifters, sand blasters, jolt squeeze machines, etc. Compressed air requirements are, therefore, highly variable from foundry to foundry and demand may range from under one hundred to many thousands of ft^3/min, according to the size of the foundryand chosen tool types. As regards the latter, pneumatic tools are, to a large extent, directly competitive with electric powered tools and actual compressed air demand is largely a matter of the balance arrived at between compressed and electric power.

Some tools are specifically designed for foundry purposes – eg sand rammers. Others, such as grinders, are general purpose workshop tools which find equal application in foundries. Some typical tools and their uses are summarized briefly:-

Sand rammers – are used specifically for compacting moulds and may be either bench type or floor mounted.

Core box vibrators – designed to clamp to a core box (often by an integral pneumatic vice, but alternatively by manual clamping). These are used to loosen moulds so that cores can be withdrawn free of flaws, etc, and also avoid damage to the boxes. Core box vibrators may also be used during the initial stages of making the cores for settling the sand to give a satisfactory core.

Fettling – chipping scaling hammers and grinders are widely used for fettling, with a wide variety of designs and weights available for different duties.

Grinding – for general fettling high speed wheels are used for rapid stock removal and may run at surface speeds up to 10 000 ft/min. Softer and finer wheels are required for surfacing and finishing, speed depending on wheel diameter and type. All pneumatic grinders can normally take a wide variety of different grinding wheel shapes, as well as wire brushes, etc. The grinders themselves may be straight or angled, choice being largely dictated by the type of work involved and corresponding ease of manipulation of the two types of grinder.

Jolt squeeze machines – used for consolidating sand moulds. The moulding box containing the sand is placed on the machine table. In the case of a shallow mould the sand can be consolidated by squeezing it between the table and an air operated platen ('squeeze'). For deeper moulds, more satisfactory consolidation is achieved by raising the table and then allowing it to drop suddenly ('jolt').

Fettling grips – normally comprise a piston operated clamp which is designed to facilitate removal of fins or flash from hot castings, drop forgings, etc. Fettling grips can accommodate a variety of sizes and shapes of castings, and hold them rigidly against a solid steel plate. Excess material can then be removed by any suitable tool, such as a pneumatic chisel or grinder, with the casting or workpiece securely located all the time.

Sand blasting machines – these are widely used in foundries for cleaning up and finishing castings, forgings, etc. Such machines are simple devices for blowing a concentration of solid abrasive particles (shot, sand, etc), through a nozzle onto the surface of the workpiece at high velocities by means of compressed air.

Percussive Tools

Chipping hammers range in size from about 3–15 lb (1.4–2.3 kg) weight and are normally fast-hitting hammers carrying a chisel blade in place of a hammer head. They are generally grouped in three categories – 'light', 'medium' and 'heavy'. Light chippers are also used for scaling, caulking, beading and similar duties. The heavier tools are used mainly for chipping duties, such as chipping castings, removing flaws in steel billets, etc, but may also be used for riveting. Since the duty can readily be varied by fitting a chisel blade or tool steel of a suitable size and shape to suit different operations, it is difficult to distinguish between smaller chipping hammers and caulking hammers, chisels and scalers, etc, although the latter may be produced as specific types. Typical air consumption figures for chipping hammers are: light 10–15 ft^3/min (0.28–0.42 m^3/min); medium 18–20 ft^3/min (0.5–0.6 m^3/min); heavy 24–28 ft^3/min (0.7–0.8 m^3/min); at 80 lb/in^2 (5.6 bar). Blows per minute range from about 4 000–45 000 for light hammers; to 1 500 –1 650 for heavy hammers. Typical accessories for chipping hammers include various chisels with hexagonal or round shanks.

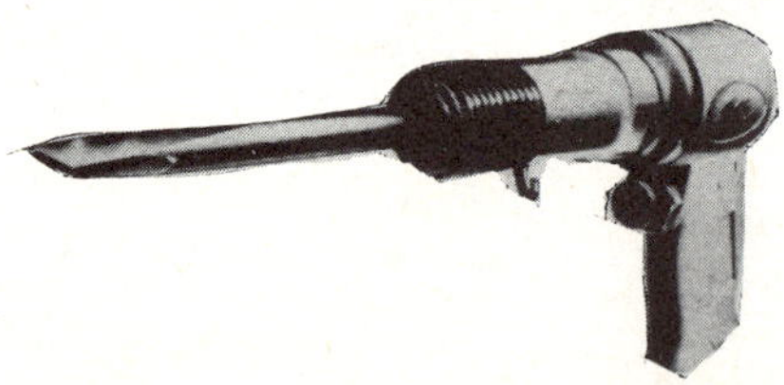

Light duty chipping hammer (Thor Tools).

Scaling Hammers

Scaling hammers are usually the smallest and lightest of the hammer tools and are commonly produced with valveless construction for high speed of operation of up to 6 000 blows per minute. In such cases the piston itself may be used as the hammer with the tip or working face hard faced to resist wear. The blows are relatively light, although rapidly delivered, and so brinelling of the surface being treated is usually no problem. Multiple piston types are also produced, either as hand held tools or mounted on a frame for multiple tool work. Other scaling hammers are designed with detachable or interchangeable steels with positive retention for the steel when fitted, and with mechanical valving. There are also various forms of handles and controls favoured by different manufacturers. Most scalers are designed so that exhaust air is directed onto the work surface to assist in cleaning, whilst some may also incorporate a specific blower arrangement rather on the lines of the rock drills.

Scaling hammers – or more correctly, scalers, since not all such tools are fitted with specific hammer heads – are normally made in sizes ranging from about 3–8 lb (1.4–3.6 kg) weight, with strokes ranging from about 9/16–1¼ in (14–30 mm). They are intended primarily for cleaning off paint or scale from large expanses of sheet metalwork, etc, but may also be suitable for other duties. Thus a small scaler, being lighter than a chipping hammer, may be more convenient to use for light metalwork operations, such as peening, beading, etc. Typical air consumption figures for scaling hammers are 5–7 ft^3/min (0.19–0.20 m^3/min) for the smaller valveless types, up to 20–25 ft^3/min (0.6–0.7 m^3/min) for the largest scalers employed for boiler tube work, etc.

Weld flux scalers are simply smaller sizes of chipping hammer intended specifically for the treatment of welds. They may also be used for other scaling duties within their work capacity.

BroomWade power tools are used extensively on this assembly line for Motorola car radios. Magnetic screw-driving bits help making driving each screw a one-handed operation, also note balanced suspension of tools.

Sand Rammers

The sand rammer is a more or less conventional percussive tool adapted to take a rod attached to the piston to which may be fitted ramming butts or peens (usually of rubber). In some cases the rod is threaded so that extension can be fitted for ramming into deep pipe moulds. Hammering rate is usually of the order of 650 to 800 blows per minute.

Sand rammers are used for foundry work and their design is specific to such use (ie differs from the contractors' rammer). Sizes range from about 4–32 lb (1.8–14.5 kg) in weight, the smaller models being used for bench ramming or light core or mould work and the heavier type for heavy ramming and building up large moulds, etc. Physical sizes range from 5/8 in (16 mm) bore and 2 in (50 mm) stroke up to 1½ in (37.5 mm) bore and 6 in (150 mm) stroke (the latter being equivalent to a back-fill tamper in work capacity). A balanced design is desirable so that the tool 'walks' over the work.

Bench type sand rammers have air consumption figures ranging from about 5 to 15 ft^3/min (0.14 to 0.4 m^3/min) at 80 lb/in^2 (5.6 bar). Floor type (heavier) rammers have a typical consumption of 17–20 ft^3/min (0.48–0.56 m^3/min) in the medium sizes (1 in bore) (25 mm), and 23–30 ft^3/min (0.65–0.85 m^3/min) in the heavy sizes.

Riveting Hammers

Riveting hammers probably represent the largest in numbers of all the percussive tools and range in size from about 2½ lb (1.1 kg) weight for 1/8 in (3 mm) diameter rivets up to 25 lb (11.3 kg) in weight for hot rivets up to 1¼ in (31 mm) or 1½ in (37.5 mm) diameter. They are also designed for specific industries. Thus aircraft or lightweight riveting hammers generally range in piston diameter from 3/8 in (10 mm) for dural rivets of 1/16 in (1.5 mm) to 3/32 in (2.4 mm) diameter up to ½ in (12.5 mm) piston diameter for rivets of 3/16 in – ¼ in (5–6.4 mm) diameter in the softer materials. Slightly heavier models may be used for general metal working duties, and heavy-duty models for the heavier engineering industries, such as shipbuilding. In addition riveting hammers may be adapted for dimpling, scaling, caulking, beading, light shipping, etc.

Riveter performance varies with the particular class or application for which the tool is designed. Thus fast action riveters with short strokes are normally used for working ferrous metal rivets, with speeds up to 5 000 blows per minute. For riveting other metals which would work-harden under repeated hammer blows, slower hitting longer stroke riveters are used. For other work a 'one-shot' riveter may be used, where the tool delivers only a single blow at a time on manipulation of the control lever.

The impact achieved is governed both by the piston diameter and the stroke, the longer the stroke (for a given piston size) the greater the impact. Moderate to long strokes are normally chosen for lightweight riveters, together with spool valves for 'timing'. Construction follows that of chipping hammers, except that the cylinder is usually hardened steel with the bore ground and honed. With larger sizes of riveters strokes may be shorter (in proportion to size), although again there are numerous variants. Some models may incorporate adjustment of force and number of blows per minute to adapt to various work requirements.

Various types of rivet sets can be accommodated, both for driving different rivet heads and for different rivet materials. In addition a 'holder' or 'holder-on' may be incorporated to back up the rivet so that the head is not mutilated and completely fills the rivet cup. Other forms of head may be fitted to adapt riveters as nail and spike drivers, etc, also for pressing, punchwork and stamping. Thus the basic riveter may be made to perform a whole range of duties.

Special types include compression riveters for squeeze-riveting, pop-riveting, eyeletting, etc. Complementary tools include rivet cutters and rivet busters.

It is virtually impossible to give specific figures for air consumption with air riveters since sizes and types vary enormously. As a generalization, however, the heavier tools for hot riveting usually have consumption figures of the order of:-

10–20 ft^3/min (0.3–0.6 m^3/min) for hot rivets up to ½ in (12.5 mm) diameter
25 ft^3/min (0.7 m^3/min) for hot rivets up to ¾ in (19 mm) diameter
25–28 ft^3/min (0.7–0.8 m^3/min) for hot rivets up to 1 in (25 mm) diameter
28–30 ft^3/min (0.8–0.85 m^3/min) for hot rivets up to 1¼ in (31 mm) diameter
35 ft^3/min (1 m^3/min) for hot rivets up to 1.3/8 in (35 mm) diameter
40 ft^3/min (1.15 m^3/min) for hot rivets up to 1½ in (37.5 mm) diameter

One-shot riveters for aircraft type rivets usually have a capacity of 10 to 20 rivets per ft^3/min and rivet busters a consumption of 2 to 5 ft^3/min (0.5 to 0.15 m^3/min) per rivet, depending on size.

Air Starters

AIR MOTORS can offer a number of advantages over electric motors for starting large capacity i/c engines and gas turbines. Air motor weight is only about 50% of that of an electric motor for the same power and an even more substantial reduction in battery weight is possible. Initial cost is also lower, with substantial savings in depreciation and maintenance. A further advantage is that starter operation is independent of battery condition. The air starter is powered from an air tank, either charged from the existing compressor (on vehicles fitted with air brakes, for example) or from a separate engine-driven compressor.

A vane-type air motor is normally used for starter duties, coupled to a conventional Bendix drive via reduction gearing. Bendix pinion speed is usually of the order of 1 000–2 000 rev/min. Operating air pressure for the starter motor is usually of the order of 7 bar (100 lb/in^2) which is readily available from the air brake system compressor charging the separate air starting tank via a one-way check valve. A safety valve is also incorporated, so that the air tank can be kept permanently topped up at full charge when the engine is running, the check valve preventing air bleeding back into the brake system.

The size of tank needed is somewhat arbitrary, but is normally based on sufficient reservoir capacity to give some five or six starts from a full charge. Some weight saving is possible by making the tank from aluminium rather than steel, although this will increase the initial cost.

A typical starter circuit is shown in Fig 1. The starter control is a quick-opening valve, normally manually operated, both for simplicity and to divorce the starter system entirely from the electrics.

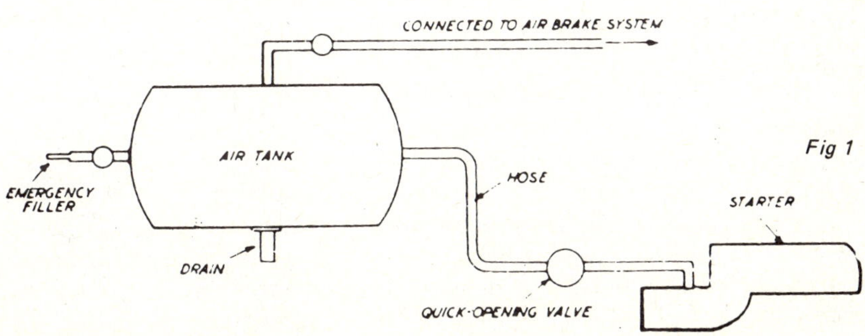

Fig 1

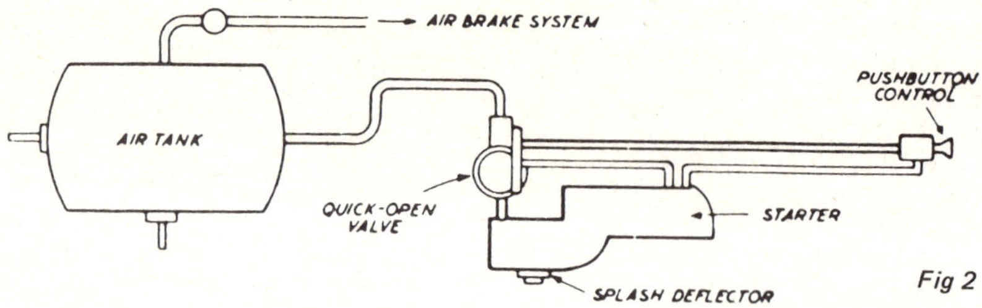

Fig 2

It could, however, be solenoid operated via a push-button if preferred, or pneumatically operated via a remote control position as in Fig 2. Besides the two connection points, the tank is fitted with a drain to enable condensation to be drained off at regular intervals, and a second valve for emergency filling. This enables the tank to be charged from an independent source if necessary — eg a service station air supply, another vehicle's air brake system, or even a gas bottle.

With a stationary system the tank can be eliminated entirely, the motor being connected directly to an existing air supply. Similarly, no separate electrical system is needed in the case of a spark-ignition engine, as magneto ignition can be employed.

The complete elimination of electrics on diesel and spark-ignition driver installations can be particularly advantageous in certain applications — eg for the savings possible in initial costs, installation weight, reduced maintenance, and so on. If auxiliary electric power is still required, this can be provided by an engine driven generator or alternator, with or without a battery as necessary. If a battery still has to be used, then air starting renders the system independent of battery condition and all services are always available as long as there is sufficient compressed air in the tank for starting — ie even with a flat battery, all battery powered services are restored as soon as the driver is started by the air motor.

Air motors are known for their reliability, and their consistency of performance is generally superior to that of electric motors since they are less affected by surrounding conditions. Thus the air motor is dependent only on the availability of a sufficient quantity of air to generate the speed required, and at a suitable pressure to generate the torque required. This is largely independent of the ambient temperature and thus the air motor retains good power characteristics over a very wide range of temperatures. Electric motors, on the other hand, can suffer from a loss of battery power as high as 50% in extremes of low temperature. A further advantage of an air motor is that it can be stalled under full power without damage, and does not place an additional demand on the air supply in a stalled or near-stalled condition. A heavily loaded dc electric motor, on the other hand, will draw increasing current with increasing load, eg when starting a very cold engine, and if it is stalled there is the very real risk of burning out the armature, or at least flattening the battery.

Because they are independent of battery performance and battery condition, air starters are attractive for emergency and standby sets. These can be independent units with their own precharged air tanks, or can draw their air from a normal air line. The system is flexible enough to adapt to a variety of situations. Thus,with a precharged starter unit driving a portable compressor, the air tank can be recharged from the compressor unit when running. If isolated from any source of compressed air, the air tank could be recharged as necessary from gas bottles, or even from a manual pump.

Air starters are finding increasing application in the field of mobile and portable engine powered units, and even on portable and emergency electric generating sets, to remove their reliance on battery condition. In the case of standby units automatically controlled by line loads, the air valve controlling the air starter can be of solenoid-operated type.

The possibility of operating the air starter on a medium other than compressed air may also exist, eg any high pressure gas that may be available. An example here is the use of high pressure natural gas to operate the air starters on gas turbines installed to pump that particular gas. There is no reason why such a 'free' source of compressed gas should not be utilized, providing the pressure available is adequate for operating the air motor and the use of the gas in this manner does not represent any hazard. If the available gas supply pressure exceeds the rated limit for the starter, then it is a simple matter to drop this to a suitable level by interposing a reducing valve in the supply line.

Air Springs

IN ITS BASIC form an air spring is a closed pneumatic cylinder, the 'spring' action resulting from the elasticity of the trapped air when a force is applied to the piston – Fig 1. The basic relationship which applies is:–

$$F = AP_1 \left(\frac{S}{S - \delta}\right)^{1.4} - P_o$$

where A = area of the piston
P_1 = initial pressure in cylinder
P_o = atmospheric or ambient pressure
S = stroke
δ = deflection of travel of the piston under force F.

The spring *rate* (or load divided by deflection) is thus quite different from that of a mechanical spring, being related to an exponential function of the volume/area change rather than directly

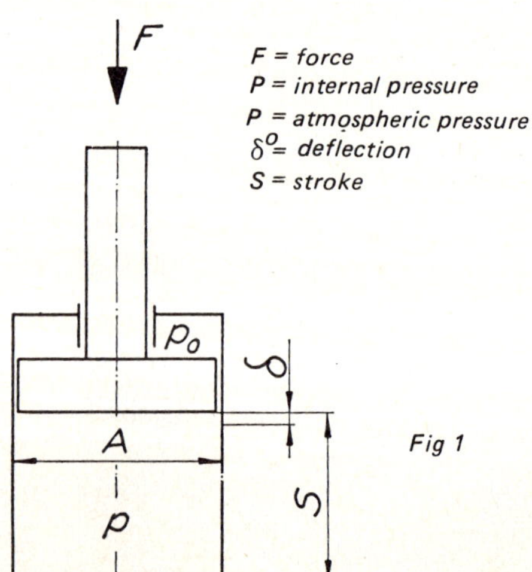

Fig 1

proportional to deflection. It is also widely variable by altering the geometry of the spring, making it possible to design air springs with a wide range of dynamic characteristics.

Another advantage of pneumatic springs is that they are cheap to produce, easy to install and can be designed to cover a wide range of performance requirements. The spring is efficient, since it contains no sliding or mechanical parts, and has no breakout or static friction. A secondary characteristic is the excellent isolation provided by pneumatic springs, which can be of particular advantage for shock isolation mounts on stationary machinery, or vibration control on vibration machines. For continuous performance calculations a figure of about 5% can be assumed to take into account piston seal function and piston rod friction.

Constructional problems are largely reduced to the choice of a suitable material for the hermetically sealed chamber and a form which will provide the necessary contraction and expansion in the direction of the applied load. A bellows form, sometimes with more than one convulution, can be provided by wire rings or a wire spiral. Alternative forms of suitable bellows are rolling sleeve and rolling diaphragm types.

The usual choice of material for the bellows is an elastomer, reinforced as necessary, to withstand the likely pressure changes handled by the chamber. Elastomeric bellows generally offer simpler fabrication and freedom from fatigue, as well as excellent resistance to corrosion, etc. The material most commonly used is nylon reinforced rubber.

With elastomeric bellows the maximum static air pressure in the spring is normally restricted to about 7 bar (100 lb/in^2). Actual working pressure range within the chamber would then be of the order of 0.7–7 bar (10–100 lb/in^2). Higher maximum pressures of up to 12.3 bar (175 lb/in^2) can be accommodated by nylon-reinforced neoprene bellows. If working at higher air pressures is required, metal bellows would normally be used.

The static load range for a typical single-convoluted bellows with a maximum design working pressure of 7 bar (100 lb/in^2) could extend from 227–25 000 kg (500–55 000 lb) depending on physical size – see Table 1. Springs of this type are particularly suitable for anti-vibration or shock-insulation mounts.

TABLE I – TYPICAL AIR SPRING DATA *

Type	Natural Frequency cycles/min	Stroke Range mm (in)	Static Load Range kg (lb)	Axial Space Required Min. mm (in)	Axial Space Required Max. mm (in)	Lateral Space Required mm (in)	Static Pressure bar (lb/in^2)
Bellows:							
Single Convolution	150–300	2.5–150 (0.1–6.0)	225–25 000 (500–55 000)	4.5 (1.8)	75 (3.0)	150–740 (6.0–29.0)	0.7–7 (10–100)
Two Convolution	110–150	2.5–300 (0.1–11.75)	70–25 000 (150–55 000)	70 (2.75)	83 (3.25)	180–740 (7.0–29.0)	0.7–7 (10–100)
Three Convolution	90–110	2.5–400 (0.1–16.0)	180–25 000 (400–55 000)	115 (4.5)	125 (5.0)	270–740 (10.5–29.0)	0.7–7 (10–100)
Rolling Sleeve	60–100	2.5–460 (0.1–18.0)	45–3 200 (100–7 000)	120 (4.75)	180 (7.29)	175–380 (5.0–15.0)	0.7–7 (10–100)
Rolling Diaphragm	20–90	2.0–230 (0.1–9.0)	25–1 150 (50–250 000)	127 (5.0)	200 (8.0)	140–635 (5.5–25.0)	0.7–7 (10–100)

*Firestone Industrial Products

Fig 2

Fig 3

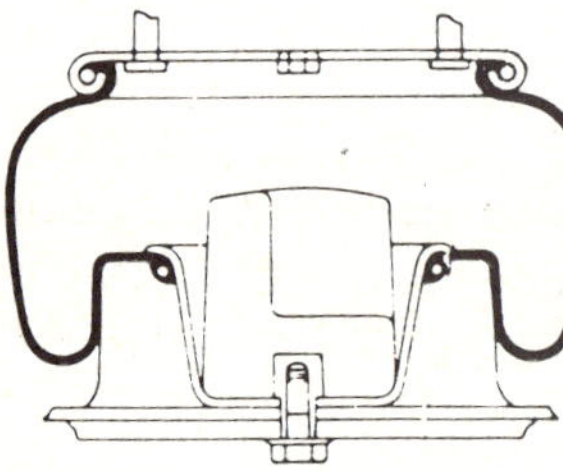

Fig 4

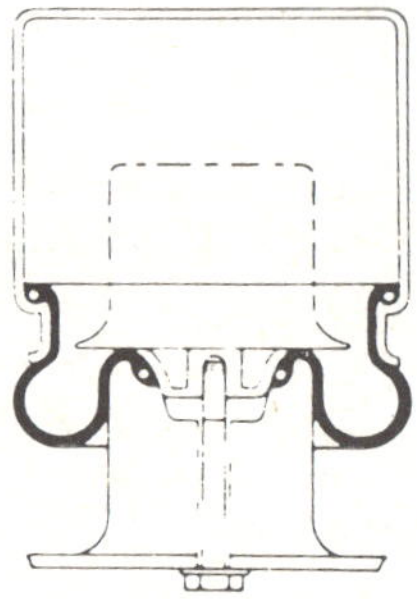

Fig 5

Where greater deflections have to be accommodated, the stroke range can be increased roughly in direct proportion to the number of convolutions. Thus Fig 2 shows a two-convolution pneumatic spring with approximately double the stroke of a single-convolution spring of the same diameter, and Fig 3 a three-convolution pneumatic spring. In this case the stroke is about 2.7 times that offered by a single-convolution pneumatic spring of the same diameter.

The rolling sleeve bellows – Fig 4 – offers an even greater capacity, but at the expense of a much reduced static load range. This also produces the softest form of pneumatic spring. The rolling diaphragm pneumatic spring – Fig 5 – is appreciably stiffer, but has only about half the stroke range for the same spring diameter.

Pneumatic springs, particularly those of the multi-convolution type and rolling sleeve type, are suitable for direct application as vehicle springs in conjunction with such damping devices as may be necessary. Springs of these two types can generally be used on trailers without damping.

Single convolution bellows and rolling diaphragm configurations produce stiffer springs which are more suitable for shock and vibration absorption.

The load capacity of a pneumatic spring can be as high as 227–450 kg (500–1 000 lb) per 0.45 kg (1 lb) weight of spring, depending on the design size and type. This figure compares favourably with most other types of conventional spring. The natural frequency can also readily be selected to come within prescribed limits by suitable attention to geometry. This normally falls within the range of 1–2.5 Hz, or appreciably lower than that of most other spring elements. The natural frequency is a constant factor with a pneumatic spring, since any given design produces essentially the same deflection or natural frequency under all static load conditions.

The performance of pneumatic springs can be modified by connecting the (normally) sealed cylinder to a suitable external circuit.

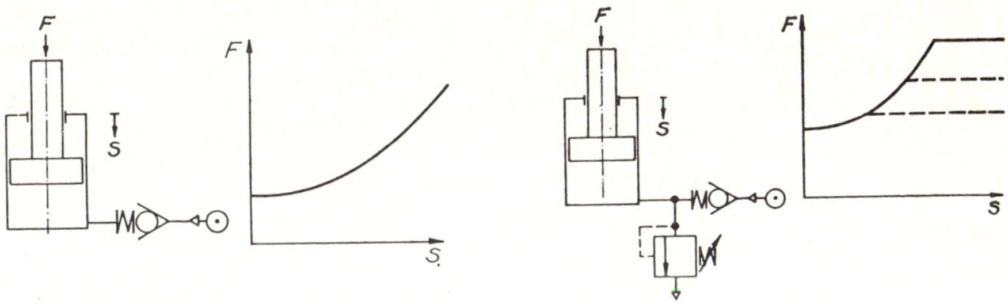

Fig 6 Spring force diagrams — cylinder with non-return valve (left) and with relief valve limiting spring force (right).

In such applications a short pneumatic cylinder is normally employed as the spring, pressurized by an external supply and with the exhaust port omitted (or closed) — Fig 6. A non-return valve in the supply line ensures that the spring is maintained at the desired air pressure.

With such a circuit the spring force can be limited by inserting a relief valve between the non-return valve and the cylinder. Additional characteristics can be imparted by including other pneumatic control valves in the system to give almost any form of Force-Stroke characteristics required.

If a constant force pneumatic spring is required then this characteristic can be achieved by inserting a reservoir between the supply and the spring cylinder — Fig 7. The reservoir should have a capacity two or three times that of the cylinder; the larger the capacity the more constant the spring force achieved. Alternatively a constant force spring may be produced by interposing a pressure regulator between the supply and cylinder.

Fig 7 Force diagram with air spring connected to a large reservoir.

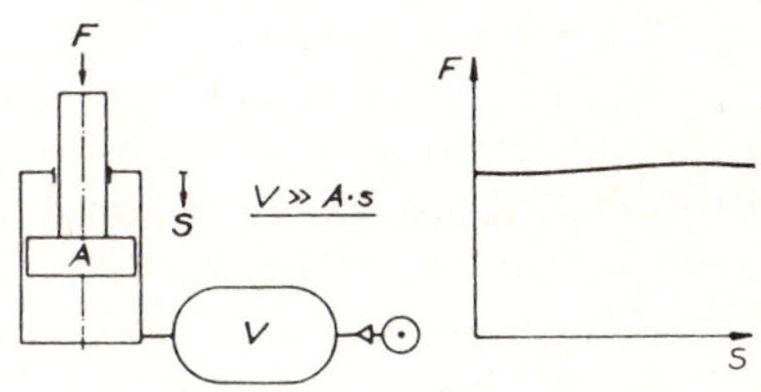

TABLE II — SELECTION GUIDE TO PNEUMATIC SPRINGS

Requirement	Type
Maximum load	Bellows
Maximum load and minimum deflection	Single-convolution bellows
Maximum load and maximum deflection	Three-convolution bellows
Maximum stroke	Rolling sleeve or three-convolution bellows
Minimum stroke	Any type
Minimum axial space	Single-convolution bellows
Minimum lateral space	Rolling sleeve
Lowest natural frequency	Rolling diaphragm
Highest natural frequency	Single-convolution bellows
Greatest range of natural frequency	Rolling diaphragm

Blowers and Fans

A *blower* is defined as a low pressure (eg 2.5 bar) compressor delivering a small volume of air at a moderate pressure. A *fan* is defined as an air mover for displacing or delivering a large volume of air at a very low pressure (eg under 0.1 bar). In practice, the main difference is in the size and form of the impeller. Single-stage blowers may, however, have fan-type impellers generating very low pressures, when they are correctly described as fan-blowers.

Propeller Fans

The propeller type fan consists of a number of blades mounted at a pitch angle and assembled on (or integral with) a hub, mounted on a driven shaft. Rotation generates a slip stream in a similar manner to an airscrew, although the efficiency is very low compared with a true airscrew, since no attempt is made to provide a constant pitch along the length of the blades (the blades may even have a constant pitch angle). Further, a relatively large number of blades may be used, with the resulting interference again reducing efficiency. At best, therefore, a simple propeller type of fan can only generate a very low pressure, although it may be capable of moving a reasonable volume of air.

The following fan laws apply:-

i) For a constant blade size:

(a) volume varies as rev/min
(b) pressure varies as $(\text{rev/min})^2$
(c) horsepower input varies as $(\text{rev/min})^3$

ii) For a constant speed:

(a) volume varies as $(\text{diameter})^3$
(b) pressure varies as $(\text{diameter})^2$
(c) horsepower input varies as $(\text{diameter})^5$

Performance can be expressed quantitatively by the following formulas:

Volume = k_v x rev/min x (diameter)^3
Pressure = k_p x (rev/min)^2 x (diameter)^2
Horsepower = K_{hp} x (rev/min)^3 x (diameter)^5

where k_v = volume coefficient. For volume in ft^3/min and diameter in feet, typical values of k_v lie between 0.65 and 0.9.

k_p = pressure coefficient. For pressure In Inches water gauge and diameter in feet, typical values of k_p lie between 0.08×10^{-6} and 0.18×10^{-6}. For simplicity of working the 10^{-6} factor can be eliminated by entering

$$\frac{\text{rev/min}^2}{1\,000} \quad \text{in the formula}$$

K_{hp} = horsepower coefficient. For diameter in feet typical values for K_{hp} lie between 0.01×10^{-9} and 0.03×10^{-9}. Again the 10^{-9} factor can be eliminated by entering

$$\frac{\text{rev/min}^3}{1\,000} \quad \text{in the formula}$$

The same fan laws and basic formulas also apply in the case of axial flow and centrifugal fans, only in these cases the values of the coefficients vary widely, depending on the specific design of the impeller and other empirical factors.

The performance of a simple propeller-type fan be considerably improved by enclosing it within a close fitting open-ended cylindrical casing, producing a ducted fan – Fig 1. The design of the ducting is critical. It may be parallel sided or convergent-divergent (with a fan at the throat). Diffuser vanes may also be added. Much depends on the individual installation as to how efficient the fan is, although quite high pressure can be realized with suitably proportioned ducted fans.

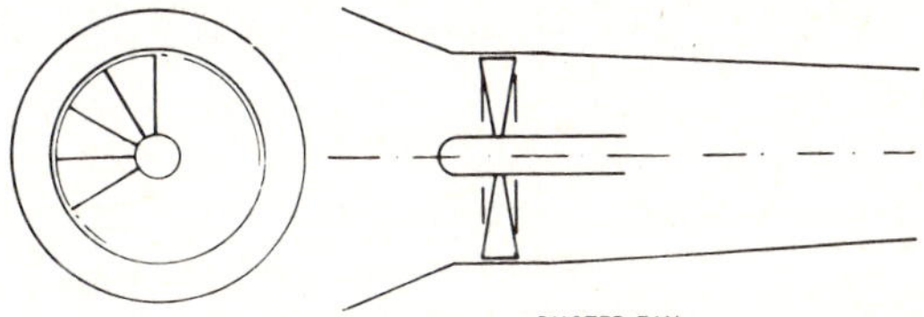

DUCTED FAN

Fig 1

Thus with peripheral velocities of the order of 120 m/s (400 ft/sec) a pressure rise of 0.07 bar (1 lb/in^2) may be achieved although in practice this is seldom realizable because of the high noise level generated. Maximum peripheral velocities of the order of 75 m/s (250 ft/sec) are more usual, when if higher pressure is required a multi-stage unit can be employed

Particular advantages of propeller type fans are relatively simple construction, low cost and the ability to handle contaminated and even gritty air. Their short axial length, even with motor fitted, makes them suitable for installation directly in walls, or for compact installation elsewhere (with or without ducting). Their performance capabilities are limited to those of a true fan rather than a blower, although blower pressure levels can be realized with multi-stage units. They are widely used for straightforward ventilation, forced draught and cooling purposes and are normally quiet operating, running at low to moderate speeds. They may, however, pose both noise and vibration problems, run at higher speeds and higher efficiencies, or where the fan blades pass close to a stationary member. The latter problems can sometimes be overcome by adopting an asymmetrical fan form.

Axial flow fans are essentially a development of the simple propeller-type fan, developing the advantages of the latter but eliminating the disadvantages as far as possible. They are invariably ducted with the smallest possible clearance between the fan blades and casing consistent with safe operation. The smaller the clearance, the more effective is the casing in preventing recirculation of the air at the blade tips and loss of efficiency. The sides of the casing are normally parallel, although entry may be convergent, and relatively long to completely enclose both the fan and motor. Particular attention is given to the aerodynamic form of the impeller blades and to streamlining the through-flow. As a consequence an axial fan, can generate a higher pressure than a propeller-type fan and has a higher efficiency. It is still, however, essentially a low pressure machine, with high volumetric capability (Fig 2).

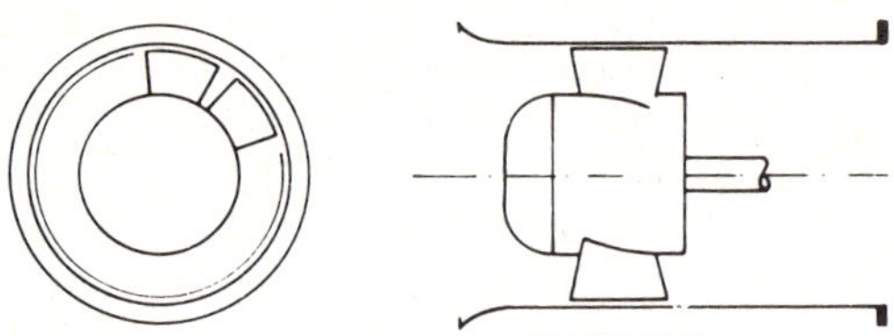

Fig 2

Critical parameters are the aerofoil form of the blades, number of blades, blade angle and area, overall diameter and hub ratio – in addition to tip clearance and through-flow design as already mentioned. For low pressure working, the hub ratio is relatively small (ie a small diameter hub with long blades). At the other end of the scale a higher pressure axial fan (blower) would have a large hub ratio and short blades, possibly of quite different aerofoil section. More attention may also be paid to the design of the fairings to streamline the flow. Guide or diffuser vanes may also be fitted downstream to straighten out the flow and improve pressure development.

Noise can be a problem and can limit the practical maximum speed of the fan. In this case the critical parameter is the entry velocity of the air, a nominal limit being about 60 m/s (200 ft/s) above which velocity the noise can be intolerable. The presence of guide vanes can also cause noise problems by setting up local oscillatory flow, which can generate intense high pitched noise.

To increase the pressure available from an axial flow fan a two-stage unit may be employed. The stages may be driven by a common shaft (eg by the same motor with a shaft at each end), or be contra-rotating. In the former case, fixed flow straightening vanes are necessary between the two impellers, so that the second impeller can operate efficiently. No inter-vaning is necessary with contra-rotation, although the mechanics of the drive are far more complicated. In practice, the simplest solution is often to drive the two fans by separate motors. The impeller form may also be modified slightly to provide forced vortex flow.

A further development is the bifurcated coaxial fan where the motor is isolated from the downstream flow by taking this flow through two separate ducts – Fig 3. This can be particularly useful where the air being handled is contaminated, corrosive or excessively hot, and could cause damage to a motor exposed to it. It also places the motor in a very accessible position for inspection and maintenance. Alternative methods of isolating the drive motor include the use of V-belt drives, suitably faired, and mounting the fan at a bend in the ducting system, where a straight drive shaft can be taken through the wall of the bend to an externally mounted motor.

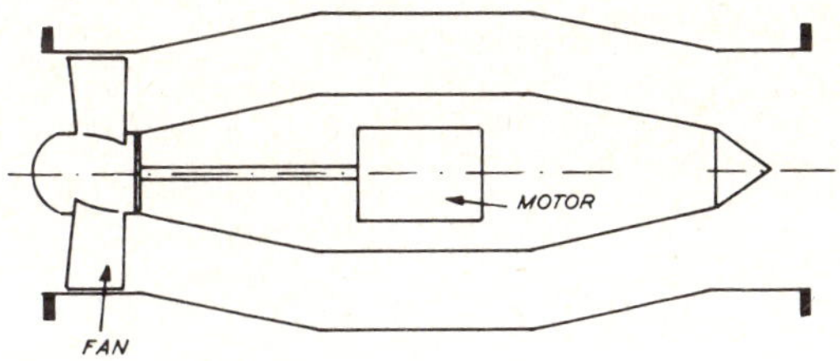

Fig 3

Radial Flow Fans

Radial flow impellers have similar characteristics to centrifugal compressors. The inducted air is turned through a right angle and ejected radially from the impeller. This provides greater pressure generation. The volume of air moved depends largely on the form of the impeller. Three basic blade forms are employed on blowers:-

(i) forward-curved blades (squirrel cage impeller)

(ii) backward-curved blades

(iii) straight radial blades

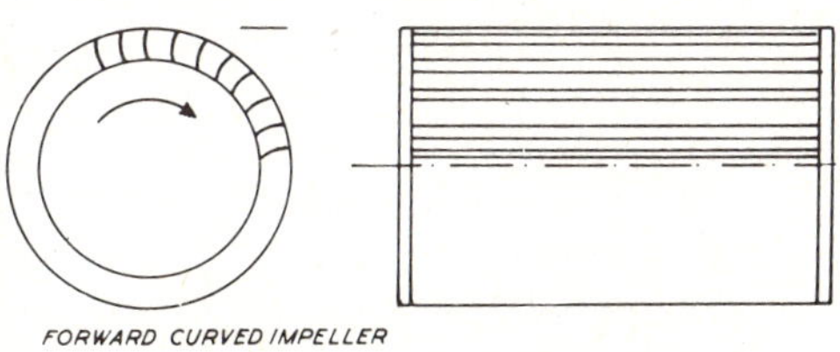

Fig 4

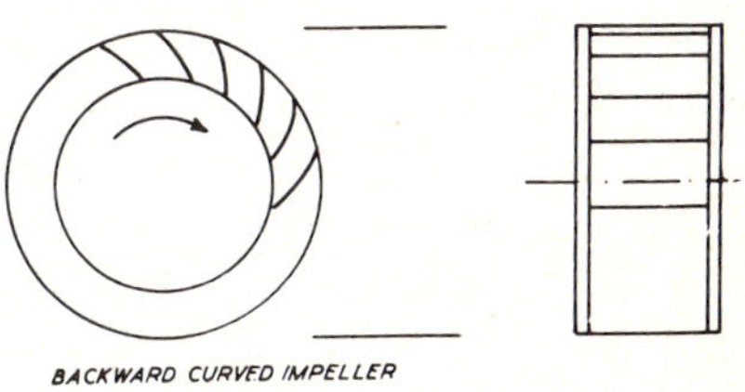

Fig 5

The forward-curved bladed impeller — Fig 4 — has a pressure-generating potential some five times greater than that of an axial flow fan and because of its form, with a large number of scoop-shaped blades usually of generous length, also has a high volume capacity. It is thus particularly suitable for handling a large volume of air where higher pressures are required than could be given by an axial flow fan. However, for the same capacity it will be much larger in diameter and more costly. Also its efficiency tends to be relatively low and restrictions may be placed on operating speed because of the noise level generated. This in turn can reduce the effective or usable capacity of the blower. They are used only when the discharge duct is short, since the high velocity at the discharge of the volute should not be converted to pressure.

The backward-curved bladed impeller is also a high volume machine, capable of generating high flows at lower speeds than the forward-curved fan. Size for size, it could handle the same volume of air at an appreciably lower speed, and thus operate with a lower noise level. The latter advantage is further enhanced by the fact that the flow through the blades remains relatively streamlined, although much depends on the aerofoil form of the blades. True aerofoil blades produce the smoothest flow and highest efficiencies, but result in the highest cost impellers. Flat plate blades may be used on simpler, less expensive designs, still capable of giving a good performance with high efficiency. The basic configuration is still sufficiently versatile to enable it to be suitable for working over a wide range of sizes and specific speeds and so is a widely favoured type for blowers. Pressure capabilities are also high.

The pressure generated by either type of centrifugal impeller are generally sufficient for all blower duties, thus more than one stage will not usually be required. Multi-stage units are, however, used as compressors. Impellers with forward-curved blades are generally unsuitable for parallel operation.

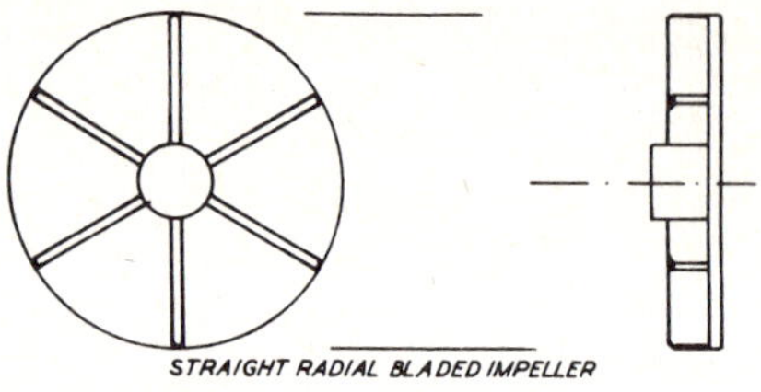

Fig 6

The radial bladed impeller – Fig 6 – is essentially a high pressure low capacity form (virtually the opposite characteristics to an axial flow fan). It has a relatively low efficiency. Its chief advantages are the simple construction and the open form of the impeller, which make it particularly suitable for working with dusty or gritty air – hence its common application as an exhaust fan in heavily contaminated atmospheres. If necessary the impeller can be made even more non-clogging by fitting a backplate to prevent fibrous material collecting at the back of the hub, or be protected against a corrosive atmosphere by simple coating treatment.

Some performance comparisons between axial flow fans and the three basic forms of centrifugal blowers are summarized in Figs 7 and 8. Fig 7 shows the characteristic horsepower and efficiency curves, and Fig 8 static pressure curves. It should be noted that these are related to 100% performance of the individual machines and that no quantitative comparison between the different types can be drawn for pressure or volume.

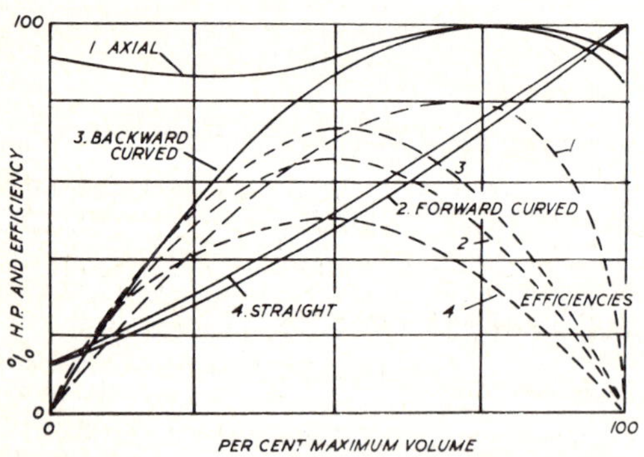

Fig 7

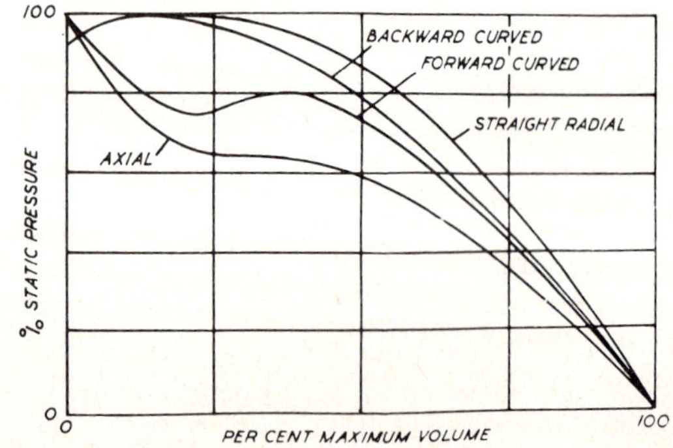

Fig 8

Choice of Type

For any application it is obviously desirable that the pressure and volume requirements be met with the highest possible machine efficiency. A convenient parameter for relating machine performance parameters and efficiency is *specific speed,* calculated as:-

$$\text{Specific speed } N_s = \frac{\text{rev/min x (volume)}^{0.5}}{\text{(static pressure)}^{0.75}}$$

TABLE I – OPTIMUM SPECIFIC SPEEDS FOR FANS, BLOWERS AND POSITIVE DISPLACEMENT MACHINES

Machine	OPTIMUM SPECIFIC SPEED RANGE	
	Metric Units	English Units
Propeller-type fans	600 000 – 2 500 000	100 000 – 400 000
Axial flow fans	300 000 – 2 000 000	50 000 – 300 000
Centrifugal impellers:		
Forward-curved, single suction	55 000 – 25 000	9 000 – 40 000
Forward-curved, double suction	185 000 – 425 000	30 000 – 70 000
Backward-curved	60 000 – 500 000	60 000 – 500 000
Straight radial blade	25 000 – 120 000	4 000 – 20 000
Positive displacement (compressor)	1 000 – 8 000	6 000 – 50 000

Each type and size of machine will have a specific speed range over which its efficiency will be around a maximum, and initial selection of a machine type can be made accordingly – see Table I.

The performance characteristics of the selected type must then be analysed to see if it can provide the actual pressure and volume required, at a practical number of revolutions per minute.

Propeller-type fans are most efficient operating at high specific speeds, consistent with high rotational speeds, and very low pressures. Axial flow fans can operate efficiently at lower specific speeds, but maintain good efficiency into the region also covered by propeller-type fans. They are essentially high volume/low pressure devices, with again the emphasis on high rotational speeds.

The specific speed range of centrifugal blowers is less clearly defined, since this is influenced by individual design geometry, particularly in the case of forward-curved bladed impellers. Here two separate ranges are shown, one for impellers with single suction and the other for impellers with double suction. The two together cover a similar range to that of the backward-curved bladed impeller. The straight radial bladed impeller achieves its best efficiency at much lower specific speeds, although it overlaps the lower part of the other centrifugal types.

A generalization which can be drawn from this is that a high specific speed is consistent with high volume, low pressure delivery; a low specific speed is consistent with high pressure, low volume delivery. Positive displacement machines also fall into the latter category, with specific speeds extending down even lower than the straight radial bladed centrifugal blower.

Lifts and Hoists

LIFTING and hoisting machines can be powered by either air cylinders or air motors. When produced as integral units they are normally described as air hoists. Winching actions are most commonly provided by air motors (air winches), but cylinders are not completely excluded from this duty.

Cylinder hoists have a more limited field of application than motor hoists for working at extreme or variable heights. Also cylinders tend to be bulky by comparison with air motors of the same lifting capacity. The usefulness of cylinders can, however, be extended by compounding the mechanism, using two or more cylinders.

Another basic difference between the two types is that cylinder hoists are best suited to fixed installations (although many are trolley-mounted). They lend themselves particularly to siting alongside or on the device to be operated, such as a pneumatic loader. For specific lifting applications they have the advantage over air motors that they can provide greater load stability and a more rigid system.

Examples of hoists with power trolleys 150–2000 kp (330–22000 lbs). (Deutsche Gardner-Denver).

Motor hoists and winches are basically conventional winching mechanisms powered by an air motor. They can be used to advantage where other types of driver are less economic to operate or are excluded because of the conditions involved (eg air motors can operate successfully at high temperatures or in an explosive atmosphere which could preclude the use of an i/c engine or an electric motor as a driver). An air motor has the further advantage of being very simple to control. Motor hoists are equally suitable for fixed installations or for mounting inside hoist installations running on an elevated I-section rail.

The specific advantages normally provided by motor-powered air hoists include:-

(i) small, lightweight compact construction, which can be allied to general solidness.

(ii) low maintenance requirements.

(iii) variable lift speed.

(iv) safe (ie undamaged) if stalled or overloaded.

(v) high working cycle (as much as 100% if necessary).

The characteristics of a motor hoist or winch are determined by the type of motor fitted. Thus a piston motor would be used where high starting torque is required. Heavy duty or general purpose air hoists are usually fitted with piston motors. Vane motors are, however, also widely used for air hoists and winches, particularly where a compact lightweight unit is required.

The basic hoist unit can be equipped in different ways for lifting – eg with roller or link chain, rope or pendant control. Rail-mounted units can also incorporate traction drive to provide traverse motions both along a horizontal rail and up gradients. An ingenious extension of this is the mounting of the hoist on a flexible wire rope suspended from a safety booth above the working area. The hoist then runs up and down the wire rope which can also be swung to one side to provide

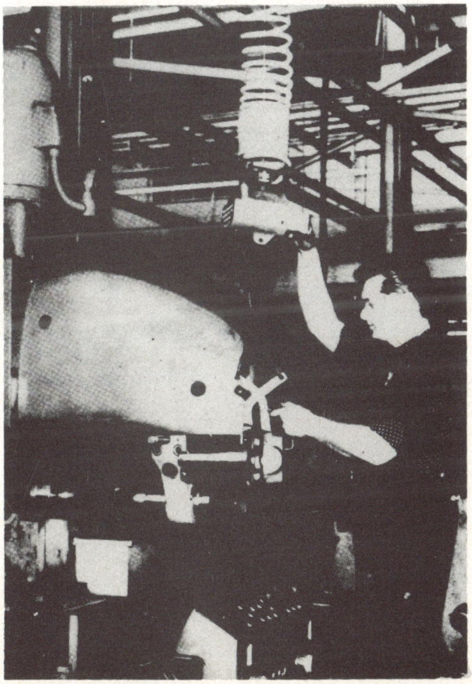

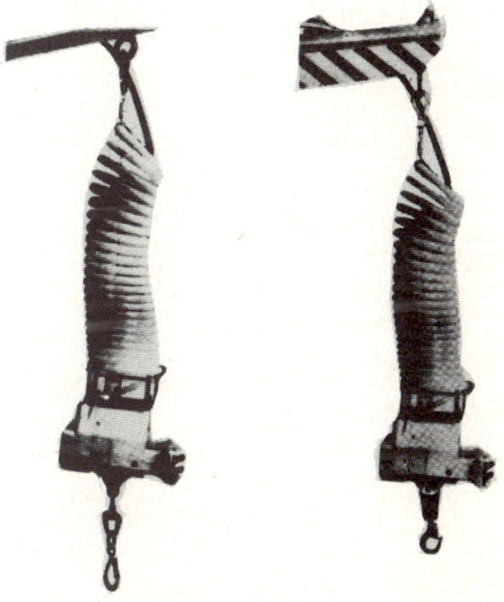

Deutsche Gardner-Denver hoist mounted on a flexible wire rope.

pick-up over a wide area. This form of suspension is limited to smaller capacity hoists – ie with a typical lifting capacity of up to 100 kg (220 lb). The general range of compressed air (motor driven) hoists is from 10–10000 kg (220 lb–10 tons). The all-up weight of an air hoist can be as low as 1.5% of its lifting capacity (in larger sizes).

Lifting speed is largely governed by the reduction gear ratio employed in the design of the hoisting unit. The actual speed achieved is also dependent on load. Typically the full load speed is 50% of the free (unloaded) speed. Some representative performance figures are given in Table I.

TABLE I – TYPICAL PERFORMANCE OF AIR HOISTS

CAPACITY		LIFTING SPEED			
		Free		Full Load	
kg	lb	m/min	ft/min	m/min	ft/min
250	550	50.0	165	27.0	88
500	1 100	25.0	82	14.0	46
1 000	2 200	12.0	40	7.0	23
2 000	4 400	6.0	20	4.0	13
3 000	6 600	4.5	15	3.0	10
6 000	13 200	2.2	7	1.5	55
10 000	22 000	1.9	6	1.0	3

Capacity	AIR CONSUMPTION			
	Free		Under Load	
	m^3/min	ft^3/min	m^3/min	ft^3/min
250	1.7	60	1.60	56
500	1.7	60	1.60	56
1 000	1.7	60	1.60	56
2 000	1.4	50	1.15	40
3 000	2.4	85	2.20	78
6 000	2.4	85	2.20	78
10 000	2.4	85	2.20	78

One of the range of Gardner-Denver axial piston motor powered air hoists.

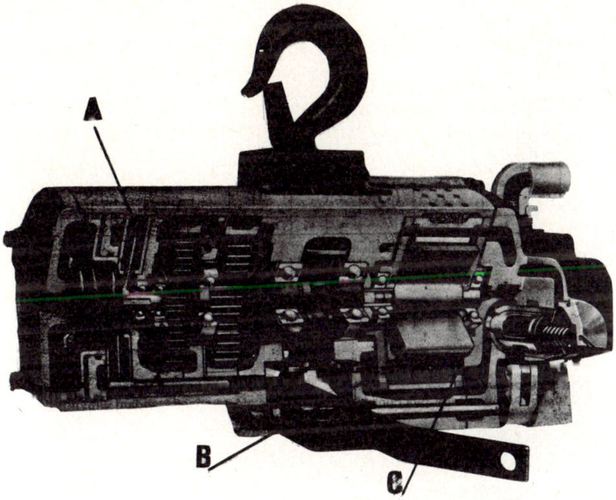

Thor rotary hoist. A – disc brake. B – unitized motor unit. C – vane motor.

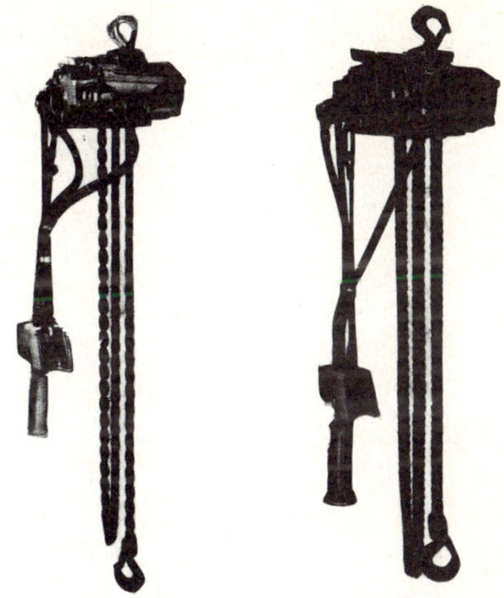

Hook mounted hoists with pendant control (Thor Tools).

TABLE II – TYPICAL DATA – PENDANT CYLINDER AIR HOISTS*

Lifting Capacity tons (metric tonnes)	Average Speed ft/min (m/min)	Lift ft (m)	Rope Length ft (m)	Rope Diameter ins (mm)	Chain Length ft (m)	Chain Line Diameter in (mm)	Weight, inc Lip Hook lb (kg)	Hose Size in (mm)
1.0	65.5 (20)	21.6 (6.6)	54 (16.5)	¼ (6.4)			109 (95)	¾ (20)
1.5	36 (11)	10 (3)			21.3 (6.5)	3/8 (10)	176 (80)	¾ (20)
2.0	36 (11)	21.6 (6.6)	56 (17)	3/8 (10)			308 (138)	¾ (20)
3.0	21.3 (6.5)	10 (3)			21.3 (6.5)	½ (12.5)	275 (124)	¾ (20)
3.2	29.5 (9)	23 (7)	59 (18)	½ (12.5)			594 (270)	1 (25)
5.0	23 (7)	23 (7)	61 (18.6)	5/8 (16)			835 (378)	1 (25)
5.0	18.4 (5.6)	10 (3)			21.3 (6.5)	5/8 (16)	715 (325)	1 (25)
10.0	11.2 (3.4)					1 (25)	945 (428)	1 (25)

*Atlas Copco Ltd.

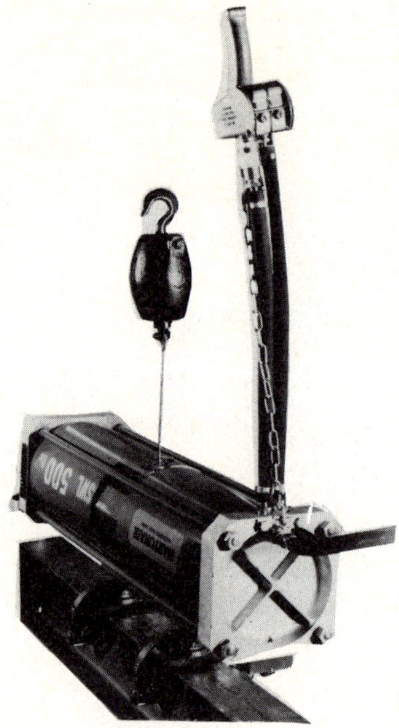

Martonair air hoist

Air winches are similar in configuration and design to other types of power winches, with similar rugged construction. The usual choice is a piston motor for high starting torque and maximum 'feel' in the controls. As with motor-driven hoists, braking must be incorporated in the units as an air motor has no self-braking capabilities when unpressurized.

In addition to air motors, pendant cylinders may be used for air hoists with an appreciable saving in cost and complication. Such cylinders are normally direct acting and lift capacity is directly related to the cylinder size and line pressure — see Table II. Pendant type cylinder hoists, of course, take up more vertical height than air-motor hoists or winches and so can only be used where space permits.

Pneumatic Timers

IN its simplest form the pneumatic timer comprises an elementary piston/cylinder combination with an air bleed at the blind end of the cylinder. The piston is spring loaded so as to advance towards the blind end, the rate of travel, and thus the 'timing' interval, being adjusted by the rate of air bleed — see Fig 1. To reset, the piston rod is simply pulled outwards, allowing air at atmospheric pressure to leak past the piston to fill the cylinder. Release of the piston then starts its motion down the cylinder once more, trapping and compressing the air in the blind end.

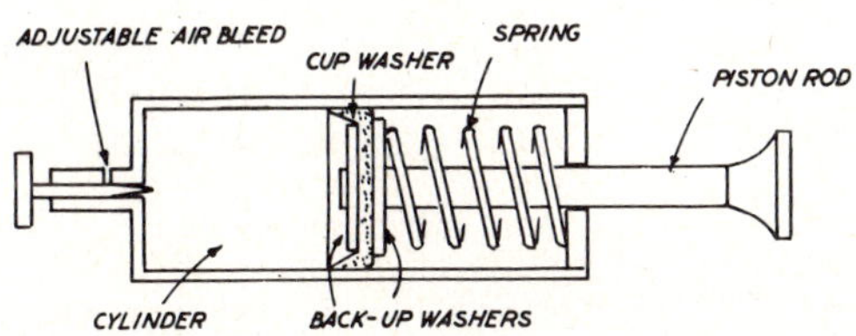

Fig 1

The performance of such a simple unit has distinct limitations. Positive timing is dependent on a perfect piston seal on the closing stroke and positive adjustment of the air bleed. Any setting of bleed rate timing (actual time of stroke) will tend to be variable with air density and temperature. Nevertheless, quite consistent results can be achieved from such timers if they are well designed and made, particularly in the small sizes. The attraction is that they can be made extremely small and light — eg the cylinder may be as small as ¼ in diameter and the total weight of the timer ¼ oz or less. They are not, however, generally suitable for precision timing.

With the mode of working described, the piston rod movement is used to operate a mechanical trip or close/open electrical contacts, the available activating force being that of the return spring. Substantially higher forces can be achieved by adopting the reverse mode of working, ie a fixed or variable orifice. In this case the force available is equal to the product of the pressure of the air supply (less pressure drop through the orifice) and the piston area. The range of timing control is more restricted, however, particularly if reasonably high output forces are required.

In order to produce longer time delays and rotary movement the piston can be varied and improved and used to operate a ratchet. The air supply is arranged so that, when the ratchet is operated, the air supply is removed from the piston, allowing it to return by means of an integral spring. In such a system the air flow to the piston controls the speed at which the ratchet rotates. Such a stepping mechanism can be used to provide longer time delays or to operate any rotating equipment.

A more refined method of producing time delays is by the use of a mechanical escapement timer. It is possible to wind up the spring using an air-operated piston and when the escapement timer has completed its timing, it can release a lever which will mechanically operate a pneumatic valve. The output from this valve will produce the timing signal at the end of the time delay set on the escapement timer.

In general piston/air bleed timers tend to be suitable only for relatively short timing periods (eg up to 1 minute), with their accuracy largely dependent on ambient temperature. They are also rather prone to malfunction due to their need for free movement of the piston and for clean air to prevent the orifice becoming clogged. In spite of these drawbacks this type of timing equipment is quite commonly used in applications where cost is important but accurancy is not paramount. The timers are generally available with a delay function, ie timing is initiated by connection of the air supply and the output is made available after the time delay has elapsed, or with interval function, ie the output is made available at the same instant as the air input supply is turned on and is then removed after the time delay has elapsed. There are a number of other variations on these timing duties which can be achieved with this type of equipment.

Various forms of camshaft timers driven by a pneumatic piston/ratchet drive system are commercially available. Depending upon the gear ratios utilized, they can provide cam-shaft speeds as slow or fast as is required. This type of drive system, however, suffers from poor accuracy and from a fairly high ambient noise level.

Timers which work from a definite compressed air supply rather than by a simple piston are more generally referred to as timing valves although similar principles are involved – eg a basic time-delay volume is incorporated with timing rate or timing action controlled by an air bleed. Frequently the air bleed path is thread clearance to provide a convenient and practical method of adjustment.

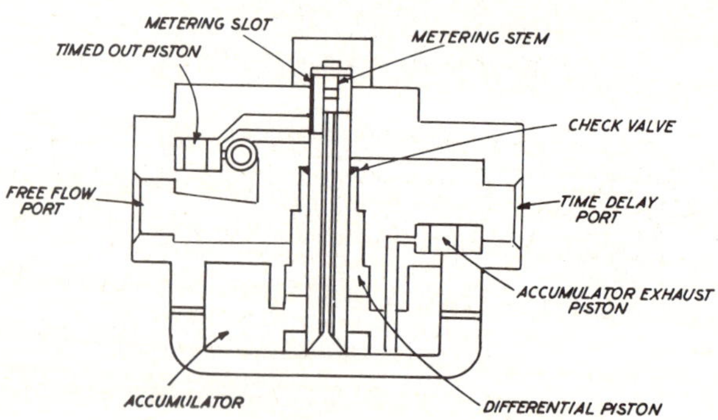

Fig 2

A design of timing valve is shown in Fig 2 where the timing volume is metered by a sharp edged orifice produced by a slot cut through the side of a hollow stem in which a close fitting spindle travels. Flow to the time delay volume is adjusted by altering the position of the spindle along the length of the slot, the time required to change the volume and actuate the valve being determined by the length of slot left open by the spindle.

A particular advantage offered by this system is the infinitely variable adjustment of orifice area possible giving fine resolution, whilst the flow characteristics are those of a single sharp-edged orifice rather than the more usual thread leakage. This gives high repeatability of the order of ±2% at fixed supply pressure and ±5% with a 10 lb/in^2 variation in supply pressure. These figures refer to a specific design, as accuracy of working and repeatability will depend very largely on actual construction and working conditions. For long-term accuracy it would be necessary to ensure that the air supply was clean and free from moisture.

In the particular design shown, three snap-action valves are incorporated in the valve envelope. The main valve is a check valve which controls the flow of working fluid actuated by a differential area piston exposed to the pressure in the timing volume or accumulator chamber. It can be arranged to be either normally closed or normally open, giving either timed-in or timed-out control of flow respectively.

The other two valves are differential area pistons operated by pressure in the inlet and outlet ports. These pistons control flow in to and out of the timing circuit.

All-pneumatic timers can also be operated on air motor drivers with standard (clock timers) or specific mechanisms as required. These can rate as true precision instruments with an accuracy comparable with that of mechanical or electrical timing equipment. Such all-pneumatic timers have a wide application in hazardous locations, pneumatic control systems and installations where the provision of electrical timing controlling electro-pneumatic components may be uneconomic.

Motors for such timers can be of miniature type, eg vane or turbine, with a suitable governor. A low speed high torque output can be derived by suitable reduction gearing. The actual output speed achieved may range from one revolution per second to one revolution in several hours according to the duty required.

Such timers are intended to be operated direct from the normal compressed air supply. If the motor is rated for low pressure they can be used in conjunction with a pressure reducing valve.

Fig 3 Pneumatic camshaft timer type D63/P.

Air motor timers can be associated with quite sophisticated pneumatic crankshaft timing equipment, the motor driving the camshaft via a gearbox (see Fig 3). In this example the camshaft carries five 180° individually adjustable cam pairs. Each cam pair operates a simple plastic three port valve. The timer is provided with a calibrated knob at the right hand side which can be utilized for setting duties. The cams are of the non slip ratchet type which can be adjusted to give a dwell at any position from 0–180° of mechanical rotation of the camshaft. Dwell times from

180–360° of mechanical rotation are effected by changing the pivot position on the air valve to convert it from normally open to normally closed. When a duty has been set, the ratchet cams can be locked into position by the locking screws. This very versatile camshaft timer has an aluminium extruded chassis and can be built with virtually any number of cams as required. More complicated duties can be achieved by utilizing an intermediate gearbox between separate camshafts on the same unit. With this technique it is possible to operate, for example, one camshaft at one revolution per minute and a second camshaft at one revolution per hour. Camshaft speeds from one revolution per second to one revolution per day are very readily available.

A number of pneumatic timers are available with mechanical escapements for the time element. An example of such a timer is shown in Fig 4. In this unit, the required time can be set using the pointer and dial on the front of the unit. This mechanism can achieve various timing functions, including interval and delay functions. The pneumatic circuit for this unit is shown in Fig 5. The control air is fed to the piston to start the timing function. It can be arranged for the output PV air valve to be operated immediately for the time period to provide an interval function or at the end of the time period to provide a delay function. Although time delays in excess of fifteen minutes are not practical using escapement timers, the timers provide extremely good repeat accuracy.

Fig 4 Pneumatic escapement timer type SP.

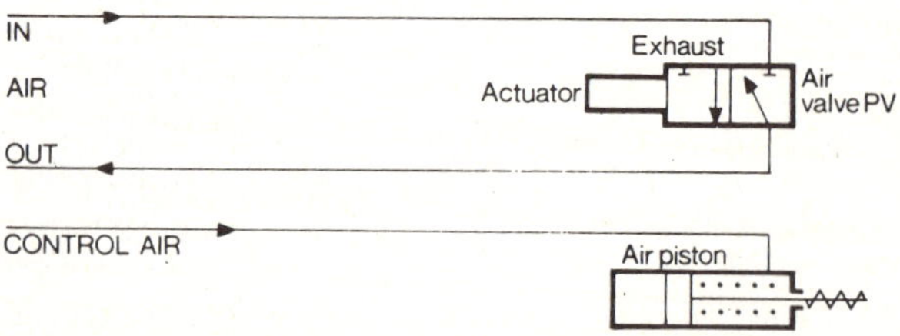

Fig 5 Pneumatic circuit for SP pneumatic escapement timer.

Dial timers utilizing air motors are produced in various forms. These, in general, enjoy the advantages of the escapement timer with the added advantage that there is virtually no limit to the length of time interval or delay which can be achieved. A typical motor driven pneumatic dial timer is shown in Fig 6. This timer is set to the desired time by rotating the external setting bezel. The time set can be read in the top window, the pointer always remaining at zero on the dial, which rotates with the bezel. When the timer is operated the pointer will rotate until it reaches the set time.

Fig 6 Pneumatic dial timer type Air Star.

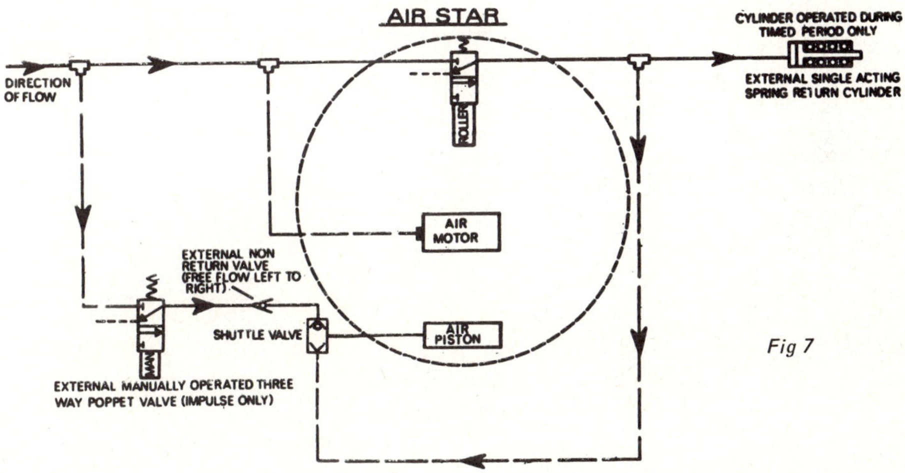

Fig 7

Fig 7 shows this timer used in a typical circuit for producing an interval action. In order to initiate the circuit, the manually operated spring return three port valve is operated for an impulse. This causes the piston and clutch in the timer to actuate. The actuation of the piston lowers the timing valve roller arm into the dwell of the cam thus operating the three port control valve within the timer. This action causes air to be fed to the piston via the timer's controlled valve. The externally operated push button can then be released and the piston within the timer will be maintained. When the timer is in this condition it is timing the interval and the shuttle valve will be changed over so that the shuttle is in the upper position. The timing pointer will rotate clockwise until it reaches the vertical position. At this point the pip on the timing cam will operate the control valve removing the air supply from the timer's piston and from the external load. The removal of the air from the timer's piston will cause the clutch to disengage and the timer will reset to zero. No further action will take place until the push button is again operated. It should be noted that, in this circuit, the motor is left running continuously. This eliminates starting errors due to the initial transitory speed build up on the motor.

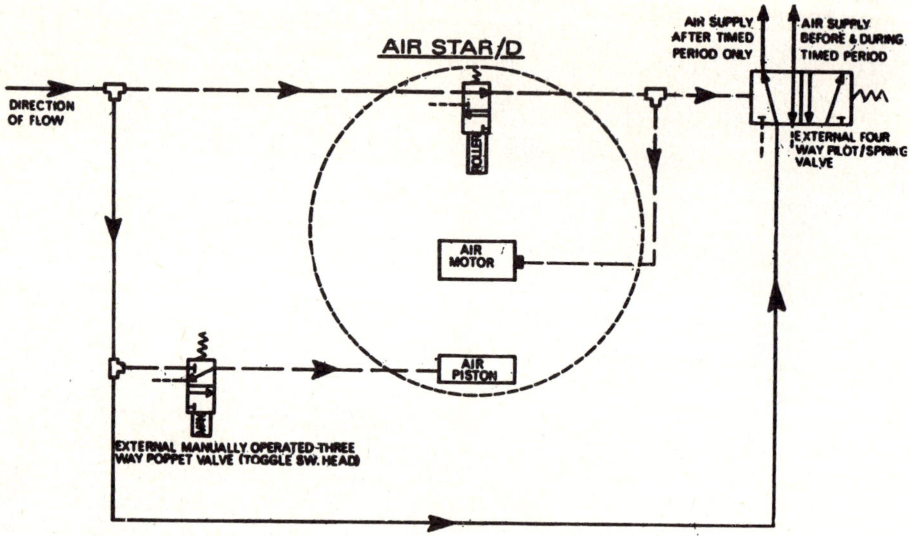

Typical delay-to-off circuit timer. (Electrical Remote Control Ltd).

Time Delay Circuits

Diaphragm valves used in conjunction with flow control valves and an air reservoir can also provide time delays. The three working examples shown in Fig 8 are by Schrader.

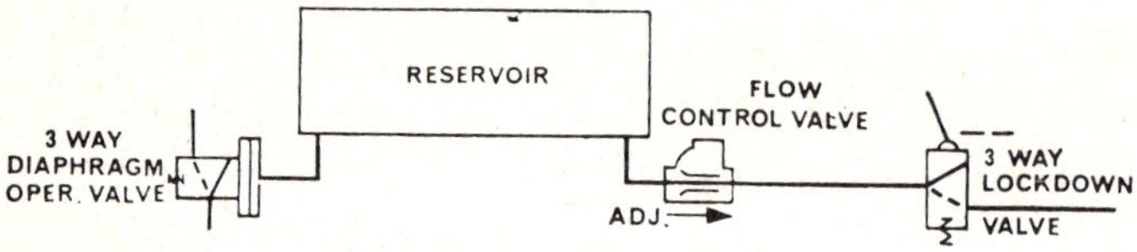

Basic time-delay circuit.

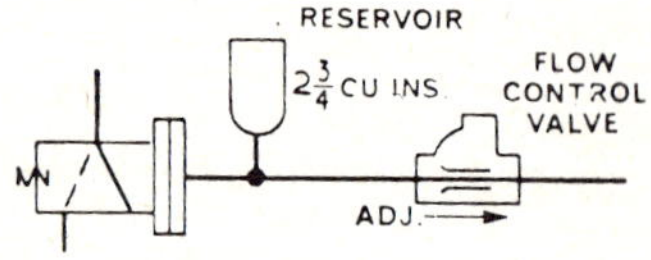

Delay-time up to 12 seconds (at 80 psig).

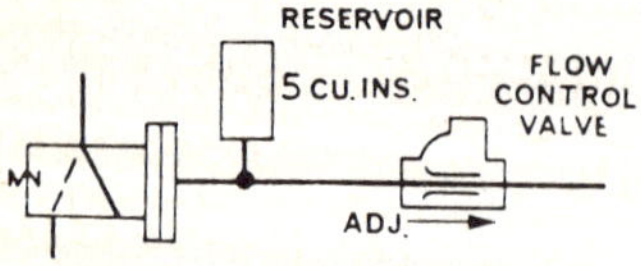

Delay-time up to 16 seconds (at 80 psig).

Fig 8

Vacuum Pumps

VACUUM pumps can be classified under two general headings:-

(i) *Backing pumps* which are capable of pumping from atmospheric pressure downwards.

(ii) Pumps which only begin to operate below a certain limiting pressure level and thus require the use of a forepump for working from atmospheric pressure down to that level — ie must be used in combination with a backing pump.

The majority of backing pumps are of the mechanical type, although fluid jet pumps, condensers, liquid air traps and cryogenic pumps will work from atmospheric pressure down. Mechanical pumps

TABLE I – TYPICAL PERFORMANCES OF MECHANICAL VACUUM PUMPS

Type	Design	Minimum Partial Pressure (approx.)
Reciprocating	Dry-single-stage	10 Torr
	double-stage	1 Torr
	Wet	20 Torr
Rotary Vane	Single blade — single-stage*	5×10^{1} Torr
	double-stage	2×10^{-5} Torr
	Double-blade — single-stage	5×10^{-5} Torr
	double-stage	2×10^{-5} Torr
	Multi-blade	5×10^{-1} Torr
Rotary Plunger	Single-stage	5×10^{-3} Torr
	Double-stage	2×10^{-5} Torr
Twin Rotor	Roots — single-stage	100 Torr
	double-stage	10 Torr
	Roots type booster	10 Torr down to 10^{-3} Torr
	Lysholm	2×10^{-1} Torr
	Northey	2 Torr

are widely used for producing coarse vacuums, and can extend into the medium vacuum range by adopting a second stage. The first stage is then effectively acting as a forepump for the second stage in pumping off the quantities of gas removed by the second stage against the forepressure provided by the first stage.

Typical performance figures for mechanical vacuum pumps are given in Table I.

The two major parameters in the design of a vacuum system are the rate of evacuation required and the ultimate pressure to be realized and held. These parameters determine the size and number of pumps required.

In the case of mechanical pumps, the following is an empirical formula which can be used when initial evacuation time is not critical:-

$$S_m = 4\,V^{2/3}$$

where S_m = the minimum mean speed of the pump between atmospheric pressure and operating vacuum in L^3 per minute,

V = the volume of the system in L^3

Where evacuation time and the vacuum level to be attained are critical, the formula:-

$$S_e = \frac{2.3V}{t} \log_{10} \frac{p_1}{p_2}$$

should be applied, where S_e = mean effective speed in pump in L^3 per minute

V = volume in L^3

t = permissible evacuation time in minutes

p_1 = initial system pressure in torr

p_2 = final system pressure in torr

F = service factor

Pump capacity necessary to maintain the required vacuum level against air leakage only can be estimated from the following empirical formula:

$$S_e = \frac{Vp_i}{6p}$$

where S_e = effective speed of pump at operating pressure in L^3/minute

V = volume of system in L^3

p = operating pressure in torr

p_i = pressure rise due to air in-leakage in torr/hour (measured by pressure rise test and/or specified as leak-tightness standard).

In a practical system account must also be taken of gases evolving from the process or system.

It is normally recommended that in selecting $\propto$ vacuum pumps, the minimum 'blank-off' pressure capability of a pump should be of the order of a half to one decade lower than the ultimate operating pressure to be attained by the pump. Thus a rotary piston pump with a 'blank-off' capability of 0.01 torr would be suitable for use on a system requiring an ultimate operating pressure above 0.05 to 0.1 torr. Increasing the margin between 'blank-off' and operating pressure is advantageous but decreasing it below the recommended figure will result in reduction of system reliability.

Many single stage rotary piston pumps are capable of reaching 'blank-off' pressures of 0.010 torr or lower, which represents a compression ratio capability of the order of 100 000 to 1. In terms of operating pressure range, such pumps perform efficiently between 75 torr to 0.075 torr and thus have operational compression ratios ranging from 10 to 1 to 10 000 to 1.

The efficiency of the pump can be estimated from the following formula:

$$F_c = (F_1 \times F_2)^{0.5}$$

where F_c = combined pump efficiency factor

$$F_1 = \frac{Sav}{41.63\ kW}$$

$$F_2 = \frac{2.88 \log_{10} p}{4.88}$$

where Sav = mean speed of pump from atmosphere to pressure p.

kW = installed power in kilowatts of driving motor(s).

p = lowest operating pressure obtained in torrs.

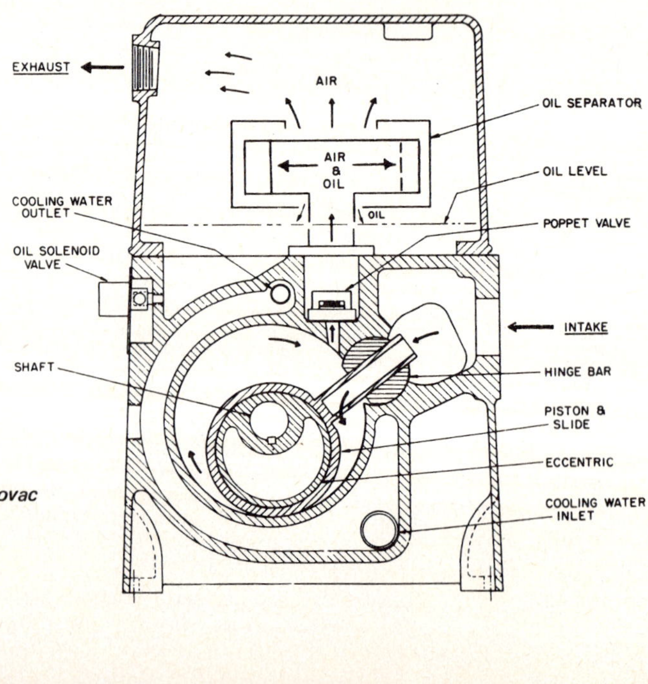

Sectional drawing of Stokes Microvac Pump.

Two giant pumping units designed to evacuate and dry large electrical transformers. The units are designed to be connected to a transformer by a pipeline up to 24 inches in diameter, and can pump down to a pressure of less than 10^{-2} torr. (Edwards High Vacuum).

Gas Ballast

If water vapour (or other vapour) is likely to be present in the system it will cause trouble by condensing during the compression cycle of mechanical pumps, contaminating the oil circulating in them. On entering the high vacuum side, this condensation will re-evaporate, reducing the ultimate pressure and pumping speed.

This can be overcome by using the technique of gas ballasting on the locking pump in the system.

Gas ballasting involves the ingestion of a small quantity of gas (usually air) at a constant rate on the compression side of the pump. This effectively reduces the partial pressure of the vapour in the mixture, preventing condensation during compression provided the sealing oil temperature is above the saturation temperature of water vapour at the reduced partial pressure. Normally, for effective gas ballast, the oil temperature should be maintained at $60^{\circ}C$ to $70^{\circ}C$ ($140^{\circ}F$ to $160^{\circ}F$) where its lubricating and sealing properties are not impaired due to reduction in viscosity.

The use of gas ballast limits 'blank-off' and operating pressure capability of the pump. Depending on the particular design and the characteristics of the pump, there is an upper limit to the pressure at which gas ballast is effective. This is normally in the region of 25–30 torr which limits the quantity of water vapour which can be handled by a particular capacity pump. Where large quantities of water vapour have to be handled, the mechanical pumps should be protected by a water cooled or refrigerated condenser or trap, as condensing is the most efficient means of pumping water vapour.

TABLE II – VACUUM PUMPS – TYPES AND CHARACTERISTICS

Type	Normal or Typical Operating Range	Remarks
Reciprocating, Dry type	Single-stage – down to 10 torr Two-stage – down to 1 torr	Vertical or horizontal Single-, twin- or multi-cylinder. Slow running speeds normal
Wet type	Single-stage – down to 5 torr	
Vane, Dry type	Down to 5×10^{-1} torr	With or without gas ballast
Oil sealed	Single-stage – down to 5×10^{-2} torr Two-stage – down to 5×10^{-5} torr	
Meshing rotor	Single-stage, down to 1 torr, depending on type Two-stage, down to 10^{-1} torr	With or without gas ballast
Rotary plunger	Down to 10^{-2} torr	With or without gas ballast
Molecular Drag, finned rotor	10^{-1} to 10^{-5} torr	Require fore-pump
plain rotor	10^{-1} to 10^{-6} torr	
Steam Ejector	Single-stage – down to 26 in Hg Two-stage – down to 29½ in Hg Three-stage – down to 29¾ in Hg Four-stage – down to 1 torr Five-stage – down to 10^{-1} torr	Normally employed with intermediate condensers
Liquid Ring	Down to 50 torr	Single- or two-stage. Liquid used is generally water. Remove condensable vapours.
Water Ejector	Down to 50 torr	Larger than steam ejector for similar duty.
Vapour-Booster	From 10 to 10^{-2} torr	Combines diffusion pump backed by ejector pump.
Diffusion, Oil	Non-fractionating – from 10^{-3} to 10^{-6} torr Semi-fractionating – from 10^{-3} to 10^{-7} torr Fractionating – from 10^{-2} to 5×10^{-8} torr	Require fore-pump for backing pressure
Mercury	From 10^{-2} to 10^{-8} torr	Regular freezing traps may be preferred for oil-free systems
Cryogenic	Down to 10^{-10} torr	Condensation pumps employing liquefied gas refrigerant
Chemical (Absorption)	Down to 2×10^{-2} torr	Alumino-silicate chemical compound at -200°C (liquid nitrogen coolant)
Sorption	Down to 10^{-3} torr	Molecular sieves at *circa* -200°C (liquid nitrogen coolant).
Getter-ion (evaporative ion)	5×10^{-4} to 10^{-11} torr	Usually titanium or zirconium getter. Outgassing reduced by uniform baking prior to pumping (*eg* at 450°C).
Cold Cathode (Penning pump)	5×10^{-2} to 10^{-11} torr	Outgassing reduced by uniform baking prior to pumping (*eg* at 500°C).

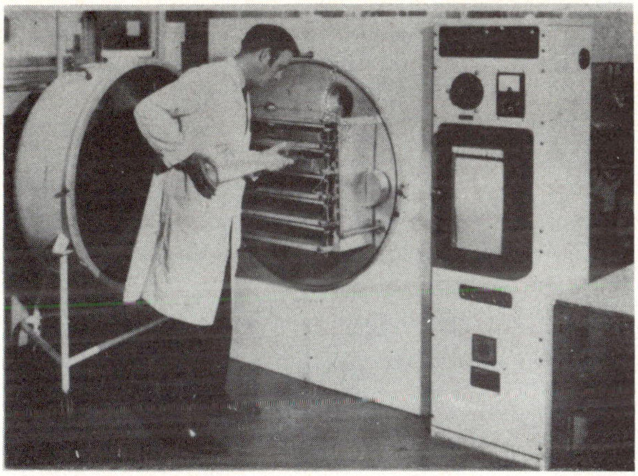

An Edwards Type EF 10 Freeze Dryer. The drying chamber with removable cover is on the left, and the instrumentation which includes a chart recorder is contained in the free-standing cubicle. (Edwards High Vacuum)

Condensers

Condensers usually consist of a vacuum envelope containing tubes or parallel plates, which may be either water cooled or refrigerated. The most efficient vapour trapping path is achieved by correct positioning of the vapour inlet and outlet.

Condenser performance is independent of the capacity of the backing pump, which merely maintains the required low partial pressure. However, high permanent gas pressures must be avoided as they will blanket the cooled surface and reduce the condenser efficiency.

If very high water vapour pumping speeds are required, it is necessary to use a water cooled surface condenser. A condenser consists of a suitably designed tube unit which has a high cooling water velocity, and pumping speeds as high as 2–2.5 kg water vapour per m^2 condenser area per hour per degC mean temperature difference can be achieved. For lower partial vapour pressures, refrigerated condensers can be used. However, there is always the possibility that the vapour fluids may freeze.

Vacuum pump/condenser system for a large drying plant. (Edwards High Vacuum)

Therefore, space must be left between the condenser tubes to prevent ice bridging across them and causing too great a temperature difference. The coefficient of heat transfer depends mainly on the thickness of the frozen layer; for water vapour which has formed an ice layer 6 mm thick, the figure is about 320 kcal/m^{2}hdegC. However, the temperature difference between the evaporating refrigerant and the ice also affects the coefficient, as this affects the turbulence and hence the rate of heat transfer.

A disadvantage of refrigerated condensers is that they must be defrosted periodically. Therefore, if continuous operation is required, two units must be installed in parallel ; one condensing, the other defrosting.

An ideal pump/condenser combination which has excellent vapour handling characteristics, and will operate efficiently in a variety of conditions, is a two-stage mechanical system consisting of a rotary pump and a mechanical booster pump with an interstage condenser. Very fast vapour pumping is possible due to the fact that the interstage pressure is high, and vapour can be condensed and collected in a suitable receptacle.

Contamination of the primary low pressure pump is avoided by maintaining an interstage vapour pressure below that at which condensation occurs in the pump. The backing pump, which has a much smaller capacity, is protected by gas ballasting. Using this system, it is possible to cope with water vapour with a partial pressure of up to 100 torr. Furthermore, solvents, particularly those with a high vapour pressure, can be easily pumped, trapped and recovered.

Fluid Jet Pumps

Fluid jet pumps work on the principle of introducing a high-velocity fluid jet (liquid or vapour) into a mixing section where it meets the gas to be evacuated and entrains that gas by turbulent mixing. Some of the momentum of the jet is thus imparted to the gas which is carried down into a diffuser section and thence to a condenser if necessary, for separation. Where the operating fluid is a liquid, separation can be achieved in the diffuser, and where the operating fluid is a gas usually in a condenser, although the complete system may comprise two or more stages, with or without intercooling, depending upon the performance required. The usual types may be grouped as water jet pumps (fluid), steam ejector pumps (vapour) and vapour-booster pumps. The two former are essentially similar in principle and application whilst the vapour-booster pump comprises a two-stage system with a condenser — Fig 1.

Fig 1 Vane pump working without gas ballast (condensed water vapour remaining over a certain level if partial evacuation is needed).

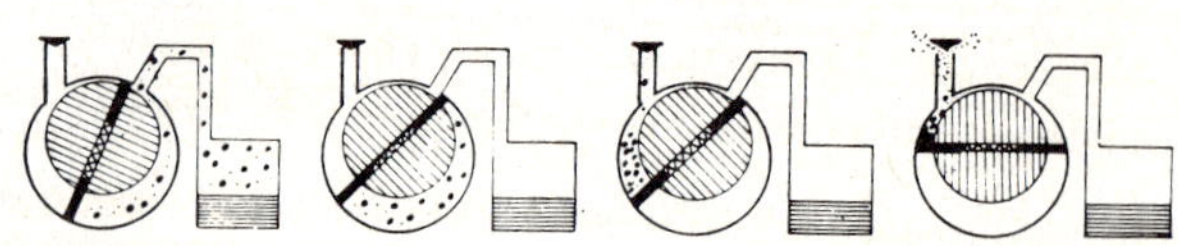

The steam ejector pump has the most wide-spread industrial application and has, in fact, largely replaced other types of vacuum equipment in certain fields (eg for use with power plant condensers) because of its simplicity, reliability of operation and low maintenance costs. It can also be used to handle corrosive or extremely hot gases, etc, since many materials can be adapted for its construction — eg the major pump elements can be made from carbon or refractory materials, glazed ceramics, etc, to resist corrosive attack. It may be termed variously a jet pump, ejector, augmenter (or vacuum augmenter), or thermo-compressor, depending on the duty performed. When used to produce and maintain a condition of vacuum by the removal of incondensable gases it is correctly termed an *ejector.* If used to compress vapours from a condition of high vacuum to a lower vacuum at which condensation can take place, it is correctly termed a vacuum augmenter.

Fig 2 Vane pump working with gas ballast (water vapour ejected with gas ballast).

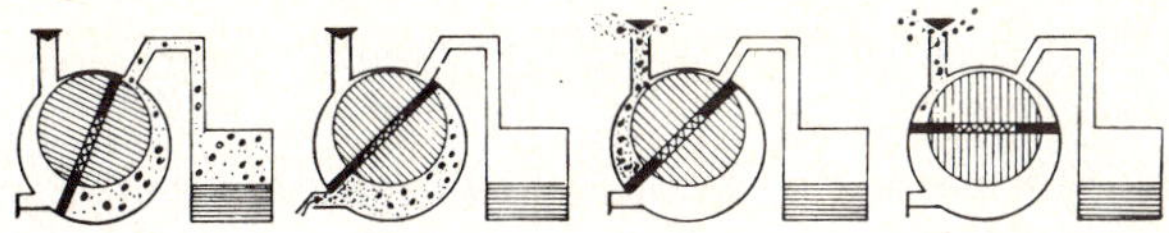

The design of all steam jet pumps is essentially similar and follows the form shown in Fig 2. Steam is introduced via a divergent nozzle into a converging section of the body. The body section also incorporates a suction branch connected to the vessel or system to be evacuated and gases drawn in through this opening are entrained with the steam jet and carried down with it into the diffuser section or compression tube. During this process of entrainment the steam jet loses a proportion of its energy which is imparted to the aspired gases. In the diffuser section the kinetic energy of the mixture of steam and gases is re-converted into pressure energy. The degree of compression which can be achieved is proportional to the amount of kinetic energy each unit mass of mixture has when it enters the diffuser and thus, basically, to the quantity of steam passing through the jet nozzle. The normal range of single-stage ejectors embraces a steam consumption ranging from about 50–60 up to 8 000–9 000 lb per hour.

The compression ratio of the ejector is defined as the ratio of the absolute pressure of the steam-gas mixture at the discharge end to that of the absolute pressure of the aspired gas at the suction branch. Compression ratios of up to about 15:1 are practical with simple ejectors, although it is more usual to work to a lower figure in order to achieve better economy in steam consumption. Thus the single-stage ejector produces economic operation down to vacua of about 26 in. For higher vacua a two-stage or multi-stage unit is normally preferred.

When two or more stages are employed it is usual to employ a condenser between each stage so that the steam of the first stage is condensed out and withdrawn before reaching the next stage. Thus each succeeding stage has only to deal with the incondensable vapours remaining. The usual method of cooling is to employ a cold water spray when the steam is condensed by direct contact with the cooling water, whilst at the same time the incondensable gases are cooled and reduced in volume before being drawn into the suction branch of the following stage ejector.

Three-stage ejectors are usually provided with an intermediate condenser for the first stage (although this may be omitted on smaller units) and a further condenser following the second stage. Four-stage units seldom require a condenser following the first stage since the amount of steam consumed from this stage is small and can be efficiently compressed by the following stages. Five- and six-stage units follow on the lines of a four-stage unit, with the addition of one or two non-condensing stages. The bulk of industrial applications are covered by one-, two- or three-stage injectors.

Water-operated air ejector pumps operate on a similar principle to the steam ejector except that the velocity of the water jet is due to pressure alone rather than expansion and is therefore usually much lower (steam jet velocities of over 4 000 ft per second are commonplace in steam ejectors designed for high vacua). For a similar duty, the water ejector pump is considerably larger in order to deal with a similar volume of gas (normally air). Initially the air is in contact with the outer surface of the water jet only, until subsequently entrained in the cone.

Diffusion Pumps

Diffusion pumps as a type require a forepump for backing pressure, when they are capable of working from 10^{-8} down to 19^{-8} torr, depending on type. They are basically condensation pumps, using mercury or oil as the working fluid. Oil-diffusion pumps are generally preferred for industrial applications, and may be of either fractionating or non-fractionating type.

Diffusion pumps are manufactured both as individual pumps and pump sets incorporating a suitable forepump.

Oil diffusion pumps are capable of much greater pumping speeds than mercury pumps of similar size, because of the difference in molecular weight and molecular volume of the operating fluid. Oils also have a low vapour pressure compared with mercury and thus the pumping speed is relatively unrestricted. Also the use of a freezing trap can be avoided (usually necessary with a mercury diffusion pump to prevent mercury vapour from diffusing back into the vessel being exhausted), except in specific applications where it is absolutely necessary to ensure complete absence of oil vapour in the work-piece (eg as in the evacuation of envelopes of thermionic valves and similar devices).

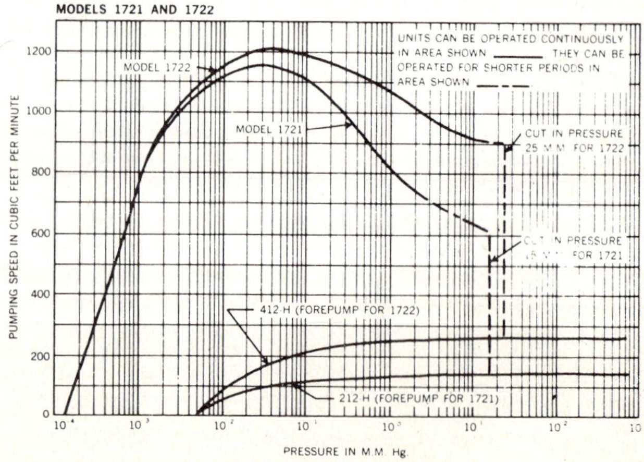

Mechanical booster and representative performance curves. (Stokes Pumps).

The typical modern oil diffusion pump is capable of attaining pressures of the order of 10^{-6} torr without the aid of cold traps or baffle systems and maintains a high pumping speed over the majority of the working range. Apiezon oils are commonly employed as the working fluid, having a low vapour pressure and being extremely reliable in use. Construction is usually of metal with either water- or air-cooling. The jet system may be fractionating, semi-fractionating or non-fractionating, according to which is best suited to the particular application.

A certain amount of back-streaming is inevitable with a diffusion pump, although this may be largely reduced by fitting a baffle plate directly above the pump inlet. Some baffle plates incorporate a valve as well as a baffle so that the pump can be completely isolated from the vacuum system if necessary. The use of a baffle plate, however, automatically means a reduction in pumping speed and so where one is employed it is most important that the design offers the best possible compromise between reduction of back-streaming and loss of pump speed. No simple baffle is completely effective in eliminating back-streaming entirely and where it is necessary to reduce back-streaming to an absolute minimum refrigerated chevron baffles or cold traps have to be employed.

For applications where the slightest trace of oil vapour in the vacuum system must be avoided, mercury diffusion pumps may be used. To eliminate mercury vapour from the vacuum system (although this may be acceptable in some cases, where oil vapour would not), very low temperature traps are employed, using liquefied gases as coolants.

Vapour Booster Pumps

Vapour booster pumps fall between the conventional diffusion pump and a rotary pump as far as general performance is concerned. They comprise, usually, a diffuser pump (one or more stages) backed by an ejector pump or ejector stages.

SECTION 6

Techniques

System Maintenance

THE GENERAL high reliability of a pneumatic system — and the fact that it will usually continue to function, with increasing loss of power, even when allowed to degenerate — tends to obscure the basic truth that *preventative maintenance* is both a necessary and economic requirement. Preventative maintenance will normally repay its cost in keeping the system at the highest possible efficiency, as well as reducing or eliminating costly shut downs due to system failure.

Maintenance requirements fall under three main headings, in the following order:—

(i) Maintenance of tools and appliances.
(ii) System maintenance.
(iii) Compressor maintenance.

All are equally important, but require different maintenance intervals. The life of pneumatic tools and appliances is generally appreciably lower than that of a compressor, hence the need for more regular preventative routine maintenance.

Air tool maintenance is primarily concerned with:—

(i) Ensuring that the air supply is maintained at the required pressure and with the required delivery.
(ii) Regular checking and refilling of lubricators in order to maintain adequate lubrication.
(iii) Checking individual tools for handling or manipulative damage.
(iv) Individual tool servicing as required.
(v) Checking that tools are handled in a reasonable manner (eg cleaned after use) and used in the proper manner (eg not abused).

Item (v) is a disciplinary rather than a maintenance factor, but can be very important in its influence on the efficiency of working and the life of the tool.

The most common cause of early troubles with workshop tools is lack of lubrication. The next is abuse, such as tools being dropped or dragged around by the hose, or generally ill-used. The design of the work area itself, and particularly the allocation of racks for holding tools when not in use, can in itself be a form of preventative maintenance.

Scheduled maintenance should, preferably, be the responsibility of a single individual rather than part of the tool operator's job. With adequate records and experience, tools can then be withdrawn at specified intervals for stripping and inspection and the replacement of worn or damaged parts, to maintain a high operating efficiency. This applies generally to all forms of pneumatic tool. Records may also provide a valuable pointer as to where (which operator) or under what conditions individual tools are suffering abuse, as shown by higher wear or damage rates, and remedial action should be taken as appropriate. This may be an operator fault or a tool fault (ie the tool not adequate for the job at a particular station).

Stocks of spares should be carried covering the normal regular replacement items likely to be required at scheduled maintenance intervals. Stocking an extensive range of spares is, however, not always an economic proposition, even though a large number of individual tools may be involved. This is because a considerable amount of capital may be tied up in spares which are seldom required. On the other hand, lack of a particular spare may put a wanted tool out of action until the necessary replacement part has been ordered and delivered.

An alternative solution, which can prove most effective, is to carry only a minimum stock of necessary spares — ie those normally expected to be required at short notice during scheduled maintenance — together with a spare stock of complete tools. A faulty tool can then be replaced immediately from the surplus stock should it fail before its scheduled maintenance period; or, if a necessary replacement part is not available from stock at the scheduled maintenance period, the tool can be replaced and the part ordered. When delivered, the original tool is fully serviced and becomes part of the stand-by stock.

TABLE I – AIR TOOL MAINTENANCE POINTS

Check or Fault	Likely Cause	Action
Dirt or gummy deposits on motor	Insufficient or wrong oil; dirty air	Check lubricator. Check line filter.
High blade wear on motor	Insufficient oil; normal wear	Reduce periods between scheduled maintenance.
Bearing play	Excessive load; normal wear	Check tool capacity for job; replace bearing
Bearing rough	Dirty air Lack of lubrication Normal wear	Check filter Check lubricator Replace bearing
Governor action sluggish	Air leakage	Check governor bushings for wear
Loss of power	Low air pressure Air leakage Governor fault	Check line pressure and filter Check and rectify Check governor condition; adjust action if necessary
Vibration	Breakage	Replace faulty part(s)
Loss of speed	Low air pressure Air leakage Governor fault	Check line pressure and filter Check and rectify Check

TABLE IA – ACTION WITH ROCK DRILLS

Before Use	During Use
Blow out air lines Check hoses Check pressure Check for leaks Check oil reservoir or lubricator Check drill shanks Check bits	Start hole at reduced throttle Apply consistent pressure to drill Blow out hole frequently Tighten side rod nuts, oil plug, nuts and hose connections periodically After use : clean off and wipe over with oily rag

Such a system normally demands a minimum of 10% of the number of tools in regular, continuous use as stand-by stock, but may reduce the value of the spares stock which would otherwise be necessary by a similar capital amount. Having the bulk of the 'spares capital' in tools will show further benefits in minimizing shut-down time, particularly when dealing with emergency breakdowns.

In the case of major repairs, tools would normally be returned to the manufacturer. With adequate preventative maintenance and efficient scheduled maintenance, a major breakdown is unlikely to occur until the tool is nearing the end of its useful life. The true value of the tool is represented by its initial cost less the amortization rate and if the major repair cost exceeds the value of the tool at that time it would be better replaced by a new one. This applies particularly in the case of workshop tools.

Table I summarizes air tool maintenance points.

Air Cylinder Maintenance

Air cylinder maintenance requirements should, in general, be minimal, except for the periodic replacement of expendable parts. These are normally confined to piston seals, end cover seals (where applicable), rod seals and rod bearings. The latter may be in the form of a replaceable cartridge assembly. The period required between servicing will depend on the frequency with which the cylinder is used and the total number of operations, and the working conditions. Scraper rings and, if necessary, an external gaiter can protect the inside of the cylinder against abrasive atmospheric contaminants. The provision of adequate lubrication by an air line lubricator can ensure good working conditions for the seals as well as give protection against corrosion, provided the lubricant used is resistant to water-washing. Most modern cylinders, too, are pre-coated during assembly with a special lubricating compound which normally adheres to the surfaces for a considerable period. Thus, even in the absence of air line lubrication, good internal working conditions are established between servicing periods with a minimum of wear. Separate, continuous lubrication is, however, always recommended for the highest degree of reliability and minimum wear. This can considerably extend the period between servicing.

Seal wear, and thus seal life and performance, is very largely dependent on the bore condition. Corrosion, pitting, abrasion or damage to the bore can result in high seal wear and premature failure. Intermittent use, with long periods of idleness during which seals may dry out completely, is also damaging. Most seal failure, however, can usually be traced to bad initial fitting of the seal, the use of the wrong type of seal, or wrong material specification.

Piston seal leak is not always easy to detect, since the first evidence may appear as leakage at the exhaust ports or a control valve, when a valve seal rather than the piston seal may be suspect.

A simple check is to disconnect the exhaust line from the cylinder to the valve. If there is leakage of air through this line, the cylinder piston seals are faulty. If there is no leakage from this line, the presence of leakage at the exhaust port of the valve indicates internal valve leakage.

Tie-rod construction generally makes cylinder disassembly easy. A point to watch with single-acting spring-return cylinders is that one end cover will be under spring load and needs backing off carefully until the compression of the spring is relieved. On longer stroke spring-return cylinders it may be advantageous to extend the length of the tie rods to accommodate the full (uncompressed) spring length with the cover still on the tie rods. Nuts can also be welded onto the opposite ends of the tie rods so that disassembly can only be attempted from the extended tie rod end.

Replacement of piston seals may or may not require disassembly of the piston, depending on the type of seal fitted and the manner in which they are retained. Every precaution should be observed in refitting seals to avoid stretching, nicking or otherwise damaging them. End cover seals (where applicable) should always be replaced with new ones when a cylinder is dismantled and reassembled. A general lubrication of the cylinder before assembly is also recommended.

Table II details cylinder maintenance features.

TABLE II – TROUBLE-SHOOTING – AIR CYLINDERS

Fault	Cause	Action
Loss of thrust*	(i) leaking piston seals	Replace seals
	(ii) corrosion on bore	Refinish bore as necessary to eliminate wear on seals
	(iii) dirt trapped in seals	Replace seals. Clean cylinder. Check condition of rod wiper seal. Fit external gaiter if necessary for better protection.
	(iv) excessive friction	Check condition of hose, seals, wear rings and rod bearing
	(v) low pressure	Check supply pressure at inlet
Leakage past rod	Rod seals faulty	Replace seals unless compression type seals are fitted in an adjustable gland In that case, tighten gland.
Loss of cushioning	(i) cushion valve blocked	Remove and clean
	(ii) cushion seals failed	Replace cushion seal(s)

*Loss of thrust can also result from substantially increasing the speed of working with no fault present in the cylinder or system.

System Maintenance

Scheduled periods for the inspection of drains, moisture traps and separators, should be based on experience with the particular components involved (manufacturer's recommendations should be used as an initial guide).

Filters and filter/separators fitted at individual tool stations must receive regular attention, particularly if they are of the manual drain type. The level of contaminants will normally be visible in the filter bowl or an excessively high level indicated by a tell-tale. Both types – manual drain or automatic – will also require periodic cleaning of the filter element; thus a regular period must be set down, depending largely on the 'dirtiness' of the system, for cleaning or replacement of the filter elements as appropriate. Some types of filter have an element which can be cleaned very simply by blowing out; others have an element which can only be cleaned by special techniques (and which thus need replacement whilst the element is returned to the manufacturers for cleaning); and other have elements which are discarded and replaced after a specified period of use. A filter element accidentally damaged during cleaning should always be replaced by a new element.

Clogged filters are the most common source of pressure drop at tool points. Air leaks or constrictions in the flexible line are other obvious sources of pressure drop at tool points. The former can normally be heard with the tool connected but switched off. Kinks or damage to hose that has not produced puncturing may show up in the form of variable performance as the tool position is changed. In this case end fittings should first be checked for air tightness, then the hose length itself checked for damage.

Less obvious sources of pressure loss may occur farther back in the system. With adequate line sizing and everything in good condition, pressure drop between the air receiver and the tool station take-off point should not normally exceed 0.2–0.35 bar (3–5 lb/in^2) (or in some installations 10% of the nominal system pressure). Additional losses at the tool point itself should not normally exceed 0.35 bar (5 lb/in^2), although again a maximum figure of 15% total pressure drop to the tool is sometimes accepted.

Regardless of the actual system pressure drop as installed and originally worked, any further loss of pressure at take-off points or at individual tools is a subject for immediate attention so that the fault can be rectified. Loss of pressure not only reduces tool working efficiency, but may also be expensive in terms of compressed air wastage if due to a leak. Leakage may develop at joints, connections, etc, or at traps where the valve has been jammed open by contaminants, apart from such obvious sources as mechanical damage to the lines.

It is not commonly appreciated that an excessive line pressure can be as bad as lack of line pressure. Whilst the latter reduces the working efficiency of the tool [a tool designed to operate on 5.6–7 bar (80–100 lb/in^2), for example, may suffer a loss of efficiency of 25% operating on 5 bar (70 lb/in^2)], excessive pressure results in accelerated wear and possible over-stressing of parts, which could lead to early failure.

Lubricators at tool points also need regular attention to ensure that an adequate supply of oil is maintained – and the rate of oil injection can also be checked if necessary. It is particularly important that only the specified oil be used to refill the lubricators as an oil of different viscosity may not only upset the lubrication rate but lack the qualities necessary for a pneumatic tool lubricant. Light oils of approximately SAE 10 viscosity are normally specified, specially compounded for pneumatic system use (eg normally containing emulsifying agents and rust inhibitors).

Pipework should require only minimum attention – eg inspection for possible leaks. The following may, however, show up as *installation faults* needing correction:–

(i) Sagging or displaced pipes – requiring proper support with pipe clips, etc.
(ii) Evidence of whipping at bends – again due to inadequate pipe support under pressure surges.
(iii) Distortion of runs – evidence of expansion/contraction movement which should be accommodated by bends or expansion joints.

Compressor Maintenance

Compressor maintenance is usually specific to the type of compressor and the installation. Basic requirements in all cases are cleanliness (eg regular cleaning of intake filters), regular attention to drain points as appropriate (eg on the compressor itself if any, intercoolers and receiver) and maintenance of an adequate supply of lubricant. Other basic requirements involve maintaining the correct tensioning of intercooler and main driving belts, and routine oil changes. Some compressors may also be fitted with intake silencers which may need either periodic cleaning or replacement of felt layers at regular intervals in order to maintain volumetric efficiency. The valves of a reciprocating compressor should be inspected and cleaned in accordance with the maker's instructions and as frequently as the duty necessitates.

Compressor manufacturers normally specify routine maintenance requirements in detail in a manual appropriate to the machine. They may also list recommended spares to be carried, with an indication of which may be regarded as replaceable components. Servicing and attention required will also vary widely with different types of machine. Thus rotating machines normally need more detailed maintenance and inspection than reciprocating machines.

Check charts for compressor maintenance and fault finding are given in Tables III, IIIA and IV for reciprocating compressors, and in Table V for rotary compressors.

TABLE III – MAINTENANCE POINTS – RECIPROCATING COMPRESSORS

Component	Interval*	Action
Air filter (inlet)	Fortnightly	Clean as necessary
Cooling system	Continuous	Check water temperatures
	As necessary	Clean water side on water-cooled systems
Lubrication	Periodically*	(i) Check cylinder oil feed rate (ii) Check oil levels (iii) Change crankcase oil
Bearings	Periodically*	Check wear. Adjust for wear or replace as necessary
Drains	Hourly*	Intercooler drains
	Daily*	Other drains or all automatic drains
Valves	Periodically*	Check for condition and replace plates or springs as necessary
Safety valves	Periodically*	Clean and inspect
Piston rings	Annually	Inspect for condition and replace as necessary
Packing glands	Periodically*	Adjust if applicable; otherwise inspect for condition and replace if necessary

*Suitable intervals can only be established by experience or on manufacturer's recommendations

TABLE IIIA – OIL CHANGES – RECIPROCATING COMPRESSORS

Compressor Type	Working Conditions	Change Interval
All stationary	Running in	After 100 hours
All portable	Running in	After 50 hours
Stationary	Clean atmosphere	6 months or 2 000 hours
	Dirty atmosphere	3 months or 1 000 hours*
Portable	Average	1 month or 500 hours
	Dirty atmosphere	2 weeks or 250 hours*
	Very dirty atmosphere	1 week or 100 hours*

*May be adjusted by experience

TABLE IV – TROUBLE-SHOOTING – RECIPROCATING COMPRESSORS

Fault or Symptom	Possible Causes	Action
'Knocking' or heavy vibration	(i) drive or coupling elements loose	Check and rectify
	(ii) unstable oil-relief valve	Causing hydraulic hammer
	(iii) excessive working clearances on crosshead or bearings	(a) Check for wear (b) Check for mis-alignment
	(iv) cylinder clearance volume too small	Check for accumulation of water, oil or other deposits
Abnormal air temperature		See OVERHEATING (check at inter-stages as well as at delivery)
Abnormal air pressure	(i) leakage	Find and correct source of leakage
	(ii) loss of compression efficiency	See EXCESSIVE CYLINDER WEAR
	(iii) seal failure	Check compression seals
Whistling noise	(i) valve leakage (ii) gland leakage	Identified as a high pitched whistle varying in intensity with pressure
	(iii) leakage in connections	Identified as a hiss
Rumbling noise	(i) air leakage past piston rings	
	(ii) lack of lubrication	
Valve chatter	Defective or broken valve springs	Replace springs

TABLE V – MAINTENANCE POINTS – ROTARY COMPRESSORS

Component	Inspection	Further Action
Electric motor	Clean motor; inspect for wear; check bearings; check clearances; check general condition	As necessary
Couplings (where appropriate)	Clean and examine	Replace if necessary
Casing	Open and examine for corrosion and wear	
Journal bearings	Examine for wear	Replace, reline, adjust with shims } as necessary
Shaft seals	Check condition of clearances	Replace if necessary
Rotor	Examine for corrosion, pitting and wear; check clearances; check balance	As necessary
Gear case (where appropriate)	Examine for corrosion and wear; check damper	As necessary
Other parts	Examine for corrosion, erosion and wear	As necessary
Governor/Regulator	Clean and inspect; check for wear; check adjustment	Replace worn parts; adjust for proper operation
Auxiliary drives	Check	Replace worn parts
Instrumentation	Check for operation and correct reading	Recalibrate or adjust as necessary

TABLE VI – MAINTENANCE POINTS – PIPEWORK

Type of Line	Possible Cause of Trouble	Action
Rigid line system	(i) dead weight of piping	Support with pipe clips, closer spaced on horizontal runs
	(ii) expansion and contraction	(a) use clips which allow lateral movement of pipes (b) incorporate bends to accommodate movement where there is a large difference in ambient temperatures
	(iii) internal pressure	Provide adequate support as necessary to prevent movement and flexing
	(iv) leakage	(a) all pipe joints to be properly made (b) replace faulty valves or fittings (c) if caused by damage check environmental conditions; re-site or protect vulnerable fittings
	(v) water entrainment	Check that suitable water traps are provided and that their location is satisfactory. Establish suitable draining period and check period for manual traps. All loop bends should be upward facing, not downward.
Flexible lines	(i) leakage	(a) check for wear or deterioration at end couplings (b) re-site or protect hoses subjected to bad environmental conditions (c) lay down standard procedure for manipulation of flexible hoses (d) consider possible use of automatic close coil flexible lines
	(ii) excessive pressure drop	(a) check hose for bore condition (b) check that suitable size of hose is being used

Low Temperature Pneumatics

SINCE a conventional supply of compressed air is normally saturated with water vapour, utilization of such a supply at constant temperatures below 0°C is inhibited by ice formation unless special precautions are taken.

The basic methods of approach are either to dry the compressed air before delivery, thus reducing the water content, or to treat the water content of the compressed air with 'anti-freeze'. Both are practical methods for working compressed air systems at sub-zero temperatures, but components themselves may need separate treatment (eg with local heaters) to ensure freedom from freezing or icing up.

Air Drying

Using absorption-type dryers removal of some 99.9% of the water vapour present is possible, when it is generally possible to maintain satisfactory operation of a compressed air system down to about −70°C. The actual minimum permissible 'working' temperature will, however, depend on the performance and capacity of the dryer and the limit to which it is worked. For economic reasons it may be desirable to work a dryer at considerably below its potential water-removal performance.

For continuous drying, two drying cylinders are used, worked alternately (see Chapter on *Air Conditioning and Treatment*). These normally have an oil filter between the receiver and dryers.

Air Dosage

'Anti-freeze' treatment by dosage of the compressed air with alcohol or vaporizing spirit is usually simpler and cheaper, particularly in simple systems. Its effectiveness is, however, limited by the maximum dosage that can be applied without adverse dilution of lubricants present in cylinders and valves, etc or of oil-mist lubrication where used. Permissible dosage will, therefore, depend somewhat on the type of system and its components. A typical *maximum* dosage rate is ten drops per cubic metre (one drop to 3.5 ft^3) of free air. This can be calculated to maintain satisfactory working of the system down to between −15 and −20°C.

The optimum point for introducing the alcohol or vaporizing oil 'dose' is immediately following the compressor, where the air is relatively warm and thus promotes more rapid and complete vaporization of the spirit – eg between the compressor and aftercooler – Fig 1. In a small system, however, adequate performance can often be obtained by introducing the spirit dose in the downstream line or lines which are operating in a cold zone — Fig 2. When the supply feeds both cold and normal temperature service lines only those in the cold zone need dosage. Thus downstream dosage can be more economic where only part of the system has to operate in freezing temperatures. It also has the merit that the disadvantages associated with dosage are avoided in the 'normal temperature' service lines.

A — compressor
B — oil filter
C — alcohol dosage apparatus
D — after-cooler
E — moisture trap
F — compressed air reservoir
G — condensate and dust separator
H — oil mist lubricating device

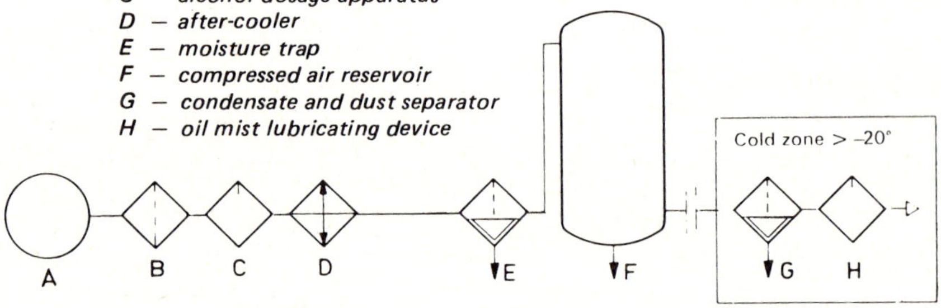

Fig 1 Alcohol dosage for large compressed air system (Mecman)

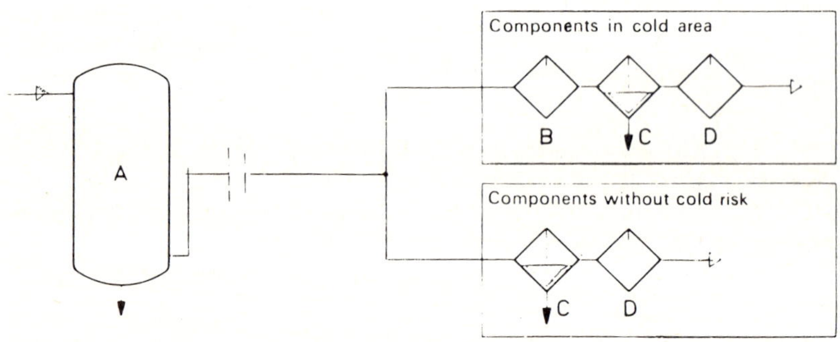

Fig 2 Alcohol mist dosage for small compressed air system (Mecman)

A — compressed air reservoir
B — alcohol dosage apparatus (oil mist lubricating device)
C — condensate and dust separator
D — oil mist lubricating device

System Requirements

Air dosage or air drying provides protection to systems down to the temperatures indicated. This does not necessarily mean that the complete system will continue to function satisfactorily at these low temperatures. Some water vapour will still be present (particularly with spirit dosage), and ice may still form at points where local expansion of the compressed air produces a further drop in temperature. Also water vapour present in the surrounding air can turn into ice in contact with low temperature parts in the system — eg outlet ports. Localized heating of components may, therefore, be necessary, to maintain satisfactory minimum temperature working.

In less severe applications, localized heating may prove satisfactory on its own. Thus, where an air line has to pass through a cold zone, heating by a jacket or similar means may be sufficient to maintain the air supply above dew point. If valves are located in cold zones, these can be protected from freezing by containing them within a thermally insulated enclosure with simple internal heating (eg provided by a low or moderate wattage light bulb). If localized heating is considered too complicated or unjustified, then alcohol dosage, applied at the critical point of the system where freezing troubles occur, may offer a simpler solution.

The main risk of ice formation in pipes occurs just after throttling points. If the system is intermittent in operation the quantity of ice formed may be insignificant and melt during the off times without giving any trouble. Icing problems may well show up on such a system operated at a faster rate or for longer periods of time because of restricted melting times and consequent ice build-up.

Operating at low ambient temperatures, but above freezing point, these critical areas subject to freezing may well show up by the formation of ice on the outside of the pipe. This in itself is not necessarily harmful in a continuously operated system; but with intermittent use can cause ice formation inside the pipe unless the compressed air supply is dosed or dry.

Outlet ports are a particular source of ice formation because of the sudden drop in air temperature at this expansion point. Ice can build up quickly, blocking the port. Dosed or dry air cannot guarantee protection from icing because the port is exposed to outside air. If protection from exhaust port icing proves troublesome then two possible solutions are:—

(i) apply alcohol dosage at or immediately preceding the port (mostly applicable to a continuously operated system);

(ii) apply localized heating to the port.

High Pressure Pneumatics

THE COMPRESSIBLE nature of air limits the degree of compression which can be achieved with optimum overall efficiency to compression ratios of the order of 7:1 to 8:1 (ie to compressed air pressures of up to 7–8 bar). Very much higher pressures can, of course, be achieved but the increase in power required for compression progressively decreases the overall efficiency of compressed air as a working medium, thus limiting its use to more specialized applications (see Table I).

TABLE I – HIGH PRESSURE PNEUMATICS

Category	Pressure Range bar	Typical Applications	Working Pressures bar	Working Pressures lb/in^2
Moderate	17.5–35	(i) Engine starting	25–42	350–600
		(ii) Air blast switchgear	17.5 up	250 up
		(iii) Deep sea diving	10	150
High	35–210	(i) Sea diving (air cylinder charge)	up to 210	up to 3 000
		(ii) Deep sea diving (air cylinder charge)	105–140	1 500–2 000
		(iii) Secondary/emergency aircraft control services	35–210	500 3 000
Very high	210–450	(i) For specialized process plant	210–350	3 000–5 000
		(ii) Supersonic wind tunnels	210–420	3 000–6 000
Ultra high	700–1 400	(i) Air-blasting	700–840	10 000–12 000
		(ii) Metal forming	up to 14 000	up to 20 000

The only straightforward advantage of higher compression ratios is the reduced storage capacity necessary, which can be particularly useful for compressed air storage systems for use as energy, power, etc. For working purposes, the compressed air can be 'let down' to conventional delivery pressures, although specific systems may demand the use of high pressure air.

High pressure pneumatics can be considered to embrace three distinct pressure ranges, viz:–

(i) *Medium pressure, 17.5–35 bar* (250–500 lb/in^2), normally derived directly from a compressor, or reservoir charged by a compressor.

(ii) *High pressure, 210–350 bar* (3 000–5 000 lb/in^2), normally derived from a charged air bottle.
(iii) *Very high pressure, 210–350 bar* (3 000–5 000 lb/in^2), either derived from high pressure reservoirs charged by multi-stage compressions for large volume applications, or from intensifiers for 'single slot' or small volume application.
(iv) *Ultra-high pressure, 700–1 400 bar* (10 000–20 000 lb/in^2), for highly specialized applications with supply from multi-stage reciprocating compressors.

Typical application of these pressure ranges are summarized in Table II.

TABLE II – THEORETICAL POWER TO COMPRESS 1 CUBIC METRE OF AIR AT ATMOSPHERIC PRESSURE

Power in Kilowatts

	Delivery Pressure (bar)					
	7	70	140	210	280	350
Two-stage	–	10.0	–	–	–	–
Three stage	–	8.9	10.7	11.8	–	–
Four-stage	–	8.4	10.4	11.0	11.8	12.4
Five-stage	–	–	–	10.5	11.2	11.8
Isothermal compression	3.6	7.1	8.3	9.1	9.5	9.7

TABLE IIA – THEORETICAL POWER TO BOOST 1 CUBIC METRE OF FREE AIR PER MINUTE

Power in Kilowatts

Inlet Pressure	Boost	Delivery Pressure – bar				
		70	140	210	280	350
7	Multi-stage	4.5	6.0	6.70	7.25	7.82
70	Single-stage	–	1.3	2.20	3.00	3.35
140	Single-stage	–	–	0.75	1.30	1.86
210	Single-stage	–	–	–	0.56	1.00

The production of high pressures direct demand the use of multi-stage compressors of the reciprocating type. Because output is necessarily restricted, a compressor (or batch of compressors) can be used to charge up a reservoir to the required pressure, the reservoir volume being selected on the basis of the working capacity required. Alternatively, if the demand is relatively low, it can probably be met by a gas bottle where the initial pressure available is usually of the order of 125 bar (1 800 lb/in^2). If the working pressure required is very much lower, this again may feed a reservoir, through a pressure reducing valve. The third alternative is to obtain the high pressure supply indirectly by boosting an available low pressure supply. This may be a normal air line supply, eg 7 bar (100 lb/in^2), or a supply from a gas bottle (say at nominal 105 bar [1 500 lb/in^2]) according to the final pressure required and the pressure intensification possible from the booster.

All three methods tend to have the common limitation that the volume of high pressure air available is restricted, either in terms of free air equivalent, or in the time over which the required working pressure is available. The higher the working pressure level, the more marked this restriction becomes. Thus high pressure pneumatics tends to be limited to specialized systems or requirements which only need short term or non-continuous operation.

This convention may, however, be subject to change. In certain fields, the use of higher pressures under exact control may provide the higher output forces normally associated with hydraulic power, particularly for 'impulse' applications, and in servo mechanism working. Even the methods of obtaining air pressure may change, such as by using liquid air for storage, or even generation of pressurized air by chemical processes.

Currently, practically all the compressed air (and other high pressure gases) for industrial use are derived from conventional multi-stage compressors. Only the lower pressures are normally supplied direct from compressors. For higher pressures up to about 140 bar (2 000 lb/in^2) the most convenient form of supply is usually a charged gas bottle. For pressures of 105 bar (1 500 lb/in^2) up to about 1 400 bar (20 000 lb/in^2) a gas booster is used.

A *gas booster* is simply a pressure intensifier or air pump accepting low pressure air in a low pressure cylinder, to drive a smaller piston in a high pressure cylinder. This tandem configuration is associated with a valve mechanism which acts at the end of each stroke to produce continuous reciprocation. The high pressure piston and cylinder form the upper portion of the intensifier or pump and the inlet and outlet valves are normally mounted on the end of this cylinder where they are readily accessible.

The inlet is compressed air at normal line pressures (eg 7 bar or 100 lb/in^2 gauge). The outlet pressure is determined by the ratio of the areas of the low and high pressure pistons offering, in theory at least, a direct method of boosting the supply to almost any pressure required. In practice, however, performance is limited by the unswept volume present (which is kept to a practical minimum), and that optimum performance is achieved with a pressure ratio of outlet pressure to inlet pressure between 4:1 and 6:1.

To obtain higher outlet pressures with economic and efficient booster working, a charged gas bottle can be used to provide the initial supply. For still higher pressures two boosters can be used, operating in series. To reduce pressure fluctuations in the supply to the second booster a receiver can be incorporated between the two.

Basically, a booster continues to pump until a preset pressure is reached. It will then continue to deliver air at this higher pressure, or will remain stalled with no demand, until the pressure falls below the preset level. Since the output pressure is directly related to the input pressure, any variations in the latter will affect the output pressure. Thus, when the supply is from a gas bottle, pressure from the supply will fall as the bottle is discharged. The maximum pressure available from the booster is then based on the lowest pressure to which it is considered economic to let the bottle pressure fall. This may be further modified by the output volume required.

Let-Down Systems

Compressed air stored in bottles or high pressure receivers provides a cheap, compact source of emergency power. It can also be used for portable systems, divorcing the actuator or appliance from a compressor. The supply pressure required may be low – eg a conventional 7 bar (100 lb/in^2) supply – when it can be advantageous to store the compressed air at a much higher pressure in order to increase the volume of free air available. The storage vessel can then be fitted with a reducing valve to supply air at the required pressure. A shut-off valve necessary to isolate the pressurized supply vessel, cannot provide satisfactory pressure control. If the pressure level to be maintained is critical, then further control may be provided by a back-pressure or pressure-maintaining valve.

A *back-pressure valve* is shown in Fig 1. This is of the 'inverted' type, with inlet pressure tending to close the valve rather than open it. The valve is opened by inlet pressure lifting a spring-loaded piston to which the poppet is loosely connected. The poppet valve is seated by a light spring, the inlet pressure acting on the small sealed area. Before the valve can open, the spring-loaded plunger must lift sufficiently to take up the slack between it and the poppet.

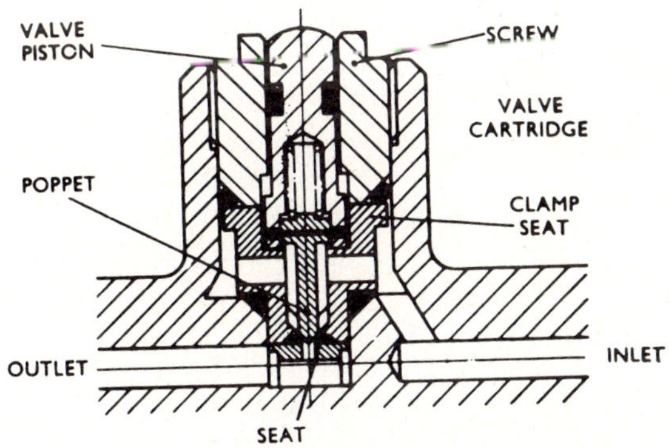

Fig 1

Unlike a liquid, high pressure air (or gas) is highly compressed and, on release of restraint, will tend to revert instantaneously to its original free air volume, within limits imposed by the nature of compressible gas flow. The very high compression ratios which may be present (see Table III) indicate the very high energy which may be stored in even a small volume of high pressure air, and the potential explosive hazard if suddenly released. Special care and precautions are, therefore, necessary with all high pressure systems and appliances.

TABLE III – COMPRESSION RATIOS

Final Pressure		Compression Ratio
bar	lb/in^2	
7	100	7.8
14	200	14.6
21	300	21.4
28	400	28.2
35	500	35.0
70	1 000	69.0
140	2 000	136.5
210	3 000	204.6
280	4 000	273.0
350	5 000	340.0
700	10 000	680.0
1 055	15 000	1 020.0
1 406	20 000	1 364.0

Conventional copper piping is generally unsuitable, and nylon tubing quite unsuitable, for high pressure lines. All pipework should be of hydraulic quality steel, or preferably stainless steel, for pressures in excess of 350 bar (5 000 lb/in^2). Fittings and couplings must be of compatible design and rated for the pressures involved for use with air (gas pressure ratings differ from hydraulic pressure ratings, with most types of compression couplings).

Pressure vessels should be the Safety Gas Pressure type, and free from oil. No pressure vessels, eg receivers and gas bottles, should be used unless they have a current test certificate. By law, all receivers and gas bottles must be insured, and the insurance company is responsible for carrying out the necessary certification tests.

Oil in any form is not tolerated in a high pressure gas system since its presence can cause 'dieseling' when the gas is introduced.

Engine Starters

Compressed air motors used for engine starting are normally powered from a receiver charged to conventional low air pressures. Much higher air pressures are, however, used on starters for marine engines, when the receiver pressure may be of the order of 25–35 bar (300–500 lb/in^2). The receiver in this case is charged directly by a compressor either from the main engine power take-off, or from its own electric motor. A reducing valve is used between the receiver and the starter motor or engine, as necessary.

Two distinct methods of using the compressed air are, in fact, utilized. In the one case, the air drives a conventional air motor or rotary actuator working as a starter motor. In the other, compressed air may be directly injected into a selected number of engine cylinders to achieve rotation for starting. In this case higher air pressures may be used directly.

Air Blasting

The 'explosive' nature of highly compressed air is used in air blasters or airbreakers, this technique having replaced the use of conventional explosives in coal mines to some extent. The working principle involved is that air at very high pressure (eg circa 850 bar [1 200 lb/in^2]) is suddenly released in a conventional bored shot hole, the resulting release of stored energy producing first, rupture of the coal in the locality, and then further break-up by the expansion of the escaping air into the fissures so formed. The action is not so violent as that produced by explosives, but is generally more far reaching. Thus, a greater proportion of coal is dislodged per blast. The amount of coal dislodged depends primarily on the thickness of the seam and can vary from three to fifteen tons per shot.

High pressure air is introduced into the borehole via a special breaker assembly – eg Fig 2. Basically, this consists of a steel shell, sealed at one end by a shear disc inserted between two shear rings and retained by a shell head. The shell head also contains the backward facing discharge ports. A supply of high pressure air is connected via a reinforced hose to a blasting valve. When this valve is opened the pressure in the shell rises until the shear disc ruptures. Air is then free to expand through the discharge ports into the bore hole, producing a jet force causing fracture of the coal. At the same time, the fact that the ports are backward facing drives the shell further into the hold, increasing the effectiveness of the blast.

Air blasters of this type used in mining are generally supplied by multi-stage reciprocating compressors, located at some convenient central point. Hose lengths may run to many miles. Similar types of air breakers have also been used for blasting concrete and other materials. The rupture action is, however, somewhat different in materials having higher ruputure strengths and different fracture characteristics.

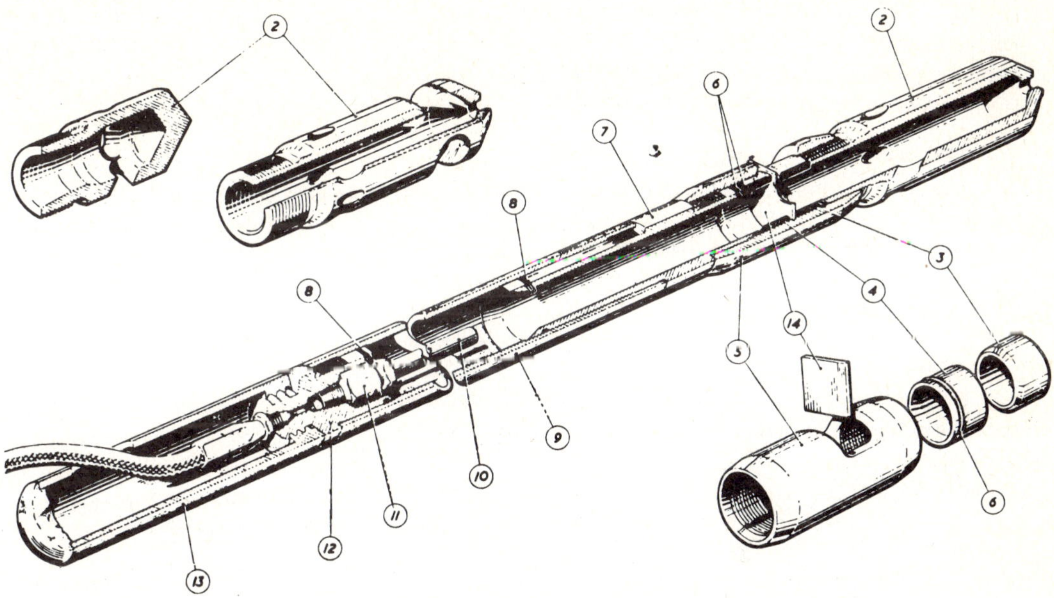

Fig 2 Bursting shell for air blasting coal. 1. 2½ in Breaker assembly. 2. Retainer head. 3. Shear ring. 4.Seal ring. 5. Body. 6. 'O' ring. 7. Adaptor. 8. 'O' ring. 9. Tube. 10. Cooling tube. 11. Inlet fitting. 12. End cap. 13. End guard. 14. Slugs. (Olin Mathieson Ltd).

Wind Tunnels

Wind tunnels, designed for working with airstreams of high Mach number, are normally of the intermittent type. That is, the high flow velocities required are achieved by a sudden discharge of high pressure air from a reservoir. Reservoir storage pressure may be of the order of 210–420 bar (3 000–4 000 lb/in^2). Multi-stage compressors are used to produce this charge, the number and capacity determined by the charge-up time required. Commonly it may take the compressors several hours to charge the reservoir to a pressure level sufficient for a single test lasting a few seconds.

Whilst a relatively simple method of obtaining high air velocities, a basic problem with this application of high pressure air is the necessity for the air to be completely dry. The sudden expansion of the compressed air causes a severe drop in temperature, which would mean that any moisture present in the air would immediately be turned into ice particles, projected through the tunnel at high velocities. For similar reasons it is also desirable to remove all traces of oil vapour and any solid particles from the air.

High Energy Rate Forming

High pressure air has been used successfully for high energy rate forming of sheet metals, the main requirement being that for maximum effectiveness the high pressure air should be applied instantaneously. Again, this is commonly applied through a bursting disc, although it may be further controlled by a quick-opening valve. The method is essentially similar to explosive forming but avoids the use (and storage) of potentially unstable materials.

Air Blast Switchgear

Compressed air at a pressure of 17 bar (250 lb/in^2) upwards is employed on modern switchgear for cooling and extinguishing the arc formed by opening contacts breaking a high voltage circuit. The resulting improvement in contact life can be substantial.

Aquasports and Skin Diving

Air supply for underwater swimming is provided by lightweight air bottles, charged with air pressures up to 210 bar (3 000 lb/in^2). The purity of the air is very important in such cases. It must be oil free and clean. If the compressor used for charging the bottles is not an oil-free type, it must be followed by an oil separator. The air is also filtered before filling the gas bottle, the filter often being charged with alumina, or silica gel, and activated charcoal.

The total quantity of air contained in a charged bottle will be dependent on the temperature, as well as the pressure, and so it is an advantage to deliver the air to the cylinder at as low a temperature as possible. Rapid filling of the bottle can cause a rise in temperature, and in such cases further topping up may be possible when the bottle has cooled down.

TABLE IV – PRESSURE OF WATER AT DEPTH

Depth of Salt Water		Positive Pressure		Absolute Pressure
m	ft	bar	lb/in^2	bar
0.000	0	0.00	0.0	1.0
0.686	2¼	0.07	1.0	–
10.000	33	1.03	14.7	2.0
40.230	132	4.13	58.8	5.0
91.440	300	10.10	133.3	10.0
104.850	344	10.70	152.8	11.4
121.900	400	12.40	177.0	12.0

Deep Sea Diving

In diving with a helmet and waterproof diving dress, the diver is supplied with air at a pressure sufficient to keep the dress inflated so that whatever depth the diver may be in, provided a proper air supply is maintained, the pressure on the outside of the body cannot be more than a pound or two over that inside the body. There is also the fact that much less exertion is involved when breathing air from an inflated dress than from within a rigid space, such as the helmet alone.

The deepest practical depth for 'dress diving' is about 90 m (300 ft), at which air pressures of the order of 9 bar (130 lb/in^2) are required – see Table III. Quite commonly, however, it is advantageous to dispense with pumps or compressors on the diving vessel, and provide the diver with air from high pressure cylinders, charged to the order of 105–130 bar (1 500–2 000 lb/in^2). This is particularly the case when operating from small boats and requiring deep dives. The pressure capabilities of hand pumps normally reach their limit at about 30 to 40 fathoms depth.

When high pressure air supplies are used, these are arranged to deliver through a reducing valve into a low pressure reservoir. The diver is then supplied from this reservoir through a throttle valve. Table IV lists free air requirements of divers at various depths. A rough estimate of the time a supply would last is given by dividing the amount of free air available by the specific air requirement. In practice, however, part of the free air in the high pressure cylinders is not available, as some is lost in filling the low pressure reservoir, and another part is lost when the pressure in the cylinder becomes too low to force the air down to the diver.

.ow Cost Automation

ECHANIZATION implies the replacement of manual effort by mechanical power — with ›vious advantages. In all degrees of mechanization, however, the operator remains an essential ›rt of the system, although with changing demands on physical input as the degree of mechaniza- ›n is increased. Table I illustrates the basic 'degrees' of mechanization, taking as an example, a nple machining operation, such as drilling.

The point at which 'mechanization' becomes 'automation' is where the machine operator's role reduced to that of a supervisory nature, or even eliminated entirely. In other words, the essential ature of command is transferred to a timed operation of the system or semi- or fully-automatic ntrol utilizing a suitable detector generating feedback signals. Extension of the previous example rough for basic degrees of automation is summarized in Table II.

Pneumatics is an attractive medium for low cost mechanization and automation, particularly r sequential or repetitive operations. Many factories and plants already have a compressed air stem installed, and adaption of manual machines to mechanization/automation can be done with ly moderate capital expenditure. Where the compressed air services are extended, however, ditional compressors may have to be installed to meet the demand.

More elaborate automation schemes may be based on logic element controls and circuits, where *eumatics* again offers the advantage of an all-fluid system, using the same fluid medium roughout (compressed air). This can offer distinct advantages and savings over electronic ntrols and electro-hydraulic systems.

In this latter respect, fluidic controls are directly competitive with electronics. In the field of id power application, *pneumatics* is directly competitive with *hydraulics.* In the mechanization/ tomation field, pneumatics is generally superior to hydraulics in taking over direct from the nual operator. The superiority of hydraulics is largely limited to the higher output forces and wer available, and to the greater rigidity of control movements directly available.

In the broad field of conversion to mechanization/automation, pneumatics can cater for ›bably 90% of possible requirements, and in the control field for virtually 100% requirements.

Also, of course, compressed air is capable of providing both the power or energy requirements d the control system (although equally pneumatic control systems may be economic and can be vantageously applied to other forms of power). The main advantages of an all-pneumatic system usually economy and simplicity, the latter reducing maintenance requirements to a low level. It also have outstanding advantages in term of safety.

TABLE I – BASIC DEGREES OF MECHANIZATION

Work Process		Work Effort and Work Rate	Operator's Responsibility	Comments
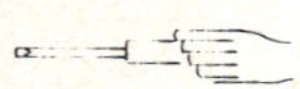	HAND ONLY	Very limited	Effort, speed and accuracy	Also limited by hardness and thickness of material, etc. Accuracy depends on operator's skill rate
1. 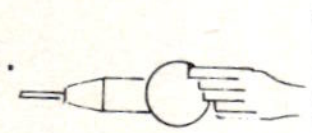	HAND DRILL	Limited	Effort, speed and accuracy	Improved work capability rate and effort, also scope of operation. Accuracy still depends on operator's skill
2. 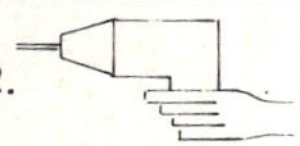	POWER DRILL	Work effort limited by acceptable weight of power tool	Part effort and accuracy, also work rate	Greatly improved work rate and scope of operation. Accuracy still depends on operator's skill
3. 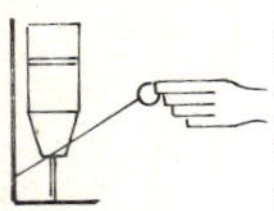	SIMPLE DRILL PRESS	Highest work effort needed. Work rate related to materials and subject	Mainly judgment, but also work rate	Similar to 2. with substantially less operator fatigue. Accuracy far less dependent on operator's skill
4. 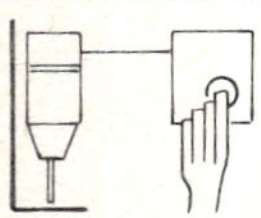	SEMI-AUTOMATIC DRILL PRESS	As (3). Simultaneous multiple operation possible		Operator's role is largely supervisory

Category	Method		Basic System	Operator's Responsibility
Semi-automatic	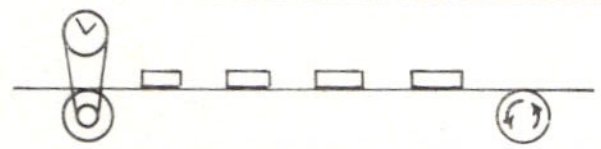	Mechanically timed	Mechanical	Start/Stop, inspection and supervision
Semi-automation	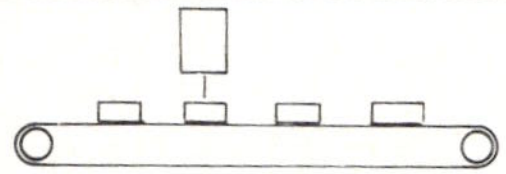	Sequential control		Start/Stop, inspection and supervision
Automation	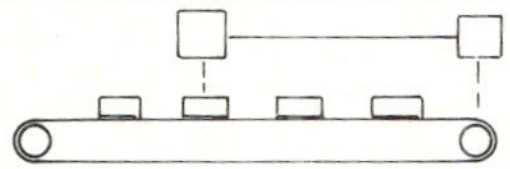	Sensing with feedback	Electro-mechanical	Start/Stop. Action on alarm signal
Full Automation	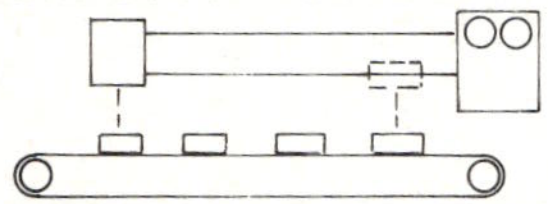	Programmed control	Programmed control by punched tape or computer	None required
Fully Programmed Automation	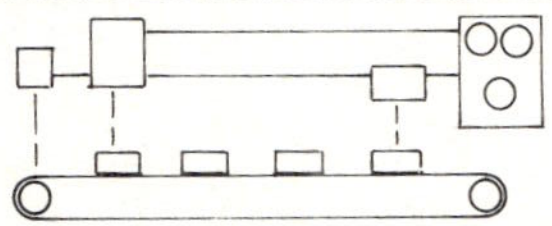	Fully programmed control	Computer controlled and capable of self-analysis, correction, etc.	None required

Special purpose drilling/tapping machine designed by Tecalemit Ltd, Plymouth. Incorporates the 22a and 5e Bellows-Locke drill units and Bellows rotary feed table.

Every application is different to some extent, although all may employ the same basic element Optimum design solutions must therefore be based on a detailed study of the economics and con petitive performance of alternative systems.

Power Units

The most widely used pneumatic component for power units is the *air cylinder.* The majority c air cylinders are 100 mm (4 in) diameter or smaller,giving a maximum thrust of the order of 450 k (1 000 lb) and a line pressure of 6 bar. Speed range usually allowed is from 0.5–2.5 m/sec (1.6 8 ft/sec), with speed control if required allowed by throttling; more precise control can b obtained by contraction with an oil-hydraulic cylinder.

Cylinders can be single- or double-acting; and are also capable of 'positioning' movements. A cylinders are basically linear actuators of limited power, it is essential to have methods c converting linear motion into semi-rotary motion, or of magnifying the force for actuating clamp cutters, etc;

See chapters on *Air Cylinders* and *Air-Hydraulic Cylinders*.

Limited Rotary Movements

Where rotary outfits are required to work for not more than one or two revolutions, the use c an air-cylinder is again a logical choice. A simple method of producing semi-rotary motion for on or two revolutions is by rack and pinion, as shown in Fig 1. The rack can be of the cut gear type as shown, or formed by stretching roller chain along a flat support. Roller chain is also used in th method shown in Figs 2 and 3.

Fig 1 Rotating spindle is driven by a cylinder (3) through a limited number of revolutions by rack (2) and pinion(1).

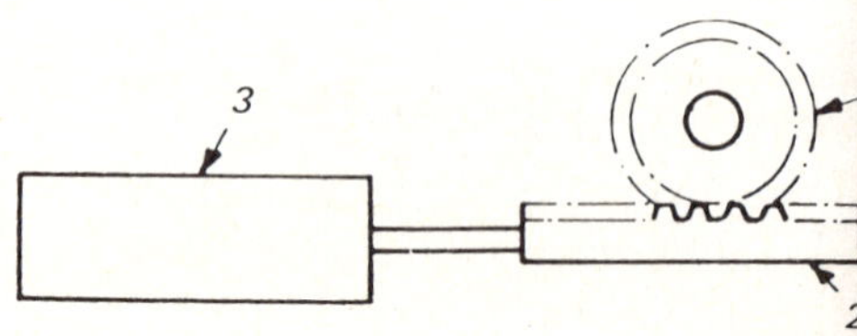

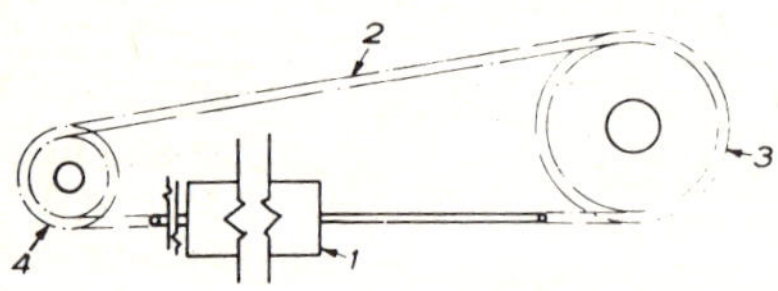

Fig 2 Rotary motion from a double rodded cylinder with chain drive. 1. Cylinder. 2. Chain 3. Pinion. 4. Idler pinion.

Fig 3 Rotary motion from two cylinders driving a chain pinion. 1. Single acting cylinder. 2. Chain. 3. Pinion.

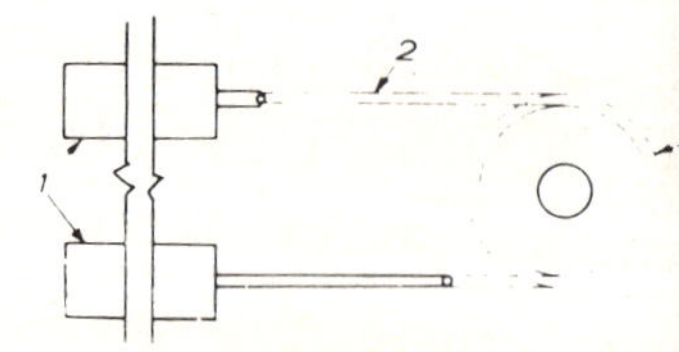

The method adopted will normally depend on the local conditions. An angular movement c 90° or less can also be obtained by a crank mechanism and this, by a ratchet and pawl, can b further extended to give an intermittently full rotary motion (Fig 4).

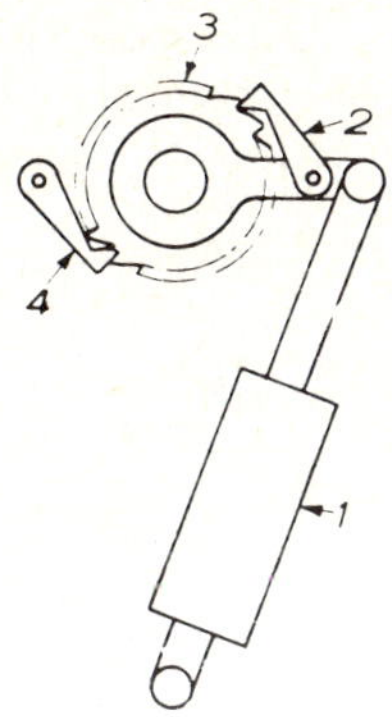

Fig 4 Ratchet drive giving intermittent uni-directional motion. 1. Cylinder. 2. Pawl. 3. Ratchet wheel. 4. Stop pawl.

Fig 5 Small riveting machine with lever to multiply the force exerted by the cylinder. About 15 tons can be exerted over a short stroke. 1. Cylinder and frame. 2. Piston. 3. Lever. 4. Port. 5. Plunger. 6. Anvil. 7. Frame.

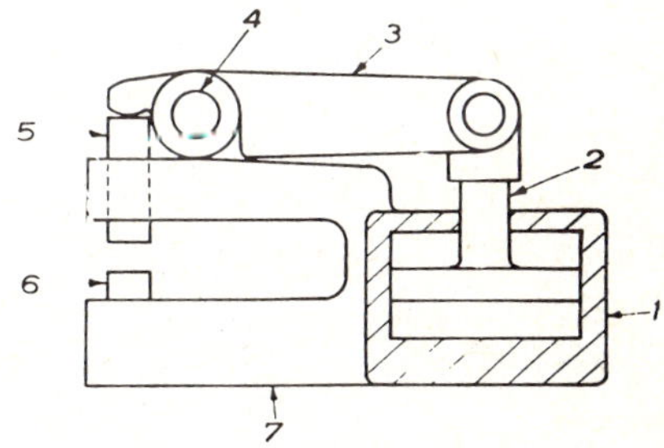

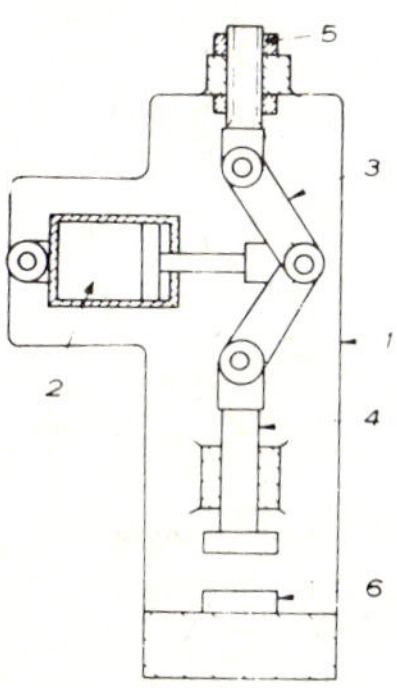

Fig 6 Toggle press. The force exerted by the plunger increases considerably as the toggle levers straighten out. By arranging the cylinder to 'pull' slightly more thrust is available for unlocking the toggle when commencing the upstroke. 1. Frame. 2. Cylinder. 3. Toggle. 4. Plunger. 5. Adjusting screw. 6. Anvil.

Force Magnification

This is best used when the final large force has only to be applied over a short distance. Then simple mechanical methods will give a limited degree of magnification. Fig 5 shows a simple squeeze riveting machine with a leverage of about 10:1. A toggle mechanism (Fig 6) has the advantage that a greater 'daylight' can be provided. The force exerted by a toggle mechanism at the end of the stroke is considerably influenced by the stiffness of the framework and links.

Fig 7 PowRlock clamping head for mounting on bolt or stud. 1. & 2. O-rings. 3. Spigot. 4. Seal. 5. Body. 6. Cap. 7. Piston. 8. Spring. 9. Wiper.

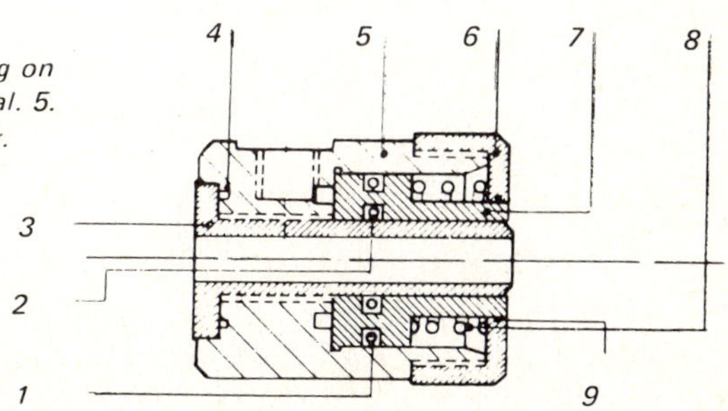

An air-hydraulic intensifier system has the advantage that line air pressure can be boosted to 175.7 bar (2 500 lb/in^2) or so, thus allowing the use of very small actuators whilst retaining the advantages of pneumatic operation. Such a system can employ single shot air hydraulic intensifiers of various types, which are connected directly to one or more of the clamping heads. A typical head is shown in Fig 7. It is intended for fitting over a bolt or stud and provides an instant clamping action, so making spanners or screwed handles unnecessary. Up to 24 clamping heads can be operated simultaneously from one booster.

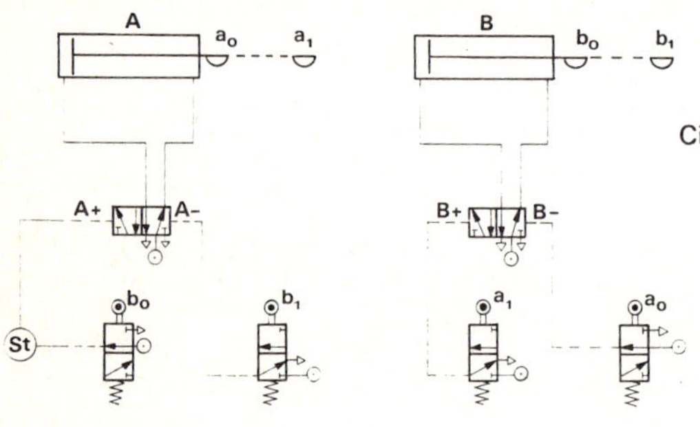

Circuit diagram for cylinders in the transport system

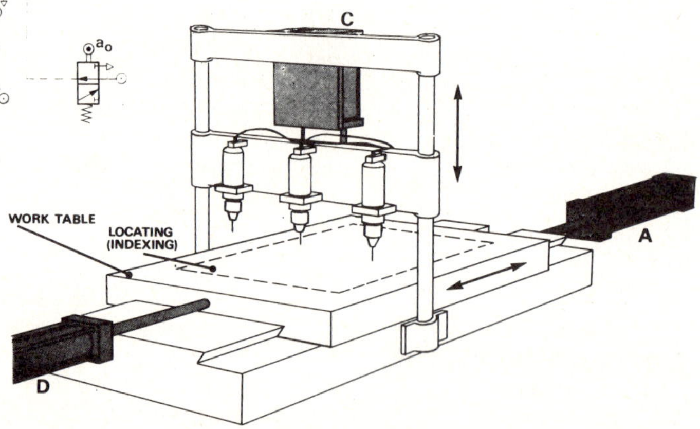

Example of mechanized drilling operation (Atlas Copco)

Linear Movements

The normal method is to arrange the cylinder so that it is co-axial with the part being moved, but this may not always be convenient and the use of levers and other devices may be justified. In this connection it may be pointed out that for a given length of stroke, a cylinder with piston extended will measure twice the stroke length plus 101–203 mm (4–8 in) to allow for seals, housings, etc. By using a shorter, larger diameter cylinder the stiffness is also greatly increased.

A simple lever (Fig 8) makes it possible to triple the cylinder stroke, but the length of movement depends on the space available for the lever. An alternative, using a bell crank, enables the cylinder to be placed at any angle. The traditional method of doubling the stroke using two racks

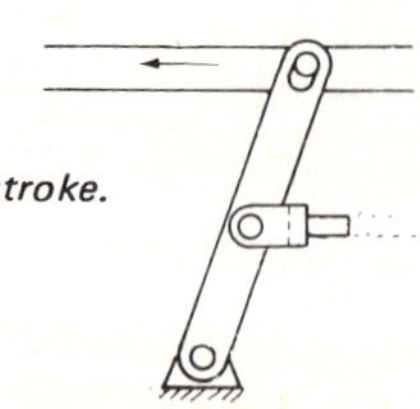

Fig 8 Simple lever to increase stroke.

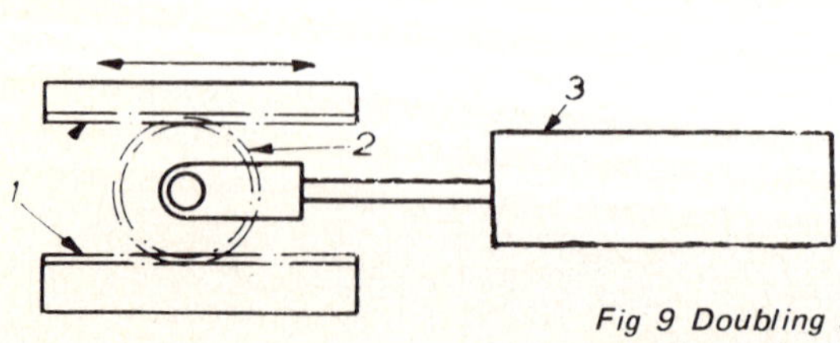

Fig 9 Doubling stroke length with stationary and moving racks. 1. Rack. 2. Pinion. 3. Cylinder.

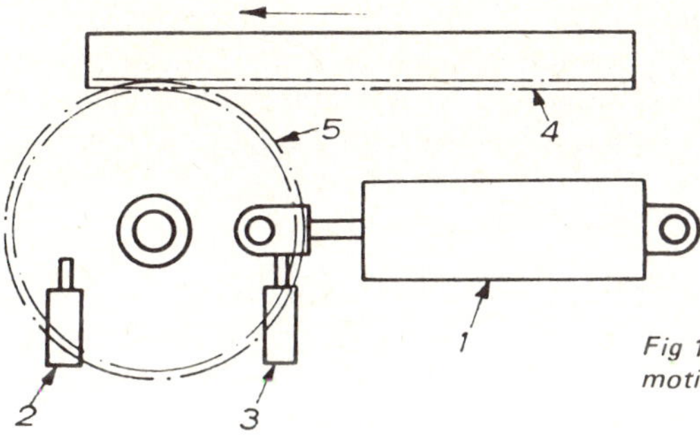

Fig 10 Gear and rack to give approximate harmonic motion. 1. Cylinder. 2. & 3. Subsidiary cylinder. 4. Rack. 5. Pinion.

(Fig 9) can also be useful. Racks of any length can be made by stretching roller chain along suitable supports.

To minimize the shock involved in accelerating and decelerating heavy slides, a method which involves a slight degree of magnification with approximate simple harmonic motion of the mass is shown in Fig 10. As the cylinder is dead centre at the ends of the stroke, two subsidiary cylinders are provided to give an initial impetus and if they are provided with suitable valving they can also help the final deceleration.

Rotary Movements

Where continuous rotary movements are required at low to moderate power levels, the *air motor* becomes the logical choice and it is being used to an increasing extent in modern industry. *Vane-type* motors, characterized by a smooth power output are generally available in sizes up to about 1.5 horsepower. *Piston-type* motors are most widely used in the power range 1.5 to 15 horse-power. They work with lower rotary speeds than vane motors (usually under 5 000 rev/min). Piston-type motors also have better starting and low speed characteristics than vane-type motors.

The normal working pressure for air motors is usually 6–7 bar (80–100 lb/in^2). See also chapters on *Air Motors.*

Controls

The simplest control method is a manually operated control valve, which, even in an automated system, is normally retained for manual override, or as an emergency control. The manual control valve remains the usual solution when the control required is too simple to justify other control methods, although valves of this type, triggered by machine movements, may readily be adapted for 'automated' working. See chapter on *Valves.*

Time Control

A time control usually takes the form of a camshaft running at a steady speed which actuates small valves or electric switches at predetermined intervals. Normally it is used where the governing factor is the time taken by the process rather than that taken for the pneumatic cylinder to complete its travel. Sometimes an interlock with a limit detecting device may be necessary.

Sequential Control

This is the normal method when two or more phases occur in an operation or series of operations. The end of each phase triggers off the next. Because of the difficulty of fixing the exact time taken for each phase this method is usually preferred.

Sensing Devices

The choice of sensing device is bound up with the type of control system used, but the correct choice can, nevertheless, be very important and may, indeed, influence the control system.

Where great precision is required for a point of reversal, this can sometimes be done by using an air jet, as in pneumatic gauging.

Electrical Control

Practically everything that can be done with electrical controls can be done pneumatically. However electrical controls may be preferred because of their compatibility with signals coming from electrical sources (eg timers); because wiring is neater than piping; or because of the convenience of microswitches to initiate control signals.

Using microswitches, proximity switches etc, with appropriate ancillary devices, in conjunction with double solenoid valves, very simple circuits can often be obtained.

Control System Design

Control system design must start with an analysis of the movements which must be performed. Thus analysis is usually done by means of a distance-time diagram. It is then necessary to allocate power components (eg cylinders) to achieve these movements and select suitable control components (eg valves). The major problem is then to plot the circuit, connecting the components to achieve correct working. This circuit plotting demands considerable skill and experience on the part of the designer.

One method is described in the Atlas Copco publication, 'Mechanization with Compressed Air' — Part 3 — 'Solving Methods for Sequence Problems'.

Noise Control

AS FAR AS generalizations are acceptable, it is now recognized that continual exposure to sound pressure levels above 70 dB is particularly damaging, although *short-term* exposure to sound pressure levels up to 100 dB is likely only to lead to temporary loss of hearing, generally with full recovery. Sound pressure level and time of exposure are thus primary parameters in assessing the risk or damage potential of noise. Permissible exposure noise level figures established on this basis are:—

Maximum Permissible Exposure (Per Day)	Sound Level dB(A)	Band
8 hours or more	under 90	
6 hours	90—92	A
4 hours	92—95	B
3 hours	95—97	C
2 hours	97—100	D
1½ hours	100—102	E
1 hour	102—105	F
½ hour	108—110	G
¼ hour	110—115	H
none	above115	

On this basis, personal protection should be provided in any working area where the noise level is in excess of 90 dB(A) for the whole of the working day. This level may be increased when the noise exposure is less frequent to the extent of an extra 3 dB for each halving of the total time of exposure during the day; thus for an exposure of 4 hours per day, 93 dB(A) is acceptable; for 2 hours per day 96 dB(A) is acceptable, and so on. There are three provisions: firstly that there is no isolated frequency with an exceptionally high peak level; secondly that no exposure at all greater than 120 dB(A) is permitted to the unprotected ear; thirdly that no exposure, of any part of the body, is allowable above 150 dB(A).

These levels are intended to apply to reasonably young, healthy workers and are framed to apply in mines, building sites, demolition areas and the like. Much better conditions ought to be reached in factories and other permanent establishments, where every effort should be made to reduce the ambient noise level to that acceptable to the unprotected ear.

In practice, exposure is likely to embrace various noise levels for various durations. The combined effects can be assessed as

$$\frac{t_A}{6} + \frac{t_B}{4} + \frac{t_C}{3} + \frac{t_D}{2} + \frac{t_E}{1\frac{1}{2}} + \frac{t_F}{1} + \frac{t_G}{\frac{1}{2}} + \frac{t_H}{\frac{1}{4}}$$

where t_A = actual exposure time (hours) to band A noise level, t_B = exposure to band B noise level, etc. For hearing conservation the sum of all the fractions must be less than unity.

A third parameter to be considered is the frequency content of noise. High frequency noise of the same level is potentially more damaging than lower frequencies. Also, for the same overall noise level, a concentration of frequency in one narrow band or 'spot' frequency is more damaging than if spread over a band of frequencies. Various attempts have been made to establish risk damage criteria embracing all three parameters.

Two such examples are shown in Figs 1 and 2, intended as a general guide to where noise reduction should be applied, or hearing protectors used, so that continuous exposure can be tolerated without risk. In these examples exposure time is excluded as a parameter.

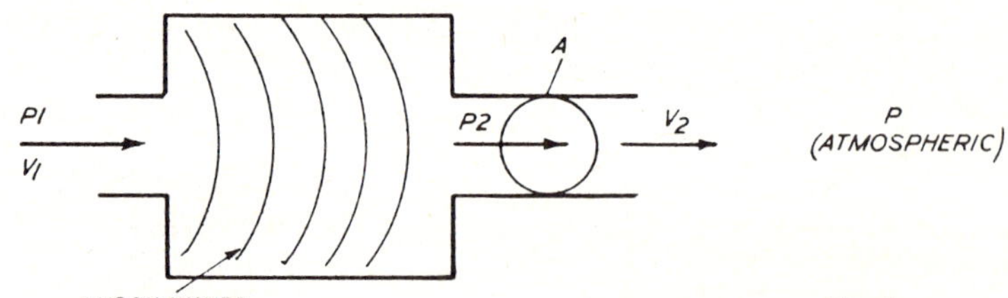

Fig 1

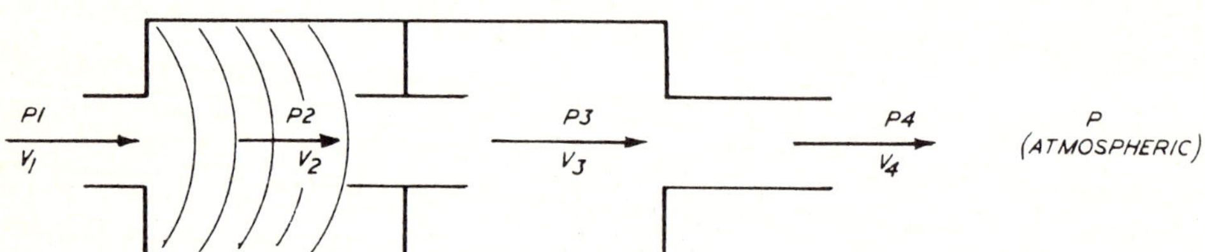

Fig 2

Compressed Air Noise

Compressed air as a working fluid tends, unfortunately, to be a noisy medium, both in its generation and utilization. The problem of noise control can be exaggerated by locating the generator (compressor) and air-operated equipment close together (eg as with roadworking equipment). In an industrial application the compressor can be separated from the remainder of the factory in a compressor house, which lends itself to effective noise reduction and control. However, this is not without its disadvantages, notably the cost of soundproofing the complete compressor house, and the extra length of piping which may be required, resulting in increased pressure drop and increased possibility of leakage losses.

In the case of compressors, the predominant noise is generated by pressure fluctuations at the intake and/or discharge, particularly in the case of reciprocating machines. A rotary compressor is inherently less noisy and also has fewer moving parts, simplifying balance and vibration control requirements.

At the other end of the system, pneumatic tools and devices tend to be noisy because relatively high pressure air is exhausted direct to the atmosphere, producing a choked jet stream. Noise level may further be raised by the mechanical noise generated by the device itself (particularly in the case of percussive tools). Basic solutions are to silence the exhaust and apply some form of sound deadening or sound absorption to the tool itself.

TABLE I – COMPRESSOR NOISE : SOURCES AND TREATMENTS

Source	Treatment
1. Driver vibration	Isolation mounting
2. Compressor vibration	(i) isolation mounting (ii) decoupling
3. Driver running noise	(i) intake silencing (ii) exhaust silencing (iii) enclosure
4. Compressor running noise	(i) intake silencing (ii) enclosure
5. Cooling fan*	Fan should be designed for minimum noise
6. Gear noise	Complete enclosure
7. Relief valve operation	Complete enclosure
8. Resonance – whole chassis* canopy*	Isolation mounting and decoupling Isolation Suitable damping and isolation mounting

*Applicable to portable compressors

Noise Control of Compressors

Where noise reduction is essential with a stationary compressor, the following treatment can be considered (see also Table I):

(i) Complete enclosure of the compressor and driver in a properly soundproofed room or chamber. Isolation treatment may also have to be applied to ensure that no vibrations are transmitted through the walls or floor.

(ii) Choice of a vane-type rotary compressor, which is inherently quieter than a reciprocating machine. The rotary compressor runs with absence of the vibration and freedom from the pulsations inherent in reciprocating machines. Provided effective intake silencing is employed, the bulk of the noise will be that generated by the driver, which should be amenable to direct treatment.

(iii) Good alignment and rigidity of coupling without transmitting vibrations between driver and compressor are also important factors in noise control.

A number of other points affecting design in relation to noise suppression can be considered. There is, for example, the use of an oil sump within the compressor housing – a feature which further helps to muffle the noise. In the cooling area, attention to fan design, air speeds and construction of the after-cooler all minimize noise. Air intake is often a source of considerable noise and must be controlled. This means that attention has to be given to the pattern of air flow, its velocity and the position of the cleaner in the system.

Portable Compressors

Mobile compressors are commonly designed as truck or trolley mounted units with a diesel engine driver. Smaller portable compressors may have spark-ignition drivers (or possibly electric motors in some cases). Likely noise sources are summarized in Table I.

The modern approach to silencing a portable compressor is to identify each noise source systematically and to reduce its level. The total sum of all noise levels does not add up, however, to the total annoyance factor to the human ear. There are some low frequencies whose annoyance factor is out of all proportion to their magnitude. Using the right shape and size of canopy has the effect of reducing resonance and hence these low frequency vibrations.

The ducting required on a portable compressor to allow the free passage of cooling air conflicts with the silencing concept, because air passing through each duct opening is a potential noise source; hence the need to consider the total noise source and to balance the outcome.

Whilst portable compressors are silenced to 75 dB(A) at 7 metres (23 feet) with little extra cost, silencing thereafter becomes progressively more expensive, but generally feasible.

More and more attention is being given to producing mobile compressors for road breaking operations, etc, where the noise level conforms to the recommendations of the Wilson Report (or the Pneurop Test Code):

70 dB(A) in rural, suburban and urban areas away from main traffic and industrial noise.
75 dB(A) in urban areas near main roads in heavy industrial areas.

Rather different conditions apply in the case of stationary compressors in industrial applications where each case normally has to be considered on its own merits. Complete enclosure is usually the best answer where appreciable reduction in noise has to be realized.

Appliance Noise

Pneumatic tools and appliances basically exhaust compressed air directly to atmosphere. Where the pressure ratio at the exhaust exceeds 2:1 the exhaust flow will be supersonic. Shock waves are generated and a considerable degree of turbulence with the sudden breakdown of the shock waves and mixture of high velocity air with surrounding static air. Noise reduction can only be effected by (i) eliminating shock wave formation and (ii) reducing the exit velocity as much as possible to minimize the degree of turbulence. Both depend on the pressure ratio involved.

The most effective way of reducing the noise is to allow the air to escape into an expansion chamber in order to reduce the pressure ratio to below 2:1 before it finally escapes into the air. Exhaust flow will then be subsonic and restricted to pulses of flow and turbulent mixing.

Three further problems then exist. The flow into the expansion chamber will still be supersonic, with the formation of shock waves. Suitable attention must, therefore, be given to the expansion shocks so formed. Noise from the final outlet will be primarily dependent on the escape velocity, which is in turn dependent on the final pressure ratio and the area of outflow. Further reduction in noise is thus possible by increasing the outflow area, so that the flow is diffused. The final requirement is that the complete device, or silencer, does not impart excessive back pressure to the outflow, which could affect the operating efficiency of the device.

The simplest type of silencer consists of a simple expansion chamber attached to the exhaust port of the device – Fig 1. This provides a volume in which the pressure ratio can be reduced to a suitable figure, ie P2 less than 1.9 times atmospheric pressure, when the exit flow will be subsonic. The final noise energy will be proportional to $AV_2{}^8$, from which it is obvious that anything reducing the value of V_2 will have a substantial effect on the noise energy level. Some improvement can often be realized by the simple practical expedient of attaching a length of hose to the silencer outlet to act as a second expansion chamber, further reducing the value of V_2.

Silencers of this type may be in the form of simple attachments to fit on the exhaust port of a cylinder or valve, or in specific cases may be incorporated in the design of the tool or device. Noise reduction of up to 10 dB(A) is generally possible, with a suitable chamber volume. Further improvement can be realized by using a two-port chamber, and/or combining an expansion chamber with a diffuser. Here savings of up to 25 dB(A) are sometimes possible, although the figure achieved is usually materially less (Fig 2).

Two-chamber expansion provides better control of expansion shocks, particularly if the first chamber can be enclosed. This will prevent any expansion shocks formed from passing directly out of the final port. Also final expansion through the exhaust port can take place isothermally, or nearly so, because of the low pressure ratio involved. This is particularly useful, since it eliminates the possibility of ice build-up in the exhaust chamber or exhaust port – a problem which can arise with simple single-chamber exhaust silencers. This, in fact, is the main limitation of simple silencers. The sudden drop in temperature of the air on expansion deposits water vapour which can freeze around the exhaust port or exhaust passages, or even block the silencer completely. In practice, however, complete blockage is not likely because of the scouring action of the high velocity air passing through the chamber, provided the flow path is simple. Icing up of the first chamber is also a possibility in the case of two-chamber expansion. Here, however, the design can allow for the fact that should the first chamber become iced up, flow to the second chamber will not be restricted. In that case the exhaust cannot become restricted by icing up, only the efficiency of the silencer reduced.

Percussive tools are inherently noisy in operation, since the majority of the expansive energy of the air is dissipated in the exhaust at each end of the stroke; a piston impacts against an anvil block at a frequency of the order of 20 times per second and a high intensity stress wave is transmitted down the drill rod at a similar frequency. The typical noise spectrum comprises a number of discrete frequency peaks spread over a wide range with a high overall level of the order of 100 dB – see Figs 3 and 4. Basically, however, about 80% of the noise comes from the exhaust, and most of the remainder from the rod. Simple jacketing, without any attempt to silence the exhaust, would thus be singularly ineffective.

Fitted jackets or mufflers provide a means of silencing an existing (unsilenced) tool without any structural modification. Jacket-type mufflers (eg Pfister and similar) are also generally lighter and comprise, basically, a lightweight jacket of PVC coated terylene lined with profiled foam rubber, which is reinforced adjacent to the exhaust positions. They are held in place with cords and straps.

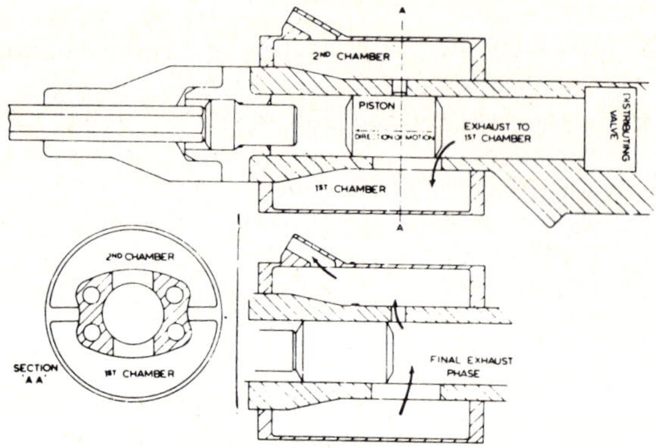

Fig 3

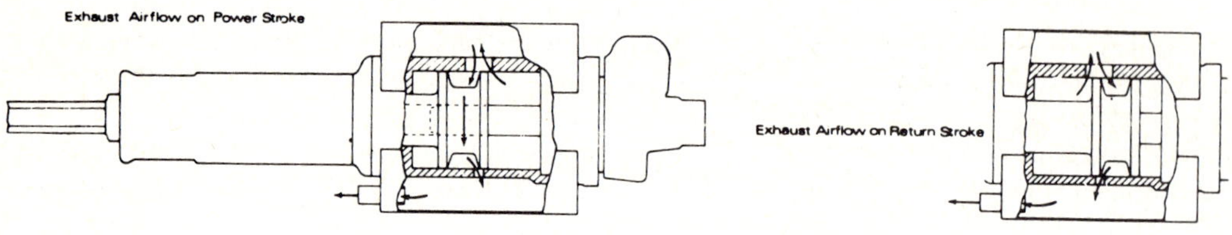

Fig 4

Rigid, detachable mufflers (basically exhaust silencers) may also be produced by manufacturers to fit individual standard tools, although in this case the tool must be designed initially with provision for accommodating a matching silencer. The simpler solution of producing an exhaust silencer to fit on to the exhaust in the form of an extension piece is less attractive because of its vulnerability. Also the efficiency of simple single-chamber exhausts can be relatively low, and variable, if icing-up is experienced.

Particularly good results have been obtained on road breakers with both exhaust and steel muffling, and the combination of a silenced portable compressor with a fully muffled breaker is likely to meet most of the objections from the public when they are used in a built-up or residential area. It is unlikely that any further reductions of appreciable magnitude can be obtained in this equipment in the near future. It has been suggested that hydraulic breakers which are coming onto the market now are much quieter than the corresponding pneumatic ones, but tests show this not to be so.

Simple exhaust systems applied to pneumatic tools are generally based on the addition of a collector ring around the body of the tool and encircling the original exhaust openings. This is

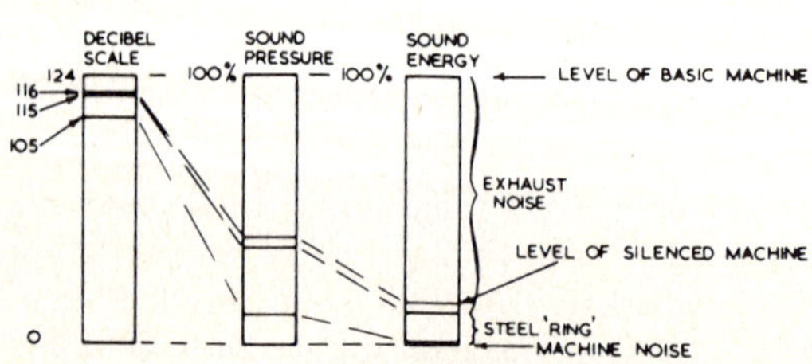

Fig 5 Comparison of the decibel scale with sound pressure and energy for an experimental silenced rock drill. (Holman Bros.)

suitably sealed to the body to form a complete expansion chamber. Some further silencing may be achieved by the addition of a length of hose to the exhaust outlet of the silencer, the collector ring then acting as a first expansion chamber with subsonic flow through the hose and to the final exit. The noise reduction achieved by these methods on a pneumatic hammer is shown in Fig 5.

Drill Silencers

Results obtained on exhaust noise tests on a representative rock drill fitted with kidney-shaped silencers and conducted by the US Steel Research Laboratory are summarized in Table II. The basic design of silencer is shown in Fig 6 (Muffler I) and the testing arrangements in Fig 7. Mufflers II and III were modifications of the geometry and size of the internal chambers of Muffler I.

The drill was mounted in Room B and was pressurized with a portable compressor. The air discharge is transmitted through a double wall by an exhaust pipe into Room A. Through this installation, the mechanical noise of the drill is separated acoustically from the air exhaust noise.

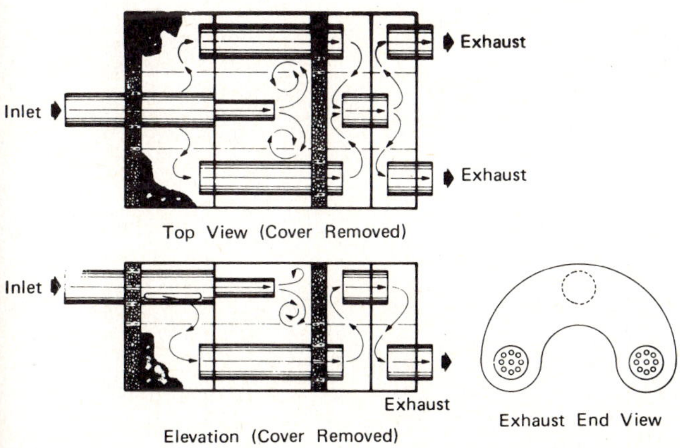

Fig 6

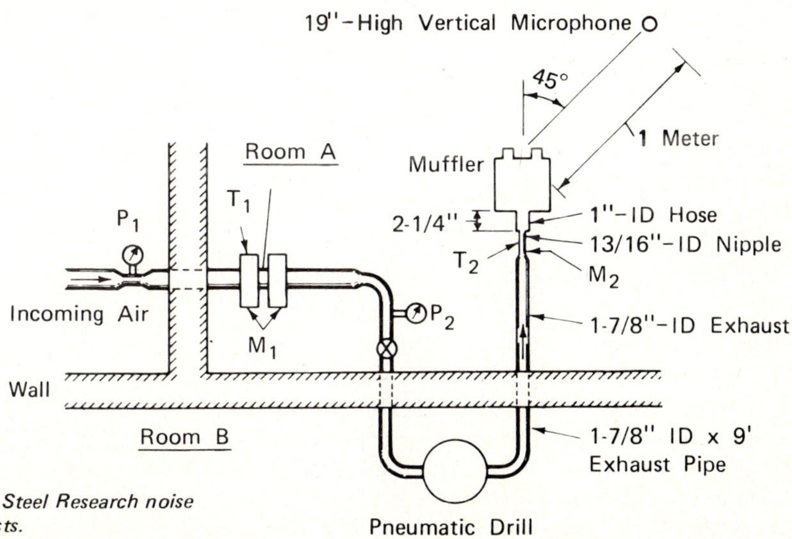

Fig 7 Set-up for US Steel Research noise tests.

The exhaust noise measurements were made for the drill when used without a muffler and when used with three variations of the muffler (designated as Muffler I, II and III).

TABLE II – SUMMARY OF EXHAUST NOISE TEST RESULTS FOR STOPER DRILL RP38E*

Muffler Condition	Flow Rate (ft/min)	Inlet Drill Pressure (lb/in^2)	Drill Back Pressure (lb/in^2)	Overall Noise Level [dB(A)]
Exhaust Noise Tests				
1. No muffler; microphone positioned in Room A as shown in Fig 1	103 94	96–102 73–77	–0.58 –0.22	113 110
2. Kidney Muffler 1	99 97	97–101 73–77	0.73 0.40	84 84
3A. Kidney Muffler II	104 99	100 75	1.56 1.00	83 80
3B. Muffler II (½ in i.d. hole to second chamber blocked)	105 100	100 75	1.61 0.89	84 80
3C. Muffler II (no discs on two outlets of last chamber)	104 102	100 75	0.15 0.07	88 86
3D. Muffler II (new second chamber)	98 90	100 75	1.35 1.20	79 77
3E. Muffler II (new second chamber and no discs on outlet of last chamber)	96 90	100 75	0.29 0.19	86 84
4A. Muffler III	103 100	100 75	1.69 0.88	84 83
4B. Muffler III (no discs on outlet of last chamber)	105 102	100 74–79	0.00 0.00	94 93
Total Noise Tests				
5. No muffler	99 98	100 75	– –	121 118

*US Steel Research

Tool Steels

Sound reduction treatment of steels remains a particularly difficult problem. Piston and anvil noise is, to some extent, alleviated by the use of a padded jacket, when a noise reduction of up to 10 dB(A) may be achieved, although the figure is usually less. There is also user-objection to the greater bulk of a padded tool.

The direct approach is the use of a high damping material for the rod, although unless the material has directional properties this will inevitably lead to a loss of effectiveness of the tool point.

An alternative solution which has been developed by BroomWade is the fitting of an elastomeric damping device located at a critical point on the steel. The object of this is to reduce mechanical vibration and recoil, and hence the noise radiated. A comparison under similar conditions, between a standard breaker using a standard steel and a muffled breaker with elastomerically damped steel has shown a noise reduction in the latter case as high as 10 dB(A).

Fan Noise

The noise generated by a fan tends to be directly proportional to the static pressure developed and the bhp absorbed by the fan. Other major parameters covered are:—

(i) Air velocity — responsible for self-noise.
(ii) Turbulence — responsible for frequency content of noise.
(iii) Operating rev/min — responsible for mechanical noise (bearing and motor noise).

An empirical formula for estimating the fan sound power level direct from pressure, volume and input power is:—

$$SWL = 90 + \log_{10} hp + 10 \log_{10} P$$

or

$$SWL = 55 + 10 \log_{10} q + 20 \log_{10} P$$

or

$$SWL = 125 + 20 \log_{10} hp - 10 \log_{10} q$$

where SWL = overall sound power level of the total noise spectrum (31.5 Hz to 8 kHz) transmitted to the associated ductwork in either upstream or downstream direction (measured in dB re 10^{-12} watts)
hp = rated motor horsepower
P = static pressure (inches wg)
q = supply volume ft^3/min

Noise level and frequency content is also influenced by the number of blades in the fan. General characteristics of different fan types are summarized in Table III.

The noise spectra of fans vary with the type of fan, but in general they are typical of all aerodynamic noise in that the acoustic power generated over a wide spectrum rises to a shallow peak at a frequency determined by the type, size and duty of the fan. Centrifugal fans having either forward or backward curved blades tend to produce a spectrum which fails at approximately 5 dB per octave with increasing frequency, although there is a tendency for backward curved impellers to generate more energy at high frequencies and correspondingly less at low frequencies.

TABLE III — FAN NOISE PARAMETERS

Type of Fan	Air Velocity	Turbulence	Speed	Remarks
Propeller	Low	Low	Low	Inherently low noise level
Axial flow	Low* High†	Low* High†	High	*Low pressure types †High pressure types
Centrifugal				
Forward curved	High	High	High	Large number of blades
Backward curved	Low	Low	Low	
Straight radial blade	High	High	Medium	

Axial-flow fans have a much flatter spectrum and contain a much greater proportion of high frequency energy. Typical generalized spectra are given in Fig 8. The use of this chart can give octave band sound power levels of standard fan configuration to an accuracy of ±4 dB.

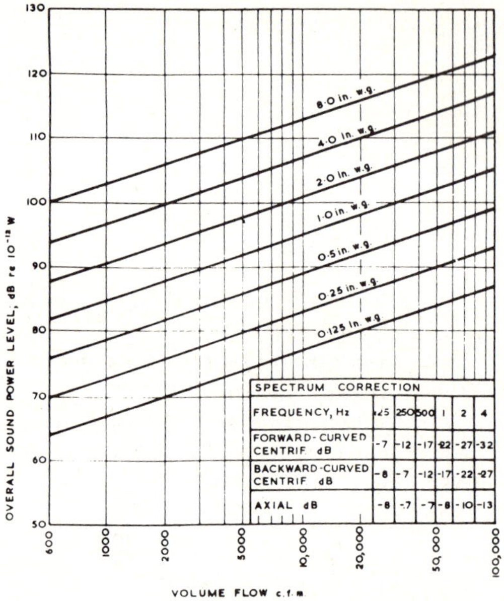

Fig 8 Chart for estimating octave band sound power generation by fans[7]

Not all noise generated by fans and transmitted along the ductwork comes from aerodynamic sources at the impeller. Some of the noise may be attributable to fan unbalance, bearing noise, structural resonance of the casing, or to noise generated in the motor, its couplings or belts. The fan noise may also be affected by its mode of installation and whilst the sound power generation of a centrifugal fan does not seem to be greatly affected by variations in inlet and discharge arrangements, extreme care should be taken to ensure that axial flow fans are installed in accordance with the manufacturers' recommendations. Quoted sound power levels are generally applicable to the case where the axial flow fan is coupled to inlet and discharge ducts having the same diameter as the fan casing and a minimum straight duct-run upstream and downstream of the fan equal in length to two duct diameters. The rated sound power levels are also applicable to fans having coned inlets and inlet transformation sections, provided the included angle does not exceed 60°.

The octave band sound power level and the spectrum of fan noise vary not only with the speed and duty of the fan, but also with its type and individual manufacturer. A standardized procedure for the measurement of fan sound power output is laid down by British Standard 848, Part II, and most fan makers will be found to publish full details of noise generation by their products in accordance with this Standard.

Fluidics

FLUIDICS may be defined as the study and application of elemental devices working on aerodynamic or hydrodynamic principles and having the capability of performing logic functions. The main application of fluidics is in providing control signals, and in this respect it can offer a degree of sophistication previously only provided by electronic circuits. The two forms of logic – fluid logic and electronic logic – are to some extent competitive, both performed by conventional fluid valves and related components. The main limitations in this field are the bulk and weight of the components and the complex external circuitry which may be involved. The former disadvantage can be reduced by miniaturization, although this can introduce manufacturing difficulties in establishing the very close tolerances necessary to obtain tightness and sensitivity. This tends to increase the cost of the miniaturized unit; also, performance can suffer from increased friction and wear.

Pure fluidic elements offer a theoretically complete answer to such problems in that they have no moving mechanical parts and rely for their action and response on flow phenomena and, in the absence of wear problems, have a very high degree of reliability and virtually infinite life. Also, they can be readily manufactured in miniature sizes with a high degree of accuracy and repeatability by straight-forward techniques (eg printed circuit or photographic reproduction and chemical milling).

Pure fluidic elements, however, have their own practical limitations. They have no shut-off capability, which means there is a constant loss of power through them. To reduce this to a minimum (usually of the order of milliwatts only), they operate at very low pressures (of the order of 0.07–0.35 bar) and utilize very small orifices. This means that they must be operated with absolutely clean, dry air at closely controlled pressures, and the performance of each element in a system must be closely matched by the use of bleeds and restrictors.

For these reasons miniaturized moving-part devices continue to be developed, and in fact are preferred by some manufacturers for use as the basic element in their logic systems. They include both original devices and specially developed miniaturized versions of spool valves. Where they are more or less directly comparable in bulk to pure fluidic elements they are direct alternatives, and as such, classified as fluidic elements or moving-part devices. Elements with no moving parts, and working by flow phenomena, are correspondingly known as pure fluidic devices.

Fluidics offers an all-fluid system, working on aerodynamic or hydraulic principles (normally the former), whereas electronic controls are normally associated with a two-media system (electricity for controls and pneumatics or hydraulics for the working medium). Practical fluidics, however, is almost invariably employed with pneumatic power systems to provide an all-air system.

Fluidics has certain inherent advantages over electronics, notably an ability to work in environments which would be highly unfavourable to electronics without extensive shielding and protection – eg at high temperatures and in hazardous surroundings. Electronics, on the other hand, has the advantage of higher speed of response (ie the speed of light compared with the speed of sound, or a speed difference of the order of 10^6). This difference is not normally significant, for the majority of control requirements, particularly where a human operator is involved, but can be important in particular fields of application (eg computer working).

Miniaturized Spool Valves

Miniaturized spool valves have the specific advantage that they can operate at much higher pressures than pure fluidic devices, even up to the full pressure normally associated with working pneumatic circuits. Large flow paths are also possible, so that cylinders can often be operated direct from a fluid logic system. Pure fluidic devices, operating at much lower pressures, need the output signals boosted, eg using step-up relays, before useful work can be done. Also no constant power loss is present, since such moving-part devices can have shut-off capabilities.

Apart from the practical difficulties of manufacture associated with the extremely fine tolerances involved, the main limitation inherent in a miniaturized spool valve is high friction, which reduces the operating pressure band and amplification available, as well as the speed and sensitivity of response. Various methods of reducing friction to a minimum have been tried, but probably the most successful is the valve designed on air bearing principles, using a sealless spool. Whilst shut-off is not complete, leakage is limited to the relatively minute amount of air needed for the correct functioning of the air bearing. This results in a spool valve which can be operated over an extremely high pressure band with high flow output, and with a high step ratio and low pilot chamber volume, giving extremely rapid switching times. Pilot pressures of 0.07–0.35 bar are adequate for switching a 7 bar supply, the higher figure generally being necessary if the air is not dry (eg contains lubricant), as the presence of the lubricant adds viscous friction, which slightly increases the pilot pressure necessary to reverse the spool.

Pure Fluidic Devices

Pure fluidic devices are based on some dozen different flow phenomena, although practical applications as yet are largely confined to just three. These are–

(i) Wall attachment devices (Coanda effect)

(ii) Jet interaction devices (momentum exchange)

(iii) Critical flow devices (turbulence amplifier)

All such devices are true amplifiers in that a low pressure control jet is capable of switching a higher pressure power jet, thus providing a power gain in terms of input power to output power. This may be a pressure gain, a flow gain or both.

Detecting Devices

Simple fluidic elements are also used as detecting devices, the interruptible air jet being a typical example. In the interruptible jet circuit shown in Fig 1, airflow from tube A impinges on the outlet of tube B, fed in a contrary direction from the other side of the loop via a restrictor. The resulting jet impingement prevents escape of air through B, forcing it to escape through the outlet C. If the airflow from A to B is interrupted, however, say by the presence of a solid object, the restraint against flow is removed and so air escapes from B, shutting off the original output flow through C.

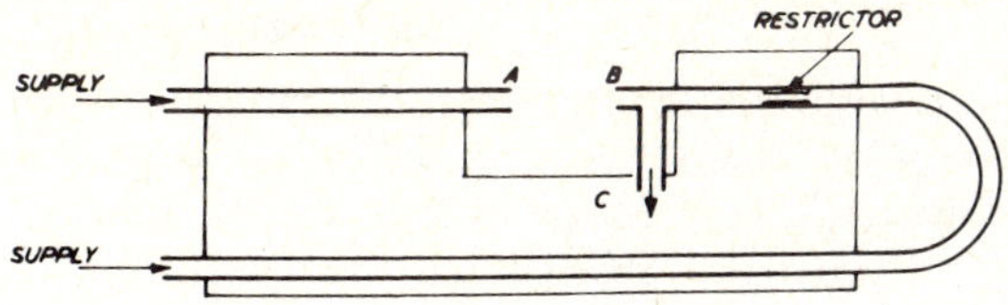

Fig 1

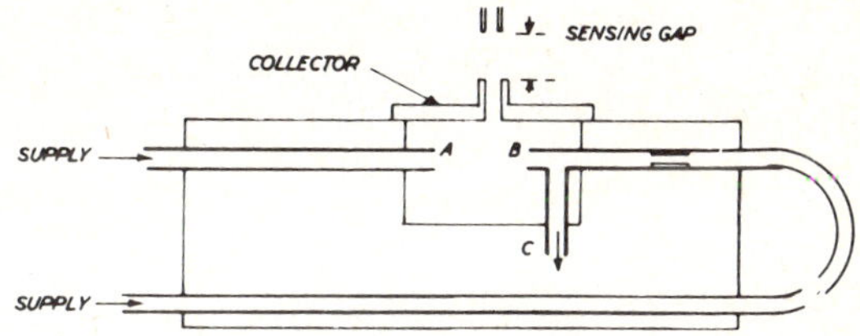

Fig 2

A simple detector of this type is normally limited to a maximum gap of about 20 mm (¾ in) between A and B, working off a supply pressure of the order of 0.07–0.2 bar. To accommodate larger gaps, a collector can be fitted, as shown in Fig 2. This configuration is more specifically described as an airstream detector. Both work on the principle of being responsive to backpressure.

Backpressure devices are also used as input units for actuating fluidic circuits, eg they are designed to give an output signal any time the exhaust flow is restricted. Thus, with the outlet port connected to a blocked load, the device responds to contact, providing an output pressure equal to the supply pressure.

More recent practical systems using back pressure or jet sensing may work on higher pressure air (eg a low pressure supply of circa 1 bar). An example is the Martonair Bi-Selector system shown in Fig 3 together with a functional diagram.

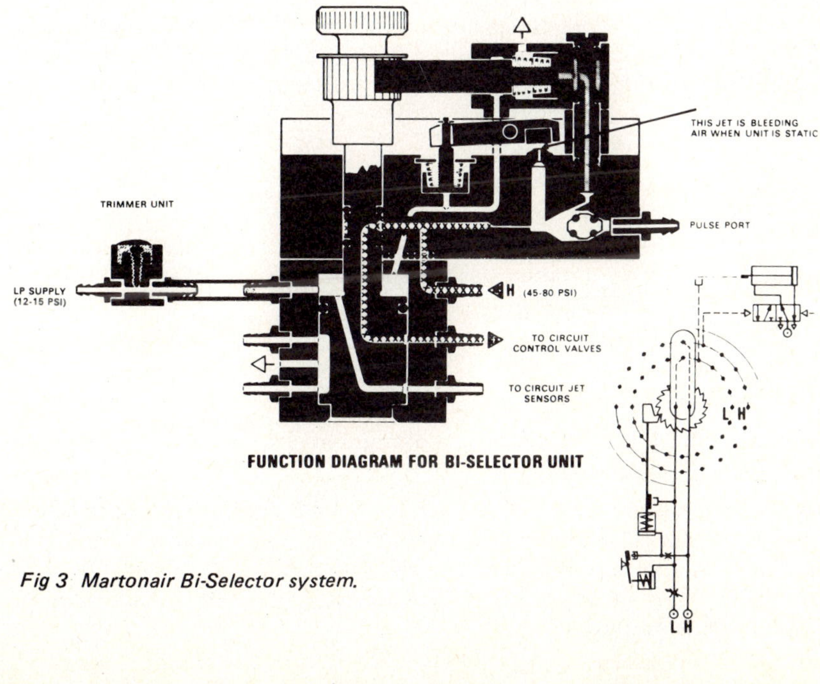

Fig 3 Martonair Bi-Selector system.

Filters

Pure logic elements, operating at low air pressures and with very low consumption, demand the use of dry filtered air. High efficiency filters with a 5 micron rating are often specified and can normally provide adequate protection, if properly located. However, a much higher degree of protection may be required for some devices, particularly as regards the separation and removal of oil mist, which is commonly present in any compressed air supply. Sub-micron particles of oil in the 0.01 to 1 micron range are often the most troublesome, hence special oil-removing filters may be required.

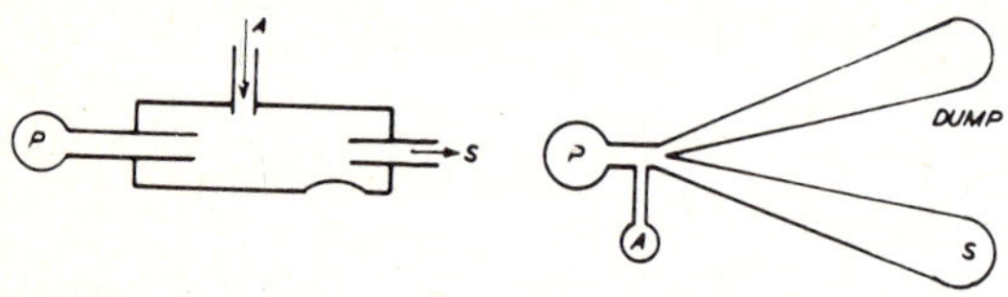

Fig 4

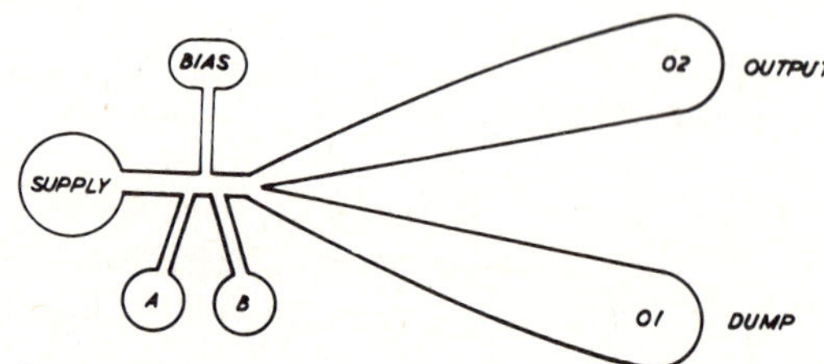

Fig 5

Logic Functions

Logic function or behaviour can be reduced to NOT, OR and AND response, and the inverse functions NOR (NOT-OR) and NAND (NOT-AND).

The NOT function implies a single input (A) and single output (S) controlling a power supply (power jet) (P). This can be provided by a turbulence amplifier or a wall attachment device, as shown in Fig 4, or by suitably constructed moving-part devices. The logic involved is essentially straightforward, in the absence of any input signal (which can be written $\overline{A}$, the bar standing for inversion or opposite) there is output. In the presence of a signal, there is no output.

$$\overline{A} = S$$

$$A = \overline{S}$$

There is NOT an output ($\overline{S}$) in the presence of a control signal. This function can also be written in the form of a truth table, where '1' designates an 'on' condition and '0' an 'off' condition.

A	S
0	1
1	0

There is no inverse of NOT as such, although a detector may work on this inverted principle if it is detecting the *removal* of an obstruction.

The OR function embodies the principle that signal A, or B (or C, etc) will give an output, and thus implies at least two inputs. It can be given directly by a wall attachment device, as shown in Fig 5, where the power flow is normally biased to 01. The application of signal A, or signal B, will transfer the flow to 02. Thus to operate as an OR device, output 01 would normally be dump, with 02 becoming the actual output.

The same function can be performed by two turbulence amplifiers, connected as shown in Fig 6. Any input signal, A, B, C, etc, applied to the first amplifier, removes the input to the second amplifier and gives an output.

In both cases the function is known as an inclusive OR, since A or B or C, etc, or all inputs together, or in any combination, will give an output.

The *inverse* function of OR is NOR, which in simple logic means that neither signal A NOR signal B, etc, will produce an output.

$$\bar{A} + \bar{B} + \bar{C} \ldots\ldots = S$$

or

$$A + B + C \ldots\ldots = \bar{S}$$

The wall attachment device of Fig 5 will perform this function if the outputs are switched, ie 01 is made the output and 02 dump. This is, therefore, an OR/NOR device, the mode of working depending only on the utilization of the outputs.

In the case of turbulence amplifiers, a *single* unit with multiple inlets will perform the NOR function – Fig 7. Thus a multiple-inlet turbulence amplifier is essentially a NOR device.

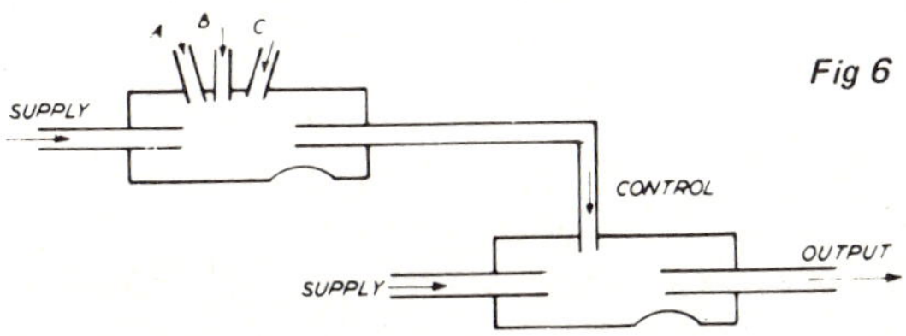

Fig 6

Fig 7

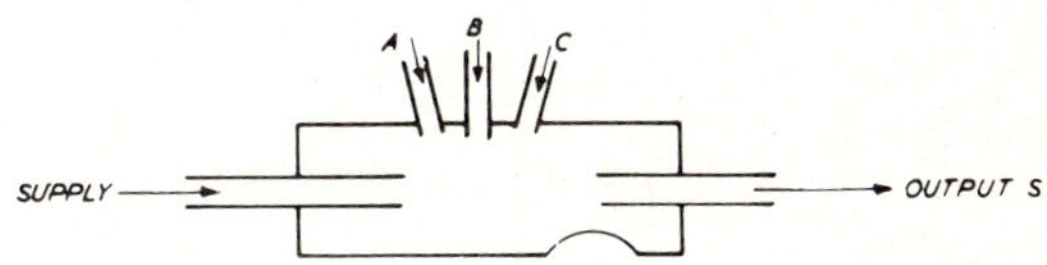

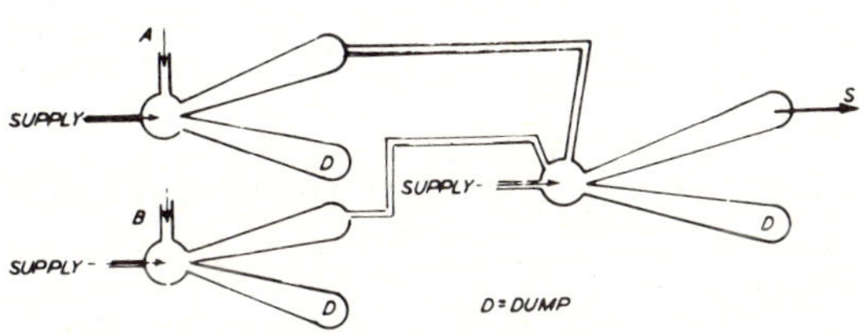

Fig 8

NOT logic can also be used as an inverter. Just as NOR is the inverse of OR, then OR must be the inverse of NOR. Referring to Fig 6, the first turbulence amplifier is working in an identical manner to Fig 4, ie as a NOR device. The second turbulence amplifier has a single input, and so is working as a NOT device (Fig 4). Thus NOR followed by NOT means simply an inversion of NOR to OR.

The AND function implies that signals A and B (and C, etc for more than two inputs) are necessary to give an output. Fig 8 shows how this function can be provided by three wall attachment devices, and Fig 10 shows the AND function performed by three turbulence amplifiers. Note that the solution applies only for two inputs; each additional input would require an additional device feeding the last device.

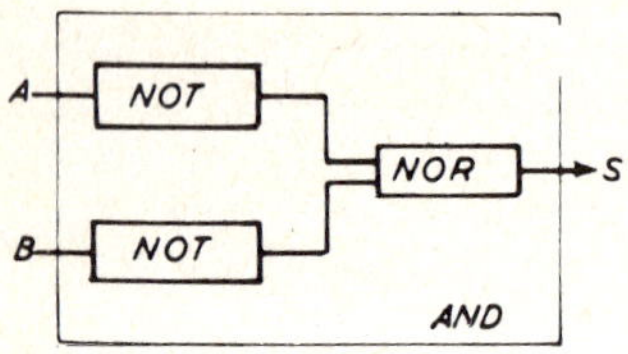

Fig 9

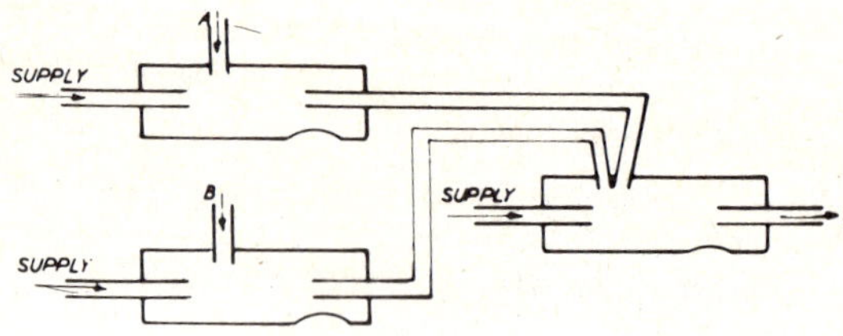

Fig 10

The mathematical sign for 'and' is . and so the mathematical equation is

$$A.B = S$$

The AND function can also be built up from NOT and NOR devices – Fig 9. This, in fact, is an exact description of the functions of the three turbulence amplifiers in Fig 10. It could equally well be built up from NOT, OR and NOR devices, as shown in Fig 11 (the NOT following the OR merely inverting the OR function to NOR output).

The inverse function of AND is NAND. Again this can be derived in various ways. Fig 12 shows suitable combinations of both wall attachment devices and turbulence amplifiers. Fig 13 shows derivation of the NAND function in block form.

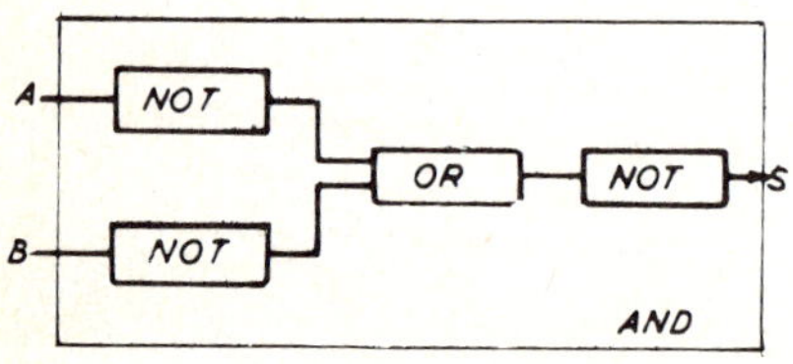

Fig 11

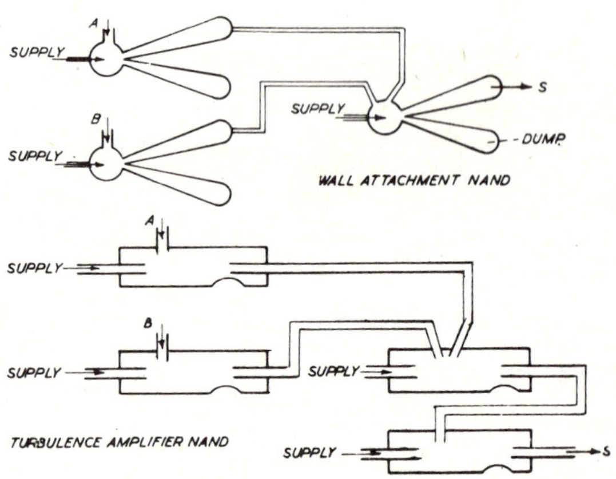

Fig 12

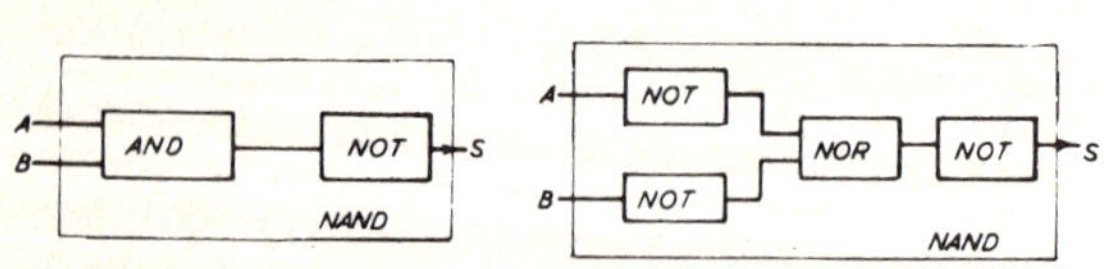

Fig 13

The *exclusive* OR is a special case of the OR function, normally restricted to two inputs. It implies that output is obtained with either signal A OR signal B present, but not with both present. Fig 14 shows how this can be obtained from NOT and NOR devices (or an AND device coupled to two NORs).

The aforementioned embrace all the basic logic functions, from which it will be appreciated that all logic workings can be derived from NOT and NOR (or OR) devices, in suitable combinations. These functions or logic gates can be drawn in block form for convenience (eg in Figs 8 and 10), or represented by symbols. Fig 15 shows preferred symbols, where it should be noted that the presence of a vertical bar on the output side indicates an inversion.

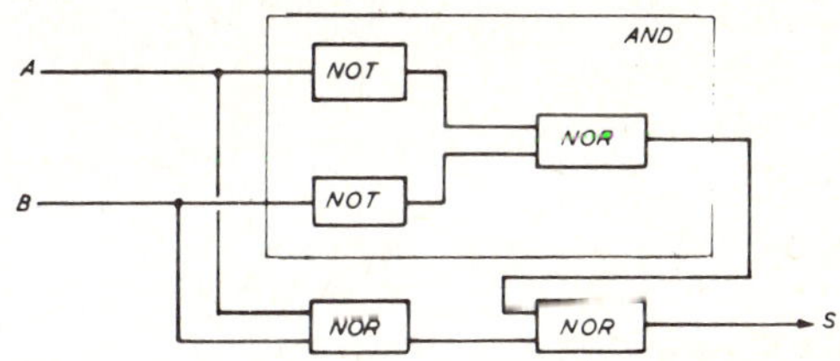

Fig 14

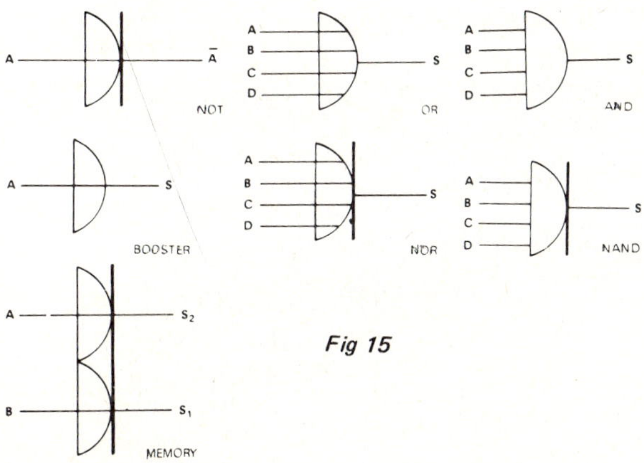

Fig 15

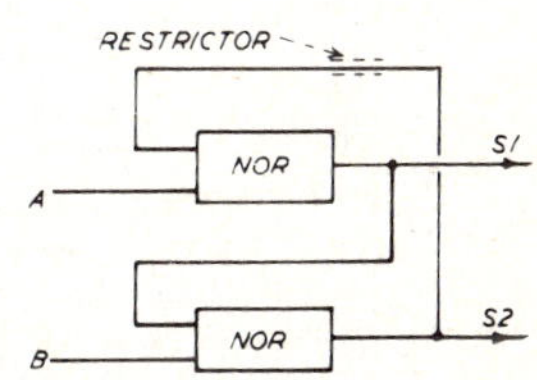

Fig 16

Memory (Flip-Flop)

A memory device, or flip-flop comprises two interconnected NOR units – see Fig 16. Since the output of one device is connected to the input of the other, only one device can give an output at any one time. The circuit is bi-stable so that a signal pulse only is required, not a continuous input.

$$A.\bar{B} = S1.\overline{S2}$$
$$\bar{A}.B = \overline{S1}.S2$$
$$A.\bar{B} = S1.\overline{S2}$$
$$\bar{A}.B = \overline{S1}.S2$$

In practice, bias is usually necessary to ensure a determined state of the flip-flop. This can be done quite simply by fitting a restrictor in one of the cross feeds. In the event of simultaneous application of pulsed signals A and B, the flip-flop will now have a preferred or predetermined output. In the absence of bias, simultaneous signals would give no outputs.

Pneumaid moving part logic provides a useful bridge to span the gap. between conventional pneumatic circuitry and fluidics. Pneumaid logic elements can be used equally effectively where circuit design is carried out by intuition, ladder, cascade or logic methods. To produce a circuit some basic knowledge of logic is an advantage, but not essential. A range of units is available, so that each unit performs a separate discrete function, thus enabling direct substitution in a logic circuit. (CompAir Maxam Ltd).

Buffering

Flip-flops derived from NOR devices may also require buffering, where several outputs are required to control other units. In this case, the two outputs may be taken to individual NOR elements, to ensure an equal load on each of the flip-flop elements, and balanced operation. Outputs are then taken off the second pair of NOR devices – Fig 17.

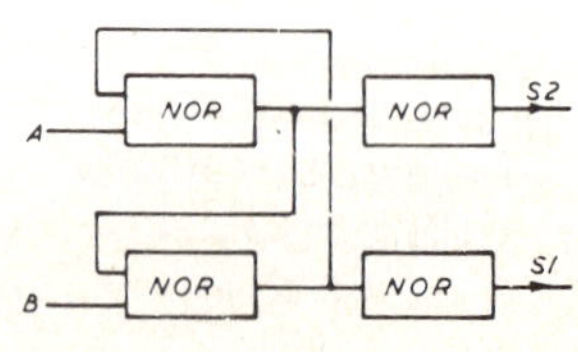

Fig 17

Booster

This is basically a NOT-NOT device, which produces double inversion, and double amplification. Double inversion restores the output to the original form, but the final output power is increased. A buffer may be used between a memory, or flip-flop, to increase the number of effective outputs (fan-out), or generally to increase output power (eg for feeding to a step-up relay).

The Half Adder

The half adder performs the function of adding two binary numbers. Fig 18 shows the geometry of a wall attachment device capable of performing this function, A being normally biased to 02 in the absence of B.

Mathematically: Sum	(S)	=	$\bar{A}.B + A.\bar{B}$
Inverting again	$\bar{S}$	=	$\underline{\overline{A.B.} + A.B}$
Inverting	S	=	$\overline{A.B} + A.B$
Carry	(C)	=	$A.B$
Hence	S	=	$\overline{\bar{A}.\bar{B} + (Carry)}$

Fig 18

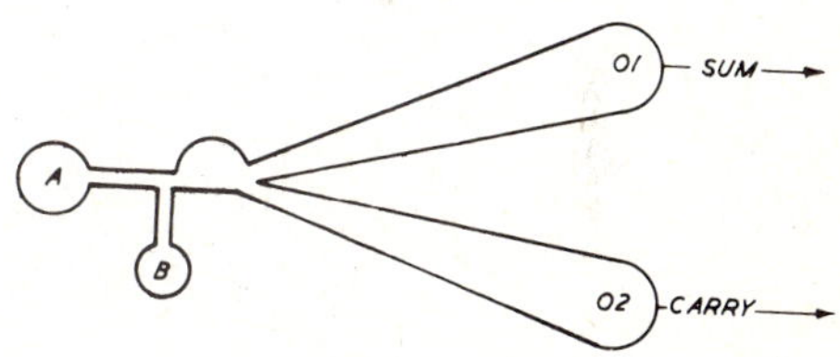

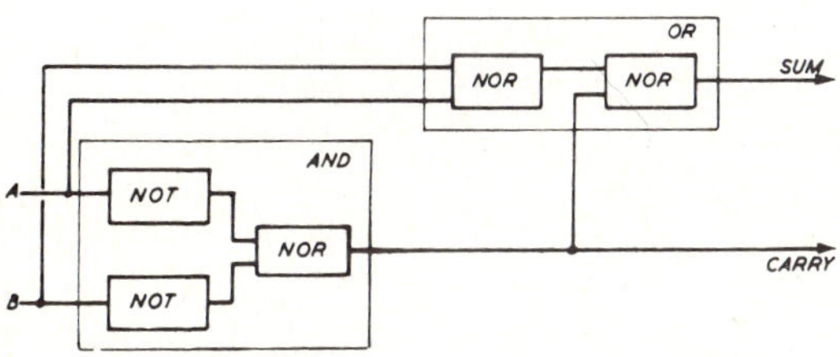

Fig 19

Fig 20

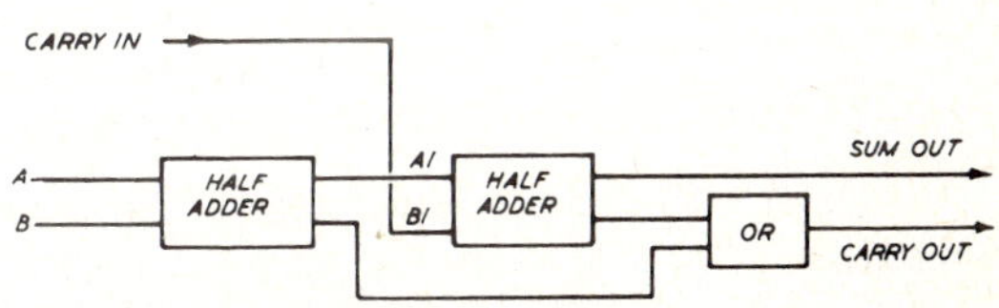

The function for S can be performed by NOR logic, as shown in Fig 19. The 'carry' appears in the equation for the sum, and can thus be extracted from a suitable part of the circuit without requiring any separate treatment.

To perform the complete function of binary addition, input carry must be accommodated at every successive stage. This demands the use of two half adders, one for the numbers A and B and the other to add the sum of these to the carry from the previous stage, as shown in Fig 20. This circuit is known as a *full adder*.

Low Vacuum Techniques

IN MODERN industrial processes the most common use of vacuum is to reduce the boiling point of water contained in a product in order to achieve one or more of the following ends:

(i) to prevent thermal damage to the product when heat is applied to vaporize and to effect mass transfer of the water and thus dry the product,

(ii) to speed up the process, and

(iii) to ensure very low residual moisture content in products difficult to dry by other means. Examples of this application are distillation, evaporation, vacuum drying of liquids and solids and freeze drying processes.

Vacuum cooling is closely allied to vacuum drying and is a simple mass transfer process which results in cooling of the product by the removal of a controlled quantity of moisture.

Depression of the vaporization point is also the objective of vacuum coating processes in which metals and other materials are melted, evaporated and eventually sublimated on prepared substrates. The high vacuum condition prevents oxidation of the material evaporated at high temperature.

Simple removal of atmospheric air and of moisture so that they can be replaced by another gas or liquid is the reason for the use of vacuum in impregnation; in filling of capillaries and containers with small orifices; in some vacuum investment and die casting processes and in refrigeration, cryogenic and gas filled cooling system evacuation; electric lamp and switch gear evacuation and inert gas welding systems.

In vacuum packaging of food, vacuum is applied to remove atmospheric air and its oxygen from the pack so that the contained product will have an increased shelf life, once it is sealed with an impervious plastic pack. An important aspect of vacuum packaging for some products is that the differential pressure between the exterior and the interior causes the flexible pack material to collapse and cling to the product, providing increased structual strength.

'Mechanical' Processes

Almost as numerous are the various ways in which a low vacuum can be used for mechanical handling and similar low-cost automation. Some examples are:-

High speed paper feeding for printing, processing, etc.

Vacuum forming of short thermoplastics

Liquid filling
Dry powder filling
Continuous liquid/solid separation
Mobile materials handling
Air sampling
Printing
Vacuum feeding sheet steel, glass, etc.
Machine tool chucks and vices
Miscellaneous applications include:

Automatic Addressing, Assembling, Book Binding, Shoe Making, Bread Wrapping, Bottle Inspection, Carton Opening and Feeding, Cigar Banding, Collating, Etching, Paper Folding, Automatic Grinding and Polishing, Labelling, Money Counting and Banding, Automatic Bread Handling, Papermill Sheeting, Photo-Copying, Ruling, Tag Handling, Tile Cutting, Perforating, X-Y Plotting, Punch Press Feeding, Check Sorting, Automatic Welding, etc, etc.

Selection of Pumps

Pressures from 760 down to 25 torr are generally referred to as *low vacuum,* and pressures from 25 down to 10^{-3} torr as *medium vacuum.* Alternatively, within this range, the term *rough vacuum* may be used for pressures from 760 down to 1 torr, and *fine vacuum* for pressures from 1 down to 10^{-3} torr. The remaining stages of vacuum are then generally described as:-

High vacuum – from 10^{-3} down to 10^{-6} torr

Very high vacuum – from 10^{-6} down to 10^{-10} torr

Ultra-high vacuum – less than 10^{-10}

Vacuum pressures in the low and medium range may continue to be quoted in other units such as inches of mercury, pounds per square inch, millimetres of mercury or millibars, etc.

The production of vacuum pressure implies exhaustion of a given vessel by suitable pumping action. Certain types of compressors can also work as exhausters, but in general the machines used for producing vacuum pressures are known as pumps (vacuum pumps).

The pump or pump combination is selected on the capacity needed for the range to be covered, with due consideration to pump down time and duration of cycle, the latter factors determining the most suitable pump size. The latter is usually a straightforward matter in the case of coarse vacuums, for the pump size is related solely to the size of the container. In the medium and medium-high vacuum range the shape as well as the volume of the container must be taken into account, the lower the final pressure, the greater the proportionate effect of the surface area factor. Selection of pump capacity in this range tends to become somewhat complicated and is normally best based on empirical data.

In the high and ultra-high vacuum range the pump size required is related almost entirely to the size and state of the surface of the container. Thus at pressures below 10^{-3} torr there are more adsorbed or absorbed molecules in the system than free gas molecules. The performance of the pump must thus ensure that at the required working pressure a rate of evacuation is maintained equivalent to the rate at which adsorbed and absorbed gas is emitted from the surfaces of the system. This can vary widely with different surfaces.

Paper feeding for data processing, printing presses etc.

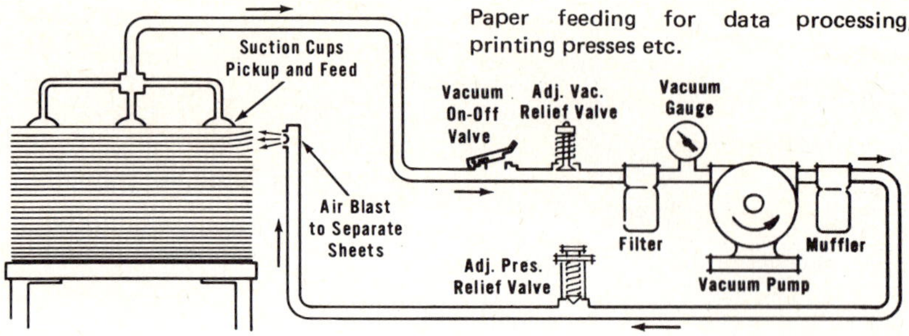

Dry powder vacuum filler

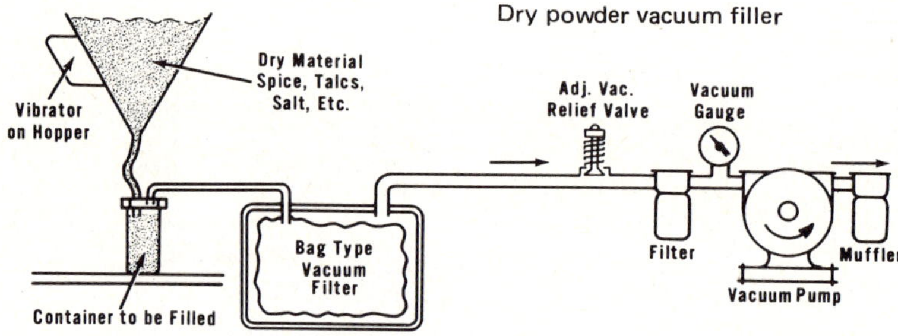

Core box duplicator equipment

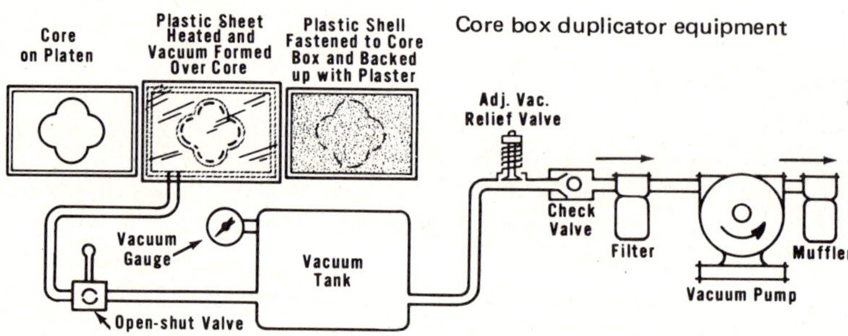

Vacuum forming plastics

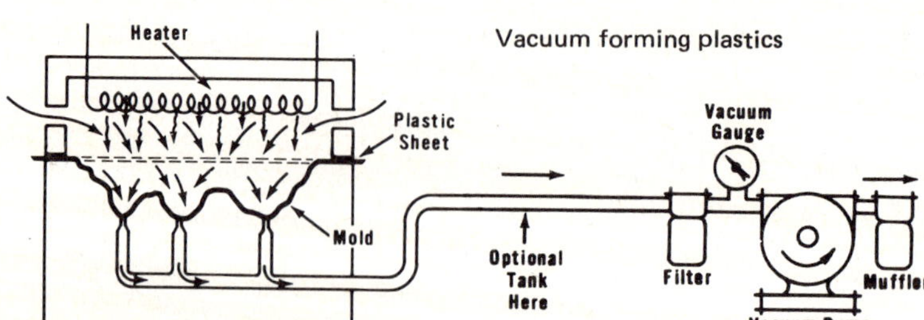

Examples of low vacuum application (Gast Manufacturing Co Ltd)

Silk screening printed circuits

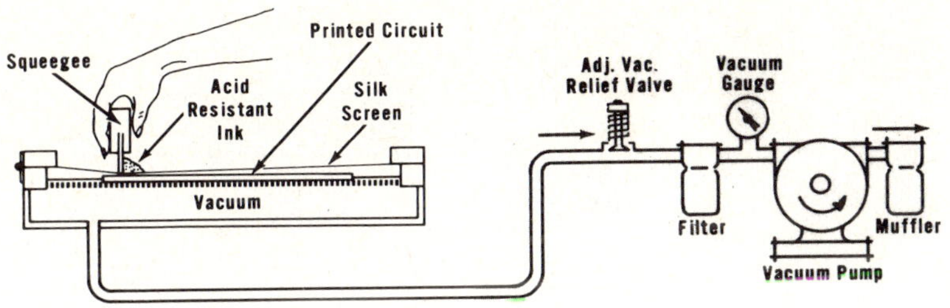

Book page turning machines

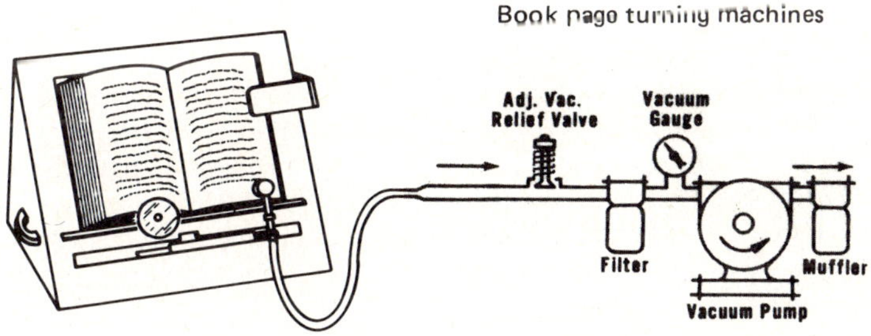

Oil purifying equipment

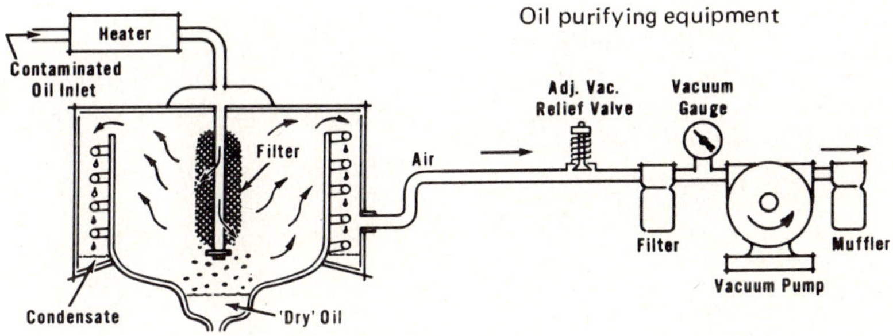

Automatic liquid sampling device

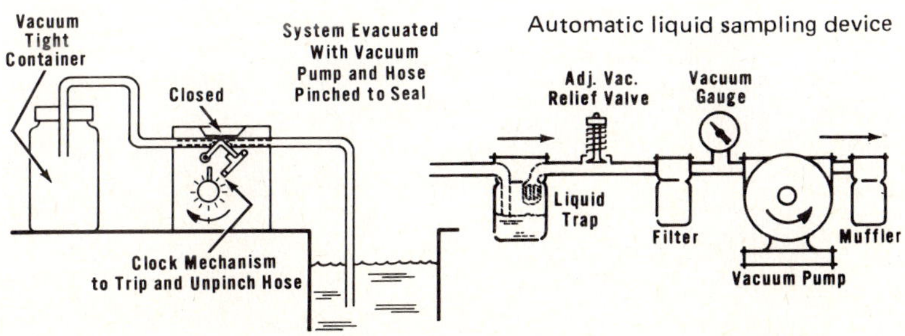

Examples of low vacuum application (Gast Manufacturing Co Ltd)

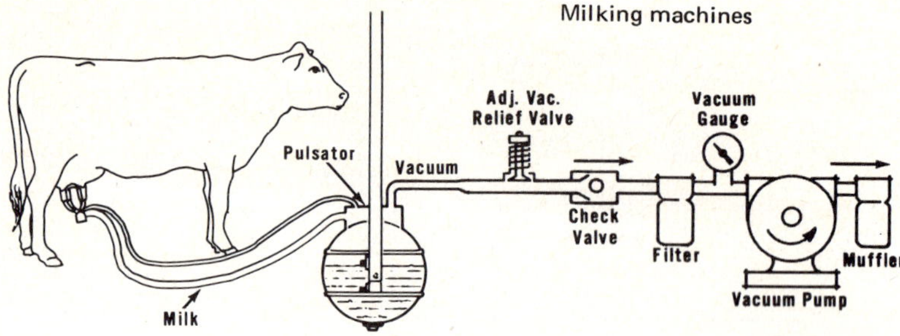

Example of low vacuum application (Gast Manufacturing Co Ltd)

This condition may be further complicated in practical systems by the fact that processes take place within the container being evacuated, whereby vapours are given off — eg in drying, degassing, distilling, etc. In such cases it is necessary to form an accurate estimate of the amount of vapour likely to be released or accumulated during the process so that the pump can be sized accordingly, bearing in mind also the nature of the vapour. Where large quantities of vapour have to be dealt with it is often necessary to install an efficient condenser system before the pump.

The working pressure to be produced or maintained in the system thus determines the type of pump required (or pump combination, where a fore-pump is necessary), and also indicates whether the pump size is dependent primarily on volume or total vacuum surface area. Where additional gases or vapours are present as a result of processing, their volume and properties also have to be taken into account. Finally the pumping down time has to be considered in arriving at a suitable pump size, balanced against the operating characteristics involved — eg whether the system exhausts as rapidly as possible down to the working pressure as part of definite work cycles, or whether the process is a continuous one with the pumps continuing to operate after pumping down to the working pressure.

See also chapter on *Vacuum Pumps.*

SECTION 7

Ancillaries

Pressure Gauges

PRESSURE under the SI system of units is expressed in Newtons per square metre (N/m^2). This yields low quantitive values for most practical pressure measurements and the unit now normally adopted is the bar, of 10^5 N/m^2. The quantitive value of 1 bar corresponds approximately to standard atmospheric pressure or 14.5 lbf/in^2. 1 bar also equals 10^6 dynes per square centimetre ($dyne/cm^2$) which was the basic pressure unit in the CGS system.

1 $dyne/cm^2$	=	1 microbar
10 $dyne/cm^2$	=	1 N/m^2
10^3 $dyne/cm^2$	=	1 milibar
10^6 $dyne/cm^2$	=	1 bar

The English pressure unit pounds per square inch (lb/in^2) may continue to be used in various literature. The original metric counterpart, kilogrammes per square centimetre (kg/cm^2) is no longer used in the pneumatic industry, being replaced by the bar. These two units do not yield identical synthetic values.

1 bar	=	1.02 kg/cm^2	(= 14.5 lb/in^2)
1 kg/cm^2	=	0.98 bar	(= 14.2 lb/in^2)

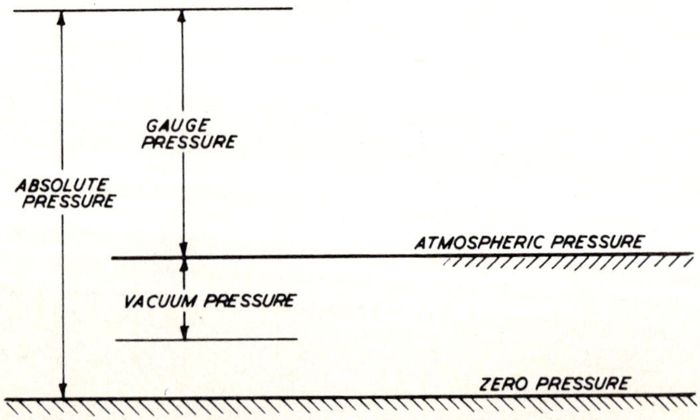

Fig 1

Pressure values may be related either to zero pressure or atmospheric pressure as the datum. Pressure above zero pressure is termed absolute pressure. Pressure related to atmospheric pressure as datum is termed gauge pressure if positive, or vacuum pressure if negative – see Fig 1. Pressure-measuring devices normally measure the difference between applied pressure and atmospheric pressure and thus indicate gauge pressure (or vacuum pressure). Since pneumatics is commonly concerned with relatively low numerical pressure values (eg of the order of 7 atmospheres) the difference between gauge pressure and absolute pressure can be quite significant.

Manometers

The classic form of manometer is a glass tube bent in the shape of a U and filled with water or mercury. One limb of the tube is open to atmosphere, the other connected to the source of pressure. Pressure value is derived directly as the difference in levels between the liquid in the limbs – Fig 2. Pressure is traditionally quoted as head in mm (or inches) of water or mercury.

where P_1 = the higher pressure

P_2 = atmospheric pressure (usually)

p = mass density of the fluid.

The simple U-tube manometer is thus a differential pressure measuring device which, if P_2 is atmospheric pressure, reads gauge pressure.

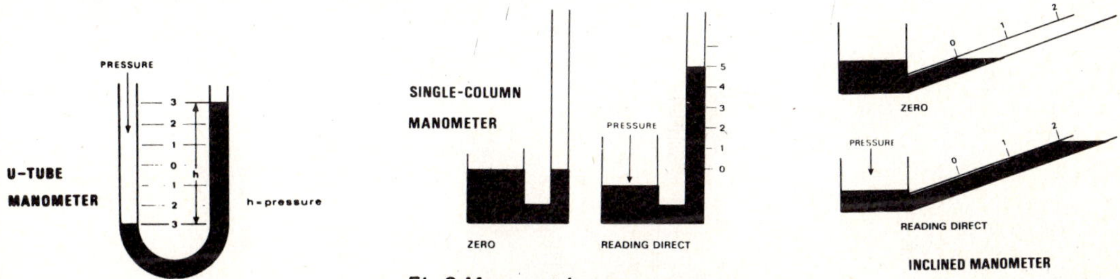

Fig 2 Manometric pressure gauges.

There are many variations on the U-tube manometer – eg single column, well, inclined tube, float, etc – see Fig 3. The inclined tube manometer has greater sensitivity being directly related to the angle of the inclined tube. The limit to this angle is that at which a definite meniscus is formed by the filling liquid. At the other extreme the float type manometer, using mercury as fluid, enables much higher pressures to be measured than is possible with U-tube manometers.

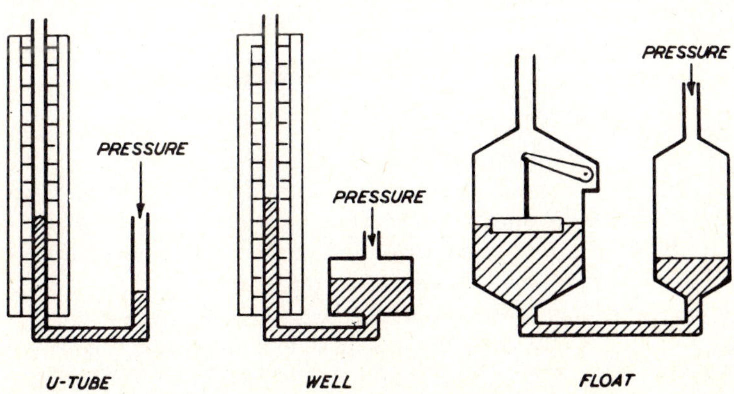

Fig 3

Inclined Manometers

Inclined manometers frequently have scales graduated to 0.25mm or 0.001 in. The calibration accuracy of an inclined manometer depends upon the angle of inclination, so the inclined manometer must be levelled very precisely before use to prevent calibration errors. Like its upright counterpart, an inclined manometer can be operated at moderate line pressure, it is equally unwieldy and is not ordinarily used in the field.

Bourdon Gauge

The Bourdon tube gauge is the most widely used type for general pressure movement. Typical construction is shown in Fig 4. The basic element is a thin walled tube, slightly flattened in section and bent into a 'C' shape. One end is fixed and the free end is sealed and attached to a mechanical linkage (Fig 7). Application of internal pressure to the tube causes it to straighten by an amount directly proportional to the internal pressure. The resulting movement of the free end is transmitted via the mechanical linkage to rotate a pointer associated with a graduated dial. This mechanical movement involves multiplication so that something like 300° of pointer rotation is achieved over the working pressure range of the gauge.

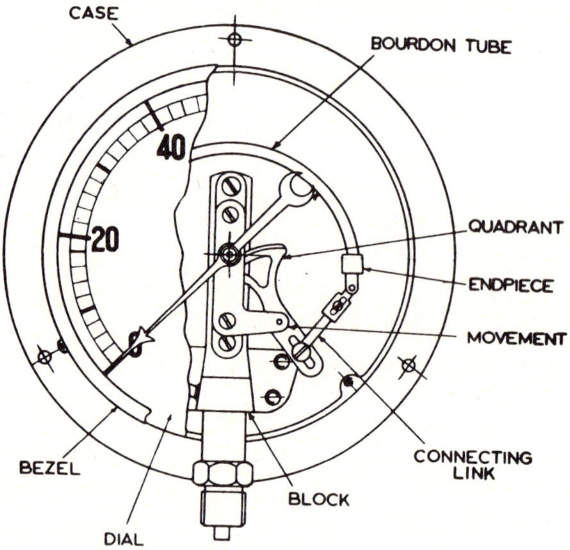

Fig 4 Typical arrangement of Bourdon gauge.

Bourdon tube gauges can be designed for pressure ranges of about 0.35 bar (5 lb/in^2) up to about 700 bar (10000 lb/in^2). With modified geometry and materials this range can be extended up to 7000 bar (100000 lb/in^2). For working pressures up to 70 bar (1000 lb/in^2) the Bourdon tube is usually made of steel or phosphor bronze. For pressures in the range of 70–250 bar 1000–3000 lb/in^2), beryllium copper is the preferred material. To accommodate corrosive fluids (or for even higher pressure working), stainless steel tubes can be used.

Free movement of the end (for any given material) is dependent on the length – the greater the movement for a given pressure. For low pressure working an extra long tube may be used, arranged in the form of a flat spiral.

Accuracy achieved with a Bourdon tube gauge can be of the order of plus or minus 2½% of the full scale deflection. Greatest accuracy is normally achieved at the upper end of the scale. Thus selection of a Bourdon tube gauge should be based on a pressure range where the normal reading point will be toward the upper end of the scale. Consistency of performance, however, is related to the fatigue characteristics of the tube material and some drift is inevitable with time. As a general rule, therefore, Bourdon gauges need regular checking and recalibration as necessary when in use.

Bourdon gauges are susceptible to mechanical shock and pressure surges and should be protected against both forms of disturbance. A simple method of protection against pressure surges is to fit a snubber or restrictor in the feed line. In severe cases, mechanical throttling, or a shut-off device, responding directly to any sudden increase in pressure, may be employed.

Another possible source of trouble is blocking of the tube by contaminants in the fluid. To obviate this a diaphragm type seal can be incorporated in the inlet or feed. The Bourdon tube is then filled with clean fluid (usually oil) which is isolated from the working fluid, the pressure of the latter being transferred to the Bourdon tube through the flexible seal. This principle can be extended by employing a diaphragm unit or capsule as a transmitter, separate from the Bourdon gauge and connected to it by a length of tubing. A distinct advantage offered here is that the gauge itself can be distant from the transmitter without affecting the accuracy of the reading.

Modifications to the original form of element includes the use of helical tubes, some examples being shown in Fig 5.

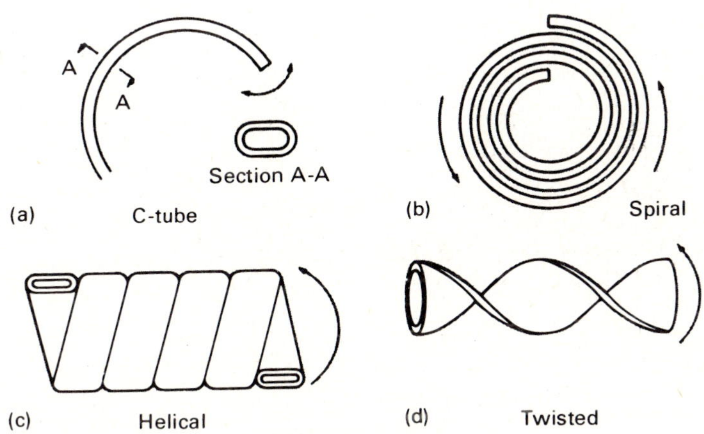

Fig 5 Bourdon tubes (Schaevitz)

The spiral tube is similar to the C-tube, except that its radius of curvature changes along the length of the tube, creating a greater length of tube for a given circumference. The number of turns in the spiral vary. Because the active length of tube is larger, the arc length of stroke is greater. A straight-line pickoff from this larger angle of sweep causes a larger non-linearity in output displacement than with a C-tube.

The helical tube permits a long tube length for a given radius of curvature. The average tube has from two to six turns. The movement of the tip is greater than that for the C-tube and is more nearly circular. The increase in length of the helical tube while pressurized is negligible. The travel of the tip lies in a plane perpendicular to the centreline of the helix, increasing as the active tube length increases. The tip travel is in the same order of magnitude (per unit length of tube) as an equivalent C-tube.

The twisted tube maintains a straight centreline along the length of the tube but is twisted about the centreline at a uniform rate. The expansion of the tube by internal pressure causes the tube to unwind angularly about the tube centreline. The angle of rotation is proportional to applied pressure. The change in tube length when the tube unwinds is negligible. Different angles of twist alter the angular rotation for any given pressure and length of tube.

Low Range Pressure Elements

Axially convoluted bellows, radially convoluted diaphragms, flat (unconvoluted) diaphragms, and welded pressure capsules are alternatives to Bourdon tubes as sensing elements for low range pressure measurement. All of these elements react to applied pressure by deforming elastically in an axial direction.

The basic difference between bellows, diaphragms, and capsules is the relative deflection of each for a given pressure and element size. Under similar conditions, a convoluted diaphragm can deflect much more than a flat diaphragm, and a bellows can deflect more than either a capsule or a convoluted diaphragm.

Bellows, capsules and convoluted diaphragm assemblies are ordinarily used in low pressure transducers, because these elements can provide sufficient deflection to operate an LVDT even at pressures as low as a fraction of a pound per square inch. Conversly, flat diaphragms require substantial pressures to yield significant deflections and so are generally found in high pressure transducers. The greater stiffness of a flat diaphragm results in significantly higher dynamic response characteristics in transducers using them as sensing elements.

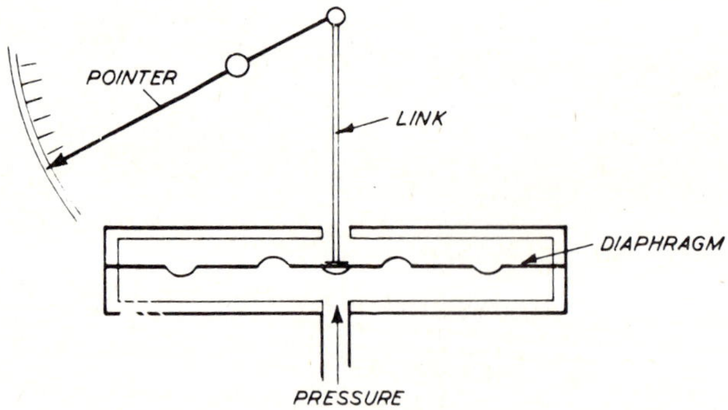

Fig 6

Diaphragm Gauges

Simple diaphragm pressure gauges are suitable only for the measurement of relatively low pressures, eg typically up to 1 bar (14.5 lb/in^2). They are, however, particularly sensitive devices and well suited to the measurement or detection of small differential pressures fed to either side of the diaphragm. A simple diaphragm gauge of this type is shown in Fig 6. The travel of the diaphragm, and thus the stroke available to operate an indicating movement, is dependent on the cross sectional form of the diaphragm and its diameter. Extreme movement may only be obtained at the expense of loss of linearity of movement.

The stroke of a single diaphragm gauge can be extended considerably by employing a slack diaphragm, made from a suitable limp material such as nylon, oiled silk or thin leather. Such a diaphragm requires biasing with spring pressure opposing the direction of applied pressure (or of the highest applied pressure) and stops to limit the movement. The particular advantage offered by a slack diaphragm is the large diaphragm area which can be exposed to pressure, making the unit particularly sensitive for the measurement of very low pressures.

Metallic diaphragms offer the possibility of working at higher pressure ratings, although the movement of stroke is largely dependent on the extent of corrugation, for a given diameter. Performance can generally be improved, both as regards stroke and pressure sensitivity, by fabricating a series of diaphragms in the form of a bellows, although here the contour adopted can have a bearing on linearity of stroke and force output with pressure, the two not necessarily being the same. Thus, some contours can give good linearity of force output with poor linearity of stroke, and vice versa.

Welded metallic diaphragm gauges, based on a built-up bellows shape or series of capsules, are mostly used for pressure measurements within the range 25 mm wg to 0.35 bar (1 inch wg to 5 lb/in^2). In the case of capsule types, the capsules form hermetically sealed chambers which may be partially evacuated to render the element more sensitive – eg the type of element used in an aneroid barometer.

Bellows type elements formed from thin-walled tubing have a higher pressure rating, and such elements may be used in several ways. For pressure measurement within the range 0.35–2.50 bar (4–50 lb/in^2), or vacuum pressures up to 762 mm Hg (30 in Hg) an open ended bellows is commonly used, enclosed within a chamber. Pressure is then applied to the chamber, with mechanical movement derived from the free end of the bellows. This basic layout may be further modified by the application of spring loading, opposing the movement produced by the pressure force, ie expanding the bellows where the applied pressure to be measured is positive, and contracting the bellows in the case of measurement of vacuums.

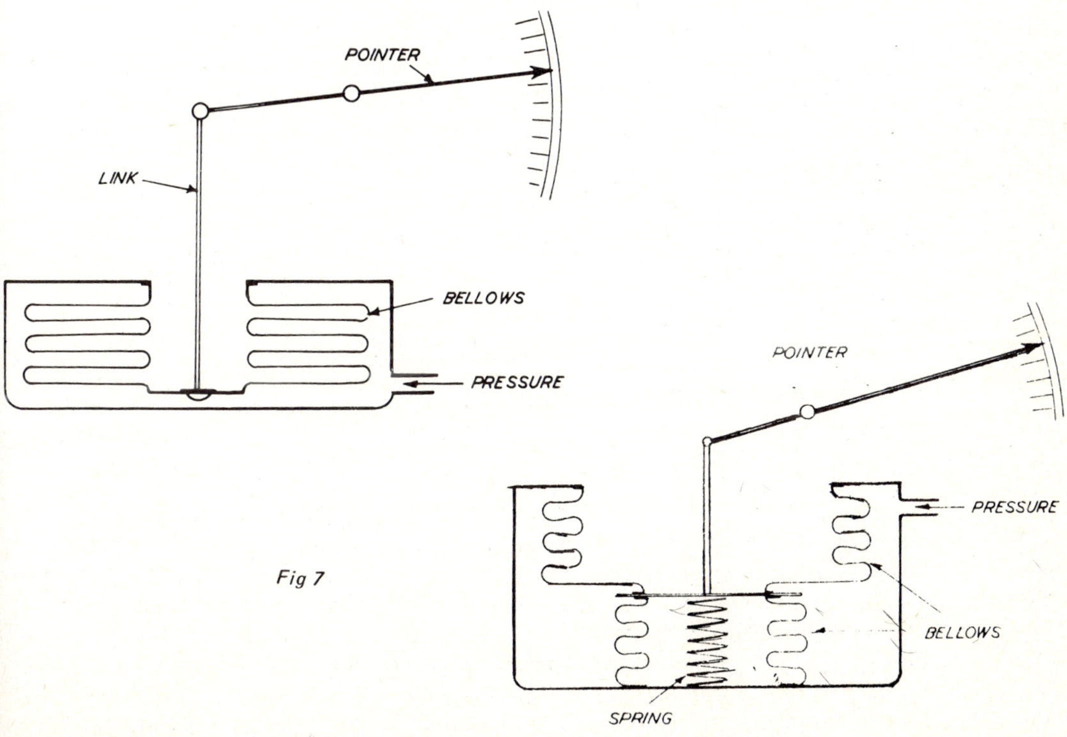

Fig 7

Sensitivity may be improved by the use of additional bellows to apply spring bias. alternatively, matched bellows may be used, each in a separate sealed chamber. The interiors of the two bellows are then interconnected and filled with a liquid with a high bulk modulus, such as oil or water. Pressure applied to one chamber, with the other chamber vented, will produce a change in bellows length in the first chamber, and a contrary movement in the other bellows due to liquid displacement, thus doubling the mechanical movement available from the applied pressure. Alternatively, if one of a pair of matched bellows is evacuated and sealed, and the other is connected to the pressure source, the movement will read absolute pressure directly.

Examples of bellows-type diaphragm gauges are shown in Fig 7.

Electronic Pressure Gauges

Electronic pressure gauges employ a pressure transducer normally connected into a resistance bridge network, followed by an amplifier. Pressure may be indicated by a meter or, in modern designs, through a complementary circuit with digital read-out via light emitting diodes. The pressure range depends on the type of transducer used, but the general range covered by instruments of this type is 0–700 bar (0–10000 lb/in^2). Accuracy again depends mainly on the transducer and normally compares favourably with, or is better than, a comparable Bourdon gauge. Accuracy is normally maintained to within ±1% of the maximum pressure rating of the transducer.

Checking Pressure Gauges

The accuracy achieved by various types of pressure gauges involving mechanical movements can vary considerably. Accuracy of any one gauge is also likely to vary with the length of time it has been in use, hence, for critical or important services, pressure gauges should be checked, at regular intervals, and recalibrated if necessary.

Compressed air is the most convenient medium for testing pressure gauges removed from the system, their accuracy being established by comparison with the reading of a suitable standard gauge of known accuracy. Alternatively, a deadweight test apparatus may be used directly for calibration or recalibration.

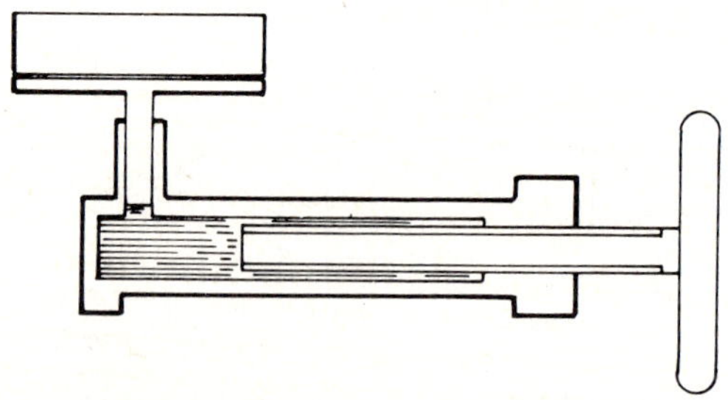

Fig 8

This takes the form of a horizontal screwed ram and vertical piston assembly, the latter terminating in a platform which can be loaded with a set of known weights – Fig 8. The interior volume is filled with oil. A weight corresponding to the required pressure is placed in position and the ram screwed in until sufficient pressure is produced on the oil fluid to 'float' the piston, thus establishing the oil pressure at the predetermined level in the connection to the pressure gauge under test. Test pressures within the range 1 to 15000 bar can readily be covered with this type of apparatus.

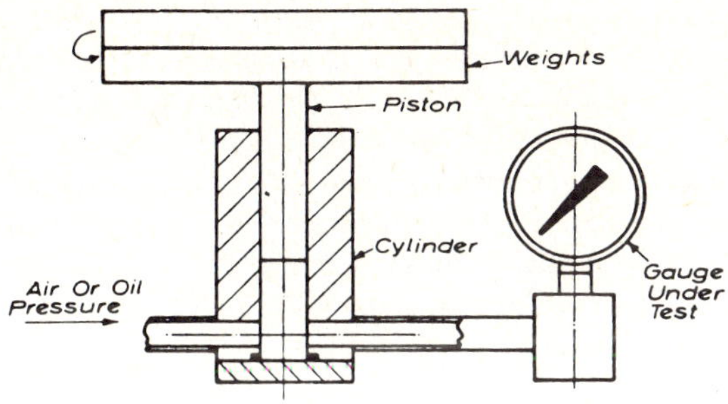

Fig 9 Deadweight pressure gauge tester for calibrating and checking dial gauges for accurate testing.

Where the pressure gauge is of a type which cannot be actuated directly by oil pressure, compressed air may be used as the working fluid, fed to the piston assembly through a pressure regulator, as shown in Fig 9. The same principle of working applies — ie the supply air pressure is adjusted sufficiently to just lift the piston with the predetermined weight added to the unit.

Vacuum Gauges

The vacuum pressure range covers pressures from atmospheric down to absolute zero. The most convenient (and generally accepted) pressure unit is the torr, corresponding to an absolute pressure of 1 millimetre of mercury. Normal atmospheric pressure is 760 torr. Various other units may be quoted, however, eg —

Microns or Micromillimetres — where 1 micron equals 10^{-3} torr

Millibars — where 1 torr equals 1.33 millibars, or 1 millibar equals 0.75 torr

Microbars — equivalent to dyne/cm^2 or millibars x 10^{-3}

Per cent vacuum — which may be defined as —

$$\frac{760 - \text{pressure in torrs}}{7.6}$$

Empirical units — such as inches of mercury (in Hg) lb/in^2, kg/cm^2, etc.

Conventional pressure gauges of the capsule or bellows type may be suitable for pressure measurements down to 25 torr. This pressure range is generally referred to as low vacuum. For higher vacua, ie lower absolute pressures, special vacuum gauges are employed.

The Mcleod gauge is a special form of mercury manometer with three limbs, one of which is sealed at the top. Pressure is determined by measurement of the degree of compression of the gas trapped in the sealed limb. The normal pressure range of such a gauge is from 10 torr down to 10^{-4} torr. It is more suitable for laboratory work rather than as an industrial instrument, but more robust and easier to use gauges working on a similar principle have been developed for general use.

The Pirani gauge is based on the behaviour of a heater wire element enclosed within a suitable gas chamber and connected to a Wheatstone bridge circuit. The principle involved is that the thermal conductivity of the gas, and thus the temperature of the wire supplied with a constant heater voltage, is pressure dependent. This change is indicated as a change in resistance of the wire element, measured on the bridge circuit. Correction for ambient temperature can, if necessary, be arranged by installing a second gauge in the opposite arm of the bridge. Normal range of a Pirani gauge is from 5 torr down to about 10^{-4} torr.

Thermocouple gauges have also proved suitable for pressure measurement in the range 10^{-1} to 10^{-4} torr. These work on a similar principle to the Pirani gauge, except that the temperature of the element is measured directly (by thermocouple reaction) rather than indirectly as a change in resistance.

For measurement in the higher ranges ionic or radionic gauges are normally employed. These are specialized devices giving meticulous calibration and control (eg the use of stabilized circuits). Degassing can also be a problem, upsetting the calibration and sensitivity of such gauges at high vacua. Measurement range is, however, readily extended down to 10^{-6} or 10^{17} torr.

Seals and Packings

SEALS for pneumatic services fall into the following general categories:-

Static seals — eg cover gaskets for valves or reservoirs, end cover seals for cylinders, gasket seals for line accessories.

Piston seals — eg for cylinders, spool valves, piston motors, reciprocating pumps and compressors.

Rod seals — for cylinders, valve stems.

Rotary seals — for rotating shafts or rotating assemblies (eg distribution chests on rotating cylinders).

Whilst such duties can be met by conventional fluid seals the modern trend is to develop specialized designs of pneumatic seal, either based on conventional sections or utilizing original sections. This is more than justified by the fact that pneumatic seals often have to work under more arduous conditions than in liquid systems. In particular, pneumatic seals often have to work in 'dry' rather than 'wet' conditions, with minimum lubrication, and often at higher temperatures. This can place a premium on material selection for satisfactory performance from conventional fluid seal forms, particularly as elastomers are less resistant to dry than wet heat. Further adverse operating factors are the relatively low pressure difference which may be involved to 'energize' a flexible seal, the deteriorating effect of oxygen on seal materials in general and the high speed of operation normally associated with pneumatic cylinders, etc.

In the case of a seal operating dry, or substantially dry, the apparent permanent set of an elastomeric seal may appear much higher (typically twice as high) than that of the same seal used as an oil seal. This may call for greater initial interference when assembling the seal, and subsequently higher initial friction during the first period of running when the permanent set is taken up. Selection of a suitable interference is more difficult, and errors may cause the seal either to continue to bind and develop high friction and wear or to run in a loose fit with consequent leakage. A permanently binding seal is more likely than a 'loose' seal, particularly if the seal is proportioned as a 'dry' type and subsequently lubricated by oil-mist or oil-fog. Only a proportion of the lubricant will operate under oil-mist or oil-fog conditions, unless provision is made for the amount of swell the seal is likely to undergo during its 'conditioning' period. This period may, in some cases, extend to several weeks of operation.

A period of immersion in fluid may also leach out a proportion of the plasticizer in the seal. If the seal is subsequently allowed to dry out it will shrink. For a dry seal, therefore, the material may have to be chosen specifically for a small amount of permanent set so that interference is not reduced too drastically in service. Seal sets and packings which can be adjusted can thus offer very real advantages in certain applications, although adjustment of packing pressure is not necessarily a satisfactory answer where consistency of performance is required. It may provide a means of adjusting sealing, but the friction will be modified both by the varying seal pressure and a variable coefficient of friction.

Specialized pneumatic seals are usually designed for simple assembly. In general the aim is to provide increased flexibility in the section design, with particular attention to conformation at the seal lip to provide maximum retention of the small amount of lubricant generally found in pneumatic applications. Increased flexibility of the lip can also reduce friction. At the same time the section must be stable enough to prevent spiralling or 'roll over' when sealing reciprocating motions.

Further improvements can be realized by development of specialized elastomers, 'tailored' for optimum hardness, flexibility and resistance to compression set, and for maximum compatibility with the temperature, pressure and lubricants normally involved in pneumatic systems.

O-Rings

O-rings are simple, low-cast seals with a particular application to smaller cylinders and light duty working. They are also equally suited to both static and dynamic seals. Properly assembled and adequately lubricated, an O-ring should not normally tend to roll or twist under reciprocating motions since the contact area with the groove is far greater than the contact area with the sliding surface, the torsional stiffness of the ring itself resisting twisting. The fact that the groove friction is static friction, and thus higher than the rubbing friction in any case, further tends to hold the ring in its groove. Anything which modifies these conditions, however, can allow some parts of the ring to slide and other parts to roll, simultaneously, resulting in an excessive amount of twist being developed in the ring. This can ultimately lead to ring failure, the break appearing as a tight spiral fracture around the ring, known as spiral failure.

Metric O-ring sizes are covered in BS4518 : 1969, based on the Swedish Standard SMS.1588 which has been generally used by European countries for many years. Inch sizes of O-rings are covered by BS1806 : 1962. Groove dimensional data are given in Tables IA and IB.

Duties for which O-rings are not generally suitable are operations involving:-

(i) high acceleration and high rubbing speeds
(ii) poor lubrication
(iii) long strokes
(iv) high unit loads
(v) side and/or bending loads

Simple Rings

Various other simple sections have been developed as alternatives to the O-ring, many of which match standard O-ring grooves. These include rectangular, Quad-ring, H-ring, Delta-ring, Nu-lip, D-ring, etc. Of these rectangular rings many have specific advantages for pneumatic applications, being less subject to spiral failure because of the stiffer section torsionally. Such rings may also be pressure-cut for close control of dimensions allowing a smaller radial interference to be applied with consequent reduction in friction. In general, however, the use of O-rings far exceeds any alternative simple ring sections for application as simple double-acting reciprocating seals. Where the duty involved is such that O-rings prove unsatisfactory (or are considered unsatisfactory), a heavier ring section would normally be selected.

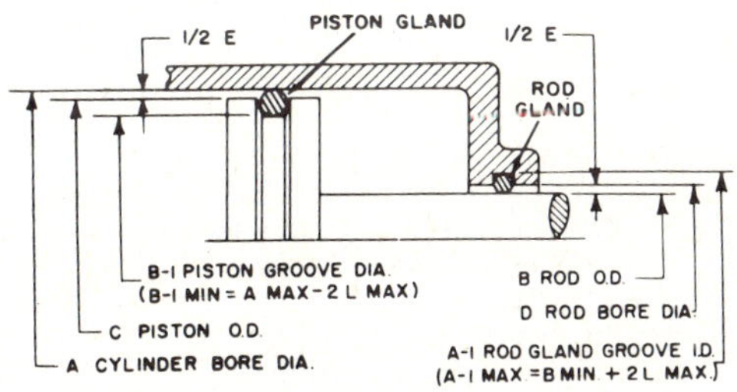

TABLE IA – O-RING GROOVE DESIGN DATA FOR RECIPROCATING APPLICATIONS

W Cross Section		L	Squeeze		E †	G Groove Width			R	
Nominal	Actual	Gland Depth	Actual	%	Diametrical Clearance	No Back-Up Rings	One Back-Up Ring	Two Back-Up Rings	Groove Radius	Eccentricity Maximum T.I.R. *
1/16	0.070 ±.003	0.055 to 0.057	0.010 to 0.018	15 to 25	0.002 to 0.005	0.093 to 0.098	0.138 to 0.143	0.205 to 0.210	0.005 to 0.015	0.002
3/32	0.103 ±.003	0.088 to 0.090	0.010 to 0.018	10 to 17	0.002 to 0.005	0.140 to 0.145	0.171 to 0.176	0.238 to 0.243	0.005 to 0.015	0.002
1/8	0.139 ±.004	0.121 to 0.123	0.012 to 0.022	9 to 16	0.003 to 0.006	0.187 to 0.192	0.208 to 0.213	0.275 to 0.280	0.010 to 0.025	0.003
3/16	0.210 ±.005	0.185 to 0.188	0.017 to 0.030	8 to 14	0.003 to 0.006	0.281 to 0.286	0.311 to 0.316	0.410 to 0.415	0.020 to 0.035	0.004
1/4	0.275 ±.006	0.237 to 0.240	0.029 to 0.044	11 to 16	0.004 to 0.007	0.375 to 0.380	0.408 to 0.413	0.538 to 0.543	0.020 to 0.035	0.005

† It is desirable to hold this to a minimum consistent with maintaining positive clearance over full working temperature range.

* Total Indicator Reading between groove and adjacent bearing surface.

Note : for 1 500 lb/in^2 maximum without back-up rings

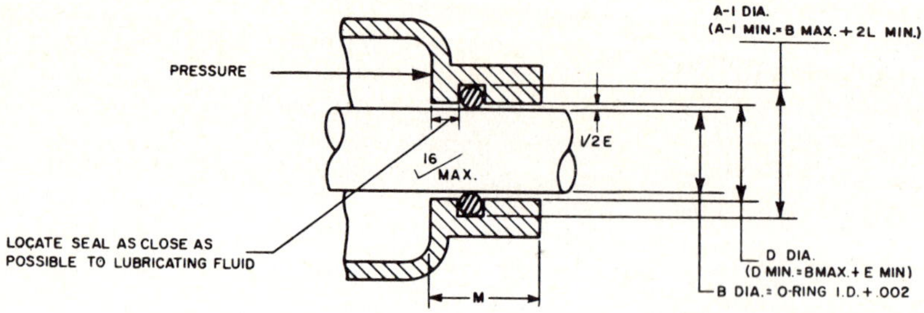

TABLE IB – O-RING GROOVE DESIGN DATA FOR ROTARY APPLICATIONS

W Cross Section			L	G	E†		M	R
Nominal	Actual	Maximum Rubbing Speed ft/min	Gland Depth	Groove Width	Diametrical Clearance	Eccentricity Maximum T.I.R.*	Bearing Length Minimum	Groove Radius
1/16	0.070 ±.003	1 500	0.065 to 0.067	0.075 to 0.079	0.012 to 0.016	0.002	0.700	0.005 to 0.015
3/32	0.103 ±.003	600	0.097 to 0.099	0.108 to 0.112	0.012 to 0.016	0.002	1.030	0.005 to 0.015
1/8	0.139 ±0.004	400	0.133 to 0.135	0.144 to 0.148	0.016 to 0.020	0.003	1.390	0.010 to 0.025

† For maximum working pressure of 500–750 lb/in^2. For higher pressures clearances would have to be reduced.

* Total Indicator Reading between the groove o.d., shaft and adjacent bearing surface.

Note : For standard sizes of O-rings and grooves, consult BS1806

O-ring seal performance can also be influenced by the use of an O-ring as the elastomeric element in a combination seal, the actual seal being a ring of harder material with good anti-wear and anti-friction properties. Such seals are made either as external or internal seals – Fig 1. O-ring and seal ring may be separately assembled in a matching groove, or the two elements bonded together to form a single (composite) seal ring.

Automatic-type piston rings may be employed as piston seals for use at elevated temperatures. These may be of metal or 'hard' plastic, depending on the service temperature involved. Hollow metallic O-rings may also be used for high temperature seals.

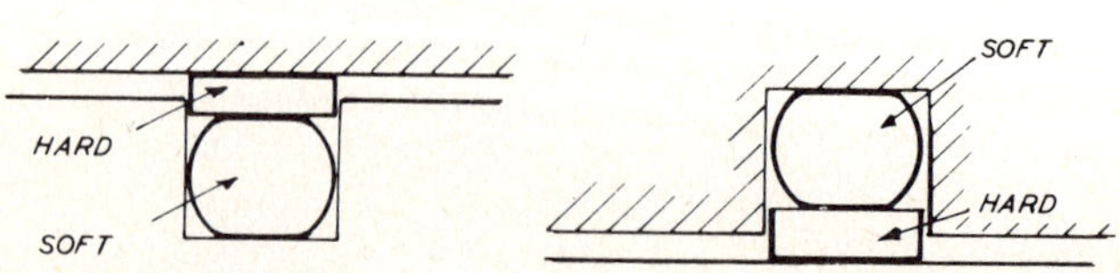

Fig 1 External and internal combination seals.

U-Rings

A U-ring is essentially a lip seal and is the main general-purpose type. At the lower pressures normally encountered in pneumatic applications, U-ring sections can be thin and flexible, using relatively soft rubbers. Vitrile rubber is the usual choice.

An inherent limitation of the basic U-ring section, particularly in the case of unreinforced elastomeric rings, is the tendency for the heel to develop high wear and damage which, in turn, can affect the proper function of the flexible lip by modifying the stress distribution through the section. The heel section is, therefore, commonly modified to eliminate or reduce localized stressing and wear.

A modified form of U-ring developed specifically as a pneumatic seal is the Gaco H-7, shown in Fig 2. This has long, thin lips (for good flexibility), rounded lips for maximum retention of lubricant, and generous interference for good sealing.

A considerably more elaborate section, which in effect is a rocking U-ring, is the Trist 'Pneuflex' seal, shown in Fig 3. This incorporates a Y-shaped section which can 'rock' about the base of the main U-section when assembled.

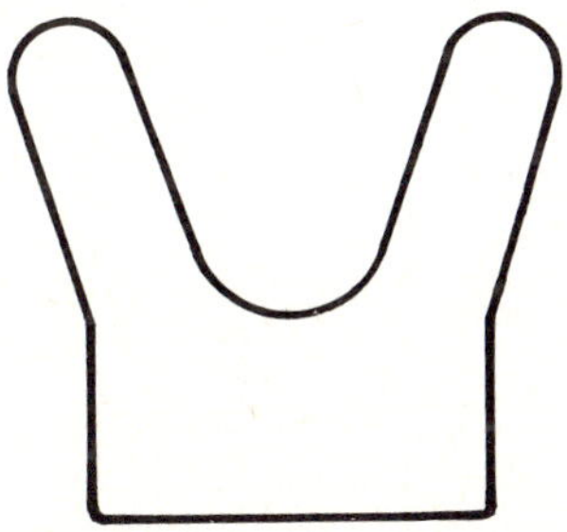

Fig 2

(c) *(d)*

(c) Pneumatic packing in H7 material shortly after release.

(d) Pneumatic packing in conventional material shortly after release (strictly comparative time).

Example of superior recovery of 'specialized' pneumatic-duty elastomer.

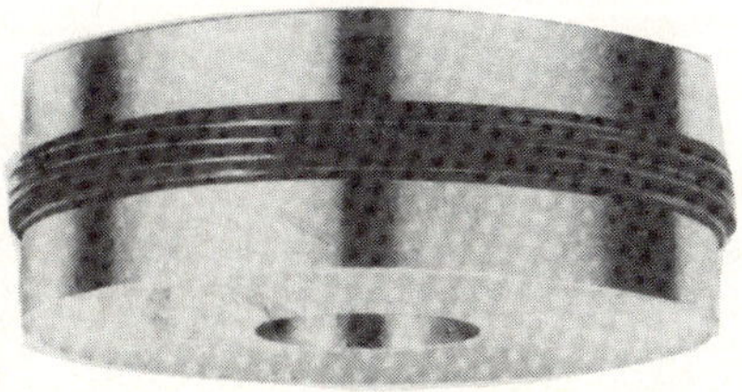

Trist 'Pneuflex' pneumatic seal.

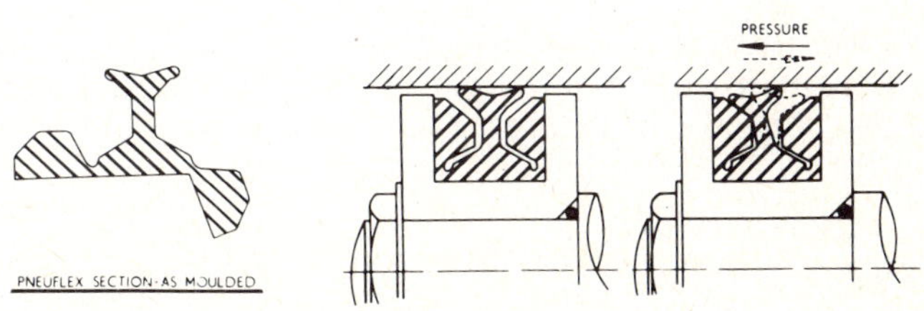

Fig 3 'Pneuflex' pneumatic cylinder seal (Trist).

C-Rings

C-rings employ similar sections to U-rings, except that one leg is shortened and the other extended into a more flexible form of lip seal. This leg only provides the sealing action, hence C-rings are made in either external form (for rod seals) or internal form (for friction seals).

A number of proprietary rings have been developed from this section form, including more rigid sections for heavy duty working where the lip is formed by a V-shaped notch in the section and on the shorter limb. These are sometimes called automatic sections or distributor seals. They can be used singly or in sets. In the latter case their rigidity may be sufficient for them to be fitted without headers. True automatic sections (distributor seals) work on the principle of a compression seal, the sealing face being the solid limb opposite to the lip, although the section is also pressure energized – eg see Fig 4.

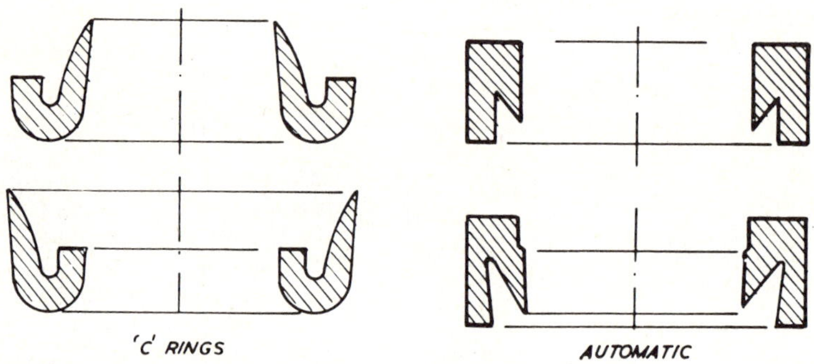

Fig 4

Cups

The cup is essentially a piston seal with the characteristic of very low friction in the reversed direction of motion. Two cups, mounted back-to-back, are required to seal a double-acting piston.

A cup section is relatively limp and requires back-up support and front support (eg with a junk plate). The cup shoulder, in particular, needs close support to prevent nibbling, and also to prevent the bottom of the cup being tightened up too much by excessive tightening of the junk plate. Too much compression on the bottom of the cup will cause the cup to bind, generating excessive friction and wear.

Hats and Collars

Hat seals (also known as flange seals) are virtually an internal form of cup, used as rod seals. The hat section is more compact than a U-ring but less efficient on a seal. The collar seal is a variation of the hat section with tapered lips and more rounded heel. It has improved resistance to scuffing and wear.

Piston Seals

Piston seals are essentially reciprocating seals. The main types of seals used in this category are summarized in Table II. For low pressure services choice is, in fact, fairly open.

TABLE II – PISTON SEALS

Type of Seal	Fitting	Remarks
O-ring	Simple groove	Suitable for small cylinder sizes and light duties
O-ring composite	Simple groove	O-ring is elastic element backing hard seal ring; integral or separate constructions. Low friction, good pressure sealing
Cup	Needs special backing and assembly	Single-acting; can be assembled back-to-back for double-acting. A basic form of piston seal
U-ring	In groove, with or without support ring	Single-acting; can be assembled back-to-back for double-acting. Widely used form of piston seal
Modified U-ring	In groove, without support rings	Numerous proprietary fluid seals have been developed around this basic form, including special types for pneumatic services
C-ring	In matching groove, or in sets	More flexible than U-ring with only one sealing lip
Modified C-ring	In matching groove, or in sets	Basis of proprietary 'automatic' seals
V-ring	In sets with headers	Limited application for pneumatic services. U-rings are generally superior for lower pressures
Solid rings	In matching groove(s)	Automatic type piston rings in metal and non-metals. Metallic piston rings are logical low-cost choice for high temperature service. Limited sealing at low pressures.
Compression packings	In glands	Largely replaced by modern seals or seal sets

Rod Seals

Rod seals are summarized in Table III. The range is similar, although it does not follow that a cylinder manufacturer will use the same type of seals on the rod as for the piston. Equally, an alternative choice of seals may be available with a given cylinder design.

It will be noticed that traditional type compression packings (and other types of compression seal) are not excluded in modern practice. There are applications where relatively high friction on the rod can be an advantage, eg to add stiffness to the movement in the case of a vertical air cylinder used to raise a weight.

An additional seal on the outside of the rod seal is also commonly used. This is a wiper seal, the duty of which is to scrape over the entering rod surface and remove dirt, etc, which may have collected on it, thus preventing it being drawn into the main rod seal where it could be retained to cause wear. Wiper-seals can take the form of simple single lip seals, metallic scraper rings bonded to an elastomeric section, thin elastomeric sections backed by metal to provide support and rigidity, or even simple compression seals (eg felt rings). Polyurethane rubber wiper rings are finding increased favour because of their good wear and abrasion resistance properties. A gaiter may also be added for positive protection where the cylinder is used in particularly dirty conditions.

TABLE III – ROD SEALS

Seal Type	Assembly	Remarks
O-rings	In matching groove	Small, light duty cylinders
Quad rings	In matching groove	Small, light duty cylinders
Hat	Suitably supported	Susceptible to heel scuffing
Collar	Suitably supported	Modified hat section with rounded heel and better resistance to scuffing
U-ring	In groove with or without support ring	Widely used
Modified U-ring	In groove with or without support ring	Basis of numerous proprietary seals
C-Ring	In matching groove or in gland in sets	May offer advantages over U-rings
Modified C-ring	In matching groove or in gland in sets	Basis of many proprietary seals
V-ring (chevron)	In gland in sets with headers	For heavier duties and larger sizes
Scraper rings	On outside of main seal or gland	To protect main seal against dirt, etc.
Compression packings	In glands	May be used in some heavy duty cylinders, for high temperature duties or special applications

Materials

Nitrile rubber is the usual choice for elastomeric seals, except when extremes of temperature may be encountered. Both the high temperature and low temperature performance of nitrile rubber can, however, be improved by compounding, although not usually both together. Thus in order to obtain good low temperature performance with a nitrile rubber it is usually necessary to sacrifice some higher temperature characteristics, such as oil resistance.

Hardness values for rubber seal materials may range from soft (40–45°) to very hard (90–95°). Choice of suitable hardness is largely a compromise. A low hardness improves conformability and reduces break-out friction with dynamic seals, and is particularly desirable with low-pressure, pressure-energized seals (ie pneumatic seals). A high hardness, however, shows lower rubbing friction and reduces susceptibility to wear, abrasion and extrusion. Rubbing friction, however, will also be dependent on the degree of interference or 'squeeze' present, which for adequate sealing may need to be appreciably greater in the case of a hard rubber than a soft rubber.

TABLE IV – WORKING TEMPERATURE RANGE OF TYPICAL ELASTOMERS

Elastomer	MINIMUM °F	MINIMUM °C	MAXIMUM °F	MAXIMUM °C
Butyl rubber	–60	–50	200	95
Ethylene propylene	–65	–54	300	150
Nitrile – *65°	–65	–56	225	110
Nitrile – *70°	–40	–40	200–250	95–120
Nitrile – *80°	–40	–40	225–300	110–150
Nitrile – *90°	–30	–35	300	150
Neoprene	–65	–55	300	150
Polyurethane	–65	–55	300	150
Silicone *60°	–70	–60	450	230
Silicone *70°	–70	–60	450	230
Fluorosilicone	–65	–55	450	230
Viton	–65	–55	450	230

*Shore hardness

Hardness can be measured in terms of BS degrees, Industrial Rubber Hardness Degrees (I.R.H.D.) or Shore Durometer degrees. All three empirical scales yield essentially similar values. BS and I.R.H.D. hardness can be measured to plus or minus 1 degree, but is usually quoted to the nearest 5 degrees. Shore Durometer hardness can be measured to plus or minus 5 degrees and is normally quoted to the nearest 10 degrees.

The working temperature range of typical elastomers used for seals is given in Table IV.

Diaphragms and Bellows

A DIAPHRAGM is a single-layered flexible or elastomeric member constrained around its periphery It is capable of flexural movement substantially at right angles to the plane of restraint, such movement being initiated by differential pressure or mechanically. Diaphragms are specifically described by their material of construction (eg metallic or elastomeric) and geometry — see Table I.

TABLE I — BASIC TYPES OF DIAPHRAGMS

Type	Movement by	Travel	Remarks
Flat disc	Stretching of material	Strictly limited with metallic diaphragm. Limited with elastomeric diaphragm	Easiest to make
Dish or saucer-shape	Flexing of material	Approximately twice the height of dishing	Preformed or moulded. Performance difficult to predict
Convoluted	Flexing of material	Slightly more than twice the height of convolutions	Generally superior performance but more difficult to make
Slack	Taking-up of slack	Limited by amount of slack which can be accommodated without wrinkling occurring	Limited to elastomeric element. Simple to make but has performance limitations

Single diaphragms are usually located between flange areas closed by clamping, bolting, crimping or, less usually, riveting. The seating areas may be further shaped or treated to provide an adequate gripping surface.

Bolted flanges are usually to be preferred as they are cheaper than clamps and readily capable of providing an adequate and consistent seal, provided the bolts are torqued evenly to ensure equal closed pressure over the whole of the sealing area. Bolted flanges are also easy to disassemble to replace a diaphragm.

Crimping is a low cost production method, particularly suitable for lighter assemblies and involves rolling and turning one flange up over the others. The actual crimping pressure applied can, however, be critical. Excess crimping pressure can wrinkle, cut or otherwise damage the diaphragm material. Insufficient crimping pressure can result in leakage at the joint. In either case the fault cannot be rectified since a crimped assembly is a permanent unit which cannot be dis-assembled and remade.

Attachments to the centre of the diaphragm invariably demand the use of centre plates. Cementing the centre plate in position avoids the necessity of piercing the diaphragm and causing a possible source of leakage, but it is not suitable where high stresses may be involved. Spinning or riveting are the more usual methods of centre plate mounting. Bolting is mainly restricted to larger units, where the ability to disassemble the unit and reassemble with the same, or a new, diaphragm may be an advantage. If springs are incorporated in the assembly they are invariably made to ride on the centre plate or on a buffer plate riding on the centre plate.

Rolling Diaphragms

A *rolling diaphragm* is essentially a very slack diaphragm contained within a cylinder and capable of piston-type movement. The movement is normally pressure energized in one direction only, the return motion being by spring action. Rolling diaphragms are normally employed as sealing devices, rather than pressure sensitive elements. A variant, the rolling sleeve, is employed as a pneumatic spring.

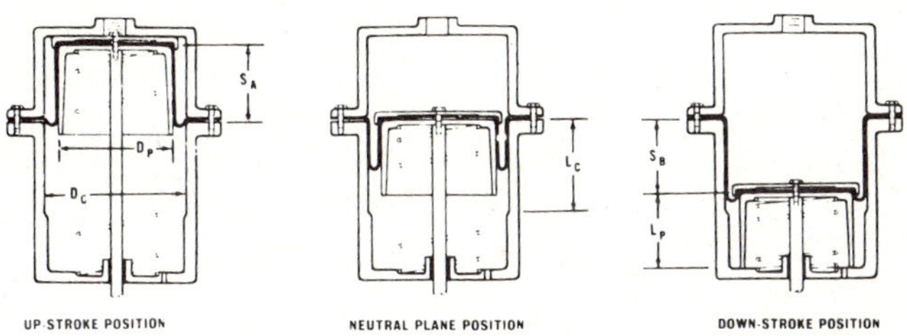

D_p = piston diameter
D_c = cylinder bore
L_c = cylinder bore length
L_p = piston skirt length
S_A = maximum upstroke
S_B = maximum downstroke

Fig 1 'Bellofram' rolling diaphragm (Angus Fluid Sealing Products).

A proprietary type of rolling diaphragm is shown in Fig 1. The slack diaphragm is moulded in a suitable elastomer, usually reinforced with high tenacity rayon, PTFE or glass fibre. Reinforcement is necessary to permit free circumferential elongation of the diaphragm with no axial elongation and so give a true rolling action. A plain elastomeric bag will not 'roll true' and will also develop friction.

The rolling diaphragm is effective only when pressurized in one direction. The bulk of the pressure load is carried by a piston over which the diaphragm drapes, the slack in the diaphragm then forming a convolution of U-shaped section in the coaxial clearance space. Pressure on the diaphragm maintains it in intimate contact with the piston and cylinder bore. Piston movement merely causes the convolution to roll up or down the axial clearance space, travel being limited only by the amount of slack available. Thus a diaphragm of this type can have a very large travel, determined by the depth of the original bag shape. Diaphragm motion is smooth and continuous and free from friction (because the motion is rolling) and hysteresis effect (because the diaphragm is not stretched).

Rolling diaphragms have a particular application, as frictionless and non-leaking seals, for sealing pressurized fluids (eg certain types of pneumatic actuators). Working pressure up to 35 bar (500 lb/in^2) can be accommodated, using readily available fabric materials, since the fabric stress is relatively low. This stress can be calculated directly as

$$S = \frac{P \times C}{2} \quad \text{(in consistent units)}$$

where P is the applied pressure
and C is the radial clearance or convolution width

The rolling diaphragm is a special case and is normally considered as part of a whole unit design. About the only mechanical requirements are the avoidance of sharp edges on the piston and at the clamping point of the diaphragm, and excess slack to ensure that the convolution is never pulled out straight.

Diaphragm-Type Bellows

Diaphragm-type bellows normally called welded diaphragms, are fabricated from a series of diaphragms stacked together. They are invariably made of metal and are in the form of contoured shapes welded together. Four basic contours are shown in Fig 2. Each shape has particular advantages and disadvantages. On the basis of pressure rating, or resistance to pressure, the torus shape is superior, followed by the single ripple. Both the other shapes are comparatively poor and thus only suited for lower pressure ratings.

For stroke rating, the nesting ripple configuration is far superior to the other types and also gives minimum nested length. At the other extreme, the flat plate bellows has only limited stroke capability.

Spring characteristics are also widely affected by the contour. Thus, the flat plate type has excellent linearity of stroke with pressure, and also excellent linearity of force output with pressure, making it particularly attractive for pressure sensing and proportional response. The spring rate is low. The torus also has excellent linearity of force output with stroke, but the linearity of stroke with pressure is not so good. The spring rate is high in this case. The nesting ripple has a reasonable linearity of stroke with pressure, but poor linearity of force output with pressure. Spring rate is low. The single ripple contour has poor linearity and a high spring rate. Table II summarizes typical applications of each type, based on these inherent characteristics.

Fig 2 Metallic bellows forms

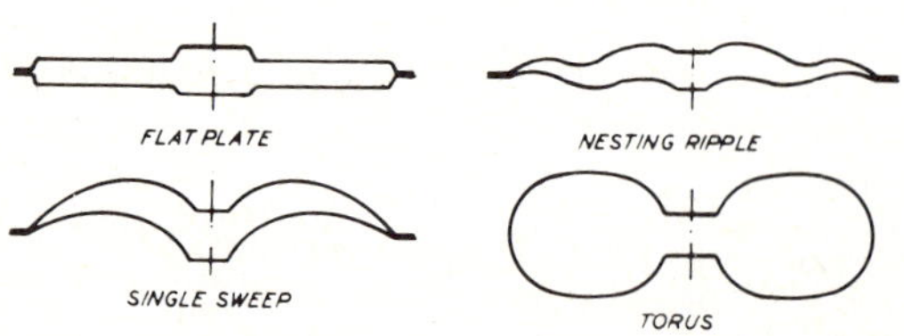

TABLE II – APPLICATIONS OF WELDED DIAPHRAGM CONTOURS

Application	CONTOUR			
	Flat Plate	Single Ripple	Nesting Ripple	Torus
Pressure sensing (null devices)	E	VG		
Pressure-sensitive (actuators)	VG	VG	E	VG
Low spring rate device	VG		VG	
High spring rate device		G		G
Variable volume pressure vessel (accumulator)			VG	
Flexible connector			VG	
Shaft seal		VG	VG	VG
Rotary seal	VG	G		G

Key: E = Excellent, VG = Very good, G = Good. Absence of any entry implies that the contour is not used for such an application.

Materials used for diaphragm-type bellows include stainless steels and high duty alloys for corrosion resistant or high temperature duties, phosphor bronze and beryllium copper, down to brass. The choice may be dictated by performance requirements, such as spring properties, drift, etc, as well as environmental requirements, although as described previously, the mechanical properties of welded diaphragm bellows can also be adjusted by means of the contours used.

Travel is limited by suitable stops to prevent over-stressing and possible rupture of the diaphragm. To ensure maximum diaphragm life, not more than 80–90% of the diaphragm travel available should be used in normal working – ie the diaphragm should be sized and proportioned accordingly.

With dish-type diaphragms, an optimum height of dishing is about 25% of the diameter, although this figure can be varied somewhat depending on the material. Excessive dishing is to be avoided. With convolute diaphragms, the optimum height of the convolution is equal to the convolution width, although lower values may be preferred from the point of view of ease of fabrication. The variation in effective area with travel can be reduced by increasing the size or number of convolutions.

Welded diaphragms are widely used as pressure-sensitive devices.

Tubular Bellows

Tubular bellows is the description given to bellows constructed in the form of a fabricated or moulded corrugated tube with suitable end covers. Deflection and spring rate are related to the number of convolutions and outside diameter, and also to the wall thickness and modular elasticity of the wall material.

In the case of convoluted metallic tubes, deflection is directly proportional to the number of convolutions and inversely proportional to the wall thickness and modulus of elasticity of the material. Spring rate varies directly with the modulus of elasticity of the material and with the cube of the wall thickness, and inversely with the number of convolutions and the square of the outside diameter – see also Table III. The required performance can thus be derived from these parameters.

If a higher spring rate is required than that which can be obtained from acceptable gometry and material, then the rate can be increased by incorporating an internal or external spring, arranged to oppose any force tending to compress the bellows. In such cases the spring used must have a higher rating than that of the bellows.

TABLE III – PERFORMANCE PARAMETERS FOR CORRUGATED TUBE BELLOWS

Parameter	PROPORTIONALITY	
	Deflection	Spring Rate
Number of convolutions (N)	$\propto N$	$\propto {}^{1}/N$
Wall thickness (t)	$\propto {}^{1}/t$	$\propto t^3$
Modulus of elasticity (E)	$\propto {}^{1}/E$	$\propto E$
Outside diameter of bellows (D)	$\propto {}^{1}/E$	$\propto {}^{1}/D^2$

'Pneuride' bellows have a wide range of applications both as linear actuators and vibration isolation devices, are available with nominal diameters ranging from 6 to 16 ins. and either 2 or 3 convolutions to provide a variety of working movements. Moulded in flexible reinforced neoprene rubber, the units can be operated on air pressure up to 8 bar (120 lb/in^2). Mounting is by means of 4 M.8 metric threaded studs in the steel end plates and there is a single ½'' BSP inlet port, actuated by Martonair Ltd by agreement with Dunlop Suspension Division.

Flexibility or deflection characteristics can be improved by deepening the convolutions. as well as reducing the maximum pressure rating, this can have other practical limitations. Deeply convoluted bellows are more prone to permanent distortion and fatigue effects.

For high pressure working a fairly substantial thickness is required, which has the effect of producing a very stiff bellows. This can be overcome by using multiple construction where the bellows is formed from two or more thin-walled co-axial tubes, yielding the total wall thickness required. The individual thin-walled tubes forming the layered construction are, however, free moving and thus the complete bellows retains good flexibility.

The spring rate characteristics, however, may be considerably modified and depart markedly from the linear characteristics associated with homogeneous corrugated tubes. This makes them far less suitable as pressure sensing devices.

Bellows materials include those already given for welded diaphragm assemblies, and also aluminium (for expansion joints). An 80:20 brass is commonly used for instrument bellows provided pressures are moderate and a continuous ambient temperature of 150°C (300°F) is not exceeded. For somewhat higher temperatures, or for use in mildly corrosive surroundings, phosphor bronze is generally preferred. For ease of fabrication, the more ductile metals are always preferred for corrugated tube bellows, which are preformed by 'ridging' under internal hydraulic pressure in a suitable die, and then compressed in length in another die to complete formation of the corrugation.

The maximum deflection to which corrugated tube bellows can be subjected is strictly limited, and is normally specified by the manufacturers. In general, under theoretical maximum deflection, the bellows is closed virtually to its solid length and the working range of stroke is represented by the travel from the free length. A figure of 80% of this length is normally taken as the practical limit for deflection. Such bellows should not be extended past their free length, ie all the deflection should be accommodated as a compression from the free length, never any extension. This particular limitation does not necessarily apply to welded diaphragm type bellows where a 10% extension beyond the free length is commonly (but not invariably) permitted.

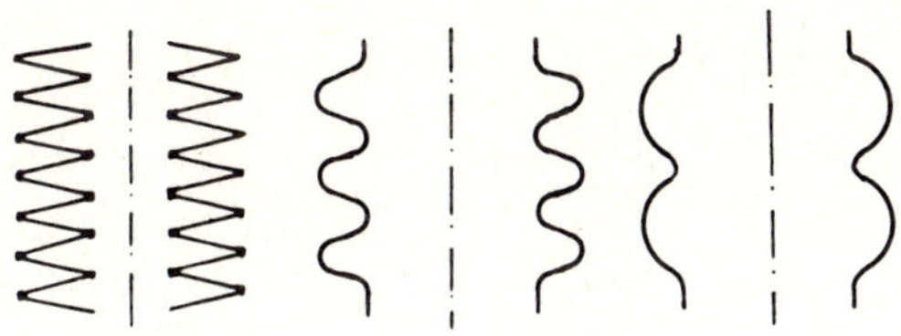

Fig 3 Moulded elastomeric bellows forms.

Elastomeric bellows may be formed by welding or moulding, the latter technique normally being preferred. Here the optimum contours are largely affected by the pressure rating involved. For low pressures, deep but narrow convolutions are best. Rounded and more widely spaced convolutions are best for higher pressures — see Fig 3. For higher pressures still, laminated construction is necessary, with the elastomer forming an inner liner, reinforced with a fabric sleeve.

Elastomeric bellows are not normally used as pressure-sensitive elements or spring devices. Their main application is for expansion joints or flexible pipe couplings, and for variable volume pressure vessels. In the latter case maximum pressure rating is usually of the order of 100 lb/in^2 (7 bar), although figures of twice this value or more may be realized in specific constructions.

SECTION 8

Surveys

Survey of Valves

Air Automation

Air Automation air control valves are manufactured in port sizes of from 1/8 in to ½ in BSP, and are designed to give long trouble free service at an economical price. Operating ends are generally interchangeable within the same range of valves, and valves are supplied with manual, air pilot, solenoid, foot, mechanical (roller or one way trip) mechanisms. Many are ideal for panel mounting, being provided with switch, lever, or twist knob actuators similar to those used on electrical controls. All sizes of valves are available in 2-way, 3-way and 4-way 5-port versions, normally open or normally closed and with closed or neutral centre positions with or without spring over-ride. The 1/8 in BSP range is constructed with compact brass one piece bodies provided with a honed bore in which operates a one piece spool and the larger sizes employ stainless steel spools floating in nitrile seals.

Also produced are shuttle, non-return and stop, flow control, binary dividing, and single pulse valves. The latter type is used to overcome held-on signals by exhausting its output after it has signalled the main valve and remaining exhausted while holding input pressure until the input is removed and re-applied.

524LL ¼ BSP 4-way lever operated valve.

Bank of 421LL toggle lever operated 4-way valves. 3-way version available.

The Aro Corporation

The range of air valves produced under the 'Modernair' brand name covers virtually all valve functions. The majority are either of poppet or spool type. Diaphragm valves are used for quick-exhaust functions, and needle valves for fine flow control.

Aro miniature 3-way control valves are offered in sixteen actuator styles – five manual, ten mechanical and one pneumatic – of which eight may be panel-mounted. They are ideally suitable for the operation of cylinders having up to 1.1/8 in bore diameters, and for actuating pilot valves controlling large-bore cylinders. Miniature valves operate at air pressures from 25 to 150 lb/in^2, with an actuating force of from 10 to 60 ounces, depending upon the actuator. They have a flow capacity of 10 ft^3/min at 100 lb/in^2 (gauge) and may be used normally closed or normally open. As many as six valves can be stacked without need of a manifold.

They are designed for quick easy overhaul and have zinc die-cast bodies and stainless steel valve spool with floating Viton O ring seals. They have 1/8 in NPTF dry-seal pipe threads. The range is supported by a wide selection of ancillary fitments including snap-together flush faced toggle and slide panel switches.

Other standard valves:-

Series 50 – 1/8 in ports. Available with manual, pilot and mechanical operation.
Series 51 – 3-way and 4-way, ¼ in side ports. Two- and three-position models.
Circuitair Series 52 – 3- or 4-way poppet valves.
Series 53 – 4-way 5-port spool valves. Sizes 3/8 in to 1in. Two- and three-position models.
Series 54 – solenoid valves – 2- and 3-wav. N/C, N/O and directional control models.

The Aro Corporation also produces a wide variety of logic elements and components for pneumatic logic controls and a selection of air cylinders.

Bestobell Mobrey Ltd

This company markets the 'Herion' range of valves, which is claimed to be the largest range of pneumatic valves produced in Europe.

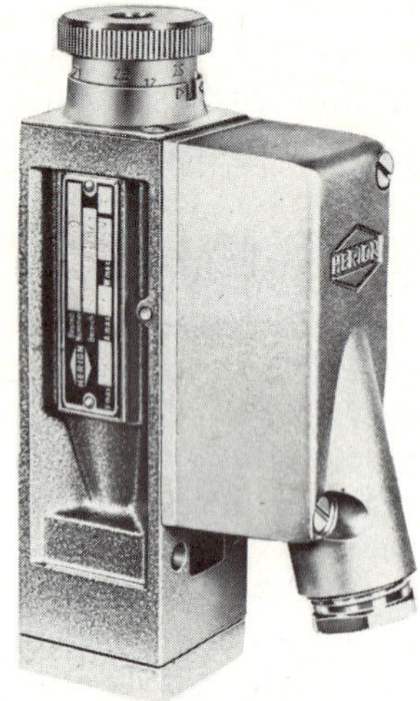

Over 30000 permutations of the standard equipment are available to control air, water, oil, corrosive and non-corrosive liquids, gases, steam and vacuum. There is also a comprehensive range of pneumatic cylinders and ancillary equipment. Herion products cover a wide range of voltages and many different materials, including PTFE and Teflon. For difficult applications, these can also be supplied in weatherproof and explosionproof housings.

Two of the pneumatic range, Models 5/2 and 3/2, contain a unique T seal on the spool in place of the standard O-ring. When clamped between the two surfaces, its shape prevents the usual rolling and chaffing between the valve ports. Special air holes allow the T seal to expand, thereby creating a better seal. The Herion range has numerous applications in most industries including pneumatic process control and pneumatic and hydraulic press control.

The British Steam Specialities Ltd

This company manufactures 'Bossmatic' diaphragm type control valves, which are actuated by compressed air, gas or liquid at control pressures of 3–15 lb/in^2. Standard sizes are from ½ in to 12 in (15–300 mm). Also ON/OFF control valves from ½ in to 6 in with actuating pressures from 30–150 lb/in^2.

The company also supplies 'Claudo' valves, based on an eccentric cam movement enclosed in a rubber cylinder (diaphragm) filled with grease, only the valve body and rubber cylinder contacting the medium. 'Claudo' valves are claimed to be 100% leakproof and are suitable for use with compressed air, vacuum, water, high and low viscosity oils and many other fluids. Sizes are ¼ in to 2 in BSP.

The 'Butterball' range of butterfly valves can be used to perform the duties of conventional ball, gate or globe valves. Special features include: blow-out proof stainless steel stem; adjustable torque; permanently aligned stainless steel disc with a Viton seal, which wipes the seat area clean with every operation and ensures tight shut-off as well as being completely flushable, eliminating all traps for contaminants.

Pressure rating: 847 mb to 12 bar (25 in hg to 175 lbf/in^2). Temperature range: –34°C (–30°F to +350°F).

One-piece bronze valves can be supplied screwed from ½ in to 2 in (15 mm to 50 mm) with adjustable handle torque.

The stainless steel range of valves is available screwed from ¼ in to 2 in (8 mm to 50 mm).

Burkert Contromatic Ltd

The range of solenoid valves offered by this company includes 2-, 3-, 4- and 5-way types.

3-way types are available in sizes from miniature (1.2 mm orifice) up to 40 mm orifice (1½ in BSP). Of special interest in this group is the 303/304 series, of 1/8 in BSP size, which is a further development of the well-known 311/312 series and contains all three ports in the base of the valve. The sub-base 304 is designed such that N/C and N/O versions can be combined on the same manifold, so that extracts from a bank of valves can be routed to a single location.

The 4-way 420 series of solenoid valves is of moulded-plastic construction and employs as a pilot a low-wattage version of the 312 series. Standard features are manual override and separate adjustable flow restrictors for each exhaust line. The 420 series is of extremely compact design and is offered with a choice of either 1/8 in BSP ports or integrally moulded connectors for direct connection without fittings of 6/4 mm tubing. Manifolds are available to accommodate from 2 to 6 valves.

Type 420 four-way solenoid valve. (Burkert Contromatic Ltd).

The 330/331 series of pivoted-armature solenoid valves are supplied in the full range of 2- and 3-way circuit functions (ie 2-way N/C and N/O, 3-way N/C, N/O and diverter). Manual override is standard and optical and electrical position indicators are now offered as optional features.

Impulse versions of this series are also supplied, designated 332/333. Spool valves with lapped steel-to-steel seals for long life are available in ¼ in and ½ in sizes with a variety of pneumatic and electro-pneumatic actuators.

The 411/413 range of 3- and 4-way solenoid valves are of ¼ in BSP size and of poppet-type design, with moulded plastic bodies.

The largest pneumatic valves in the Burkert range are the 340 series, available in sizes up to 1½ in BSP. N/C and N/O 3-way versions are offered, with position indicators (optical and electrical) now being optional.

Several of the products outlined above can be supplied with special approvals, eg CSA-approved, GL-approved, explosion-proof (PTB and BASEEFA).

Circlex Ltd.

The range of Circlex low and high pressure valves includes a wide variety of needle and ball types in brass and hardened steel construction. In the case of needle valves, neoprene and other seal materials are used. PTFE seats are standard in the ball valves.

Valve sizes range up to 3 inch (75 mm) bore.

CompAir Maxam Ltd

The Maxam 'Flo-Master' range of valves is capable of withstanding working temperatures up to 90°C (194°F), except the solenoid operated types which are continuously rated at an ambient temperature of 70°C (158°F) maximum.

They are design tested at a pressure of 200 lb/in^2 (140 bar) and unless otherwise stated are suitable for a maximum working pressure of 150 lb/in^2 (10.5 bar) air.

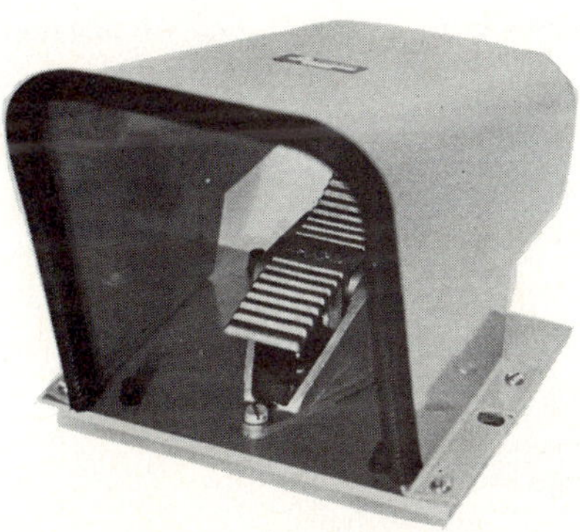

CompAir Maxam shroud assembly for foot operated valves.

CompAir Maxam 'Flo-Master— valve.

Maximum air flow for ¼ in BSP valves at 80 lb/in^2 (5.62 bar) is 50 ft^3/min (1.41 m^3/min).

Maximum air flow for ½ in BSP valves at 80 lb/in^2 (5.62 bar) is 150 ft^3/min (4.23 m^3/min).

Electrical data for solenoid operated valves — all solenoid coils are fully capsulated, tropicalized and flash tested, and can be supplied in all recognized standard voltages conveying 40, 50 and 60 cycles AC and DC currents. Voltages outside the standard range are available if required.

Mean power rating of coil ∓ 10% nominal volts — AC 13VA — DC 9 watts

All coils are rated for continuous energization.

All valves are available for three- and four-way operation. All three-way valves can be arranged for two-way operation by plugging the appropriate port.

Compressor Accessories Ltd.

This company manufactures the 'Comprax' range of automatic plate valves for all types of reciprocating compressors; also the following regulation equipment:

Non-return valves for pipelines

Pilot Valves, Diaphragm actuators for suction valve unloading.

Complete capacity regulation systems for start/stop, 2-step, 3-step or stepless control.

Pneumatic regulation instruments and systems for screw compressors, comprising suction throttle control unit, (for on/off and modulating control, including butterfly valve, actuator and regulator), oil stop valve, minimum pressure valve, non-return valves for air and oil circuits, blow-down and pressure reducing valves for receiver.

Crouzet

Pneumatic control valves manufactured by this company include 4/5 air-pilot operated and solenoid-pilot operated models; also 2/3 two-position valves. All are of modular construction with CETOP standard sub-base mounting, the sub-bases containing instant connections for 6, 8 or 12 mm nylon tubing. Other types include:

Microvalves — miniature and reinforced types, available with a wide range of mechanical and manual actuators.

Pilot valves — particularly suitable for the direct drive of 5 port control valves and cylinders.

Lourde series pilot valves — heavy duty types available with ¼ in or ½ in BSP ports.

Crouzet VM-directional in-line restrictor

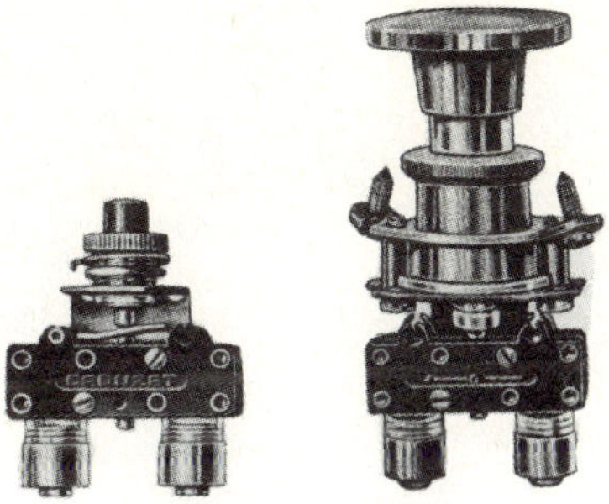

Miniature and reinforced microvalve

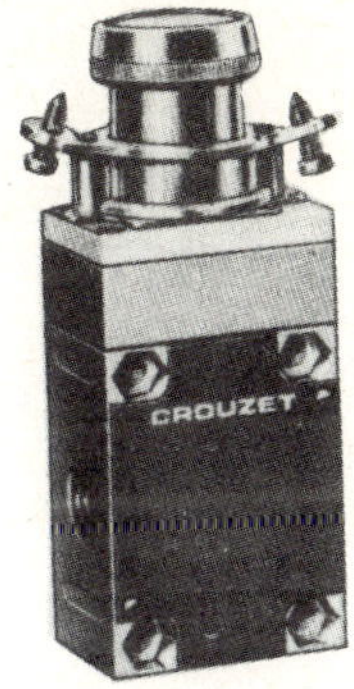

Pilot valve

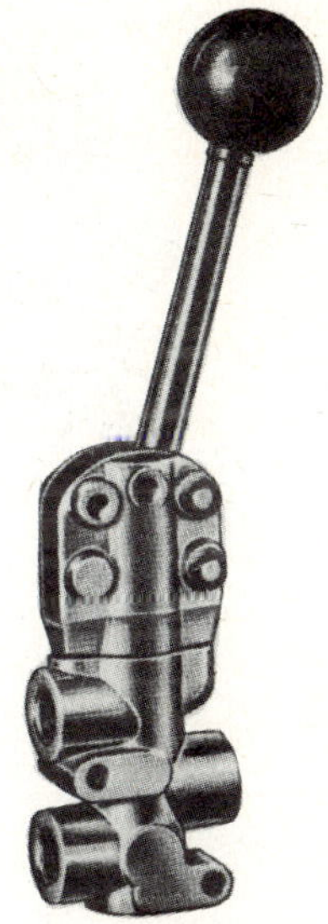

Lourde series pilot valve

Recent additions include a new range of uni-directional in-line restrictors with various fixed orifice diameters ranging from 0.3 mm to 1.0 mm.

Davis Pneumatic Systems Ltd

Valves produced by this company are of 1/8 in, ¼ in and ½ in BSP sizes in 3- or 5-way types, and are provided with all the usual types of operator, ie solenoid, lever, pilot etc.

Sub-base or manifold mounted versions of these are also available.

Airflow regulators in 1/8 in, ¼ in and ½ in BSP sizes.

Impulse valves and "light touch" trip valves and signal boosters.

Drallim Controls Ltd

This company produces the unique Drallim Rotary Switch Valve (RSV). This is a rotary selector or distributor for use in small bore piping systems (1/8 in to 3/8 in BSP) based on a square or hexagonal section body with four or six ports respectively, each port containing a spring-loaded poppet valve which is actuated by the rotary movement of a cam on a spindle passing through the body.

An important feature of the RSV is that it can have a permanently-open connection in the base giving direct access to the peripheral ports via the central chamber. With such a connection, the RSV can switch clock-fashion. Simultaneous operation of more than one poppet valve can also be provided by suitable shaping of the cam (actually the number of 'flats' on the spindle). Numerous permutations are also possible on this basis, thus offering extreme versatility in switching sequences for different applications. A number of RSVs can also be ganged to form a multibank valve, all operated by the same spindle.

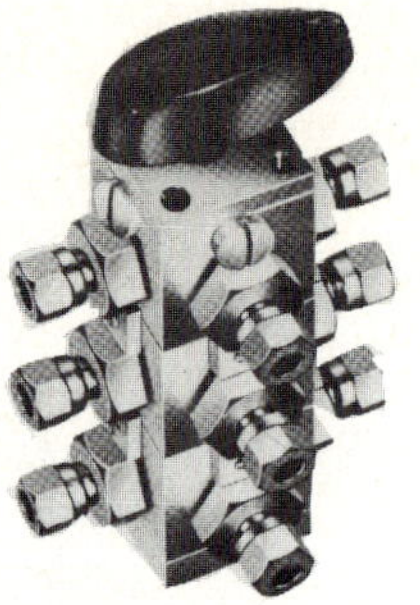

Three-bank RSV with square body.

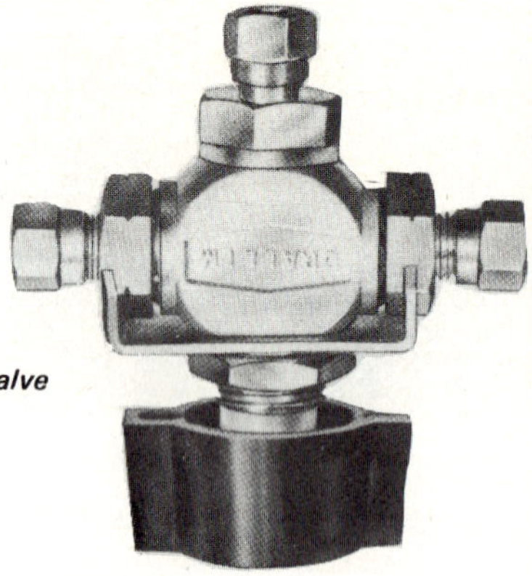

Series 31 rotary switch valve

Series 91 valves

Other types manufactured by this company include:

Large Bore Rotary Switch Valves — similar in operating principle to the RSV with clearflow stems. Nominal bore is 3/8 in with tube connections for ½ in or 5/8 in o.d. tubing. The valve can have one or two banks and be fitted with an electrical rotary switch base.

Series 20 Control Valves — instrument-type valve with one-piece acetal plastic screwed spindle. Open bore size 3/16 in with interchangeable external connectors to take 1/8 in to ½ in o.d. tubing.

Series 31 Rotary Switch Valves — stem-type valve with O-ring seal, operated by flats machined on the spindle. Sizes 1/8 in to 3/8 in.

Series 50 Miniature Manifold Valve — incorporating three single 1/8 in valves in a square section body.

Series 91 — 1/8 in BSP miniature balanced poppet in-line valves in 2- and 3-port versions.

Series 175—180 Ball Valves — ¼ in, 3/8 in and ½ in BSP.

Series 1000 Control Valves — ¼ in bore with interchangeable connectors to take 1/8 in to 3/8 in o.d. tubing.

Series 1010, 1020 and 1030 Push-Button Valves — press-to-open, press-to-close and changeover.

Series 1100, 1200 and 1300 Fine Control Valves — made in 1/8 in, ¼ in and ½ in port sizes. These leak-free valves employ a diaphragm seal isolating the operating mechanism from the fluid control element. Connectors are Drallim Captive Seal type or conventional metal compression couplings.

Dyna-Quip Ltd

Shut-off valves produced by this company are of compact ball type with quick on/off action (quarter-turn), non-rising blowout-proof stem and minimum restriction to flow in the open position. These valves are made in pipe sizes up to 2½in in bronze, brass, stainless steel and carbon steel with a choice of seat and seat materials to suit the fluid being handled. New additions to the VC Series include a 1¼ in model in carbon steel and high pressure models for use at pressures up to 3000 lb/in^2.

Dyna-Quip also have a new Series VC exhaust valve. This bleeds down downstream residual pressure when in the shut position to safeguard connected equipment. The type VP are of sandwich construction which permit the centre section to swing away for easy installation and maintenance inspections without disturbing the connected pipework. All models are available with safety locking handle if required.

Electrical Remote Control Ltd.

The pneumatic valve Type PV manufactured by this company features single die-cast body construction. It will operate in a normally closed or normally open mode at pressures up to 80 lb/in^2. Working data are:

Operating pressure	10-80 lb/in^2 (gauge)
Operating force on lever arm roller	10-150 lb/in^2 (gauge) when used with connection securing collars and increased spring pressure.
Flow rate typical	2 s.c. ft^3/min at 40 lb/in^2 (gauge)
	3.8s.c.ft^3/min at 80 lb/in^2 (gauge)
Operating time	25 ms typical
Operating temperature	-5 +90°C
Weight	170 grms

Electrical Remote Control type PV valve

Fluid Automation Systems S.A.

Fluid Automation Systems S.A. are manufacturers of automation components and a wide range of solenoid valves. They have specialized, in particular, in miniaturization of solenoid valves and claim to have produced the smallest solenoid valve with high performance. The complete series of miniaturized solenoid valves includes those specifically adapted for pilot duties for directional control valves and also a whole series of 2/3, 3/2 and 5/2 solenoid valves with direct connection and sub-base mounting.

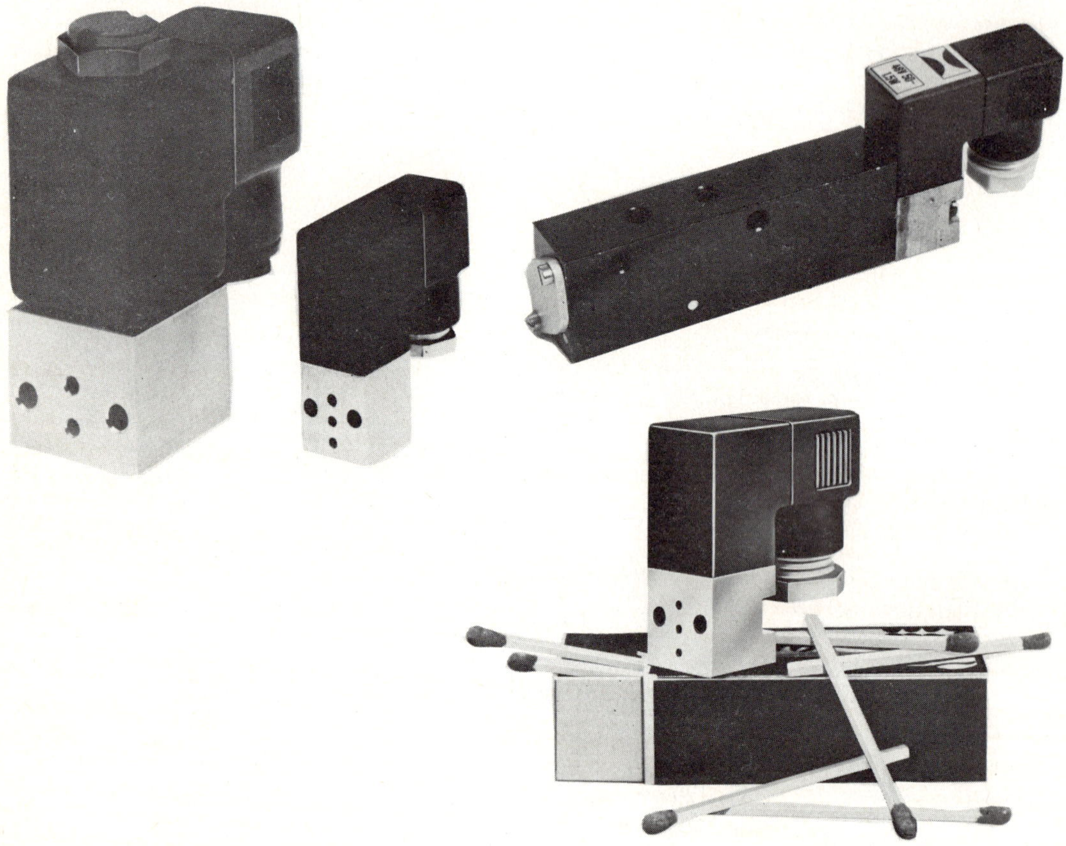

Joucomatic Controls Ltd.

This company supplies pneumatic control valves, solenoid-operated, air-operated and mechanically-operated in sizes 1/8 in to 1½ in BSP.

Spool Valves – 1/8 in to ¾ in BSP – the new 540/541 series added to this range feature a sealing system with a life of 10 million cycles and a multifunction selection system allowing valves to be changed from air-operated to solenoid-operated.

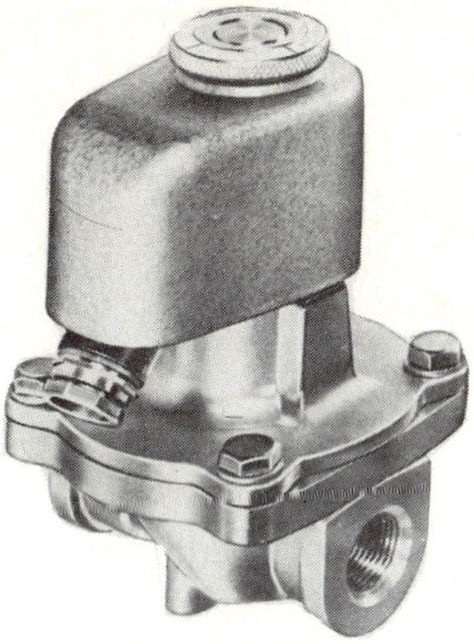

Type 138 Solenoid valve

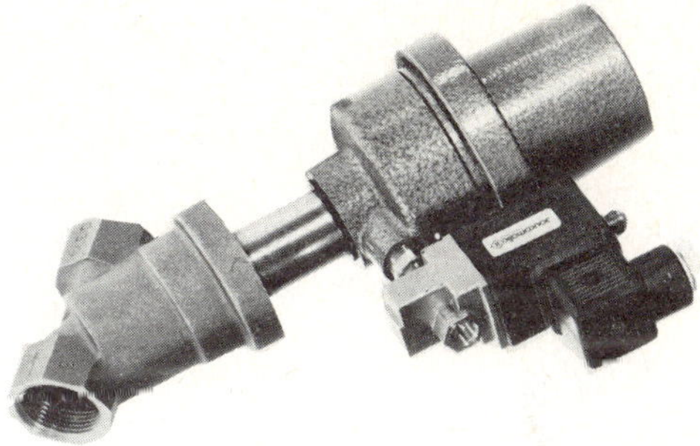

Type 212 valve

Poppet Valves — 1/8 in BSP to 1½ in BSP.

Valves are available line mounted and sub-base mounted 3-way and 4-way.

Joucomatic also supply a complete range of solenoid valves 1/8 in BSP to 2 in BSP and flanged to 80 mm. These are available weatherproof IP65 or flameproof.

Globe valves in bronze, cast iron, steel or stainless steel are also available in sizes ½ in BSP to 2 in BSP and flanged to 150 mm. New piston valves have also been introduced in sizes ½″ BSP to 2″ BSP.

Solenoid spool valve

DP Solenoid valve

Kay Pneumatics Ltd.

Kay Pneumatics produce two ranges of high-flow spool valves which are sub-base mounted, together with a full support programme of ancillary valves and timers.

Power Logic 43/45 Series

4 mm through bore: 3 port (3/2 way) and 5 port (4/2 way)
Flow 430 N1/min: Response time < 4 milliseconds: for 4 mm or 6 mm o.d. tubing.

Power 95 Series

9 mm through bore: 5 port (4/2 way)
Flow 3400 N1/min: Response time < 10 milliseconds: for 8 mm or 12 mm o.d. tubing.

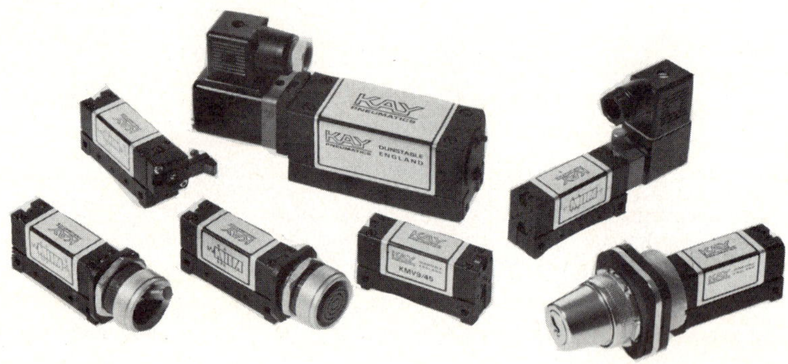

New metric valve range

Valve bases are ordered separately to simplify stockholding. Bases are either bottom entry, ideal for gang-mounting, or BSP ported side entry to CETOP specifications.

Martonair Ltd.

Abridged details of the Martonair range of valves are given in the Table.

CONTROL VALVES

Type	Range	Pressure bar	Remarks
2/2	S/534 S/728 S/738 M/1328 M/1338	0 – 10	simple on/off valves; also suitable for hydraulic services
3/2	S/666-7 M/21 S/560 S/509 S/1340	2 – 10 4 – 10 1 – 7 2 – 10 1 – 10	midget type poppet valves highly sensitive valves requiring only very light operating effort general purpose free standing and panel-mounted models normally closed (S/2509 normally open) heavy duty valve
3/2 & 3/3	S/441-2	1 – 7	simple valves with good flow capacity
5/2 & 5/3	M/1555 M/1701 M/1721 CM/1701 CM/1721 S/556 M/1702 M/1712 M/1722 M/1732 CM/1702 etc P/1702 etc	2 – 10 2 – 10 2 – 10 1 – 7 2 – 10 2 – 10 2 – 10	miniature valves for double-acting cylinder control sub-base mounted valves similar to M series, intended for manifold mounting similar to S/560, but suitable for double-acting cylinder control and pilot systems larger versions of the M/1701 similar to M series, intended for manifold mounting virtually identical to M series but with integral push-on connectors for 8mm nylon tubing.
Miscellaneous Valves	S/562 M/1704 etc CM/1704 etc S/563-5 M/1718	1 – 7 2 – 10 2 – 10 1 – 7 2 – 10	available with a wide choice of operating mechanisms in the same style as M1701 and M1702 similar to M series, intended for manifold mounting available with a wide choice of operating mechanisms sub-base mounted, incorporate a silencer as standard
	M/805-6 BM/805-6 PBM/806 S/808	1 – 10	4/2, 4/3 and 5/2 plate valves M/805 and M/806 are sub-base mounted
	S/594 M/649	1.5 – 7 2 – 10	manually operated valves
	DM/1400 DM/1402	0 – 10	'Minijet' low power solenoid valves for pilot operation. Suitable for manifold mounting
	DM/1500 DM/1502	0 – 10	'Electro-jet' range of solenoid valves. Suitable for manifold mounting.
	S/557 S/657 S/757	0 – 16, depending on type	Micro-valves for pilot control of directional control valves. S/657 is flameproof version.
Non-return valves	S/520-524	0.5 – 17	1/8in to 1in BSP (5 models)
Shuttle valves	S/575 S/592	1 – 10 1 – 10	1/8in BSP 1/4in BSP
Relief valves	S/611 S/612	3 – 10 2 – 7	1/4in BSP 1/4in BSP

Midland Pneumatic Ltd

This company manufactures valves from 1/8 in to 1½ in BSP including the patented 'Fluoslide' range of slide valves with PTFE slides. The 3-way poppet valves have recently been redesigned and are available in metric form, the range covering from ¼ in Gaz up to 1½ in Gaz.

Midland Pneumatic 'poppet valve.

Midland Pneumatic semi-rotary valve.

C. A. Norgren Ltd.

The Olympian Plug-in (Industrial) Pressure Regulator is the foremost general purpose type available in ¼,3/8, ½, ¾, 1 and 1¼ in port sizes. Plug-in design enables the regulator to be removed

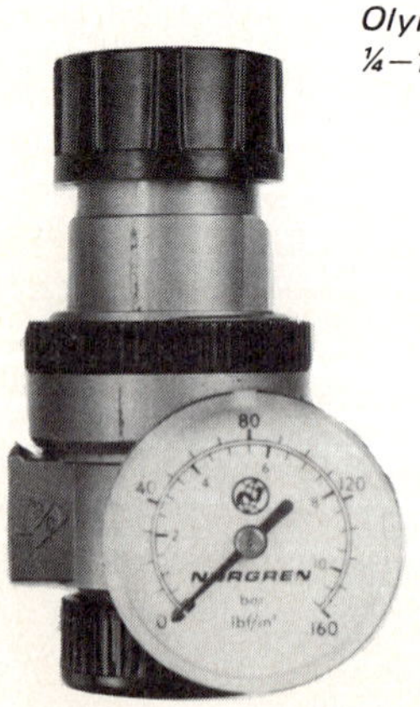

Designer ¼, 3/8 in sizes.

Olympian 'Plug-In' ¼–1¼ in sizes.

Precision controller. ¼ in size.

from the line quickly without disturbing connected pipework. They are offered with 'relieving' or 'non-relieving' operation and there is a wide range of regulating springs to choose from.

Other main industrial ranges include miniature models for low flow applications, the Designer Type in ¼ and 3/8 sizes and Pilot Operated Models in ¾, 1, 1¼, 1½ and 2in sizes.

Special purpose models produced include:—

Precision Controllers and Precision Regulators for air gauging, precision pilot operation, fluid logic circuits, instrumentation and similar applications where the highest degree of accuracy, sensitivity, fast response and precise repeatability are required.

In addition a small but important unit is the new R16 Safety Blow Gun regulator, completely tamperproofed and pre-set at 2 bar.

They also produce a range of integral Filter-Regulators (See *Survey of Air Line Fittings*).

Parker-Hannifin

The Parker Hannifin range of air valves includes the following types:

4-way valves (3/8 in, ½ in, ¾ in and 1 in)
electrically operated, single solenoid pilot
single solenoid direct
double solenoid pilot
double solenoid direct

4-way valves, (3/8 in, ½ in, ¾ in and 1 in)
pressure actuated

3-way valves, (1/8 in, 3/8 in, ½ in, ¾ in and 1 in)
pilot operated
electrically actuated, single solenoid direct
electrically actuated, single solenoid pilot

4-way valves (¼ in and 3/8 in)
foot operated
hand actuated
cam actuated

3-way valves (¼ in and 3/8 in)
cam actuated
hand actuated
foot actuated

Speed control valves (¼ in, 3/8 in, ½ in and ¾ in)
Needle valves (1/8 in, ¼ in, 3/8 in, ½ in and ¾ in)
Check valves (1/8 in, ¼ in, 3/8 in, ½ in and ¾ in)
Quick-exhaust valves (¼ in, 3/8 in, ½ in and ¾ in)

R.G.S. Electro-Pneumatics Ltd.

This company manufactures directional control static seal spool valves up to 1 in BSP and 1/8 in BSP and ¼ in BSP direct solenoid valves. Many valves are arranged for ganging, sub-base or direct face mounting. The valves incorporate low friction twin lip seals. The solenoid valves are low power units and cover intrinsic safety CSA and mining specifications.

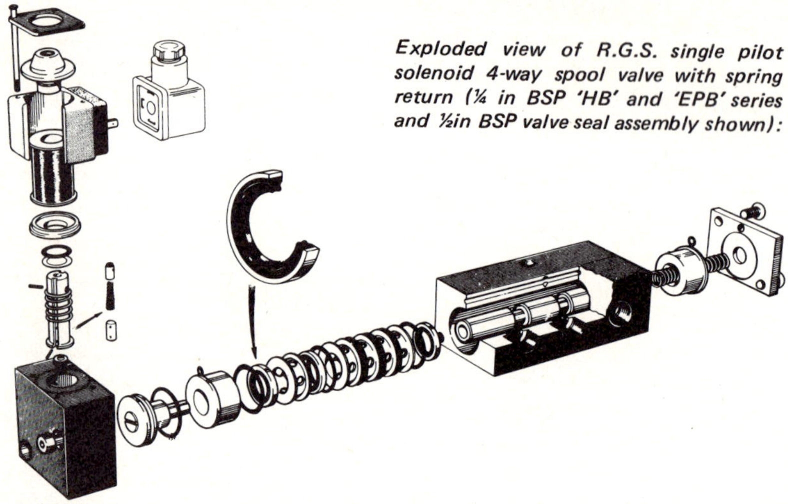

Exploded view of R.G.S. single pilot solenoid 4-way spool valve with spring return (¼ in BSP 'HB' and 'EPB' series and ½in BSP valve seal assembly shown):

Ross James Engineering Ltd

This company is the sole UK distributors of Numatic valves of patented spool and sleeve construction. The range includes direct solenoid, solenoid-pilot, air-pilot and numerous limit, manual, mechanical and special valves. Numatic valves are corrosion free (being of hardened stainless steel construction) and self-cleaning. They can also operate without lubricant.

Numatics also manufacture Numatrol II, a fully integrated system of all-air moving part logic control. This consists of a range of logic components, supplied either as individual units or as complete ready-built circuits.

A. Schrader's Son

Schrader control valves use spool, poppet, disc, and their unique, balanced, double poppet mechanisms and provide a range of port sizes from 5 mm to ¾ inch. Most of the valves are available ported, sub-base or ganged manifold with a wide choice of actuation and return methods in manual, mechanical and automatic configurations. Ancillary valves comprise such types as quick exhaust/shuttle valves, flow controls, non-return valves plus pressure switches, restrictors and port silencers.

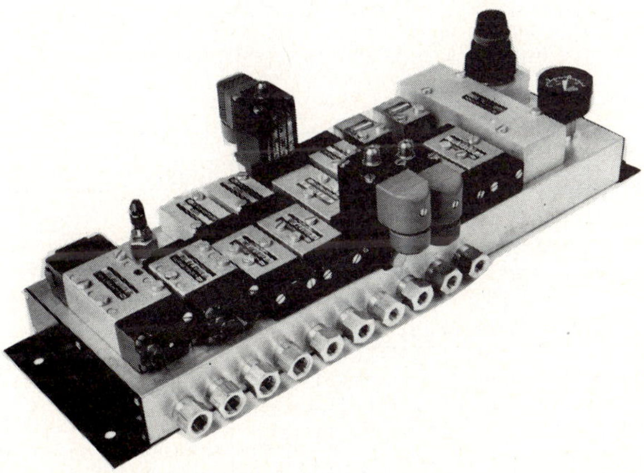

Sperry Vickers Lang Pneumatic

Types of valves manufactured by this company are summarized in the Table.

Series	General Description	Port Sizes	Remarks
VA170	3-way or, 4-way heavy duty pilot slide valves	1/8 in BSP	spring return; various methods of actuation
VL13	4-way heavy duty linear slide valves	1/8 in BSP	various methods of actuation, including air
VL25	4-way heavy duty linear slide valves, port block mounted	¼ in BSP (1/8 in pilot)	air, solenoid or spring return operated
VL50	4-way heavy duty linear slide valves, port block mounted	½ in BSP	air, solenoid or spring return operated
VL75	4-way heavy duty linear slide valve sub-base mounted	¾ in BSP	air or solenoid operated
VL100	4-way heavy duty linear slide valve, sub-base mounted	1 in BSP	air or solenoid operated
VR13) VR25)	2- and 3-position semi-rotary slide valves	1/8 in BSP ¼ in BSP	pressure release or pressure held neutral (3-way valves)
VR50	4-way double acting semi-rotary slide valve	½ in BSP	lever switch or foot operated
VP	miniature pilot valves 3-way poppet type	1/8 in	various methods of actuation
PV	2- and 3-way poppet valves	¼ in, ½ in, ¾ in and 1 in BSP	air or solenoid operated spring return
VE25	3-way solenoid pilot valve	¼ in BSP	metric design
VE26	3-way solenoid pilot valve		manifold mounting
F1113 series	3-way sleeve valve S	1/8 in, ¼ in, and ½ in BSP	pneumatic isolators
F1141 series	shuttle valves	1/8 in and ¼ in BSP	
F1123 series	relief valves	¼ in and ½ in BSP	60–130 lb/in^2, 10–60 lb/in^2 130–250 lb/in^2
F1116 series	non return valves	1/8 in, ¼ in and ½ in BSP	0–145 lb/in^2
EV3042 series	quick-exhaust valves	¼ in to 1 in BSP	10–145 lb/in^2

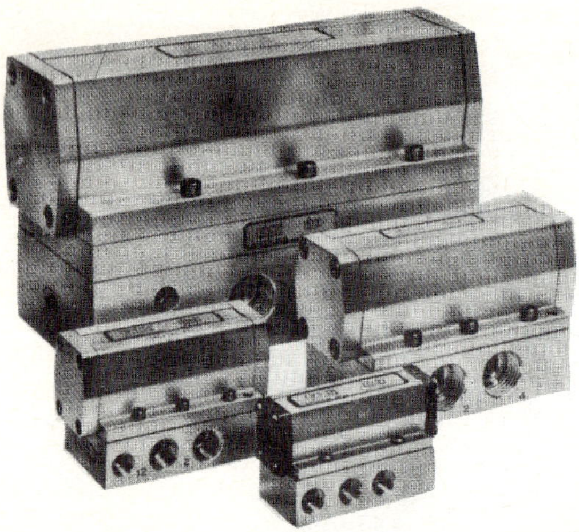

¼ in, ½ in, ¾ in and 1 in.

Two typical miniature pilot poppet valves 1/8in BSP.

Telektron Ltd.

The design of valves manufactured by this company is based on the use of a flexible diaphragm and an annular poppet in axial contact to give friction-free operation. Solenoid operators available with these valves are also intrinsically safe.

The basic valve is a 3-port type. The 4-port valve is made up of two 3-port valve modules assembled face to face. These are available as 2-position or 3-position types. The basic valve size has a ¼ in bore.

Wade Couplings Ltd

The following is a summary of the type and sizes of pneumatic valve manufactured by this company.

Ball Valves. Female x female and female x male ball valves in brass or gunmetal for use with fuel, heating oil, lubricating oil, gas (including natural gas), liquid gas, water and air. Also diesel oil, paint, lacquer, solvents, weak alkalis and acids. Temperature range –20 to +200°C. Nominal pressure up to 40 bar, pneumatic applications up to 10 bar. Vacuum tight to 10^{-3} torr.

Wade Mini Valves. Brass mini valves to suit 3/16in, ¼in and 5/16in o.d. tube. Stainless steel, aluminium bronze, or brass spindles.

Wade Straight and Angle Valves. Brass valves, with brass or aluminium bronze spindles to suit ¼in, 5/16in and 3/8in o.d. tube.

Wade Foot Mounted Valves. Brass for 3/16in or ¼in o.d. tube to tube and chromium plated brass for ¼in o.d. tube to hoze nozzle.

Wade Brass Plug Cocks. Brass 1/8in o.d. to ½in o.d. tube to tube type, or tube to BSP type.

GUNMETAL BALL VALVES

Pattern No	Nominal Size inch	Nominal Size mm	LW	W	t	L	H	R	Sw	Sw4kt	Nominal pressure in bars
							Sizes in mm				
509/3	1/4"	6	6	1/4"	10	46	54	90	22	7	40
509/8	3/8"	10	6	3/8"	10	46	54	90	22	7	40
509/11	1/2"	15	10	1/2"	11	50	56	90	26	7	25
509/14	3/4"	20	13	3/4"	13	58	58	90	32	7	25
509/17	1"	25	19	1"	15	72	72	110	40	8	25
509/1-3	1¼"	32	24	1¼"	17	82	75	110	50	8	16
509/1-11	1½"	40	30	1½"	18	92	90	130	56	10	16
509/1-17	2"	50	35	2"	20	102	95	130	68	10	16

BRASS BALL VALVES

Pattern No	Nominal Size inch	Nominal Size mm	LW	W	t	L	H	R	Sw	Sw4kt	Nominal pressure in bars
							Sizes in mm				
GM 509/8	3/8"	10	10	3/8"	10	48	56	90	22	7	40
GM 509/11	1/2"	15	13	1/2"	11	54	58	90	26	7	25
GM 509/14	3/4"	20	19	3/4"	13	68	72	110	32	8	25
GM 509/17	1"	25	24	1"	15	78	75	110	40	8	25
GM 509/1-3	1¼"	32	30	1¼"	17	90	90	130	50	10	16
GM 509/1-11	1½"	40	35	1½"	18	99	95	130	56	10	16
GM 509/1-17	2"	50	46	2"	20	114	100	130	68	10	16

Williams & James Ltd

Series 400 pressure reducing valves manufactured by this company are suitable for steam, air, gas and liquids (including corrosive fluids). They are of diaphragm type.

Survey of Air Motors

The Aro Corporation

Aro air motors are of the rotary-vane type with enclosed planetary gearing, giving a wide range of speeds and torques from each basic motor in a series embracing ten different sizes. All planet gears run on needle bearings and planet cages have ball bearings at each end to withstand moderate end thrust and radial loads. Individual models are available with forward rotation or as reversible motors, with single and double gearing reduction in both cases.

Leading performance details of the range are summarized in the Table. Load speed and load torque performance is based on an air inlet pressure of 90 lb/in^2 (6.5 bar) gauge. The motors are not generally recommended for running at air pressures of less than 50 lb/in^2 (3.5 bar) gauge.

ARO AIR MOTORS

Series	Maximum Horsepower	Speed Range rev/min	Torque Range lb.ft	Remarks
0000	0.10	1 000 – 25 000	0.10–1.94	(6) forward and (3) reversible models
000	0.20	550 – 20 000	0.22–7.80	(10) forward and (5) reversible models
0	0.60 (0.50 reversible)	475 – 19 000	0.52–24.00	(36) forward and (18) reversible models
0–2200	0.60 (0.50 reversible)	40 – 5 700	1.70–226.00	double and triple gear reduction; forward and reversible models
2200	0.90 (0.65 reversible)	250 – 18 000	1.07–56.00	forward and reversible models

cont...

ARO AIR MOTORS (contd).

Series	Maximum Horsepower	Speed Range rev/min	Torque Range lb.ft	Remarks
2200–44	0.90–0.65* (0.80–0.33 reversible)	30 – 1 200	10.00–355.00	*unbalanced cylinder available on reversible models
33–2200	0.90	160 – 14 000	1.30 – 73.00	(18) forward rotating models with governor control.
33–44	0.90	25–875	21.00–480.00	(21) forward rotating models
44	2.0 (1.85 reversible)	275–14 000	2.70–124.00	(6) forward and (6) reversible models
7700	2.75 (2.0–1.50 reversible)	260–6 000	7.00–135.00	(4) forward and (12) reversible models, the latter available with unbalanced cylinders

Electrical Remote Control Ltd.

The Gnome Mk II air motor is a miniature tube designed to be interchangeable with miniature electrical synchronous motors for timer duties. It has two inputs, each with its own needle valve for speed regulation. A reduction gearbox is supplied as standard. Leading dimensions are: overall width 63 mm, overall height 70 mm; length 51 mm.

Gnome MkII miniature air motor connected to a timer gear train.

Gast Manufacturing Co. Ltd.

Gast produce a range of seven different sizes of vane-type air motors covering horsepowers of 1/20 up to 9. Foot, hub, face and NEMA-C flange mountings are standard for most sizes. Four vanes are normally standard, but 8-vane versions are available where precise low speed control and minimum installed blow-by is required.

Globe Pneumatic Engineering Co. Ltd.

This company manufactures piston type air motors from ¼ to 22 hp. These motors can be supplied with hand or pilot operated control valves, automatic brakes and reduction gearboxes. In addition, a range of Vane type motors are made, these motors being available with gearboxes if required.

Other products manufactured are ½ to 5 ton pneumatic wire rope hoists as standard units and larger hoists to customers specification. 1 ton winches are also available, with larger winches again to customers specification.

Williams & James Ltd

Series 350 air starter motors manufactured by this company have been designed for starting high speed diesels of 10–250 hp.

Survey of Air Line Fittings

Adflow Engineering Ltd.

This company manufactures 'Snap-Tite' quick-disconnect couplings, available in both valves and straight-through types. On the valved units, flow is automatically shut off when the coupling is disconnected, and automatically opened when the coupling is connected.

Sizes (I.D.): 1/8 in, ¼in, 3/8 in, ½ in, ¾ in, 1 in, 1¼ in, 1½ in, 2½ in, 3 in, 4 in, 6 in.

Metals: Steel, brass, aluminium, 303 and 316 stainless, electroless nickel plate.

Finishes: Cadmium plate – chromate conversion treated, electroless nickel, chrome, nickel, anodized, or passivation depending on metal selection.

There are 6 different Snap-Tite models for specific service, pressure, temperature, or fluid applications. There are from 4 to 13 sizes per model, up to 3 different types per model, 9 choices of end connections, over 15 different seal materials, 5 types of metal and 6 finishes.

Seals: Available in all commercial compounds (bunas, butyls, neoprenes, vitons, silicons, teflon, rubber, etc.) to accommodate the most troublesome fluids.

End Fittings: Female npt, male npt, hose shank, AND-10050, MS-33515, MS-33657, MS-33656, and SAE/JIC.

Airtools and Compressors Ltd

The ATC range of equipment includes Aftercoolers and Air Dryers for moisture removal; also Watts Lubro-Control Equipment: filters, lubricators and pressure regulators. PLC quick-release couplers, flexible hoses, and Wade fittings are stocked in large quantities.

Compressed air cooler.

A team of ATC pipe fitters are well equipped to deal with extensive pipework installations up to 76 mm (3 in) bore.

The new ATC air cooler is designed to reduce the problem of damage and malfunctioning of pneumatic equipment caused when moisture condenses out of compressed air.

Designed by Airtools and Compressors Limited, the 'ATC Mini Cooler' meets the need of the small 20 ft^3/min compressor user who hasn't the demand for one of the larger aftercoolers already on the market. It can also be used on the branch line of a large compressed air system where hot air is causing a water problem.

The compressed air is cooled by a flow of air from an electrically driven fan as it passes through a copper tube system which is fitted with numerous aluminium cooling fins. The 18 gauge steel casing supports the whole assembly which, being light in weight can be wall mounted.

Bestobell Mobrey Ltd

This company markets the 'Herion' range of pneumatic components — see under *Survey of Valves*.

Crouzet

The Crouzet LF 3000 range of Instant fittings is designed for use with plastic tubing, providing positive connection by simply pushing the tubing into the fitting. Tubing can be connected or

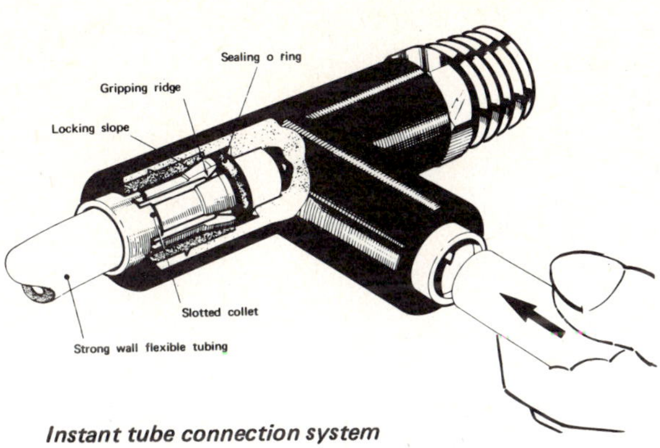

Instant tube connection system

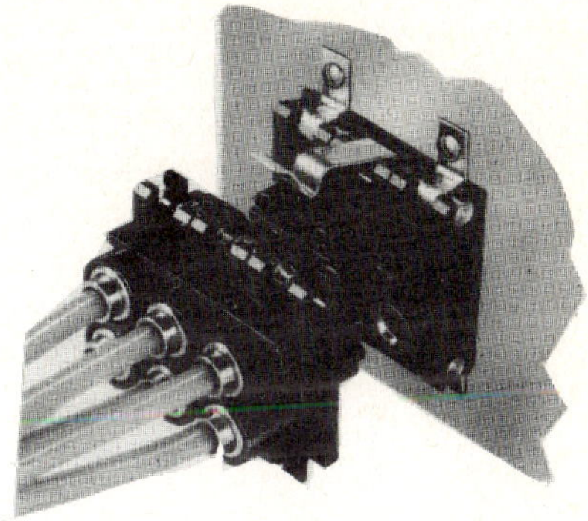

Crouzet modular plug-in connector.

disconnected instantly without any tool. The range of fittings includes male studs, stud elbows, stud branches and stud tees, straight and bulkhead fittings, tees and elbows. Accessories include reducers, plugs and barb connectors.

The Crouzet range also includes modular plug-in connectors for 1/8 in, 5/32 in and 4 mm o.d. tubes.

Domnick Hunter Engineers Ltd.

Designed to overcome problems associated with the removal of acidic mineral oils and synthetic lubricants from compressed air, OIL-X filters from Domnick Hunter Engineers Ltd have been fitted with SAFE-T-PORE, a new type of anti re-entrainment barrier having exceptional chemical resistance. Acidic mineral oils created by oxidation – a natural phenomena – and synthetic lubricants, such as phosphate ester and phosphate diester, are compatible with all materials of construction of OIL-X elements.

Four grades of OIL-X are being manufactured, 5 micron, 1 micron, 0.01 micron and vapour adsorption. Complete line filters having flow rates from 1 ft^3/m to 10 000 ft^3/min for use at pressures up to 17 bar, are available.

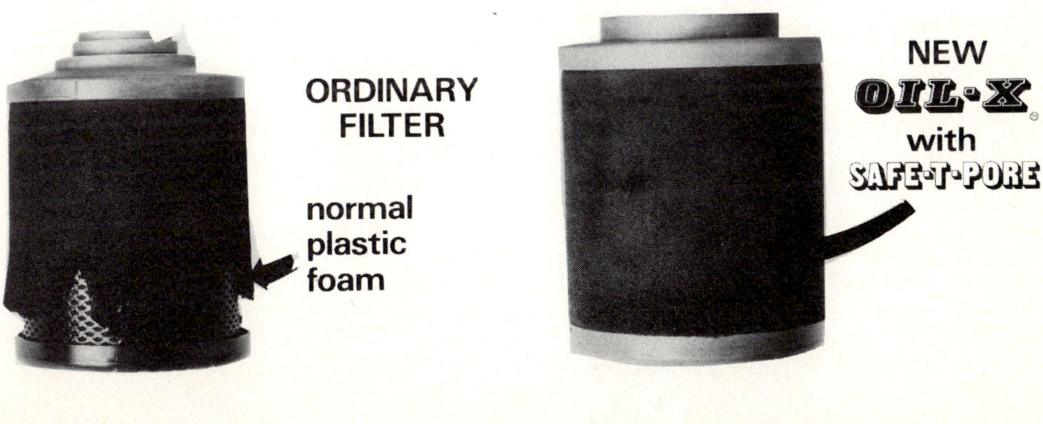

Drallim Tube Couplings Ltd.

Drallim couplings have the unique feature of being smaller and lighter than those of any other manufacturer. Drallim brass tube couplings are of captive seal type, with a resilient rubber seal, and connection can be broken and re-made without damage to the tube. They are designed for use with both metal and piastic tubes and are suitable for pressures up to 240 lb/in^2 (17 bar). For higher pressures, conventional metal compression couplings are available with brass olives in the smaller sizes and annealed copper rings in the larger sizes. All types are available as straights, elbows, tees and crosses, together with stud and bulkhead types. Sizes covered range from 1/8 in to ¾ in BSP.

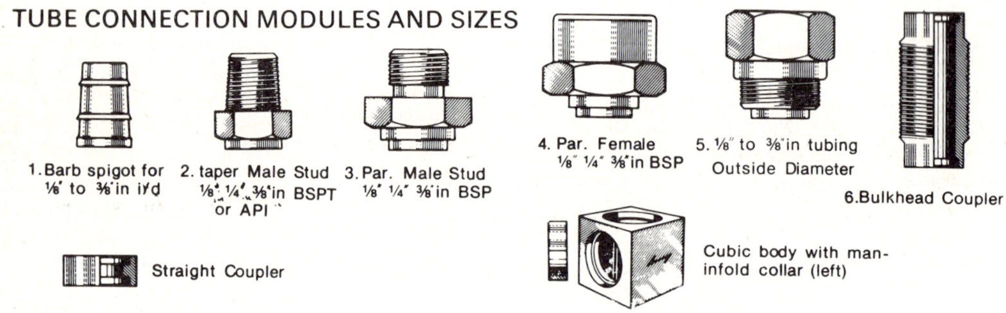

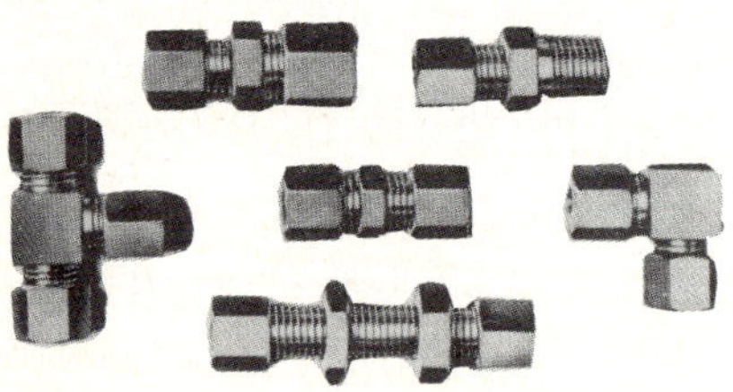

Drallim stainless steel couplings.

Resiliant steel coupling.

Drallim stainless steel tube couplings are available with two types of union nuts, one containing a stainless steel olive and the other a resilient captive seal in Viton. Sizes available are 1/8 in, 3/16 in and ¼ in and the range includes straights, tees, elbows, stud couplings, reducers and bulkhead fittings.

Drallim 'Interface' couplings are built from six types of modular tube connectors in a range of sizes, straight couplers for in-line couplings, and cubic bodies for multiway couplings. Parts are available in kit form or in bulk for assembly by simple soldering technique.

For plastic tubing systems, the Interface barbed spigot provides a positive instant push-on connection without tools. The tubing cannot be removed without cutting.

Dyna-Quip Ltd.

The several ranges of Dyna-Quip Quick Release Couplings include the Series DC with slide action and integral shut-off valve. It can be coupled with one hand and without effort. It provides for safe operation because downstream air is automatically vented to atmosphere before disconnection is possible, eliminating the possibility of flying plugs and hoses causing a serious accident. Series DM and D are 'Push Type' and 'Ball Type' quick couplings respectively. All plugs are to specification MIL-C-4109D.

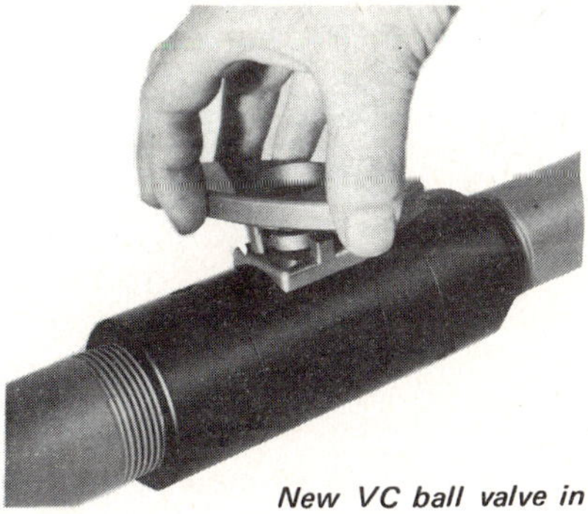

New VC ball valve in carbon steel 1¼in size.

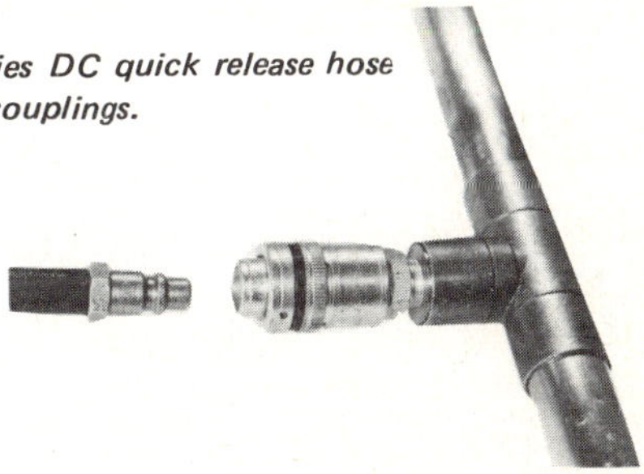

Dyna-Quip series DC quick release hose couplings.

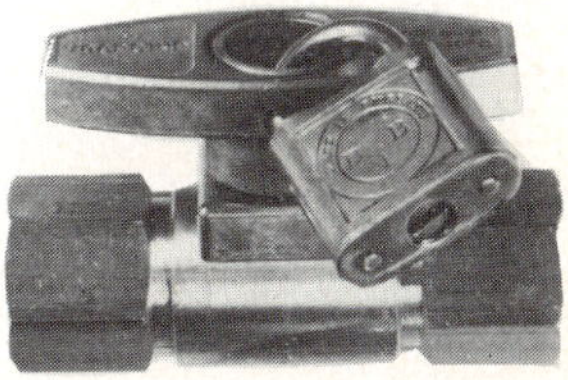

Dyna-Quip ball type shut-off valve with safety lock.

Series DB in brass one-way shut-off couplings for liquids, air and non-corrosive gases. Series DH are double shut-off couplings for hydraulics, liquids and gases — they are in brass or steel for pressures up to 5000 lb/in^2. Seal options are Buna-N, Viton or EPR.

Economatics Ltd

Economatics are distributors for Schrader, Broom-Wade, Nycoil, Gast and Rutter Templair products, and stockists of an extensive range of hoses and tubing.

The company also operates projects and installations divisions, the former offering advice on the design of pneumatic circuitry, whilst the installation division offers a comprehensive service for the supply of compressors and their accessories, together with complete air lines, etc.

Legris (UM) Ltd.

The Legris LF 3000 Push-in Fittings System is designed for nylon or other hard plastic tubes used in pneumatic systems. Components range from fittings, including swivel elbows and tees,

specially designed for easy installation in tight spaces, to multiple plug-in connectors, DIN Rail Terminal Connections and a range of Plug-in Accessories and Multi-Way connectors. Recent additions to the system are multiple banio outlets and stacking banjos.

This equipment is offered in a full range of Imperial sizes from 1/8 in to ½ in o.d. tubing and in a full metric range of 4 mm to 14 mm o.d. tubing.

Legris also supply LF Cartridge assemblies for implantation in valve bodies or sub-bases and also complete sub-bases with LF cartridges in the ports.

Martonair Ltd.

The wide range of air line fittings and ancillary equipment produced by this company includes the following:

Filters — eight models 1/8 in to ½ in BSP, designed to meet the requirements of most applications.

Lubricators — six models, 1/8 in to 1 in BSP, design pressure 0-10 bar.

Drain Valves — 1/8 in BSP (two models) with automatic drain action.

Pressure Regulators — **Miniature** type (0-7 bar), particularly for use with the Martonair Bi-Selector system; **Standard** (0.5—10 bar); and **Precision type** (0—2 bar).

Compression Fittings — for use with copper, Bundy and nylon tubing. Also Instant Couplers and Flat Unions.

Silencers and Exhaust Filters — 1/8 in to 1 in BSP.

Relief Valves — ¼ in BSP, 3—10 bar and 0—10 bar.

Bi-Selector control system specially designed to simplify the design and installation of sequential circuitry, together with a range of ancillary equipment.

C. A. Norgren Ltd

Norgren manufacture a very complete range of air filters, air line lubricators, etc; also precision air sets (filter-regulators with high efficiency oil-removing filters) for use where extremely clean, dry oil-free air is required at precise operating pressures.

The Olympian Plug-in system features units with plain unthreaded outlet ports with O-ring seals which can be plugged in to adaptors installed in an air line. Basic units in this range are a filter, regulator, filter-regulator or oil-fog or Micro-fog lubricator. Abridged details of Olympian and other units are given in the following tables. Combination Units are also available, ie any combination of filters, regulators, filter-regulators and lubricators can be assembled together.

INDUSTRIAL RANGE FILTERS

Model	Filtration	Size(s)	Notes
Olympian Plug-in	50 micron	¼, 3/8, ½	Automatic or manual drain
General Purpose	50 micron	½, ¾, 1	25 micron element optional Automatic or manual drain
	25 micron	1¼, 1½, 2	Automatic drain
Designer – Intermediate	50 micron	¼, 3/8	Automatic or manual drain
Miniature	25 micron	1/8, ¼	Automatic or manual drain

ULTRA-HIGH EFFICIENCY FILTERS

Model	Sizes	Notes
Olympian Plug-in	¼, 3/8	Automatic or manual drain
Puraire	¼, 3/8	Two-stage filtration. Automatic or manual drain
Ultraire	¼, 3/8	Three-stage filtration. Automatic drain

INTEGRAL FILTER-REGULATORS

Model	Filtration	Pressure Range bar	Pressure Range lb/in^2	Sizes	Notes
Olympian Plug-in	50 micron	0.4–10	5–150	¼, 3/8, ½	Automatic or manual drain
Designer Intermediate	50 micron	0.4–10	5–150	¼, 3/8	Automatic or manual drain. 25 micron element optional
Miniature	25 micron	0.16–7	2–100	1/8, ¼	Automatic or manual drain.

OIL-FOG LUBRICATORS

Model	Sizes *	Capacity	Notes
Olympian Plug-in	¼, 3/8,½	0.2 litres 1/3 pint	Fully automatic with built-in flow senser. Constant density type.
Designer Intermediate	¼, 3/8	0.15 litres ¼ pint	Fully automatic with built-in flow senser. Constant density type.
'E' Range	½,¾,1	0.2 litres 1/3 pint	General purpose utility lubricators with thumbscrew setting of oil-feed rate. Can be filled under pressure.

* All dimensions in inches

MICRO-FOG LUBRICATORS

Model	Sizes *	Capacity	Notes
Olympian Plug-in	¼,3/8,½	0.2 litres 1/3 pint	Fully automatic with built-in flow senser. Constant density type.
Designer Intermediate	¼, 3/8	0.15 litres ¼ pint	Fully autromatic with built-in flow senser. Constant density type.
Miniature	1/8,¼	17 cc ½ ounce	Fully automatic with built-in flow senser. For small air tools, etc.
'L' Range	½,¾,1	0.2 litres 1/3 pint	General application lubricator with adjustable vane on bypass.

* All dimensions in inches

Parker-Hannifin

A wide variety of connectors, couplers and clamps are produced by Parker-Hannifin for pneumatic equipment. Latest connectors are the 'Prestolok' brass fittings for nylon tube; the tube only having to be pushed into the fitting. The Parker PL is another brass fitting for plastic tube. Both types are re-usable. PL fittings are available with taper gas threads to ISO R7 and parallel gas threads to ISO R228.

Metru-Lok fittings are another range for copper and plastic tubes,being very compact and quick to assemble. They are re-usable. p to 15 times.

Parker 'Push-lok' is a low cost, low pressure hose with push-on barbed end fittings. No additional clamping is required.

For rubber or plastic tubing which *does* require clamping there is the Parker range of stainless steel worm drive clamps, available in diameter sizes from 6 mm to 178 mm (¼ in to 7 in).

A.Schrader's Son

This company, a division of Scovill Manufacturing Company, manufactures a range of airline filters, lubricators and air pressure regulators to meet all air line requirements.

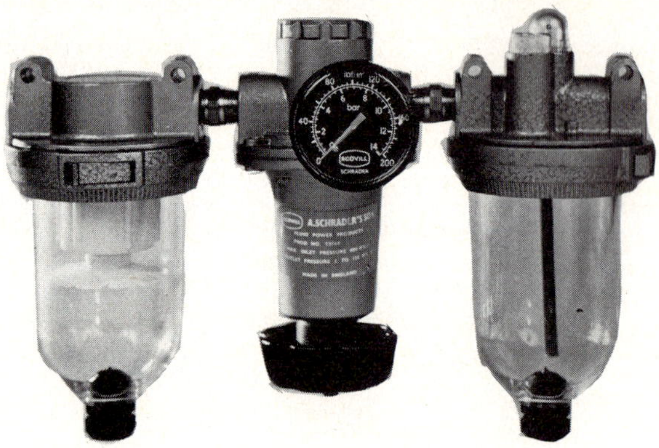

Schrader air line fittings.

Standard 'Prep-air', and high capacity types provide a spread of miniature sizes from 1/8 in to 2 in; of particular note are the coalescing type oil removing filters and the compact design, combined filter-regulator units.

A feature of the Schrader range is the wide spread of airline ancillary equipment, covering such products as blow guns, quick release airline couplers, airlines and airline fittings plus hydraulic gauges, rotating pressure joints and the 'Aquaject' automatic airline drain.

Spirax-Sarco Ltd

Air line fittings and ancillary equipment manufactured by this company include:

A comprehensive range of pneumatic filters with connections from 1/8in to 1½in with a choice of 25 or 5 micron rated filter cartridges, both manual and automatic drain models.

¼ in/3/8in/½in high efficiency coalescing filters capable of removing both sub micron solid and liquid particles with efficiencies in excess of 99.95%, when tested to BS 3928 and the American D.O.P. test. Both manual and automatic drain models are available.

A range of pressure regulators from 1/8in to 1in pipe sizes, both relieving and non-relieving pattern.

A range of combined filter/regulators, pipe sizes 1/8in to 3/8in.

A range of lubricators, pipe sizes, 1/8in to 1 in, with inbuilt venturi valve so that once set the venturi valve will also adjust the oiling rate to compensate for varying oil flow rates.

SCIA Coalescing filter with auto-drain. (Spirax Sarco Ltd).

MPI Combined miniature filter/regulator (Spirax Sarco Ltd.).

Standard FRL, filter, regulator and lubricator. (Spirax Sarco Ltd.).

A comprehensive range of compressed air drain traps for use on compressed air mains, receivers, aftercoolers and intercoolers.

Pipeline strainers in sizes from ½ in to 8 in.

A range of separators from 3/8in to 8 in.

Pressure gauges in various sizes and ranges, including some for panel mounting.

A range of temperature controls to regulate cooling water flow through water cooled compressors, intercoolers and aftercoolers.

Wade Couplings Ltd

Couplings and fittings manufactured by this company fall into the following two categories:

Brass Medium Pressure — Full Imperial and metric ranges of compression couplings using copper or universal brass ferrules for copper tube or various metal and plastic tubes. Also WADE 'N' ferrule for use with nylon tube, and 'P' fitting for polythene tube.

Complete range of pieces with alternative stud threads available as standard — BSP parallel and taper, API, NPT, others on demand.

Reinforced Nylon

"Serkit Systems" fittings in most standard forms, studs, straights, elbows, tees, plus nozzle adaptors and manifold blocks.

One piece compression nut and ferrule for ease of assembly.

Suitable for wide range of applications with various fluids in addition to standard pneumatics.

Watts Regulator (UK) Ltd.

The Watts 'MAXI' range of air line fittings include one size each for lubricators, filters and regulators with a kit of connection inserts to fit air lines of ¼ in, 3/8 in, ½ in or 1 in diameter. Interchangeable inserts make up any combination of units to connect into lines. Units can be removed for servicing and replaced in seconds, using no tools and without breaking the pipe connections.

All units handle air flows up to 250 ft^3/min. Two-stage filter action to 5 micron is available, if required. The Regulator provides precise control from 225 lb/in^2 down to zero, and is practically unaffected by variations in supply pressure or flow rate.

The Watts 'Perma-Fog' lubrication system is fitted with a bypass and is self-adjusting to any pressure condition. It is suitable for continuous or rapid cycling applications.

Survey of Compressors

Airtools and Compressors Limited

Although powered by only a ¾ horsepower motor which will work off a domestic electrical supply, the ATC N3 air compressor will operate a ¼ in air drill drilling through ¼ in plate, power a ½ in impact wrench to remove four wheel nuts in quick succession from a family saloon car and work a spraygun — but not all at the same time!

Designed initially for the small garage which may need its 10.5 bar (150 lb/in^2) pressure for the type inflation of lorry tyres, the N3 compressor will find a ready use in shoe and upholstery workshops for operating pneumatic staplers, in dental surgeries and in buildings using pneumatic door opening and air conditioning control mechanisms.

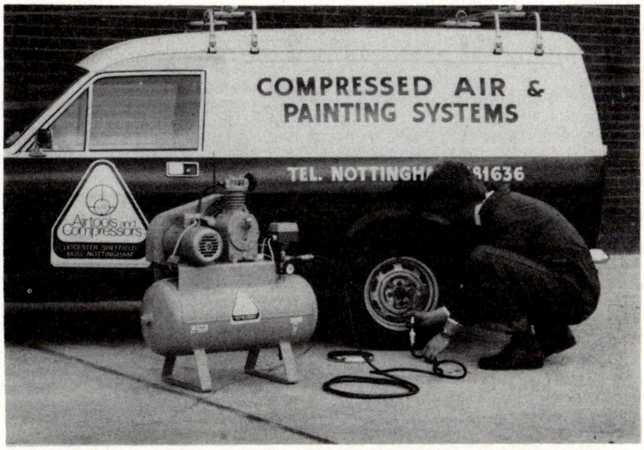

Versatile ATC compressor.

Airtools and Compressors Ltd (ATC) was formed in 1965 to give a comprehensive service to the users of compressed air power and paint finishing.

Compressors – ATC are a major distributor of the Hydrovane oil sealed, rotary air compressor. It is compact, quiet in operation and delivers a continuous smooth flow of cool, clean air at 7 bar (100 lb/in^2). The range covers 2-60 hp.

Reciprocating Compressors are supplied for pressures of 7 bar (100 lb/in^2) and above and for oil free duties. Many can be receiver mounted and supplied for automatic stop/start operation. The range extends up to 600 ft^3/min.

Rotary Screw Compressors together with certain Hydrovanes and reciprocating types can be supplied in sound proofed cases when low noise levels are required.

APE-Bellis Ltd

APE-Bellis Limited (formerly Belliss & Morcom Limited), a subsidiary of Amalgamated Power Engineering Limited produces a wide range of gas and air reciprocating compressors of various cylinder configurations meeting an overall capacity range from 200 to 3000 ft^3/min (5.7–84 Nm^3/min). Air or water-cooled machines to supply lubricated or oil-free air are available, with minimum installation requirements or alternatively as self-contained, free standing totally-engineered package modules in acoustically controlled enclosures.

Commander range

The Commander compressor is a medium-duty compressed air package, designed for instant installation, being air-cooled and requiring no special foundation or complicated electrical supplies.

Three (90 degree) V-twin and two V-four cylinder models are available in capacities of 200, 300 and 420 ft^3/min (5.7, 8.5 and 11.9 Nm^3/min) in V-twin form and 400 and 600 ft^3/min (11.3 and 17.0 Nm^3/min) in V-four form. All models feature a large diameter, duplex piston design for dynamic balance.

V- and W-type Compressors

V-twin cylinder machines are available in single stage from (VL) for pressures up to 4.0 barg (58lb/in^2) and capacities up to 56Nm^3/min (2 000 ft^3/min). Two-stage compressors (VH) are suitable for pressures up to 10.5 barg (150 lb/in^2) and capacities up to 28 Nm^3/min (1000 ft^3/min). Boosting duties up to 41.5 barg (600 lb/in^2) are also catered for.

The APE-Belliss W-type three cylinder machines are available with three first-stage cylinders (WL) for pressures up to 4.0 barg (58 lb/in^2) and capacities up to 84 Nm^3/min (3 000 ft^3/min). The series also includes two first-stage and one central, second-stage cylinder (WH) for pressures up to 10.5 barg (150 lb/in^2) and capacities up to 56 Nm^3/min (2000 ft^3/min). A three-stage system (WH3) caters for pressures up to 41.5 barg (600 lb/in^2).

V-type compressor (APE-Bellis)

Packaged Compressors

The new PBM 30/42/60 series of medium duty packaged air compressors provides for outputs of 300, 420 and 600 ft^3/min (8.5, 11.9 & 17.0 Nm^3/min) respectively. Each unit is supplied completely housed in a ventilated, acoustically-lined enclosure to give quiet operation and lowest possible operating temperatures. The packages are complete, requiring no special foundations or complex electrical supplies.

The same principles of 'packaging', have been applied to their larger VH series of heavy duty V-twin, single and two-stage compressors for capacities from 21 to 28 m^3/min (750 – 1000 ft^3/min) at pressures up to 10.5 barg (150 lb/in^2) and for lubricated or oil-free air delivery.

H25 Series Compressor

This range of horizontal, water-cooled, double acting crosshead compressors is suitable for duty with a wide range of gases and air. They are available in lubricated or oil-free form. In two-stage form the output range is from 2 500 to 5 000 ft^3/min at pressures up to 150 lb/in^2.
In single stage form the range is up to 1 000 ft^3/min at pressures up to 35 lb/in^2.

Dryaire – Air & Gas Dryers

These dryers use the refrigeration method which has proved cheaper to run than those which use electrical heaters to evaporate the absorbed moisture. The unit is fully automatic.

Atlas Copco AB

The following Tables summarize the range of Atlas Copco compressors.

PORTABLE OIL-FLOODED SCREW COMPRESSORS TYPE XA

Compressor	Engine or motor	Actual FAD at norm. oper. pr.		Norm. oper. eff. pressure		Max. oper. eff. pressure		Vol. of compr. lubr. system	
		l/s	ft^3/min	bar	lb/in^2g	bar	lb/in^2g	l	gal(US)
XA60Dd	F2L912	58	123	7	102	8.0	116	25	6.6
XA60Pd	D3.1521	62	130	7	102	8.5	123	25	6.6
XAS60Dd	F2L912	58	123	7	102	8.0	116	25	6.6
XAS60Pd	D3.1521	62	130	7	102	8.5	123	25	6.6
XASS60Dd	F2L912	58	123	7	102	8.0	116	25	6.6
XASS60Pd	D3.1521	62	130	7	102	8.5	123	25	6.6
XA80Dd	F3L912	80	170	7	102	8.5	123	25	6.6
XA80Pd	4.2361	80	170	7	102	8.5	123	25	6.6
XAS80Dd	F3L912	80	170	7	102	8.5	123	25	6.6
XAS80Pd	4.2361	80	170	7	102	8.5	123	25	6.6
XASS80Dd	F3L912	80	170	7	102	8.5	123	25	6.6
XASS80Pd	4.2361	80	170	7	102	8.5	123	25	6.6
XA120Dd	F4L912	118	250	7	102	8.5	123	46	12.2
XA120Pd	4.2481	118	250	7	102	8.5	123	46	12.2
XAS120Dd	F4L912	118	250	7	102	8.5	123	46	12.2
XAS120Pd	4.2481	118	250	7	102	8.5	123	46	12.2
XASS120Dd	F4L912	118	250	7	102	8.5	123	46	12.2
XASS120Pd	4.2481	118	250	7	102	8.5	123	46	12.2
XA160Dd	F6L912	158	335	7	102	8.5	123	46	12.2
XA160Pd	4.3541	158	335	7	102	8.5	123	46	12.2
XAS160Dd	F6L912	158	335	7	102	8.5	123	46	12.2
XAS160Pd	4.3541	158	335	7	102	8.5	123	46	12.2
XASS160Dd	F6L912	158	335	7	102	8.5	123	46	12.2
XASS160Pd	4.3541	158	335	7	102	8.5	123	46	12.2
XAS50E	Siemens 1RA6167	52	110	7	102	8.0	116	25	6.6
XAS60E	Siemens 1RA6183	67.5	143	7	102	8.0	116	25	6.6
XAS80E	Siemens 1RA6186	81	171	7	102	8.0	116	25	6.6
XAS100E	Siemens 1RA6203	99	210	7	102	8.0	116	43	11.3
XAS120E	Siemens 1RA6206	120	254	7	102	8.0	116	43	11.3

PORTABLE COMPRESSORS (Summary of Types)

Maximum effective pressure		Nominal capacity range		Compressor type	Model design
bar	lb/in^{2}g	l/s	ft^3/min		
7.0	100	75	159	reciprocating	VT*
		425	900	dry screw	PT
8.0	115	40 – 42	85 – 88	reciprocating	UT
		52 – 120	110 – 254	oil injected screw	XA
		188 – 312	398 – 661	oil injected screw	PR
8.5	125	62 – 158	130 – 335	oil injected screw	XA
8.8	130	34 – 42	72 – 00	reciprocating	UT
		53	113	reciprocating	VT*
		58 – 155	124 – 328	reciprocating	ST
		200 – 330	424 – 700	oil injected screw	PR
		425 – 708	900 – 1500	dry screw	PT
10.5	150	193 – 321	409 – 680	oil injected screw	PRH
		413 – 553	825 – 1172	dry screw	PTH
20.0	290	567	1200	dry screw	PN

PORTABLE TWO-STAGE PISTON COMPRESSORS

Compressor	Engine or motor	Actual FAD at normal operating pressure		Normal operating effective pressure		Max. operating effective pressure		Volume of compr. lubr. system		Max. loaded speed
		l/s	ft^3/min	bar	lb/in^{2}g	bar	lb/in^{2}g	l	gal(US)	rev/min
UT85Dd	F2L912	41.7	88.3	7	102	8.0	116	3.5	0.92	1800
US85Dd	F2L912	40.0	84.7	7	102	8.0	116	3.5	0.92	1800
UT75DdA	F2L912	35.3	74.8	7	102	8.8	128	3.5	0.92	1800
UT75E	Siemens 1LA3166-4	34.1	72.2	7	102	8.8	128	3.5	0.92	1480
UT85Fd	2504E	41.7	88.3	7	102	8.8	128	3.5	0.92	1800
US85Fd	2504E	40.0	84.7	7	102	8.8	128	3.5	0.92	1800
VT115H/F	–	53.3	113	7	102	8.8	128	6.5	1.72	1600
VT160H/F	–	75.0	159	7	102	7.0	102	6.5	1.72	1600
ST35Dd	F3L912	58.3	124	7	102	8.8	128	5	1.32	1750
STS35Dd	F3L912	58.3	124	7	102	8.8	128	5	1.32	1750
ST48Dd	F4L912	75.0	159	7	102	8.8	128	5	1.32	1750
STS48Dd	F4L912	75.0	159	7	102	8.8	128	5	1.32	1750
ST71Dd	F6L912	117	248	7	102	8.8	128	10	2.64	1750
STS71Dd	F6L912	117	248	7	102	8.8	128	10	2.64	1750
ST95Dd	BF6L913	155	328	7	102	8.8	128	10	2.64	1750
STS95Dd	BF6L913	155	328	7	102	8.8	128	10	2.64	1750

PORTABLE TWO-STAGE SCREW COMPRESSORS

Compressor	Engine or motor	Actual FAD at normal operating pressure		Normal operating effective pressure		Max. operating effective pressure		Volume of compr. lubr. system		Max. loaded speed
		l/s	ft^3/min	bar	lb/in^{2}g	bar	lb/in^{2}g	l	gal(US)	rev/min
PR365E	Asea MCK250S	188	398	7.0.	102	8.0	116	60	15.9	2925 (50Hz)
PR425Dd	F6L413	200	424	7.0	102	8.8	128	60	15.9	2100
PR425DdA	F8L413	200	424	7.0	102	8.8	128	62	16.4	2100
PR425Vod	D100B	200	424	7.0	102	8.8	128	60	15.9	2100
PRH425Dd	F8L413	193	409	10.5	152	10.5	152	62	16.4	2100
PRSS425Dd	F6L413	200	424	7.0	102	8.8	128	90	23.8	2100
PR600E	Asea MCK280M	312	661	7.0	102	8.0	116	90	23.8	2925 (50Hz)
PR700Dd	F10L413	330	699	7.0	102	8.8	128	90	23.8	2100
PR700DdA	F12L413	330	699	7.0	102	8.8	128	100	26.4	2100
PR700Vod	TD100A	330	699	7.0	102	8.8	128	90	23.8	2100
PR700Cud	N855C250	330	699	7.0	102	8.8	128	90	23.8	2100
PRH700Dd	F12L413	321	680	10.5	152	10.5	152	100	26.4	2100
PRH700Gd	8V71N60	321	680	10.5	152	10.5	152	95	25.1	2100
PRSS700Dd	F10L413	330	699	7.0	102	8.8	128	92	24.3	2100
PRSS700Vod	TD100A	330	699	7.0	102	8.8	128	92	24.3	2100
PRHSS700Dd	F12L413	321	680	10.5	152	10.5	152	–	–	2100
PT900Gd	8V71N65	425	900	7.0	102	8.8	128	65	17.2	2100
PT900Cud	NT855C280	425	900	7.0	102	8.8	128	65	17.2	2100
PT1050Gd	8V71N65	500	1059	7.0	102	8.8	128	65	17.2	2100
PT1200Gd	12V71N55	567	1201	7.0	102	8.8	128	65	17.2	2100
PT1200Cud	NTA855C360	567	1201	7.0	102	8.8	128	65	17.2	2100
PT1200E	Asea MCK355S	567	1201	7.0	102	8.8	128	65	17.2	1470
PTS900Dd	F10L413F	425	900	7.0	102	8.8	128	65	17.2	2500
PTS900Gd	6V-92	425	900	7.0	102	8.8	128	65	17.2	2100
PTHS900Dd	BFIOL413	413	825	10.5	152	10.5	152	65	17.2	2100
PTHS900Gd	8V92N75	413	875	10.5	152	10.5	152	65	17.2	2100
PTS1200Dd	BF10L413	567	1200	7.0	102	8.8	128	65	17.2	2500
PTS1200Gd	8V92N75	567	1200	7.0	102	8.8	128	65	17.2	2100
PTHS1200Dd	BF12L413	553	1172	10.5	152	10.5	152	65	17.2	2500
PTHS1200Gd	12V71N60	553	1172	10.5	152	10.5	152	65	17.2	2100
PNS1200Gd	16V-71N60	567	1200	20.0	290	20.0	290	65	17.2	2100
PTS1500Dd	BF12L413	708	1500	7.0	102	8.8	128	65	17.2	2500
PTS1500Gd	12V71N60	708	1500	7.0	102	8.8	128	65	17.2	2100

PNS COMPRESSORS

	Engine	Actual FAD at normal oper. pressure		Normal oper. eff. pressure		Max. oper. effective pressure		Volume of compr. lub system		Max. loaded speed	Total mass without tools		Overall length		Overall width		Overall height	
		l/s	ft^3/min	bar	lb/in^{2}g.	bar	lb/in^{2}g.	l	gal(US)	rev/min	kg	lb	mm	in	mm	in	mm	in
PNS1200Gd	16V-71N60	567	1200	20	29C	20	290	65	17.2	2100	8360	18430	5550	219	2100	82.7	2400	94.5

	Maximum effective pressure		Nominal capacity range		Coolant		Stages	Model designation
	bar	lb/in²g.	l/s	ft³/min	air	water		
Reciprocating, lubricated	2.5	35	150	318	●	–	1	IE
			980	2076	–	●	1	EA
	3.0	45	1 – 2	3 – 4	●	–	1	FE
			99 – 136	210 – 189	●	–	1	IE
	3.5	50	63 – 83	134 – 176	●	–	1	IE
			63 – 247	134 – 525	●	–	1	BF
			363	769	–	●	1	DA
			728 – 1243	1542 – 2633	–	●	1	EA
	4.5	65	592	1254	–	●	1	EA
	7.0	100	5 – 17	10 – 36	●	–	1	KE
	8.8	130	34 – 133	72 – 282	●	–	2	BT
			182 – 267	386 – 566	●	–	2	DT
			187 – 270	396 – 572	–	●	2	DR
			403 – 662	854 – 1403	●	–	2	ET
			415 – 1640	883 – 3492	–	●	2	ER
	10.0	145	0.5 – 2	1 – 4	●	–	1	FE
	14.0	200	2 – 13	5 – 28	●	–	2	KT
			25 – 118	52 – 250	●	–	2	BT
	18.0	260	165	350	–	●	2	DR
			250 – 432	534 – 917	–	●	2	ER
	21.0	300	322	684	–	●	2	ER
	30.0	435	2 – 7	5 – 14	●	–	2	KT
	300	4350	6 – 10	12 – 21	●	–	4	BP
Reciprocating, non-lubricated	7.0	100	38	81	–	●	1	AA-KT
	8.8	130	143	303	–	●	2	AR-KT
	16.0	230	93 – 170	196 – 350	–	●	2	VK
Dry screw	2.5	35	9457	20030	–	●	1	UA
	3.0	45	7556	16010	–	●	1	UA
	3.5	50	173 – 1669	367 – 3540	–	●	1	ZA
			834 – 5137	1765 – 10900	–	●	1	UA
	8.8	130	2635 – 7589	5580 – 16100	–	●	2	UR
	10.5	150	164 – 1663	348 – 3530	–	●	2	ZR

BT COMPRESSORS

Type	Max. work press.		Normal work pres.		Piston displacement		FAD at normal work press.		Power consumption		Speed
	bar	lb/in^{2}g.	bar	lb/in^{2}g.	l/s	ft^3/min	l/s	ft^3/min	kW	hp(US)	rev/min
Normal press.											
BT2	8.8	128	7	102	45.9	97.2	34.0	72.0	14.2	19.0	1460
BT3	8.8	128	7	102	72.1	153.0	50.0	106.0	18.6	24.9	1460
BT4	8.8	128	7	102	92.5	196.0	66.7	141.0	25.2	33.8	1460
BT5	8.8	128	7	102	144.0	305.0	100.0	212.0	35.5	47.6	1460
BT6	8.8	128	7	102	185.0	392.0	133.0	282.0	49.2	66.0	1460
High press.											
BT114	14	203	12	174	36.7	77.8	24.7	52.3	11.9	16.0	1460
BT314	14	203	12	174	72.1	153.0	42.8	90.7	20.6	27.6	1460
BT414	14	203	12	174	92.5	196.0	58.3	123.0	28.0	37.6	1460
BT614	14	203	12	174	185.0	392.0	118.0	250.0	55.5	74.4	1460

B-PACK UNITS

Type	Max. work press.		Normal work pres.		Mean sound pressure level	FAD at normal work press.		Power consumption		Speed
	bar	lb/in^2/g.	bar	lb/in^2/g.	dB(A)*	l/s	ft^3/min	kW	hp(US)	rev/min.
B Pack 2E	8.8	128	7	102	75.0	33	71	14.2	19.0	1460
B Pack 3E	8.8	128	7	102	77.0	48	102	18.5	23.8	1460
B Pack 4E	8.8	128	7	102	78.0	64	136	24.9	33.4	1460
B Pack 5E	8.8	128	7	102	77.5	98	207	35.2	47.1	1460
B Pack 6E	8.8	128	7	102	77.5	130	275	48.6	65.1	1460
B Pack 114E	14.0	203	12	174	75.0	24	50	12.0	16.0	1460
B Pack 314E	14.0	203	12	174	77.0	41	88	20.4	27.2	1460
B Pack 414E	14.0	203	12	174	78.0	56	119	27.6	37.0	1460
B Pack 614E	14.0	203	12	174	78.5	115	245	54.8	73.2	1460

**According to CAGI-PNEUROP test code (measured at 1m distance).*

BE/IE COMPRESSORS

Type	Max. work press.		Normal work pres.		Piston Displacement		FAD at normal work press.		Power consumption		Speed
	bar	$lb/in^2/g.$	bar	$lb/in^2/g.$	l/s	ft^3/min	l/s	ft^3/min	kW	hp(US)	rev/min
BE 23	3.5	51	3.0	43	92	194	62.5	134	16.5	22.1	1460
BE 33	3.5	51	3.0	43	144	306	93.3	203	22.5	30.2	1460
BE 43	3.5	51	3.0	43	185	392	122.0	269	29.6	39.7	1460
BE 63	3.5	51	3.0	43	370	784	246.7	531	57.9	77.7	1460
IE 23	3.5	51	3.0	43	91	193	62.5	134	16.7	22.4	1450
IE 23	3.5	51	3.0	43	113	240	75.8	160	21.0	28.2	1800
IE 23	3.5	51	3.0	43	126	266	83.3	176	23.7	31.8	2000
IE 43	3.0	43	2.0	30	139	295	99.2	210	19.4	26.0	1100
IE 43	3.0	43	2.0	30	190	403	135.8	285	28.5	38.3	1500
IE 43	2.5	36	2.0	30	222	470	150.0	316	34.2	45.9	1750

DR/DA COMPRESSORS

Type	Max. work press.		Normal work press.		Piston displacement		FAD at normal work press.		Power consumption		Speed
	bar	$lb/in^2/g.$	bar	$lb/in^2/g.$	l/s	ft^3/min	l/s	ft^3/min	kW	hp(US)	rev/min
DR 2	8.8	128	7	102	244	517	187	396	58.4	78.3	970
DR 4	8.8	128	7	102	357	756	270	572	83.7	112.0	970
DR 418	18.0	203	16	232	244	517	165	350	73.1	98.0	970
DA 23	3.5	51	3	43	488	1034	363	769	82.9	111.0	970

DR-PACK UNITS

Type	Max. work press.		Normal work press.		Mean sound pressure level dB(A)*	FAD at normal work press.		Power consumption		Speed
	bar	$lb/in^2/g.$	bar	$lb/in^2/g.$		l/s	ft^3/min	kW	hp(US)	rev/min
DR2 Pack E	8.8	128	7	102	74	185	392	59.1	79.3	970
DR4 Pack E	8.8	128	7	102	74	268	568	84.4	113.0	970
DR418 Pack E	18.0	261	16	232	73	163	345	73.8	99.0	970
DA23 Pack E	3.5	51	3	43	73	360	763	83.5	112.0	970

**According to CAGI-PNEUROP test code (measured at 1m distance).*

ER-COMPRESSORS

Type	Max. eff. working press.		Piston displacement			FAD at 7 bar (102 lb/in^2/g.			Recommended motor size for max. eff. pressure				Speed	Cooling water consumption (serial system 15°C (60°F) inlet temp)		
									direct coupling		V-belt drive					
	bar	lb/in^2/g.	l/s	m^3/min	ft^3/min	l/s	m^3/min	ft^3/min	kW	hp(US)	kW	hp(US)	rev/min	l/s	m^3/h	gal/US)/h
ER5	8.8	128	487	29.2	1032	415	24.9	883	140	190	147	200	730	0.445	1.6	423
ER6	8.8	128	622	37.3	1318	510	30.6	1087	169	230	176	240	490	0.639	2.3	608
ER7	8.8	128	800	48.0	1695	672	40.3	1430	220	300	235	320	590	0.695	2.5	661
ER8	8.8	128	1228	73.7	2602	1040	62.4	2225	331	450	–	–	500 (50 Hz)	1.111	4.0	1057
	8.8	128	1263	75.8	2676	1070	64.2	2288	342	465	–	–	514 (60 Hz)	1.111	4.0	1057
ER9	8.8	128	1850	111.0	3919	1563	93.8	3331	478	650	–	–	429 (50 Hz)	1.667	6.0	1585
	8.8	128	1933	116.0	4095	1640	98.4	3492	507	690	–	–	450 (60 Hz)	1.667	6.0	1585
ER518	18.0	261	363	21.8	769	250*	15.0*	534*	121	165	128	175	730	0.417	1.5	397
ER618	21.0	305	448	26.9	949	322*	19.3*	684*	147	200	154	210	490	0.639	2.3	608
ER718	18.0	261	622	37.3	1318	432*	25.9*	917*	220	300	235	320	590	0.695	2.5	661

at 18 bar (261 lb/in^2/gauge)

EA-COMPRESSORS

Type	Max. eff. working press.		Piston displacement			FAD at max. eff. working pressure			Recommended motor size for max. eff. pressure				Speed	Cooling water consumption (15°C (60°F) in temp.).		
									direct coupling		V-belt drive					
	bar	lb/in^2/g.	l/s	m^3/min	ft^3/min	l/s	m^3/min	ft^3/min	kW	hp(US)	kW	hp(US)	rev/min	l/s	m^3/h	gal/US)/h
EA53	3.5	51	728	43.7	1542	513	30.8	1087	125	175	136	185	730	0.167	0.6	159
EA62	2.5	36	1243	74.6	2633	980	58.8	2076	176	240	185	250	490	0.222	0.8	211
EA64	4.5	65	887	53.2	1879	592	35.5	1254	158	215	165	225	490	0.194	0.7	185
EA73	3.5	51	1243	74.6	2633	915	54.9	1939	220	300	235	320	590	0.278	1.0	264

DT COMPRESSORS

Type	Max. work press.		Normal work press.		Piston displacement		FAD at normal work press.		Power consumption		Speed rev/min
	bar	lb/in²/g.	bar	lb/in²/g.	l/s	ft³/min	l/s	ft³/min	kW	hp(US)	
DT 2	8.8	128	7	102	244	517	182	386	59.3	79.5	970
DT 4	8.8	128	7	102	357	756	267	566	84.3	114.0	970

DT-PACK UNITS

Type	Max. work press.		Normal work press.		Mean sound pressure level	FAD at normal work press.		Power consumption		Speed
	bar	lb/in²/g.	bar	lb/in²/g.	dB(A)*	l/s	ft³/min	kW	hp(US)	rev/min
DT2 Pack E	8.8	128	7	102	75	180	381	60.0	80.5	970
DT4 Pack E	8.8	128	7	102	75	263	558	85.5	115.0	970

**According to CAGI-PNEUROP test code (measured at 1m distance)*

ET-COMPRESSORS

Type	Max. eff. working press.		Piston displacement			FAD at 7 bar (102 lb/in²/g)			Recommended motor size for max. eff. pressure				Speed
									Direct coupling		V-belt drive		
	bar	lb/in²/g.	l/s	m³/min	ft³/min	l/s	m³/min	ft³/min	kW	hp(US)	kW	hp(US)	rev/min
ET5	8.8	128	487	29.2	1032	403	24.2	854	140	190	147	200	730
ET6	8.8	128	622	37.3	1318	503	30.2	1066	169	230	176	240	490
ET7	8.8	128	800	48.0	1695	662	39.7	1377	220	300	235	320	590

DATA – ZR COMPRESSORS, TWO-STAGE, MAX. EFFECTIVE WORKING PRESSURE: 10.5 bar (155 lb/in²/g.)

Type	Step-up gear	Electric motor speed		FAD (1) at 7 bar*		Power (1) consumption at 7 bar*	Cooling water (2) consumption		Net mass (approx) (4)
		rev/min	Hz	l/s	ft³/min	kW	l/s	gal(US)/h	kg
ZR3	A	2950	50	303	642	113	0.92	870	2300
	B	2950	50	250	530	94	0.75	710	2200
	B	3550	60	309	655	116	0.95	900	2200
	C	1470	50	158	335	63	0.55	520	2200
	C	1760	60	203	430	76	0.61	580	2200
ZR4	A	1470	50	583	1235	206	1.67	1590	3600
	B	1470	50	478	1012	169	1.36	1290	3600
	B	1760	60	592	1254	209	1.67	1590	3600
	C	1470	50	380	805	136	1.05	1000	3600
	C	1760	60	477	1010	168	1.36	1290	3600
ZR5	A	1470	50	943	1998	320	2.52	2400	5500
	A	1170	60	727	1540	244	1.91	1810	5500
	B	1470	50	763	1617	256	2.03	1920	5500
	B	1760	60	935	1981	317	2.49	2360	5500
ZR6	A	1470	50	1663	3523	559	4.37	4150	7500
	A	1170	60	1282	2716	430	3.55	3370	7500
	B	1470	50	1327	2811	443	3.75	3570	7200
	B	1760	60	1630	3453	540	4.37	4150	7500

**effective working pressure*

Charles Austen Pumps Ltd

The company's products include a range of 'oil free' diaphragm pumps for flow rates from 0 to 70 l/min, pressure from 0 to 4 kg/cm^2, or vacuum of 0 to 750 mm Hg. These pumps are used extensively to sample air and gas for process control instruments. Diaphragm liquid pumps capable of 50 to 800 cc/min and centrifugal liquid pumps covering flow rates up to 1600 litres per hour are also available.

"OIL FREE" DIAPHRAGM PUMPS, FOR AIR AND GAS, PRESSURE OR VACUUM

TYPE	Height mm	Width mm	Length mm	Weight kg.	Flow L/Min	Vacuum mm Hg	Pressure Kg/cm^2	Motor Standard Pump	Motor Consumption Standard Pumps in Amps
LVM 8	83	53	68	0.18	1.2	150	0.3	A	0.85 (6V)
LVM 9	83	53	51	0.2	0.2	200	0.3	E	0.017
DYMAX Mk. 1	108	82	168	1.6	4.0	300	0.75	A	2.20 (12V)
DYMAX Mk. 2	127	82	168	2.0	3.0	300	0.65	B	0.18
DYMAX Mk. 2A	127	82	168	2.0	1.5	150	0.30	B	0.18
CAPEX Mk. 1	143	86	149	1.6	15.0	400	1.00	A	2.20 (12V)
CAPEX Mk. 2	142	92	139	2.0	6.0	400	0.80	B	0.25
CAPEX Mk.2C	160	180	230	3.7	6.0	400	0.80	B	0.25
M 361	157	111	168	3.6	16.0	530	1.5	C	0.20
M 361 C	180	210	280	6.4	16.0	530	1.5	C	0.20
M 371	157	108	188	3.8	16.0	530	1.5	A	4.5 (12V)
F 65 D.E.	192	120	280	6.8	30.0	*650	2.00	D	1.00
F 65 D.E.P.	240	120	280	7.0	30.0	*650	2.00	D	1.00
F 65 D.E.R.	192	120	280	7.0	20.0	--	4.00	D	1.00
DUPLEX 2F	127	239	232	8.0	40.0	600	2.00	D	1.00
DUPLEX 4N	156	235	362	13.1	70.0	*740	2.00	C	1.60
DUPLEX 4NP	156	235	430	13.3	70.0	*740	2.00	C	1.60

* The vacuum figure stated is with the pump heads connected in series

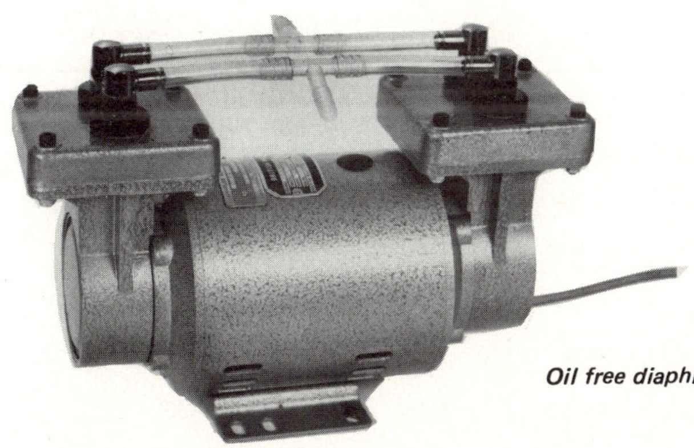

Oil free diaphragm pump

Bristol Pneumatic Ltd

Details of all models produced are summarized in the following Tables.

LIGHT INDUSTRIAL RANGE

Model	Displacement		FAD		Max. Pressure		Motor	Air REceiver		
	ft^3/min	m^3/min	ft^3/min	m^3/min	lb/in^2	bar	kW	ft^3	gals	ltrs
307/HL6	6	0.167	4.25	0.12	150	10.5	1.1	4.0	25.2	115
307/J10	10	0.283	7.3	0.20	200	14.0	2.2	4.0	25.2	115
307/B15	15	0.425	12.5	0.35	200	14.0	3.0	7.9	49.2	222
307/B20	20	0.566	16.0	0.45	200	14.0	4.0	7.9	49.2	222
307/D30	30	0.850	23.6	0.67	200	14.0	5.5	7.9	49.2	222
307/D40	40	1.130	30.5	0.86	200	14.0	7.5	7.9	49.2	222
307/ATS/60	57.3	1.620	47.0	1.33	200	14.0	11.0	20.0	125.0	567
309/HL6	6	0.167	4.25	0.12	150	10.5	1.1	4.0	25.2	115
309/J10	10	0.283	7.3	0.20	200	14.0	2.2	4.0	25.2	115
309/B15	15	0.425	12.5	0.35	200	14.0	3.0	7.9	49.2	222
309/B20	20	0.566	16.0	0.45	200	14.0	4.0	7.9	49.2	222

HEAVY DUTY INDUSTRIAL MODELS

Model	Displacement ft^3/min	Displacement m^3/min	FAD ft^3/min	FAD m^3/min	Maximum Pressure lb/in^2	Maximum Pressure bar	Motor kW
308/ATS/60	57	1.61	48	1.35	100	7	11.0
308/ATS/70	67	1.89	57	1.61	100	7	11.0
308/ATS/115	115	3.25	98	2.77	100	7	18.5
308/ATS/135	135	3.82	115	3.25	100	7	22.0
308/ATS/182	182	5.15	155	4.38	100	7	30.0
308/ATS/240	246	6.97	200	5.66	100	7	45.0
308/ATS/325	323	9.15	275	7.78	100	7	55.0

PORTABLE DIESEL-DRIVEN MODELS

Model	Displacement ft³/min	Displacement m³/min	FAD ft³/min	FAD m³/min	Max. Pressure lb/in²	Max. Pressure bar	Speed rev/min	Engine bhp	Diesel engine
70P2	85	2.40	70	1.98	100	7	1500	19.5	Perkins 4 cylinder
144F2	171	4.84	144	4.08	100	7	1500	40.0	Ford 4 cylinder
170P2	200	5.66	170	4.82	100	7	1750	52.0	Perkins 4 cylinder
250P2	313	8.86	250	7.08	100	7	1750	75.0	Perkins 6 cylinder
TD140	162	4.58	140	3.96	100	7	1400	35.0	Ford Tractor

Bristol pneumatic type BPT 2026 compressor

Bristol pneumatic type BP 2156 compressors

HIGH PRESSURE COMPRESSORS

Model	Displacement		FAD		Max. Pressure		Motor
	ft^3/min	m^3/min	ft^3/min	m^3/min	lb/in^2	bar	hp
308/KL6	6	0.170	4.25	0.120	450	31.5	2.0
308/A11H	11	0.311	7.5	0.212	400	28	4.0
308/C18H	18	0.510	12.0	0.340	300	21	5.0
308/BM3S-16	16	0.453	12.0	0.340	500	35	7.5
308/BM3S-20	20	0.566	15.5	0.439	500	35	7.5
308/BM3S-24	24	0.679	20.0	0.566	500	35	10.0
308/BM4S-16	16	0.453	12.0	0.340	2000	140	7.5
BM3S-4	4	0.114	2.0	0.057	4000	281	2.5
BM4S-7J	7	0.198	5.0	0.141	4000	281	5.0
BM4S-7K	7	0.198	5.0	0.141	6000	422	6.0
BM4S-16HP	16	0.453	11.0	0.312	4000	281	10.0
BM4S-24	23	0.651	18.0	0.509	4000	281	15.0

Compair Construction and Mining Ltd

Leading details of Holman screw and Broomwade rotary vane portable air compressors are summarized in these Tables.

BROOMWADE ROTARY VANE

Model	Output (FAD)		Pressure		Engine	rev/min
	ft^3/min	litres/sec	lbf/in^2	bar		
CA1	85	40.0	100	7	Perkins	2000
CA1S Silent	85	40.0	100	7	Perkins	2000
CA1SS Super silent	85	40.0	100	7	Perkins	2000
CV125	125	59.0	100	7	Perkins, Ford, Fiat	1900
CV125S Silent	125	59.0	100	7	Perkins, Ford, Fiat	1900
CV125SS Super Silent	125	59.0	100	7	Perkins, Ford	1900
CV160	160	75.0	100	7	Perkins, Ford, Fiat	1900
CV160S Silent	160	75.0	100	7	Perkins, Ford, Fiat	1900
CV160SS Super Silent	160	75.0	100	7	Perkins, Ford	1900
CV250	250	117.5	100	7	Perkins, Ford, Fiat	2210
CV250S Silent	250	117.5	100	7	Perkins, Ford	2210
CV400	400	188	100	7	Perkins, Cummins	2350
WR600	600	282	100	7	Rolls Royce, Cummins	1800

Broomwade CV160SS Super Silent portable compressor.

HOLMAN SCREW

Model	Output (FAD) ft³/min	Output (FAD) litres/sec	Pressure lbf/in²	Pressure bar	Engine	rev/min
140	140	66.0	100	7.0	Ford, Perkins, Fiat	1650
140 Silent	140	66.0	100	7.0	Ford, Perkins, Fiat	1650
140 Super Silent	140	66.0	100	7.0	Ford, Perkins	1650
175	175	82.6	100	7.0	Ford, Perkins, Fiat	2050
175 Silent	175	82.6	100	7.0	Ford, Perkins, Fiat	2050
175 Super Silent	175	82.6	100	7.0	Ford, Perkins, Fiat	2050
275	275	130	100	7.0	Ford, Perkins, Fiat	2050
275 Silent	275	130	100	7.0	Ford, Perkins	2050
275 Super Silent	275	130	100	7.0	Ford, Perkins	2050
CR400	400	189	100	7.0	Perkins, Cummins	2400
RO45-170	450	212	170	11.7	Rolls Royce, Cummins	1800
RO60-100	600	283	100	7.0	Rolls Royce, Cummins	1975
6060-100 Silent	600	283	100	7.0	Rolls Royce, Cummins	1975
RO60-100 Super Silent	600	283	100	7.0	Rolls Royce, Cummins	1975
RO60-170	600	283	170	11.7	Rolls Royce, Cummins	2000
RO70VHP	700	329	250	17.2	General Motors	1900
RO75-100	750	354	100	7.0	Rolls Royce, Cummins	2010
RO90	900	423	100	7.0	General Motors	2100
RO100HP	1000	470	150	10.3	General Motors	1950
RO120	1200	564	100	7	General Motors	1800

Holman R045 — 170 high pressure air compressor.

Consolidated Pneumatic Tool Company Ltd

Consolidated Pneumatic Class RO portable compressors are of oil-flooded rotary vane type and are available in 2 or 4 wheel versions with alternative engines. 'Hushed Power' models with acceptable noise levels for operating in built-up areas are available up to 260 ft^3/min. Leading data are summarized in the Table.

Model		125-R0-1	160-R0-2	210-R0-2	210-R0-2	260-R0-2	365-R0-2	600-R0-2	900-R0-2
Speed rev/min	full load	1850	1200	2300	1900	2000	1825	1750	1775
	ft^3/min	85	125	160	210	260	365	800	900
Capacity	m^3/min	2.4	3.5	4.5	6	7.36	10.62	17	25.5
Normal working pressure (gauge)	lb/in^2	100	100	100	100	100	100	100	100
	bar	7	7	7	7	7	7	7	7
Maximum working pressure (gauge)*	lb/in^2	125	110	135	135	135	135	125	120
	bar	8.5	7.6	9.3	9.3	9.3	9.3	8.8	8.3
Engine make and type (standard)		Perkins D3-152	Ford 2711E	Ford 2711E	Ford 2713E	Ford 2714E	Cummins VS04	Rolls Royce C6NFL	Rolls Royce C6TFL

** Under idling conditions the pressure will be 15 lb/in^2 or 1 bar higher than these figures.*

New silenced version of Consolidated Pneumatic's 600 ft^3/m rotary power vane air compressor.

Class R stationary compressors capacity 100 ft^3/m to 500 ft^3/m, from Consolidated Pneumatic

Class R stationary compressors are completely self-contained units available as air-cooled models (designated REA) or water-cooled models (designated REW). Leading data relative to the five models in this range are summarized in the Table.

CLASS R STATIONARY COMPRESSORS

Speed	**Model No.**	**100**	**160**	**200**	**300**	**500**
Speed	rev/min	1460	1460	1460	1460	1460
Actual delivery	ft^3/min	100	160	200	300	500
	m^3/min	2.85	4.5	5.7	8.5	14
Normal discharge pressure (max with std. motor)	lb/in^2 (gauge)	100	100	100	100	100
	bar	7	7	7	7	7
Motor	Horsepower	30	50	60	80	130
Oil capacity	Imp galls	5	7	7	9.25	13
	litres	30	32	32	42	60
Cooling water requirements (REW)	Imp galls/min	3	5	6	8	13
	litres/min	14	23	27	37	60
Additional cooling water for aftercooler	Imp galls/min	1.5	2	3	3.5	6
	litres/min	7	9	14	16	27
Alternative motor horse power for operation at 150 lb/in^2 (10 kg/cm^2)		37	54	64	91	161

Dawson, McDonald and Dawson Ltd

Compton compressors are reciprocating machines designed to provide oil-free delivery, utilizing the diaphragm principle for pumping. All the machines are suitable for vacuum duties as well as air or gas compression.

Type No.	Pressure lb/in^2	FAD (delivery) at max pressure) ft^3/m	HP	Remarks
D180	20	1.8	¼	Single head, air-cooled, with motor
D296	20	1.8	¼	As D180 for coupling to flame-proof motor
D245S	30	2.25	¼ or ½	
D497	20	1.8	*	* 12 or 24 V DC motor
D351VM	30	0.3	1/12	
D189	60	0.58		Also mounted on receiver with automatic pressure switch (D274)
D617	30	0.7		Lightweight portable version also available
2D/245	20	5	1	Double head
2D/197	60	1.7	1	Double head; also as automatic installation
2D/657	100	1.56	1	Two-stage; also as automatic installation
4D/245	20	10		Multiple units (4 heads)
4D/197	40	5.2		
4D/657	100	3.1		
6D/245	20	15		Multiple units (6 heads)
6D/197	40	7.8		
6D/657	100	4.68		
8D/245	20	20		Multiple units (8 heads)
8D/197	40	10.4		
8D/657	100	6.2		
20/351VM	30	0.6	1/8	Can be arranged as two stages vacuum up to 28 in Hg

Bryan Donkin Co Ltd

The Bryan Donkin range of reciprocating and turbo compressors encompasses single and multi-stage machines suitable for a wide range of applications.

The reciprocating machines, including both vertical and horizontal opposed designs, in oil free, mini-lube and lubricated forms offer capacities up to 26700 Nm^3/h, (10^6 scfh) and pressures up to 492 bar (7000 lbf/in^2) in 1, 2, 4 and 6 throw crank configurations. These high pressure machines lend themselves well to high pressure gas storage systems and numerous applications in the chemical, petro-chemical and process industries.

Bryan Donkin manufacture a comprehensive range of single stage centrifugal SD and SG type blowers suitable for a wide variety of applications such as pneumatic conveying, sewage aeration and any industrial processes where large volumes of air need to be handled at medium pressure viz. Volume flows up to 20 m^3/sec (42000 scfm) with 700 mbar (10 lbf/in^2) pressure rise. One of the features of this range of machines is that inlet guide vane control can be fitted as standard to provide volume flow control with a constant speed motor without recourse to a variable speed prime mover. This type of control provides facilities for manual or most forms of automatic control to deal efficiently and economically with variations in throughput or discharge pressure.

The Bryan Donkin Mk V Roots type compressor/exhauster is an improved range of positive displacement machines. Operation is both simple and effective, two rotors mounted on parallel shafts within a casing rotate in opposite directions through a pair of timing gears. The profile and machine finish of the rotors is such that accurate clearance is maintained between the two, and between the rotors and casing. The Mk V Roots type blowers are designed to give a pressure rise of 1.03 bar from atmosphere on all sizes of machine with a capacity range of .28 m^3/s, to 9.5 m^3/s, (592 scfm to 18000 scfm).

As an exhauster vacuums up to 380 mm Hg can be achieved under normal dry operation and 560 mm Hg with water injection to the inlet side.

These machines are suitable for sewage aeration, pneumatic handling, furnace air blowing, gas boosting, coke oven gas exhausting, steam boosting and circulation for de-salination plant and vacuum pump applications.

Gast Manufacturing Co Ltd

The Gast range of oilless piston air compressors comprises fourteen different motor-mounted models offering FAD outputs ranging up to 7.5 ft^3/min at 50 lb/in^2, and 5 ft^3/min at 100 lb/in^2. All units are noted for their compact form, light weight and general versatility. A further range of three separate-drive compressors is available (suitable for V-belt or coupling drive). All the motor-mounted compressors can also be supplied as tank-mounted compressors.

A particular feature of the piston construction is the use of lap joint PTFE rings backed by stainless steel springs in a patented configuration specially developed to overcome progressive loss of efficiency caused by ring wear. The large piston skirt is also made of filled PTFE to give extreme wear resistance, whilst retaining low rubbing friction.

Howden Compressors Limited

Howden compressors are of the screw type, giving oil-free deliveries in the volume range 500–10000 ft^3/min (236–4720 l/sec). Single-stage compressors are designed for discharge pressures up to 3.5 bar gauge. Two-stage compressors cover the volume range 1000–10000 ft^3/min (470–4720 l/sec) with discharge pressures up to 13 bar (190 lb/in^2).

Howden oil-injected screw compressor.

Oil-injected screw compressors are also produced by this company. These are built into 'Howdenair' package air compressors which cover a range of 75–1000 ft^3/min (35–470 l/sec) with outlet pressures up to 10.3 bar gauge. These self-contained sets may be air- or water-cooled. No special foundations are needed.

Oil-injected screw compressors can also be offered as bare units for building into customers' sets.

HPC Engineering Ltd

HPC Engineering produce a range of horizontal receiver-mounted compressors driven by electric motors, four models of which are also available with petrol engine drivers (see Table). Other models include the BLC Series (four models) and F Series (two models) base-mounted 2-stage (series) compressors for delivery at 150 lb/in^2 and 170 lb/in^2 respectively. In addition there are two small portable compressors (one an oil-free diaphragm type for spray gun working) and three small mobile compressors. All the latter are particularly suitable for spray work.

BLC Series (150 lb/in^2)

Model	Design	Displacement ft^3/min	lit/min	Delivery at 100 lb/in^2 ft^3/min	lit/min	Delivery at 150 lb/in^2 ft/min	lit/min
BLC80	2-stage	105.0	2974.0	80.0	2266.0	70.0	1983.0
BLC110	2-stage	140.0	3962.0	110.0	3115.0	100.0	2832.0
BLC145	2-stage	175.0	4956.0	145.0	4107.0	135.0	3823.0
BLC215	2-stage	274.0	7760.0	215.0	6089.0	195.0	5523.0

MWP Series (150 lb/in^2)

Model	Design	Displacement ft^3/min	Displacement lit/min	Delivery at 100lb/in^2 ft^3/min	Delivery at 100lb/in^2 lit/min	Motor kW	Motor hp
SH4	1-stage	3.2	90.6	2.0	59.4	0.37	0.5
SH6	1-stage	6.0	169.8	3.7	104.7	1.1	1.5
SH10	2-stage	8.4	237.7	6.0	169.8	1.5	2.0
SH15	1-stage	15.5	438.0	10.3	283.0	3.0	4.0
SH23	1-stage	22.6	639.6	16.2	458.5	4.0	5.5
SH26	1-stage	26.0	735.8	18.0	509.4	4.0	5.5
SH42*	2-stage	32.0	905.6	25.5	721.6	7.5	10.0
SH65*	1-stage	64.7	1831.0	46.0	1301.8	11.0	15.0
SH105	2-stage	105.0	2971.5	80.0	2265.3	15.0	20.0
SH140	2-stage	140.0	3962.0	110.0	3115.0	22.0	30.0
SH12	1-stage	12.9	365.0	8.1	229.2	2.2	3.0

**also with petrol engine driver*

MWP Series (200 lb/in^2)

Model	Design	Displacement ft^3/min	Displacement lit/min	Delivery at 100lb/in^2 ft^3/min	Delivery at 100lb/in^2 lit/min	Motor kW	Motor hp
TSH9	2-stage	7.1	200.9	5.5	155.6	1.5	2.0
TSH13	2-stage	12.7	359.4	10.5	297.1	2.2	3.0
TSH18	2-stage	17.1	483.9	14.1	399.0	4.0	5.5
TSH26	2-stage	25.0	707.5	20.0	566.0	5.5	7.5
TSH56	2-stage	56.0	1595.0	39.0	1100.0	11.0	15.0
TSH75	2-stage	75.0	2122.5	50.0	1415.0	15.0	20.0

Design

TSH26

TSH13–TSH18

TSH 9

TSH56

The Hydrovane Compressor Co Ltd

Hydrovane compressors are of oil-sealed vane design; the injection of oil using the pressure of delivered air for oil circulation also provides air cooling. Volumetric efficiencies up to 90% are claimed; oil separation is automatic with air delivery particularly clean. The ranges available include both stationary and portable units.

PORTABLE COMPRESSORS

Model	FAD		Working Pressure		Power Unit
	ft^3/min	m^3/min	lb/in^2	bar	
90D	76	2.1	100	7	Perkins 4.108 Diesel
85DS	85	2.4	100	7	Perkins D3.152 Diesel
100DS	100	2.8	100	7	Perkins D3.152 Diesel
152DS	152	4.3	100	7	Ford 2711E Diesel
180DS	180	5.1	100	7	Ford 2711E Diesel
280D	250	7.1	100	7	Perkins 6.354 Diesel

SELF-CONTAINED STATIONARY UNITS

Model	Displacement		FAD		Working Pressure		Electric Motor Drive
	ft^3/min	l/min	ft^3/min	l/min	lb/in^2	bar	
6PUM	6	185	5.2	147	100	7	1.5 KW 2HP Direct
13PUM	13	368	11.4	322	100	7	2.5 KW 3HP Direct
23PUM	23	651	21.5	608	100	7	4.1 KW 5HP Direct
33PUM	33	934	30	849	100	7	5.5 KW 7.5 HP Direct
43PUM	46	1322	39	1104	100	7	7.5 KW 10 HP Direct
66PUM	64	1812	54	1529	100	7	11.2 KW 15HP Direct
120PUM	109	3087	95	2690	100	7	18.5 KW 25HP Direct
170PUM	170	4800	150	4247	100	7	30 KW 40HP Direct
250 PUM	250	7075	205	5805	100	7	45 KW 60HP Direct
140 AIR CENTRE	140	3964	112	3171	100	7	22KW 30HP Direct
170 AIR CENTRE	170	4800	150	4247	100	7	30 KW 40 HP Direct

Ingersoll-Rand Co Ltd

Ingersoll-Rand, pioneer of portable compressors, currently produces the following portable compressor units:-

Model DR85 – 2407l/m at 7 bar (85ft^3/m at 100lb/in^2)
With Perkins 3–152 diesel engine and two wheel running gear for speed towing.
Spring starter is not available on this unit.
With Ford 2504E diesel engine Simms spring starter, two wheel running gear for speed towing.
The above units can be supplied silenced.

Model DRAF 125 – 3539lit/m at 7 bar (125ft^3/m at 100lb/in^2)
With Ford 2711E diesel engine, two or four wheel running gear for speed towing.
This unit can be supplied silenced and can also be supplied less running gear for truck mounting.
It can also be supplied with Simms spring starter.

Model DRF 175 – 4954lit/m at 7 bar (175ft^3/m at 100lb/in^2)
With Ford 2711E diesel engine and two wheel running gear for speed towing.
This unit can be supplied less running gear.

Model DRF 250 – 7000l/m at 7 bar (250ft^3/m at 100lb/in^2).
With Ford 2714E diesel engine and four wheel running gear for speed towing.
This unit can be supplied with two wheel running gear for speed towing or less running gear suitable for truck mounting.

Model DR 365 – 10335l/m at 7 bar (365ft^3/m at 100 lb/in^2)
With Cummins V 504 or Rolls Royce C4FNL diesel engines and mounted on four pneumatic-tyred wheels.
The above can be supplied less running gear suitable for truck mounting.

Model DRCH 480 – 13591l/m at 100 lb/in^2 bar (480ft^3/m at 150lb/in^2)
With Cummins NH220-Bi or Rolls Royce C6FNL diesel engines mounted on four pneumatic-tyred wheels.

Model DR 600 – 17590l/m at 7 bar (600ft^3/m at 10.5
With Cummins NH220-Bi or Rolls Royce C6FNL diesel engines mounted on four pneumatic-tyred wheels.
The above can be supplied less running gear suitable for truck mounting.

TYPE 40 COMPRESSORS

Standard Two-Stage Type 40 Compressors for operation at pressures up to 7 bar (100lb/in^2). Assembled packaged units complete with motor and receiver.

Model	Free Air Delivered at 7 bar (100 lb/in^2) lit/m (cu ft/min)	Speed rev/min	Motor HP
25	3623 (128)	970	30
40H	5663 (201)	970	50
50	7248 (256)	970	60
75H	11326 (401)	970	96
100	14158 (509)	970	120

ITT Fluid Handling Limited

Compressors available from this company include Bell & Gossett oil-free reciprocating compressors and vacuum pumps in 1-,2-,3-,4- and 6- cylinder models for pressures of 35–175 lb/in^2 (2.5–12 bar) and deliveries up to 12 ft^3/min (0.34 m^3/min). Other models include lubricated or oil-free rotary compressors and vacuum pumps up to 40 lb/in^2 (2.66 bar) and 250 ft^3/min (7.08 m^3/min). Vacuum up to 25in mercury (635 mm/hg).

Also available is the 'Duraire' industrial grade diaphragm air pump capable of pressures up to 30 lb/in^2 (2 bar). Delivery ranges from 1.2 ft^3/min (0.034 m^3/min) at zero pressure to 0.4 ft^3/min (0.011 m^3/min) at 30 lb/in^2 (2 bar).

Joy Manufacturing Co (UK) Ltd

Joy reciprocating air compressors are made in two basic models, the WG and the WN series, with several variations.

WG Series

The Joy WG series are vertical single-stage, single cylinder, double-acting, water-cooled units designed for V belt drive. Their capacities range from 80 to 350 ft^3/min at 100 lb/in^2 (gauge) (2.3 to 9.9 m^3/min at 7 bar) the main frames designed for 5 in, 7 in or 9 in stroke. Operating pressures vary from 15 to 130 lb/in^2 (gauge) (1 to 9 bar) depending upon capacity. They may be used as main power sources, auxiliary power sources, as booster units where reduced quantities of air are required at higher than normal working pressure, or as vacuum pumps.

Details of other compressor types manufactured by this company are summarized in the abridged notes and tables.

Centac Centrifugal Compressors

The I-R Centac compressor is a heavy duty, vertically split, packaged, multi-stage intercooled centrifugal machine, designed for 24 hour operations. Each pinion and impeller is designed to run at its own optimum speed for maximum efficiency and minimum bearing loads. The range available is from 42–311 metre^3/m (1500 to 11000 Aft^3/m) at 7–8.7 bar (100–125 lb/in^2).

TYPE 30 COMPRESSORS (AIR COOLED, RECIPROCATING)

Receiver Mounted Compressor Model	Piston Displacement lit/min (ft ^{3}/m)	Maximum Pressure bar (lb/in^2)	HP of Motor	Size of Air Receiver litres (gal)
TWO-STAGE MODELS 7 bar (100 lb/in^2) Rating:				
L5	141 (5)	7 (100)	1	63 (24)
L6	184 (6.5)	7 (100)	1½	104 (33)
L10	283 (10)	7 (100)	2	104 (33)
L15	453 (15.8)	7 (100)	3	104 (33)
L24	680 (24)	7 (100)	5	104 (33)
7TD1	1161 (41)	7 (100)	7½	345 (76)
71TD	1150 (53)	7 (100)	12½	345 (76)
15TE1	2208 (77.5)	7 (100)*	15	500 (110)
20TE	2690 (95)	7 (100)*	20	500 (110)
25TE	3539 (125.5)	7 (100)*	30	500 (110)
TWO-STAGE MODELS 10–14 bar (150–200 lb/in^2) Rating:				
T65	184 (6.5)	10.5 (150)	1½	109 (24)
9#	254 (9)	14 (200)	2	218 (48)
13#	370 (13.2)	14 (200)	3	218 (48)
22#	623 (22)	14 (200)	5	218 (48)
7TD	1217 (43.2)	14 (200)	10	345 (76)
15TE	1925 (68)	14 (200)	15	500 (110)
20T2E	2831 (100)	14 (200)	25	500 (110)
SINGLE-STAGE MODELS 10 bar (150 lb/in^2) Rating:				
VA	70 (2.7)	10.5 (150)	½	77 (17)
VB5	141 (5.0)	10.5 (150)	1	77 (17)

* Also available for pressures up to 8.7 bar (125 lb/in^2)

Models V9, V13 and V22 available mounted on 18 in x 36 in, 38 gallon, vertical receiver

WN Series

The Joy WN series are V configuration, two-stage, two cylinder, double-acting, water-cooled units with a highly efficient water cooled heat exchanger fitted between the first and second stages. They are designed for flange mounted motors for compactness. Capacities vary from 425 to 1110 ft^3/min FAD at 100 lb/in^2 (gauge) (12.0 to 31.1 m^3/min at 7 bar) using a main frame designed for a 7 in stroke. Operating pressures vary from 60 to 150 lb/in^2 (gauge) (4.2 to 10.5 bar) depending upon capacity. They are also available as direct coupled or belt driven units.

Oil-free WGOL and WNOL

The Joy oil-free series runs parallel with the WG and WN, the running gear and main frames being common to both ranges. They are similar in all respects except where no lubricating oil is used in the cylinders.

WN 112 Two-Stage Lubricated

Cylinder size x stroke	Free Air Del'd at 100 lb/in^2 (gauge) (7 bar)		Compressor Speed	Motor Size at 100 lb/in^2 (gauge) (7 bar)
in	ft^3/min	m^3/min	rev/min	hp
11¼–7 x 7	425	12.0	585	90
	540	15.3	735	120
13½–8 x 7	615	17.4	585	130
	775	21.9	735	170
16–10 x 7	880	24.9	585	170
	1110	31.1	735	220

WNOL112 Two-Stage Oil-Free Models

11¼–7 x 7	410	11.6	585	90
	520	14.7	735	120
13½–8 x 7	590	16.7	585	130
	745	21.1	735	180
16–10 x 7	850	24.1	585	170
	1070	30.3	735	230

WG9 Single-Stage Lubricated

7 x 5	85	2.4	600	25
8 x 5	125	3.5	600	30
8 x 7	170	4.8	600	40
9 x 7	235	6.7	600	55
10 x 9	270	7.6	450	60
11¼ x 9	350	9.9	450	80

WGOL9 Single-Stage Oil-Free Models

7 x 5	80	2.3	600	25
8 x 5	110	3.1	600	35
8 x 7	155	4.4	600	45
9 x 7	210	5.9	600	55
10 x 9	250	7.1	450	65
11¼ x 9	330	9.3	450	85

MPRESSORS

Air Compressors

WN Series

PTFE piston rings

JOY RECIPROCATING Air Compressor

WG Series—Vertical, single-stage, water-cooled units, designed for 'V' belt drive and having capacities from 80 to 350 ft³/min FAD at 100 psig (2.3 to 9.9 m³/min at 7 bars).

WG Series

JOY TURBO AIR Compressors

Completely packaged centrifugal compressed air plant—inherently oil-free—ready for fast installation, with compressor drive and all components arranged on a frame. Easy access for routine servicing is an important feature of this range of machines. Two to six stage models span a capacity range from 1750 to 35000 ft³/min at pressures ranging from 20 to above 600 psig (2975 to 59500 m³/ hr at 1.4 to 41.4 bars).

JOY MANUFACTURING COMPANY (UK) LIMITED
Peel Park Place, East Kilbride, Glasgow, Scotland G74 5LN.
Telephone East Kilbride (03552) 39121. Telex 777281.

Joy WN112 two-stage reciprocating air compressor.

Joy 'Turbo Air' packaged centrifugal compressor.

Joy Turbo Air Compressors

This is a range of centrifugal compressor packages, compiled from matched impellers, pinions, scrolls and diffusers and built into standard compressor frames.

The design is compact and allows for the mounting on a common frame of all the essential items – compressor, intercoolers, lubrication system, driver and coupling and the control panel.

The whole forms a "package" for a wide range of plant and process air duties within a span of 1750 to 35000 ft^3/min at pressures ranging from 20 to above 600 lb/in^2 (gauge) (2975 to 59500 m^3/hr at 1.4 to 41.4 bar). The design provides oil-free compressed air.

Joy Turbo Air 20V Series

Model	Nominal motor power at 100 lb/in^2 (gauge)(7 bar) bhp	Inlet Capacity ft^3/min	Inlet Capacity m^3/hr
TA–17V	450	1750	2975
TA–20V	500	2100	3570
TA–25V	600	2600	4420

Two-stage models to 65 lb/in^2 (gauge) (4.5 bar)
Three-stage models to 135 lb/in^2 (gauge) (9.3 bar)

Joy Turbo Air 26/85 series

TA–26	600	2600	4420
TA–30	700	3000	5100
TA–35	800	3500	5950
TA–40	900	4000	6800
TA–50	1100	5000	8500
TA–60	1300	6000	10200
TA–70	1500	7000	11900
TA–85	1850	8500	14450

Two-stage Models to 65 lb/in^2 (gauge) (4.5 bar)
Three-stage Models to 135 lb/in^2 (gauge) (9.3 bar)
Four-stage models to 200 lb/in^2 (gauge) (13.8 bar)
Five-stage models to 400 lb/in^2 (gauge) (27.6 bar)
Six-stage models to above 600 lb/in^2 (gauge) (41.4 bar)

Joy Turbo Air 100/150 Series

Model	Nominal motor power at 100 lb/in^2(gauge)(7 bar) bhp	Inlet Capacity	
		ft^3/min	m^3/h
TA–100	2100	10000	17000
TA–150	3100	15000	25500

Two-stage models to 65 lb/in^2 (gauge) (4.5 bar)
Three-stage models to 135 lb/in^2 (gauge) (9.3 bar)
Four-stage models to 200 lb/in^2 (gauge) (13.8 bar)
Five-stage models to 400 lb/in^2 (gauge) (27.6 bar)
Six-stage models to above 600 lb/in^2 (gauge) (41.4 bar)

Joy 'Twistair' Screw Compressors

A range of packaged compressors which utilize an oil-injected rotary screw unit with electric motor drive and are supplied complete and ready for start-up without requiring foundations. The TA 100 model is air cooled and the TA 200 and TA 300 models offer air or water cooling and have the compressor and all components housed within an acoustically designed cabinet.

Model	Capacity at 100 lb/in^2 (gauge) (7 bar)	Motor hp
TA 100	105 ft^3/min FAD	30
TA 200	216 ft^3/min FAD	50
TA 300	294 ft^3/min FAD	75

K. P. Pneumatic Services Ltd.

K. P. Pneumatic Services Ltd, produce a comprehensive range of oil-free and lubricated rotary sliding vane compressors and exhausters which vary in capacity from 1 ft^3/m to 1 000 ft^3/m.

The machines in the lubricated form require a minimum of lubrication and use less H.P. than conventional rotary machines. In their air cooled form they will produce 20 lb/in^2 as compressors or 26 in Hg as exhausters, whereas in the water cooled form the ratings are 50 lb/in^2 and 27 in Hg.

The machines can be supplied as bare shaft units or directly mounted to electric motor, engine or as customer requires.

Lacy-Hulbert & Co Ltd

Lacy-Hulbert offer an extremely comprehensive range of reciprocating air compressors covering displacements from 2.6 ft^3/min up to 60 ft^3/min. The designs include single-stage, two-stage and three-stage compressors. Lubricated or oil-free versions are available. Lacy-Hulbert have made a speciality of supplying compressors for breathing quality air for centralized Hospital installations and for charging compressed air bottles.

In addition to the reciprocating range of compressors, rotary sliding vane compressors and vacuum pumps operating up to 1 atm pressure are also available. These are offered in displacements up to 60 ft^3/min in oil-free and lubricated versions. The following table summarizes the types available.

Range	General Description	Model	Displacement ft^3/min	ft^3/min	Pressure FAD at lb/in^2
'S'	Air Compressors: single stage, 1–2 cylinders, air cooled; lubricated	S3–S42		1–27	120
Grasso	Air compressors: single-stage: 1-4 cylinders; air-cooled, lubricated	L 810 L 820 L 830 L 840		11 23 34 45	105 105 105 105
	Air compressors: two-stage; multiple-cylinder air-cooled; lubricated	L 811 L 821 L 831		13 25 34	205 205 205
	Air compressors; single-stage; 1-4 cylinders air-cooled; oil-free	LO 810 LO 820 LO 830 LO 840		7 14 22 29	105 105 105 105

cont...

Range	General Description	Model	Displacement ft³/min	ft³/min	Pressure FAD at lb/in²
YB	Air compressors; single-stage	YB 1 YB 2 YB 3 YB 4		12 25 35 50	50 50 50 50
GO 6	Air-compressor: three-stage; integral water; radiator-cooling	G 06		7	4000
RB	Sliding-vane type vacuum pumps and air compressors: oil-free or lubricated	RB 1201 RB 1361	1.7 5.2	1.0 3.7	15 13
		RB 1–5 RB 1–6	5.3 7.5	3.0 4.9	10 10
		RB 2	8.5	7.2	10
		RB 2–5 RB 2–8	12.0 17.0	8.0 12.0	15 15
		RB 3 RB 4	24.5 35.0	20.0 30.0	10 10
		RB 5 RB 6	45.0 52.0	38.0 40.0	10 10

The Lead Wool Co. Ltd.

This company manufactures portable compressors, two-stage V-form two-cylinder air cooled configuration, in two sizes:—

ACD 80 — with a rated delivery of 80 ft³/min

ACD 140 — with a rated delivery of 140 ft³/min

Diesel drive is standard in each case. Both units are also available mounted on the Ford Transit, the combinations being known as the *Truckair 80 and Truckair 140*, respectively.

Northey Rotary Compressors Ltd

Northey positive rotary compressors are based on an original form of intermeshing rotor design. The twin rotors operate with positive clearance, thus no lubrication is required and delivery is oil-free. Gas flow is controlled by movement of the rotor across ports machined in the end covers, thus no inlet or exhaust valves are required.

Typical arrangement of large Northey two-stage 'oil free' air or gas compressor or vacuum pump with 100 HP motor.

Typical Northey watercooled single-stage 'oil free' air or gas compressor.

Whittaker Hall Ltd

The "Fluidair" range of rotary vane compressors covers free air deliveries from 17 to 600 ft^3/min and pressure ranges from 100 to 170 lb/in^2 according to the compressors selected. In addition Whittaker Hall Ltd manufacture other rotary compressors for deliveries of 15–1805 ft^3/min in the lower pressure ranges of 5 to 40 lb/in^2. They also manufacture Vacuum Pumps for a wide range of displacement at vacua up to 29.50 in Hg.

'Fluidair' compressors can be supplied with full acoustic enclosures if required, and both water cooled and air cooled oil coolers are available through the whole range except for the smallest unit, which is air cooled. (See Table on page 444).

FLUIDAIR OIL-SEALED ROTARY COMPRESSORS

Model	Displacement		Output FAD		Max. Pressure		Max.	Cooling	Remarks
	ft^3/min	l/m	ft^3/min	l/m	lb/in^2	bar	hp		
S2R20	20	570	17	480	170	12	6.5	Air	Two-stage
S2R66L	66	(m^3/m) 1.87	56	(m^3/m) 1.58	100	7	15	Air or Water	
S2R66H	66	1.87	52	1.47	150	10.5	20	Air or Water	Two-stage
S2R110L	110	3.14	95	2.7	100	7	25	Air or Water	
S2R110H	110	3.14	88	2.5	140	10	30	Air or Water	
SR 177	177	5.0	150	4.3	100	7	40	Air or Water	Single-stage
SR250	250	7.1	210	6.0	100	7	60	Air or Water	
SR350	350	10.0	300	8.5	100	7	75	Air or Water	Single-stage
SR460	460	13.0	400	11.4	100	7	100	Air or Water	
SR570	570	16.2	500	14.2	100	7	125	Air or Water	
SR680	680	19.3	600	17.0	100	7	150	Air or Water	

Williams & James Ltd

Series 170 compressors are a heavy duty range suitable for either air or gas duties, pressure range is 400–6000 lb/in^2 (28–280 bar) and capacities 20–40 ft^3/min (9.4–19 dm^3/s). Model 175 is a vee-form four-cylinder three-stage compressor giving 35 ft^3/min (16.5 dm^3/s) at 3800 lb/in^2 (262 bar).

Williams & James vee drive type 695 oil free, air cooled compressor with heavy duty filtration for desert use.

Series 200 compressors are air-cooled multi-stage machines, available with optional heavy-duty filtration systems with pressures up to 6000 lb/in^2 (240 bar). Capacity is 12 ft^3/min (5.7 dm^3/s).

Series 140/150 compressors offer pressures up to 125 lb/in^2 (8.6 bar) with a capacity of 140 ft^3/min (66 dm^3/s). The 140 series are air-blast cooled with integral aftercoolers. Model 156 is a water-cooled four-cylinder compressor capable of pressures up to 125 lb/in^2 (8.6 bar) at outputs of 140 ft^3/min (66 dm^3/s).

Oil-free compressors are available in three ranges: Series 700 models, pressures up to 125 lb/in^2 (8.6 bar), capacity 5–10 ft^3/min (2.4–4.7 dm^3/s); Series 190 models, pressures up to 125 lb/in^2 (8.6 bar), capacity 20–120 ft^3/min (9.4–56.6 dm^3/s); and Series 600 models, pressures up to 150 lb/in^2 (10.3 bar) and capacities 100–200 ft^3/min (47.2–94.4 dm^3/s). Model 695 gives 200 ft^3/min (94.4 dm^3/s) at pressures up to 150 lb/in^2 (10.3bar) and is complete with integral air-cooled interstage and aftercoolers.

Survey of Cylinders

Air Automation

Cylinders

Air Automation produce pneumatic cylinders in bore sizes from ½ in to 6 in in single acting, double acting, cushioned and non-cushioned, as well as through rod models.

Rods are of stainless steel and piston and rod seals are self conforming distributor type seals. Most cylinders are provided with a universal mounting facility with screwed nose, rear trunnion and often base mounting provision on the same basic cylinder. Most piston assemblies are of nylon, so avoiding possible metal to metal contact with the bore. Imperial or metric rod threads are available.

2¼in bore universally mounted air cylinder also available in 1½in and 3in bore.

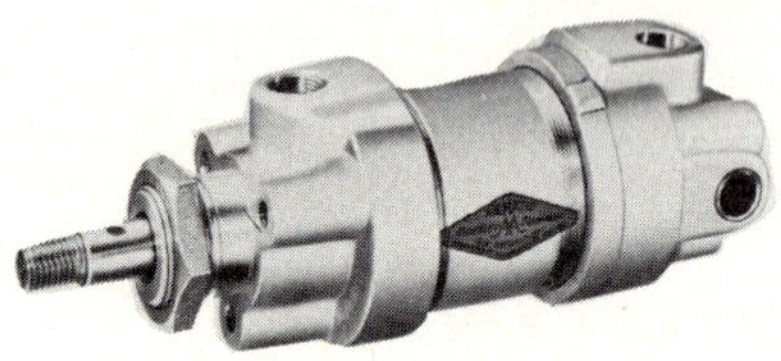

Bestobell Mobrey Ltd

This company markets the 'Herion' range of pneumatic components — see entry under Survey of Valves.

CompAir Maxam Ltd

Maxam air cylinders are made in bore sizes from 11 mm (7/16 in) up to 330 mm (13 in) bore with a wide variety of variants and mountings. International CETOP cylinders of 25 mm to 100 mm bore are manufactured from extruded aluminium. Round section barrel with tie rods are used for all sizes above 100 mm.

The CETOP International cylinder range can be supplied with two additional features.

(1) Non-lube types — specially coated cylinder bores to fulfil the requirements for cylinders to be used without lubrication. The cylinders are constructed such that no metal moving parts are in contact with the bore.

CETOP Standard International Cylinder. (CompAir Maxam Ltd.).

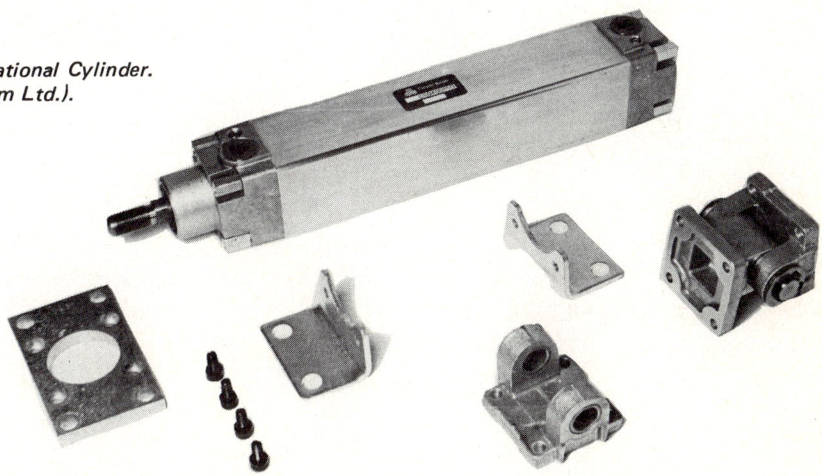

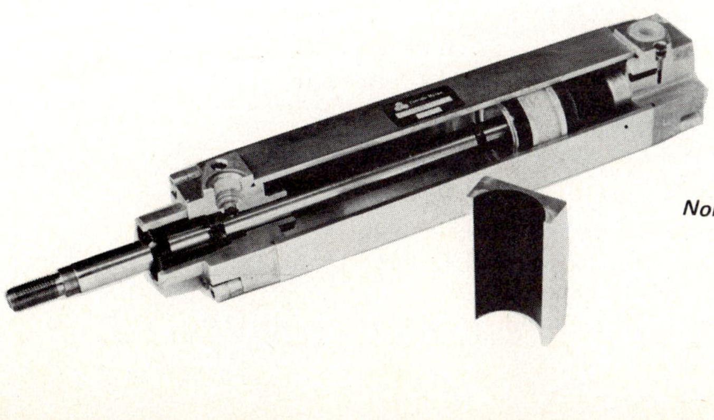

Non-lube specially coated cylinder bores for use without lubrication. (CompAir Maxam Ltd.).

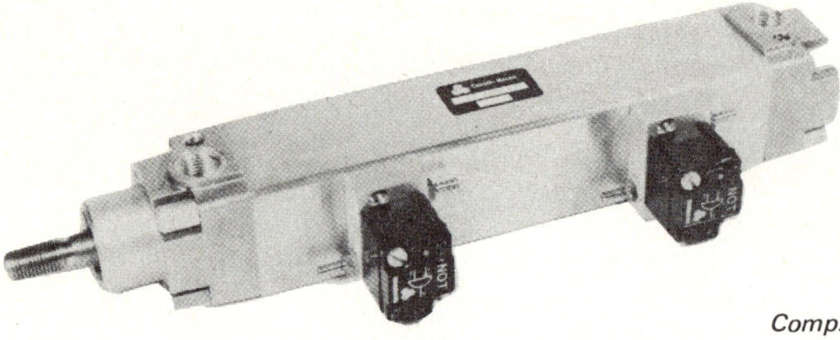

CompAir Maxam signalmaster cylinder.

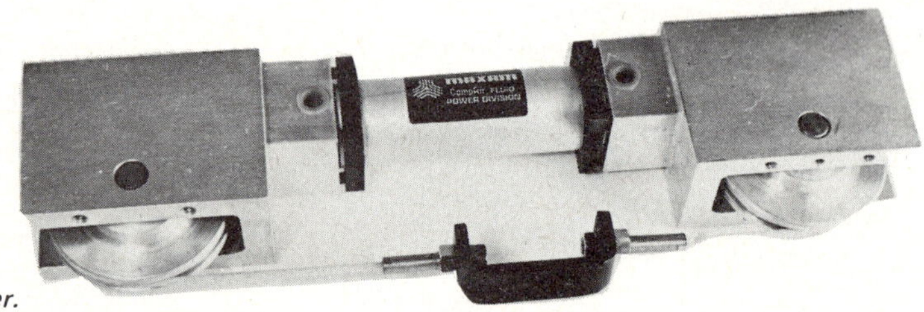

CompAir Maxam cable cylinder.

(ii) Pneumatic signalmaster — these units provide switching within the cylinders, replacing externally operated conventional pilot valves and the accompanying trip mechanisms and brackets. Basically the signalmaster is a NOT unit fitted to a standard cylinder to provide a pneumatic signal at the end of stroke. Over-travel of 3 mm nominal is provided as standard.

A new addition to the Maxam range is the double acting — fixed cushion cable cylinder. This unit provides a compact unit ideal for use in areas where space is limited. The cable cylinder will operate in approximately half the space required for a conventional cylinder of similar stroke.

Davis Pneumatic Systems Ltd

A comprehensive range of air cylinders is offered and design and manufacture of "specials" is welcomed.

The standard range embraces:—
Mini-metric series // 12.20 & 25mm bore.
Standard air cylinders to CETOP & BSS // 32 – 320 mm bore.
Imperial cylinders, both cushioned and standard // 5 in — 14 in bore.
Hydro-pneumatic cylinders // 1¾ in — 6 in bore.
Tolomatic cable cylinders // 1 in — 8 in bore with stroke lengths up to 30 ft.

Kinetrol Ltd

This company manufactures a series of ¼-turn pneumatic actuators, each model giving approximately twice the torque of the preceding model. Latest models are the 080 and 090, giving operating torques of 1000 lb-in and 2200 lb-in respectively. Features include 100 degree stroke with adjustable end stops.

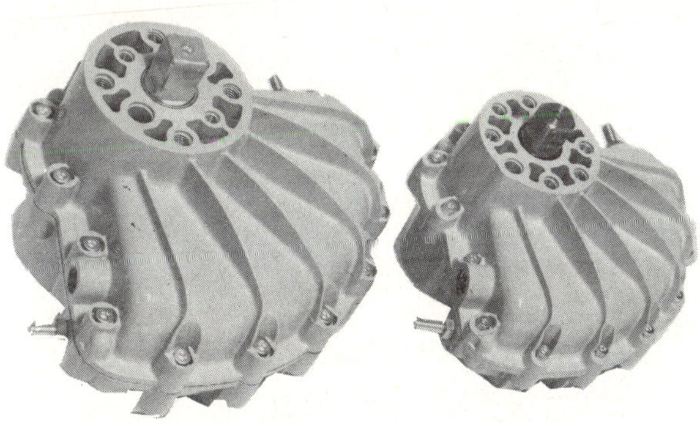

Martonair Ltd

Brief details of the various types and ranges of 'Martonair' air cylinders are given in the Table.

Type	Range	Pressure bar	Size(s)	Remarks
Midget	M/6000	2 – 10	10–25mm	fixed cushions
	M/7000	3 – 10	10–25mm	" "
	M/778	2 – 10	¾in	non-cushioned
	M/775	2 – 10	¾in	" "
Minor	M/544-9 M/4125, M/4175 M/4250	2 – 10	1¼in, 1¾in, 2½in	single, double-acting, cushioned or non-cushioned
Standard	M/900	2 – 10	1¼in–14in	wide range of stock strokes and mountings
	M/1900	2 – 10	3in – 12in	non-cushioned version of M/900
	M/13000	1 – 10	32–100mm	new series of metric designs in square styling. Extra fast operation with advanced cushion design
Heavy duty	M/1000	2 – 17	2in – 12in	designed for use in arduous conditions
Cable cylinders	M/5900	2 – 10	2in – 4in	For long stroke action in a confined space

cont...

Type	Range	Pressure bar	Size(s)	Remarks
Impact cylinders	M/3000	4 – 10	2in – 8in	
Air-hydraulic cylinders	M/3220 M/3320 M/3340	2.5–10 2.5–10 2.5–10	2in 3in 4in	air cylinder integrally coupled to a hydraulic checking cylinder
	M/428			hydraulic dashpot cylinder
	M/153- M/157	0 – 7	1¾in–8in	air/oil reservoirs for precision speed control or positive locking

Martonair M/13010 series air cylinder.

Midland Pneumatic Ltd

Midland Pneumatic manufacture a range of metric cylinders from 40 to 200 mm bore conforming to CETOP and BSI standards. These cylinders are rated for a maximum working pressure of 10 bar. (cylinders for higher pressure are available to special order). Construction features stainless steel piston rods, brass or hard anodized aluminium barrels, high tensile light alloy end covers and cartridge type bronze bearing including seal.

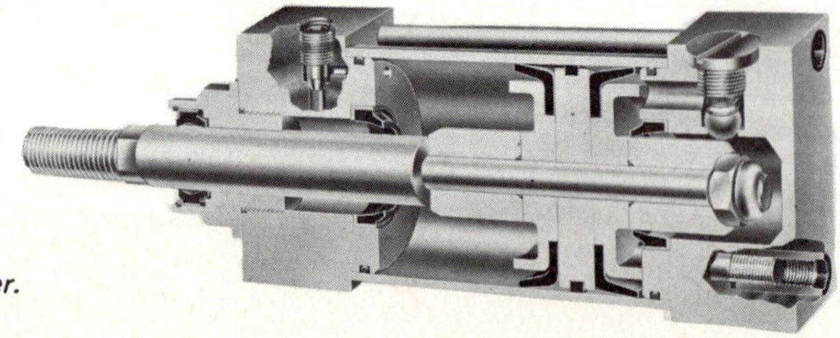

Midland Pneumatics CETOP cylinder.

The original Imperial range of cylinders produced by this company are still available, but are being phased out. The Imperial sizes which remain in production are non-cushioned versions in 1 to 3 inch bore, also 10 in, 12 in and 14 in bore cylinders which can be supplied with either BSF or metric rod threads.

R.G.S. Electro-Pneumatics Ltd

This company manufactures air cylinders in a variety of material specifications in sizes from 12 mm to 300 mm bore. They also produce a complementary range of cylinders for positioner duty. These cylinders have low friction characteristics for smooth modulation.

A. Schrader's Son

A. Schrader's Son, a division of Scovill Manufacturing Company, produces a range of air cylinders from 12 mm to 14 inch bore sizes, encompassing square endplate tie rod designs in both imperial and metric configurations, midget cylinders with screwed end plates and the "all-round" non-tie rod construction variety. Many special purpose cylinders are featured, including air clamps, air presses, reciprocating cylinders, rotary actuators and positioner units.

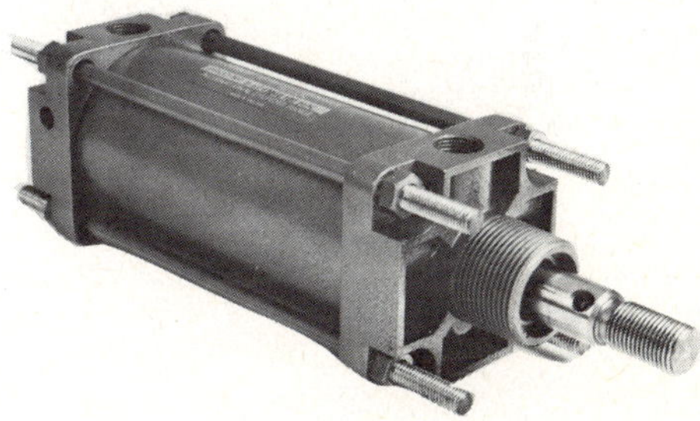

Schrader air cylinder

Sperry Vickers Lang Pneumatic

Air cylinders manufactured by this company are summarized in the Table on page 452.

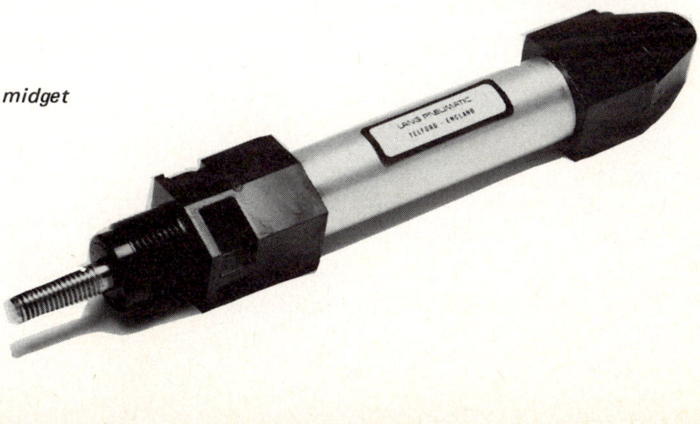

Sperry Vickers Lang Pneumatic midget cylinder.

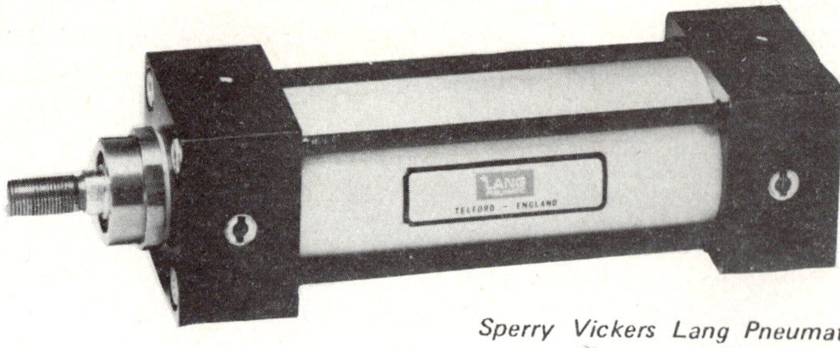

Sperry Vickers Lang Pneumatic metric cylinder.

Type Description	Type No.	Bore	Standard Strokes	Ports	Remarks
Midget	MC2	½ in	1 in only	¼ in BSP	Single-acting, spring return
Miniature 'CM' series	CM075	¾ in	½ in 1 in 1½ in 2 in 3 in 4 in	1/8 in BSP	Double-acting, non-cushioned
	CM100	1 in	½ in 1 in 1½ in 2 in 3 in & 4 in	1/8 in BSP	
Heavy duty 'S' series	S150	1½ in		¼ in BSP	
	S225	2¼ in	as required	¼ in BSP	Neck mounting as standard
	S300	3 in		¼ in BSP	
	S400	4 in			
	S500	5 in	3 in 6 in 9 in & 12 in		Double acting
	S600	6 in		BSP	Double acting through rod
	S800	8 in			also twin cylinders (front &
	S1000	10 in	as required		rear connected)
	S1200	12 in			All cylinders available with
					front or rear cushioning or both.
Metric 'M' series	M0401 DA	40 mm		¼ in BSP	
	M0501 DA	50 mm	25 mm 50 mm 80 mm 100 mm 160 mm	¼ in BSP	
	M0631 DA	63 mm	200 mm 320 mm	½ in BSP	Various mountings available.
	M0801 DA	80 mm		½ in BSP	Cushioning as standard
	M1001 DA	100 mm	–80 mm 160 mm 200 mm 320 mm	½ in BSP	To CETOP standard
	M1251 DA	125 mm		½ in BSP	RP43P & BS4862 : 1973
	M1601 DA	160 mm		¾ in BSP	
	M2001 DA	200 mm	To order	¾ in BSP	
	M2501 DA	250 mm		1 in BSP	

Telektron Ltd.

This company markets the 'Bettis' range of torque cylinders, specifically designed for use as valve actuators. The thirteen models in the range offer opening/closing torques from 300 to 17 500 lb-in at 100 lb/in^2 (7 bar) line pressure. Maximum shaft rotation is 102°, the torque arm being of Scotch yoke configuration to provide maximum torque at the beginning and end of the stroke.

Section 9

Technical Data

USE OF PRESSURE UNITS

The metric unit adopted by the UK pneumatic industry for air pressure is the *bar* rather than kg/cm^2.

$$1 \text{ bar} = 14.5038 \text{ lbf/in}^2$$
$$1 \text{ lbf/in}^2 = 0.0689474 \text{ bar}$$
$$1 \text{ bar} = 1.01972 \text{ kg/cm}^2$$
$$1 \text{ kg/cm}^2 = 0.98066 \text{ bar}$$

This is consistent with the SI system where the derived unit for pressure is newtons per square metre (N/m^2). The N/m^2 unit, however, yields very low quantitative values for pressure values normally used in fluid engineering. Thus the bar, being equal to 10^5 N/m^2, is essentially a practical unit.

Some confusion can arise from the fact that 1 bar is also a 'nominal' equivalent of standard atmospheric pressure. Compression ratios, etc, calculated on this basis will yield lower values than those calculated on the standard value normally adopted for atmospheric pressure of 14.7 lbf/in^2.

By SI standard definition, a standard atmosphere = 1 013 250 dyn/cm^2 exactly, ie

$$1 \text{ atm} = 1.013250 \text{ bar}$$
$$= 14.695975 \text{ lbf/in}^2$$
$$= 1.0332299 \text{ kg/cm}^2$$

Also remaining in use in Continental Europe is the *technical atmosphere,* equivalent to a standard atmospheric pressure of 1.0 kilopond per square centimetre (kp/cm^2). This leads to the following (closely approximate) equivalents:

$$1 \text{ at} = 1.0 \text{ kp/cm}^2$$
$$= 14.2 \text{ lbf/in}^2$$
$$= 0.998 \text{ kg/cm}^2$$
$$1 \text{ kg/cm}^2 = 1.002 \text{ kp/cm}^2$$
$$1 \text{ lbf/in}^2 = 0.070 \text{ kp/cm}^2$$
$$1 \text{ bar} = 1.0214 \text{ kg/cm}^2$$

PRINCIPAL SI UNITS

bar	Standard unit of pressure (1 bar = 10^5 N/m^2 = 10^5 Pa)
C	Celsius (= centigrade). A temperature interval of 1^oC = 1K.
cm	centimetre (100 cm = 1 m)
cSt	centistoke (1 cSt = 10^{-6} m^2/s)
dm	decimetre (10 dm = 1m)
g	gramme (1g = 10^{-3} kg)
Hz	hertz (1 Hz = 1 cycle per second)
h	hour
J	joule (j = Nm = Ws)
K	kelvin. The ^{o} sign is not used when quoting temperatures in degrees kelvin. The kelvin is $\frac{1}{273.16}$ of the thermodynamic temperature of the triple point of water (9th CGPM 1948)
kg	kilogramme (1 kg = 1 000 g)
km	kilometre (1 000 m)
kN	kilonewton = N x 10^3
kW	kilowatt (1 000W)
l	litre (originally 1 kg of water). In 1964 it was redefined by the CGPM so as to be equal to 10^{-3} m^3 = 1 dm^3
MN	meganewton = N x 10^6
m	metre
mbar	millibar (1 000 mbar = 1 bar = 10^5 N/m^2)
ml	millilitre (1 000 ml = 1l)
mm	millimetre (1 000 mm = 1 m)
N	newton. The force which will accelerate a freely movable mass of 1 kg by 1 m/s^2
Pa	pascal = N/m^2 (Note : N/mm^2 is the preferred unit for stress)
s	second
t	tonne (= 1 000 kg). Note: this abbreviation is not widely used and to avoid possible misunderstanding the full spelling of the unit is preferred
W	watt

EEC DIRECTIVE ON METRICATION

Directive 76/770 of the European Economic Community has been affirmed as United Kingdom Law by means of the Weights and Measures 'Units of Measurement Regulation 1976' No 1674 and became effective on November 1st, 1976. Its effect on units of concern to the Pneumatic industry are summarized below.

S.I. and other units which must be made mandatory from 21st April 1978 at the latest:—

Contains all the usual S.I. Units and includes the following units of concern to the Pneumatics Industry:—

bar	pressure	bar
metric ton	mass	t
litre	liquid volume	l
minute, hour, day	for time	min h d

CHAPTER B:

Units which shall not be authorized for use after the 31st December 1977 at the latest (include):—

kilogramme force	kgf
kilopond	kp
torr	torr
atmosphere	at
metre of water O	mH_2O
metric horsepower	CV
inches of water	in WG

CHAPTER C:

Units which shall not be authorized for use after the 31st December 1979 at the latest (include):—

square inch	in^2
cubic inch	in^3
cubic foot	ft^3
ton	ton
pound force	lbf
British Thermal Unit	BTU
foot pound force	ft.lbf
horsepower	HP
degree Fahrenheit	°F
poise and centipoise	P cP
stoke and centistoke	St cSt

Units whose status is to be reviewed before the 31st December 1979 (include):—

inch	in
foot	ft
fluid ounce	fl. oz
pint	pt
quart	qt
gallon	gal
pount (mass)	lb

TEMPERATURE CONVERSION TABLE

Negative Values											
C		F	C		F	C		F	C		F
−273	−459.4		−17.8	0	32.0	10.0	50	122.0	38	100	212
−268	−450		−17.2	1	33.8	10.6	51	213.8	43	110	230
−262	−440		−16.7	2	35.6	11.1	52	125.6	49	120	248
−257	−430		−16.1	3	37.4	11.7	53	127.4	54	130	266
−251	−420		−15.6	4	39.2	12.2	54	129.2	60	140	284
−246	−410		−15.0	5	41.0	12.8	55	131.0	66	150	302
−240	−400		−14.4	6	42.8	13.3	56	132.8	71	160	320
−234	−390		−13.9	7	44.6	13.9	57	134.6	77	170	338
−229	−380		−13.3	8	46.4	14.4	58	136.4	82	180	356
−223	−370		−12.8	9	48.2	15.0	59	138.2	88	190	374
−218	−360		−12.2	10	50.0	15.6	60	140.0	93	200	392
−212	−350		−11.7	11	51.8	16.1	61	141.8	99	210	410
−207	−340		−11.1	12	53.6	16.7	62	143.6	100	212	413.6
−201	−330		−10.6	13	55.4	17.2	63	145.4	104	220	428
−196	−320		−10.0	14	57.2	17.8	64	147.2	110	230	446
−190	−310		− 9.4	15	59.0	18.3	65	149.0	116	240	464
−184	−300		− 8.9	16	60.8	18.9	66	150.8	121	250	482
−179	−290		− 8.3	17	62.6	19.4	67	152.6	127	260	500
−173	−280		− 7.8	18	64.4	20.0	68	154.4	132	270	518
−169	−273	−459.4	− 7.2	19	66.2	20.6	69	156.2	138	280	536
−168	−270	−454	− 6.7	20	68.0	21.1	70	158.0	143	290	554
−162	−260	−436	− 6.1	21	69.8	21.7	71	159.8	149	300	572
−157	−250	−418	− 5.6	22	71.6	22.2	72	161.6	154	310	590
−151	−240	−400	− 5.0	23	73.4	22.8	73	163.4	160	320	608
−146	−230	−382	− 4.4	24	75.2	23.3	74	165.3	166	330	626
−140	−220	−364	− 3.9	25	77.0	23.9	75	167.0	171	340	644
−134	−210	−346	− 3.3	26	78.8	24.4	76	168.8	177	350	662
−129	−200	−328	− 2.8	27	80.6	25.0	77	170.6	182	360	680
−123	−190	−310	− 2.2	28	82.4	25.6	78	172.4	188	370	698
−118	−180	−292	− 1.7	29	84.2	26.1	79	174.2	193	380	716
−112	−170	−274	− 1.1	30	86.0	26.7	80	176.0	199	390	734
−107	−160	−256	− 0.6	31	87.8	27.2	81	177.8	204	400	752
−101	−150	−238	0.0	32	89.6	27.8	82	179.6	210	410	770
− 96	−140	−220	0.6	33	91.4	28.3	83	181.4	216	420	788
− 90	−130	−202	1.1	34	93.2	28.9	84	183.2	221	430	806
− 84	−120	−184	1.7	35	95.0	29.4	85	185.0	227	440	824
− 79	−110	−166	2.2	36	96.8	30.0	86	186.8	232	450	842
− 73	−100	−148	2.8	37	98.6	30.6	87	188.6	238	460	860
− 68	− 90	−130	3.3	38	100.4	31.1	88	190.4	243	470	878
− 62	− 80	−112	3.9	39	102.2	31.7	89	192.2	249	480	896
− 57	− 70	− 94	4.4	40	104.0	32.2	90	194.0	254	490	914
− 51	− 60	− 76	5.0	41	105.8	32.8	91	195.8	260	500	932
− 46	− 50	− 58	5.6	42	107.6	33.3	92	197.6	288	550	1 022
− 40	− 40	− 40	6.1	43	109.4	33.9	93	199.4	316	600	1 112
− 34	− 30	− 22	6.7	44	111.2	34.4	94	201.2	343	650	1 202
− 29	− 20	− 4	7.2	45	113.0	35.0	95	203.0	371	700	1 292
− 23	− 10	14	7.8	46	114.8	35.6	96	204.8	399	750	1 382
− 17.8	0	32	8.3	47	116.6	36.1	97	205.6	427	800	1 472
			8.9	48	118.4	36.7	98	208.4	454	850	1 562
			9.4	49	120.2	37.2	99	210.2	482	900	1 692
			10.0	50	122.0	37.8	100	212.0	538	1 000	1 832

METRICATION

THE principal SI units chosen by the Metrication Committee of the British Compressed Air Society for use by the compressed air industry are outlined in the following tables. CETOP and PNEUROP are in general agreement with the adoption of the multiples and submultiples recommended for general use in the last column of each table. These are based on ISO 1 000 and BSI PD 5686, which are equivalent. In practice, the choice of the appropriate multiple is governed by convenience – eg so that numerical values will be between 0.1 and 1 000.

TABLE I – LENGTH, AREA, VOLUME

Description	SI Unit in Accordance with ISO 1 000	Common Units in Current UK Use	Multiples and Sub-Multiples of Metric Units Recommended for General Use
Linear	m	in, ft, yd	m, mm
Area	m^2	in^2, ft^2, yd^2	m^2, cm^2, mm^2
Volume	m^3	in^3, ft^3, yd^3	m^3, dm^3, cm^3* ($1\ dm^3$ = 1 litre, $1\ cm^3$ = 1 ml)
Receiver capacity (usual sizes)	m^3	ft^3	m^3
Crankcase oil capacity and other capacities involving liquids (usual sizes)	m^3	gals	l (1 litre = $1\ dm^3$)

* The following units of volume are generally applicable:

m^3 for large compressors

dm^3 for pneumatic controls, air tool consumptions, and medium size compressors

cm^3 for fluidics.

(Note: $1\ dm^3/s$ = $2.1\ ft^3/min$ [approx])

TABLE II – DIMENSIONS RELATED TO TIME

Description	SI Unit in Accordance with ISO 1 000	Common Units in Current UK Use	Multiples and Sub-Multiples of Metric Units Recommended for General Use
Portable compressor towing speed	m/s	mls/hr	km/h (1 m/s = 3.6 km/h)
Velocity, eg piston speed, gas velocity	m/s	ft/s	m/s
Air or gas flow, normally quoted as Free Air at specified conditions†	m^3/s	ft^3/m cffam	m^3/s, dm^3/s, cm^3/s* (1 dm^3/s = 1 l/s, 1 cm^3/s = 1 ml/s)
Coolant or lubricant flow	m^3/s	gal/m	l/s, ml/s

† The ISO Standard Reference Atmosphere is 20°C, 65% relative humidity, 1 013 mbar, designated by ANR (atmosphere normale de reference). In general the compressor and pneumatic tool industries do not employ this standard because of the high humidity. Their preferred standard atmosphere is that of the British Gas Industry Proposal of 15°C, dry, 1 013.5 mbar.

TABLE III – PERIODIC, NON-DIMENSIONAL

Description	SI Unit in Accordance with ISO 1 000	Common Units in Current UK Use	Multiples and Sub-Multiples of Metric Units Recommended for General Use
Non-dimensional periodic phenomena			
Speed of rotation	rev/s	rev/min	r/s (now specified in place of rev/s)
Frequency	Hz	Hz	Hz

TABLE IV – MASS AND MASS-RELATED UNITS

Description	SI Unit in Accordance with ISO 1 000	Common Units in Current UK Use	Multiples and Sub-Multiples of Metric Units Recommended for General Use
Mass	kg	lb, cwt, ton	kg, t, g
Lifting capacity	N	lb, cwt, ton	kg*, t
Thrust	N	lb, lbf	N, kN
Mass related to time and/or other dimensions			
Mass flow	kg/s	lb/min	kg/s, g/s
Moisture content of air or gases	kg/m^3	$lb/1\,000\ ft^3$ $fl\ oz/1\,000\ ft^3$ and others	g/m^3
Pressure (indicating whether gauge or absolute)	$\frac{N}{m^2} = Pa$	lbf/in^2 lb/in^2 psia, psig, etc	bar, mbar †
Stress (eg metal, receiver shells)	N/m^2	lb/in^2, ton/in^2	$N/mm^2 = MN/m^2 = MPa$ ⊕
Torque (moment of force)	Nm	lb ft	Nm
Impact energy	J	ft lb	J
Work = energy	J	ft lb	J
Power	W	HP, hp	W, kW
Specific power consumption (compressor)	$\frac{Ws}{m^3} = \frac{J}{m^3}$	$HP\ min/100\ ft^3$ $(HP/100\ ft^3/min)$	$kWs/m^3 = Ws/dm^3 = J/dm^3$ (ie $kW/m^3/s$ but avoid double solidus)

* The kilogram (kg) is preferred as a more practical unit than the Newton.

† The millibar is used for fluidics and very low pressures. It replaces 'inches water gauge' or 'mm of water'. (Note: 1 mbar = 10 mm of water [approx]).

⊕ The Pascal (Pa) is recommended for use by the high vacuum industry. The Pa is reckoned to be equally acceptable for fluid pressure and stress. The unit N/mm^2 is, however, generally preferred for stress.

TABLE V – TEMPERATURE

Description	SI Unit in Accordance with ISO 1 000	Common Units in Current UK Use	Multiples and Sub-Multiples of Metric Units Recommended for General Use
Absolute temperature	K	^{o}C abs	K
Gauge temperature		^{o}C	$^{o}C = K - 273.16$

TABLE VI – VISCOSITY

Description	SI Unit in Accordance with ISO 1 000	Common Units in Current UK Use	Multiples and Sub-Multiples of Metric Units Recommended for General Use
Kinematic viscosity (oil or gas)	m^2/s	Redwood seconds Saybolt seconds Engler degrees	cSt, preferably quoted at (1 cSt = $m^2/s \times 10^{-6}$)

The subject of a standard temperature to be used when quoting oil viscosities in centistokes is still under discussion at ISO level. There is support for 50°C, a temperature previously used for viscosities quoted in Engler degrees, and 37.8^{o}C corresponding to 100°F. Viscosity figures at the latter temperature are required by the lubricant suppliers in relation to blending and it is one of the standard temperatures used in the Saybolt system of viscosities.

In the pneumatic field it is, however, frequently important to have viscosity figures at a temperature close to ambient atmosphere temperature. Hence the choice of 20°C which corresponds to 68°F and nearly corresponds to a standard Redwood viscosity temperature of 70°F.

GRAPHICAL SYMBOLS FOR PNEUMATIC SYSTEMS AND COMPONENTS

THE following symbols are specified in ISO 1219-1976 (E/F). They are virtually identical to those specified in BS2917. The other standard source of reference for Europe, CETOP RP3, is no longer available and superceded by ISO 1219.

General (Basic and Functional Symbols)

The symbols for hydraulic and pneumatic equipment and accessories are *functional* and consist of one or more *basic symbols* and in general of one or more *functional symbols.* The symbols are neither to scale nor in general orientated in any particular direction. The relative sizes of symbols in combination should correspond approximately to those in clauses 11 and 12.

BASIC SYMBOLS

Description	Application	Symbol
Line: — continuous	flow lines	
— long dashes	flow lines	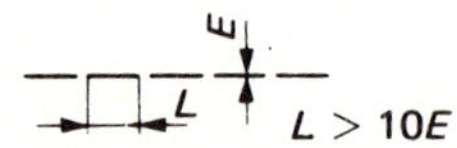
— short dashes	flow lines	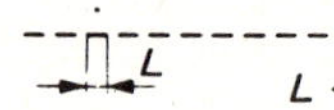
— double	Mechanical connections (shafts, levers, piston-rods)	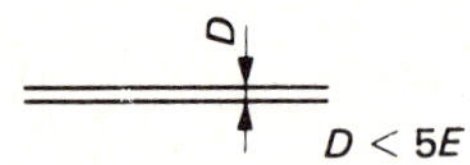
— long chain thin (optional use)	Enclosure for several components assembled in one unit	 *L* = length of dash *E* = thickness of line *D* = space between lines
Circle, semi-circle	As a rule, energy conversion units (pump, compressor, motor)	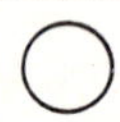
	Measuring instruments	
	Non-return valve, rotary connection, etc	
	Mechanical link, roller, etc	
	Semi-rotary actuator	
Square, rectangle	As a rule, control valves (valve) except for non-return valves	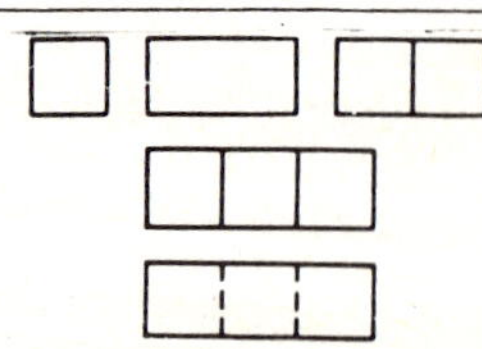
Diamond	Conditioning apparatus (filter, separator, lubricator, heat exchanger)	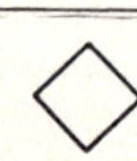

BASIC SYMBOLS

Description	Application	Symbol
	Flow line connection	 $d \approx 5E$ E = thickness of line
	Spring	
	Restriction: — affected by viscosity	
	— unaffected by viscosity	

FUNCTIONAL SYMBOLS

Description	Application	Symbol
Triangle	The direction of flow and the nature of the fluid	
— solid	Hydraulic flow	
— in outline only	Pneumatic flow or exhaust to atmosphere	
Arrow	Indication of: — direction	
	— direction of rotation	
	— path and direction of flow through valves	
	For regulating apparatus as in *Pressure Control Valves* both representations with or without a tail to the end of the arrow are used without distinction. As a general rule the line perpendicular to the head of the arrow indicates that when the arrow moves the interior path always remains connected to the corresponding exterior path.	
Sloping arrow	Indication of the possibility of a regulation or of a progressive variability	

PUMPS AND COMPRESSORS

Description	Remarks	Symbol
Fixed capacity compressor		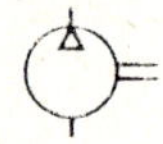
Fixed capacity pneumatic motor:		
– with one direction of flow		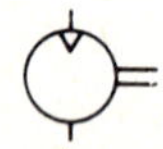
– with two directions of flow		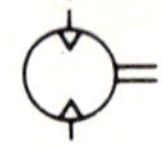
Variable capacity pneumatic motor:		
– with one direction of flow		
– with two directions of flow		
Oscillating motor:		
– hydraulic		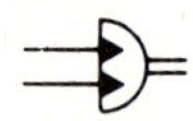
– pneumatic		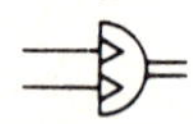
Pump/motor units	Unit with two functions, either as pump or as rotary motor	
Fixed capacity pump/motor unit:		
– with reversal of the direction of flow	Functioning as pump or motor according to direction of flow	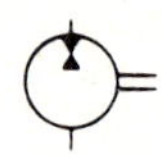
– with one single direction of flow	Functioning as pump or motor without change of direction of flow	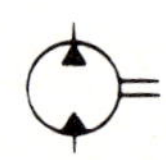
– with two directions of flow	Functioning as pump or motor with either direction of flow	
Variable capacity pump/motor unit:		
– with reversal of the direction of flow		
– with one single direction of flow		
– with two directions of flow		

CYLINDERS

Description	Remarks	Symbol	
		Detailed	Simplified
Single-acting cylinder:	Cylinder in which the fluid pressure always acts in one and the same direction (on the forward stroke)		
– returned by an unspecified force	General symbol when the method of return is not specified		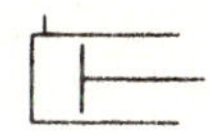
– returned by spring		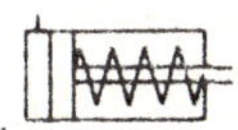	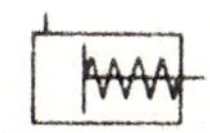
Double acting cylinder	Cylinder in which the fluid pressure operates alternately in both directions (forward and backward strokes)		
– with single piston rod			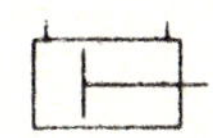
– with double-ended piston rod			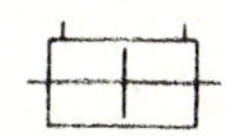
Differential cylinder	The action is dependent on the difference between the effective areas on each side of the piston		
Cylinder with cushion:			
– with single fixed cushion	Cylinder incorporating fixed cushion acting in one direction only		
– with double fixed cushion	Cylinder with fixed cushion acting in both directions	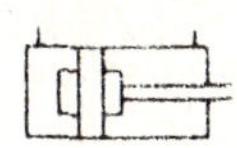	
– with single adjustable cushion		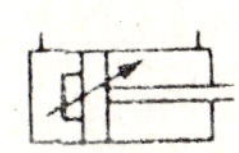	
– with double adjustable cushion		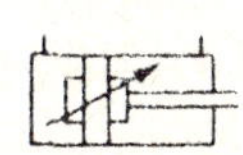	
Telescopic cylinder:			
– single acting	The fluid pressure always acts in one and the same direction (on the forward stroke)	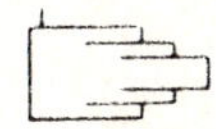	
– double acting	The fluid pressure operates alternately in both directions (forward and backward strokes)	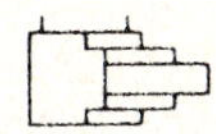	

PRESSURE INTENSIFIERS

Description	Remarks	Symbol	
		Detailed	Simplified
— for one type of fluid	eg a pneumatic pressure x is transformed into a higher pneumatic pressure y	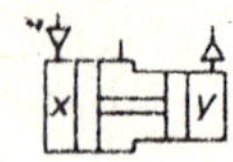	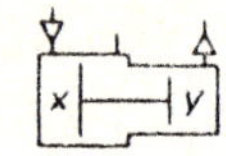
— for two types of fluid	eg a pneumatic pressure x is transformed into a higher hydraulic pressure y	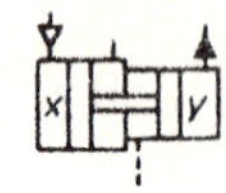	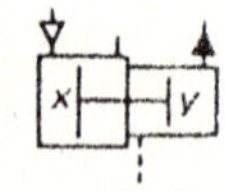
Air-oil actuator	Equipment transforming a pneumatic pressure into a substantially equal hydraulic pressure or vice versa		

CONTROL VALVES

Description	Remarks	Symbol
Method of representation of valves	Made up of one or more squares and arrows *In circuit diagrams hydraulic and pneumatic units are normally shown in the unoperated condition*	
One single square	Indicates unit for controlling flow or pressure, having in operation an infinite number of possible positions between its end positions so as to vary the conditions of flow across one or more or its ports, thus ensuring the chosen pressure and/or flow with regard to the operating conditions of the circuit	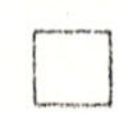
Two or more squares	Indicate a directional control valve having as many distinct positions as there are squares. The pipe connections are normally represented as connected to the box representing the unoperated condition. The operating positions are deduced by imagining the boxes to be displaced so that the pipe connections correspond with the ports of the box in question	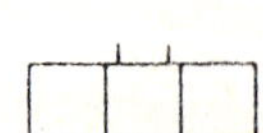
Simplified symbol for valves in cases of multiple repetition	The number refers to a note on the diagram in which the symbol for the valve is given in full	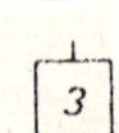

DIRECTIONAL CONTROL VALVES

Description	Remarks	Symbol
Flow paths:	Square containing interior lines	
– one flow path		
– two closed ports		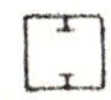
– two flow paths		 
– two flow paths and one closed port		
– two flow paths with cross connection		
– one flow path in a by-pass position, two closed ports		
Non-throttling directional control valve	The unit provides distinct circuit conditions each depicted by a square	
	Basic symbol for 2-position directional control valve	
	Basic symbol for 3-position directional control valve	
	A transitory but significant condition between two distinct positions is optionally represented by a square with dashed ends	
	A basic symbol for a directional control valve with two distinct positions and one transitory intermediate condition	
Designation: The first figure in the *designation* shows the number of ports (excluding pilot ports) and the second figure the number of distinct positions		

cont..........

DIRECTIONAL CONTROL VALVES (contd.)

Description	Remarks	Symbol
Directional control valve 2/2: – with manual control – controlled by pressure operating against a return spring (eg on air unloading valve)	Directional control valve with 2 ports and 2 distinct positions	 
Directional control valve 3/2: – controlled by pressure in both directions – controlled by solenoid with return spring	Directional control valve with 3 ports and 2 distinct positions	 
Directional control valve 4/2: – controlled by pressure in both directions by means of a pilot valve (with a single solenoid and spring return)	Directional control valve with 4 ports and 2 distinct positions	Detailed 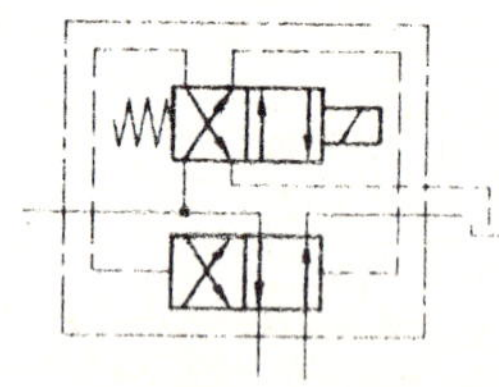 Simplified 
Directional control valve 5/2: – controlled by pressure in both directions	Directional control valve with 5 ports and 2 distinct positions	
Throttling directional control	The unit has 2 extreme positions and an infinite number of intermediate conditions with varying degrees of throttling All the symbols have parallel lines along the length of the boxes. Showing the extreme positions Showing the extreme positions and a central (neutral) position	 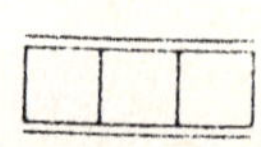

cont..........

DIRECTIONAL CONTROL VALVES (contd.)

Description	Remarks	Symbol
Throttling directional control (contd.)		
– with 2 ports (one throttling orifice)	For example: Tracer valve plunger operated against a return spring	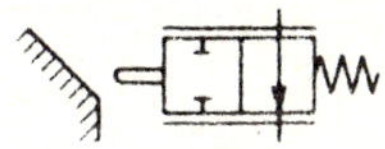
– with 3 ports (two throttling orifices)	For example: Directional control valve controlled by pressure against a return spring	
– with 4 ports (four throttling orifices)	For example: Tracer valve, plunger operated against a return spring	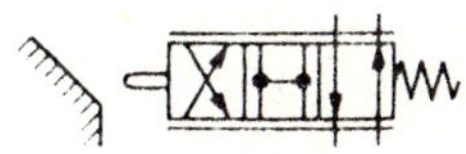
Electro-hydraulic servo valve: Electro-pneumatic servo valve:	A unit which accepts an analogue electrical signal and provides a similar analogue fluid power output	
– single stage	– with direct operation	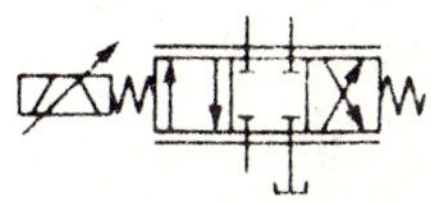
– two stage with mechanical feedback	– with indirect pilot operation	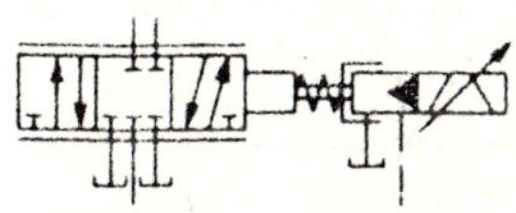
– two stage with hydraulic feedback	– with indirect pilot operation	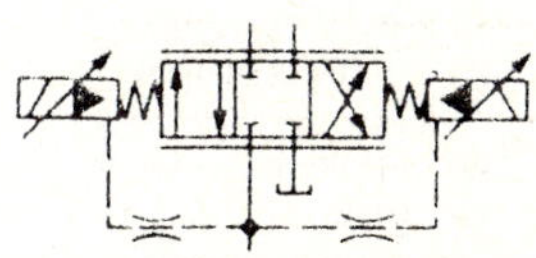

NON-RETURN VALVES, SHUTTLE VALVE, RAPID EXHAUST VALVE

Description	Remarks	Symbol
Non-return valve:		
— free	Opens if the inlet pressure is higher than the outlet pressure	
— spring loaded	Opens if the inlet pressure is greater than the outlet pressure plus the spring pressure	
— pilot controlled	With pilot control it is possible to prevent: — closing of the valve	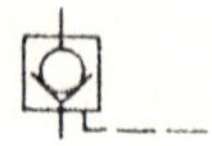
	— opening of the valve	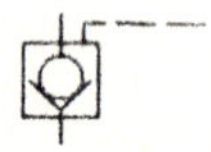
— with restriction	Unit allowing free flow in one direction but restricted flow in the other	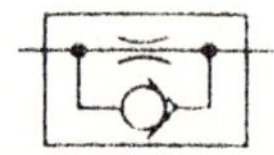
Shuttle valve	The inlet port connected to the higher pressure is automatically connected to the outlet port while the other inlet port is closed	
Rapid exhaust valve	When the inlet port is unloaded the outlet port is freely exhausted	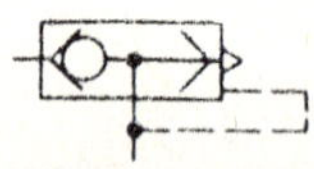

PRESSURE CONTROL VALVES

Description	Remarks	Symbol
Pressure control valve: — 1 throttling orifice normally closed		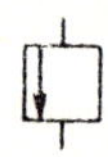 or 
— 1 throttling orifice normally open		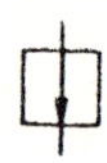 or 
— 2 throttling orifices, normally closed		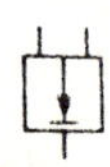
Pressure relief valve (safety valve):	Inlet pressure is controlled by opening the exhaust port to the reservoir or to atmosphere against an opposing force (for example a spring)	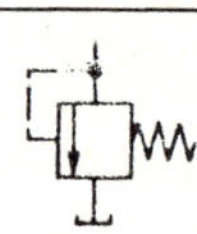 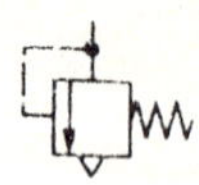
— with remote pilot control	The pressure at the inlet port is limited, or to that corresponding to the setting of a pilot control	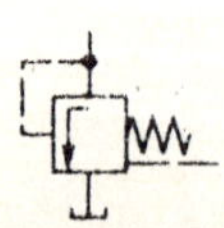 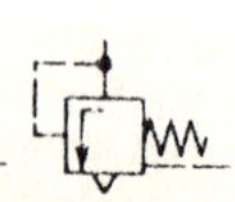

cont..........

PRESSURE CONTROL VALVES (contd.)

Description	Remarks	Symbol
Proportional pressure relief	Inlet pressure is limited to a value proportional to the pilot pressure	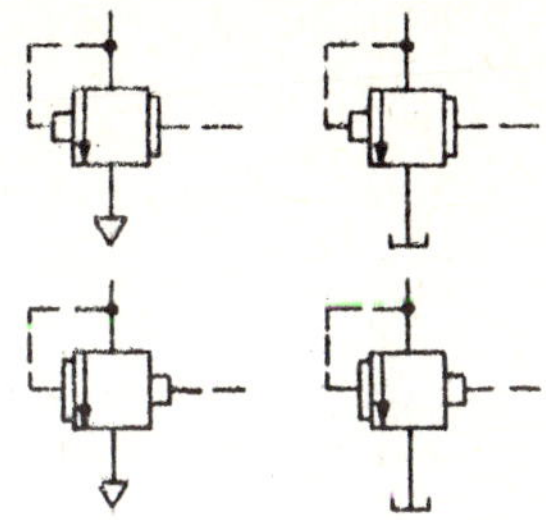
Sequence valve	When the inlet pressure overcomes the opposing force of the spring, the valve opens permitting flow from the outlet port	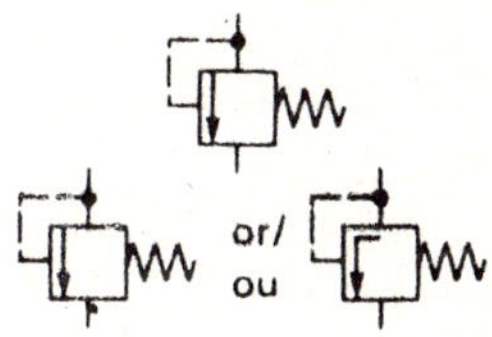
Pressure regulator or reducing valve (reducer of pressure):	A unit which, with a variable inlet pressure, gives substantially constant output pressure provided that the inlet pressure remains higher than the required outlet pressure	
– without relief port		
– without relief port with remote control	Outlet pressure is dependent on the control pressure	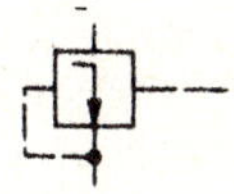
– with relief port		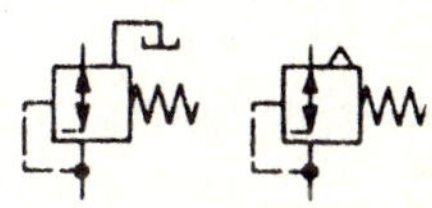
– with relief port, with remote control	Outlet pressure is dependent on the control pressure	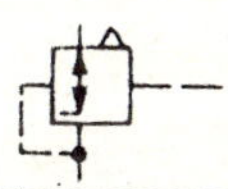
Differential pressure regulator	The outlet pressure is reduced by a fixed amount with respect to the inlet pressure	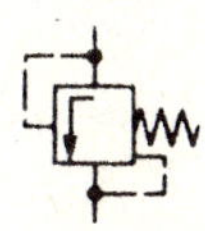
Proportional pressure regulator	The outlet pressure is reduced by a fixed ratio with respect to the inlet pressure	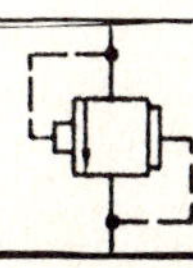

FLOW CONTROL VALVES

Description	Remarks	Symbol	
Throttle valve:	Simplified symbol (Does not indicate the control method or the state of the valve)		
– with manual control	Detailed symbol (indicates the control method or the state of the valve)	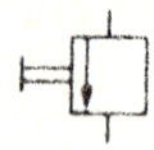	
– with mechanical control against a return spring (braking valve)		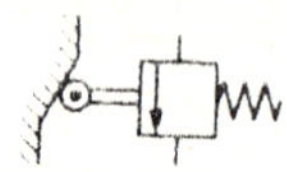	
Flow control valve:	Variations in inlet pressure do not affect the rate of flow	Detailed	Simplified
– with fixed output		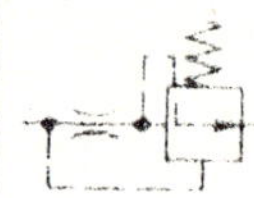	
– with fixed output and relief port to reservoir	With relief for excess flow	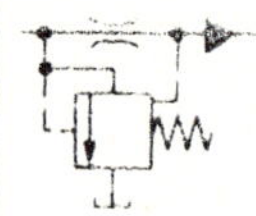	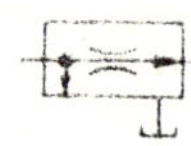
– with variable output		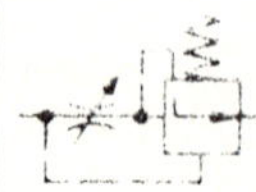	
– with variable output and relief port to reservoir	With relief for excess flow	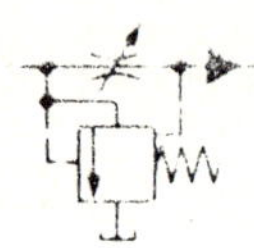	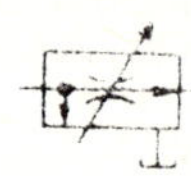
Flow dividing valve	The flow is divided into two flows in a fixed ratio substantially independent of pressure variations	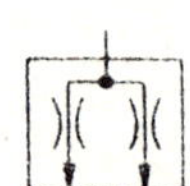	
Shut-off valve	Simplified symbol		

SOURCES OF ENERGY

Description	Remarks	Symbol
Pressure source	Simplified general symbol	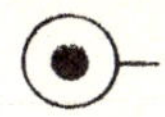
Hydraulic pressure source	Symbols to be used when the nature of the source should be indicated	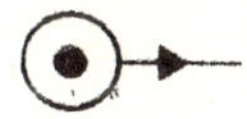
Pneumatic pressure source		
Electric motor	Symbol 113 in IEC Publication 117.2	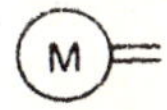
Heat engine		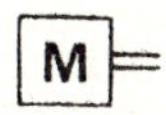
Flow lines and connections		
Flow line:		
– working line, return line and feed line		
– pilot control line		
– drain or bleed line		
– flexible pipe	Flexible hose, usually connecting moving parts	
– electric line		
Pipeline junction		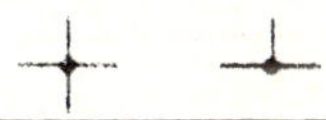
Crossed pipelines	Not connected	
Air bleed		
Exhaust port:		
– plain with no provision for connection		
– threaded for connection		

cont..........

SOURCES OF ENERGY (contd.)

Description	Remarks	Symbol
Power take-off:	On equipment or lines, for energy take-off or measurement	
– plugged		
– with take-off line		
Quick-acting coupling:		
– connected, without mechanically opened non-return valve		
– connected, with mechanically opened non-return valves		
– uncoupled, with open end		
– uncoupled, closed by free non-return valve		
Rotary connection:	Line junction allowing angular movement in service	
– one way		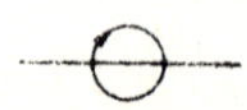
– three way		
Silencer		

RESERVOIRS

Description	Remarks	Symbol
Reservoir open to atmosphere:		
– with inlet pipe above fluid level		
– with inlet pipe below fluid level		
– with a header line		
Pressurized reservoir		
Accumulators	The fluid is maintained under pressure by a spring, weight or compressed gas (air, nitrogen, etc)	
Filter or strainer		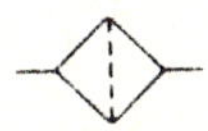
Water trap: – with manual control		
– automatically drained		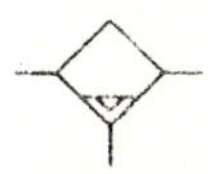
Filter with water trap: – with manual control		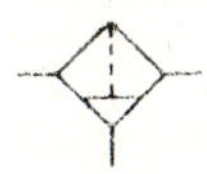
– automatically drained		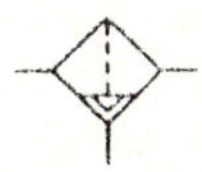
Air dryer	A unit drying air (for example by chemical means)	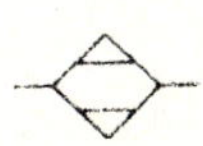

cont..........

RESERVOIRS (contd.)

Description	Remarks	Symbol
Lubricator	Small quantities of oil are added to the air passing through the unit, in order to lubricate equipment receiving the air	
Conditioning unit	Consisting of filter, pressure regulator, pressure gauge and lubricator – Detailed symbol – Simplified symbol	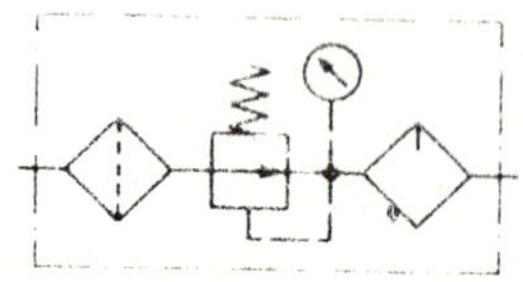 
Heat exchangers	Apparatus for heating or cooling the circulating fluid	
Temperature controller	The fluid temperature is maintained between two pre-determined values. The arrows indicate that heat may be either introduced or dissipated	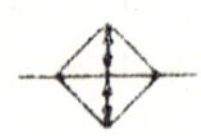
Cooler	The arrows in the diamond indicate the extraction of heat – without representation of the flow lines of the coolant – indicating the flow lines of the coolant	 
Heater	The arrows in the diamond indicate the introduction of heat	

CONTROL MECHANISMS

Description	Remarks	Symbol
Rotating shaft:	The arrow indicates rotation	
– in one direction		
– in either direction		
Detent	A device for maintaining a given position	
Locking device	*The symbol for unlocking control is inserted in the square	
Over-centre device	Prevents the mechanism stopping in a dead centre position	
Pivoting devices:		
– simple		
– with traversing lever		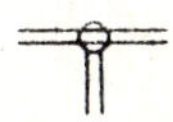
– with fixed fulcrum		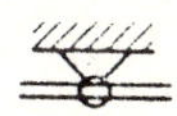

CONTROL METHODS

Description	Remarks	Symbol
Muscular control:	General symbol (without indication of control type)	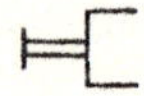
– by push button		
– by lever		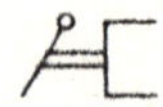
– by pedal		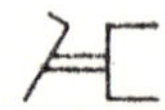
Mechanical control:		
– by plunger or tracer		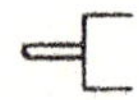
– by spring		
– by roller		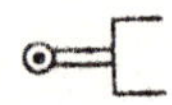
– by roller, operating in one direction only		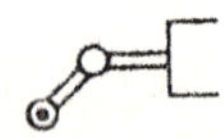
Electrical control:		
– by solenoid	– with 1 winding	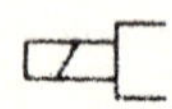
	– with 2 windings operating in opposite directions	
	– with 2 windings operating in a variable way progressively, operating in opposite directions	
– by electric motor		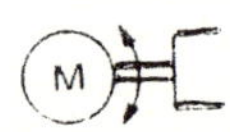
Control by application or release of pressure		
Direct acting control:		
– by application of pressure		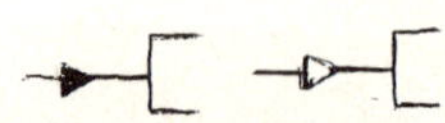

cont.........

CONTROL METHODS (contd.)

Description	Remarks	Symbol
– by release of pressure		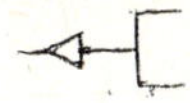
– by different control areas	In the symbol the larger rectangle represents the larger control area, ie the priority phase	
Indirect control, pilot actuated: – by application of pressure	General symbol for pilot directional control valve	
– by release of pressure		
Interior control paths	The control paths are inside the unit	
Combined control: – by solenoid and pilot directional valve –	The pilot directional valve is actuated by the solenoid	
– by solenoid or pilot directional valve	Either may actuate the control independently	
Mechanical feedback	The mechanical connection of a control apparatus moving part to a controlled apparatus moving part is represented by the symbol which joins the two parts connected	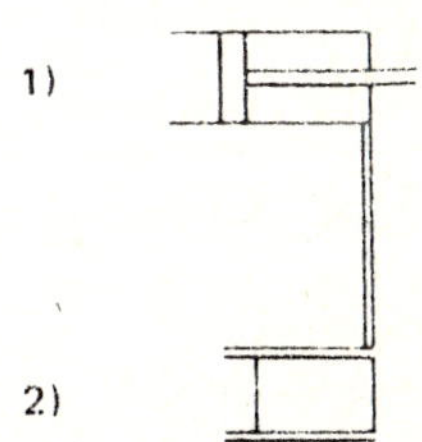 1) Controlled apparatus 2) Control apparatus

SUPPLEMENTARY EQUIPMENT

Description	Remarks	Symbol
Pressure measurement: — pressure gauge	The point on the circle at which the connection joins the symbol is immaterial	
Temperature measurement: —Thermometer	The point on the circle at which the connection joins the symbol is immaterial	
Measurement of flow: — Flow meter — Integrating flow meter		 
Other apparatus Pressure electric switch		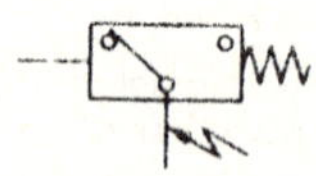

VDMA SYMBOLS

VDMA symbols for valve actuation are shown below. This graphic system was originated by the Organisation of German Machine Manufacturing Industry specifically for hydraulic valves, but has been widely used in Europe for pneumatic as well as hydraulic components.

Actuation Method	Symbol	Examples	
Hand lever			Direct actuated hand valve
Push Button			Direct actuated push-button (spring return)
Treadle			Direct actuated foot valve (spring return)
Cam Follower		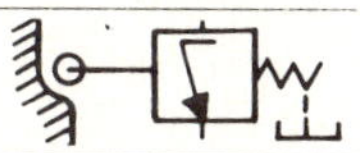	Cam valve
Air or Fluid Actuator	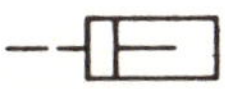		3-position, air actuated valve
Solenoid	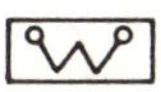		Solenoid (direct) actuated valve
Other Method			The asterisk in the first box refers to a footnote. For example, this footnote would say "photo electric cell"
Indirect actuation (basic symbol)	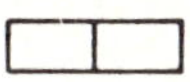	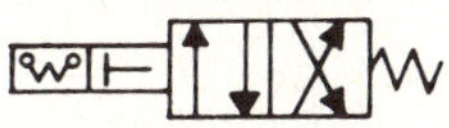	Solenoid actuated, pilot operated valve

Auxiliary Terms

Spring		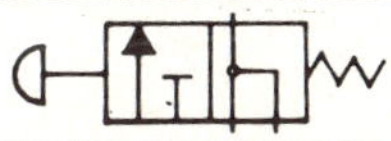	Foot actuated valve with spring return
Adjustment			Adjustable spring load
Detent			Direct operated hand valve, spring returned with detent
Ratchet (Pawl)			The asterisk means a footnote will explain the ratchet.

JIC GRAPHICAL SYMBOLS

The Joint Industry Conference (JIC) graphical pneumatic symbols are widely used by the American pneumatic industry (and also elsewhere). The system is based on graphic symbols emphasizing both the function of a component and its construction.

The basic symbols are divided into three main categories – supply, control and work element. The basic valve symbol consists of one or more square envelopes, with lines to show flow direction between ports. Direction of flow paths is indicated by arrows. The valve is drawn to show normal or neutral position. A single envelope symbol is used for valves with only one flow path. Multiple envelopes are used where there is more than one flow path. Envelopes move to show how flow paths change when the valve elements are shifted to various positions.

Actuation method is indicated in a small rectangular box attached to the end of the envelope opposite the flow side. When there are two actuators, one box is shown at each end.

External air lines are designated by solid lines. Crossover of air lines is represented by a half circle. Pilot lines are shown by dashed lines.

MISCELLANEOUS UNITS	Symbol
Motor, Drive, Electric	
Lubricator	
Intensifier	
Accumulator	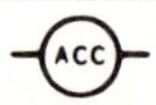
Filter	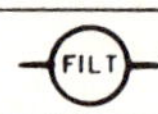
Separator	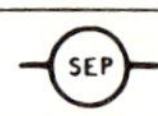
Pressure Switch	
Pressure Gauge	
Spring	
Shaft, Rotating (Arrow, Front of Shaft)	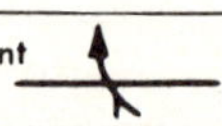
Component Enclosure	
Storage or Surge Tank	
Heat Exchanger (Intercooler or Aftercooler)	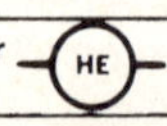
Muffler, Exhaust	

VALVES	Symbol
Valve, Check	
Valve, Restriction, Variable	
Valve, Basic Symbol (Insert Model No. for Special Valves	
Method of Indicating Internal Flow	
VALVE EXAMPLES	
Valve, Manual Shut Off	
Valve, Maximum Pressure	
Valve, Relief Remotely Operated	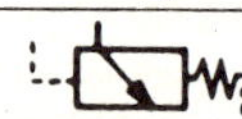
Valve, Sequence Directly Operated	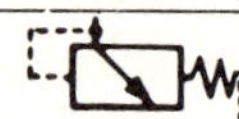
Valve, Pressure Regulator	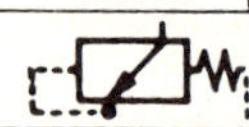
Valve, Flow Control Pressure Compensated	
Valve, Shut Off 2 Pos.— 2 Connection	
Valve, Directional 2 Pos.— 3 Connection	
Valve, Directional 2 Pos.— 4 Connection	
Valve, Directional 3 Pos.— 4 Connection Closed Center	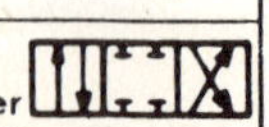

LINES	
Line, Working	
Line, Pilot	
Exhaust	
Line, Flexible	
Connector (Dot to be 3 x width of Associated Line)	
Direction of Flow	
Line, Passing	
Line, Joining (Tee Cross, etc.) (Dot to be 3 x w)	
Plug or Plugged Connection	
Testing Station (Gauge Connection)	T S
Power Take-Off	P T O
Restriction, Fixed	

MOTORS AND CYLINDERS	
Motor, Rotary Fixed Displacement	MF
Motor, Rotary Variable Displacement	MV
Motor, Oscillating	MO
Cylinder, Single Acting Plunger Type	
Piston Type	
Cylinder, Double Acting Single End Rod	
Double End Rod	
Cylinder, Rotating, Air	R

COMPRESSORS	
Compressor, Single Fixed Displacement	CF
Compressor, Single Variable Displacement	CV

METHODS OF OPERATION	
Control, Basic Symbol	
Control, Centrifugal	CENT
Control, Compensator	COMP
Control, Compensator Pressure	PR COMP
Control, Compensator Temperature	TEMP CONT
Control, Cylinder	CYL
Control, Detent	DET
Control, Manual	MAN
Control, Mechanical	MECH
Control, Motor Electric	ELEC MTR
Control, Motor Air	AIR MTR
Control, Pilot Hydraulic	HYD
Control, Pilot Air	AIR
Control, Servo	SERVO
Control, Solenoid	SOL
Control, Solenoid Hyd. Pilot Operated	SOL HYD
Control, Solenoid Air Pilot Operated	SOL AIR
Control, Thermal	THERM
Control, Pilot Hyd. Differential Area	HYD DIFF

CETOP RECOMMENDATIONS FOR CIRCUIT DIAGRAMS

The following rules and recommendations are extracted from CETOP Provisional Recommendations RP41 – Hydraulic and Pneumatic Systems Circuit Diagrams – published by the Comite European des Transmissions Oleohydrauliques et Pneumatiques.

The purpose of RP41 is to facilitate the design, construction, description and maintenance of fluid power systems by aiding communication through uniform presentation and by preventing confusion. The recommendation is based on the applicable ISO standards and CETOP recommendations. It is not necessary for a circuit diagram to incorporate all the requirements listed but anything which is shown must be in accordance with the appropriate sections of the document. The recommendation is generally applicable to the representation of all fluid power systems.

Design of Circuit Diagrams

The circuit diagram must show fluid circulation for all the control and motion functions. All hydraulic and pneumatic components and flow line junctions in the plant will be represented functionally. In addition all control components that affect the operational processes of the fluid power system will be shown.

The circuit diagram need not be constructed with reference to the physical arrangement of the equipment in the system. Interdependent circuits should be shown on one diagram. For electro-hydraulic or electro-pneumatic controls the circuit diagram will, however, be divided into one hydraulic or pneumatic circuit diagram and, if necessary, a separate electrical circuit.

The following circuit diagram formats should be used:–

ISO-A-Format (ISO Recommendation R216)
Length: preferably 297 mm
exceptionally 420 or 594 mm
Width: up to 1189 mm
Diagrams will be folded to ISO A4 format with binding margin (except for diagrams for reproduction).

The layout of the circuit diagram must be clear. Care should be taken to see that the circuit paths are easily followed. Pneumatic circuits should be drawn if possible in the sequence of the operational process of function. The components of the individual control chains and groups should be drawn wherever possible in the direction of energy flow.

Cylinders and directional control valves should preferably be shown horizontally. Fluid lines should be represented wherever possible by straight lines and without intersections.

The location of control components such as sequence switch cams and simple cams, etc should be indicated. If a signalling element is to be actuated by a one way trip, then the actuating direction will be indicated by an arrow. In the case of electro-hydraulic and electro-pneumatic control systems, signalling and final control elements such as limit switches and electromagnetic valves will be shown in both circuit diagrams.

Individual sub-circuits may be identified. The function of each operating element should be indicated beside it (eg clamping, lifting, etc). Where necessary for better understanding a function chart can be drawn up.

Components should be represented by symbols in accordance with the applicable standards, eg ISO R1219 for fluid power equipment. Where no symbols exist, the operation of the part should be clearly shown, eg by means of a simplified sectional representation or the like.

Symbols that are repeated in the diagram can be represented by a numbered rectangle and the corresponding symbols shown separately.

Data to go on Circuit Diagram

Every component must be clearly identified, eg by item number. Components in a given sub-circuit may be given a common prime identification plus a running sub-number, eg sub-circuit 2 component 3 = 2.3.

Component Operating Positions

In the case of valves with discrete switching positions these positions may be identified by letters, the letter 'O' being used only for the spring returned position (eg springloaded valves).

Identification of Ports

Ports will be identified on the circuit diagram by the characters marked on the components or the connection plate. This applies to individual components and sub-assemblies.

Identification of Oil or Air Lines

The pipelines are to be drawn according to their function in line with ISO Recommendations ISO R1219 (5.2.1).

In cases where hydraulic and pneumatic circuits are represented on one diagram they should be differentiated in accordance with ISO R1219.

Should it be necessary to identify the pipelines more precisely than in ISO R1219, the following code applies:—

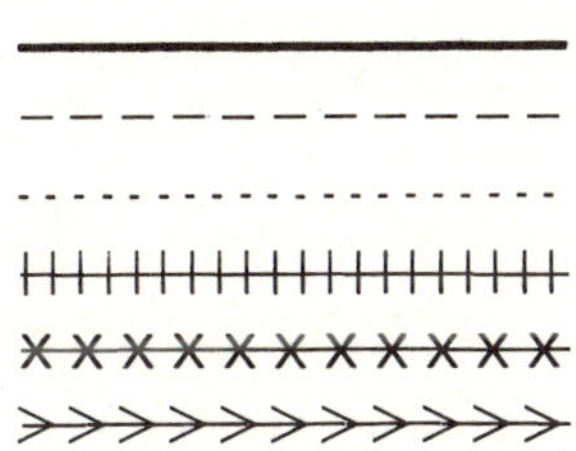

full line	pressure line
dashed line	pilot control line
short dotted line	drain or bleed line
full line with short cross lines	return or exhaust line
full line with crosses	replenishing line
full line with arrows	pump inlet line

If pipelines are identified by colour, the following colour code should be used:—

red line	pressure line
red dashed line	pilot control line
blue line	return or exhaust line
blue dashed line	drain or bleed line
green line	replenishing line
yellow line	pump inlet line

Any deviations from the above or additional colours should be explained on the circuit diagram.

The following technical data should be given beside each individual component or on a component list.

For reservoirs, give the volume represented by the maximum quantity of fluid in the system and, if required, the minimum quantity, the latter figure indicating the point at which replenishment is necessary. Show also the type and viscosity grade of the fluid. For pneumatic reservoirs indicate the volume.

For constant delivery pumps, indicate nominal delivery rate for the application and, for variable delivery pumps, minimum and maximum delivery rate. In both cases show speed of rotation and nominal power of the prime movers. For prime movers indicate nominal power at stated speed and speed range if applicable. For pressure valves and pressure switches, show pressure setting or permissible pressure range for the system.

For cylinders, the internal diameter of cylinder, the diameter of piston rod, maximum stroke in mm, eg 100/50 x 500, and if necessary speed and force should be given. For each cylinder, describe the function, eg clamping, lifting, transverse feed etc. For telescopic cylinders this information should be given for each stage. Where necessary the circuit should be supplemented by a stroke-time diagram showing the speed.

For semi rotary actuators show torque in relation to pressure, if necessary in each direction, the angular movement, the speed and the function.

For fixed displacement motors, the nominal capacity and, for variable displacement motors, the maximum capacity per revolution should be indicated. For both give the torque in relation to pressure for the application, speed range, direction of rotation and function.

For gas loaded hydraulic accumulators indicate the precharge pressure, type of gas, effective gas volume, working pressure range and corresponding draw off capacity. See RP62 H when available.

For pipes, give the external diameter in millimetres, wall thickness in millimetres and pipe material if necessary. For flexible hose, indicate the nominal diameter and type.

For delaying devices the time delay in seconds or range of adjustment should be shown. Any temporary connection ports such as bleed and test points should be identified by size, type and as marked on component.

For filters give the nominal filtration rating, flow rate and nominal pressure.

Conversion Factors

The following Table (by Norgren) gives practical conversion factors for both approximate conversions (suitable for normal working) and more exact working.

To convert from the OLD UNIT	and find the SI UNIT	use as the APPROXIMATE conversion	which gives an accuracy of	use as the more EXACT conversion	which gives an accuracy of
PRESSURE					
lbf/in^2 or p.s.i.g.	bar*	× 7 then ÷ 100	1.5%	÷ 14.5	0.025%
lbf/in^2 or p.s.i.g.	kilopascal (kPa)	× 7	1.5%	× 6.9	0.08%
lbf/in^2 or p.s.i.g.	megapascal (MPa)	× 7 then ÷ 1000	1.5%	× 6.9 then ÷ 1000	0.08%
inches Water Gauge (inWG)	millibar (mbar)	× 10 then ÷ 4	0.6%	× 2.49	0.16%
kgf/cm^2 or kp/cm^2	bar*	× 1	2.0%	× 0.98	0.07%
kgf/cm^2 or kp/cm^2	kilopascal (kPa)	× 100	2.0%	× 98	0.07%
kgf/cm^2 or kp/cm^2	megapascal (MPa)	÷ 10	2.0%	× 0.098	0.07%
millimetres Water Gauge (mmH_2O)	millibar (mbar)	÷ 10	2.0%	× 0.098	0.07%
Atmosphere	bar*	× 1	1.3%	× 1.013	0.02%
Atmosphere	kilopascal (kPa)	× 100	1.3%	× 101.3	0.02%
Atmosphere	megapascal (MPa)	÷ 10	1.3%	× 0.101	0.3%
Tons/in^2	bar*	× 1000 then ÷ 7	7.5%	× 154	0.3%
Tons/ft^2	bar*	× 1	1.5%	× 1.07	0.24%
FORCE (WEIGHT)					
lbf (pound force)	newton (N)	× 4	10%	× 9 then ÷ 2	1.2%
kp (kilopond)	newton (N)	× 10	2%	× 9.8	0.1%
FLOW					
c.f.m. or ft^3/min	cubic decimetres/second (dm^3/s)*	÷ 2	5.9%	× 0.472	0.01%
c.f.m. or ft^3/min	litres/s (l/s)	÷ 2	5.9%	× 0.472	0.01%
c.f.m. or ft^3/min	cubic metres/second (m^3/s)	÷ 2 then ÷ 1000	5.9%	× 0.472 then ÷ 1000	0.01%
ft^3/hour	dm^3/s*	× 8 then ÷ 1000	1.7%	× 7.9 then ÷ 1000	0.4%
Nl/m litres/min	dm^3/s*	× 2 then ÷ 100	20%	÷ 60	exact
Nm^3/hour	dm^3/s*	÷ 4	10%	× 0.28	0.8%
TORQUE					
Pounds. feet (lbf.ft)	newton metres (N.m)*	× 3 then ÷ 2	10%	× 1.36	0.3%
Pounds. inches (lbf.in)	newton metres (N.m)*	÷ 10	11%	× 0.11	2.6%
POWER					
Horsepower (hp)	watts (W)	× 3 then ÷ 4 then × 1000	0.6%	× 746	0.04%
Horsepower (hp)	kilowatts (kW)*	× 3 then ÷ 4	0.6%	× 0.746	0.04%
TEMPERATURE					
Fahrenheit (°F)	Celcius (°C)	Subtract 32 then ÷ 2	10% between 0°F and 400°F	Add 40 then × 5 then ÷ 9 then subtract 40	exact
MASS					
Pound (lb)	gram (g)	× 1000 then ÷ 2	10%	× 9 then × 100 then ÷ 2	0.8%
Pound (lb)	kilogram (kg)*	÷ 2	10%	× 9 then ÷ 2 then ÷ 10	0.8%
Ton (UK = 2240 lb)	tonne (t)*	× 1	1.6%	× 1.02	0.4%
Ton (USA = 2000 lb)	tonne (t)*	× 9 then ÷ 10	0.8%	× 0.91	0.3%
ounce (oz)	gram (g)*	× 3 then × 10	6%	× 28.4	0.2%
VOLUME					
Gallon (UK)	litre (l)*	× 5	10%	× 4.5	1%
Gallon (USA)	litre (l)*	× 4	5.7%	× 3.8	0.4%
Pint (UK)	litre (l)*	× 6 then ÷ 10	5.6%	× 0.57	0.3%
Pint (USA)	litre (l)*	÷ 2	5.7%	× 0.47	0.7%
Fluid ounce (UK)	cubic centimetres (cm^3)*	× 3 then × 10	5.6%	× 28.4	0.05%
Fluid ounce (USA)	cubic centimetres (cm^3)*	× 3 then × 10	1.4%	× 29.6	0.09%
MEASUREMENT					
inches	millimetres (mm)*	÷ 4 then × 100	1.6%	× 25.4	exact
$y/_{16}$ in	millimetres (mm)*	'y' × 3 then ÷ 2	5.5%	× 1.6	0.8%
$y/_{1000}$ in	millimetres (mm)*	'y' ÷ 4 then ÷ 10	1.6%	× 0.0254	exact
feet	metres (m)	× 3 then ÷ 10	1.6%	× 0.305	0.1%

*Preferred unit

PRESSURE CONVERSION TABLE – LOW VACUUM

Torr (mm of mercury)	Micrometres (microns) of mercury	Inches of mercury	lb/in^2
1.0×10^{-3}	1	0.00003937	0.00001934
2.0	2	0.00007874	0.00003868
3.0	3	0.0001181	0.00005802
4.0	4	0.0001575	0.00007736
5.0	5	0.0001969	0.00009670
6.0	6	0.0002362	0.0001160
7.0	7	0.0002756	0.0001354
8.0	8	0.0003150	0.0001547
9.0	9	0.0003543	0.0001741
1.0×10^{-2}	10	0.0003937	0.0001934
1.2	12	0.0004724	0.0002321
1.4	14	0.0005512	0.0002708
1.6	16	0.0006299	0.0003094
1.8	18	0.0007087	0.0003481
2.0	20	0.007874	0.0003878
2.5	25	0.0009843	0.0004835
3.0	30	0.001181	0.000580
3.5	35	0.001378	0.000677
4.0	40	0.001575	0.000774
4.5	45	0.001772	0.000870
5.0	50	0.001969	0.000967
6.0	60	0.002362	0.001160
7.0	70	0.002756	0.001354
7.5	75	0.002953	0.001451
8.0	80	0.003149	0.001547
9.0	90	0.003543	0.001740
1.0×10^{-1}	100	0.003937	0.001934
2.0	200	0.007874	0.003870
2.5	250	0.009845	0.004835
3.0	300	0.01181	0.00580
4.0	400	0.01575	0.00774
5.0	500	0.01969	0.00967
6.0	600	0.02362	0.01160
7.0	700	0.02756	0.01354
7.5	750	0.02953	0.01450
8.0	800	0.03149	0.01547
9.0	900	0.03543	0.01741
1.0	1 000	0.03937	0.01934

PRESSURE CONVERSION TABLE – MEDIUM VACUUM

Torr (mm of mercury)	Inches of mercury	lb/in^2	Millibars
1	0.0394	0.01934	1.333
2	0.0787	0.03868	2.666
3	0.1181	0.05802	3.999
4	0.1575	0.07736	5.332
5	0.1970	0.09670	6.665
6	0.2360	0.1160	7.998
7	0.2756	0.1354	9.331
8	0.3150	0.1547	10.664
9	0.3543	0.1741	11.997
10	0.3940	0.1934	13.330
12	0.4724	0.2321	15.996
14	0.5512	0.2708	18.662
16	0.6299	0.3094	21.328
18	0.7087	0.3481	23.994
20	0.7870	0.3868	26.660
25	0.9843	0.4835	33.325
30	1.1811	0.580	39.990
35	1.378	0.677	46.655
40	1.575	0.774	53.320
45	1.777	0.870	59.985
50	1.969	0.967	66.650
60	2.362	1.160	79.980
70	2.756	1.354	93.310
80	3.149	1.547	106.64
90	3.543	1.740	119.97
100	3.937	1.934	133.30
150	5.906	2.900	199.95
200	7.874	3.870	266.60
250	9.843	4.830	333.25
300	11.810	5.800	399.90
350	13.780	6.760	466.55
400	15.750	7.740	533.20
450	17.720	8.700	599.85
500	19.690	9.670	666.50
550	21.650	10.630	733.15
600	23.620	11.600	799.80
650	25.590	12.560	866.45
700	27.560	13.540	933.10
750	29.530	14.500	999.75
760	29.921	14.696	1 013.10

VACUUM PRESSURE CONVERSION TABLE

Inches of Mercury	mm of Mercury	lbf/in^2	Vacuum Inches of Mercury	Millibars	Inches of Mercury	mm of Mercury	lbf/in^2	Vacuum Inches of Mercury	Millibars
0.01	0.254	0.00491	29.911	0.3386		80.0	1.547	26.772	106.64
	0.500	0.00966	29.901	0.6664		90.0	1.740	26.378	119.97
0.02	0.508	0.00982	29.901	0.6772		100.0	1.934	25.984	133.30
0.03	0.762	0.01474	29.891	1.0160	4.000	101.6	1.965	25.921	135.40
	1.000	0.01934	29.882	1 3330	5.000	127.0	2.456	24.921	169.30
0.04	1.016	0.01965	29.881	1.3540		150.0	2.900	24.015	199.95
0.05	1.270	0.02456	29.871	1.6930	6.000	152.4	2.947	23.921	203.20
0.06	1.524	0.02947	29.861	2.0320	7.000	177.8	3.438	22.921	237.00
0.07	1.778	0.03438	29.851	2.3700		200.0	3.870	22.047	266.60
	2.000	0.03868	29.000	2.6660	8.000	203.2	3.930	21.921	270.90
0.08	2.032	0.03930	29.831	2.7090	9.000	228.6	4.421	20.921	304.80
0.09	2.286	0.04421	29.831	3.0480		250.0	4.833	20.078	333.25
0.10	2.540	0.04912	29.821	3.3860	10.000	254.0	4.912	19.921	338.60
	3.000	0.05802	29.803	3.9990	11.000	279.4	5.403	18.921	372.40
	4.000	0.07736	29.763	5.3320		300.0	5.800	18.111	399.90
	5.000	0.09670	29.724	6.6650	12.000	304.8	5.894	17.921	406.30
0.20	5.080	0.09820	29.721	6.7730	13.000	330.2	6.386	16.921	440.20
	6.000	0.11600	29.685	7.9980		350.0	6.760	16.141	466.55
	7.000	0.13540	29.646	9.3310	14.000	355.6	6.877	15.921	474.00
0.30	7.620	0.14740	29.621	10.1600	15.000	381.0	7.368	14.921	507.90
	8.000	0.15740	29.606	10.6640		400.0	7.740	14.171	533.20
	9.000	0.17410	29.567	11.9970	16.000	406.4	7.859	13.921	541.70
	10.000	0.19230	29.527	13.330	17.000	431.8	8.350	12.921	575.60
0.40	10.160	0.19340	29.521	13.5400		450.0	8.700	12.201	599.85
	12.000	0.23210	29.449	15.9960	18.000	457.2	8.842	11.921	609.40
0.50	12.700	0.24560	29.421	16.9300	19.000	482.6	9.333	10.921	643.30
	14.000	0.27080	29.370	18.6620		500.0	9.667	10.131	666.50
0.60	15.240	0.29470	29.321	20.3200	20.000	508.0	9.824	9.921	677.20
	16.000	0.30940	29.290	21.3280	21.000	533.4	10.315	8.921	711.00
0.70	17.780	0.34380	29.221	23.7000		550.0	10.630	8.271	733.15
	18.000	0.34810	29.212	23.9940	22.000	558.8	10.806	7.921	744.90
	20.000	0.38680	29.134	26.6600	23.000	584.2	11.298	6.921	778.70
0.80	20.320	0.39300	29.121	27.0900		600.0	11.600	6.291	799.80
0.90	22.860	0.44210	29.021	30.4800	24.000	609.6	11.789	5.921	812.60
	25.000	0.48350	28.937	33.3250	25.000	635.0	12.280	4.921	846.50
1.00	25.900	0.49120	28.921	33.8600		650.0	12.560	4.331	866.45
	30.000	0.58000	28.740	39.9900	26.000	660.4	12.771	3.921	880.30
	35.000	0.67700	28.543	46.6650	27.000	685.8	13.262	2.921	914.20
	40.000	0.77400	28.346	53.3200		700.0	13.540	2.361	933.10
	45.000	0.87000	38.149	59.9850	28.000	711.2	13.754	1.921	948.00
	50.000	0.96700	27.952	66.6500	29.000	736.6	14.245	0.921	981.90
2.00	50.800	0.98200	27.921	67.7200		750.0	14.500	0.391	999.75
	60.000	1.16000	27.559	79.9800		755.0	14.597	0.191	1 006.42
	70.000	1.35400	27.165	93.3200	29.921	760.0	14.696	0.000	1 013.10
3.00	76.200	1.47400	26.921	101.6000					

'RATIONALIZED' EQUIVALENTS OF STANDARD INCH FRACTIONS

Inch Fraction	mm (nominal)	Remarks
1/64	0.4	May also be quoted as 0.5
1/32	0.8	
3/64	1.0	
1/16	1.5	
3/32	2.5	
1/8	3.0	
5/32	4.0	
3/16	5.0	
1/4	6.0	Alternatively, 6.5 mm nominal
5/16	7.5	
3/8	10.0	
1/2	12.5	
5/8	15.0	Rounded for consistent progression
3/4	20.0	
7/8	22.5	

MILLIMETRE EQUIVALENTS OF FRACTIONS OF AN INCH

Inch		Millimetre	Inch		Millimetre
1/64		0.3969	33/64		13.0969
	1/32	0.7937		17/32	13.4937
3/64		1.1906	35/64		13.8906
	1/16	1.5875		9/16	14.2875
5/64		1.9844	37/64		14.6843
	3/32	2.3812		19/32	15.0812
7/64		2.7781	39/64		15.4781
	1/8	3.1750		5/8	15.8750
9/64		3.5719	41/64		16.2718
	5/32	3.9687		21/32	16.6687
11/64		4.3656	43/64		17.0656
	3/16	4.7625		11/16	17.4625
13/64		5.1594	45/64		17.8593
	7/32	5.5562		23/32	18.2562
15/64		5.9531	47/64		18.6531
	1/4	6.3500		3/4	19.0500
17/64		6.7469	49/64		19.4468
	9/32	7.1437		25/32	19.8437
19/64		7.5406	51/64		20.2406
	5/16	7.9375		13/16	20.6375
21/64		8.3344	53/64		21.0343
	11/32	8.7312		27/32	21.4312
23/64		9.1281	55/64		21.8281
	3/8	9.5250		7/8	22.2250
25/64		9.9219	57/64		22.6218
	13/32	10.3187		29/32	23.0187
27/64		10.7156	59/64		23.4156
	7/16	11.1125		15/16	23.8125
29/64		11.5094	61/64		24.2093
	15/32	11.9062		31/32	24.6062
31/64		12.3031	63/64		25.0031
	1/2	12.7000		1	25.4000

MILLIMETRE EQUIVALENTS OF INCHES AND INCH FRACTIONS

Inches	0.0	1/8	1/4	3/8	1/2	5/8	3/4	7/8
1	25.400	28.575	31.750	34.925	38.100	41.275	44.450	47.625
2	50.800	53.975	57.150	60.325	63.500	66.675	69.850	73.025
3	76.200	79.375	82.550	85.725	88.900	92.075	95.250	98.425
4	101.60	104.77	107.95	111.12	114.30	117.47	120.65	123.82
5	127.00	130.17	133.35	136.52	139.70	142.87	146.05	149.22
6	152.40	155.57	158.75	161.92	165.10	168.27	171.45	174.62
7	177.80	180.97	184.15	187.32	190.50	193.67	196.85	200.02
8	203.20	206.37	209.55	212.72	215.90	219.07	222.25	225.42
9	228.60	231.77	234.95	238.12	241.30	244.47	247.65	250.82
10	254.00	257.17	260.35	263.52	266.70	269.87	273.05	276.22
11	279.40	282.57	285.75	288.92	292.10	295.27	298.45	301.62
12	304.80	307.97	311.15	314.32	317.50	320.67	323.85	327.02
13	330.20	333.37	336.55	339.72	342.90	346.07	349.25	352.42
14	355.60	358.77	361.95	365.12	368.30	371.47	374.65	377.82
15	381.00	384.17	387.35	390.52	393.70	396.87	400.05	403.22
16	406.40	409.57	412.75	415.92	419.10	422.27	425.45	428.62
17	431.80	434.97	438.15	441.32	444.50	447.67	450.85	454.02
18	457.20	460.37	463.55	466.72	469.90	473.07	476.25	479.42
19	482.60	485.77	488.95	492.12	495.30	498.47	501.65	504.82
20	508.00	511.17	514.34	517.52	520.70	523.87	527.05	530.22
21	533.40	536.57	539.75	542.92	546.10	549.27	552.45	555.62
22	558.80	561.97	565.15	568.32	571.50	574.67	577.85	581.02
23	584.20	587.37	590.55	593.72	596.90	600.07	603.25	606.42
24	609.60	612.77	615.95	619.12	622.30	625.47	628.65	631.82
25	635.00	638.17	641.35	644.52	647.70	650.87	654.05	657.22
26	660.40	663.57	666.75	669.92	673.10	676.27	679.45	682.62
27	685.80	688.97	692.15	695.32	698.50	701.67	704.85	708.02
28	711.20	714.37	717.55	720.72	723.90	727.07	730.25	733.42
29	736.60	739.77	742.95	746.12	749.30	752.47	755.65	758.82
30	762.00	765.17	768.35	771.52	774.70	777.87	781.05	784.22
31	787.40	790.57	793.75	796.92	800.10	803.27	806.45	809.62
32	812.80	815.97	819.15	822.32	825.50	828.67	831.85	835.02
33	838.20	841.37	844.55	847.72	850.90	854.07	857.25	860.42
34	863.60	866.77	869.95	873.12	876.30	879.47	882.65	885.82
35	889.00	892.17	895.35	898.52	901.70	904.87	908.05	911.22
Inches	0.0	1/8	1/4	3/8	1/2	5/8	3/4	7/8

EQUIVALENT FREE AIR VOLUME litres^3 (dm^3) PER MINUTE

Volume of Compressed Air litres^3/min	LINE PRESSURE BAR (GAUGE)									
	1	2	3	4	5	6	7	8	9	10
0.1	0.2	0.3	0.4	0.5	0.6	0.7	0.8	0.9	1.0	1.1
0.2	0.4	0.6	0.8	1.0	1.2	1.4	1.6	1.8	2.0	2.2
0.3	0.6	0.9	1.2	1.5	1.8	2.1	2.4	2.7	3.0	3.3
0.4	0.8	1.2	1.6	2.0	2.4	2.8	3.2	3.6	4.0	4.4
0.5	1.0	1.5	2.0	2.5	3.0	3.5	4.0	4.5	5.0	5.5
0.6	1.2	1.8	2.4	3.0	3.6	4.2	4.8	5.4	6.0	6.6
0.7	1.4	2.1	2.8	3.5	4.2	4.9	5.6	6.3	7.0	7.7
0.8	1.6	2.4	3.2	4.0	4.8	5.6	6.4	7.2	8.0	8.8
0.9	1.8	2.7	3.6	4.5	5.4	6.3	7.2	8.1	9.0	9.9
1	2	3	4	5	6	7	8	9	10	11
2	4	6	8	10	12	14	16	18	20	22
3	6	9	12	15	18	21	24	27	30	33
4	8	12	16	20	24	28	32	36	40	44
5	10	15	20	25	30	35	40	45	50	55
6	12	18	24	30	36	42	48	54	60	66
7	14	21	28	35	42	49	56	63	70	77
8	16	24	32	40	48	56	64	72	80	88
9	18	27	36	45	54	63	72	81	90	99
10	20	30	40	50	60	70	80	90	100	110

EQUIVALENT COMPRESSED AIR VOLUME litres^3 (dm^3)

Free Air Volume litres^3	COMPRESSION PRESSURE – BAR (GAUGE)									
	1	2	3	4	5	6	7	8	9	10
0.1	0.05	0.033	0.025	0.02	0.017	0.014	0.0125	0.011	0.01	0.009
0.2	0.10	0.067	0.050	0.04	0.033	0.029	0.025	0.022	0.02	0.018
0.3	0.15	0.10	0.075	0.06	0.050	0.043	0.038	0.033	0.03	0.027
0.4	0.20	0.13	0.100	0.08	0.066	0.057	0.050	0.044	0.04	0.036
0.5	0.25	0.17	0.125	0.10	0.083	0.071	0.063	0.056	0.05	0.045
0.6	0.30	0.20	0.150	0.12	0.100	0.088	0.075	0.067	0.06	0.055
0.7	0.35	0.23	0.175	0.14	0.120	0.100	0.088	0.078	0.07	0.064
0.8	0.40	0.27	0.200	0.16	0.130	0.114	0.100	0.089	0.08	0.073
0.9	0.45	0.30	0.225	0.18	0.150	0.129	0.113	0.100	0.09	0.082
1	0.5	0.33	0.25	0.20	0.17	0.14	0.125	0.112	0.10	0.090
2	1.0	0.67	0.50	0.40	0.33	0.29	0.75	0.22	0.20	0.180
3	1.5	1.00	0.75	0.60	0.50	0.43	0.38	0.33	0.30	0.270
4	2.0	1.33	1.00	0.80	0.67	0.57	0.50	0.44	0.40	0.360
5	2.5	1.67	1.25	1.00	0.83	0.71	0.63	0.56	0.50	0.450
6	3.0	2.00	1.50	1.20	1.00	0.89	0.25	0.67	0.60	0.550
7	3.5	2.33	1.75	1.40	1.17	1.00	0.88	0.78	0.70	0.640
8	4.0	2.67	2.00	1.60	1.33	1.14	1.00	0.89	0.80	0.730
9	4.5	3.00	2.25	1.80	1.50	1.29	1.13	1.00	0.90	0.820
10	5.0	3.33	2.50	2.00	1.67	1.43	1.25	1.11	1.00	0.910

PERMEABILITY OF ELASTOMERS

Elastomers do not provide complete barriers to air, thus the permeability of such materials used as seals, bellows, diaphragms, etc, may be of significance in particular applications. This can influence the choice of elastomer for a specific duty (assuming the other characteristics of that elastomer are suitable).

For general duties the air permeability of most common elastomers is negligible at normal ambient temperatures (with silicone rubber as a notable exception). Rate of permeability does, however, increase with increasing temperatures. This effect is more marked with some elastomers than others. Butyl and nitrile rubbers are generally the most favourable.

Useful data are summarized in the tables.

MAXIMUM SERVICE TEMPERATURE AND AIR PERMEABILITY OF ELASTOMERS

Elastomer	Maximum Service Temp °C (°F)	Air Permeability $cm^3/sec/cm^2/cm \times 10^{-6}$ at			
		21 (70)	80 (175)	110 (230)	175 (350)
Butyl	95 (200)	0.2	3.2	13	
Low nitrile	120 (250)	1.3	8.0	22	
High nitrile	150 (300)	negligible	4.1	15	
Buna S	95 (200)	2.5	29.0	47	
Natural rubber	85 (185)	4.9	44.0	71	
Neoprene	150 (300)	1.0	9.8	26	
Polyacrylate	120 (250)	1.9	18.0	48	94
Polyurethane	120 (250)	0.5	9.7	31	
Fluorosilicone	230 (450)		128.0		
Silicone	250 (480)	115.0	350.0		690
Viton A	230 (450)		8.8	36	146

PERMEABILITY OF ELASTOMERS TO OTHER GASES

Elastomer	Permeability $cm^3/sec/cm^2/cm \times 10^{-8}$				
	Oxygen	Nitrogen	Carbon Dioxide	Hydrogen	Methane
Butyl	0.3–1.0	0.11–0.35	0.03–0.10	0.03–0.10	0.01–0.33
SBR	13.0–34.5	4.8–14.5	94.0–195.0	30.5–74.0	16.0–43.0
Neoprene	3.0–10.0	0.9–3.55	19.5–56.5	10.3–28.5	2.5–9.8
Nitrile	3.2–10.5	0.9–3.7	23.0–66.0	11.5–31.5	2.4–10.1
Natural rubber	18.0–19.5	6.6–22.5	102.0–220.0	39.0–97.0	22.0–64.0

Note: Values refer to permeability at 26°C (78°F) and 50°C (122°F) respectively.

PRESSURE EQUIVALENTS

	lb/in²	kN/m²	N/m²	in of H_2O	cm of Hg	bar	m bar	atm
lb/in²	1	0.15	14.5×10^{-5}	36×10^{-3}	0.2	14.5	14.5×10^{-3}	14.7
kN/m²	6.9	1	10^{-3}	0.25	1.3	100	0.1	101
N/m²	6.9×10^{3}	10^{3}	1	249	1.3×10^{3}	10^{5}	100	101×10^{3}
in of H_2O	27.7	4.1	4.1×10^{-3}	1	5.5	400	0.4	406
cm of Hg	5.2	75	75×10^{-3}	0.18	1	75×10^{-3}	75×10^{-6}	76
bar	69×10^{-3}	10^{-2}	10^{-5}	2.5×10^{-3}	13.3	1	10^{-3}	1.01
m bar	69	10	10^{-2}	2.5	13.3×10^{3}	10^{3}	1	1010
atm	68×10^{-3}	9.9×10^{-3}	9.9×10^{-6}	2.5×10^{-3}	13.2×10^{-3}	0.99	0.99×10^{-3}	1

VOLUMETRIC EQUIVALENTS

	m^3	cm^3	mm^3	yd^3	ft^3	in^3	litre	Imp.gall.	US gall.
m^3	1	10^{-6}	10^{-9}	76.5×10^{-2}	28.3×10^{-3}	16.4×10^{-6}	10^{-3}	4.5×10^{-3}	3.8×10^{-3}
cm^3	10^{6}	1	10^{-3}	76.5×10^{4}	28.3×10^{3}	16.4	10^{3}	4.5×10^{3}	3.8×10^{3}
mm^3	10^{9}	10^{3}	1	76.5×10^{7}	28.3×10^{6}	16.4×10^{3}	10^{6}	4.5×10^{6}	3.8×10^{6}
yd^3	1.3	1.3×10^{-6}	1.3×10^{-9}	1	37×10^{-3}	21.4×10^{-6}	1.3×10^{-3}	5.9×10^{-3}	5×10^{-3}
ft^3	35.3	35.3×10^{-6}	35.3×10^{-9}	27	1	5.8×10^{-4}	35.3×10^{-3}	0.16	0.13
in^3	61×10^{3}	61×10^{-3}	61×10^{-6}	46700	1728	1	61	276	231
Litre	10^{3}	10^{-3}	10^{-6}	765	28.3	16.4×10^{-3}	1	4.5	3.8
Imp.gall	220	22×10^{-5}	22×10^{-8}	169	6.25	3.6×10^{-3}	0.22	1	0.82
US gall	263	26×10^{-5}	26×10^{-8}	200	7.4	4.3×10^{-3}	0.26	1.20	1

CONVERSION – ATMOSPHERES* TO lb/in^2

	At									
	0	0.1	0.2	0.3	0.4	0.5	0.6	0.7	0.8	0.9
0	–	1.47	2.92	4.41	5.88	7.35	8.82	10.3	11.8	13.2
1	14.7	16.2	17.6	19.1	20.6	22.1	23.5	25.0	26.5	27.9
2	29.4	30.9	32.3	33.8	35.3	36.8	38.2	29.7	41.2	42.6
3	44.1	45.6	47.0	48.5	50.0	51.5	52.9	54.4	55.9	57.3
4	58.8	60.3	61.7	63.2	64.7	66.2	67.6	69.1	70.6	72.0
5	73.5	75.0	76.4	77.9	79.4	80.9	82.3	83.8	85.3	86.7
6	88.2	89.7	91.1	92.6	94.1	95.6	97.0	98.5	100.0	101.4
7	102.9	104.4	105.8	107.3	108.8	11.03	111.7	113.2	114.7	116.1
8	117.6	119.1	120.5	122.0	123.5	125.0	126.4	127.9	129.4	130.8
9	132.3	133.8	135.2	136.7	138.2	139.7	141.1	142.6	144.1	145.5
10	147.0	–	–	–	–	–	–	–	–	–

Atmospheres to lb/in^2: *These tables are based on an atmosphere of 14.7 lbf/in^2, as used in conventional English practice. This differs from the bar by approximately 1.35%, since a bar is approximately 14.5 lbf/in^2. Note also that the value of the technical atmosphere is approximately 14.2 lbf/in^2, on which the kolopond per square centimetre (kp/cm^2) is based.

CONVERSION – lb/in^2 TO ATMOSPHERES*

lb/in^2	lb/in^2									
	0	1	2	3	4	5	6	7	8	9
0	–	0.068	0.136	0.204	0.272	0.340	0.408	0.476	0.544	0.612
10	0.68	0.75	0.82	0.88	0.95	1.02	1.09	1.16	1.22	1.29
20	1.36	1.43	1.50	1.56	1.63	1.70	1.77	1.84	1.90	1.97
30	2.04	2.11	2.18	2.24	2.31	2.38	2.48	2.52	2.58	2.65
40	2.72	2.79	2.86	2.92	2.99	3.06	3.13	3.20	3.26	3.33
50	3.40	3.47	3.54	3.60	3.67	3.74	3.81	3.88	3.94	4.01
60	4.08	4.15	4.22	4.28	4.35	4.42	4.49	4.56	4.62	4.69
70	4.76	4.83	4.90	4.96	5.03	5.10	5.17	5.24	5.30	5.37
80	5.44	5.51	5.58	5.64	5.71	5.78	5.85	5.92	5.98	6.05
90	6.12	6.19	6.26	6.32	6.39	6.46	6.53	6.60	6.66	6.73
100	6.80	–	–	–	–	–	–	–	–	–

Atmospheres to lb/in^2: *These tables are based on an atmosphere of 14.7 lbf/in^2, as used in conventional English practice. This differs from the bar by approximately 1.35%, since a bar is approximately 14.5 lbf/in^2. Note also that the value of the technical atmosphere is approximately 14.2 lbf/in^2, on which the kolopond per square centimetre (kp/cm^2) is based.

CONVERSION : JOULES TO ft/lbf

Joules	0	0.1	0.2	0.3	0.4	0.5	0.6	0.7	0.8	0.9
0	—	0.074	0.148	0.221	0.295	0.369	0.443	0.516	0.590	0.664
1	0.738	0.811	0.885	0.959	1.033	1.106	1.180	1.254	1.328	1.401
2	1.475	1.549	1.623	1.696	1.770	1.844	1.918	1.991	2.065	2.139
3	2.213	2.286	2.360	2.434	2.508	2.581	2.655	2.729	2.803	2.876
4	2.950	3.024	3.098	3.171	3.245	3.319	3.393	3.466	3.540	3.614
5	3.688	3.761	3.835	3.909	3.983	4.056	4.130	4.204	4.278	4.351
6	4.425	4.499	4.573	4.646	4.720	4.794	4.868	4.941	5.015	5.089
7	5.163	5.236	5.310	5.384	5.458	5.531	5.605	5.679	5.753	5.826
8	5.900	5.974	6.048	6.121	6.195	6.269	6.343	6.416	6.490	6.564
9	6.638	6.711	6.785	6.859	6.933	7.006	7.080	7.154	7.228	7.301
10	7.375	—	—	—	—	—	—	—	—	—

Joules	0	1	2	3	4	5	6	7	8	9
0	—	0.738	1.475	2.213	2.950	3.688	4.425	5.163	5.900	6.638
10	7.375	8.112	8.850	9.587	10.325	11.062	11.800	12.537	13.275	14.012
20	14.750	15.487	16.225	16.962	17.700	18.437	19.175	19.912	20.650	21.387
30	22.125	22.862	23.600	24.337	25.075	25.812	26.550	27.287	28.025	28.762
40	29.500	30.237	30.975	31.712	32.450	33.187	33.925	34.662	35.400	36.137
50	36.875	37.612	38.350	39.078	39.825	40.562	41.300	42.037	42.775	43.512
60	44.250	44.987	45.725	46.462	47.200	47.937	48.675	49.412	50.150	50.887
70	51.625	52.362	53.100	53.837	54.575	55.312	56.050	56.787	57.525	58.262
80	59.000	59.737	60.475	61.212	61.950	62.687	63.425	64.162	64.900	65.637
90	66.375	67.112	67.850	68.587	69.325	70.062	70.800	71.537	72.275	73.012
100	73.750	—	—	—	—	—	—	—	—	—

Joules	0	10	20	30	40	50	60	70	80	90
0	—	7.38	14.75	22.13	29.50	36.88	44.25	51.63	59.00	66.38
100	73.75	81.12	88.50	95.87	102.25	110.62	118.00	125.37	132.75	140.12
200	147.50	154.87	162.25	169.62	177.00	184.37	191.75	199.12	206.50	213.87
300	221.25	228.62	236.00	243.37	250.75	258.12	265.50	272.87	280.25	287.62
400	295.00	302.37	309.75	317.12	324.50	331.87	339.25	346.62	354.00	361.37
500	368.75	376.12	383.50	390.87	398.25	405.62	413.00	420.37	427.75	435.12
600	442.50	449.87	457.25	464.62	472.00	479.37	486.75	494.12	501.50	508.87
700	516.25	523.62	531.00	538.37	545.75	553.12	560.50	567.87	575.25	582.62
800	590.00	597.37	604.75	612.12	619.50	626.87	634.25	641.62	649.00	656.37
900	663.75	671.12	678.50	685.87	693.25	700.62	708.00	715.37	722.75	730.12
1 000	737.50	—	—	—	—	—	—	—	—	—

CONVERSION : ft lbf TO JOULES

ft lbf	0	0.1	0.2	0.3	0.4	0.5	0.6	0.7	0.8	0.9
0	–	0.136	0.271	0.407	0.542	0.678	0.814	0.949	1.085	1.220
1	1.356	1.491	1.627	1.762	1.898	2.034	2.169	2.305	2.440	2.576
2	2.712	2.847	2.983	3.118	3.254	3.390	3.525	3.661	3.796	3.932
3	4.067	4.203	4.339	4.474	4.610	4.745	4.881	5.017	5.152	5.288
4	5.423	5.559	5.694	5.830	5.966	6.101	6.237	6.372	6.508	6.644
5	6.779	6.915	7.050	7.186	7.321	7.457	7.593	7.728	7.864	7.999
6	8.135	8.271	8.406	8.542	8.677	8.813	8.948	9.084	9.200	9.355
7	9.491	9.626	9.762	9.898	10.033	10.169	10.304	10.440	10.575	10.711
8	10.847	10.982	11.118	11.253	11.389	11.524	11.660	11.796	11.931	12.067
9	12.202	12.338	12.474	12.609	12.745	12.880	13.016	13.151	13.287	13.423
10	13.558	–	–	–	–	–	–	–	–	–

ft lbf	0	1	2	3	4	5	6	7	8	9
0	–	1.36	2.71	4.07	5.42	6.78	8.14	9.49	10.85	12.20
10	13.56	14.91	16.27	17.62	18.98	20.34	21.69	23.05	24.40	25.76
20	27.12	28.47	29.83	31.18	32.54	33.90	35.25	36.61	37.96	39.32
30	40.67	42.03	43.39	44.74	46.10	47.45	48.81	50.17	51.52	52.88
40	54.23	55.59	56.94	58.30	59.66	61.01	62.37	63.72	65.08	66.44
50	67.79	69.15	70.50	71.86	73.21	74.57	75.93	77.28	78.64	79.99
60	81.35	82.71	84.06	85.42	86.77	88.13	89.48	90.84	92.20	93.55
70	94.91	96.26	97.62	98.98	100.33	101.69	103.04	104.40	105.75	107.11
80	108.47	109.82	111.18	112.53	113.89	115.24	116.60	117.96	119.31	120.67
90	122.02	123.38	124.74	126.09	127.45	128.80	130.16	131.51	132.87	134.23
100	135.58	–	–	–	–	–	–	–	–	–

ft lbf	0	10	20	30	40	50	60	70	80	90
0	–	13.56	27.12	40.67	54.23	67.79	81.35	94.91	108.47	122.02
100	135.56	149.12	162.68	176.24	189.80	203.36	216.92	230.48	244.04	257.60
200	271.16	284.72	298.28	311.83	325.39	338.95	352.51	366.06	379.62	393.18
300	406.74	420.30	433.86	447.42	460.97	474.53	488.09	501.65	515.21	528.77
400	542.32	555.88	569.44	583.00	596.55	610.11	623.67	637.23	650.79	664.35
500	677.90	691.46	705.02	718.58	732.13	745.69	759.26	772.82	786.38	799.94
600	813.49	827.05	840.61	854.17	867.72	881.28	894.84	908.40	921.96	935.52
700	949.07	962.63	976.19	989.75	1 003.30	1 016.86	1 030.42	1 043.98	1 057.54	1 071.10
800	1 084.65	1 098.21	1 111.77	1 125.33	1 138.88	1 152.44	1 166.00	1 179.56	1 193.12	1 206.68
900	1 220.23	1 233.79	1 247.35	1 260.91	1 274.46	1 288.02	1 301.58	1 315.14	1 328.70	1 342.26
1 000	1 355.82	–	–	–	–	–	–	–	–	–

CONVERSION : HORSEPOWER–KILOWATTS

Horse-power	0	0.1	0.2	0.3	0.4	0.5	0.6	0.7	0.8	0.9
0	–	0.0746	0.1492	0.2238	0.2984	0.3730	0.4477	0.5223	0.5969	0.6715
1	0.7461	0.8207	0.8953	0.9699	1.0445	1.1191	1.1937	1.2683	1.3429	1.4175
2	1.4921	1.5667	1.6413	1.7159	1.7905	1.8651	1.9397	2.0143	2.0889	2.1636
3	2.2383	2.3129	2.3875	2.4621	2.5367	2.6113	2.6859	2.7608	2.8351	2.9097
4	2.9843	3.0589	3.1335	3.2087	3.2827	3.3573	3.4319	3.5068	3.5822	3.6558
5	3.7304	3.8050	3.8796	3.9542	4.0288	4.1034	4.1780	4.2526	4.3272	4.4018
6	4.4765	4.5511	4.6257	4.7003	4.7749	4.8495	4.9241	4.9987	5.0733	5.1479
7	5.2226	5.2972	5.3718	5.4464	5.5210	5.5956	5.6702	5.7448	5.8194	5.8940
8	5.9687	6.0433	6.1179	6.1925	6.2671	6.3417	6.4163	6.4909	6.5655	6.6401
9	6.7148	6.7894	6.8640	6.9386	7.0132	7.0878	7.1624	7.2370	7.3116	7.3002
10	7.4608	–	–		–	–	–	–	–	–

CONVERSION : KILOWATTS–HORSEPOWER

Kilo-watts	0	0.1	0.2	0.3	0.4	0.5	0.6	0.7	0.8	0.9
0	–	0.1340	0.2681	0.4021	0.5361	0.6702	0.8042	0.9382	1.0723	1.2063
1	1.3403	1.4743	1.6083	1.7424	1.8764	2.0104	2.1445	2.2785	2.4125	2.5466
2	2.6807	2.8147	2.9487	3.0828	3.2168	3.3508	3.4849	3.6189	3.7529	3.8870
3	4.0210	4.1550	4.2890	4.4231	4.5571	4.6911	4.8252	4.9592	5.0932	5.2273
4	5.3613	5.4953	5.6293	5.7634	5.8974	6.0314	6.1658	6.2995	6.4335	6.5676
5	6.7017	6.8357	6.9697	7.1038	7.2378	7.3718	7.5059	7.6399	7.7639	7.8980
6	8.0420	8.1760	8.3100	8.4441	8.5781	8.7121	8.8462	8.9802	9.1142	9.2483
7	9.3823	9.5163	9.6503	9.7844	9.9184	10.0524	10.1865	10.3205	10.4545	10.5886
8	10.7227	10.8567	10.9907	11.1248	11.2588	11.3928	11.5269	11.6609	11.7949	11.9290
9	12.0630	12.1970	12.3310	12.4651	12.5991	12.7331	12.8672	13.0012	13.1352	13.2693
10	13.4033	–	–	–	–	–	–	–	–	–

CONVERSION : Kgfm/s–WATTS

Kgfm/s	0	0.1	0.2	0.3	0.4	0.5	0.6	0.7	0.8	0.9
0	–	0.981	1.961	2.942	3.923	4.903	5.884	6.865	7.845	8.826
1	9.807	10.788	11.769	12.749	13.730	14.711	15.691	16.674	17.653	18.634
2	19.614	20.595	21.576	22.556	23.537	24.518	25.498	26.479	27.460	28.441
3	29.420	30.401	31.382	32.362	33.343	34.324	35.304	36.285	37.266	38.247
4	39.227	40.208	41.189	42.169	43.150	44.131	45.111	46.092	47.073	48.054
5	49 034	50.015	50.996	51.976	52.957	53.938	54.918	55.899	56.880	57.880
6	58.840	59.821	60.802	61.782	62.763	63.744	64.724	65.705	66.686	67.667
7	68.647	69.628	70.609	71.589	72.570	73.551	74.531	75.512	76.493	77.474
8	78.454	79.435	80.416	81.396	82.377	83.358	84.338	85.319	86.300	87.281
9	88.261	89.242	90.223	91.203	92.184	93.165	94.145	95.126	96.107	97.088
10	98.067	–	–	–	–	–	–	–	–	–

CONVERSION : NEWTONS TO POUNDS FORCE

Newtons	0	0.1	0.2	0.3	0.4	0.5	0.6	0.7	0.8	0.9
0	–	0.022	0.045	0.067	0.090	0.112	0.135	0.157	0.180	0.202
1	0.225	0.247	0.270	0.292	0l315	0.337	0.360	0.382	0.405	0.427
2	0.450	0.472	0.495	0.517	0.540	0.562	0.585	0.607	0.630	0.652
3	0.674	0.697	0.719	0.742	0.764	0.787	0.809	0.832	0.854	0.877
4	0.899	0.922	0.944	0.967	0.989	1.011	1.034	1.056	1.079	1.101
5	1.124	1.147	1.169	1.192	1.214	1.236	1.259	1.281	1.304	1.326
6	1.349	1.371	1.394	1.416	1.439	1.461	1.484	1.506	1.529	1.551
7	1.574	1.596	1.619	1.641	1.664	1.686	1.709	1.731	1.754	1.776
8	1.798	1.821	1.843	1.866	1.888	1.911	1.933	1.956	1.978	2.001
9	2.023	2.046	2.068	2.091	2.113	2.136	2.158	2.181	2.203	2.226
10	2.248	–	–	–	–	–	–	–	–	–

Newtons	0	1	2	3	4	5	6	7	8	9
0	–	0.225	0.450	0.674	0.899	1.124	1.349	1.574	1.798	2.023
10	2.248	2.473	2.698	2.922	3.147	3.372	3.597	3.822	4.046	4.271
20	4.496	4.721	4.946	5.171	5.396	5.620	5.845	6.070	6.295	6.520
30	6.744	6.969	7.194	7.419	7.644	7.868	8.093	8.318	8.543	8.768
40	8.992	9.215	9.440	9.665	9.890	10.114	10.339	10.564	10.789	11.014
50	11.240	11.465	11.690	11.915	12.140	12.364	12.589	12.814	13.039	13.264
60	13.488	13.713	13.938	14.163	14.388	14.612	14.837	15.062	15.287	15.512
70	15.736	15.961	16.186	16.411	16.636	16.860	17.086	17.310	17.535	17.760
80	17.984	18.209	18.434	18.659	18.884	19.108	19.333	19.558	19.783	20.008
90	20.232	20.457	20.682	20.907	21.132	21.356	21.581	21.806	22.031	22.256
100	22.480	–	–	–	–	–	–	–	–	–

Newtons	0	10	20	30	40	50	60	70	80	90
0	–	2.25	4.50	6.74	8.99	11.24	13.49	15.74	17.98	20.23
100	22.48	24.73	26.98	29.22	31.47	33.72	35.97	38.22	40.46	42.71
200	44.96	47.21	49.46	51.71	53.96	56.20	58.45	60.70	62.95	65.20
300	67.44	69.69	71.94	74.19	76.44	78.68	80.93	83.18	85.43	87.68
400	89.92	92.15	94.40	96.65	98.90	101.14	103.39	105.64	107.89	110.14
500	112.40	114.65	116.90	119.15	121.40	123.64	125.89	128.14	130.39	132.64
600	134.88	137.13	139.38	141.63	143.88	146.12	148.37	150.62	152.87	155.12
700	157.36	159.61	161.86	164.11	166.36	168.60	170.85	173.10	175.35	177.60
800	179.84	182.09	184.34	186.59	188.84	191.08	193.33	195.58	197.83	200.08
900	202.32	204.57	206.82	209.07	211.32	213.56	215.81	218.06	220.31	222.56
1 000	224.80	–	–	–	–	–	–	–	–	–

CONVERSION : POUND FORCE TO NEWTONS

lbf	0	0.1	0.2	0.3	0.4	0.5	0.6	0.7	0.8	0.9
0	–	0.45	0.89	1.34	1.78	2.22	2.67	3.11	3.56	4.03
1	4.45	4.89	5.34	5.78	6.23	6.67	7.12	7.56	8.01	8.45
2	8.90	9.34	9.79	10.23	10.68	11.12	11.57	12.01	12.46	12.90
3	13.35	13.79	14.24	14.68	15.12	15.57	16.01	16.46	16.90	17.35
4	17.79	18.24	18.68	19.13	19.57	20.12	20.46	20.91	21.35	21.80
5	22.24	22.69	23.13	23.58	24.02	24.47	24.91	25.36	25.80	26.25
6	26.69	27.13	27.58	28.02	28.47	28.91	29.36	29.80	30.25	30.69
7	31.14	31.58	32.03	32.47	32.92	33.36	33.81	34.25	34.70	35.14
8	35.59	36.03	36.48	36.92	37.37	37.81	38.26	38.70	39.14	39.59
9	40.03	40.48	40.92	41.37	41.81	42.26	42.70	43.15	43.59	44.04
10	44.48	–	–	–	–	–	–	–	–	–

lbf	0	1	2	3	4	5	6	7	8	9
0	–	4.45	8.90	13.35	17.79	22.24	26.69	31.14	35.59	40.33
10	44.48	48.93	53.38	57.83	62.28	66.72	71.17	75.62	80.07	84.52
20	88.96	93.41	97.86	102.31	106.76	111.21	115.65	120.10	124.55	129.00
30	133.45	137.90	142.35	146.79	151.24	155.69	160.14	164.58	169.03	173.48
40	177.93	182.38	186.83	191.27	195.72	201.17	204.62	209.07	213.52	217.96
50	222.41	226.86	231.31	235.76	240.20	244.65	249.10	253.55	258.00	262.45
60	266.89	271.34	275.79	280.24	284.69	289.13	293.58	298.03	302.48	306.93
70	311.38	315.82	320.27	324.72	329.17	333.62	338.06	342.51	346.96	351.41
80	355.86	360.31	364.75	369.20	373.65	378.10	382.55	387.00	391.44	395.89
90	400.34	404.79	409.24	413.69	418.13	422.58	427.03	431.48	435.93	440.37
100	444.82	–	–	–	–	–	–	–	–	–

lbf	0	10	20	30	40	50	60	70	80	90
0	–	44.5	89.0	133.5	177.9	222.4	266.9	311.4	355.9	403.3
100	444.8	489.3	533.8	578.3	622.8	667.2	711.7	756.2	800.7	845.2
200	889.6	934.1	978.6	1 023.1	1 067.6	1 112.1	1 156.5	1 201.0	1 245.5	1 290.0
300	1 334.5	1379.0	1 423.5	1 467.9	1 512.4	1 556.9	1 601.4	1 645.9	1 690.3	1 734.8
400	1 779.3	1 823.8	1 868.3	1 912.7	1 957.2	2 011.7	2 046.2	2 090.7	2 135.2	2 179.6
500	2 224.1	2 268.6	2 313.1	2 357.6	2 402.0	2 446.5	2 491.0	2 535.5	2 580.0	2 624.5
600	2 668.9	2 713.4	2 757.9	2 802.4	2 846.9	2 891.3	2 935.8	2 980.3	3 024.8	3 069.3
700	3 113.8	3 158.2	3 202.7	3 247.2	3 291.7	3 336.2	3 380.6	3 425.1	3 469.6	3 514.1
800	3 558.6	3 603.1	3 647.5	3 692.0	3 736.5	3 781.0	3 825.5	3 870.0	3 914.4	3 958.9
900	4 003.4	4 047.9	4 092.4	4 136.9	4 181.3	4 225.8	4 270.3	4 314.8	4 359.3	4 403.7
1 000	4 448.2	–	–	–	–	–	–	–	–	–

QUANTITIES FOR COMPRESSION CALCULATIONS

Compression Ratio r	Isothermal $\frac{1}{r}$	Isothermal $\frac{57.17}{\log_e r}$	$\left(\frac{1}{r}\right)^{\frac{1}{1.3}}$	$\left(\frac{1}{r}\right)^{\frac{1}{1.41}}$	$(r)^{\frac{n-1}{n}}$ $n = 1.3$*	$(r)^{\frac{n-1}{n}}$ $n = 1.41$†	$\frac{53.17n}{n-1}\left(r^{\frac{n-1}{n}}-1\right)$ $n = 1.3$*	$\frac{53.17n}{n-1}\left(r^{\frac{n-1}{n}}-1\right)$ $n = 1.41$†
1.0	1.0000	0.000	1.0000	1.0000	1.0000	1.0000	0.000	0.000
1.1	0.9091	5.068	0.9393	0.9346	1.0222	1.0281	5.124	5.138
1.2	0.8333	9.694	0.8691	0.8787	1.0430	1.0544	9.900	9.954
1.3	0.7692	13.950	0.8172	0.8316	1.0625	1.0793	14.382	14.495
1.4	0.7143	17.890	0.7719	0.7877	1.0807	1.1028	18.603	18.794
1.5	0.6667	21.560	0.7371	0.7501	1.0981	1.1251	22.600	22.880
1.6	0.6250	24.990	0.6966	0.7165	1.1146	1.1464	26.398	26.778
1.7	0.5882	28.210	0.6649	0.6864	1.1303	1.1668	30.012	30.507
1.8	0.5556	31.250	0.6362	0.6591	1.1453	1.1864	33.471	34.078
1.9	0.5263	34.130	0.6103	0.6343	1.1596	1.2052	36.784	37.519
2.0	0.5000	36.860	0.5867	0.6116	1.1738	1.2233	39.966	40.830
2.1	0.4762	39.450	0.5651	0.5908	1.1867	1.2408	43.028	44.027
2.2	0.4545	41.920	0.5452	0.5717	1.1996	1.2577	45.979	47.118
2.3	0.4348	44.290	0.5269	0.5539	1.2119	1.2740	48.827	50.109
2.4	0.4167	46.550	0.5099	0.5375	1.2249	1.2899	51.814	52.980
2.5	0.4000	48.720	0.4942	0.5221	1.2355	1.3053	54.253	55.827
2.6	0.3846	50.800	0.4795	0.5078	1.2467	1.3203	56.840	58.564
2.7	0.3704	52.810	0.4658	0.4944	1.2576	1.3348	59.354	61.228
2.8	0.3571	54.750	0.4529	0.4817	1.2682	1.3490	61.796	63.823
2.9	0.3448	56.610	0.4409	0.4699	1.2785	1.3629	64.172	66.354
3.0	0.3333	58.410	0.4295	0.4588	1.2886	1.3764	66.485	68.822
3.1	0.3226	60.160	0.4188	0.4482	1.2953	1.3896	68.741	71.233
3.2	0.3125	61.850	0.4087	0.4383	1.3078	1.4024	70.918	73.589
3.3	0.3030	63.480	0.3991	0.4288	1.3172	1.4150	73.089	75.895
3.4	0.2941	65.070	0.3901	0.4198	1.3263	1.4274	75.185	78.149
3.5	0.2857	66.610	0.3815	0.4113	1.3352	1.4395	77.233	80.359
3.6	0.2778	68.110	0.3733	0.4031	1.3439	1.4513	79.243	82.523
3.7	0.2703	69.560	0.3655	0.3954	1.3525	1.4629	81.208	84.647
3.8	0.2632	70.980	0.3581	0.3880	1.3608	1.4748	83.131	86.817
3.9	0.2564	72.360	0.3510	0.3809	1.3690	1.4855	85.016	88.773
4.0	0.2500	73.710	0.3442	0.3741	1.3770	1.4965	86.864	90.781
4.1	0.2439	75.020	0.3378	0.3676	1.3848	1.5072	88.670	92.752
4.2	0.2381	76.300	0.3316	0.3614	1.3926	1.5179	90.456	94.696
4.3	0.2326	77.550	0.3256	0.3554	1.4002	1.5283	92.204	96.596
4.4	0.2273	78.780	0.3199	0.3497	1.4076	1.5385	93.919	98.468
4.5	0.2222	79.970	0.3144	0.3441	1.4149	1.5486	95.606	100.320
4.6	0.2174	81.140	0.3092	0.3388	1.4221	1.5585	97.265	102.130
4.7	0.2128	82.280	0.3041	0.3337	1.4292	1.5683	98.894	103.920
4.8	0.2083	83.400	0.2992	0.3288	1.4362	1.5779	100.500	105.680
4.9	0.2041	84.500	0.2944	0.3240	1.4430	1.5874	102.080	107.410
5.0	0.2000	85.570	0.2900	0.3193	1.4498	1.5968	103.630	109.120
5.1	0.1961	86.630	0.2856	0.3149	1.4564	1.6060	105.160	110.810
5.2	0.1923	87.660	0.2813	0.3106	1.4629	1.6151	106.670	112.470
5.3	0.1887	88.670	0.2772	0.3062	1.4694	1.6241	108.150	114.110
5.4	0.1852	89.670	0.2733	0.3024	1.4758	1.6329	109.620	115.730

*True Isothermal

†True Adiabatic

QUANTITIES FOR COMPRESSION CALCULATIONS (contd.)

Compression Ratio r	Isothermal $\frac{1}{r}$	Isothermal $\frac{57.17}{\log_e r}$	$\left(\frac{1}{r}\right)^{\frac{1}{1.3}}$	$\left(\frac{1}{r}\right)^{\frac{1}{1.41}}$	$(r)^{\frac{n-1}{n}}$ n = 1.3*	$(r)^{\frac{n-1}{n}}$ n = 1.41†	$\frac{53.17n}{n-1}\left(r^{\frac{n-1}{n}}-1\right)$ n = 1.3*	$\frac{53.17n}{n-1}\left(r^{\frac{n-1}{n}}-1\right)$ n = 1.41†
5.5	0.1818	90.64	0.2695	0.2985	1.4820	1.6417	111.06	117.33
5.6	0.1786	91.60	0.2657	0.2947	1.4882	1.6503	112.48	118.91
5.7	0.1754	92.54	0.2622	0.2910	1.4943	1.6588	113.88	120.46
5.8	0.1724	93.47	0.2587	0.2874	1.5003	1.6672	115.27	122.00
5.9	0.1695	94.37	0.2553	0.2840	1.5062	1.6755	116.64	123.52
6.0	0.1667	95.27	0.2520	0.2806	1.5121	1.6835	117.99	124.98
6.1	0.1639	96.15	0.2488	0.2773	1.5178	1.6918	119.31	126.51
6.2	0.1613	97.01	0.2457	0.2742	1.5236	1.6998	120.63	127.97
6.3	0.1587	97.86	0.2427	0.2711	1.5292	1.7078	121.93	129.42
6.4	0.1563	98.70	0.2398	0.2681	1.5348	1.7156	123.21	130.85
6.5	0.1539	99.52	0.2370	0.2651	1.5403	1.7232	124.48	132.23
6.6	0.1515	100.34	0.2342	0.2623	1.5457	1.7310	125.73	133.67
6.7	0.1493	101.14	0.2315	0.2595	1.5511	1.7386	126.97	135.06
6.8	0.1471	101.92	0.2289	0.2568	1.5564	1.7461	128.19	136.43
6.9	0.1449	102.70	0.2263	0.2541	1.5616	1.7536	129.40	137.79
7.0	0.1429	103.46	0.2238	0.2516	1.5668	1.7609	130.60	139.14
7.1	0.1408	104.22	0.2214	0.2490	1.5720	1.7682	131.78	140.47
7.2	0.1389	104.96	0.2190	0.2466	1.5770	1.7754	132.95	141.78
7.3	0.1370	105.70	0.2167	0.2442	1.5821	1.7825	134.11	143.09
7.4	0.1351	106.42	0.2145	0.2418	1.5871	1.7896	135.26	144.38
7.5	0.1333	107.13	0.2123	0.2395	1.5920	1.7966	136.39	145.66
7.6	0.1316	107.84	0.2101	0.2373	1.5968	1.8035	137.52	146.93
7.7	0.1299	108.53	0.2080	0.2351	1.6017	1.8104	138.63	148.18
7.8	0.1282	109.22	0.2059	0.2330	1.6064	1.8172	139.73	149.43
7.9	0.1266	109.90	0.2039	0.2309	1.6112	1.8239	140.82	150.66
8.0	0.1250	110.56	0.2020	0.2287	1.6159	1.8306	141.90	151.88
8.1	0.1235	111.22	0.2001	0.2268	1.6205	1.8373	142.97	153.10
8.2	0.1220	111.88	0.1982	0.2249	1.6251	1.8438	144.03	154.30
8.3	0.1205	112.52	0.1963	0.2229	1.6296	1.8504	145.07	155.49
8.4	0.1191	113.16	0.1945	0.2210	1.6343	1.8568	146.11	156.67
8.5	0.1177	113.78	0.1928	0.2192	1.6386	1.8632	147.14	157.84
8.6	0.1163	114.40	0.1910	0.2174	1.6431	1.8695	148.17	159.00
8.7	0.1149	115.02	0.1893	0.2156	1.6474	1.8758	149.18	160.15
8.8	0.1136	115.63	0.1877	0.2139	1.6518	1.8821	150.18	161.29
8.9	0.1124	116.23	0.1861	0.2122	1.6561	1.8883	151.17	162.41
9.0	0.1111	116.83	0.1845	0.2105	1.6604	1.8944	152.15	163.55
9.1	0.1099	117.41	0.1829	0.2085	1.6646	1.9005	153.13	164.66
9.2	0.1087	118.00	0.1814	0.2072	1.6688	1.9066	154.10	165.77
9.3	0.1075	118.57	0.1799	0.2056	1.6730	1.9126	155.06	166.86
9.4	0.1064	119.14	0.1784	0.2041	1.6771	1.9185	156.01	167.95
9.5	0.1053	119.70	0.1770	0.2026	1.6812	1.9244	156.96	169.04
9.6	0.1042	120.26	0.1758	0.2011	1.6853	1.9303	157.90	170.11
9.7	0.1031	120.81	0.1742	0.1996	1.6893	1.9361	158.83	171.17
9.8	0.1020	121.35	0.1728	0.1981	1.6933	1.9419	159.75	172.23
9.9	0.1010	121.89	0.1714	0.1967	1.6963	1.9476	160.66	173.28
10.0	0.1000	122.43	0.1701	0.1953	1.7012	1.9533	161.56	174.32

*True Isothermal †True Adiabatic

PRESSURE CONVERSION: lbf/in^2 TO BARS

lbf/in^2	0	0.1	0.2	0.3	0.4	0.5	0.6	0.7	0.8	0.9
0	–	0.007	0.014	0.021	0.028	0.035	0.041	0.048	0.055	0.062
1	0.069	0.076	0.083	0.090	0.097	0.103	0.110	0.117	0.124	0.131
2	0.138	0.145	0.152	0.159	0.166	0.172	0.179	0.186	0.193	0.200
3	0.207	0.214	0.221	0.228	0.234	0.241	0.248	0.255	0.262	0.269
4	0.276	0.283	0.290	0.297	0.303	0.310	0.317	0.324	0.331	0.338
5	0.345	0.352	0.359	0.365	0.372	0.379	0.386	0.393	0.400	0.407
6	0.414	0.421	0.428	0.434	0.441	0.448	0.455	0.462	0.469	0.476
7	0.483	0.490	0.496	0.503	0.510	0.517	0.524	0.531	0.538	0.545
8	0.552	0.559	0.565	0.572	0.579	0.586	0.593	0.600	0.607	0.614
9	0.621	0.627	0.634	0.641	0.648	0.655	0.662	0.669	0.676	0.683
10	0.689	–	–	–	–	–	–	–	–	–

lbf/in^2	0	1	2	3	4	5	6	7	8	9
0	–	0.069	0.138	0.207	0.276	0.345	0.414	0.483	0.552	0.621
10	0.689	0.758	0.827	0.896	0.965	1.034	1.103	1.172	1.241	1.310
20	1.379	1.448	1.517	1.586	1.655	1.724	1.793	1.862	1.931	2.000
30	2.069	2.137	2.206	2.275	2.344	2.413	2.482	2.551	2.620	2.689
40	2.758	2.827	2.896	2.965	3.034	3.103	3.172	3.241	3.310	3.379
50	3.448	3.516	3.585	3.654	3.723	3.792	3.861	3.930	3.999	4.068
60	4.137	4.206	4.275	4.344	4.413	4.482	4.551	4.620	4.689	4.758
70	4.827	4.895	4.964	5.033	5.102	5.171	5.240	5.309	5.378	5.447
80	5.516	5.585	5.654	5.723	5.792	5.861	5.930	5.999	6.068	6.137
90	6.206	6.274	6.343	6.412	6.481	6.550	6.619	6.688	6.757	6.826
100	6.895	–	–	–	–	–	–	–	–	–

lbf/in^2	0	10	20	30	40	50	60	70	80	90
0	–	0.689	1.379	2.069	2.758	3.448	4.137	4.827	5.516	6.206
100	6.895	7.584	8.274	8.963	9.653	10.342	11.032	11.721	12.411	13.100
200	13.790	14.479	15.169	15.858	16.548	17.237	17.927	18.616	19.306	19.995
300	20.685	21.374	22.064	22.753	23.443	24.132	24.822	25.511	26.201	26.890
400	27.580	28.269	28.959	29.648	30.338	31.028	31.717	32.406	33.096	33.785
500	34.475	35.164	35.854	36.543	37.233	37.922	38.612	39.301	39.991	40.680
600	41.370	42.059	42.749	43.438	44.128	44.817	45.507	46.196	46.886	47.575
700	48.265	48.954	49.644	50.333	51.023	51.712	52.402	53.091	53.781	54.470
800	55.160	55.849	56.539	57.228	57.918	58.607	59.297	59.986	60.676	61.365
900	62.055	62.744	63.434	64.123	64.813	65.502	66.192	66.881	67.571	68.260
1 000	68.950	–	–	–	–	–	–	–	–	–

PRESSURE CONVERSION: BARS TO lbf/in^2

Bars	0	0.1	0.2	0.3	0.4	0.5	0.6	0.7	0.8	0.9
0	–	1.45	2.90	4.35	5.80	7.25	8.70	10.15	11.60	13.05
1	14.50	14.95	17.40	18.85	20.30	21.75	23.20	24.65	26.10	27.55
2	29.00	30.45	31.90	33.35	34.80	36.25	37.70	39.15	40.60	42.05
3	43.50	44.95	46.40	47.85	49.30	50.75	52.20	53.65	55.10	56.55
4	58.00	59.45	60.90	62.35	63.80	65.25	66.70	68.15	69.60	71.05
5	72.50	73.95	75.40	67.85	78.30	79.75	81.20	82.65	84.10	85.55
6	87.00	88.45	89.90	91.35	92.80	94.25	95.70	97.15	98.60	100.05
7	101.50	102.95	104.40	105.85	107.30	108.75	110.20	111.65	113.10	114.55
8	116.00	117.45	118.90	120.35	121.80	123.25	124.70	126.15	127.60	129.05
9	130.50	131.95	133.40	134.85	136.30	137.75	139.20	140.65	142.10	143.55
10	145.00	–	–	–	–	–	–	–	–	–

Millibars	0	1	2	3	4	5	6	7	8	9
0	–	0.015	0.029	0.044	0.058	0.073	0.087	0.102	0.116	0.131
10	0.145	0.160	0.174	0.189	0.203	0.218	0.232	0.247	0.261	0.276
20	0.290	0.305	0.319	0.334	0.348	0.363	0.377	0.392	0.406	0.421
30	0.435	0.450	0.464	0.479	0.493	0.508	0.522	0.537	0.551	0.566
40	0.580	0.595	0.609	0.624	0.638	0.653	0.667	0.682	0.696	0.711
50	0.725	0.740	0.754	0.769	0.783	0.798	0.812	0.827	0.841	0.856
60	0.870	0.885	0.899	0.914	0.928	0.943	0.957	0.972	0.986	1.001
70	1.015	1.030	1.044	1.059	1.073	1.088	1.102	1.117	1.131	1.146
80	1.160	1.175	1.189	1.204	1.218	1.233	1.247	1.262	1.276	1.291
90	1.305	1.320	1.334	1.349	1.363	1.378	1.392	1.407	1.421	1.436
100	1.450	–	–	–	–	–	–	–	–	–

Millibars	0	10	20	30	40	50	60	70	80	90
0	–	0.145	0.290	0.435	0.580	0.725	0.870	1.015	1.160	1.305
100	1.450	1.600	1.740	1.890	2.030	2.180	2.320	2.470	2.610	2.760
200	2.900	3.050	3.190	3.340	3.480	3.630	3.770	3.920	4.060	4.210
300	4.350	4.500	4.640	4.790	4.930	5.080	5.220	5.370	5.510	5.660
400	5.800	5.950	6.090	6.240	6.380	6.530	6.670	6.820	6.960	7.110
500	7.250	7.400	7.540	7.690	7.830	7.980	8.120	8.270	8.410	8.560
600	8.700	8.850	8.990	9.140	9.280	9.430	9.570	9.720	9.860	10.010
700	10.150	10.300	10.440	10.590	10.730	10.880	11.020	11.170	11.310	11.460
800	11.600	11.750	11.890	12.040	12.180	12.330	12.470	12.620	12.760	12.910
900	13.050	13.200	13.340	13.490	13.630	13.780	13.920	14.070	14.210	14.360
1 000	14.500	–	–	–	–	–	–	–	–	–

EQUIVALENT COMPRESSED AIR VOLUME
(Cubic feet)

Volume of Free Air ft^3	Compressed Air Pressure — lbf/in^2 Gauge								
	40	50	60	70	80	90	100	110	120
1	0.27	0.23	0.20	0.17	0.16	0.14	0.13	0.12	0.11
2	0.54	0.46	0.39	0.35	0.31	0.28	0.26	0.24	0.22
3	0.81	0.68	0.59	0.52	0.47	0.42	0.39	0.35	0.33
4	1.08	0.91	0.79	0.69	0.62	0.56	0.51	0.47	0.44
5	1.34	1.14	0.99	0.87	0.78	0.70	0.64	0.59	0.55
6	1.61	1.36	1.18	1.04	0.93	0.84	0.77	0.71	0.66
7	1.88	1.59	1.38	1.22	1.09	0.98	0.90	0.83	0.77
8	2.15	1.82	1.58	1.39	1.24	1.12	1.03	0.95	0.87
9	2.42	2.05	1.77	1.56	1.40	1.26	1.16	1.06	0.98
10	2.69	2.27	1.97	1.74	1.55	1.40	1.28	1.18	1.09
11	2.96	2.50	2.17	1.91	1.71	1.54	1.41	1.30	1.20
12	3.23	2.73	2.36	2.08	1.86	1.69	1.54	1.42	1.31
13	3.50	2.95	2.56	2.26	2.02	1.83	1.67	1.54	1.42
14	3.76	3.18	2.76	2.43	2.17	1.97	1.80	1.65	1.53
15	4.03	3.41	2.95	2.60	2.33	2.01	1.93	1.77	1.64
16	4.30	3.64	3.15	2.78	2.49	2.25	2.05	1.89	1.75
17	4.57	3.86	3.35	2.95	2.64	2.39	2.18	2.01	1.86
18	4.84	4.09	3.55	3.12	2.80	2.53	2.31	2.13	1.97
19	5.11	4.32	3.74	3.30	2.95	2.67	2.44	2.24	2.08
20	5.38	4.55	3.94	3.47	3.11	2.81	2.57	2.36	2.19
21	5.65	4.77	4.14	3.65	3.26	2.95	2.70	2.48	2.30
22	5.91	5.00	4.33	3.82	3.42	3.09	2.83	2.60	2.40
23	6.18	5.23	4.53	3.99	3.57	3.23	2.95	2.72	2.51
24	6.45	5.46	4.73	4.17	3.73	3.37	3.08	2.83	2.62
25	6.72	5.68	4.92	4.34	3.88	3.51	3.21	2.95	2.73
26	6.99	5.91	5.12	4.51	4.04	3.65	3.34	3.07	2.84
27	7.26	6.14	5.32	4.69	4.19	3.79	3.47	3.19	2.95
28	7.53	6.36	5.51	4.86	4.35	3.93	3.60	3.31	3.06
29	7.80	6.59	5.71	5.03	4.50	4.07	3.72	3.42	3.17
30	8.06	6.82	5.91	5.21	4.66	4.21	3.85	3.54	3.28
31	8.33	7.05	6.10	5.38	4.81	4.35	3.98	3.66	3.39
32	8.60	7.27	6.30	5.56	4.97	4.49	4.11	3.78	3.50
33	8.87	7.50	6.50	5.72	5.13	4.63	4.24	3.90	3.61
34	9.14	7.73	6.70	5.90	5.28	4.77	4.37	4.02	3.72
35	9.41	8.06	6.89	6.07	5.44	4.91	4.49	4.13	3.83
36	9.68	8.18	7.09	6.24	5.59	5.05	4.62	4.25	3.94
37	9.95	8.41	7.29	6.42	5.75	5.19	4.75	4.37	4.05
38	10.21	8.64	7.48	6.59	5.90	5.34	4.88	4.49	4.15
39	10.48	8.87	7.68	6.76	6.06	5.48	5.01	4.61	4.26
40	10.75	9.09	7.88	6.94	6.21	5.62	5.14	4.72	4.37
41	11.02	9.32	8.07	7.12	6.37	5.76	5.26	4.84	4.48
42	11.29	9.55	8.27	7.29	6.52	5.90	5.39	4.96	4.59
43	11.56	9.77	8.47	7.47	6.68	6.04	5.52	5.08	4.70
44	11.83	10.00	8.66	7.64	6.83	6.18	5.65	5.20	4.81
45	12.10	10.23	8.86	7.81	6.99	6.32	5.78	5.31	4.92
46	12.37	10.46	9.06	7.99	7.14	6.46	5.91	5.43	5.03
47	12.63	10.68	9.26	8.16	7.30	6.60	6.04	5.55	5.14
48	12.90	10.91	9.45	8.33	7.45	6.74	6.16	5.67	5.25
49	13.17	11.14	9.65	8.51	7.61	6.88	6.29	5.79	5.36
50	13.44	11.37	9.85	8.68	7.77	7.02	6.42	5.91	5.47

cont...........

EQUIVALENT COMPRESSED AIR VOLUME (contd.)
(Cubic feet)

Volume of Free Air ft^3	Compressed Air Pressure – lbf/in^2 Gauge								
	40	50	60	70	80	90	100	110	120
60	16.1	13.6	11.8	10.4	9.3	8.4	7.7	7.1	6.6
70	18.8	15.9	13.8	12.2	10.9	9.8	9.0	8.3	7.7
80	21.5	18.2	15.8	13.9	12.4	11.2	10.3	9.5	8.7
90	24.2	20.5	17.7	15.6	14.0	12.6	11.6	10.6	9.8
100	26.9	22.7	19.7	17.4	15.5	14.0	12.8	11.8	10.9
110	29.6	25.0	21.7	19.1	17.1	15.4	14.1	13.0	12.0
120	32.3	27.3	23.6	20.8	18.6	16.9	15.4	14.2	13.1
130	35.0	29.5	25.6	22.6	20.2	18.3	16.7	15.4	14.2
140	37.6	31.8	27.6	24.3	21.7	19.7	18.0	16.5	15.3
150	40.3	34.1	29.5	26.0	23.3	20.1	19.3	17.7	16.4
160	43.0	36.4	31.5	27.8	24.9	22.5	20.5	18.9	17.5
170	45.7	38.6	33.5	29.5	26.4	23.9	21.8	20.1	18.6
180	48.4	40.9	35.5	31.2	28.0	25.3	23.1	21.3	19.7
190	51.1	43.2	37.4	33.0	29.5	26.7	24.4	22.4	20.8
200	53.8	45.5	39.4	34.7	31.1	28.1	25.7	23.6	21.9
210	56.5	47.7	41.4	36.5	32.6	29.5	27.0	24.8	23.0
220	59.1	50.0	43.3	38.2	34.2	30.9	28.3	26.0	24.0
230	61.8	52.3	45.3	39.9	35.7	32.3	29.5	27.2	25.1
240	64.5	54.6	47.3	41.7	37.3	33.7	30.8	28.3	26.2
250	67.2	56.8	49.2	43.4	38.8	35.8	32.1	29.5	27.3
260	69.9	59.1	51.2	45.1	40.4	36.5	33.4	30.7	28.4
270	72.6	61.4	53.2	46.9	41.9	37.9	34.7	31.9	29.5
280	75.3	63.6	55.1	48.6	43.5	39.3	36.0	33.1	30.6
290	78.0	65.9	57.1	50.3	45.0	40.7	37.2	34.2	31.7
300	80.6	67.2	59.1	52.1	46.6	42.1	38.5	35.4	32.8
310	83.3	70.5	61.0	53.8	48.1	43.5	39.8	36.6	33.9
320	86.0	72.7	63.0	55.6	49.7	44.9	41.1	37.8	35.0
330	88.7	75.0	65.0	57.2	51.3	46.3	42.4	39.0	36.1
340	91.4	77.3	67.0	59.0	52.8	47.7	43.7	40.2	37.2
350	94.1	79.6	68.9	60.7	54.4	49.1	44.9	41.3	38.3
360	96.8	81.8	70.9	62.4	55.9	50.5	46.2	42.5	39.4
370	99.5	84.1	72.9	64.2	57.5	51.9	47.5	43.7	40.4
380	102.1	86.4	74.8	65.9	59.0	53.4	48.8	44.9	41.5
390	104.8	88.7	76.8	67.6	60.6	54.8	50.1	46.1	42.6
400	107.5	90.9	78.8	69.4	62.1	56.2	51.4	47.2	43.7
410	110.2	93.2	80.7	71.2	63.7	57.6	52.6	48.4	44.8
420	112.9	95.5	82.7	72.9	65.2	59.0	53.9	49.6	45.9
430	115.6	97.7	84.7	74.7	66.8	60.4	55.2	50.8	47.0
440	118.3	100.0	86.6	76.4	68.3	61.8	56.5	52.0	48.1
450	121.0	102.3	88.6	78.1	69.9	63.2	57.8	53.1	49.2
460	123.7	104.6	90.6	79.9	71.4	64.6	59.1	54.3	50.3
470	126.3	106.8	92.6	81.6	73.0	66.0	60.4	55.5	51.4
480	129.0	109.1	94.5	83.3	74.5	67.4	61.6	56.7	52.5
490	131.7	111.4	96.5	85.1	76.1	68.8	62.9	57.9	53.6
500	134.4	113.7	98.5	86.8	77.7	70.2	64.2	59.1	54.7

cont.......

EQUIVALENT COMPRESSED AIR VOLUME (contd.)
(Cubic Feet)

Volume of Free Air ft^3	Compressed Air Pressure – lbf/in^2 Gauge								
	40	50	60	70	80	90	100	110	120
600	161.3	136.4	118.1	104.2	93.2	84.2	77.0	70.9	65.6
700	188.2	159.1	137.8	121.5	108.7	98.3	89.9	82.7	76.5
800	215.0	181.8	157.5	138.9	124.2	112.3	102.7	94.5	87.4
900	241.9	204.6	177.2	156.2	139.8	126.3	115.6	106.3	98.4
1 000	268.8	227.3	196.9	173.6	155.3	140.4	128.4	118.1	109.3
1 100	295.7	250.0	216.6	191.0	170.8	154.4	141.2	129.9	120.2
1 200	322.6	272.7	236.3	208.3	186.3	168.5	154.1	141.7	131.1
1 300	349.5	295.4	256.0	225.7	201.9	182.5	166.9	153.5	142.0
1 400	376.4	318.1	275.7	243.0	217.4	196.5	179.7	165.3	152.9
1 500	403.3	340.8	295.4	260.4	232.9	210.6	192.5	177.1	163.8
1 600	430.2	363.5	315.1	277.7	248.9	224.6	205.3	188.9	174.7
1 700	457.1	386.3	334.8	295.1	264.0	238.6	218.2	200.7	185.7
1 800	484.0	409.1	359.5	312.4	279.5	252.7	231.0	212.5	196.6
1 900	510.8	431.8	374.2	329.8	295.1	266.7	243.9	224.3	207.6
2 000	537.6	454.6	393.8	347.2	310.6	280.8	256.8	236.2	218.6
2 100	564.5	477.3	413.5	364.6	326.1	294.8	269.6	248.0	229.5
2 200	591.4	500.0	433.2	381.9	341.6	308.8	282.5	259.8	240.4
2 300	618.3	522.8	452.9	399.3	357.2	322.9	295.3	271.6	251.4
2 400	645.2	545.5	472.6	416.6	372.7	336.9	308.2	283.4	262.3
2 500	672.1	568.2	492.3	434.0	388.2	351.0	321.0	295.2	273.3
2 600	699.0	591.0	512.0	451.3	403.8	365.1	333.9	307.0	284.2
2 700	725.8	613.7	531.7	468.7	419.3	379.1	346.7	318.8	295.1
2 800	752.7	636.4	551.4	486.0	434.8	393.1	359.5	330.6	306.0
2 900	779.5	659.2	571.1	503.4	450.3	407.2	372.4	342.4	316.0
3 000	806.4	681.9	590.7	520.8	465.9	421.2	385.2	354.3	327.9
3 100	833.3	704.6	610.4	538.2	481.4	435.2	398.0	366.1	338.8
3 200	860.2	727.3	630.1	555.5	496.9	449.3	410.9	377.9	349.7
3 300	887.1	750.1	649.8	572.4	512.5	463.3	423.7	389.7	360.7
3 400	914.0	772.8	669.5	589.7	528.0	477.3	436.6	401.5	371.6
3 500	940.9	795.5	689.2	607.1	543.5	491.4	449.4	413.3	382.5
3 600	967.7	818.2	708.9	624.4	559.1	505.4	462.3	425.1	393.5
3 700	994.6	841.0	728.6	641.7	574.6	519.4	475.2	436.9	404.4
3 800	1 021.0	863.7	748.3	659.1	590.1	533.5	488.0	448.7	415.3
3 900	1 048.0	886.5	768.0	676.4	605.7	547.5	500.8	460.5	426.3
4 000	1 075.0	909.2	787.6	694.4	621.2	561.6	513.6	472.4	437.2
4 100	1 102.0	931.9	807.3	711.8	636.7	575.6	526.4	484.2	448.1
4 200	1 129.0	954.6	827.0	729.1	652.2	589.6	539.3	496.0	459.0
4 300	1 156.0	977.4	846.7	746.5	667.8	603.7	552.1	507.8	470.0
4 400	1 183.0	1 000.0	866.4	763.8	683.8	617.7	565.0	519.6	480.9
4 500	1 210.0	1 023.0	886.1	781.2	698.8	631.8	577.8	531.4	491.8
4 600	1 237.0	1 046.0	905.8	798.5	714.4	645.8	590.7	543.2	502.8
4 700	1 263.0	1 068.0	925.5	815.9	729.9	659.8	603.5	555.0	513.7
4 800	1 290.0	1 091.0	945.2	833.2	745.4	673.9	616.4	566.8	524.6
4 900	1 317.0	1 114.0	964.9	850.6	761.0	687.9	629.2	578.6	535.6
5 000	1 344.0	1 137.0	984.5	868.0	776.5	702.0	642.0	590.5	546.5
6 000	1 613.0	1 364.0	1 181.0	1 042.0	931.8	842.4	770.4	708.6	655.8
7 000	1 882.0	1 591.0	1 378.0	1 215.0	1 087.0	982.8	898.8	826.7	765.1
8 000	2 150.0	1 818.0	1 575.0	1 389.0	1 242.0	1 123.0	1 027.0	944.8	874.4
9 000	2 419.0	2 046.0	1 772.0	1 562.0	1 398.0	1 264.0	1 156.0	1 063.0	983.7
10 000	2 688.0	2 273.0	1 969.0	1 736.0	1 553.0	1 404.0	1 284.0	1 181.0	1 093.0

TORQUE/HORSEPOWER EQUIVALENTS
(Torque in Kilogramme-metres)

Torque kg/m	Rev/Min 1 000	2 000	3 000	4 000	5 000	6 000	7 000	8 000	9 000	10 000
0.01	0.014	0.028	0.041	0.055	0.069	0.083	0.096	0.110	0.124	0.138
0.02	0.028	0.055	0.083	0.110	0.138	0.165	0.193	0.220	0.248	0.275
0.03	0.041	0.083	0.124	0.164	0.207	0.248	0.289	0.330	0.372	0.413
0.04	0.055	0.110	0.164	0.220	0.275	0.330	0.386	0.441	0.506	0.551
0.05	0.069	0.138	0.207	0.276	0.344	0.413	0.482	0.551	0.620	0.689
0.06	0.082	0.165	0.248	0.330	0.413	0.496	0.578	0.661	0.744	0.826
0.07	0.096	0.193	0.289	0.386	0.482	0.578	0.675	0.771	0.868	0.964
0.08	0.110	0.220	0.331	0.441	0.551	0.661	0.771	0.881	0.991	1.102
0.09	0.124	0.248	0.372	0.496	0.620	0.743	0.868	0.992	1.115	1.239
0.10	0.138	0.275	0.413	0.551	0.689	0.826	0.964	1.102	1.239	1.377
0.20	0.276	0.550	0.826	1.100	1.377	1.652	1.928	2.203	2.479	2.759
0.30	0.413	0.836	1.239	1.652	2.066	2.478	2.892	3.306	3.718	4.131
0.40	0.551	1.102	1.692	2.204	2.754	3.304	3.856	4.408	5.057	5.508
0.50	0.689	1.378	2.066	2.754	3.442	4.132	4.820	5.512	6.195	6.885
0.60	0.826	1.652	2.478	3.304	4.131	4.956	5.783	6.612	7.436	8.262
0.70	0.964	1.928	2.891	3.856	4.820	5.782	6.747	7.712	8.675	9.639
0.80	1.102	2.204	3.306	4.408	5.508	6.612	7.711	8.814	9.914	11.016
0.90	1.239	2.479	3.717	4.958	6.197	7.434	8.675	9.916	11.150	12.393
1.00	1.377	2.754	4.131	5.508	6.885	8.262	9.639	11.020	12.390	13.770
1.10	1.515	3.029	4.544	6.059	7.573	9.088	10.603	12.120	13.630	15.147
1.20	1.652	3.304	4.956	6.608	8.262	9.912	11.566	13.220	14.870	16.524
1.30	1.790	3.579	5.369	7.159	8.950	10.738	12.530	14.320	16.110	17.901
1.40	1.928	3.856	5.782	7.712	9.639	11.564	13.490	15.420	17.350	19.278
1.50	2.066	4.131	6.195	8.263	10.328	12.390	14.450	16.520	18.590	20.655
1.60	2.203	4.408	6.612	8.816	11.016	13.224	15.420	17.630	19.830	22.032
1.70	2.341	4.683	7.025	9.367	11.705	14.050	16.380	18.730	21.070	23.409
1.80	2.479	4.958	7.434	9.916	12.393	14.868	17.350	19.830	22.300	24.786
1.90	2.616	5.233	7.847	10.467	13.082	15.694	18.310	20.930	23.540	26.163
2.00	2.754	5.508	8.262	11.020	13.770	16.520	19.280	22.030	24.790	27.540
2.50	3.443	6.885	10.330	13.770	17.220	20.660	24.100	27.550	30.980	34.430
3.00	4.131	8.260	12.390	16.520	20.660	24.780	28.920	33.060	37.180	41.310
3.50	4.820	9.640	14.460	19.280	24.100	38.910	33.740	38.560	43.380	48.200
4.00	5.508	11.020	16.520	22.040	27.540	33.040	38.560	44.080	50.570	55.080
4.50	6.197	12.400	18.590	24.790	30.990	37.170	43.380	49.580	55.750	61.970
5.00	6.885	13.770	20.660	27.540	34.430	41.310	48.200	55.120	61.950	68.850
6.00	8.262	16.520	24.780	33.040	41.310	49.560	57.830	66.120	74.360	82.620
7.00	9.639	19.280	28.910	38.560	48.200	57.820	67.470	77.120	86.750	96.390
8.00	11.020	22.040	33.060	44.080	55.100	66.120	77.110	88.140	99.140	110.200
9.00	12.390	24.790	37.170	49.580	61.950	74.340	86.750	99.160	111.500	123.900
10.00	13.770	27.540	41.310	55.080	68.850	82.620	96.390	110.200	123.900	137.700

NOTE: To obtain equivalent horsepower for torque in gramme-centimetres, divide table figures by 100 000.

It is recommended by the B.C.A.S. that the term 'Brake Power' or 'Shaft Power' should be used as standard practice in describing power outputs previously quoted as 'Brake Horse Power' or 'Shaft Horse Power'. 746 W = 1 hp.

TORQUE/HORSEPOWER EQUIVALENTS
(Torque in Pound-feet)

Torque lb/ft	Rev/Min 1 000	2 000	3 000	4 000	5 000	6 000	7 000	8 000	9 000	10 000
0.1	0.02	0.04	0.06	0.08	0.10	0.11	0.13	0.15	0.17	0.19
0.2	0.04	0.08	0.11	0.15	0.19	0.23	0.27	0.31	0.34	0.38
0.3	0.06	0.11	0.17	0.23	0.29	0.34	0.40	0.46	0.51	0.57
0.4	0.08	0.15	0.23	0.30	0.39	0.46	0.53	0.61	0.69	0.76
0.5	0.10	0.19	0.29	0.38	0.48	0.57	0.67	0.76	0.86	0.95
0.6	0.11	0.23	034	0.46	0.58	0.69	0.80	0.91	1.03	1.14
0.7	0.13	0.27	0.40	0.53	0.67	0.80	0.93	1.00	1.20	1.33
0.8	0.15	0.31	0.46	0.61	0.76	0.91	1.07	1.22	1.37	1.52
0.9	0.17	0.34	0.51	0.69	0.86	1.03	1.20	1.37	1.54	1.72
1.0	0.19	0.38	0.57	0.76	0.95	1.14	1.33	1.52	1.72	1.91
2.0	0.39	0.76	1.14	1.52	1.91	2.29	2.67	3.05	3.43	3.81
3.0	0.57	1.14	1.72	2.29	2.86	3.43	4.00	4.57	5.14	5.72
4.0	0.76	1.53	2.29	3.05	3.81	4.57	5.33	6.10	6.86	7.62
5.0	0.95	1.91	2.86	3.81	4.76	5.72	6.67	7.62	8.57	9.53
6.0	1.14	2.29	3.43	4.57	5.72	6.86	8.00	9.14	10.29	11.43
7.0	1.33	2.67	4.00	5.33	6.67	8.00	9.33	10.66	12.00	13.34
8.0	1.52	3.05	4.57	6.10	7.62	9.14	10.67	12.20	13.72	15.24
9.0	1.72	3.43	5.14	6.86	8.57	10.28	12.00	13.72	15.43	17.15
10.0	1.91	3.81	5.72	7.62	9.53	11.43	13.34	15.24	17.15	19.05
11.0	2.10	4.19	6.29	8.38	10.46	12.58	14.67	16.76	18.86	20.96
12.0	2.29	4.57	6.86	9.14	11.44	13.72	16.00	18.28	20.57	22.87
13.0	2.48	4.95	7.43	9.90	12.39	14.86	17.33	19.80	22.29	24.78
14.0	2.67	5.33	8.00	10.66	13.35	16.00	18.67	21.32	24.00	26.69
15.0	2.86	5.72	8.57	11.43	14.30	17.14	20.00	22.84	25.72	28.60
16.0	3.05	6.10	9.14	12.20	15.26	18.28	21.33	24.40	27.43	30.51
17.0	3.24	6.48	9.72	12.96	16.21	19.44	22.67	25.92	29.15	32.42
18.0	3.43	6.86	10.29	13.72	17.16	20.58	24.00	27.44	30.86	34.32
19.0	3.62	7.24	10.86	14.48	18.12	21.72	25.33	28.96	32.58	36.23
20.0	3.81	7.62	11.43	15.24	19.05	22.86	26.67	30.48	34.29	38.10
25.0	4.76	9.53	14.29	19.05	23.82	28.58	33.33	38.10	42.86	47.63
30.0	5.72	11.43	17.15	22.86	28.58	34.30	40.01	45.72	51.44	57.15
35.0	6.67	13.34	20.00	26.67	33.34	40.00	46.67	53.34	60.01	66.68
40.0	7.62	15.24	22.86	30.48	38.10	45.72	53.34	60.96	68.58	76.20
45.0	8.57	17.15	25.72	34.29	42.87	51.44	60.00	68.58	77.15	85.73
50.0	9.53	19.05	28.58	38.10	47.63	57.15	66.67	76.20	85.73	95.25
60.0	11.43	22.86	34.29	45.72	57.15	68.58	80.00	91.44	102.87	114.30
70.0	13.34	26.67	40.01	53.34	66.68	80.02	93.35	106.68	120.02	133.35
80.0	15.24	30.48	45.72	60.96	76.20	91.44	106.68	121.92	137.16	152.40
90.0	17.15	36.74	51.44	73.58	85.73	102.80	120.01	147.16	154.31	171.45
100.0	19.05	38.10	57.15	76.20	95.25	114.30	133.35	152.40	171.45	190.50

Table figures are equivalent horsepower.

NOTE: It is recommended by the B.C.A.S. that the term 'Brake Power' or 'Shaft Power' be used as standard practice in describing power outputs previously quoted as 'Brake Horse Power' or 'Shaft Horse Power' 746 W = 1 hp.

TYPICAL MAXIMUM AIR CONSUMPTION OF PNEUMATIC TOOLS

	m^3 free air per minute at 6 bar (gauge)	Cubic feet free air per minute on full throttle 80 lb/in^2 (gauge)
Rock Drills		
Drifter drill (cradle mounted), 3 in	4–4.8	140–170
Drifter drill (cradle mounted), 3½ in	4.2–6	150–210
Drifter drill (cradle mounted), 4 in	5.7–8	200–280
Wagon drill with 4 in drifter	6	210
Wagon drill with 3½ in drifter	5	175
Hand hammer drill (jack hammer), 28 lb	1.5	50
Hand hammer drill (jack hammer), 35 lb	1.7	60
Hand hammer drill (jack hammer), 45 lb	2.25	80
Hand hammer drill (jack hammer), 60 lb	4.25	150
Plug drill	0.85	30
Sinker drill, 3 in	3.5	125
Sinker drill, 3½ in	4.5	160
Stoping drill, light	2	70
Stoping drill, heavy	3.5	125
Dustless dry drill, 3½ in	5.1	180
Dustless dry drill, 3 in	3.4	120
Underwater drill	2.4–2.6	85–90
Pneumatic Tools (Percussive)		
Concrete breaker, 75–80 lb	1.8–2	65–75
Concrete breaker, 50 lb	1.4	50
Concrete breaker, 30 lb	1.2	40
Pile driver	1.7	60
Chipping and caulking hammer, light	0.3–0.35	10–13
Chipping and caulking hammer, medium	0.5	18
Chipping and caulking hammer, heavy (suitable also for ½ in hot rivets)	0.6	23
Riveting hammer, ¾ in, hot rivets	0.7	24
Riveting hammer, 1 in hot rivets	0.8	26
Riveting hammer, 1¼ in hot rivets	0.9	29
Riveting hammer, 1⅜ in hot rivets	1.1	35
Rivet buster, heavy single blow type	0.09 per rivet	3 per rivet
Rivet cutter, rapid blow type for rivets up to ¾ in	1.1	35
Single blow riveter (one shot for aircraft rivets)	0.002 per rivet	1/16 per rivet
Staybolt riveter for 1 in copper staybolts	0.9	33
Stand riveter for ½ in hot rivets	0.7	24
Stand riveter for ¾ in hot rivets	0.8	26
Stand riveter for 1 in hot rivets	0.85	28
Jam riveters as for stand riveters		
Yoke riveters as for stand riveters		
Hammer holder-on	0.85	28
Holder-on	0.009 per rivet	1/8 per rivet
Wood deck caulker	0.6	22

cont.........

TYPICAL MAXIMUM AIR CONSUMPTION OF PNEUMATIC TOOLS

	m^3 free air per minute at 6 bar (gauge)	Cubic feet free air per minute on full throttle 80 lb/in^2 (gauge)
Pneumatic Tools (Percussive) ...continued		
Pick, light	0.6–0.85	20–30
Pick, medium	0.85–1.2	30–40
Pick, heavy	0.7–1.1	25–35
Spader, light	0.6–0.85	20–30
Spader, medium	0.85–1.2	30–40
Tie tamper	0.6	20
Rammer, foundry, bench type	0.3–0.5	12–16
Rammer, foundry, floor type, medium	0.5–0.6	17–20
Rammer, foundry, floor type, heavy	0.6–0.85	23–30
Rammer, back fill trench type	1	35
Spike driver	1.7	60
Scaling hammer, valveless, for surface work	0.2	7
Scaling hammer for large boiler tubes	0.6	21
Paint scraper	0.2	7
Stone tool, for lettering and light carving	0.2	6
Stone tool, for medium dressing	0.3	10
Stone tool, for roughing and bushing	0.4	14
Weld flux chipper	0.75	24
Pneumatic Tools (Rotary)		
Drilling machine for ¼ in holes in steel	0.3–0.5	10–16
Drilling machine for 3/8 in holes in steel	0.6–0.85	20–30
Drilling machine for ½ in–¾ in holes in steel	0.6–1.15	20–40
Drilling machine for 7/8 in – 1 in holes in steel	1.15–2.3	40–80
Drilling machine for 1¼ in holes in steel	1.25–1.4	45–50
Drilling machine for l½ in holes in steel	1.7–3.4	60–120
Drilling machine for 2 in holes in steel	2.3–3.4	80–120
Drilling machine for 3 in holes in steel	2.85–3.7	100–130
(For wood boring take next size smaller)		
(For reaming and tapping in steel, take next size larger)		
Corner drills as above		
Grinders with ½ in – ¾ in diameter wheels	0.3–0.6	10–20
Grinders with 2 in diameter wheels	0.3–0.85	10–30
Grinders with 4 in diameter wheels	1.15–1.6	40–55
Grinders with 6 in diameter wheels	1.2–2	45–70
Grinders with 8 in diameter wheels	2–2.5	70–85
Saw, 6 in diameter	0.6	20
Auger drill for coal	2.1	74
Sander (reciprocating or orbital)	0.275	9
Tube cutter, for tubes up to 2½ in	1.25	45
Tube cutter, for tubes up to 2½ in – 4 in	1.6	55
Tube expander, for tubes up to 2½ in	1.8–2.0	65
Tube expander, for tubes up to 3 in	2.125	75
Tube expander, for tubes up to 4 in	2.7	95
Wrenches (rotary torque) for nuts ¼ in	0.15–0.25	5–8
Wrenches (rotary torque) for nuts 3/8 in	4.3–6	15–20
Wrenches (rotary torque) for nuts ½ in – ¾ in	0.75–0.85	25–30
Wrenches (rotary torque) for nuts 7/8 in – 1 in	0.95–1.5	35–40

cont.........

TYPICAL MAXIMUM AIR CONSUMPTION OF PNEUMATIC TOOLS (contd.)

	m^3 free air per minute at 6 bar (gauge)	Cubic feet free air per minute on full throttle 80 lb/in^2 (gauge)
Impact wrench for nuts up to ¾ in	0.75–0.85	25–30
Impact wrench for nuts up to 1¼ in	0.85–1.7	30–60
Spike puller, per spike	0.085	3
Bath enameller	0.25	8
Spray gun, small	0.03–1	1–3
Spray gun, medium	0.1–0.2 } at 3.5 bar	3–7 } at 50 lb/in^2
Spray gun, large	0.2–0.6 } at 3.5 bar	7–20 } at 50 lb/in^2
Cement gun, small	2.85	100
Cement gun, medium	4.55	160
Air arbor	0.15	5
	0.004 up	1/8 upwards, per operation according to size
Air chuck, as for air arbor		
Air lift pump. Considerable variation according to prevailing conditions. Individual problems should be submitted to members of the Society who specialize in this work.		
Sump pump, 90–250 gal/m	1.5–3.4	50–120
Air press	0.008 up	¼ upwards per operation according to size
Air motors, up to 1 bhp	0.85–0.9 per bhp	30–35 per bhp
Air motors 1–5 bhp	0.85 per bhp	30 per bhp
Air motors over 5 bhp	0.75 per bhp	25 per bhp
Jigger conveyor engine 11 in	3.4	120
Jigger conveyor engine 13½ in	4.8	170
Jigger conveyor engine 15¾ in	7.1	250
Jigger conveyor engine 17¾ in	10.5	370
Coal cutter, percussive 2½ in	2.8	98
Coal cutter,percussive 3½ in	4.25	144
Coal cutter,percussive 4 in	5.1	180
Coal cutter, Longwall, 50 hp	22.75	800
Drill sharpener, small	1.45	50
Drill sharpener, large	3.4	120
Forging hammer (power hammer) 1 cwt	1.7	60
Forging hammer (power hammer) 3 cwt	3.7	130
Forging hammer (power hammer) 5 cwt	5.4	190
Forging hammer (power hammer) 10 cwt	9.1	320
Forging hammer (power hammer) 20 cwt	15.3	540
Pedestal grinder with wheel, 8 in dia x 1¼ in face	1.3	45
Grouting machine	up to 2.85 at 3 bar	Variable up to 100 ft^3/min at 40 lb/in^2
Rivet forge	0.17	6
Sand blast (shot) 5/16 in jet	2.85	105
Sand blast (shot) 3/8 in jet	4.25 } at 60 bar	151 } at 60 bar
Sand blast (shot) 7/16 in jet	5.7 } at 60 bar	205 } at 60 bar
Sand blast (shot) ½ in jet	7.65 } at 60 bar	268 } at 60 bar
Sand riddler up to 3 ft x 2 ft	0.6	22

cont.........

TYPICAL MAXIMUM AIR CONSUMPTION OF PNEUMATIC TOOLS

	m^3 free air per minute at 6 bar (gauge)	Cubic feet free air per minute on full throttle 80 lb/in^2 (gauge)
Squeeze riveter, for 1/8 in x 3/16 in aircraft rivets	0.002 per rivet	1/16 per rivet
Squeeze riveter, for ¼ in cold rivets	0.03–0.1	1–3 per rivet
Squeeze riveter, for 3/8 in cold rivets	0.015 per river	7/8 per rivet
Squeeze riveter, for 7/16 in cold rivets	0.04 per rivet	1½ per rivet
Squeeze riveter (for girder work) for ½ in rivets	0.04 per rivet	1.1/3 per rivet
Squeeze riveter (for girder work) for ¾ in rivets	0.06 per rivet	2 per rivet
Squeeze riveter (for girder work) for 7/8 in rivets	0.08 per rivet	2¾ per rivet
Squeeze riveter (for girder work) for 1 in rivets	0.011 per rivet	4 per rivet
Squeeze riveter (for girder work) for 1¼ in rivets	0.017 per rivet	6 per rivet
Squeeze riveter (for girder work) for 1½ in rivets	0.0275 per rivet	9½ per rivet
Squeeze riveter (boiler work) 1.3/8 in	0.03 per rivet	11½ per rivet
Squeeze riveter (boiler work)1.5/8 in	0.07 per rivet	24½ per rivet
Haulage or winch hoists, pendant wire rope, motor type		3 per foot-ton of lift
Hoists, straight lift, direct acting cylinder type		1½ per foot-ton of lift
Air cylinders		1½ per foot-ton of lift
Stone surfacer, for large blocks		30 per foot-ton of lift
Hydro-pneumatic riveter (as for squeeze riveters)		
Fettling grip or vice	0.004 up	1/8 upwards per operation according to size
Hot miller (1 or 2 cutters)	0.85 per cutter	30 per cutter
Pneumatic drill leg	0.85–0.95	2 above consumption of rock drill
Internal vibrator o/d 2½ in	0.85–0.95	30–35
Internal vibrator o/d 3 in	1.75–2	65–70
Internal vibrator o/d 4½ in	2.4	85
Internal vibrator, o/d 4½ in	1.75	65
Internal vibrator, o/d 5½ in	2.4	85
Shuttering vibrators, 7 lb	0.15–0.3	5–10
Shuttering vibrators, 10 lb	0.3–0.45	10–15
Shuttering vibrators, 14 lb	0.45–0.6	15–20
Concrete compactor	0.17–0.3	6–10

APPROXIMATE CONSUMPTION OF AIR MOTORS – FT^3/MIN (M^3/MIN)

hp	ft^3/min per hp Figure								m^3/min per hp Figure							hp
	25	30	35	40	45	50	55	60	0.8	1.0	1.2	1.4	1.6	1.8	2.0	
0.1	2.5	3	3.5	4	4.5	5	5.5	6	2.5	3	3.6	4.2	4.8	5.4	6	0.1
0.2	5.0	6	7.0	8	9.0	10	11.0	12	5.0	6	7.2	8.4	9.6	10.8	12	0.2
0.3	7.5	9	10.5	12	13.5	15	16.5	18	7.5	9	10.8	12.6	14.4	16.2	18	0.3
0.4	10.0	12	14.0	16	18.0	20	22.0	24	10.0	12	14.4	16.8	19.2	21.6	24	0.4
0.5	12.5	15	17.5	20	22.5	25	27.5	30	12.5	15	18.0	21.0	24.0	27.0	30	0.5
0.6	15.0	18	21.0	24	27.0	30	33.0	36	15.0	18	21.6	25.7	28.8	32.4	36	0.6
0.7	17.5	21	24.5	28	31.5	35	38.5	42	17.5	21	25.2	29.4	33.6	37.8	42	0.7
0.8	20.0	24	28.0	32	36.0	40	44.0	48	20.0	24	28.8	33.6	38.4	43.2	48	0.8
0.9	22.5	27	31.5	36	40.5	45	49.5	54	22.5	27	32.4	37.8	43.2	48.6	54	0.9
1.0	25.0	30	35.0	40	45.5	50	55.0	60	25.0	30	36.0	42.0	48.0	54.0	60	1.0
2.0	50.0	60	70.0	80	90.0	100	110.0	120	50.0	60	72.0	84.0	96.0	108.0	120	2.0
3.0	75.0	90	105.0	120	135.0	150	165.0	180	75.0	90	108.0	126.0	144.0	162.0	180	3.0
4.0	100.0	120	140.0	150	180.0	200	220.0	240	100.0	120	144.0	168.0	192.0	216.0	240	4.0
5.0	125.0	150	175.0	200	225.0	250	275.0	300	125.0	150	180.0	210.0	240.0	270.0	300	5.0
6.0	150.0	180	210.0	240	270.0	300	330.0	360	150.0	180	216.0	252.0	288.0	324.0	360	6.0
7.0	175.0	210	245.0	280	315.0	350	385.0	420	175.0	210	252.0	294.0	336.0	378.0	420	7.0
8.0	200.0	240	280.0	320	360.0	400	440.0	480	200.0	240	288.0	336.0	384.0	432.0	480	8.0
9.0	225.0	270	315.0	360	405.0	450	495.0	540	225.0	270	324.0	378.0	432.0	486.0	540	9.0
10.0	250.0	300	350.0	400	450.0	500	550.0	600	250.0	300	360.0	420.0	480.0	540.0	600	10.0
11.0	275.0	330	385.0	440	495.0	550	605.0	660	275.0	330	396.0	462.0	528.0	594.0	660	11.0
12.0	300.0	360	420.0	480	540.0	600	660.0	720	300.0	360	432.0	504.0	576.0	648.0	720	12.0
13.0	325.0	390	455.0	520	585.0	650	715.0	780	325.0	390	468.0	546.0	624.0	702.0	780	13.0
14.0	350.0	420	490.0	560	630.0	700	770.0	840	350.0	420	504.0	588.0	672.0	756.0	840	14.0
15.0	375.0	450	525.0	600	675.0	750	825.0	900	375.0	450	540.0	630.0	720.0	810.0	900	15.0
16.0	400.0	480	560.0	720	800.0	880	960.0	960	400.0	480	576.0	672.0	768.0	864.0	960	16.0
17.0	425.0	510	595.0	680	765.0	850	935.0	1 020	425.0	510	612.0	714.0	816.0	918.0	1 020	17.0
18.0	450.0	540	630.0	720	810.0	900	990.0	1 080	450.0	540	648.0	756.0	864.0	972.0	1 080	18.0
19.0	475.0	570	665.0	760	855.0	950	1 045.0	1 140	475.0	570	684.0	798.0	912.0	1 026.0	1 140	19.0
20.0	500.0	600	700.0	800	900.0	1 000	1 100.0	1 200	500.0	600	720.0	840.0	960.0	1 080.0	1 200	20.0

	25–35	40–45	50–60	0.8–1.2	1.4–1.6	1.8–2.0
Motor type	Vane and Diaphragm Motors	Large	Small	Vane and Diaphragm Motors	Large	Small
		Medium Piston Motors			Medium Piston Motors	

CYLINDER DISPLACEMENT (ENGLISH UNITS) DISPLACEMENT IN CUBIC INCHES

Bore		Stroke											
in	mm	in 1 mm 25.4	2 50.8	3 76.2	4 101.6	5 127.0	6 152.4	7 177.8	8 203.2	9 228.6	10 254.0	11 279.4	12 304.8
½	12.70	0.200	0.39	0.59	0.79	0.97	1.16	1.375	1.57	1.77	1.96	2;16	2.36
¾	19.05	0.440	0.88	1.33	1.77	2.21	2.65	3.09	3.54	3.98	4.42	4.86	5.31
1	25.40	0.785	1.57	2.36	3.14	3.93	4.71	5.49	6.28	7.07	7.85	8.64	9.42
1¼	31.75	1.23	2.45	3.68	4.91	6.14	7.36	8.59	9.82	11.04	12.27	13.50	14.72
1½	38.10	1.77	3.53	5.30	7.07	8.83	10.60	12.37	14.14	15.90	17.67	19.36	21.12
1¾	44.45	2.41	4.81	7.22	9.62	12.03	14.43	16.84	19.24	21.65	24.05	26.45	28.86
2	50.80	3.14	6.28	9.43	12.57	15.71	18.85	21.99	25.14	28.28	31.42	34.56	37.70
2¼	57.15	3.98	7.95	11.93	15.90	19.83	23.86	27.83	31.81	35.78	39.76	43.74	47.71
2½	63.50	4.91	9.82	14.73	19.64	24.55	29.45	34.36	39.27	44.18	49.09	54.00	58.90
2¾	69.85	5.94	11.88	17.82	23.76	29.70	35.64	41.58	47.52	53.46	59.40	63.54	71.28
3	76.20	7.07	14.14	21.21	28.28	35.36	42.41	49.48	56.56	63.62	70.67	77.74	84.83
3½	88.90	9.62	19.24	28.86	38.48	48.11	57.73	67.35	71.97	86.59	96.21	105.83	115.45
4	101.60	12.57	25.13	37.70	50.26	62.83	75.40	87.96	100.53	113.09	125.66	138.23	150.80
4½	114.3	15.90	31.81	47.71	63.62	79.52	95.42	111.33	127.23	143.14	159.00	174.90	190.80
5	127.0	19.64	39.27	58.91	78.54	98.18	117.81	137.45	157.08	176.72	196.35	216.00	235.62
5½	139.7	23.76	47.52	71.27	95.03	118.79	142.55	166.31	190.06	213.82	237.58	261.34	285.10
6	152.4	28.27	56.55	84.82	113.10	141.37	169.64	197.89	226.20	254.57	282.74	311.01	339.28
7	177.8	38.49	76.97	115.46	153.94	192.43	230.92	269.39	307.88	346.38	384.85	423.34	461.84
8	203.2	50.27	100.53	150.80	201.06	251.33	301.60	351.86	402.12	452.40	502.7	553.0	603.2
9	228.6	63.62	127.23	190.83	254.46	318.09	381.66	445.32	508.9	572.5	636.2	699.8	763.3
10	254.0	78.54	157.08	235.62	314.16	392.70	471.24	549.8	628.3	706.9	785.4	863.9	942.5
12	304.8	113.10	226.2	339.3	452.4	565.5	678.6	791.7	904.8	1017.9	1131.0	1244.1	1357.2
14	355.6	153.94	307.9	461.8	615.8	769.7	923.6	1077.5	1231.5	1385.4	1539.4	1693.3	1847.2
16	406.4	201.06	402.1	603.2	804.3	1005.3	1206.4	1407.4	1608.5	1809.5	2010.6	2211.7	2412.8
18	457.2	254.47	509.0	763.5	1018.0	1272.4	1526.9	1781.3	2035.8	2290.3	2544.7	2799.2	3053.8
20	508.0	314.16	628.4	942.6	1256.8	1570.8	1885.2	2199.4	2513.6	2827.8	3141.6	3455.8	3770.4

This table gives the displacement or swept volume of cylinders for strokes of 1 to 12 inches. Displacement for intermediate values of stroke can be interpolated. Displacement for longer strokes can be obtained by factoring (eg displacement for 18 inches stroke is 2 x displacement for 9 inches stroke); or by addition, (eg displacement for 18 inches stroke equals displacement for 10 inches stroke plus displacement for 8 inches stroke).

Total air consumption is given by multiplying the displacement per stroke by the number of strokes per minute.

CYLINDER DISPLACEMENT (METRIC UNITS)
DISPLACEMENT IN CUBIC METRES

Bore		Stroke											
mm	in	mm 25 in 0.98	50 1.97	75 2.95	100 3.94	125 4.92	150 5.91	175 6.89	200 7.87	225 8.86	250 9.84	275 10.83	300 11.81
10	0.40	0.0020	0.0039	0.0059	0.0078	0.0098	0.012	0.014	0.016	0.018	0.020	0.021	0.023
20	0.8	0.0079	0.0157	0.0236	0.0314	0.039	0.047	0.055	0.063	0.071	0.079	0.086	0.094
30	1.18	0.0177	0.0353	0.0530	0.0707	0.088	0.106	0.124	0.141	0.160	0.178	0.195	0.212
40	1.57	0.0314	0.0628	0.0942	0.1257	0.157	0.189	0.220	0.251	0.283	0.314	0.346	0.377
50	1.97	0.0491	0.0982	0.1473	0.1964	0.245	0.294	0.344	0.393	0.442	0.491	0.540	0.589
60	2.36	0.0707	0.1413	0.2120	0.2827	0.353	0.424	0.494	0.565	0.636	0.707	0.778	0.848
70	2.76	0.0962	0.1924	0.2886	0.3848	0.481	0.577	0.673	0.770	0.866	0.962	1.058	1.154
80	3.15	0.1257	0.2514	0.3771	0.5027	0.628	0.754	0.880	1.005	1.131	1.257	1.383	1.508
90	3.54	0.1590	0.3181	0.4771	0.6362	0.795	0.954	1.113	1.272	1.431	1.590	1.749	1.909
100	3.94	0.1964	0.3927	0.5892	0.7854	0.982	1.178	1.374	1.571	1.767	1.964	2.160	2.356
125	4.92	0.3068	0.6136	0.9204	1.2272	1.534	1.841	2.148	2.454	2.761	3.068	3.375	3.682
150	5.91	0.4418	0.8836	1.3254	1.7672	2.209	2.651	3.093	3.534	3.976	4.418	4.860	5.302
175	6.89	0.6014	1.2027	1.8041	2.4053	3.007	3.608	4.209	4.811	5.412	6.014	6.615	7.216
200	7.87	0.7854	1.5708	2.3562	3.1416	3.927	4.712	5.497	6.283	7.068	7.854	8.639	9.425
250	9.84	1.2272	2.4544	3.6816	4.9088	6.136	7.363	8.590	9.818	11.045	12.272	13.499	14.726
300	11.81	1.7671	3.5343	5.3012	7.0686	8.835	10.602	12.369	14.137	15.904	17.671	19.438	21.206
350	13.78	2.4053	4.8106	7.2159	9.6211	12.027	14.432	16.837	19.242	21.647	24.053	26.458	28.863
400	15.75	3.1416	6.2832	9.4248	12.5664	15.708	18.850	21.992	25.133	28.275	31.416	34.558	37.699
450	17.72	3.9761	7.9522	11.928	15.904	19.880	23.856	27.832	31.809	35.785	39.761	43.737	47.713
500	19.69	4.9088	9.8175	14.726	19.635	24.544	29.453	34.362	39.270	44.179	49.088	53.997	58.905

This table gives the swept volume of cylinders for strokes from 25 mm to 300 mm.
Displacements for intermediate values of stroke can be interpolated.
Displacements for longer strokes can be obtained by factoring (eg displacement for a 450 mm stroke would be 2 x that for a 225 mm stroke); or by addition (eg displacement for a 450 mm stroke equals displacement for 300 mm stroke plus displacement for 150 mm stroke).

Total air consumption is given by multiplying the displacement per stroke by the number of strokes per minute.

UNIVERSAL DISPLACEMENT TABLES

Bore	Cross Sectional Area	Displacement per Unit Stroke*	
		ft^3	m^3
1	0.7854	0.00045	0.00079
1.1	0.9503	0.00055	0.00095
1.2	1.1310	0.00065	0.00113
1.25	1.2272	0.00071	0.00123
1.3	1.3273	0.00077	0.00133
1.4	1.5394	0.00089	0.00154
1.5	1.7672	0.00102	0.00177
1.6	2.0106	0.00112	0.00201
1.7	2.2698	0.00131	0.00227
1.75	2.4053	0.00139	0.00241
1.8	2.5447	0.00147	0.00254
1.9	2.8353	0.00164	0.00284
2.0	3.1416	0.00182	0.00314
2.1	3.4636	0.00205	0.00346
2.2	3.8013	0.00220	0.00380
2.25	3.9761	0.00230	0.00398
2.3	4.1548	0.00241	0.00415
2.4	4.5239	0.00262	0.00452
2.5	4.9087	0.00284	0.00491
2.6	5.3093	0.00307	0.00531
2.7	5.7256	0.00331	0.00573
2.75	5.9396	0.00343	0.00594
2.8	6.1575	0.00356	0.00616
2.9	6.6052	0.00382	0.00661
3.0	7.0686	0.00409	0.00707
3.1	7.5477	0.00437	0.00755
3.2	8.0425	0.00465	0.00804
3.25	8.2958	0.00480	0.00830
3.3	7.5530	0.00495	0.00855
3.4	9.0792	0.00525	0.00908
3.5	9.6211	0.00557	0.00962

Bore	Cross Sectional Area	Displacement per Unit Stroke*	
		ft^3	m^3
3.6	10.1788	0.00589	0.01018
3.7	10.7521	0.00622	0.01075
3.75	11.0450	0.00640	0.01105
3.8	11.3412	0.00656	0.01134
3.9	11.9459	0.00691	0.01195
4.0	12.566	0.00727	0.01257
4.1	13.203	0.00764	0.01320
4.2	13.855	0.00802	0.01386
4.25	14.186	0.00820	0.01419
4.3	14.522	0.00840	0.01452
4.4	15.205	0.00880	0.01521
4.5	15.904	0.00920	0.01590
4.6	16.619	0.00962	0.01662
4.7	17.350	0.01004	0.01735
4.75	17.721	0.0103	0.01772
4.8	18.096	0.0105	0.01810
4.9	18.858	0.0109	0.01886
5.0	19.635	0.0114	0.01964
5.1	20.428	0.0118	0.02043
5.2	21.237	0.0123	0.02124
5.25	21.648	0.0125	0.02165
5.3	22.062	0.0128	0.02206
5.4	22.902	0.0133	0.02290
5.5	23.758	0.0138	0.02376
5.6	24.630	0.0143	0.02463
5.7	25.518	0.0147	0.02552
5.75	25.967	0.0150	0.02597
5.8	26.421	0.0153	0.02642
5.9	27.340	0.0158	0.02734
6.0	28.274	0.0164	0.02827
6.1	29.225	0.0169	0.02923

UNIVERSAL DISPLACEMENT TABLES

Bore	Cross Sectional Area	Displacement per Unit Stroke*	
		ft^3	m^3
6.2	30.191	0.0174	0.03019
6.25	30.680	0.0178	0.03068
6.3	31.173	0.0180	0.03117
6.4	32.170	0.0186	0.03217
6.5	33.183	0.0192	0.03318
6.6	34.212	0.0198	0.03421
6.7	35.257	0.0204	0.03526
6.75	35.785	0.0207	0.03579
6.8	36.317	0.0210	0.03632
6.9	37.393	0.0216	0.03739
7.0	38.485	0.0223	0.03849
7.1	39.592	0.0229	0.03959
7.2	40.715	0.0236	0.04072
7.3	41.854	0.0242	0.04185
7.4	43.009	0.0248	0.04301
7.5	44.179	0.0256	0.04418
7.6	45.365	0.0262	0.04537
7.7	46.566	0.0270	0.04657
7.8	47.784	0.0276	0.04778
7.9	49.017	0.0283	0.04902
8.0	50.266	0.0291	0.05027
8.1	51.530	0.0299	0.05153
8.2	52.810	0.0306	0.05281
8.3	54.106	0.0314	0.05411
8.4	55.418	0.0321	0.05542
8.5	56.745	0.0329	0.05675
8.6	58.088	0.0336	0.05809
8.7	59.447	0.0344	0.05945
8.8	60.821	0.0352	0.06082
8.9	62.222	0.0360	0.06222

Bore	Cross Sectional Area	Displacement per Unit Stroke*	
		ft^3	m^3
9.0	63.617	0.0369	0.06362
9.1	65.039	0.0377	0.06504
9.2	66.476	0.0384	0.06648
9.3	67.929	0.0393	0.06793
9.4	69.398	0.0402	0.06940
9.5	70.882	0.0412	0.07088
9.6	72.383	0.0419	0.07238
9.7	73.898	0.0427	0.07390
9.8	75.430	0.0436	0.07543
9.9	76.927	0.0445	0.07698
10	78.540	0.0454	0.07854
11	95.033	0.0550	0.09503
12	113.097	0.0650	0.11310
13	132.732	0.0769	0.13273
14	153.938	0.0892	0.15394
15	176.715	0.1023	0.17672
16	210.062	0.1124	0.20106
17	226.980	0.1314	0.22698
18	254.469	0.1473	0.25447
19	283.529	0.1642	0.28353
20	314.159	0.1818	0.31416
21	346.361	0.2005	0.34636
22	380.134	0.2200	0.38013
23	415.477	0.2405	0.41548
24	452.389	0.2619	0.45239
25	490.874	0.2840	0.49087
26	530.929	0.3072	0.53093
27	572.556	0.3313	0.57256
28	615.753	0.3563	0.61575
29	660.521	0.3822	0.66052

UNIVERSAL DISPLACEMENT TABLES

Bore	Cross Sectional Area	Displacement per Unit Stroke*	
		ft^3	m^3
30	706.860	0.4091	0.70686
31	754.769	0.4368	0.75477
32	804.249	0.4654	0.80425
33	855.300	0.4950	0.85530
34	907.921	0.5253	0.90792
35	962.114	0.5568	0.96211
36	1017.88	0.5890	1.01788
37	1075.21	0.6223	1.07521
38	1134.12	0.6563	1.13412
39	1194.59	0.6907	1.19459
40	1256.64	0.7273	1.25664
41	1320.26	0.7642	1.32026
42	1385.45	0.8020	1.38545
43	1452.20	0.8403	1.45220
44	1520.53	0.8798	1.52053
45	1590.43	0.9204	1.59043
46	1661.91	0.9619	1.66191
47	1734.95	1.0040	1.73495
48	1809.56	1.0472	1.80956
49	1885.75	1.0915	1.88575
50	1963.50	1.1368	1.96350
51	2042.8	1.18	2.043
52	2123.7	1.23	2.124
53	2206.2	1.28	2.206
54	2290.2	1.33	2.290
55	2375.8	1.38	2.376
56	2463.0	1.43	2.463
57	2551.8	1.47	2.552
58	2642.1	1.53	2.642
59	2734.0	1.58	2.734
60	2827.4	1.64	2.827

Bore	Cross Sectional Area	Displacement per Unit Stroke*	
		ft^3	m^3
61	2922.5	1.69	2.923
62	3019.1	1.74	3.019
63	3117.3	1.80	3.117
64	3217.0	1.86	3.217
65	3318.3	1.92	3.318
66	3421.2	1.98	3.421
67	3525.7	2.04	3.526
68	3631.7	2.10	3.632
69	3739.3	2.16	3.739
70	3848.5	2.23	3.849
71	3959.2	2.29	3.959
72	4071.5	2.36	4.072
73	4185.4	2.42	4.185
74	4300.9	2.48	4.301
75	4417.9	2.56	4.418
76	4536.5	2.62	4.537
77	4656.6	2.70	4.657
78	4778.4	2.76	4.778
79	4901.7	2.83	4.902
80	5026.6	2.91	5.027
81	5153.0	2.99	5.153
82	5281.0	3.06	5.281
83	5410.6	3.14	5.411
84	5541.8	3.21	5.542
85	5674.5	3.29	5.675
86	5808.8	3.36	5.809
87	5944.7	3.44	5.945
88	6082.1	3.52	6.082
89	6222.2	3.60	6.222
90	6361.7	3.69	6.362
91	6503.9	3.77	6.504

UNIVERSAL DISPLACEMENT TABLES

Bore	Cross Sectional Area	Displacement per Unit Stroke*	
		ft^3	m^3
92	6647.6	3.84	6.648
93	6792.9	3.93	6.793
94	6939.8	4.02	6.940
95	7088.2	4.12	7.088
96	7238.3	4.19	7.238
97	7389.8	4.27	7.390
98	7543.0	4.36	7.543
99	7697.7	4.45	7.698
100	7854.0	4.54	7.854

*This figure multiplied by the stroke in the same units as the bore gives displacement as under:-

For bore and stroke in inches, use ft^3 column figure to give displacement in cubic feet.

For bore and stroke in centimetres, use m^3 figure to give displacement in cubic metres.

Note that the bore column can also be entered in millimetres, if preferred, when the displacement must be adjusted accordingly. The cross sectional area column gives the equivalent cross section in any units, ie in^2 for bore in inches; mm^2 for bore in millimetres; cm^2 for bore in centimetres.

RATIONALIZED TABLE : AIR CONSUMPTION OF DOUBLE-ACTING CYLINDERS
(English Sizes)

Bore	Rod Diameter – Inches									
Ins	3/8	1/2	5/8	3/4	7/8	1	1.1/8	1.1/4	1.3/8	1.1/2
¾	0.00349	0.00310	—	—	—	—	—	—	—	—
1	0.00660	0.00621	0.00572	—	—	—	—	—	—	—
1¼	0.0105	0.0101	0.0097	—	—	—	—	—	—	—
1½	0.0154	0.0150	0.0146	0.0140	—	—	—	—	—	—
1¾	0.0212	0.0208	0.0204	0.0197	0.0191	—	—	—	—	—
2	—	0.0275	0.0271	0.0266	0.0257	0.0249	—	—	—	—
2¼	—	—	0.0346	0.0340	0.0332	0.0324	0.0314	—	—	—
2½	—	—	0.0430	0.0424	0.0416	0.0407	0.0400	0.0389	—	—
2¾	—	—	0.0522	0.0517	0.0510	0.0502	0.0492	0.0481	0.0470	—
3	—	—	0.0624	0.0619	0.0611	0.0604	0.0595	0.0584	0.0572	0.0559
3¼	—	—	0.0735	0.0729	0.0722	0.0712	0.0704	0.0694	0.0683	0.0669
3½	—	—	0.0854	0.0849	0.0841	0.0835	0.0825	0.0814	0.0803	0.0791
3¾	—	—	—	0.0978	0.0971	0.0962	0.0953	0.0943	0.0932	0.0918
4	—	—	—	0.112	0.111	0.110	0.109	0.108	0.106	0.105
4½	—	—	—	0.142	0.141	0.140	0.139	0.138	0.137	0.136
5	—	—	—	0.175	0.174	0.173	0.172	0.172	0.171	0.169
5½	—	—	—	0.212	0.211	0.211	0.210	0.209	0.208	0.207
6	—	—	—	—	—	0.252	0.251	0.250	0.249	0.248
7	—	—	—	—	—	0.343	0.343	0.342	0.341	0.340
8	—	—	—	—	—	—	—	0.448	0.447	0.445
9	—	—	—	—	—	—	—	0.568	0.567	0.566
10	—	—	—	—	—	—	—	—	—	0.701
11	—	—	—	—	—	—	—	—	—	0.850
12	—	—	—	—	—	—	—	—	—	1.014
14	—	—	—	—	—	—	—	—	—	—
16	—	—	—	—	—	—	—	—	—	—
18	—	—	—	—	—	—	—	—	—	—
20	—	—	—	—	—	—	—	—	—	—

These rationalized tables cover the normal range of proprietary air cylinder sizes and rod sizes. The table figures give the consumption for a complete inward and outward stroke in terms of cubic feet of free air per inch of stroke, based on isothermal compression and expansion, assuming that the temperature of the compressed air supplied to the cylinder is the same as that at the compressor inlet.

To arrive at the actual consumption the table figures must be multiplied by a factor determined by the gauge pressure of the compressed air. These factors are:-

Gauge Pressure lb/in^2	Factor
150	1.44
120	1.175
100	1.0
90	0.913
80	0.905
70	0.740
60	0.653
50	0.565
40	0.480

RATIONALIZED TABLE : AIR CONSUMPTION OF DOUBLE-ACTING CYLINDERS
(English Sizes)

Rod Diameter – Inches										Bore
1.3/4	**2**	**2.1/2**	**3**	**3.1/2**	**4**	**4.1/2**	**5**	**5.1/2**	**6**	**Ins**
										¾
										1
										1¼
										1½
										1¾
										2
										2¼
										2½
										2¾
										3
0.0640										3¼
0.0763										3½
0.0892	0.0856									3¾
0.103	0.0994									4
0.134	0.129									4½
0.167	0.163									5
0.204	0.200									5½
0.245	0.241	0.233	0.224							6
0.337	0.334	0.326	0.315							7
0.443	0.440	0.431	0.422	0.411						8
0.563	0.561	0.552	0.543	0.532	0.518					9
0.697	0.695	0.687	0.677	0.665	0.652	0.638				10
0.847	0.845	0.836	0.827	0.815	0.802	0.784	0.770			11
1.010	1.008	0.998	0.990	0.978	0.965	0.947	0.934	0.914		12
—	1.378	1.369	1.360	1.348	1.336	1.319	1.304	1.280	1.264	14
—	—	—	1.788	1.772	1.760	1.745	1.730	1.710	1.690	16
—	—	—	—	2.194	2.174	2.153	2.132	2.121	2.100	18
—	—	—	—	2.800	2.780	2.765	2.750	2.730	2.710	20

For any other gauge pressure the appropriate factor can be calculated from the formula

$$\text{Factor} = \frac{P + 14.7}{14.7}$$

where P is the actual gauge pressure (lb/in^2)

RATIONALIZED TABLE : AIR CONSUMPTION OF DOUBLE-ACTING CYLINDERS
(Metric)

Bore	Rod Diameter — Millimetres										
mm	13	20	25	32	40	50	65	75	90	100	120
25	0.00847	—	—	—	—	—	—	—	—	—	—
39	0.02267	0.02076	—	—	—	—	—	—	—	—	—
50	0.03793	0.03612	0.03437	—	—	—	—	—	—	—	—
64	0.0630	0.0612	0.0594	0.0563	—	—	—	—	—	—	—
76	0.0894	0.0876	0.0858	0.0827	—	—	—	—	—	—	—
100	—	0.1539	0.1522	0.1490	0.1445	0.1325	—	—	—	—	—
124	—	—	0.2351	0.2320	0.2274	0.2200	0.2068	—	—	—	—
154	—	—	—	0.3640	0.3594	0.352	0.339	0.328	—	—	—
203	—	—	—	0.638	0.633	0.626	0.613	0.602	0.583	—	—
254	—	—	—	—	1.000	0.992	0.979	0.968	0.948	0.934	—
305	—	—	—	—	—	1.440	1.427	1.416	1.376	1.382	—
350	—	—	—	—	—	1.904	1.891	1.880	1.860	1.846	1.811
400	—	—	—	—	—	2.492	2.479	2.468	2.448	2.434	2.400
450	—	—	—	—	—	—	3.146	3.136	3.116	3.102	3.067
500	—	—	—	—	—	—	—	3.882	3.863	3.848	3.813

This rationalized table covers the normal range of proprietary air cylinders and rod sizes. The table figures are based on displacement in cubic centimetres of free air per cm of stroke, based on isothermal compression and expansion and assuming that the temperature of the compressed air supplied to the cylinder is the same as at the compressor inlet.

To arrive at the actual air consumption per complete inward and outward stroke the table figure derived for cylinder size is used in the following formula:-

air consumption (litres) — table figure x (P + 1) x stroke

where P = gauge pressure in bars and the stroke is in centimetres

DYNAMIC AIR PRESSURE (lb/in^2) SPEED IN FEET/SECOND
(Atmospheric Air At Standard Density*)

V→ ↓ ft/sec	0	1	2	3	4	5	6	7	8	9
0		0.00119	0.00476	0.0107	0.0190	0.0297	0.0428	0.0583	0.0762	0.0964
10	0.119	0.143	0.171	0.201	0.233	0.268	0.304	0.343	0.390	0.429
20	0.476	0.525	0.575	0.628	0.685	0.745	0.805	0.868	0.932	1.000
30	1.070	1.14	1.22	1.29	1.38	1.458	1.54	1.63	1.72	1.81
40	1,904	2.00	2.10	2.20	2.30	2.410	2.53	2.62	2.74	2.86
50	2.971	3.09	3.22	3.34	3.47	3.600	3.73	3.86	4.00	4.14
60	4.280	4.42	4.57	4.72	4.87	5.025	5.18	5.34	5.50	5.66
70	5.831	5.99	6.16	6.34	6.52	6.70	6.86	7.05	7.23	7.41
80	7.610	7.80	8.00	8.20	8.40	8.60	8.80	9.00	9.20	9.42
90	9.639	9.85	10.08	10.30	10.50	10.70	10.90	11.20	11.40	11.70
100	11.890									

DYNAMIC AIR PRESSURE (bar) SPEED IN METRES/SECOND
(Atmospheric Air At Standard Density*)

V→ ↓ m/sec	0	0.1	0.2	0.3	0.4	0.5	0.6	0.7	0.8	0.9
0		0.001	0.002	0.003	0.004	0.005	0.005	0.006	0.007	0.008
1	0.009	0.010	0.011	0.012	0.013	0.014	0.014	0.015	0.016	0.017
2	0.018	0.019	0.020	0.021	0.022	0.023	0.023	0.024	0.025	0.026
3	0.027	0.028	0.029	0.030	0.031	0.032	0.032	0.033	0.034	0.035
4	0.036	0.037	0.038	0.039	0.040	0.041	0.041	0.042	0.043	0.044
5	0.045	0.046	0.047	0.048	0.049	0.050	0.050	0.051	0.052	0.053
6	0.054	0.055	0.056	0.057	0.058	0.059	0.059	0.060	0.061	0.062
7	0.063	0.064	0.065	0.066	0.067	0.068	0.068	0.069	0.070	0.071
8	0.072	0.073	0.074	0.075	0.076	0.077	0.077	0.078	0.079	0.080
9	0.081	0.082	0.083	0.084	0.085	0.086	0.086	0.087	0.088	0.089
10	0.090									

* The value of dynamic air pressure (q) at any other density (ρ) can be found from the following relationship

$$q = qo \times \frac{\rho}{\rho_o}$$

where q_o = table figure

ρ_o = standard air density

ISENTROPIC FLOW TABLES

Mach No	P^*/P_o	T/T_o	ρ/ρ_o	A/A^*	V/V^*
0.01	0.9999	0.9999	0.9999	57.874	0.01096
0.02	0.9997	0.9999	0.9998	28.942	0.02191
0.03	0.9993	0.9998	0.9996	19.300	0.03286
0.04	0.9989	0.9997	0.9992	14.482	0.04381
0.05	0.9983	0.9995	0.9988	11.592	0.05476
0.06	0.9974	0.9993	0.9982	9.666	0.06570
0.07	0.9966	0.9990	0.9976	8.292	0.07664
0.08	0.9955	0.9987	0.9968	7.262	0.08758
0.09	0.9945	0.9984	0.9960	6.461	0.09851
0.10	0.9930	0.9980	0.9950	5.822	0.10943
0.11	0.9916	0.9976	0.9940	5.299	0.12035
0.12	0.9900	0.9971	0.9928	4.864	0.13126
0.13	0.9883	0.9966	0.9916	4.497	0.14216
0.14	0.9864	0.9961	0.9903	4.182	0.15306
0.15	0.9844	0.9955	0.9888	3.910	0.16395
0.16	0.9823	0.9949	0.9873	3.673	0.17483
0.17	0.9800	0.9943	0.9857	3.464	0.18569
0.18	0.9777	0.9936	0.9840	3.178	0.19654
0.19	0.9751	0.9928	0.9822	3.112	0.20737
0.20	0.9725	0.9921	0.9802	2.964	0.21822
0.21	0.9697	0.9913	0.9783	2.829	0.22904
0.22	0.9669	0.9904	0.9762	2.708	0.23984
0.23	0.9638	0.9895	0.9740	2.597	0.25063
0.24	0.9607	0.9886	0.9718	2.496	0.26141
0.25	0.9575	0.9877	0.9694	2.403	0.27216
0.26	0.9541	0.9867	0.9670	2.317	0.28291
0.27	0.9506	0.9856	0.9645	2.239	0.29364
0.28	0.9470	0.9846	0.9619	2.166	0.30435
0.29	0.9433	0.9835	0.9592	2.098	0.31504
0.30	0.9395	0.9832	0.9564	2.035	0.32572
0.31	0.9355	0.9811	0.9535	1.977	0.33638
0.32	0.9315	0.9799	0.9506	1.922	0.34701
0.33	0.9274	0.9787	0.9476	1.871	0.35762
0.34	0.9231	0.9774	0.9445	1.823	0.36821
0.35	0.9188	0.9761	0.9413	1.778	0.37879
0.36	0.9143	0.9747	0.9380	1.735	0.38935
0.37	0.9098	0.9734	0.9347	1.696	0.39988
0.38	0.9052	0.9719	0.9313	1.659	0.41039
0.39	0.9004	0.9705	0.9278	1.623	0.42087
0.40	0.8956	0.9690	0.9243	1.590	0.43133
0.41	0.8907	0.9675	0.9207	1.559	0.44177
0.42	0.8857	0.9659	0.9170	1.529	0.45218
0.43	0.8807	0.9643	0.9132	1.501	0.46256
0.44	0.8755	0.9627	0.9094	1.474	0.47292
0.45	0.8703	0.9611	0.9055	1.449	0.48326
0.46	0.8650	0.9594	0.9016	1.425	0.49357
0.47	0.8596	0.9577	0.8976	1.402	0.50385
0.48	0.8541	0.9560	0.8935	1.380	0.51400
0.49	0.8486	0.9542	0.8894	1.359	0.52432
0.50	0.8430	0.9524	0.8852	1.340	0.53432

Note: A^* = critical area
P^* = critical pressure
V^* = critical volume

cont...

ISENTROPIC FLOW TABLES (contd.)

Mach No	P^*/P_o	T/T_o	ρ/ρ_o	A/A^*	V/V^*
0.51	0.8374	0.9506	0.8809	1.321	0.54469
0.52	0.8317	0.9487	0.8766	1.303	0.55482
0.53	0.8259	0.9468	0.8723	1.286	0.56493
0.54	0.8201	0.9449	0.8679	1.270	0.57501
0.55	0.8142	0.9430	0.8634	1.255	0.58506
0.56	0.8082	0.9410	0.8589	1.240	0.59508
0.57	0.8022	0.9390	0.8544	1.226	0.60506
0.58	0.7962	0.9370	0.8498	1.213	0.61500
0.59	0.7901	0.9349	0.8451	1.200	0.62491
0.60	0.7840	0.9328	0.8405	1.188	0.63480
0.61	0.7778	0.9307	0.8357	1 177	0.64466
0.62	0.7716	0.9286	0.8310	1.166	0.65448
0.63	0.7654	0.9265	0.8262	1.155	0.66427
0.64	0.7591	0.9243	0.8213	1.145	0.67402
0.65	0.7528	0.9221	0.8164	1.136	0.68374
0.66	0.7465	0.9199	0.8115	1.127	0.69342
0.67	0.7401	0.9176	0.8066	1.118	0.70307
0.68	0.7338	0.9154	0.8016	1.110	0.71268
0.69	0.7274	0.9131	0.7966	1.102	0.72225
0.70	0.7209	0.9108	0.7916	1.094	0.73179
0.71	0.7145	0.9082	0.7865	1.087	0.74129
0.72	0.7080	0.9061	0.7814	1.081	0.75076
0.73	0.7016	0.9037	0.7763	1.074	0.76019
0.74	0.6951	0.9013	0.7712	1.068	0.76958
0.75	0.6886	0.8989	0.7660	1.062	0.77893
0.76	0.6821	0.8964	0.7609	1,057	0.78825
0.77	0.6756	0 8940	0.7557	1.052	0.79753
0.78	0.6691	0.8915	0.7505	1.047	0.80677
0.79	0.6625	0.8890	0.7452	1.043	0.81597
0.80	0.6560	0.8865	0.7400	1.038	0.82514
0.81	0.6495	0.8840	0.7347	1.034	0.83426
0.82	0.6430	0.8815	0.7295	1.030	0.84334
0.83	0.6365	0.8789	0.7242	1.027	0.85239
0.84	0.6300	0.8763	0.7189	1.024	0.86140
0.85	0.6235	0.8737	0.7136	1.021	0.87037
0.86	0.6170	0.8711	0.7083	1.018	0.87929
0.87	0.6106	0.8685	0.7030	1.015	0.88817
0.88	0.6041	0.8659	0.6977	1.013	0.89702
0.89	0.5977	0.8632	0.6924	1.011	0.90583
0.90	0.5913	0.8606	0.6870	1.009	0.91460
0.91	0.5849	0.8579	0.6817	1.007	0.92333
0.92	0.5785	0.8552	0.6764	1.006	0.93201
0.93	0.5721	0.8525	0.6711	1.004	0.94065
0.94	0.5658	0.8498	0.6658	1.003	0.94925
0.95	0.5595	0.8471	0.6604	1.002	0.95780
0.96	0.5532	0.8444	0.6551	1.001	0.96633
0.97	0.5469	0.8416	0.6498	1.000	0.98325
0.98	0.5345	0.8361	0.6392	1.000	0.99165
1.00	0.5283	0.8333	0.6339	1.000	1.00000

Note: A^* = critical area
P^* = critical pressure
V^* = critical volume

cont........

ISENTROPIC FLOW TABLES (contd.)

Mach No	P^*/P_o	T/T_o	ρ/ρ_o	A/A^*	V/V^*
1.01	0.5221	0.8306	0.6287	1.0000	1.0083
1.02	0.5160	0.8278	0.6234	1.0003	1.0166
1.03	0.5099	0.8250	0.6181	1.0007	1.0248
1.04	0.5039	0.8222	0.6129	1.0013	1.0330
1.05	0.4979	0.8193	0.6077	1.0020	1.0411
1.06	0.4919	0.8165	0.6024	1.0029	1.0492
1.07	0.4860	0.8137	0.5972	1.0039	1.0573
1.08	0.4801	0.8108	0.5920	1.0051	1.0653
1.09	0.4742	0.8080	0.5869	1.0064	1.0733
1.10	0.4684	0.8052	0.5817	1.0079	1.0812
1.11	0.4626	0.8023	0.5766	1.0096	1.0891
1.12	0.4568	0.7994	0.5714	1.0111	1.0970
1.13	0.4511	0.7966	0.5663	1.0132	1.1048
1.14	0.4455	0.7937	0.5612	1.0153	1.1126
1.15	0.4398	0.7908	0.5562	1.0175	1.1203
1.16	0.4343	0.7880	0.5511	1.0198	1.1280
1.17	0.4287	0.7851	0.5461	1.0222	1.1356
1.18	0.4232	0.7822	0.5411	1.0248	1.1432
1.19	0.4178	0.7793	0.5361	1.0276	1.1508
1.20	0.4124	0.7764	0.5311	1.0304	1.1583
1.21	0.4070	0.7735	0.5262	1.0334	1.1658
1.22	0.4017	0.7706	0.5213	1.0366	1.1732
1.23	0.3965	0.7677	0.5164	1.0398	1.1806
1.24	0.3912	0.7648	0.5115	1.0432	1.1879
1.25	0.3861	0.7619	0.5067	1.0468	1.1952
1.26	0.3809	0.7590	0.5019	1.0504	1.2025
1.27	0.3759	0.7561	0.4971	1.0542	1.2097
1.28	0.3708	0.7532	0.4923	1.0581	1.2169
1.29	0.3659	0.7503	0.4876	1.0621	1.2240
1.30	0.3609	0.7474	0.4829	1.0663	1.2311
1.31	0.3560	0.7445	0.4782	1.0706	1.2382
1.32	0.3512	0.7416	0.4736	1.0750	1.2452
1.33	0.3464	0.7387	0.4690	1.0796	1.2522
1.34	0.3417	0.7358	0.4644	1.0842	1.2591
1.35	0.3370	0.7329	0.4598	1.0890	1.2660
1.36	0.3323	0.7300	0.4553	1.0940	1.2729
1.37	0.3277	0.7271	0.4508	1.0990	1.2797
1.38	0.3232	0.7242	0.4463	1.1042	1.2865
1.39	0.3187	0.7213	0.4418	1.1095	1.2932
1.40	0.3142	0.7184	0.4374	1.1149	1.2999
1.41	0.3098	0.7155	0.4330	1.1205	1.3065
0.42	0.3055	0.7126	0.4287	1.1262	1.3131
1.43	0.3012	0.7097	0.4244	1.1320	1.3197
1.44	0.2969	0.7069	0.4201	1.1379	1.3262
1.45	0.2927	0.7040	0.4158	1.1440	1.3327
1.46	0.2886	0.7011	0.4116	1.1502	1.3392
1.47	0.2845	0.6983	0.4074	1.1565	1.3456
1.48	0.2804	0.6954	0.4032	1.1629	1.3520
1.49	0.2764	0.6925	0.3991	1.1695	1.3583
1.50	0.2724	0.6897	0.3950	1.1762	1.3646
1.51	0.2685	0.6868	0.3909	1.1830	1.3708
1.52	0.2646	0.6840	0.3869	1.1899	1.3770

See Note on page 526

cont........

ISENTROPIC FLOW TABLES (contd.)

Mach No	P^*/P_o	T/T_o	ρ/ρ_o	A/A^*	V/V^*
1.53	0.2608	0.6811	0.3829	1.1970	1.3832
1.54	0.2570	0.6783	0.3789	1.2042	1.3894
1.55	0.2533	0.6755	0.3750	1.2115	1.3955
1.56	0.2496	0.6726	0.3711	1.2190	1.4016
1.57	0.2459	0.6698	0.3672	1.2266	1.4076
1.58	0.2423	0.6670	0.3633	1.2343	1.4135
1.59	0.2388	0.6642	0.3595	1.2422	1.4195
1.60	0.2353	0.6614	0.3557	1.2502	1.4254
1.61	0.2318	0.6586	0.3520	1.2583	1.4313
1.62	0.2284	0.6558	0.3483	1.2666	1.4371
1.63	0.2250	0.6530	0.3446	1.2750	1.4429
1.64	0.2217	0.6502	0.3409	1.2836	1.4487
1.65	0.2184	0.6475	0.3373	1.2922	1.4544
1.66	0.2152	0.6447	0.3337	1.3010	1.4601
1.67	0.2120	0.6419	0.3302	1.3099	1.4657
1.68	0.2088	0.6392	0.3266	1.3190	1.4713
1.69	0.2057	0.6365	0.3232	1.3282	1.4769
1.70	0.2026	0.6337	0.3197	1.3376	1.4825
1.71	0.1996	0.6310	0.3163	1.3471	1.4880
1.72	0.1966	0.6283	0.3129	1.3567	1.4935
1.73	0.1936	0.6256	0.3095	1.3665	1.4989
1.74	0.1907	0.6229	0.3062	1.3764	1.5043
1.75	0.1878	0.6202	0.3029	1.3865	1.5097
1.76	0.1850	0.6175	0.2996	1.3967	1.5150
1.77	0.1822	0.6148	0.2964	1.4071	1.5203
1.78	0.1794	0.6121	0.2931	1.4176	1.5256
1.79	0.1767	0.6095	0.2900	1.4282	1.5308
1.80	0.1740	0.6068	0.2868	1.4390	1.5360
1.81	0.1714	0.6042	0.2837	1.4499	1.5412
1.82	0.1688	0.6015	0.2806	1.4610	1.5463
1.83	0.1662	0.5989	0.2776	1.4723	1.5514
1.84	0.1637	0.5963	0.2745	1.4837	1.5564
1.85	0.1612	0.5937	0.2715	1.4952	1.5614
1.86	0.1587	0.5911	0.2686	1.5069	1.5664
1.87	0.1563	0.5885	0.2656	1.5188	1.5714
1.88	0.1539	0.5859	0.2627	1.5308	1.5763
1.89	0.1516	0.5833	0.2598	1.5429	1.5812
1.90	0.1492	0.5807	0.2570	1.5552	1.5861
1.91	0.1470	0.5782	0.2542	1.5677	1.5909
1.92	0.1447	0.5756	0.2514	1.5804	1.5957
1.93	0.1425	0.5731	0.2486	1.5932	1.6005
1.94	0.1403	0.5705	0.2459	1.6062	1.6052
1.95	0.1381	0.5680	0.2432	1.6193	1.6099
1.96	0.1360	0.5655	0.2405	1.6326	1.6146
1.97	0.1339	0.5630	0.2378	1.6461	1.6193
1.98	0.1314	0.5605	0.2352	1.6597	1.6239
1.99	0.1298	0.5580	0.2326	1.6735	1.6285
2.00	0.1278	0.5556	0.2301	1.6875	1.6330
2.01	0.1258	0.5531	0.2275	1.7017	1.6375
2.02	0.1239	0.5506	0.2250	1.7160	1.6420
2.03	0.1220	0.5482	0.2225	1.7305	1.6465
2.04	0.1201	0.5458	0.2200	1.7452	1.6509
2.05	0.1182	0.5433	0.2176	1.7600	1.6553
2.06	0.1164	0.5409	0.2152	1.7750	1.6597
2.07	0.1146	0.5385	0.2128	1.7902	1.6640
2.08	0.1128	0.5360	0.2105	1.8056	1.6683

See note on page 526

cont........

ISENTROPIC FLOW TABLES (contd.)

Mach No	P^*/P_o	T/T_o	ρ/ρ_o	A/A^*	V/V^*
2.09	0.1111	0.5337	0.2081	1.8212	1.6726
2.10	0.1094	0.5314	0.2058	1.8369	1.6769
2.11	0.1077	0.5290	0.2035	1.8529	1.6811
2.12	0.1060	0.5266	0.2013	1.8690	1.6853
2.13	0.1043	0.5248	0.1990	1.8853	1.6895
2.14	0.1027	0.5219	0.1968	1.9018	1.6936
2.15	0.1011	0.5196	0.1946	1.9185	1.6977
2.16	0.0996	0.5173	0.1925	1.9354	1.7018
2.17	0.0980	0.5150	0.1903	1.9525	1.7059
2.18	0.0965	0.5127	0.1882	1.9698	1.7099
2.19	0.0950	0.5104	0.1861	1.9873	1.7139
2.20	0.0935	0.5081	0.1841	2.0050	1.7179
2.21	0.0921	0.5059	0.1820	2.0229	1.7219
2.22	0.0906	0.5036	0.1800	2.0409	1.7258
2.23	0.0892	0.5014	0.1780	2.0592	1.7297
2.24	0.0878	0.4991	0.1760	2.0777	1.7336
2.25	0.0865	0.4969	0.1740	2.0964	1.7374
2.26	0.0851	0.4947	0.1721	2.1154	1.7412
2.27	0.0838	0.4925	0.1702	2.1345	1.7450
2.28	0.0825	0.4903	0.1683	2.1538	1.7488
2.29	0.0812	0.4881	0.1664	2.1734	1.7526
2.30	0.0800	0.4859	0.1646	2.1931	1.7563
2.31	0.0787	0.4837	0.1628	2.2131	1.7600
2.32	0.0775	0.4816	0.1610	2.2333	1.7637
2.33	0.0763	0.4794	0.1592	2.2537	1.7673
2.34	0.0751	0.4773	0.1574	2.2744	1.7709
2.35	0.0740	0.4752	0.1556	2.2953	1.7745
2.36	0.0728	0.4731	0.1539	2.3164	1.7781
2.37	0.0717	0.4710	0.1522	2.3377	1.7817
2.38	0.0706	0.4689	0.1505	2.3593	1.7852
2.39	0.0695	0.4668	0.1489	2.3811	1.7887
2.40	0.0684	0.4647	0.1472	2.4031	1.7922
2.41	0.0673	0.4626	0.1456	2.4254	1.7957
2.42	0.0663	0.4606	0.1440	2.4479	1.7991
2.43	0.0653	0.4585	0.1424	2.4706	1.8025
2.44	0.0643	0.4565	0.1408	2.4936	1.8059
2.45	0.0633	0.4544	0.1399	2.5168	1.8093
2.46	0.0623	0.4524	0.1377	2.5403	1.8126
2.47	0.0613	0.4504	0.1362	2.5640	1.8159
2.48	0.0604	0.4484	0.1347	2.5880	1.8192
2.49	0.0595	0.4464	0.1332	2.6122	1.8225
2.50	0.0583	0.4444	0.1317	2.6367	1.8259
3.00	0.0272	0.3571	0.0762	4.2346	1.9640
4.00	0.0066	0.2381	0.0277	10.7190	2.1381
5.00	0.0019	0.1667	0.0113	25.0000	2.2361

EQUIVALENT FREE AIR VOLUME (Cubic Feet)

Volume of Compressed Air ft^3/min	Line Pressure lbf/in^2 Gauge								
	30	40	50	60	70	80	90	100	120
0.1	0.311	0.372	0.440	0.508	0.576	0.644	0.712	0.780	0.916
0.2	0.621	0.744	0.880	1.016	1.152	1.288	1 424	1.560	1.833
0.3	0.931	1.116	1.320	1.525	1.729	1 922	2.137	2.341	2 749
0.4	1.242	1.488	1.760	2.033	2.305	2.577	2.849	3.121	3.666
0.5	1.553	1.861	2.201	2.541	2.881	3.221	3.561	3.901	4.682
0.6	1.863	2.233	2.641	3.049	3.457	3.865	4.273	4.681	5.498
0.7	2.173	2.605	3.081	3.557	4.033	4.509	4.985	5.461	6.415
0.8	2.484	2.977	3.521	4.066	4.610	5.154	5.698	6.242	7.331
0.9	2.795	3.349	3.961	4.574	5.186	5.798	6.410	7 022	8.248
1.0	3.105	3.721	4.401	5.082	5.762	6.442	7.122	7.802	9.164
2.0	6.210	7.442	8.802	10.164	11.524	12.884	14.244	15.604	18.328
3.0	9.315	11.163	13.203	15.246	17.286	19.326	21.366	23.406	27.492
4.0	12.420	14.884	17.604	20.328	23.048	25.768	28.488	31.208	36.656
5.0	15.530	18.605	22.005	25.410	28.810	32.210	35.610	39.010	45.820
6.0	18.630	22.326	26.406	30.492	34.572	38.652	42.732	46.812	54.984
7.0	21.740	26.047	30.807	35.574	40.334	45.094	49.854	54.614	64.148
8.0	24.840	29.768	35.208	40.656	46.096	51.536	56.976	62.416	73.312
9.0	27.950	33.489	39.609	45.738	51.858	57.978	64.098	70.218	82.476
10.0	31.050	37.210	44.010	50.820	57.620	64.420	71.220	68.020	91.640
11.0	34.155	40.930	48.410	55.900	63.380	70.860	78.340	85.820	100.800
12.0	37.260	44.650	52.810	60.980	69.140	77.300	85.460	93.620	109.970
13.0	40.365	48.370	57.210	66.060	74.900	83.740	92.580	101.420	119.130
14.0	43.470	52.090	61.610	71.140	80.660	90.180	99.700	109.220	128.290
15.0	46.575	55.810	66.010	76.230	86.420	96.620	106.820	117.020	137.460
16.0	49.680	59.530	70.410	81.310	92.180	103.060	113.940	124.820	146.620
17.0	52.785	63.250	74.810	86.390	97.940	109.500	121.060	132.620	155.790
18.0	55.890	66.970	79.210	91.470	103.700	115.940	128.180	140.420	164.950
19.0	58.995	70.690	83.610	96.550	109.460	122.380	135.310	148.230	174.110
20.0	62.100	74.420	88.020	101.640	115.240	128.840	142.440	156.040	183.280
21.0	65.205	78.140	92.420	106.720	121.000	135.280	149.560	163.840	192.440
22.0	68.310	81.860	96.820	111.800	126.760	141.720	156.680	171.640	201.610
23.0	71.415	85.580	101.220	116.880	132.520	148.160	163.800	179.440	210.780
24.0	74.520	89.300	105.620	121.960	138.280	154.600	170.930	187.240	219.940
25.0	77.625	93.020	110.020	127.040	144.040	161.040	178.050	195.040	229.100
26.0	80.730	96.740	114.420	132.120	149.800	167.480	185.170	202.840	238.260
27.0	83.835	100.460	118.820	137.200	155.560	173.920	192.290	210.640	247.430
28.0	86.940	104.180	123.220	142.280	161.320	180.360	199.420	218.440	256.590
29.0	90.045	107.900	127.620	147.360	167.080	186.800	206.540	226.240	265.760
30.0	93.150	111.630	132.030	152.460	172.860	193.260	213.660	234.060	274.920
31.0	96.255	115.350	136.430	157.540	178.620	199.700	220.780	241.860	284.080
32.0	99.360	119.070	140.830	162.620	184.380	206.140	227.900	249.660	293.240
33.0	102.465	122.790	145.230	167.700	190.140	212.580	235.030	257.460	302.410
34.0	105.570	126.510	149.630	172.780	195.900	219.020	242.150	265.260	311.570
35.0	108.675	130.230	154.030	177.860	201.660	225.460	249.270	273.060	320.740
36.0	111.780	133.950	158.430	182.940	207.420	231.900	256.390	280.860	329.900
37.0	114.885	137.670	162.830	188.020	213.180	238.340	263.510	288.660	339.060
38.0	117.990	141.390	167.230	192.100	218.950	244.780	270.630	296.460	348.220
39.0	121.095	145.110	171.630	198.180	224.710	251.230	277.750	304.260	357.390

cont..........

EQUIVALENT FREE AIR VOLUME (Cubic Feet) (contd.)

Volume of Compressed Air ft³/min	Line Pressure lbf/in² Gauge								
	30	40	50	60	70	80	90	100	120
40	124.200	148.84	176.04	203.28	230.48	257.68	284.88	312.08	366.56
41	127.305	152.56	180.44	208.36	236.24	264.12	292.00	319.88	375.72
42	130.410	156.28	184.84	213.44	242.00	270.56	299,12	327.68	384.89
43	133.515	160.00	189.24	218.52	247.76	277.00	306.24	335.48	394.05
44	136.620	163.72	193.64	223.60	253.52	283.44	313.37	343.28	403.22
45	139.725	167.44	198.04	228.68	259.28	289.88	320.49	351.08	412.38
46	142.830	171.16	202.44	233.76	265.04	296.32	327.61	358.88	421.55
47	145.935	174.88	206.84	238.84	270.80	302.76	334.73	366.68	430.71
48	149.040	178.60	211.24	243.92	276.56	309.20	341.85	374.48	439.88
49	152.145	182.32	215.64	249.00	282.33	315.64	348.97	382.28	449.04
50	155.25	186.05	220.05	254.10	288.10	322.10	356.10	390.10	458.20
52	161.46	193.49	228.85	264.26	299.62	334.98	370.34	405.70	476.53
54	167.67	200.93	237.65	274.42	311.15	347.86	384.59	421.30	494.86
56	173.88	208.37	246.45	284.58	322.67	360.74	398.83	436.90	513.18
58	180.09	215.81	255.25	294.74	334.19	373.62	413.08	452.50	531.52
60	186.30	223.26	264.06	304.92	345.72	386.52	427.32	468.12	549.84
62	192.51	230.70	272.86	315.08	357.24	399.40	441.56	483.70	568.16
64	198.72	238.14	281.66	325.24	368.76	412.28	455.81	499.30	586.49
66	204.93	245.58	290.46	335.40	380.28	425.16	470.05	514.90	604.82
68	211.14	253.02	299.26	345.56	391.80	438.04	484.29	530.50	623.15
70	217.35	260.47	308.07	355.74	403.34	450.94	498.54	546.14	641.48
72	223.56	267.91	316.87	365.90	414.86	463.82	512.78	561.70	659.81
74	229.77	275.35	325.67	376.06	426.39	476.70	527.03	577.30	678.13
76	235.98	282.79	334.47	386.22	437.91	489.59	541.27	592.90	696.46
78	242.18	290.23	343.27	396.38	449.43	502.47	555.51	608.50	714.79
80	248.40	297.68	352.08	406.56	460.96	515.36	569.76	624.16	733.12
82	254.61	305.12	360.88	416.22	472.48	528.24	584.00	639.80	751.45
84	260.82	312.56	369.68	426.88	484.01	541.12	598.25	655.40	769.78
86	267.03	320.00	378.48	437.04	495.53	554.01	612.49	671.00	788.11
88	273.24	327.44	387.28	447.21	507.05	566.89	626.74	686.60	806.44
90	779.45	334.89	396.09	457.38	518.58	579.78	640.98	702.18	824.76
92	285.67	342.33	404.89	467.54	530.10	592.66	655.22	717.8	843.09
94	291.88	349.77	413.69	477.70	541.63	605.55	669.47	733.4	861.42
96	298.09	357.21	422.49	487.86	553.15	618.43	683.72	749.0	879.75
98	304.3	364.65	431.29	498.02	564.67	631.31	697.96	764.6	898.08
100	310.5	372.1	440.1	508.2	576.2	644.2	712.2	780.2	916.4
110	341.6	409.3	484.1	559.0	633.8	708.6	783.4	858.2	1 008.0
120	372.6	446.5	528.1	609.8	691.4	773.0	854.6	936.2	1 099.7
130	403.7	483.7	572.1	660.6	749.0	837.4	925.8	1 014.2	1 191.3
140	430.5	520.9	616.1	711.4	806.6	901.8	997.0	1 092.2	1 282.9
150	460.6	558.1	660.1	762.3	864.2	966.2	1 068.2	1 170.2	1 374.6
160	496.8	595.3	704.1	813.1	921.8	1 030.6	1 139.4	1 248.2	1 466.2
170	527.9	632.5	748.1	863.9	979.4	1 095.0	1 210.6	1 326.2	1 557.9
180	558.9	669.7	792.1	914.7	1 037.0	1 159.4	1 281.8	1 404.2	1 649.5
190	590.0	706.9	836.1	965.5	1 094.6	1 223.8	1 352.1	1 482.3	1 741.1
200	621.0	744.2	880.2	1 016.4	1 152.4	1 288.4	1 424.4	1 560.4	1 832.8
210	652.1	781.4	924.2	1 067.2	1 210.0	1 352.8	1 495.6	1 638.4	1 924.4
220	683.1	818.6	968.2	1 118.0	1 267.6	1 417.2	1 566.8	1 716.4	2 016.1

cont..........

EQUIVALENT FREE AIR VOLUME (Cubic Feet) (contd.)

Volume of Compressed Air ft^3/min	Line Pressure lbf/in^2 Gauge								
	30	40	50	60	70	80	90	100	120
230	714.2	855.8	1 012.2	1 168.8	1 325.2	1 481.6	1 638.0	1 794.4	2 107.8
240	745.2	893.0	1 056.2	1 219.6	1 382.8	1 546.0	1 709.3	1 872.4	2 199.4
250	776.3	930.2	1 100.2	1 270.4	1 440.4	1 610.4	1 780.5	1 950.4	2 291.0
260	807.3	967.4	1 144.2	1 321.2	1 498.0	1 674.8	1 851.7	2 028.4	2 382.6
270	838.4	1 004.6	1 188.2	1 372.0	1 555.6	1 739.2	1 922.9	2 106.4	2 474.3
280	869.4	1 041.8	1 232.2	1 422.8	1 613.2	1 803.6	1 994.2	2 184.4	2 565.9
290	900.5	1 079.0	1 276.2	1 473.6	1 670.8	1 868.0	2 065.4	2 262.4	2 657.6
300	931.5	1 116.3	1 320.3	1 524.6	1 728.6	1 932.6	2 136.6	2 340.6	2 749.2
310	962.6	1 153.5	1 364.3	1 575.4	1 786.2	1 997.0	2 207.8	2 418.6	2 840.8
320	993.6	1 190.7	1 408.3	1 626.2	1 843.8	2 061.4	2 279.0	2 496.6	2 932.4
330	1 024.7	1 227.9	1 452.3	1 677.0	1 901.4	2 125.8	2 350.3	2 574.6	3 024.1
340	1 055.7	1 265.1	1 496.3	1 727.8	1 959.0	2 190.2	2 421.5	2 652.6	3 115.7
350	1 086.8	1 302.3	1 540.3	1 778.6	2 016.6	2 254.6	2 492.7	2 730.6	3 207.4
360	1 117.8	1 339.5	1 584.3	1 829.4	2 074.2	2 319.0	2 563.9	2 808.6	3 299.0
370	1 148.9	1 376.7	1 628.3	1 880.2	2 131.8	2 383.4	2 635.1	2 886.6	3 390.6
380	1 179.9	1 413.9	1 672.3	1 931.0	2 189.5	2 447.8	2 706.3	2 964.6	3 482.2
390	1 211.0	1 451.1	1 716.3	1 981.8	2 247.1	2 512.3	2 777.5	3 042.6	3 573.9
400	1 242.0	1 488.4	1 760.4	2 032.8	2 304.8	2 576.8	2 848.8	3 120.8	3 665.6
410	1 273.1	1 525.6	1 804.4	2 083.6	2 362.4	2 641.2	2 920.0	3 198.8	3 757.2
420	1 304.1	1 562.8	1 848.4	2 134.4	2 420.0	2 705.6	2 991.2	3 276.8	3 848.9
430	1 335.2	1 600.0	1 892.4	2 185.2	2 477.6	2 770.0	3 062.4	3 354.8	3 940.5
440	1 366.2	1 637.2	1 936.4	2 236.0	2 535.2	2 834.4	3 133.7	3 432.8	4 032.2
450	1 397.3	1 674.4	1 980.4	2 286.8	2 592.8	2 898.8	3 204.9	3 510.8	4 123.8
460	1 428.3	1 711.6	2 024.4	2 337.6	2 650.4	2 963.2	3 276.1	3 588.8	4 215.5
470	1 459.4	1 748.8	2 068.4	2 388.4	2 708.0	3 027.6	3 347.3	3 666.8	4 307.1
480	1 490.4	1 786.0	2 112.4	2 439.2	2 765.6	3 092.0	3 418.5	3 744.8	4 398.8
490	1 521.5	1 823.2	2 156.4	2 490.0	2 823.3	3 156.4	3 489.7	3 822.8	4 490.4
500	1 552.5	1 860.5	2 200.5	2 541.0	2 881.0	3 221.0	3 561.0	3 901.0	4 582.0
520	1 614.6	1 934.9	2 288.5	2 642.6	2 996.2	3 349.8	3 703.4	4 057.0	4 765.3
540	1 676.7	2 009.3	2 376.5	2 744.2	3 111.5	3 478.5	3 845.9	4 213.0	4 948.6
560	1 738.8	2 083.7	2 464.5	2 845.8	3 226.7	3 607.4	3 988.3	4 369.0	5 131.8
580	1 800.9	2 158.1	2 552.5	2 947.4	3 341.9	3 736.2	4 130.8	4 525.0	5 315.2
600	1 863.0	2 232.6	2 640.2	3 049.2	3 457.2	3 865.2	4 273.2	4 681.2	5 498.4
620	1 925.1	2 307.0	2 728.6	3 150.8	3 572.4	3 994.0	4 415.6	4 837.0	5 681.6
640	1 987.2	2 381.4	2 816.6	3 252.4	3 687.6	4 122.8	4 558.1	4 993.0	5 864.9
660	2 049.3	2 455.8	2 904.6	3 354.0	3 802.8	4 251.6	4 700.5	5 149.0	6 048.2
680	2 111.4	2 530.2	2 992.6	3 455.6	3 918.0	4 380.4	4 842.9	5 305.0	6 231.5
700	2 173.5	2 604.7	3 080.7	3 557.4	4 033.4	4 509.4	4 985.4	5 461.4	6 414.8
720	2 235.6	2 679.1	3 168.7	3 659.0	4 148.6	4 638.2	5 127.8	5 617.0	6 598.1
740	2 297.7	7 753.5	3 256.7	3 760.6	4 263.9	4 767.0	5 270.3	5 773.0	6 781.3
760	2 359.8	2 827.9	3 344.7	3 862.2	4 379.1	4 895.9	5 412.7	5 929.0	6 964.6
780	2 421.8	2 902.3	3 432.7	3 963.8	4 494.3	5 024.7	5 555.1	6 085.0	7 147.9
800	2 484.0	2 976.8	3 520.8	4 065.6	4 609.6	5 153.6	5 697.6	6 241.6	7 331.2
820	2 546.1	3 051.2	3 608.8	4 167.2	4 724.8	5 282.4	5 840.0	6 398.0	7 514.5
840	2 608.2	3 125.6	3 696.8	4 268.8	4 840.1	5 411.2	5 982.5	6 554.0	7 697.8
860	2 670.3	3 200.0	3 784.8	4 370.4	4 955.3	5 540.1	6 124.9	6 710.0	7 881.1
880	2 732.4	3 274.4	3 872.8	4 472.8	5 070.5	5 668.9	6 267.4	6 866.6	8 064.4
900	2 794.5	3 348.9	3 960.9	4 573.8	5 185.8	5 797.8	6 409.8	7 021.8	8 247.6

cont..........

EQUIVALENT FREE AIR VOLUME (Cubic Feet) (contd.)

Volume of Compressed Air ft^3/min	Line Pressure lbf/in^2 Gauge								
	30	40	50	60	70	80	90	100	120
920	2 856.7	3 423.3	4 048.9	4 675.4	5 301.0	5 926.6	6 582.2	7 178.0	8 430.9
940	2 918.8	3 497.7	4 136.9	4 777.0	5 416.3	6 055.5	6 694.7	7 334.0	8 614.2
960	2 980.9	3 572.1	4 224.9	4 878.6	5 531.5	6 184.3	6 837.2	7 490.0	8 797.5
980	3 043.0	3 646.5	4 312.9	4 980.2	5 646.7	6 313.1	6 979.6	7 646.0	8 980.8
1 000	3 105.0	3 721.0	4 401.0	5 082.0	5 762.0	6 442.0	7 122.0	7.802.0	9 164.0

This table is calculated on the basis of the standard formula

$$\text{equivalent free air volume} = Q_c \times \frac{P + 14.7}{14.7}$$

where Q_c = compressed air volume
P = gauge pressure, lbf/in^2

The table has been extended to cover all likely values needed without resorting to factoring. Intermediate values can be interpolated or found additively, eg for a compressed air volume of 22.4 ft^3/min, find values for 22 and 0.4 ft^3/min separately at the appropriate gauge pressure and add.

PROPERTIES OF GASES

Gas	Formula	Molecular weight	Gas constant J/kmol·K	P_{cr} bar	t_{cr} °C	Boiling point °C	c_p/c_v
Acetylene	C_2H_2	26.04	319.4	63	36	−89	1.26
Air	–	28.96	287.1	38	−141	−193	1.40
Ammonia	NH_3	17.03	488.2	113	132	−34	1.32
Argon	Ar	39.94	208.2	49	−122	−186	1.67
Benzene	C_6H_6	78.11	106.5	49	289	80	1.08
Butane, N	C_4H_{10}	58.12	143.1	35	153	−1	1.11
Butane Iso-	C_4H_{10}	58.12	143.1	36	134	−10	1.11
Butylene	C_4H_8	56.06	148.3	44	144	−6	1.11
Carbon dioxide	CO_2	44.01	188.9	74	31	−78	1.30
Carbon disulphide	CS_2	76.13	109.2	75	278	46	1.20
Carbon monoxide	CO	28.01	296.8	35	−140	−192	1.40
Carbon tetrachloride	CCl_4	153.84	54.0	44	283	77	1.18
Chlorine	Cl_2	70.91	117.3	82	146	−34	1.33
Cyan	$(CN)_2$	52.04	159.8	61	128	−21	1.26
Dichloromethane	CH_4Cl_2	84.93	97.9	103	216	41	1.18
Ethane	C_2H_6	30.07	276.5	48	35	−89	1.22
Ethyl chloride	C_2H_5Cl	64.51	128.9	53	185	12	1.13
Ethylene	C_2H_4	28.05	296.4	50	10	−104	1.22
Freon 12	CCl_2F_2	120.92	68.8	39	112	−30	1.13
Helium	He	4.00	2077.2	2	−268	−269	1.66
Hexane, N-	C_6H_{14}	86.17	96.5	31	235	69	1.08
Hydrogen	H_2	2.02	4124.5	20	−240	−253	1.41
Hydrogen bromide	HBr	80.92	102.8	82	90	−68	1.42
Hydrogen chloride	HCl	36.46	228.0	80	51	−85	1.41
Hydrogen cyanide	HCN	30.07	276.5	49	184	−89	1.60
Hydrogen sulphide	H_2S	34.08	244.0	88	100	−60	1.30
Hydrogen jodide	HJ	127.93	65.0	—	151	−36	1.40
Krypton	Kr	83.70	99.3	55	−63	−153	1.68
Methane	CH_4	16.04	518.3	45	−83	−162	1.32
Methyl ether	$(CH_3)_2O$	46.07	180.5	52	127	−23	1.11
Methylene chloride	CH_2Cl_2	84.94	97.9	63	238	40	1.18
Methyl chloride	CH_3Cl	50.49	164.7	65	143	−24	1.20
Neon	Ne	20.18	412.0	28	−228	−246	1.64
Nitrogen	N_2	28.02	296.8	34	−147	−196	1.41
Nitric oxide	NO	30.01	277.1	66	−94	−152	1.40
Nitrous oxide	N_2O	44.02	188.9	73	37	−89	1.31
Oxygen	O_2	32.00	259.8	49	−119	−183	1.40
Ozone	O_3	48.00	173.2	69	− 5	−112	1.29
Pentane, N-	C_5H_{12}	72.14	115.3	34	197	36	1.06
Pentane, Iso-	C_5H_{12}	72.14	115.3	34	188	28	1.06
Propane	C_3H_8	44.09	188.6	39	96	−42	1.15
Propylene	C_3H_6	42.08	197.6	44	97	−48	1.17
Sulphur dioxide	SO_2	64.07	129.8	79	157	−10	1.26
Water (superheated)	H_2O	18.02	461.4	221	374	100	1.33
Xenon	X_3	131.30	63.3	57	15	−108	1.66

CONSUMPTION OF DOUBLE-ACTING CYLINDERS AT 60 lbf/in^2 GAUGE

(Cubic Feet of Free Air per Inch or Stroke for Complete Cycle)

Cylinder Bore Inches	Rod Diameter – Inches								
	1/4	3/8	1/2	5/8	3/4	1	1.1/4	1.1/2	1.3/4
½	0.00102	0.00083	–	–	–	–	–	–	–
¾	0.00245	0.00227	0.00218	0.00169	–	–	–	–	–
1	0.00447	0.00429	0.00404	0.00372	0.00332	–	–	–	–
1¼	0.00708	0.00683	0.00656	0.00631	0.00558	0.00494	–	–	–
1½	0.0103	0.0100	0.00975	0.00950	0.00910	0.00806	0.00663	–	–
1¾	0.0141	0.0138	0.0135	0.0133	0.0128	0.0118	0.0105	0.00897	–
2	0.0184	0.0182	0.0179	0.0177	0.0173	0.0162	0.0148	0.0133	0.0115
2¼	0.0232	0.0230	0.0228	0.0225	0.0221	0.0211	0.0198	0.0182	0.0164
2½	0.0287	0.0285	0.0282	0.0280	0.0276	0.0265	0.0253	0.0236	0.0218
2¾	0.0348	0.0346	0.0343	0.0340	0.0336	0.0327	0.0313	0.0297	0.0279
3	0.0414	0.0412	0.0409	0.0406	0.0403	0.0393	0.0380	0.0364	0.0346
3½	0.0566	0.0563	0.0560	0.0556	0.0552	0.0543	0.0529	0.0515	0.0496
4	0.0754	0.0751	0.0747	0.0742	0.0738	0.0715	0.0712	0.0685	0.0670
4½	–	–	0.0930	0.0927	0.0924	0.0910	0.0897	0.0885	0.0872
5	–	–	0.1150	0.1150	0.1140	0.1130	0.1120	0.1100	0.1090
5½	–	–	0.1390	0.1380	0.1380	0.1370	0.1360	0.1350	0.1330
6	–	–	0.1660	0.1650	0.1640	0.1640	0.1630	0.1610	0.1590
7	–	–	–	–	0.2240	0.2230	0.2220	0.2210	0.2190
8	–	–	–	–	0.2940	0.2930	0.2920	0.2900	0.2870
9	–	–	–	–	0.3720	0.3710	0.3700	0.3680	0.3660
10	–	–	–	–	0.4590	0.4580	0.4570	0.4560	0.4540
11	–	–	–	–	0.5580	0.5560	0.5550	0.5540	0.5520
12	–	–	–	–	0.6630	0.6620	0.6610	0.6590	0.6570
13	–	–	–	–	–	0.7800	0.7780	0.7750	0.7720
14	–	–	–	–	–	0.9040	0.9020	0.9000	0.8980
15	–	–	–	–	–	–	–	1.0350	1.0320
16	–	–	–	–	–	–	–	1.1800	1.1760
17	–	–	–	–	–	–	–	1.3320	1.3290
18	–	–	–	–	–	–	–	1.4920	1.4900
19	–	–	–	–	–	–	–	1.6620	1.6600
20	–	–	–	–	–	–	–	1.8420	1.8390
21	–	–	–	–	–	–	–	2.0300	2.0260
22	–	–	–	–	–	–	–	2.2300	2.2260
23	–	–	–	–	–	–	–	2.4400	2.4360
24	–	–	–	–	–	–	–	–	–
25	–	–	–	–	–	–	–	–	–
26	–	–	–	–	–	–	–	–	–
27	–	–	–	–	–	–	–	–	–
28	–	–	–	–	–	–	–	–	–
29	–	–	–	–	–	–	–	–	–
30	–	–	–	–	–	–	–	–	–

cont..........

CONSUMPTION OF DOUBLE-ACTING CYLINDERS AT 60 lbf/in^2 GAUGE (contd.)

(Cubic Feet of Free Air per Inch or Stroke for Complete Cycle)

Cylinder Bore Inches	Rod Diameter — Inches								
	2	2.1/4	2.1/2	2.3/4	3	3.1/2	4	5	6
½	—	—	—	—	—	—	—	—	—
¾	—	—	—	—	—	—	—	—	—
1	—	—	—	—	—	—	—	—	—
1¼	—	—	—	—	—	—	—	—	—
1½	—	—	—	—	—	—	—	—	—
1¾	—	—	—	—	—	—	—	—	—
2	—	—	—	—	—	—	—	—	—
2¼	0.0142	—	—	—	—		—	—	—
2½	0.0197	0.0172	—	—	—	—	—	—	—
2¾	0.0257	0.0232	0.0205	—	—	—	—	—	—
3	0.0324	0.0299	0.0271	0.0241	—	—	—	—	—
3½	0.0473	0.0448	0.0422	0.0391	0.0358	—	—	—	—
4	0.0647	0.0623	0.0594	0.0565	0.0531	0.0456	—	—	—
4½	0.0838	0.0818	0.0783	0.0760	0.0721	0.0650	0.0565	—	—
5	0.1060	0.1030	0.1000	0.0975	0.0944	0.0872	0.0786	—	—
5½	0.1300	0.1270	0.1250	0.1220	0.1190	0.1110	0.1030	0.0818	—
6	0.1570	0.1540	0.1510	0.1480	0.1460	0.1380	0.1290	0.1090	—
7	0.2170	0.2140	0.2110	0.2080	0.2050	0.1980	0.1890	0.1680	0.1430
8	0.2860	0.2830	0.2800	0.2770	0.2740	0.2680	0.2580	0.2380	0.2120
9	0.3650	0.3620	0.3590	0.3560	0.3530	0.3460	0.3380	0.3160	0.2900
10	0.4520	0.4490	0.4460	0.4430	0.4410	0.4330	0.4250	0.4050	0.3880
11	0.5500	0.5470	0.5430	0.5400	0.5380	0.5300	0.5220	0.5020	0.4750
12	0.6550	0.6520	0.6500	0.6470	0.6440	0.6370	0.6280	0.6070	0.5820
13	0.7720	0.7690	0.7660	0.7630	0.7600	0.7540	0.7440	0.7300	0.6970
14	0.8960	0.8920	0.8900	0.8870	0.8850	0.8780	0.8700	0.8490	0.8230
15	1.0300	1.0270	1.0230	1.0200	1.0180	1.0120	1.0000	0.9830	0.9560
16	1.1750	1.1720	1.1690	1.1660	1.1630	1.1540	1.1460	1.1300	1.1100
17	1.3280	1.3250	1.3230	1.3200	1.3180	1.3080	1.2980	1.2780	1.2530
18	1.4880	1.4860	1.4830	1.4800	1.4780	1.4700	1.4580	1.4370	1.4120
19	1.6600	1.6570	1.6540	1.6510	1.6480	1.6410	1.6340	1.6100	1.5800
20	1.8400	1.8370	1.8340	1.8310	1.8290	1.8220	1.8100	1.7900	1.7640
21	2.0300	2.0270	2.0240	2.0220	2.0190	2.0080	1.9980	1.9840	1.9560
22	2.2280	2.2250	2.2230	2.2200	2.2180	2.2100	2.2000	2.1900	2.1580
23	2.4330	2.4300	2.4260	2.4230	2.4200	2.4140	2.4080	2.3880	2.3640
24	2.6500	2.6480	2.6450	2.6420	2.6400	2.6330	2.6220	2.6020	2.5770
25	2.8760	2.8740	2.8720	2.8700	2.8670	2.8600	2.8530	2.8300	2.8090
26	3.1130	3.1100	3.1060	3.1030	3.1000	3.0940	3.0680	3.0500	3.0300
27	3.3600	3.3560	3.3500	3.3450	3.3400	3.3320	3.3250	3.3100	3.2850
28	3.6100	3.6070	3.6050	3.6020	3.6000	3.5900	3.5800	3.5600	3.5400
29	3.8800	3.8750	3.8700	3.8660	3.8500	3.8600	3.8400	3.8200	3.8000
30	4.1450	4.1430	4.1410	4.1390	4.1380	4.1360	4.1300	4.1100	4.0850

cont..........

CONSUMPTION OF DOUBLE-ACTING CYLINDERS AT 80 lbf/in^2 GAUGE

(Cubic Feet of Free Air per Inch or Stroke for Complete Cycle)

Cylinder Bore Inches	Rod Diameter – Inches								
	1/4	3/8	1/2	5/8	3/4	1	1.1/4	1.1/2	1.3/4
½	0.00128	0.00105	–	–	–	–	–	–	–
¾	0.00312	0.00288	0.00256	0.00215	–	–	–	–	–
1	0.00567	0.00545	0.00513	0.00471	0.00421	–	–	–	–
1¼	0.00904	0.00885	0.00843	0.00802	0.00753	0.00624	–	–	–
1½	0.0130	0.0128	0.0125	0.0121	0.0116	0.0103	0.0086	–	–
1¾	0.0178	0.0175	0.0172	0.0168	0.0163	0.0150	0.0134	0.0114	–
2	0.0233	0.0230	0.0227	0.0223	0.0218	0.0205	0.0189	0.0169	0.0145
2¼	0.0295	0.0293	0.0289	0.0286	0.0281	0.0268	0.0251	0.0231	0.0207
2½	0.0366	0.0362	0.0358	0.0355	0.0350	0.0336	0.0321	0.0301	0.0277
2¾	0.0442	0.0438	0.0435	0.0432	0.0427	0.0415	0.0397	0.0378	0.0354
3	0.0525	0.0522	0.0518	0.0515	0.0512	0.0498	0.0482	0.0462	0.0438
3½	0.0719	0.0714	0.0710	0.0705	0.0701	0.0690	0.0672	0.0654	0.0630
4	0.0958	0.0954	0.0950	0.0942	0.0925	0.0910	0.0892	0.0867	0.0850
4½	–	–	0.1180	0.1170	0.1170	0.1160	0.1140	0.1130	0.1110
5	–	–	0.1470	0.1460	0.1450	0.1430	0.1420	0.1400	0.1380
5½	–	–	0.1770	0.1760	0.1750	0.1730	0.1720	0.1710	0.1690
6	–	–	0.2110	0.2100	0.2090	0.2080	0.2070	0.2050	0.2030
7	–	–	–	–	0.2850	0.2840	0.2830	0.2810	0.2790
8	–	–	–	–	0.3740	0.3720	0.3710	0.3680	0.3650
9	–	–	–	–	0.4730	0.4720	0.4700	0.4680	0.4650
10	–	–	–	–	0.5840	0.5830	0.5820	0.5800	0.5770
11	–	–	–	–	0.7080	0.7070	0.7050	0.7030	0.7010
12	–	–	–	–	0.8420	0.8410	0.8390	0.8370	0.8350
13	–	–	–	–	–	0.9920	0.9880	0.9850	0.9810
14	–	–	–	–	–	1.1500	1.1480	1.1450	1.1430
15	–	–	–	–	–	–	–	1.3150	1.3100
16	–	–	–	–	–	–	–	1.4980	1.4930
17	–	–	–	–	–	–	–	1.6920	1.6870
18	–	–	–	–	–	–	–	1.8950	1.8900
19	–	–	–	–	–	–	–	2.1130	2.1080
20	–	–	–	–	–	–	–	2.3400	2.3350
21	–	–	–	–	–	–	–	2.5800	2.5730
22	–	–	–	–	–	–	–	2.8360	2.8300
23	–	–	–	–	–	–	–	3.1000	3.0050
24	–	–	–	–	–	–	–	–	–
25	–	–	–	–	–	–	–	–	–
26	–	–	–	–	–	–	–	–	–
27	–	–	–	–	–	–	–	–	–
28	–	–	–	–	–	–	–	–	–
29	–	–	–	–	–	–	–	–	–
30	–	–	–	–	–	–	–	–	–

cont..........

CONSUMPTION OF DOUBLE-ACTING CYLINDERS AT 80 lbf/in^2 GAUGE

(Cubic Feet of Free Air per Inch or Stroke for Complete Cycle)

Cylinder Bore Inches	Rod Diameter — Inches								
	2	2.1/4	2.1/2	2.3/4	3	3.1/2	4	5	6
½	—	—	—	—	—	—	—	—	—
¾	—	—	—	—	—	—	—	—	—
1	—	—	—	—	—	—	—	—	—
1¼	—	—	—	—	—	—	—	—	—
1½	—	—	—	—	—	—	—	—	—
1¾	—	—	—	—	—	—	—	—	—
2	—	—	—	—	—	—	—	—	—
2¼	0.0180	—	—	—	—	—	—	—	—
2½	0.0250	0.0218	—	—	—	—	—	—	—
2¾	0.0327	0.0295	0.0261	—	—	—	—	—	—
3	0.0411	0.0380	0.0345	0.0306	—	—	—	—	—
3½	0.0602	0.0570	0.0535	0.0496	0.0450	—	—	—	—
4	0.0821	0.0790	0.0755	0.0717	0.0675	0.0580	—	—	—
4½	0.1071	0.1041	0.1016	0.0964	0.0924	0.0828	0.0718	—	—
5	0.1350	0.1320	0.1280	0.1240	0.1200	0.1110	0.1000	—	—
5½	0.1650	0.1630	0.1590	0.1550	0.1510	0.1410	0.1310	0.1040	—
6	0.1980	0.1970	0.1930	0.1880	0.1850	0.1750	0.1650	0.1380	—
7	0.2760	0.2720	0.2690	0.2650	0.2600	0.2520	0.2410	0.2140	0.1820
8	0.3630	0.3590	0.3560	0.3520	0.3480	0.3390	0.3280	0.3020	0.2690
9	0.4640	0.4600	0.4560	0.4500	0.4480	0.4440	0.4280	0.4020	0.3680
10	0.5740	0.5710	0.5670	0.5640	0.5560	0.5550	0.5540	0.5140	0.4800
11	0.6980	0.6950	0.6900	0.6870	0.6830	0.6730	0.6620	0.6360	0.6040
12	0.8330	0.8300	0.8250	0.8210	0.8180	0.8080	0.7970	0.7720	0.7380
13	0.9800	0.9750	0.9700	0.9670	0.9640	0.9560	0.9440	0.9180	0.8850
14	1.1380	1.1360	1.1310	1.1270	1.1240	1.1130	1.1030	1.0780	1.0450
15	1.3100	1.3050	1.3000	1.2960	1.2930	1.2880	1.2720	1.2480	1.2150
16	1.4930	1.4900	1.4860	1.4810	1.4780	1.4650	1.4550	1.4300	1.3960
17	1.6860	1.6820	1.6780	1.6740	1.6700	1.6610	1.6480	1.6240	1.5900
18	1.8900	1.8870	1.8830	1.8790	1.8750	1.8690	1.8500	1.8300	1.7980
19	2.1090	2.1050	2.1010	2.0970	2.0920	2.0810	2.0700	2.0480	2.0080
20	2.3360	2.3320	2.3280	2.3240	2.3200	2.3140	2.2980	2.2740	2.2400
21	2.5800	2.5750	2.5700	2.5650	2.5600	2.5500	2.5400	2.5200	2.4800
22	2.8250	2.8220	2.8190	2.8170	2.8150	2.8080	2.7920	2.7690	2.7380
23	3.0900	3.0860	3.0820	3.0780	3.0740	3.0680	3.0600	3.0300	3.0000
24	3.3600	3.3570	3.3540	3.3530	3.3500	3.3370	3.3240	3.3020	3.2740
25	3.6500	3.6470	3.6440	3.6430	3.6400	3.6300	3.6200	3.5980	3.5600
26	3.9500	3.9470	3.9440	3.9410	3.9380	3.9220	3.9000	3.8800	3.8500
27	4.2600	4.2570	4.2540	4.2510	4.2480	4.2440	4.2200	4.2000	4.1800
28	4.5800	4.5750	4.5710	4.5660	4.5620	4.5560	4.5510	4.5200	4.4800
29	4.9300	4.9270	4.9240	4.9200	4.9160	4.9000	4.8800	4.8600	4.8250
30	5.2600	5.2570	5.2540	5.2520	5.2500	5.2460	5.2400	5.2200	5.1800

cont..........

CONSUMPTION OF DOUBLE-ACTING CYLINDERS AT 100 lbf/in^2 GAUGE
(Cubic Feet of Free Air per Inch or Stroke for Complete Cycle)

Cylinder Bore Inches	Rod Diameter — Inches								
	1/4	3/8	1/2	5/8	3/4	1	1.1/4	1.1/2	1.3/4
½	0.00156	0.00127	—	—	—	—	—	—	—
¾	0.00377	0.00349	0.00310	0.00260	—	—	—	—	—
1	0.00687	0.0066	0.00621	0.00572	0.0051	—	—	—	—
1¼	0.0109	0.1050	0.0101	0.0097	0.0086	0.0076	—	—	—
1½	0.0158	0.0154	0.0150	0.0146	0.0140	0.0124	0.0104	—	—
1¾	0.0216	0.0212	0.0208	0.0204	0.0197	0.0182	0.0162	0.0138	—
2	0.0282	0.0279	0.0275	0.0271	0.0266	0.0248	0.0228	0.0204	0.0176
2¼	0.0357	0.0354	0.0351	0.0346	0.0340	0.0324	0.0304	0.0280	0.0251
2½	0.0442	0.0438	0.0434	0.0430	0.0424	0.0407	0.0389	0.0364	0.0335
2¾	0.0535	0.0531	0.0527	0.0522	0.0517	0.0502	0.0481	0.0457	0.0428
3	0.0636	0.0632	0.0628	0.0624	0.0619	0.0604	0.0584	0.0559	0.0531
3½	0.0870	0.0865	0.0860	0.0854	0.0849	0.0835	0.0814	0.0791	0.0763
4	0.1160	0.1150	0.1150	0.1140	0.1120	0.1100	0.1080	0.1050	0.1030
4½	—	—	0.1430	0.1420	0.1420	0.1400	0.1380	0.1360	0.1340
5	—	—	0.1770	0.1760	0.1750	0.1730	0.0172	0.1690	0.1670
5½	—	—	0.2140	0.2130	0.2120	0.2110	0.2090	0.2070	0.2040
6	—	—	0.2550	0.2540	0.2530	0.2520	0.2500	0.2480	0.2450
7	—	—	—	—	0.3440	0.3430	0.3420	0.3400	0.3370
8	—	—	—	—	0.4520	0.4500	0.4480	0.4450	0.4420
9	—	—	—	—	0.5720	0.5700	0.5680	0.5660	0.5630
10	—	—	—	—	0.7060	0.7050	0.7030	0.7010	0.6970
11	—	—	—	—	0.8560	0.8530	0.8520	0.8500	0.8470
12	—	—	—	—	1.0190	1.0180	1.0160	1.0140	1.0100
13	—	—	—	—	—	1.2000	1.1970	1.1920	1.1880
14	—	—	—	—	—	1.3900	1.3870	1.3840	1.3810
15	—	—	—	—	—	—	—	1.5900	1.5850
16	—	—	—	—	—	—	—	1.8110	1.8070
17	—	—	—	—	—	—	—	2.0430	2.0400
18	—	—	—	—	—	—	—	2.2930	2.2880
19	—	—	—	—	—	—	—	2.5530	2.5490
20	—	—	—	—	—	—	—	2.8300	2.8260
21	—	—	—	—	—	—	—	3.1200	3.1160
22	—	—	—	—	—	—	—	3.4250	3.4200
23	—	—	—	—	—	—	—	3.7440	3.7400
24	—	—	—	—	—	—	—	—	—
25	—	—	—	—	—	—	—	—	—
26	—	—	—	—	—	—	—	—	—
27	—	—	—	—	—	—	—	—	—
28	—	—	—	—	—	—	—	—	—
29	—	—	—	—	—	—	—	—	—
30	—	—	—	—	—	—	—	—	—

cont...........

CONSUMPTION OF DOUBLE-ACTING CYLINDERS AT 100 lbf/in^2 GAUGE (contd.)
(Cubic Feet of Free Air per Inch or Stroke for Complete Cycle)

Cylinder Bore Inches	Rod Diameter – Inches								
	2	2.1/4	2.1/2	2.3/4	3	3.1/2	4	5	6
½	–	–	–	–	–	–	–	–	–
¾	–	–	–	–	–	–	–	–	–
1	–	–	–	–	–	–	–	–	–
1¼	–	–	–	–	–	–	–	–	–
1½	–	–	–	–	–	–	–	–	–
1¾	–	–	–	–	–	–	–	–	–
2	–	–	–	–	–	–	–	–	–
2¼	0.0218	–	–	–	–	–	–	–	–
2½	0.0302	0.0264	–	–	–	–	–	–	–
2¾	0.0395	0.0357	0.0315	–	–	–	–	–	–
3	0.0497	0.0459	0.0417	0.0370	–	–	–	–	–
3½	0.0728	0.0690	0.0647	0.0601	0.0550	–	–	–	–
4	0.0994	0.0956	0.0914	0.0868	0.0816	0.0701	–	–	–
4½	0.1290	0.1260	0.1220	0.1170	0.1120	0.1000	0.0870	–	–
5	0.1630	0.1590	0.1550	0.1500	0.1450	0.1340	0.1210	–	–
5½	0.2000	0.1970	0.1930	0.1880	0.1830	0.1710	0.1580	0.1260	–
6	0.2410	0.2380	0.2330	0.2280	0.2240	0.2120	0.1990	0.1670	–
7	0.3340	0.3300	0.3260	0.3210	0.3150	0.3040	0.2910	0.2590	0.2200
8	0.4400	0.4350	0.4310	0.4270	0.4220	0.4110	0.3970	0.3660	0.3260
9	0.5610	0.5560	0.5520	0.5470	0.5430	0.5320	0.5180	0.4860	0.4460
10	0.6950	0.6910	0.6870	0.6830	0.6770	0.6650	0.6530	0.6210	0.5820
11	0.8450	0.8410	0.8360	0.8320	0.8270	0.8150	0.8020	0.7700	0.7310
12	1.0080	1.0040	0.9980	0.9940	0.9900	0.9780	0.9650	0.9340	0.8940
13	1.1870	1.1820	1.1770	1.1730	1.1690	1.1580	1.1430	1.1120	1.0720
14	1.3780	1.3730	1.3690	1.3640	1.3600	1.3480	1.3360	1.3040	1.2640
15	1.5840	1.5800	1.5750	1.5700	1.5640	1.5580	1.5400	1.5100	1.4700
16	1.8050	1.8010	1.7970	1.7920	1.7880	1.7720	1.7600	1.7300	1.6900
17	2.0400	2.0350	2.0300	2.0250	2.0200	2.0120	1.9940	1.9630	1.9250
18	2.2840	2.2810	2.2770	2.2730	2.2700	2.2600	2.2440	2.2160	2.1740
19	2.5480	2.5440	2.5400	2.5360	2.5320	2.5200	2.5070	2.4760	2.4340
20	2.8240	2.8210	2.8170	2.8130	2.8100	2.8000	2.7800	2.7500	2.7100
21	3.1200	2.1150	3.1100	3.1050	3.1000	3.0830	3.0720	3.0440	3.0030
22	3.4200	3.4160	3.4120	3.4080	3.4040	3.3960	3.3800	3.3470	3.3100
23	3.7400	3.7350	3.7300	3.7250	3.7200	3.7100	3.7000	3.6700	3.6300
24	4.0700	4.0670	4.0650	4.0630	4.0600	4.0450	4.0300	4.0000	3.9600
25	4.4200	4.4170	4.4150	4.4130	4.4100	4.3950	4.3800	4.3500	4.3100
26	4.78000	4.7760	4.7720	4.7680	4.7650	4.7500	4.7200	4.6900	4.6600
27	5.1600	5.1550	5.1500	5.1450	5.1400	5.1300	5.1200	5.0850	5.0500
28	5.5500	5.5450	5.5400	5.5350	5.5300	5.5200	5.5100	5.4700	5.4400
29	5.9550	5.9510	5.9470	5.9430	5.9400	5.9330	5.9260	5.8800	5.8400
30	6.3700	6.3670	6.3640	6.3610	6.3580	6.3520	6.3480	6.3080	6.2700

cont............

CONSUMPTION OF DOUBLE-ACTING CYLINDERS AT 120 lbf/in^2 GAUGE (contd.)
(Cubic Feet of Free Air per Inch or Stroke for Complete Cycle)

Cylinder Bore Inches	Rod Diameter — Inches								
	1/4	3/8	1/2	5/8	3/4	1	1.1/4	1.1/2	1.3/4
½	0.00196	0.00160	–	–	–	–	–	–	–
¾	0.00475	0.00440	0.00390	0.00328	–	–	–	–	–
1	0.00865	0.00830	0.00782	0.00720	0.00643	–	–	–	–
1¼	0.0137	0.0132	0.0127	0.0122	0.0108	0.00957	–	–	–
1½	0.0199	0.0194	0.0189	0.0184	0.0176	0.0156	0.0131	–	–
1¾	0.0272	0.0267	0.0262	0.0257	0.0248	0.0229	0.0204	0.0174	–
2	0.0355	0.0351	0.0346	0.0341	0.0335	0.0312	0.0287	0.0257	0.0222
2¼	0.0450	0.0446	0.0442	0.0436	0.0428	0.0408	0.0383	0.0353	0.0316
2½	0.0556	0.0552	0.0547	0.0541	0.0535	0.0513	0.0490	0.0458	0.0422
2¾	0.0674	0.0668	0.0663	0.0657	0.0653	0.0633	0.0605	0.0575	0.0540
3	0.0800	0.0796	0.0792	0.0786	0.0780	0.0760	0.0735	0.0705	0.0670
3½	0.1100	0.1090	0.1080	0.1070	0.1070	0.1050	0.1020	0.0996	0.0960
4	0.1460	0.1450	0.1440	0.1430	0.1410	0.1390	0.1360	0.1320	0.1300
4½	–	–	0.1800	0.1790	0.1770	0.1760	0.1740	0.1710	0.1690
5	–	–	0.2230	0.2220	0.2200	0.2180	0.2160	0.2130	0.2100
5½	–	–	0.2700	0.2680	0.2670	0.2660	0.2640	0.2610	0.2570
6	–	–	0.3210	0.3200	0.3190	0.3180	0.3150	0.3120	0.3080
7	–	–	–	–	0.4340	0.4320	0.4310	0.4280	0.4250
8	–	–	–	–	0.5690	0.5670	0.5650	0.5610	0.5560
9	–	–	–	–	0.7200	0.7180	0.7160	0.7140	0.7090
10	–	–	–	–	0.8900	0.8870	0.8850	0.8830	0.8780
11	–	–	–	–	1.0800	1.0760	1.0730	1.0700	1.0660
12	–	–	–	–	1.2840	1.2820	1.2800	1.2780	1.2720
13	–	–	–	–	–	1.5100	1.5050	1.5000	1.4960
14	–	–	–	–	–	1.7500	1.7470	1.7440	1.7400
15	–	–	–	–	–	–	–	2.0020	1.9960
16	–	–	–	–	–	–	–	2.2810	2.2750
17	–	–	–	–	–	–	–	2.5780	2.5700
18	–	–	–	–	–	–	–	2.8880	2.8800
19	–	–	–	–	–	–	–	3.2180	3.2120
20	–	–	–	–	–	–	–	3.5620	3.5560
21	–	–	–	–	–	–	–	3.9300	3.9220
22	–	–	–	–	–	–	–	4.3180	4.3100
23	–	–	–	–	–	–	–	4.7200	4.7100
24	–	–	–	–	–	–	–	–	–
25	–	–	–	–	–	–	–	–	–
26	–	–	–	–	–	–	–	–	–
27	–	–	–	–	–	–	–	–	–
28	–	–	–	–	–	–	–	–	–
29	–	–	–	–	–	–	–	–	–
30	–	–	–	–	–	–	–	–	–

cont..........

CONSUMPTION OF DOUBLE-ACTING CYLINDERS AT 120 lbf/in^2 GAUGE (contd.)

(Cubic Feet of Free Air per Inch or Stroke for Complete Cycle)

Cylinder Bore Inches	Rod Diameter – Inches								
	2	2.1/4	2.1/2	2.3/4	3	3.1/2	4	5	6
½	–	–	–	–	–	–	–	–	–
¾	–	–	–	–	–	–	–	–	–
1	–	–	–	–	–	–	–	–	–
1¼	–	–	–	–	–	–	–	–	–
1½	–	–	–	–	–	–	–	–	–
1¾	–	–	–	–	–	–	–	–	–
2	–	–	–	–	–	–	–	–	–
2¼	0.0275	–	–	–	–	–	–	–	–
2½	0.0381	0.0333	–	–	–	–	–	–	–
2¾	0.0498	0.0450	0.0397	–	–	–	–	–	–
3	0.0625	0.0578	0.0525	0.0466	–	–	–	–	–
3½	0.0917	0.0870	0.0815	0.0756	0.0694	–	–	–	–
4	0.1240	0.1200	0.1140	0.1090	0.1020	0.0884	–	–	–
4½	0.1630	0.1590	0.1540	0.1480	0.1410	0.1260	0.1100	–	–
5	0.2050	0.1990	0.1940	0.1880	0.1830	0.1690	0.1520	–	–
5½	0.2520	0.2480	0.2430	0.2370	0.2300	0.2150	0.1990	0.1590	–
6	0.3040	0.2980	0.2930	0.2870	0.2820	0.2670	0.2500	0.2100	–
7	0.4210	0.4150	0.4100	0.4040	0.3970	0.3830	0.3670	0.3260	0.2770
8	0.5540	0.5480	0.5420	0.5360	0.5310	0.5170	0.5000	0.4710	0.4100
9	0.7060	0.7000	0.6950	0.6900	0.6850	0.6700	0.6530	0.6120	0.5620
10	0.8750	0.8700	0.8650	0.8590	0.8540	0.8370	0.8220	0.7830	0.7340
11	1.0640	1.0590	1.0530	1.0480	1.0420	1.0270	0.0100	0.9700	0.9200
12	1.2700	1.2650	1.2590	1.2530	1.2480	1.2320	1.2150	1.1750	1.1250
13	1.4960	1.4900	1.4840	1.4770	1.4700	1.4590	1.4400	1.4000	1.3500
14	1.7350	1.7290	1.7240	1.7190	1.7140	1.6980	1.6850	1.6430	1.5940
15	1.9930	1.9870	1.9810	1.9760	1.9700	1.9610	1.9400	1.9000	1.8500
16	2.2700	2.2650	2.2600	2.2560	2.2520	2.2300	2.2180	2.1800	2.1300
17	2.5700	2.5640	2.5580	2.5510	2.5440	2.5340	2.5100	2.4740	2.4260
18	2.8800	2.8750	2.8700	2.8650	2.8600	2.8370	2.8200	2.7900	2.7400
19	3.2100	3.2040	3.1980	3.1920	3.1860	3.1700	3.1600	3.1200	3.0660
20	3.5600	3.5550	3.5500	3.5450	3.5400	3.5280	3.5000	3.4630	3.4180
21	3.9300	3.9230	3.9160	3.9080	3.9000	3.8850	3.8700	3.8400	3.7800
22	4.3000	4.2960	4.2910	4.2870	4.2820	4.2700	4.2550	4.2150	4.1650
23	4.7080	4.7010	4.6940	4.6870	4.6800	4.6700	4.6600	4.6200	4.5700
24	5.1260	5.1220	5.1180	5.1140	5.1100	5.0900	5.0700	5.0400	5.0000
25	5.5580	5.5550	5.5510	5.5490	5.5460	5.5400	5.5200	5.5000	5.4500
26	6.0300	6.0230	6.0160	6.0080	6.0000	5.9800	5.9600	5.9000	5.8600
27	6.5000	6.4940	6.4860	6.4780	6.4700	6.4500	6.4300	6.4000	6.3600
28	7.0000	6.9900	6.9800	6.9600	6.9700	6.9400	6.9200	6.8850	6.8500
29	7.5000	7.4950	7.4900	7.4850	7.4800	7.4700	7.4600	7.4200	7.3800
30	8.0200	8.0170	8.0140	8.0130	8.0100	8.0000	7.9900	7.9500	7.9100

PRACTICAL THRUST OF AIR CYLINDERS (POUNDS)
(Based on 80% Theoretical Thrust)

Cylinder Diameter Inches	Line Pressure lb/in^2 Gauge								
	5	10	15	20	30	40	50	60	70
¾	1.75	3.5	5	7.0	10.5	14	17.5	21	24.5
1	3	6	9	12.5	18	25	31	37	43
1¼	5	10	15	19.5	29	39	49	59	69
1½	7	14	21	28	42	56	71	85	99
1¾	10	19	29	38	58	77	96	115	134
2	12	25	38	50	75	100	126	151	176
2¼	16	32	48	63	94	127	159	191	223
2½	20	39	59	78	117	157	196	234	274
2¾	24	48	71	95	142	190	238	285	332
3	28	57	85	113	170	226	283	340	396
3¼	33	66	100	132	198	265	332	396	462
3½	38	77	115	154	231	308	385	462	539
3¾	44	88	133	177	265	354	442	531	618
4	50	101	151	201	301	402	503	603	703
4¼	57	114	170	227	340	454	567	681	794
4½	63	127	191	254	381	509	636	763	890
4¾	71	142	213	283	425	567	709	850	992
5	78	157	236	314	471	628	785	942	1 099
5¼	86	173	260	346	519	692	866	1 038	1 211
5½	95	190	285	380	570	760	950	1 140	1 330
5¾	104	208	312	415	622	831	1 039	1 245	1 453
6	113	226	339	452	678	905	1 131	1 356	1 582
6¼	122	245	368	491	736	982	1 227	1 473	1 718
6½	133	266	398	531	796	1 062	1 327	1 593	1 859
6¾	143	286	429	572	858	1 144	1 431	1 716	2 003
7	154	308	462	616	924	1 232	1 539	1 848	2 156
7¼	165	330	495	660	980	1 321	1 651	1 980	2 311
7½	176	353	530	706	1 059	1 413	1 767	2 118	2 472
7¾	188	377	566	755	1 132	1 510	1 887	2 265	2 642
8	201	402	603	804	1 206	1 608	2 011	2 412	2 814
8¼	214	428	641	855	1 282	1 710	2 138	2 565	2 993
8½	227	454	681	908	1 362	1 816	2 270	2 724	3 178
8¾	240	481	722	972	1 443	1 944	2 405	2 886	3 367
9	254	509	763	1 018	1 527	2 036	2 545	3 054	3 563
9½	283	567	851	1 134	1 702	2 268	2 835	3 404	3 969
10	314	625	942	1 256	1 884	2 513	3 142	3 769	4 397
10½	346	693	1 039	1 385	2 077	2 771	3 464	4 155	4 848
11	380	760	1 140	1 520	2 280	3 041	3 801	4 560	5 320
11½	415	831	1 246	1 662	2 493	3 324	4 155	4 986	5 817
12	452	905	1 357	1 809	2 713	3 619	4 524	5 427	6 333
12½	491	982	1 473	1 963	2 944	3 927	4 909	5 889	6 871
13	531	1 062	1 593	2 123	3 184	4 247	5 309	6 369	7 432
13½	572	1 145	1 718	2 290	3 435	4 580	5 726	6 870	8 015
14	616	1 232	1 847	2 463	3 694	4 926	6 158	7 389	8 621
14½	660	1 321	1 982	2 642	3 963	5 284	6 605	7 926	9 247
15	707	1 414	2 121	2 827	4 241	5 655	7 068	8 482	9 896
16	804	1 608	2 413	3 216	4 824	6 433	8 041	9 648	11 256
17	908	1 816	2 724	3 631	5 446	7 263	9 079	10 893	12 710
18	1 018	2 036	3 058	4 071	6 107	8 143	10 179	12 215	14 250
20	1 256	2 513	3 770	5 026	7 539	10 053	12 566	15 079	17 592

cont.........

PRACTICAL THRUST OF AIR CYLINDERS (POUNDS) (contd.)

(Based on 80% Theoretical Thrust)

Cylinder Diameter Inches	Line Pressure lb/in² Gauge								
	80	90	100	110	120	130	140	150	200
¾	28	31.5	35	38.5	42	45.5	49	52.5	70
1	49	55	63	69	75	81	87	94	126
1¼	79	89	98	108	118	128	138	148	196
1½	113	127	141	155	169	183	197	212	282
1¾	153	172	192	211	230	249	268	288	384
2	201	226	251	277	302	328	354	378	502
2¼	254	256	318	350	381	413	445	477	636
2½	314	354	393	432	470	509	549	589	786
2¾	380	426	475	521	569	616	664	712	950
3	452	510	566	623	680	737	793	848	1 132
3¼	530	596	664	728	793	868	932	996	1 324
3½	616	693	770	847	924	1 001	1 078	1 154	1 540
3¾	708	796	884	972	1 060	1 148	1 236	1 325	1 768
4	804	905	1 005	1 106	1 206	1 307	1 407	1 508	2 010
4¼	908	1 021	1 135	1 248	1 361	1 474	1 587	1 701	2 270
4½	1 018	1 145	1 272	1 399	1 526	1 653	1 780	1 908	2 544
4¾	1 134	1 276	1 418	1 560	1 700	1 842	1 984	2 126	2 836
5	1 256	1 413	1 571	1 728	1 884	2 041	2 198	2 356	3 142
5¼	1 384	1 557	1 731	1 904	2 076	2 249	2 422	2 598	3 462
5½	1 520	1 710	1 901	2 091	2 280	2 470	2 660	2 852	3 802
5¾	1 662	1 870	2 077	2 285	2 490	2 698	2 906	3 116	4 154
6	1 810	2 036	2 262	2 488	2 712	2 938	3 164	3 392	4 524
6¼	1 964	2 209	2 454	2 699	2 945	3 190	3 436	3 682	4 908
6½	2 124	2 390	2 655	2 921	3 186	3 452	3 717	3 982	5 310
6¾	2 288	2 574	2 863	3 149	3 433	3 720	4 007	4 294	5 732
7	2 462	2 770	3 080	3 388	3 696	4 004	4 312	4 618	6 160
7¼	2 642	2 972	3 303	3 633	3 962	4 292	4 623	4 954	6 606
7½	2 826	3 179	3 534	3 887	4 238	4 591	4 945	5 300	7 068
7¾	3 020	3 397	3 774	4 151	4 530	4 907	5 284	5 660	7 548
8	3 216	3 618	4 021	4 423	4 824	5 226	5 628	6 030	8 042
8¼	3 420	3 848	4 276	4 704	5 130	5 558	5 986	6 419	8 552
8½	3 632	4 086	4 540	4 994	5 448	5 902	6 356	6 810	9 080
8¾	3 888	4 369	4 810	5 291	5 772	6 253	6 734	7 216	9 620
9	4 072	4 581	5 087	5 598	6 108	6 617	7 126	7 634	10 178
9½	4 536	5 103	5 670	6 237	6 804	2 371	7 938	8 506	11 340
10	5 026	5 654	6 283	6 910	7 538	8 166	8 794	9 424	12 566
10½	5 542	6 235	6 927	7 620	8 310	9 003	9 696	10 390	12 854
11	6 082	6 842	7 602	8 362	9 120	9 880	10 640	11 400	15 204
11½	6 648	74 79	8 310	9 141	9 972	10 803	11 634	12 464	16 620
12	7 238	8 143	9 048	9 953	10 857	11 762	12 667	13 572	18 096
12½	7 854	8 836	9 817	10 798	11 779	12 761	13 743	14 726	19 634
13	8 494	9 556	10 618	11 680	12 740	13 802	14 864	15 928	21 236
13½	9 160	10 305	11 451	12 596	13 740	14 885	16 030	17 176	22 902
14	9 852	11 084	12 315	13 547	14 778	16 010	17 242	18 472	24 630
14½	10 568	11 889	13 210	14 531	15 852	17 173	18 494	19 816	26 420
15	11 310	12 724	14 137	15 551	16 964	18 378	19 792	21 206	28 274
16	12 866	14 474	16 083	17 691	19 296	20 904	22 512	24 128	32 166
17	14 526	16 342	18 158	19 974	21 788	23 604	25 420	27 238	36 316
18	16 286	18 322	20 358	22 394	24 430	26 466	28 500	30 535	40 716
20	20 106	22 619	25 133	27 646	30 158	32 671	35 184	37 700	50 266

CYLINDER OUTPUT FORCE – ROD SIDE – 100 lb/in^2

(Theoretical Force in Pounds)

Cylinder Diameter Inches	Rod Diameter – Inches								
	1/2	5/8	3/4	7/8	1	1.1/4	1.1/2	1.3/4	2
1	58.9	47.9	34.4	18.4	–	–	–	–	–
1.1/8	79.8	68.7	55.2	39.3	20.9	–	–	–	–
1.1/4	103	92	78.5	62.6	44.2	–	–	–	–
1.3/8	129	118	104	88.4	70.0	25.8	–	–	–
1.1/2	157	146	133	117	98.2	54.0	–	–	–
1.5/8	188	177	163	147	129	84.7	30.7	–	–
1.3/4	221	210	196	180	162	118	63.8	–	–
1.7/8	257	245	232	216	198	153	99.4	35.6	–
2	295	284	267	254	236	191	139	73.6	–
2.1/8	335	324	311	295	276	232	178	114	–
2.1/4	378	367	353	338	319	275	221	157	83.5
2.3/8	423	412	399	383	365	320	266	203	129
2.1/2	471	460	447	431	412	368	314	250	177
2.5/8	522	511	497	481	463	419	365	301	227
2.3/4	574	563	550	534	515	471	417	353	280
2.7/8	630	619	605	589	571	527	473	409	335
3	687	676	663	647	628	584	530	466	393
3.1/8	747	736	723	707	689	644	590	526	453
3.1/4	810	799	785	770	751	707	653	589	515
3.3/8	875	864	849	835	816	772	718	654	581
3.1/2	943	931	918	902	884	839	785	722	648
3.5/8	1 013	1 001	988	972	954	909	855	792	718
3.3/4	1 085	1 074	1 060	1 044	1 026	982	928	864	790
3.7/8	1 160	1 149	1 135	1 119	1 101	1 057	1 003	939	865
4	1 237	1 226	1 212	1 197	1 178	1 134	1 080	1 016	942
4.1/8	1 317	1 306	1 292	1 276	1 258	1 214	1 160	1 096	1 022
4.1/4	1 399	1 388	1 374	1 359	1 340	1 296	1 242	1 178	1 104
4.3/8	1 484	1 473	1 459	1 443	1 425	1 381	1 327	1 263	1 189
4.1/2	1 571	1 560	1 546	1 530	1 512	1 468	1 414	1 350	1 276
4.5/8	1 660	1 649	1 636	1 620	1 607	1 557	1 503	1 440	1 366
4.3/4	1 753	1 741	1 728	1 712	1 694	1 649	1 595	1 532	1 458
4.7/8	1 847	1 836	1 822	1 806	1 788	1 743	1 690	1 626	1 552
5	1 944	1 933	1 919	1 903	1 885	1 841	1 787	1 723	1 649
5.1/8	–	2 032	2 019	2 003	1 984	1 940	1 886	1 822	1 749
5.1/4	–	2 134	2 121	2 105	2 083	2 042	1 988	1 924	1 851
5.3/8	–	2 238	2 225	2 209	2 191	2 146	2 092	2 019	1 955
5.1/2	–	–	2 332	2 316	2 298	2 253	2 199	2 135	2 062
5.5/8	–	–	2 441	2 425	2 406	2 362	2 308	2 245	2 171
5.3/4	–	–	2 553	2 537	2 518	2 474	2 420	2 356	2 283
5.7/8	–	–	2 667	2 651	2 632	2 588	2 534	2 470	2 397
6	–	–	2 784	2 767	2 749	2 705	2 651	2 587	2 513
6.1/2	–	–	–	–	3 240	3 186	3 142	3 078	3 004
7	–	–	–	–	3 770	3 726	3 672	3 608	3 534
7.1/2	–	–	–	–	4 349	4 295	4 241	4 177	4 104
8	–	–	–	–	4 948	4 904	4 850	4 786	4 712
8.1/2	–	–	–	–	5 596	5 552	5 498	5 434	5 360
9	–	–	–	–	6 283	6 239	6 185	6 121	6 048
10	–	–	–	–	7 776	7 731	7 677	7 614	7 540
11	–	–	–	–	–	9 381	9 327	9 263	9 189
12	–	–	–	–	–	–	11 133	11 069	10 996

cont.........

CYLINDER OUTPUT FORCE – ROD SIDE – 100 lb/in^2 (contd.)
(Theoretical Force in Pounds)

Cylinder Diameter Inches	Rod Diameter – Inches								
	2.1/2	3	3.1/2	4	4.1/2	5	5.1/2	6	6.1/2
1	–	–	–	–	–	–	–	–	–
1.1/8	–	–	–	–	–	–	–	–	–
1.1/4	–	–	–	–	–	–	–	–	–
1.3/8	–	–	–	–	–	–	–	–	–
1.1/2	–	–	–	–	–	–	–	–	–
1.5/8	–	–	–	–	–	–	–	–	–
1.3/4	–	–	–	–	–	–	–	–	–
1.7/8	–	–	–	–	–	–	–	–	–
2	–	–	–	–	–	–	–	–	–
2.1/8	–	–	–	–	–	–	–	–	–
2.1/4	–	–	–	–	–	–	–	–	–
2.3/8	–	–	–	–	–	–	–	–	–
2.1/2	–	–	–	–	–	–	–	–	–
2.5/8	–	–	–	–	–	–	–	–	–
2.3/4	–	–	–	–	–	–	–	–	–
2.7/8	–	–	–	–	–	–	–	–	–
3	216	–	–	–	–	–	–	–	–
3.1/8	276	–	–	–	–	–	–	–	–
3.1/4	339	123	–	–	–	–	–	–	–
3.3/8	404	158	–	–	–	–	–	–	–
3.1/2	471	255	–	–	–	–	–	–	–
3.5/8	541	325	–	–	–	–	–	–	–
3.3/4	614	398	142	–	–	–	–	–	–
3.7/8	688	472	217	–	–	–	–	–	–
4	766	550	295	–	–	–	–	–	–
4.1/8	846	630	374	–	–	–	–	–	–
4.1/4	928	712	457	162	–	–	–	–	–
4.3/8	1 012	796	541	247	–	–	–	–	–
4.1/2	1 100	884	628	334	–	–	–	–	–
4.5/8	1 189	973	718	423	–	–	–	–	–
4.3/4	1 281	1 065	810	516	182	–	–	–	–
4.7/8	1 376	1 160	904	610	276	–	–	–	–
5	1 473	1 257	1 001	707	373	–	–	–	–
5.1/8	1 572	1 356	1 101	806	473	99.4	–	–	–
5.1/4	1 674	1 458	1 205	908	574	201	–	–	–
5.3/8	1 778	1 562	1 307	1 013	679	306	–	–	–
5.1/2	1 885	1 669	1 414	1 119	785	412	–	–	–
5.5/8	1 994	1 778	1 523	1 228	895	522	–	–	–
5.3/4	2 106	1 890	1 635	1 340	1 006	633	–	–	–
5.7/8	2 220	2 004	1 749	1 454	1 121	747	–	–	–
6	2 337	2 121	1 865	1 571	1 237	864	–	–	–
6.1/2	2 827	2 611	2 356	2 062	1 728	1 355	–	–	–
7	3 358	3 142	2 886	2 592	2 258	1 885	1 473	1 021	530
7.1/2	3 927	3 711	3 456	3 161	2 828	2 454	2 042	1 591	1 100
8	4 536	4 320	4 064	3 770	3 436	3 063	2 651	2 199	1 708
8.1/2	5 184	4 968	4 712	4 418	4 084	3 711	3 299	2 847	2 356
9	5 871	5 655	5 400	5 105	4 771	4 398	3 986	3 534	3 043
10	7 363	7 147	6 892	6 597	6 264	5 891	5 478	5 027	4 536
11	9 012	8 786	8 541	8 247	7 913	7 540	7 128	6 676	6 185
12	10 819	10 603	10348	10 053	9 720	9 346	8 934	8 483	7 992

CYLINDER OUTPUT FORCE – METRIC CYLINDERS
(Force in kilograms)

Cylinder Bore mm	Pressure kg/cm^2*									
	1	2	3	4	5	6	7	8	9	10
12	1.1	2.3	3.4	4.5	5.7	6.8	7.9	9.0	10.2	11.3
16	2.0	4.0	6.0	8.0	10.0	12.0	14.0	16.0	18.0	20.0
20	3.1	6.3	9.4	12.6	15.7	18.8	22.0	25.1	28.3	31.4
25	4.9	9.8	14.7	19.6	24.8	29.4	34.3	39.2	44.1	49.0
32	8.0	16.0	24.0	32.0	40.0	48.0	56.0	64.0	72.0	80.0
40	12.6	25.2	37.8	50.4	63.0	75.6	88.2	100.8	113.4	126.0
50	20	39	59	78	98	118	137	157	176	196
63	31	62	94	125	156	187	218	250	281	312
80	50	100	151	201	251	301	351	402	452	502
100	79	157	236	314	393	471	550	628	707	785
125	123	245	368	491	614	736	859	982	1104	1227
150	176	352	528	704	880	1056	1232	1408	1584	1760
200	314	628	942	1256	1570	1884	2218	2512	2826	3140
250	492	984	1476	1968	2460	2952	3434	3936	4428	4920
300	708	1416	2124	2832	3540	4248	4956	5664	6372	7080
350	962	1924	2886	3848	4810	5772	6734	7696	8658	9620
400	1258	2516	3774	5032	6290	7548	8806	10064	11322	12580
450	1590	3180	4770	6360	7950	9540	11130	12720	14310	15900
500	1965	3930	5895	7860	9825	11790	13755	15720	17775	19650

***Note:** This table is actually calculated for pressure in kg/cm^2 and is thus in error by 2% (too high) for pressure in bars. However, this error can be considered negligible for general use of this table.

INTEGRAL GREY CAST IRON FLANGES (BS4504)

Nominal Pressure 6 bar — All Dimensions in Millimetres

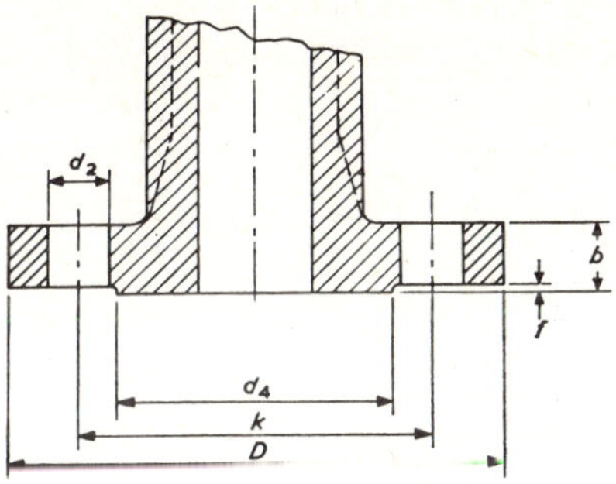

Nominal Size	Flange D	Flange b	Raised Face d_4	Raised Face f (max)	Bolting	Drilling No	Drilling d_2	Drilling k
10	75	12	35	2	M10	4	11	50
15	80	12	40	2	M10	4	11	55
20	90	14	50	2	M10	4	11	65
25	100	14	60	2	M10	4	11	75
32	120	16	70	2	M12	4	14	90
40	130	16	80	3	M12	4	14	100
50	140	16	90	3	M12	4	14	110
65	160	16	110	3	M12	4	14	130
80	190	18	128	3	M16	4	18	150
100	210	18	148	3	M16	4	18	170
125	240	20	178	3	M16	8	18	200
150	265	20	202	3	M16	8	18	225
200	320	22	258	3	M16	8	18	280
250	375	24	312	3	M16	12	18	335
300	440	24	365	4	M20	12	22	395
350	490	26	415	4	M20	12	22	445
400	540	28	465	4	M20	16	22	495
500	645	30	570	4	M20	20	22	600

Nominal Pressure 10 bar — All Dimensions in Millimetres

Nominal Size	Flange D	Flange b	Raised Face d_4	Raised Face f (max)	Bolting	Drilling No	Drilling d_2	Drilling k
200	340	26	268	3	M20	8	22	295
250	395	28	320	3	M20	12	22	350
300	445	28	370	4	M20	12	22	400
350	505	30	430	4	M20	16	22	460
400	565	32	482	4	M24	16	26	515
450	615	32	532	4	M24	20	26	565
500	670	34	585	4	M24	20	26	620

INTEGRAL COPPER ALLOY FLANGES (BS4504)

Nominal Pressure 16 bar — All Dimensions in Millimetres

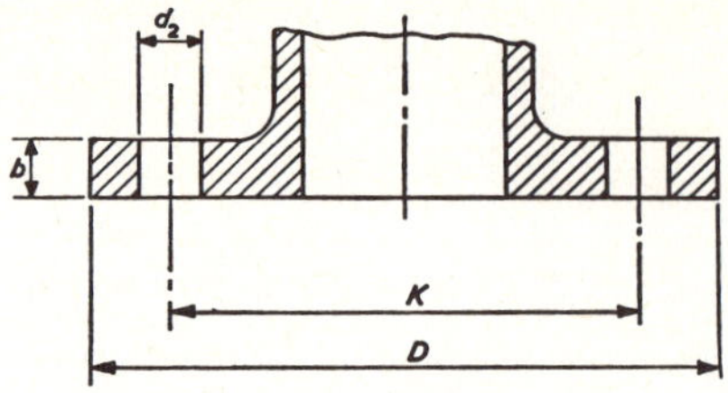

Nominal Size	Flange D	Flange b	Bolting	Drilling No	Drilling d_2	Drilling k
15	95	10	M12	4	14	65
20	105	10	M12	4	14	75
25	115	10	M12	4	14	85
32	140	10	M16	4	18	100
40	150	11	M16	4	18	110
50	165	11	M16	4	18	125
65	185	13	M16	4	18	145
80	200	15	M16	8	18	160
100	220	19	M16	8	18	180

Nominal Pressure 25 bar — All Dimensions in Millimetres

Nominal Size	Flange D	Flange b	Bolting	Drilling No	Drilling d_2	Drilling k
15	95	10	M12	4	14	65
20	105	10	M12	4	14	75
25	115	11	M12	4	14	85
32	140	11	M16	4	18	100
40	150	13	M16	4	18	110
50	165	13	M16	4	18	125
65	185	15	M16	8	18	145
80	200	17	M16	8	18	160
100	235	21	M20	8	22	190

Nominal Pressure 40 bar — All Dimensions in Millimetres

Nominal Size	Flange D	Flange b	Bolting	Drilling No	Drilling d_2	Drilling k
15	95	11	M12	4	14	65
20	105	11	M12	4	14	75
25	115	13	M12	4	14	85
32	140	13	M16	4	18	100
40	150	15	M16	4	18	110
50	165	15	M16	4	18	125
65	185	17	M16	8	18	145
80	200	19	M16	8	18	160
100	235	23	M20	8	22	190

BCAS, PNEUROP and CETOP addresses and publications

The British Compressed Air Society represents virtually all the manufacturers of every type of compressed air equipment in the U.K.

It is organized into the following product groups:

Air and Gas Compressors and Vacuum Pumps
Portable Compressors and Contractors Tools
Industrial Tools
Pneumatic Control Equipment
Ancillary Equipment

The Committees comprizing these groups meet regularly and members are thus able to take an active part in all those fields – technical and commercial – in which the Society operates to promote the best interests of the industry.

BCAS holds the General Secretariat of PNEUROP (European Committee of Manufacturers of Compressors, Vacuum Pumps and Pneumatic Tools) and is a prominent member of CETOP (European Oil Hydraulic and Pneumatic Committee) of whose Economic Commission it also holds the Secretariat. Through these two organisations the Society plays a vital role in the setting up of internationally accepted standards and codes of practice and in promoting European industrial co-operation, particularly for the elimination of technical barriers to trade.

The Society also has very close connections with the Compressed Air and Gas Institute and the National Fluid Power Association in the U.S.A.

The compressed air industry in Britain is one of the most rapidly growing sectors of the mechanical engineering industry, with an actual growth rate significantly better than the average for the industry as a whole.

Fairs and Exhibitions

The Society assists members' export endeavours by organizing joint venture schemes at exhibitions abroad with the support of the Department of Trade.

Metrication

Through its Metrication Committee the Society assists members in overcoming supply difficulties arising from this voluntary change. Agreement on S1 units preferred by the Compressed Air Industry has been established.

ALPHABETICAL LIST OF FULL MEMBERS OF THE BRITISH COMPRESSED AIR SOCIETY

Alenco Industrial Components Ltd.
Atlas Copco (G.B.) Ltd.
Austin Beech Ltd.
Baldwin Fluid Power Ltd.
Belliss & Morcom Manufacturing Ltd.
Bellows International Ltd.
Block Compressors Ltd.
Bristol Pneumatic Ltd.
Peter Brotherhood Ltd.
Bryan Donkin Co. Ltd.

CompAir Construction & Mining Ltd.
CompAir Industrial Ltd.
CompAir Maxam Ltd.
Compressor Accessories Ltd.
Consolidated Pneumatic Tool Co. Ltd.
Deltech Engineering Ltd.
Denco Miller Ltd.
Domnick Hunter (Engineers) Ltd.
Drury Engineering Co. Ltd.
Dryvent Ltd.

Durapipe Ltd.
Enots Ltd.
Gardner-Denver (U.K.) Ltd.
Garlock (G.B.) Ltd.
Halifax Tool Co. Ltd.
Howden Compressors Ltd.
Hydrovane Compressor Co. Ltd.
Ingersoll-Rand Co. Ltd.
I.T.T. Fluid Handling Ltd.
Joy Manufacturing Co. (U.K.) Ltd.
Lacy-Hulbert & Co. Ltd.
The Lead Wool Co. Ltd.
Loba (Industrial Products) Ltd.
Locker Air-Maze Ltd.
John Macdonald & Co. (P.T.) Ltd.
Martonair Ltd.
Metallic Valve Co. Ltd.
Norbro Pneumatic Ltd.

C.A. Norgren Ltd.
Northey Rotary Compressors Ltd.
Padley & Venables (Mining & Engineering) Ltd.
Pall Europe Ltd.,
Peabody Holmes Ltd.
A. Schrader's Son
Sefton Pneumatics Ltd.
Silicair Dryers Ltd.
Sperry Rand Ltd. (Lang Pneumatic Div.)
Spirax-Sarco Ltd.
Stewart-Warner Ltd. (Thor. Tools Div.)
Telektron Ltd.
Tilghman Wheelabrator Ltd.
John Trelawny Ltd.
Universal Metallic Packing Co. Ltd.
Watts Regulator (U.K.) Ltd.
Williams & James (Engineers) Ltd.
Worthington-Simpson Ltd.

ALPHABETICAL LIST OF ASSOCIATE MEMBERS OF THE BRITISH COMPRESSED AIR SOCIETY

Airtools & Compressors Ltd.
A.J. Howard Ltd.
Cotton Bros. (Longton) Ltd.
Derek Crouch (Sales) Ltd.
Economatics Ltd.
Hunts Fluid Power Ltd.
I.M.I. Pneumatics Ltd.
J. & J. Pneumatics Ltd.
Motivair Compressors Ltd.
Oscott Air Ltd.
Rodwell-Bayne Ltd.
Three Counties Pneumatics Ltd.
Thomas Wright (Bradford) Ltd.

PNEUROP – NAMES AND ADDRESSES OF MEMBERS

Austria
Fachverband der Maschinen – und Stahlbau– industrie Osterreichs,
Bauernmarkt 13,
A.1010 Vienna

Belgium
Federation des Entreprises de l'industrie des Fabrications Metalliques (FABRIMETAL),
21, rue des Drapiers,
B.1050 Brussels.

Finland
Suomen Metalliteollisuuden Keskusliitto,
Etelaranta 10,
SF.00130 Helsinki.

France
Chambre Syndicale des Constructeurs d'Outillage a Air Comprime,
10, Avenue Hoche,
F.75382 Paris.

Chambre Syndicale des Constructeurs de Compresseurs d'Air et de Gaz,
10, Avenue Hoche,
F.75382 Paris.

Germany
Fachgemeinschaft Pumpen und Verdichter im V.D.M.A.,
Postfach 710109,
(Lyonerstrasse),
D.6000 Frankfurt/M – Niederrad 71

Fachgemeinschaft Bergwerksmaschinen im V.D.M.A.,
Lueg-Alle 65,
Dusseldorf D.4000, Oberkassel.

Verband der Deutscher Feinmechanischen und Optischen Industrie E.V.,
Pipinstrasse 16,
D.5000 Cologne.

Great Britain
British Compressed Air Society (BCAS),
Leicester House,
8, Leicester Street,
London, WC2H 7BN.

Italy
Associazione Nazionale Industriali dell-Ottica Meccanica Fine e di Precisioni (ASSOTTICA),
Via Brisa 3,
Milano 20123

Associazione Nazionale Industrie Meccanica varia ed Affine (ANIMA),
Piazza Diaz 2,
Milano 20123.

Liechtenstein
Balzers A.G.,
FL – 9496 Balzers.

Netherlands
Federatie Metaal- en Electrotechnische Industrie,
Groep Fabrikanten van Compressoren,
Nassaulaan 25,
The Hague.

Sweden
Tryckluftgruppen inom Sveriges Mekanforbund,
Box 5506,
S.11485 Stockholm.

Switzerland
Verein Schweizerischer Maschinen– Industrieller, (VSM),
Kirchenweg 4,
CH.8032, Zurich.

Verein Schweizerischer Maschinen– Industrieller, (VSM),
Kirchenweg 4,
CH.8032, Zurich.

CETOP – LIST OF MEMBER ASSOCIATIONS

DISTRIBUTION

Belgium
Federations des Entreprises de l'Industrie
des Fabrications Metalliques (FABRIMETAL – Group 9),
21 rue des Drapiers,
Brussels 1050.

Denmark
Brancheforeningen for Hydraulik og Pneumatik,
Borsen BK – 1217,
Copenhagen K.

Finland
Finlands Metallindustriforening/
Suomen Metalliteollisuusyhdistys r.y,
Etelaranta 10,
Helsinki 13.

France
Syndicat des Constructeurs de Transmissions
Hydrauliques et Pneumatiques,
10 Avenue Hoche,
Paris 8e.

Germany
Fachgemeinschaft Olhydraulik und Pneumatik
im V.D.M.A.,
Postfach 710109,
Frankfurt/Main – Niederrad 1.

Great Britain
British Compressed Air Society,
Leicester House,
8 Leicester Street,
London, WC2H 7BN.

Italy
Assofluid,
Via Moscova 46/5,
20121 Milan.

Netherlands
Verenigung van Metaal-Industrieen,
Groep Fabrikanten van hydraulische en pneumatische
Apparaten (GHP),
Nassaulaan 13,
Den Haag.

Spain
Agrupacion Espanola de Fabricantes de
Transmissiones Oleo-Hidraulicas y Pneumaticas
(AEFTOP),
Via Layetana 39,
Barcelona 5.

Sweden
Sveriges Mekanforbund,
Box 5506,
S–114 85 Stockholm.

Switzerland
Verein Schweizerischer Maschinen-
Industrieller,
Kirchenweg 4,
CH.8032 Zurich.

PUBLICATIONS

B.C.A.S. Publications

BCAS Directory
(Price £1.00 plus postage)

The Selection and Installation of Compressed Air Services (Code of Practice)
(Price £3.00 plus postage to non-members)

Buyers Guide to Compressed Air Plant and Equipment
(Free)

The Directory and Code of Practice includes the Buyers Guide

The Principal SI Units to be used by the Compressed Air industry
(Price £0.25 plus postage)
This SI Unit publication is included in the "Code"

United Kingdom Export/Import Statistics of Air and Gas Compressors and Exhausters, Portable Pneumatic Tools
(Price £0.50 plus postage) (Free to Members)

General

Pneumatic Handbook (4th Edition)
(Price £21.50 plus postage in U.K. Overseas price £28.00 plus postage)

Pneumatic Technical Data, Volume II
(Price £5.00 plus postage in U.K. Overseas price £7.20 plus postage)

Principles of Pneumatics
(Price £5.00 plus postage in U.K. Overseas price £6.20 plus postage)

Pneumatic Circuits and Low Cost Automation
(Price £5.00 plus postage in U.K. Overseas price £7.20 plus postage)

Pneumatic Engineering Calculations
(Price £4.25 plus postage in U.K. Overseas price £5.80 plus postage)

Applied Hydraulics and Pneumatics in Industry
(Price £7.30 plus postage in U.K. Overseas price £9.80 plus postage)

Hydraulic and Pneumatic Cylinders
(Price £5.60 plus postage in U.K. Overseas price £8.00 plus postage)

Pneumatic Power Glossary
(Price £4.25 plus postage in U.K. Overseas price £5.80 plus postage)

PNEUROP

All the following publications are printed in English, French and German, with the exception of the first publication which is also printed in Italian, Spanish and Swedish, and the Test Code for Pneumatic Tools which is in English only.

Compressors, Classification and Glossary of Technical Terms
(Price £6.00 plus postage)

Pneumatic Tools, Classification and Glossary of Technical Terms
(Price £1.75 plus postage)

Vacuum Pumps, Classification and Glossary of Technical Terms
(Price £3.00 plus postage)

Vacuum Pumps, Rules of Acceptance, Part I (Positive Displacement, Rotary Roots Pumps)
(Price £2.00 plus postage)

Vacuum Pumps, Rules of Acceptance, Part II (Vapour Pumps)
(Price £2.00 plus postage)

Vacuum Pumps, Rules of Acceptance, Part III (Turbomolecular Pumps)
(Price £2.00 plus postage)

Flanges and Connections for High Vacuum Equipment Dimensions
(Price £1.00 plus postage)

Measurement of Sound from Pneumatic Equipment
(Price £5.00 plus postage)

Working Group Report of Sub-Committee No.8
(Price £0.60 plus postage)

PNEUROP Committee No.8. Noise Levels of Pneumatic Equipment used on building sites. Regulations in Europe and USA relating to such noise
(Price £0.50 plus postage)

CETOP

All the following documents are printed in English, French and German. The CETOP publications marked (PN) are joint PNEUROP/CETOP recommendations.

R.1 Units (Price £0.60 plus postage)

R.2 Classification of Fluid Power Terms and Documents
(Price £0.40 plus postage)

RP.3 Symbols for Hydraulic and Pneumatic Equipment and Accessories for Power Transmission by Fluid
Out of print. Use BS.2917:1969 or ISO 1219

RP.4P Pneumatic Cylinders, Suggested data for inclusion as a minimum in manufacturers' technical sales literature
(Price £0.20 plus postage)

RP.5P Specification for Pneumatic Cylinders
(Price £0.20 plus postage)

RP.6P Pipe Couplings for Pneumatic Piping — Coupling Thread
(Price £0.20 plus postage)

RP.7P. Pneumatic Cylinders, Recommended minimum relation of port size (thread) to cylinder bore (In part superseded by ISO R.1939)
(Price £0.20 plus postage)

RP.19P Characteristics, Pneumatic directional control valves
(Price £0.20 plus postage)

RP.20P Characteristics, Pneumatic flow control valves
(Price £0.20 plus postage)

RP.21P	Characteristics, Pneumatic pressure control valves (Price £0.20 plus postage)
RP.22P	Characteristics, Pneumatic shuttle, non-return and quick exhaust valves (Price £0.20 plus postage)
RP.23P	Characteristics, Pneumatic pressure intensifiers (Price £0.20 plus postage)
RP.24P	Characteristics, Pneumatic rectilinear piston type cylinders (Price £0.20 plus postage)
RP.25P	Characteristics, Pneumatic filters and water traps (Price £0.20 plus postage)
RP.26P	Characteristics, Lubricators, (Price £0.20 plus postage)
RP.27P	Characteristics, Desiccators, Air dryers (Price £0.20 plus postage)
RP.28P	Characteristics, Connections (Price £0.20 plus postage)
RP.29P	Characteristics, Quick-action couplings (Price £0.20 plus postage)
RP.30P	Characteristics, Rotating and telescopic joints (Price £0.20 plus postage)
RP.32P	Subplates for Directional Control Valves (Price £0.20 plus postage)
RP.33P	Symbols and definitions for operations of logic and related functions (Price £0.60 plus postage)
RP.34P (PN)	Couplings for Industrial Air Hoses (Price £0.15 plus postage)
RP.37P (PN)	Recommended diameters for tubes and hoses (Price £0.20 plus postage)
RP.38P (PN)	Guidance on relation between port threads and pipe/hose diameters (Price £0.20 plus postage)
RP.40P (PN)	Hose Couplings, claw type (Price £0.30 plus postage)
RP.41	Hydraulic and Pneumatic Circuits, circuit diagrams (to be published shortly)
RP.43P	Pneumatic Cylinders, Mounting dimensions (32–100 mm bore) (Price £0.40 plus postage)
RP.44P	Pneumatic Fluid Power Glossary (Price £1.75 plus postage)
RP.49P	Technological Symbols for Fluid Logic and related devices with and without moving parts (Price £0.75 plus postage)

RP.50P	Flow capacity measurement of Pneumatic components (Price £1.00 plus postage)
RP.51P	Basic data for Pneumatic cylinders (Price £0.20 plus postage)
R.52P	Pneumatic cylinders, Mounting dimensions (8—25 mm bore) (Price £0.25 plus postage)
RP.53P	Pneumatic Cylinders, Mounting dimensions (125—320 mm bore) (Price £0.20 plus postage)
RP.54P (PN)	Polyamide Tubing for Pneumatic Transmission (Price £0.30 plus postage)
RP.57P	Pressure Relief Valves (Price £0.20 plus postage)
RP.59P (PN)	Quick Action Couplings (Price £0.12 plus postage)
	Supplements Nos. 1 and 2 (Price £0.10 plus postage)
RP.68	Identification Code for Ports and Operators of Pneumatic and Hydraulic Control Valves and Other Components (Price £0.70 plus postage)
RP.71	SI Units to be used for Fluid Power (Price £0.60 plus postage)
RP.76 (PN)	Outside Diameters for Tubes in Fluid Power Applications (Price £0.25 plus postage)
RP.79P	Specification for non-ferrous pneumatic cylinder tubes. Ready to use (Price £1.25 plus postage)
RP.100	Hydraulic and Pneumatic Fluid Power Glossary (Price £5.00 plus postage)
RP.101	Lexicon of Terms from RP.100 (Price £5.00 plus postage)

SECTION 10

Buyers' Guide

*For late entries in this section, please see Contents page

Sub-section(a)

TRADE NAMES INDEX

AIRMATIC – Cylinders and valves (pneumatic) – Air Automation
"AIRVEE" – Air-cooled compressors – Tilghman Compressors, Div. of Tilghman Wheelabrator Ltd
ARO – Air tools, air hoists, fluid control equipment – The Aro Corporation (UK) Ltd
ATLAS COPCO – Compressed air equipment – Atlas Copco (Great Britain) Ltd

BIO-X – Air sterilisation filters, medical vacuum filters – Domnick Hunter Engineers Ltd
BORCAS – Vacuum pumps and compressors – Lacy-Hulbert & Co
BRISTOL DUPLEX – Air compressors – Bristol Pneumatic Ltd
BROOMWADE – Compressors and air powered tools – CompAir Industrial Ltd

COMPTON – Oil-free compressors, gas-circulating pumps and vacuum pumps – Dawson, McDonald & Dawson Ltd

DRALLIM – Pneumatic control valves – Drallim Controls Ltd
DRALLIM – Tube couplings (captive seal and metal compression) – Drallim Tube Couplings Ltd
DYNA-CON – Quick release hose coupling – Dyna-quip Ltd
DYNALOG – Logic equipment – Joucomatic Controls Ltd
DYNA-QUIP – Ball type shut-off valves, blow guns, quick release hose couplings – Dyna-quip Ltd

ECCO – Paint spraying equipment – Atlas Copco (Great Britain) Ltd
EUROLINE – Pneumatic control valves – Austin Beech Ltd

FLO-MASTER – Sub-base valves (air) – CompAir Maxam Ltd
FLUIDAIR – Rotary compressors – Whittaker Hall Ltd
FOX – Adhesives application equipment – The Aro Corporation (UK) Ltd

HOWDENAIR – Packaged screw compressors – Howden Compressors Ltd

JOUCOMATIC – Control equipment – Joucomatic Controls Ltd

KOMPEX – Rotary sliding vane compressors and exhausters – K. P. Pneumatic Services Ltd

LF 300 INSTANT FITTINGS – Push in fitting for plastic tubes – Legris (UK) Ltd

MAXAM – Pneumatic control equipment – CompAir Maxam Ltd

MECMAN – Hydraulic cylinders, pneumatic cylinders, pneumatic valves, pneumatic control systems – Mecman Ltd
MISTCOOL – Liquid spraying equipment – C. A. Norgren Ltd
MODERNAIR – Air cylinders and valves – The Aro Corporation (UK) Ltd
MODULOG – Logic equipment – Joucomatic Controls Ltd
MOTIVAIR COMPRESSORS – Air compressors – Motivair Compressors Ltd

NEAVELL – Compressors and exhausters – CompAir Industrial Ltd
NITEX – Fittings for nylon and metal tubes – Compression Joints Ltd
NORGREN – Compressed air processing equipment, fluids devices, vacuum filters, relief valves, liquid regulators and filters, lubrication systems – C. A. Norgren Ltd
NORTHEY – 'Oil-free' air and gas compressors, boosters and vacuum pumps, lubricated vacuum pumps – Northern Rotary Compressors Ltd
NUMATICS – Air valves and associated equipment – Numatics

OIL -X – Compressed air filters – Domnick Hunter Engineers Ltd

PNEUGRIP – Vices – Air Automation
PNEUMAID – High pressure logic – CompAir Maxam Ltd
PNEUMERSTAT – Constant flow bubbler unit – Williams & James (Engineers) Ltd
PROVENAIR – Air Cylinders – The Aro Corporation (UK) Ltd

RED RING – Air cylinders and pneumatic control valves – Davis Pneumatic Systems Ltd

SARCO – Temperature controllers – Spirax Sarco Ltd
SPIRAX – Pressure reducing valves, automatic drain taps and moisture separators – Spirax Sarco Ltd
SPIRAX MONNIER – Filters, regulators, lubricators – Spirax Sarco Ltd

TERITE – Pneumatic devices – Thomas Wright (Bradford) Ltd
THORITE – Air compressors – Thomas Wright (Bradford) Ltd
TURBO AIR – Turbo compressors – Joy Manufacturing Co (UK) Ltd
TWISTAIR – Screw Compressors – Joy Manufacturing Co (UK) Ltd

ULTRAIRE – High efficiency 'oil removing filters' – C. A. Norgren Ltd

WHITTAKER HALL – Rotary compressors – Whittaker Hall
W & J – Compressors, vacuum pumps, reducing valves – Williams & James (Engineers) Ltd

Sub-section(b)

CLASSIFIED INDEX OF PNEUMATIC EQUIPMENT AND COMPONENTS SUPPLIERS

Adhesives Application Equipment
The Aro Corp. (UK) Ltd

Aerators
Charles Austen Pumps Ltd

Aftercoolers, Air Cooled
Airtools & Compressors Ltd
Atlas Copco (Great Britain) Ltd
CompAir Industrial Ltd
Economatics Ltd
Motivair Compressors Ltd
Williams & James (Engineers) Ltd
Thomas Wright (Bradford) Ltd

Aftercoolers, Water Cooled
Airtools and Compressors Ltd
Atlas Copco (Great Britain) Ltd
Drury Engineering Co Ltd
Economatics Ltd
Joy Manufacturing Co (UK) Ltd
Motivair Compressors Ltd
Williams & James (Engineers) Ltd
Thomas Wright (Bradford) Ltd

Agitators
The Aro Corp. (UK) Ltd
Atlas Copco (Great Britain) Ltd

Air Actuators
The Aro Corp. (UK) Ltd
Atlas Copco (Great Britain) Ltd
CompAir Maxam Ltd
Davis Pneumatic Systems Ltd
Drallim Controls Ltd
Drury Engineering Co Ltd
Economatics Ltd
Kinetrol Ltd
Mecman Ltd
Sperry Vickers – Lang Pneumatic
Thomas Wright (Bradford) Ltd

Air Dryers
Airtools & Compressors Ltd

Air Operated Clutch Systems
Whittaker Hall Ltd

Alarm Warnings, Flow
C. A. Norgren Ltd

Alarm Warnings, Level
Crouzet Ltd
Motivair Compressors Ltd
C. A. Norgren Ltd

Alarm Warnings, Pressure
Crouzet Ltd
Drury Engineering Co Ltd
Kay Pneumatics Ltd
Motivair Compressors Ltd
C. A. Norgren Ltd

Alarm Warnings, Temperature
Drury Engineering Co Ltd
Motivair Compressors Ltd
C. A. Norgren Ltd

Alarm Warnings, Vacuum
Crouzet Ltd

Annunciator Warning Systems
C. A. Norgren Ltd

Anti-Freeze Injectors
Atlas Copco (Great Britain) Ltd
C. A. Norgren Ltd

Batch Counters (Pneumatic)
Kay Pneumatics Ltd

Bellows, Cylinder/rod protection
Davis Pneumatic Systems Ltd
Economatics Ltd
Mecman Ltd
Midland Pneumatic Ltd

Bits, Percussive
Atlas Copco (Great Britain) Ltd
Consolidated Pneumatic
Padley & Venables Ltd

Bits, Rotary
Atlas Copco (Great Britain) Ltd
Consolidated Pneumatic
Padley & Venables

Blow Guns
Dyna-Quip Ltd

Blowers, Roots
Howden Compressors Ltd

Blowers, Screw
Howden Compressors Ltd

Blowers, Sliding Vane
Williams & James (Engineers) Ltd

Boosters, Diaphragm
Charles Austen Pumps Ltd
Mecman Ltd

Boosters, Piston
Williams & James (Engineers) Ltd

Boosters, Rotary
Howden Compressors Ltd

Boosters, Turbo
Joy Manufacturing Co (UK) Ltd

Bottles, Air : Diving
Motivair Compressors Ltd

Bottles, Air : Industrial, Storage
Motivair Compressors Ltd

Bottles, Air : Tyre Inflating, Portable
Motivair Compressors Ltd

Brakes, Air Operated
Davis Pneumatic Systems Ltd
Drury Engineering Co Ltd

Breakers, Concrete and Paving
Atlas Copco (Great Britain) Ltd
Consolidated Pneumatic
Padley & Venables Ltd

Breathing Apparatus, Resuscitators
Kay Pneumatics Ltd

Chucks, Vacuum (see also Air Chucks)
Kay Pneumatics Ltd

Circulators, Air and Gas
Charles Austen Pumps Ltd
Northey Rotary Compressors Ltd
Williams & James (Engineers) Ltd

Clamps
Economatics Ltd
Parker Hannifin (UK) Ltd

Compressors, Air Diaphragm
Airtools & Compressors Ltd
Charles Austen Pumps Ltd
Dawson McDonald & Dawson Ltd
Economatics Ltd
ITT Fluid Handling Ltd
Motivair Compressors Ltd
Thomas Wright (Bradford) Ltd

Compressors, Air Reciprocating : Portable
Airtools & Compressors Ltd
Applied Pneumatics Ltd
Atlas Copco (Great Britain) Ltd
Charles Austen Pumps Ltd
Bristol Pneumatic Ltd
CompAir Industrial Ltd
Economatics Ltd
ITT Fluid Handling Ltd
Lacy-Hulbert & Co Ltd
Motivair Compressors Ltd
Williams & James (Engineers) Ltd
Thomas Wright (Bradford) Ltd

Compressors, Air Reciprocating : Stationary or Fixed
Airtools & Compressors Ltd
Applied Pneumatics Ltd
Atlas Copco (Great Britain) Ltd
Bristol Pneumatic Ltd
CompAir Industrial Ltd
Consolidated Pneumatic
Desoutter Brothers Ltd
Economatics Ltd
ITT Fluid Handling Ltd
Joy Manufacturing Co (UK) Ltd
Lacy-Hulbert & Co Ltd
Motivair Compressors Ltd
Tilghman Compressors – Div of Tilghman Wheelabrator Ltd

Controllers, Positional
Austin Beech Ltd
Kinetrol Ltd

Controllers, Pressure
Austin Beech Ltd
Drallim Controls Ltd
Mecman Ltd
C. A. Norgren Ltd
Schrader Pneumatics
Spirax Sarco Ltd
Thomas Wright (Bradford) Ltd

Controllers, Temperature
Spirax Sarco Ltd

Control Panels, Pneumatic Sequence
Applied Pneumatics Ltd
The Aro Corp (UK) Ltd
Atlas Copco (Great Britain) Ltd
Austin Beech Ltd
CompAir Maxam Ltd
Crouzet Ltd
Drallim Controls Ltd
Drury Engineering Co Ltd
Economatics Ltd
Kay Pneumatics Ltd
Mecman Ltd
Motivair Compressors Ltd
Numatics
Ross James Engineering Ltd
Schrader Pneumatics

Couplings, Air Vacuum Line : Pipe
The Aro Corp. (UK) Ltd
CompAir Industrial Ltd
CompAir Maxam Ltd
Crouzet Ltd
Drallim Tube Couplings Ltd
Economatics Ltd
Legris (UK) Ltd
Mecman Ltd
Motivair Compressors Ltd
Parker Hannifin (UK) Ltd
Schrader Pneumatics

Couplings, Air Vacuum Line : Quick Acting
The Aro Corp (UK) Ltd
Atlas Copco (Great Britain) Ltd
CompAir Industrial Ltd
CompAir Maxam Ltd
Crouzet Ltd
Economatics Ltd

Filters, High Pressure
Domnick Hunter Engineers Ltd

Filters, oil removing
C. A. Norgren Ltd

Filters and Separators, Adsorption
Domnick Hunter Engineers Ltd
Economatics Ltd
Motivair Compressors Ltd
C. A. Norgren Ltd
Numatics
Ross James Engineering Ltd
Williams & James (Engineers) Ltd

Filters and Separators, Air Intake
Economatics Ltd
Drury Engineering Co Ltd
Motivair Compressors Ltd
Parker Hannifin (UK) Ltd

Filters, and Separators, Air Pipeline (see also Desiccators)
Airtools & Compressors Ltd
The Aro Corp. (UK) Ltd
Atlas Copco (Great Britain) Ltd
Austin Beech Ltd
CompAir Maxam Ltd
Domnick Hunter Engineers Ltd
Davis Pneumatic Systems Ltd
Drury Engineering Co Ltd
Economatics Ltd
The Globe Engineering Co Ltd
Joucomatic Controls Ltd
Kay Pneumatics Ltd
Mecman Ltd
Midland Pneumatic Ltd
Motivair Compressors Ltd
C. A. Norgren Ltd
Numatics
RGS Electro-Pneumatics Ltd
Ross James Engineering Ltd
Schrader Pneumatics
Sperry Vickers – Lang Pneumatic
Spirax Sarco Ltd
Watts Regulator (UK) Ltd
Thomas Wright (Bradford) Ltd

Filters and Separators, Vacuum
Applied Pneumatics Ltd
Domnick Hunter Engineers Ltd
Economatics Ltd
C. A. Norgren Ltd
Motivair Compressors Ltd
Numatics
Parker Hannifin (UK) Ltd
Ross James Engineering Ltd

Fittings, Air Line (see also Hose): Ferrous
The Aro Corp. (UK) Ltd
Atlas Copco (Great Britain) Ltd
Consolidated Pneumatic
Crouzet Ltd
Economatics Ltd
Joucomatic Controls Ltd
Legris (UK) Ltd
Mecman Ltd
Midland Pneumatic Ltd
Motivair Compressors Ltd
C. A. Norgren Ltd
Numatics
Parker Hannifin (UK) Ltd
Ross James Engineering Ltd
Schrader Pneumatics
Thomas Wright (Bradford) Ltd

Fittings, Air Line: Non-Ferrous
Air Automation
Airtools & Compressors Ltd
Atlas Copco (Great Britain) Ltd
CompAir Maxam Ltd
Compression Joints Ltd
Consolidated Pneumatic
Crouzet Ltd
Drallim Tube Couplings Ltd
Economatics Ltd
Kay Pneumatics Ltd
Legris (UK) Ltd
Mecman Ltd
Midland Pneumatic Ltd
Motivair Compressors Ltd
Numatics
Parker Hannifin (UK) Ltd
RGS Electro-Pneumatics Ltd
Ross James Engineering Ltd
Schrader Pneumatics
Sperry Vickers – Lang Pneumatic
Thomas Wright (Bradford) Ltd

Fittings, Air Line: Plastic
Airtools & Compressors Ltd
CompAir Maxam Ltd
Crouzet Ltd
Drallim Tube Couplings Ltd
Economatics Ltd
Legris (UK) Ltd
Mecman Ltd
Motivair Compressors Ltd
C. A. Norgren Ltd

Numatics
Parker Hannifin (UK) Ltd
RGS Electro-Pneumatics Ltd
Ross James Engineering Ltd
Schrader Pneumatics

Fittings – Airline push in type
Legris (UK) Ltd

Flowmeters, Air
Drallim Controls Ltd
Economatics Ltd
Numatics
Ross James Engineering Ltd

Fluidic Devices
C. A. Norgren Ltd

Fume Removal Plant
Economatics Ltd

Gaskets
Motivair Compressors Ltd

Gauges, Inspection
Numatics
Ross James Engineering Ltd

Gauges, Pressure
Airtools & Compressors Ltd
Austin Beech Ltd
CompAir Maxam Ltd
Drallim Controls Ltd
Economatics Ltd
Joucomatic Controls Ltd
Kay Pneumatics Ltd
Midland Pneumatic Ltd
Motivair Compressors Ltd
Numatics
Parker Hannifin (UK) Ltd
RGS Electro-Pneumatics Ltd
Ross James Engineering Ltd
Sperry Vickers – Lang Pneumatic
Spirax Sarco Ltd
Thomas Wright (Bradford) Ltd

Gauges, Vacuum
Economatics Ltd
Motivair Compressors Ltd
Numatics
Ross James Engineering Ltd

Gauges, Pressure Differential
Domnick Hunter Engineers Ltd

Grinders
Airtools & Compressors Ltd
The Aro Corp. (UK) Ltd
Atlas Copco (Great Britain) Ltd
CompAir Industrial Ltd
Desoutter Brothers Ltd
Economatics Ltd
Motivair Compressors Ltd
Thomas Wright (Bradford) Ltd

Guns, Blow (or dusters)
Air Automation
Airtools & Compressors Ltd
Applied Pneumatics Ltd
The Aro Corp. (UK) Ltd
Atlas Copco (Great Britain) Ltd
Austin Beech Ltd
CompAir Industrial Ltd
Economatics Ltd
Legris (UK) Ltd
Midland Pneumatic Ltd
Motivair Compressors
Schrader Pneumatics
Thomas Wright (Bradford) Ltd

Guns, Oiling
Airtools & Compressors Ltd
The Aro Corp. (UK) Ltd
CompAir Industrial Ltd
Motivair Compressors Ltd

Guns, Spray, Automatic
Airtools & Compressors Ltd
Atlas Copco (Great Britain) Ltd
CompAir Industrial Ltd
Economatics Ltd
Motivair Compressors Ltd

Guns, Spray, Metal Coating
Economatics Ltd

Guns, Spray, Paint, Liquids
Airtools & Compressors Ltd
Atlas Copco (Great Britain) Ltd
CompAir Industrial Ltd
Dawson McDonald & Dawson Ltd
Economatics Ltd
Motivair Compressors Ltd
Thomas Wright (Bradford) Ltd

Guns, Spray, Powder and Flock
Airtools & Compressors Ltd
Atlas Copco (Great Britain) Ltd
Economatics Ltd
Motivair Compressors Ltd

Hose Couplings, Quick Release
Dyna-Quip Ltd

Hose, Fittings
Airtools & Compressors Ltd
Atlas Copco (Great Britain) Ltd
Bestobell Mobrey Ltd
CompAir Industrial Ltd
Consolidated Pneumatic
Crouzet Industrial Ltd
Desoutter Brothers Ltd
Drallim Tube Couplings Ltd
Economatics Ltd
Kay Pneumatics Ltd
Legris (UK) Ltd
Mecman Ltd
Motivair Compressors Ltd
Numatics
Parker Hannifin (UK) Ltd
Ross James Engineering Ltd
Schrader Pneumatics
Thomas Wright (Bradford) Ltd

Hose, High Pressure
Air Automation
TheAro Corp. (UK) Ltd
Atlas Copco (Great Britain) Ltd
Economatics Ltd
Motivair Compressors Ltd
Numatics
Parker Hannifin (UK) Ltd
Ross James Engineering Ltd
Thomas Wright (Bradford) Ltd

Hose, Low Pressure
Air Automation
Atlas Copco (Great Britain) Ltd
Drallim Tube Couplings Ltd
Economatics Ltd
Mecman Ltd
Midland Pneumatic Ltd
Motivair Compressors Ltd
Numatics
Parker Hannifin (UK) Ltd
Ross James Engineering Ltd
Schrader Pneumatics
Thomas Wright (Bradford) Ltd

Hose, Medium Pressure
Airtools & Compressors Ltd
Atlas Copco (Great Britain) Ltd
Consolidated Pneumatic
Desoutter Brothers Ltd
Drallim Tube Couplings Ltd
Economatics Ltd
Mecman Ltd
Motivair Compressors Ltd
Numatics
Parker Hannifin (UK) Ltd
Ross James Engineering Ltd
Schrader Pneumatics
Thomas Wright (Bradford) Ltd

Hose, Plastic
Airtools & Compressors Ltd
Atlas Copco (Great Britain) Ltd
Bestobell Mobrey Ltd
Crouzet Ltd
Desoutter Brothers Ltd
Drallim Tube Couplings Ltd
Economatics Ltd
Mecman Ltd
Midland Pneumatic Ltd
Motivair Compressors Ltd
Numatics
Parker Hannifin (UK) Ltd
Ross James Engineering Ltd
Schrader Pneumatics
Thomas Wright (Bradford) Ltd

Hose, Reels
Airtools & Compressors Ltd
The Aro Corp. (UK) Ltd
Atlas Copco (Great Britain) Ltd
Economatics Ltd
Motivair Compressors Ltd
Numatics
Ross James Engineering Ltd
Thomas Wright (Bradford) Ltd

Hose, Special types
Economatics Ltd
Motivair Compressors Ltd
Numatics
Ross James Engineering Ltd

Hose, Wiring tools
Numatics
Ross James Engineering Ltd

Hydraulic Check Units
Air Automation
The Aro Corp. (UK) Ltd
Desoutter Brothers Ltd
Motivair Compressors Ltd

Indexing Tables, Rotary
Air Automation
Economatics Ltd

Atlas Copco (Great Britain) Ltd
CompAir Industrial Ltd
Desoutter Brothers Ltd
Motivair Compressors Ltd

Nutrunners (see also Wrenches)
The Aro Corp. (UK) Ltd
Atlas Copco (Great Britain) Ltd
Desoutter Brothers Ltd
Motivair Compressors Ltd

O-rings
Parker Hannifin (UK) Ltd

Packing Glands and Seals
Parker Hannifin (UK) Ltd

Packings
Parker Hannifin (UK) Ltd

Pipe Joints and Fittings
Atlas Copco (Great Britain) Ltd
Compression Joints Ltd
Drallim Tube Couplings Ltd
Legris (UK) Ltd
Motivair Compressors Ltd
Wade Couplings Ltd

Pipes, Flexible
CompAir Maxam Ltd
Drallim Tube Couplings Ltd
Economatics Ltd
Legris (UK) Ltd
Mecman Ltd
Motivair Compressors Ltd
Wade Couplings Ltd

Pipes, Ferrous
Motivair Compressors Ltd

Pipes, Non-Ferrous
CompAir Maxam Ltd
Mecman Ltd
Midland Pneumatic Ltd
Motivair Compressors Ltd
Sperry Vickers – Lang Pneumatic
Wade Couplings Ltd

Pipes, Plastic
CompAir Maxam Ltd
Drallim Tube Couplings Ltd
Legris (UK) Ltd
Mecman Ltd
Midland Pneumatic Ltd
Motivair Compressors Ltd
Sperry Vickers – Lang Pneumatic

Pipework
Airtools & Compressors Ltd
Drury Engineering Co Ltd
Economatics Ltd
Motivair Compressors Ltd

Pneumatic Logic Control Systems, Moving Part
Numatics
Ross James Engineering Ltd

Polishers
The Aro Corp. (UK) Ltd
Atlas Copco (Great Britain) Ltd
Desoutter Brothers Ltd

Presses, Impact
Economatics Ltd
Schrader Pneumatics

Presses, Squeeze
CompAir Maxam Ltd
Economatics Ltd
Motivair Compressors Ltd
Schrader Pneumatics

Programmers
Numatics
Ross James Engineering Ltd

Pulsation Dampers
Drury Engineering Co Ltd
Motivair Compressors Ltd

Pumps (see also Vacuum Pumps): Air Lift
Atlas Copco (Great Britain) Ltd
Austin Beech Ltd
Motivair Compressors Ltd
Williams & James (Engineers) Ltd

Pumps, Reciprocating
The Aro Corp. (UK) Ltd
Parker Hanninfin (UK) Ltd

Pumps, Sludge
Atlas Copco (Great Britain) Ltd
Consolidated Pneumatic

Pumps, Sump
Atlas Copco (Great Britain) Ltd
Consolidated Pneumatic
Motivair Compressors Ltd

Rammers, Backfill
Consolidated Pneumatic
Motivair Compressors Ltd

Rammers, Foundry and Sand
Atlas Copco (Great Britain) Ltd
CompAir Industrial Ltd
Motivair Compressors Ltd

Receivers, Small
Airtools & Compressors Ltd
Atlas Copco (Great Britain) Ltd
Austin Beech Ltd
Consolidated Pneumatic
Drury Engineering Co Ltd
Economatics Ltd
Midland Pneumatic Ltd
Motivair Compressors Ltd
Schrader Pneumatics
Thomas Wright (Bradford) Ltd

Receivers, Medium
Airtools & Compressors Ltd
Atlas Copco (Great Britain) Ltd
Austin Beech Ltd
Consolidated Pneumatic
Drury Engineering Co Ltd
Economatics Ltd
Midland Pneumatic Ltd
Motivair Compressors Ltd
Schrader Pneumatics
Thomas Wright (Bradford) Ltd

Receivers, Large
Atlas Copco (Great Britain) Ltd
Drury Engineering Co Ltd
Economatics Ltd
Motivair Compressors Ltd
Thomas Wright (Bradford) Ltd

Regulators, Flow
Austin Beech Ltd
CompAir Maxam Ltd
Davis Pneumatic Systems Ltd
Drallim Controls Ltd
Economatics Ltd
Joucomatic Controls Ltd
Mecman Ltd
Midland Pneumatic Ltd
Motivair Compressors Ltd
Neumatics
Parker Hannifin (UK) Ltd
Ross James Engineering Ltd
Schrader Pneumatics
Thomas Wright (Bradford) Ltd
Sperry Vickers – Lang Pneumatic
Spirax Sarco Ltd
Watts Regulator (UK) Ltd
Thomas Wright (Bradford) Ltd

Regulators, not Self-Relieving
Austin Beech Ltd
CompAir Maxam Ltd
Davis Pneumatic Systems Ltd
Economatics Ltd
Kay Pneumatics Ltd
Mecman Ltd
Midland Pneumatic Ltd
Motivair Compressors Ltd
C. A. Norgren Ltd
Numatics
RGS Electro-Pneumatics Ltd
Ross James Engineering Ltd
Spirax Sarco Ltd
Watts Regulator (UK) Ltd
Thomas Wright (Bradford) Ltd

Regulators, Remote Operated
Austin Beech Ltd
CompAir Maxam Ltd
Economatics Ltd
Mecman Ltd
Midland Pneumatic Ltd
Motivair Compressors Ltd
C. A. Norgren Ltd
Numatics
Parker Hannifin (UK) Ltd
Ross James Engineering Ltd
Schrader Pneumatic
Watts Regulator (UK) Ltd

Regulators, Temperature
Spirax Sarco Ltd

Regulators, Vacuum
Economatics Ltd
Numatics
Ross James Engineering Ltd

Riveters, Aircraft
Airtools & Compressors Ltd
CompAir Industrial Ltd
Motivair Compressors Ltd

Riveters, Industrial
Atlas Copco (Great Britain) Ltd
CompAir Industrial Ltd
Economatics Ltd
Motivair Compressors Ltd

Schrader Pneumatics
Thomas Wright (Bradford) Ltd

Tampers, Backfill
Atlas Copco (Great Britain) Ltd
Consolidated Pneumatic

Tampers, Railway Tie
Atlas Copco (Great Britain) Ltd

Tappers, Thread
Airtools & Compressors Ltd
The Aro Corp. (UK) Ltd
Economatics Ltd

Thrusters
Atlas Copco (Great Britain) Ltd
Mecman Ltd

Timers
CompAir Maxam Ltd
Economatics Ltd
Joucomatic Controls Ltd
Kay Pneumatics Ltd
Mecman Ltd
Midland Pneumatic Ltd
Numatics
Ross James Engineering Ltd

Timers, Electrical (Dial, sequence, delay, process, automatic and hand reset)
Crouzet Ltd

Timers, Electro-Mechanical (Dial, sequence, delay, process, automatic and hand reset)
CompAir Maxam Ltd
Crouzet Ltd
Motivair Compressors Ltd

Timers, Pneumatic (Dial, sequence, delay, process, automatic and hand reset)
The Aro Corp. (UK) Ltd
CompAir Maxam Ltd
Crouzet Ltd
Davis Pneumatic Systems Ltd
Economatics Ltd
Joucomatic Controls Ltd
Kay Pneumatics Ltd
Motivair Compressors Ltd
Numatics
Ross James Engineering Ltd
Schrader Pneumatics

Tool Steels, Chisel Steels
Atlas Copco (Great Britain) Ltd
Consolidated Pneumatic
Padley & Venables Ltd

Tool Steels, Dressing Tools
Consolidated Pneumatic
Padley & Venables Ltd

Tool Steels, Digger Steels
Consolidated Pneumatic
Padley & Venables

Tool Steels, Feathers
Consolidated Pneumatic
Padley & Venables

Tool Steels, Plug and Feathers
Atlas Copco (Great Britain) Ltd
Consolidated Pneumatic
Padley & Venables Ltd

Tool Steels, Spades
Atlas Copco (Great Britain) Ltd
Consolidated Pneumatic
Padley & Venables Ltd

Tool Steels, Tamping Pads and Stems
Atlas Copco (Great Britain) Ltd
Consolidated Pneumatic
Padley & Venables Ltd

Tool Steels, Pick Steels
Atlas Copco (Great Britain) Ltd
Consolidated Pneumatic
Padley & Venables Ltd

Tool Steels, Moil Points
Atlas Copco (Great Britain) Ltd
Consolidated Pneumatic
Padley & Venables Ltd

Tool Steels, Pile Drivers
Consolidated Pneumatic

Tool Steels, Punches
Padley & Venables Ltd

Torque Actuators
Economatics Ltd
Kinetrol Ltd
Mecman Ltd

Traps, Water (see also Filters and Separators)
Airtools & Compressors Ltd

The Aro Corp. (UK) Ltd
Consolidated Pneumatic
Economatics Ltd
Motivair Compressors Ltd
C. A. Norgren Ltd
Spirax Sarco Ltd

Tubes, Pneumatic Conveying
Drallim Tube Couplings Ltd
Motivair Compressors Ltd
Parker Hannifin (UK) Ltd

Two Hand Safety Controls
Kay Pneumatics Ltd

Tyre Inflators
Airtools & Compressors Ltd
Motivair Compressors
Thomas Wright (Bradford) Ltd

Underwater Pneumatic Tools
Atlas Copco (Great Britain) Ltd
Desoutter Brothers Ltd

Vacuum Pumps
Airtools & Compressors Ltd
Applied Pneumatics Ltd
Atlas Copco (Great Britain) Ltd
Charles Austen Pumps Ltd
Austin Beech Ltd
CompAir Industrial Ltd
CompAir Maxam Ltd
Dawson McDonald & Dawson Ltd
Economatics Ltd
Howden Compressors Ltd
ITT Fluid Handling Ltd
Joy Manufacturing Co (UK) Ltd
K.P. Pneumatic Services Ltd
Lacy-Hulbert & Co Ltd
Motivair Compressors Ltd
Northey Rotary Compressors Ltd
Whittaker Hall Ltd
Williams & James (Engineers) Ltd
Thomas Wright (Bradford) Ltd

Vacuum Pumps, Oil Free
Davis Pneumatic Systems Ltd
Northey Rotary Compressors Ltd

Vacuum Tweezer Unit
Charles Austen Pumps Ltd

Valve Actuators
Atlas Copco (Great Britain) Ltd
Austin Beech Ltd
Drallim Controls Ltd
Drury Engineering Co Ltd
Dyna-Quip Ltd
Economatics Ltd
The Globe Pneumatic Eng. Co Ltd
Kinetrol Ltd
Mecman Ltd
Numatics Parker Hannifin (UK) Ltd
RGS Electro-Pneumatics Ltd
Ross James Engineering Ltd
Schrader Pneumatics

Valves Ball type
Dyna-Quip Ltd

Valves, Compressor, Suction and Discharge
Bestobell Mobrey Ltd
Drallim Controls Ltd
Drury Engineering Co Ltd
Economatics Ltd
Motivair Compressors Ltd

Valves, Directional Control (for controlling pneumatic cylinders, etc)
Air Automation
The Aro Corp. (UK) Ltd
Austin Beech Ltd
Atlas Copco (Great Britain) Ltd
Bestobell Mobrey Ltd
CompAir Maxam Ltd
Crouzet Ltd
Davis Pneumatic Systems Ltd
Drallim Controls Ltd
Economatics Ltd
The Globe Pneumatic Eng Co Ltd
Joucomatic Controls Ltd
Kay Pneumatics Ltd
Mecman Ltd
Midland Pneumatic Ltd
Motivair Compressors Ltd
Numatics
Parker Hannifin (UK) Ltd
RGS Electro-Pneumatics Ltd
Ross James Engineering Ltd
Schrader Pneumatics
Sperry Vickers – Lang Pneumatic
Thomas Wright (Bradford) Ltd

Valves, Limit
Numatics
Ross James Engineering Ltd

Valves, Manual
Air Automation

The Aro Corp. (UK) Ltd
Atlas Copco (Great Britain) Ltd
Austin Beech Ltd
Bestobell Mobrey Ltd
CompAir Maxam Ltd
Crouzet Ltd
Davis Pneumatic Systems Ltd
Drallim Controls Ltd
Dyna-Quip Ltd
Economatics Ltd
The Globe Pneumatic Engineering Co Ltd
Kay Pneumatics Ltd
Legris (UK) Ltd
Mecman Ltd
Motivair Compressors Ltd
C. A. Norgren Ltd
Numatics
RGS Electro-Pneumatics Ltd
Ross James Engineering Ltd
Schrader Pneumatics
Sperry Vickers – Lang Pneumatic
Wade Couplings Ltd

Valves, Mechanical
Air Automation
The Aro Corp. (UK) Ltd
Austin Beech Ltd
Atlas Copco (Great Britain) Ltd
Bestobell Mobrey Ltd
CompAir Maxam Ltd
Crouzet Ltd
Davis Pneumatic Systems Ltd
Economatics Ltd
Joucomatic Controls
Kay Pneumatics Ltd
Mecman Ltd
Midland Pneumatic Ltd
Motivair Compressors Ltd
Numatics
Parker Hannifin (UK) Ltd
RGS Electro-Pneumatics Ltd
Ross James Engineering Ltd
Schrader Pneumatics
Sperry Vickers – Lang Pneumatic
Wade Couplings Ltd
Thomas Wright (Bradford) Ltd

Valves, Needle (see also Regulators, flow)
Air Automation
Airtools & Compressors Ltd
Austin Beech Ltd
CompAir Maxam Ltd
Crouzet Ltd
Davis Pneumatic Systems Ltd
Drallim Controls Ltd
Economatics Ltd
Joucomatic Controls Ltd
Kay Pneumatics Ltd
Legris (UK) Ltd
Mecman Ltd
Midland Pneumatic Ltd
Motivair Compressors Ltd
Numatics
Parker Hannifin (UK) Ltd
RGS Electro-Pneumatics Ltd
Ross James Engineering Ltd
Schrader Pneumatics
Sperry Vickers – Lang Pneumatic
Wade Couplings Ltd
Thomas Wright (Bradford) Ltd

Valves, Non-Return or Check
Air Automation
Airtools & Compressors Ltd
Austin Beech Ltd
Bestobell Mobrey Ltd
CompAir Maxam Ltd
Davis Pneumatic Systems Ltd
Drallim Controls Ltd
Drury Engineering Co Ltd
Economatics Ltd
Kay Pneumatics Ltd
Mecman Ltd
Midland Pneumatic Ltd
Motivair Compressors Ltd
Numatics
Parker Hannifin (UK) Ltd
RGS Electro-Pneumatics Ltd
Ross James Engineering Ltd
Schrader Pneumatics
Sperry Vickers – Lang Pneumatic
Thomas Wright (Bradford) Ltd

Valves, Pressure Reducing (See also Regulators)
Air Automation
Airtools & Compressors Ltd
Atlas Copco (Great Britain) Ltd
Austin Beech Ltd
The Aro Corp. (UK) Ltd
Bestobell Mobrey Ltd
CompAir Maxam Ltd
Crouzet Ltd
Davis Pneumatic Systems Ltd
Economatics Ltd
Joucomatic Controls Ltd
Kay Pneumatics Ltd

Vibrators, External
Consolidated Pneumatic
Economatics Ltd
Thomas Wright (Bradford) Ltd

Vibrators, Internal
Consolidated Pneumatic
Thomas Wright (Bradford) Ltd

Vices, Air Operated
Air Automation
Economatics Ltd
Motivair Compressors Ltd

Winches
Atlas Copco (Great Britain) Ltd
The Globe Pneumatic Eng. Co Ltd
Motivair Compressors Ltd

Wire Brush Machines
The Aro Corp. (UK) Ltd

Wrenches, Impact
The Aro Corp. (UK) Ltd
Atlas Copco (Great Britain) Ltd
Desoutter Brothers Ltd
Economatics Ltd
Motivair Compressors Ltd

Wrenches, Torque
Airtools & Compressors Ltd
The Aro Corp. (UK) Ltd
Atlas Copco (Great Britain) Ltd
CompAir Industrial Ltd
Economatics Ltd

Sub-section(c)

ALPHABETICAL LIST OF MANUFACTURERS WITH ADDRESSES, TELEPHONE NUMBERS, TELEGRAM ADDRESSES OF HEAD OFFICE, WORKS AND BRANCHES

Airtools & Compressors Ltd, 89/91 Nottingham Road, Loughborough LE11 1ES
Tel: (0509) 66011/2
Branches: 58 Station Road, Brough, N.Humberside HU15 1EA Tel: (0482) 668591
Robinson Road, off Coleman Rd., Leicester LE5 4NS Tel: (0533) 764575
119/121 Canal St., Nottingham NG1 7HB Tel: (0602) 581636
Finchwell Rd., Handsworth, Sheffield SI3 9AS Tel: (0742) 446141

Air Automation, Navigation St., Wolverhampton Tel: Wolverhampton 52968

Applied Pneumatics Limited, Charlton Mead Lane, Hoddesdon, Herts. Tel: Hoddesdon 60411
Branch: Neachells Lane, Willenhall, Staffs. Tel: 0902 737357

The Aro Corporation (UK) Ltd., East Moons Moat Industrial Estate, Alders Drive, Redditch, Worcs, B98 0RH Tel: Redditch 25722 Telex 338169

Atlas Copco (Great Britain) Ltd, P.O. Box 79, Swallowdale Lane, Hemel Hempstead, Herts HP2 7HA Tel: (0442) 61201 Grams ATLASCOPCO
Branches: Coronation Rd, Park Royal, London NW10 7QL Tel: 01-965 2466
47a Zetland Road, Bristol BS6 7AJ Tel: Bristol 45036
Avondale Road, Pontrhydyrun, Cwmbran, Gwent NP4 1YD Tel: Cwmbran 4054
Chasetown Industrial Estate, Ring Rd, Walsall WS7 8XS Tel: Burntwood 2165
Kenyons Lane, Haydock, St. Helens, Lancs. Tel: Garswood Park 714561
Portobello Industrial Estate, Birtley, Co. Durham Tel: Newcastle 402747
Mansfield Rd., Aston, Nr. Sheffield, Yorks. Tel: Sheffield 872831
School Lane, Chandler's Ford, Eastleigh, Hants SO5 3DF Tel: Chandler's Ford 66551/2/3
Dunnswood Rd, Wardpark South, Cumbernauld, Glasgow Tel: Cumbernauld 22174
Atlas Copco (Ireland) Ltd, Kylemore Rd, Bluebell, Dublin 12, Ireland Tel: Dublin 505978

Charles Austen Pumps Ltd, 100 Royston Rd, Byfleet, Weybridge, Surrey KT14 7PB Tel: Byfleet 43224/5 Telex 928368

Austin Beech Ltd, P.O. Box 10, Leighton Buzzard, Beds LU7 7LJ Tel: (052-53) 2581
Branches (UK): T. J. Morgan (Barry) Ltd, Building 17, Atlantic Trading Estate, Barry CF6 6RF Tel: Barry 5333 & 740376
Astbury, Kilby & Co Ltd, 124 Hardwick Rd, Streetley, Sutton Coldfield, Warks. B74 3DP Tel: 021-353-8211/2
Air Accessories (Sheffield) Ltd, 24 Norton Park View, Norton, Sheffield, S8 8GS, Yorks. and at: 515 Abbeydale Rd, Sheffield S7 1FU Tel: 0742-582086
Mr A. Eden, 17 Pitcairn Crescent, Hairmyres Est., East Kilbride, Glasgow, G75 8TP Scotland

Bestobell Mobrey Ltd, 190–196 Bath Rd, Trading Estate, Slough SL1 4DN Tel: Slough 34646 Grams ARTEECON SLOUGH
Branches: 246 Bedford Ave, Trading Estate, Slough SL1 4SB Tel: Slough 31757/8
8A Cotham Hill, Bristol BS6 6LF Tel: Bristol 38992/3
318 Stratford Rd, Shirley, Solihull, West Midlands Tel: 021-745 1262/3/4
76 Cross St, Sale, Cheshire Tel: 061-973 2235
Northinch St, Whiteinch, Glasgow G14 0RN Tel: 041-954 8957/8 or 9506

Bradford Cylinders Limited, Allerton Rd, Bradford BD8 0BA Tel: (0274) 495611 Telex 517526
Branches: 1 Church St, Warwick Tel: (0926) 41808
Thorn Hydraulic Division, Rodley, Leeds LS13 1LG Tel: (0532) 577721 Grams ELECTRAULIC RODLEY

Bristol Pneumatic Limited, Lodge Causeway, Fishponds, Bristol BS16 3JS Tel: (0272) 657511 Grams PERCUSSION BSTL

Burkert Contromatic Ltd, London Rd, Bowbridge, Stroud, Glos. GL5 2BT Tel: Stroud 5981/2 Telex 43301

CompAir Industrial Ltd, P.O. Box 7, BroomWade Works, High Wycombe, Bucks Tel: High Wycombe (0494) 21181 Grams BROOM H/WYCOMBE
Branches: Reavell Works, P.O. Box 44, Ipswich, Suffolk IP2 0AE Tel: Ipswich 56124
19 Erdington Industrial Estate, Chester Rd, Erdington, Birmingham B24 0RD Tel: 021-382-6544
Bonville Rd, Brislington Trading Estate, Bristol BS4 5PA Tel: Bristol 711471
Rhymney River Bridge Rd, Cardiff. Tel: Cardiff 395645
Burnfield Rd, Giffnock, Glasgow Tel: Glasgow Giffnock 2336
Wellington Bridge Industrial Estate, Leeds 12, Yorkshire. Tel: Leeds 41306
83 Jesmond Rd, Newcastle upon Tyne NE2 1NH Tel: Newcastle 811896
Kingsland Grange, Warrington, Lancs. Tel: Padgate 811610
3 Garnett Close, Greycaine Industrial Estate, Bushey Mill Lane, Watford, Herts WD2 4PN Tel: Watford 33744

CompAir Maxam Limited, Camborne, Cornwall TR14 8DS, Tel: Camborne 712750 Telex 45501
Branches: Arundel Rd, Uxbridge, Middx UB8 2RR Tel: Uxbridge 52571 Telex 935197
955 Romford Rd, Manor Park, London E12 Tel: 01-478 1094
33 Monarch Parade, London Rd, Mitcham CR4 3HA Tel: 01-648 7709
Lightning Way, off Alvechurch Rd, West Heath, Birmingham Tel: 021-475-1191 Telex 33516
Queens Tower, Park Grange Rd, Sheffield S2 3PW Tel: 0742-79911 Telex 54221

Kingsland Grange, Warrington, Cheshire Tel: 0925-811610 Telex 627592
42 Milton Rd, College Milton Industrial Estate, East Kilbride Tel: E.Kilbride 35661 Telex 77780
Bonville Rd, Brislington Trading Estate, Bristol BS4 5PA Tel: 0272 774135/7 Telex 44719

Compression Joints Ltd, Oldmixon, Weston Super Mare, Tel: Weston 24171 Grams CONIFORM WESTON SUPER MARE

Consolidated Pneumatic (Equipment Div), Fraserburgh AB4 5TE, Aberdeenshire Tel: Fraserburgh 3131

Crouzet Ltd, 108 Hawley Lane, Farnborough, Hants GU14 8HY Tel: Farnborough 513211 Telex 858612
Branch: 320A Moseley Rd, Manchester M19 2LH Tel: 061-224-0414 Telex 669330

Davis Pneumatic Systems Ltd, Huxley Close, Newnham Industrial Estate, Plympton, Plymouth Tel: (0752) 336421 Grams RED RING, PLYMOUTH Telex 45657

Dawson McDonald & Dawson Ltd, Compton Works, Ashbourne, Derbys. Tel: Ashbourne 3184 Telex 377072

Desoutter Brothers Limited, The Hyde, Hendon, London NW9 6ND Tel: 01-205 7050 Grams DESPNUCO, LONDON NW9
Branches: 54 College Rd, Perry Barr, Birmingham B44 8BT Tel: 021 356 5064
Tyldesley House, Elliot St, Tyldesley, Lancs. Tel: 052 34 6117/8/9

Domnick Hunter Engineers Ltd, Durham Rd, Birtley, Co. Durham DH3 2SF Tel: 0632 405121 Telex 537282

Drallim Controls Ltd, Brett Drive, Bexhill-on-Sea, East Sussex TN40 2JR Tel: 0424 216611 Grams DRALLIMIND BEXHILL, Telex 95285
Branch: Drallim Industries International S.A., 9 Rue Sabatier, 6001 Marcinelle, Belgium Tel: 071 363024 Grams DRALLIMINT

Drallim Tube Couplings Limited, Brett Drive, Bexhill-on-Sea, East Sussex TN40 2JR Tel: 0424-216611 Grams DRALLIMIND Telex 95285
Drallim Industries International S.A., 9 Rue Sabatier, 6001 Marcinelle, Belgium Tel: 071 363024 Grams DRALLIMINT

Drury Engineering Co Ltd, Bentley Lane, Walsall, West Midlands WS2 8TN Tel: Walsall 29444/8 Grams 'DRURY' WALSALL

Dyna-Quip Ltd, Shipston-on-Stour, Warwickshire Tel: 0608-61676 Grams PHILSYM, SHIPSTON
Branch 477 Barlow Moor Rd., Chorlton-cum-Hardy, Greater Manchester M21 2AS Tel: 061 888 2358 Telex 668712

Economatics Ltd, 36 Wilkinson St, Sheffield S10 2GB, Tel: 738514/5/6 23268
Branches: Kean Chambers, 11 Mappin St, Sheffield S1 4DT Tel: (0742) 23096/7 20693
125 Marshalls Mill, Water Lane, Leeds LS11 9UD Tel: (0532) 40890 446878 454906
Blenheim Place, Dunston Industrial Estate, Newcastle-upon-Tyne NE11 9JE Tel: (0632) 606125 606150 606175
133 Jarrom St, Leicester LE2 7DX Tel: (0533) 57583
41 Great Lister St, Birmingham Tel: (021 359) 5738
7 Church Lane, West Bromwich, Staffs B71 1DD Tel: 021-553-6216/7/8
207A Swan Lane, Coventry CV2 4GE Tel: (0203) 56486

The Globe Pneumatic Engineering Co Ltd, Ashton Rd, Harold Hill, Romford RM3 8UA Tel: Ingrebourne 43851 Grams PNEUMATOID ROMFORD Telex 21381

Howden Compressors Limited, 133 Barfillan Drive, Glasgow G52 1BE,
Tel: 041 882 3346 Telex: 778711
Branches: Southern Sales Office, Howden Compressors Ltd, Howden & Godfrey House, Godfrey Way, Hounslow, Middx TW4 5PW Tel: 01-894 5500 Telex 24692
Northern Sales Office, Regent St, Rochdale, Lancashire Tel: 341138/9 Telex 635142

ITT Fluid Handling Limited, Belcon Industrial Estate, Bingley Rd, Hoddesdon, Herts.
Tel: Hoddesdon 67191 Grams JABSCO HODDESDON

Joucomatic Controls Ltd, 51 Milford Rd, Reading, Berkshire Tel: 0734 599947/9 Grams JOMATIC READING

Joy Manufacturing Company (UK) Ltd, Peel Park Place, East Kilbride, Glasgow G74 5LN
Tel: 2064 Grams JOY EK
Branches: Burlington House, Burlington St, Chesterfield S40 15B Tel: 70281 Grams JOYCFD
Capitol House, Church St, Epsom Tel: 03727 27661 Grams: JOYEPS

Kay Pneumatics Ltd, London Rd, Dunstable, Beds LU6 3DL Tel: 0582 609292
Grams STRUMATICS DUNSTABLE

Kinetrol Limited, Trading Estate, Farnham, Surrey GU9 9NU Tel: Farnham (025 13) 23346
Grams KINETROL Telex 858567

K. P. Pneumatic Services Ltd, Fen End Industrial Estate, Astwick Rd, Stotfold, Hitchin, Herts.
Tel: Hitchin 730371

Lacy-Hulbert & Co Ltd, Richmond Rd, Beddington, Croydon CR9 4EA Tel: 01-688 5504
Grams PERCUSSIVE CROYDON Telex 21598
Branch: Lacy-Hulbert & Co Ltd (W.Germany) 5204 Lohmar 1, Tannenweg 22c
Tel: 02246/2263

Legris (UK) Limited, Ashville Trading Estate, The Runnings, Cheltenham GL51 9PT
Tel: 0242-43103/4/5 Telex 43335 Grams LEGRIS G

Mecman Ltd, 206 Elgar Road, Reading Tel: 0734 84629

Midland Pneumatic Limited, Patrick Gregory Road, Wednesfield, Wolverhampton

Motivair Compressors Ltd, Twickenham Trading Estate, Rugby Rd, Twickenham, Middx
Tel: 01-891 0844
Branches: Motivair (Midlands) Ltd, Unit 17, Watery Lane, Willenhall, West Midlands
Tel: Willenhall 61241
Motivair (Leach) Ltd, Britannia Rd, Waltham Cross, Herts. Tel: Lea Valley 714 471
Motivair Compressors (Eastleigh) Ltd, Unit 6, Eastern Industrial Estate, Boyatt Wood, Eastleigh, Hants.

C. A. Norgren Ltd, Shipston-on-Stour, Warwickshire, CV36 4PX Tel: 0608 61676 Telex 83208
Branch: 477 Barlow Moor Road, Chorlton-cum-Hardy, Greater Manchester M21 2AS.
Tel: 061 881-2358 Telex: 668712

Northey Rotary Compressors Ltd, Alder Road, Parkstone, Poole, Dorset. Tel: Parkstone (0202) 741725 Telex 41444
Branch: Mr D. J. Parkinson, 26 Warwick Drive, Brierfield, Nelson, Lancs BB9 0PP
Tel: Nelson (0282) 692478

Numatics, Sole U.K. Distributors: Ross James Engineering Ltd, 5 Ravensmead, Chalfont St Peter, Bucks, SL9 0NB, England Tel: Chalfont St. Giles 3698 (STD 024 07) Telex 922424

Padley & Venables Ltd, Calleywhite Lane, Dronfield, Sheffield Tel: Dronfield 413301 Telex 54254 Grams PEANVEE, DRONFIELD, SHEFFIELD

Parker-Hannifin (UK) Ltd, Fluid Connectors Group, Haydock Park Rd, Derby DE2 8JA Tel: Derby 41515 Telex 37427

Porvair Ltd, Estuary Rd, Kings Lynn, Norfolk PE30 2HS Tel: Kings Lynn (0553) 61111 Grams PORLYN

RGS Electro Pneumatics Limited, Maudsley St, Accrington, BB5 6AE Tel: 0254 35015 Telex 63425

Ross James Engineering Ltd, Sole NUMATICS UK Distributor, 5 Ravensmead, Chalfont St Peter, Bucks. SL9 0NB, England. Tel: Chalfont St. Giles 3698 (STD 024 07) Telex 922424

Schrader Pneumatics, Walkmill Lane, Bridgetown, Cannock, Staffs WS11 3LR Tel: Cannock 2644 Grams AIRVALVE CANNOCK Telex 336150

Sperry Vickers – Lang Pneumatic, Halesfield 6, Telford, Salop TF7 4LF Tel: Telford 586000 Telex 35286
Branches: Sherbourne Rd, Balsall Heath, Birmingham B12 9DJ Tel: 021 440 3501
First Floor, 147 Whiteladies Rd, Clifton, Bristol 8 Tel: Bristol 32598
6 Washington Rd, Paisley, Glasgow Tel: 041 887 6281
Shepcote House, Shepcote Lane, Sheffield 9 Tel: Sheffield 449611

Spirax Sarco Ltd, Charlton House, Cheltenham, GL53 8ER Tel: (0242) 21361 Grams SPIRAX CHELTENHAM Telex 43123
Branches: 41 Curzon St, London W1Y 7RF Tel: 01-499 1671/2
29 Park Circus, Glasgow G3 6AP Tel: 041-332-2069
Church Rd, Lymm, Cheshire Tel: Lymm 4904/5

Thorite Pneumatics Ltd, Lowfields Road, Leeds LS12 6BS. Tel: 444554/9 Telex 557628

Thomas Wright (Bradford) Ltd, Thorite House, Laisterdyke, Bradford BD4 8BZ Tel: 663471/8 Grams THORITE, BRDFD. Telex 51674
Branch: Thomas Wright (Huddersfield) Ltd., Long Lane, Moldgreen, Huddersfield HD5 9LY. Tel: 34245/6

Wade Couplings Ltd, Argyle Street, Birmingham B7 5TN. Tel: 021-327-1971 (20 lines)
Branches: 23 Greenwich High Road, London S.E.10. Tel: 01-692 2680
Linnyshaw Trading Estate, Moss Lane, Walkden, Worsley, Nr. Manchester Tel: 061-790-9200 and 2407
Premier Street, Birmingham B7 5TP. Tel: 021-327-1971 (20 lines)

Watts Regulator (UK) Ltd, London Rd, Thrupp, Stroud, Glos GL5 2BA Tel: 045 388 3401

Whittaker Hall Ltd, Black Lane, Engineering Works, Radcliffe, Manchester Tel: 061-723-2421 Telex 667071
Branches: 173 Kings Rd, Reading Tel: 599661/2 Telex 847680
Sandford St, Birkenhead Tel: 051-647-6041 Telex 627731
Dalziel Drive, Glasgow G41 4NY Tel: 041-427-2597 Telex 778427

Williams & James (Engineers) Ltd, Chequers Bridge, Gloucester GL1 4LL Tel: Gloucester (0452) 28431 Grams: COMPRESSOR GLOUCESTER

Addenda to Buyers' Guide

TRADE NAMES INDEX

BACOSOL – Miniature valves – Fluid Automation Systems SA

CARSET – Drill bit – Ingersoll-Rand Co Ltd

THE COMPRESSED AIR CENTRE – Air compressors and pneumatic equipment – Hunts Fluid Power Ltd

CRAWL-IR – Drill rig, self-propelled – Ingersoll-Rand Co Ltd

DONKIN – Valves – Bryan Donkin Co Ltd

FULBOR – Valves – Bryan Donkin Co Ltd

GYRO-FLO – Portable compressors – Ingersoll-Rand Co Ltd

JACKDRILL – Rock drill, leg mounted – Ingersoll-Rand Co Ltd

JACKHAMMER – Rock drill, hand held – Ingersoll-Rand Co Ltd

JACKLEG – Drill feed leg – Ingersoll-Rand Co Ltd

MICROSOL – Miniature valves – Fluid Automation Systems SA

MINISOL – Miniature valves – Fluid Automation Systems SA

PNEUMETHODS OF CUMBRIA – Pneumatic control devices and systems – Systematic Design & Developments Ltd

QUARRYMASTER – Drill rig, self contained, self propelled – Ingersoll-Rand Co Ltd

RATEAU – Compressors – Bryan Donkin Co Ltd

SERCK HEAT TRANSFER – Heat exchangers – Serck Heat Transfer

THOR – Portable power tools including safety balancers, air hoists and air motors – Stewart Warner Ltd – Thor Tools Div

WHISPERIZED – Silenced compressors (portable) – Ingersoll-Rand Co Ltd

CLASSIFIED INDEX OF PNEUMATIC EQUIPMENT AND COMPONENTS SUPPLIERS

Actuators, Rotary
Fluid Automation Systems SA

Aftercoolers, Air Cooled
Hunts Fluid Power Ltd
Ingersoll-Rand Co Ltd
Serck Heat Transfer
Systematic Design & Developments Ltd

Aftercoolers, Water Cooled
Hunts Fluid Power Ltd
Ingersoll-Rand Co Ltd
Serck Heat Transfer

Air Actuators
Hunts Fluid Power Ltd
Martonair Ltd
Systematic Design & Developments Ltd

Alarm Warnings, Flow
Hunts Fluid Power Ltd
Systematic Design & Developments Ltd

Alarm Warnings, Level
Hunts Fluid Power Ltd
Systematic Design & Developments Ltd

Alarm Warnings, Pressure
Hunts Fluid Power Ltd
Systematic Design & Developments Ltd

Alarm Warnings, Temperature
Systematic Design / Developments Ltd

Alarm Warnings, Vacuum
Systematic Design & Developments Ltd

Bits, Percussive
Pneumatic Plant Services Ltd

Bits, Rotary
Pneumatic Plant Services Ltd

Blowers
Bryan Donkin Co Ltd

Blowers, Reciprocative
Pneumatic Plant Services Ltd

Blowers, Sliding Vane
Hunts Fluid Power Ltd
Pneumatic Plant Services Ltd

Boosters, Rotary
Bryan Donkin Co Ltd

Boosters, Turbo
Bryan Donkin Co Ltd

Bottles, Air: Industrial, Storage
Pneumatic Plant Services Ltd

Breakers, Concrete and Paving
Pneumatic Plant Services Ltd
Stewart Warner Ltd — Thor Tools Div

Clamps
Hunts Fluid Power Ltd
Systematic Design & Developments Ltd

Compressors, Air Diaphragm
Hunts Fluid Power Ltd
Ingersoll-Rand Co Ltd
Pneumatic Plant Services Ltd

Compressors, Air Reciprocating: Mobile
Ingersoll-Rand Co Ltd
Pneumatic Plant Services Ltd

Compressors, Air Reciprocating: Portable
Bryan Donkin Co Ltd
Hunts Fluid Power Ltd
Ingersoll-Rand Co Ltd
Pneumatic Plant Services Ltd

Compressors, Air Reciprocating: Stationary or Fixed
A.P.E.—Bellis Ltd
Hunts Fluid Power Ltd
Ingersoll-Rand Co Ltd
Pneumatic Plant Services Ltd
Systematic Design & Developments Ltd

Compressors, Air Rotary: Portable
Pneumatic Plant Services Ltd

Compressors, Air Rotary: Screw
Hunts Fluid Power Ltd

Compressors, Air Rotary: Sliding Vane
Bryan Donkin Co Ltd
Hunts Fluid Power Ltd
Ingersoll-Rand Co Ltd
Pneumatic Plant Services Ltd
Systematic Design & Development Ltd

Compressors, Air Rotary: Stationary or Fixed
Hunts Fluid Power Ltd
Ingersoll-Rand Co Ltd
Martonair Ltd

Compressors, Air,Bottle Charging
Pneumatic Plant Services Ltd

Compressors, Air, Turbo Axial Flow
Bryan Donkin Co Ltd
Ingersoll-Rand Co Ltd

Compressors, Air Turbo, Centrifugal
Bryan Donkin Co Ltd
Ingersoll-Rand Co Ltd

Compressors, Dental
Hunts, Fluid Power Ltd
Pneumatic Plant Services Ltd

Compressors, Diving
Pneumatic Plant Services Ltd

Compressors, Double Entry
Bryan Donkin Co Ltd
Pneumatic Plant Services Ltd

Compressors, Garage
Hunts Fluid Power Ltd
Ingersoll-Rand Co Ltd
Pneumatic Plant Services Ltd

Compressors, Gas
A.P.E.—Belliss Ltd
Bryan Donkin Co Ltd
Hunts Fluid Power Ltd
Ingersoll-Rand Co Ltd
Pneumatic Plant Services Ltd

Compressors, Oil Free: Diaphraghm
A.P.E.—Belliss Ltd
Hunts Fluid Power Ltd
Ingersoll-Rand Co Ltd
Pneumatic Plant Services Ltd

Compressors, Oil Free: Reciprocating
Hunts Fluid Power Ltd
Ingersoll Rand & Co Ltd
Pneumatic Plant Services Ltd

Compressors, Oil Free: Rotary
Ingersoll-Rand Co Ltd
Pneumatic Plant Services Ltd

Compressors, Paint Spraying
Hunts Fluid Power Ltd
Ingersoll-Rand Co Ltd
Pneumatic Plant Services Ltd

Compressors, Portable
Hunts Fluid Power Ltd
Ingersoll-Rand Co Ltd
Pneumatic Plant Services Ltd

Compressors, Portable silenced
Ingersoll-Rand Co Ltd

Containers, Paint: Gravity
Hunts Fluid Power Ltd

Containers, Paint: Pressure
Hunts Fluid Power Ltd

Controllers, Flow
Martonair Ltd
Systematic Design & Developments Ltd

Controllers, Level
Hunts Fluid Power Ltd
Systematic Design & Developments Ltd

Controllers, Positional
Hunts Fluid Power Ltd
Martonair Ltd
Systematic Design & Developments Ltd

Controllers, Pressure
Hunts Fluid Power Ltd
Martonair Ltd

Control Panels, Pneumatic Sequence
Hunts Fluid Power Ltd
Martonair Ltd
Systematic Design & Developments Ltd

Conveyors
Systematic Design & Developments Ltd

Coolers, Air
Serck Heat Transfer

Coolers, Oil
Serck Heat Transfer

Counters, Pneumatic
Systematic Design & Developments Ltd

Cylinders, Air: Single Acting
Fluid Automation Systems SA
Hunts Fluid Power Ltd
Martonair Ltd

Cylinders, Air: Double Acting
Fluid Automation Systems SA
Hunts Fluid Power Ltd
Martonair Ltd

Cylinders, Air: Cushioned
Hunts Fluid Power Ltd
Martonair Ltd

Cylinders, Air: Duplex
Hunts Fluid Power Ltd
Martonair Ltd

Cylinders, Air Impact
Hunts Fluid Power Ltd
Martonair Ltd

Cylinders, Air: Non-Cushioned
Hunts, Fluid Power Ltd
Martonair Ltd

Cylinders, Air: Oscillating
Hunts Fluid Power Ltd
Martonair Ltd

Cylinders, Air: Other Types
Hunts Fluid Power Ltd
Martonair Ltd

Cylinders, Air: Rotary
Martonair Ltd

Cylinders, Air: Through Rod
Hunts Fluid Power Ltd
Martonair Ltd

Diggers, Clay or Spade
Ingersoll-Rand Co Ltd
Pneumatic Plant Services Ltd
Stewart Warner Ltd — Thor Tools Div.

Drill, Feed Leg
Ingersoll-Rand Co Ltd

Drill, Legs
Ingersoll-Rand Co Ltd

Drill Rig, self-contained
Ingersoll-Rand Co Ltd

Drills, Angle
Hunts Fluid Power Ltd
Stewart Warner Ltd — Thor Tools Div.

Drill Bit
Ingersoll-Rand

Drills, Corner
Stewart Warner Ltd — Thor Tools Div.

Drills, Industrial: Auto feed
Hunts Fluid Power Ltd
Ingersoll-Rand Co Ltd

Drills, Industrial: Corner
Stewart Warner Ltd — Thor Tools Div.

Drills, Industrial: Heavy Duty
Hunts Fluid Power Ltd
Ingersoll-Rand Co Ltd
Stewart Warner Ltd — Thor Tools Div.

Drills, Industrial: Light Duty
Hunts Fluid Power Ltd
Ingersoll-Rand Co Ltd
Stewart Warner Ltd — Thor Tools Div.

Drills, Industrial, Medium Duty
Hunts Fluid Power Ltd
Ingersoll-Rand Co Ltd
Stewart Warner Ltd — Thor Tools Div.

Drills, Industrial: Multi-spindle
Ingersoll-Rand Co Ltd

Drills, Industrial: Self feed
Hunts Fluid Power Ltd
Ingersoll-Rand Co Ltd

Drills, Industrial, Ultra-lightweight
Ingersoll-Rand Co Ltd

Drill Rods
Ingersoll-Rand Co Ltd

Drills, Wagon
Ingersoll-Rand Co Ltd

Drills, Woodborers
Hunts Fluid Power Ltd

Dusters, Air
Hunts Fluid Power Ltd
Pneumatic Plant Services Ltd

Ejectors
Bryan Donkin Co Ltd

Exhausters, (see also Vacuum Pumps)
Bryan Donkin Co Ltd
Hunts Fluid Power Ltd
Pneumatic Plant Services Ltd

Fan Blowers
Bryan Donkin Co Ltd

Fans
Bryan Donkin Co Ltd

Filters and Separators, Adsorption
Pneumatic Plant Services Ltd

Filters & Separators, Air Intake
Hunts Fluid Power Ltd
Pneumatic Plant Services Ltd
Systematic Design & Developments Ltd

Filters & Separators, Air Pipeline (see also Desiccators)
Hunts Fluid Power Ltd
Martonair Ltd
Pneumatic Plant Services Ltd

Filters and Separators, Vacuum
Hunts Fluid Power Ltd
Pneumatic Plant Services Ltd

Fittings, Air Line (see also Hose): Ferrous
Hunts Fluid Power Ltd
Pneumatic Plant Services Ltd
Stewart Warner Ltd — Thor Tools Div

Fittings, Air Line: Non-Ferrous
Fluid Automation Systems SA
Hunts Fluid Power Ltd
Martonair Ltd
Pneumatic Plant Services Ltd

Fittings, Air Line: Plastic
Hunts Fluid Power Ltd
Martonair Ltd
Pneumatic Plant Services Ltd

Gauges, Pressure
Hunts Fluid Power Ltd
Martonair Ltd
Pneumatic Plant Services Ltd
Systematic Design & Developments Ltd

Gauges, Vacuum
Pneumatic Plant Services Ltd

Grinders
Hunts Fluid Power Ltd
Ingersoll-Rand Co Ltd
Pneumatic Plant Services Ltd
Stewart Warner Ltd — Thor Tools Div

Guns, Blow (or dusters)
Hunts Fluid Power Ltd
Pneumatic Plant Services Ltd

Guns, Spray, Automatic
Pneumatic Plant Services Ltd

Guns, Spray, Paint, Liquids
Pneumatic Plant Services Ltd

Hammers, Bush
Pneumatic Plant Services Ltd

Hammers, Chipping and Caulking
Pneumatic Plant Services Ltd
Stewart Warner Ltd — Thor Tools Div

Hammers, Jack
Ingersoll-Rand Co Ltd
Pneumatic Plant Services Ltd
Stewart Warner Ltd — Thor Tools Div

Hammers, Pick, Dry
Pneumatic Plant Services Ltd
Stewart Warner Ltd — Thor Tools Div

Hammers, Pick Wet
Pneumatic Plant Services Ltd

Hammers, Riveting
Ingersoll-Rand Co Ltd
Pneumatic Plant Services Ltd

Hammers, Scaling
Hunts Fluid Power Ltd
Ingersoll-Rand Co Ltd
Pneumatic Plant Services Ltd
Stewart Warner Ltd — Thor Tools Div

Hammers, Stone-working
Pneumatic Plant Services Ltd

Heat Exchangers
Serck Heat Transfer

Hoist, Air Motor Type
Hunts Fluid Power Ltd
Ingersoll-Rand Co Ltd
Stewart Warner Ltd — Thor Tools Div

Hoists, Cylinder
Martonair Ltd

Hoists, Overhead
Ingersoll-Rand Co Ltd
Martonair Ltd
Stewart Warner Ltd — Thor Tools Div

Hoists, Portable
Ingersoll-Rand Co Ltd
Stewart Warner Ltd — Thor Tools Div

Holders-on for Riveting
Ingersoll-Rand Co Ltd

Horns, Pneumatic
Systematic Design & Developments Ltd

Hose
Hunts Fluid Power Ltd
Pneumatic Plant Services Ltd
Martonair Ltd
Stewart Warner Ltd — Thor Tools Div

Hose, Clips
Hunts Fluid Power Ltd
Pneumatic Plant Services Ltd
Martonair Ltd
Stewart Warner Ltd — Thor Tools Div

Hose, Couplings
Hunts Fluid Power Ltd
Pneumatic Plant Services Ltd
Martonair Ltd
Stewart Warner Ltd — Thor Tools Div

Hose, Fittings
Hunts Fluid Power Ltd
Pneumatic Plant Services Ltd
Stewart Warner Ltd — Thor Tools Div

Hose, High Pressure
Pneumatic Plant Services Ltd

Hose, Low Pressure
Pneumatic Plant Services Ltd

Hose, Medium Pressure
Hunts Fluid Power Ltd
Pneumatic Plant Services Ltd
Stewart Warner Ltd — Thor Tools Div

Hose, Plastic
Hunts Fluid Power Ltd
Pneumatic Plant Services Ltd

Hose, Reels
Hunts Fluid Power Ltd
Pneumatic Plant Services Ltd

Hose, Special Types
Fluid Automation Systems SA

Hydro-Check Units
Fluid Automation Systems SA

Indexing Tables, Rotary
Fluid Automation Systems SA
Systematic Design & Developments Ltd

Instruments
Systematic Design & Developments Ltd

Intercoolers
Hunts Fluid Power Ltd
Ingersoll-Rand Co Ltd

Lubricators
Hunts Fluid Power Ltd
Martonair Ltd

Machines, Air Operated
Systematic Design & Developments Ltd

Moisture Separators, (see also Filters & Separators)
Hunts Fluid Power Ltd
Pneumatic Plant Services Ltd

Motors, Rotary Vane
Hunts Fluid Power Ltd
Stewart Warner Ltd — Thor Tools Div

Nibblers and Shearing Machines
Hunts Fluid Power Ltd

Nutrunners (see also Wrenches)
Hunts Fluid Power Ltd
Pneumatic Plant Services Ltd
Stewart Warner Ltd — Thor Tools Div

Pipes, Flexible
Martonair Ltd

Pipes, Non-Ferrous
Martonair Ltd

Pipes, Plastic
Martonair Ltd

Pipework
Systematic Design & Developments Ltd

Polishers
Ingersoll-Rand Co Ltd

Presses, Impact
Hunts Fluid Power Ltd
Martonair Ltd

Presses Squeeze
Hunts Fluid Power Ltd

Pumps (see also Vacuum Pumps): Air Lift
Ingersoll-Rand Co Ltd
Pneumatic Plant Services Ltd

Pumps, Displacement
Pneumatic Plant Services Ltd

Pumps, Sludge
Pneumatic Plant Services Ltd
Stewart Warner Ltd — Thor Tools Div

Pumps, Sump
Ingersoll-Rand Co Ltd
Pneumatic Plant Services Ltd
Stewart Warner Ltd — Thor Tools Div

Radiators
Serck Heat Transfer

Rammers, Backfill
Ingersoll-Rand Co Ltd
Pneumatic Plant Services Ltd
Stewart Warner Ltd — Thor Tools Div

Rammers, Foundry and Sand
Ingersoll-Rand Co Ltd
Pneumatic Plant Services Ltd
Stewart Warner Ltd — Thor Tools Div

Receivers, Large
Hunts Fluid Power Ltd
Pneumatic Plant Services Ltd

Receivers, Medium
Hunts Fluid Power Ltd
Pneumatic Plant Services Ltd

Receivers, Small
Hunts Fluid Power Ltd
Pneumatic Plant Services Ltd

Regulators, Flow
Hunts Fluid Power Ltd
Martonair Ltd
Pneumatic Plant Services Ltd

Regulators, Instrument
Pneumatic Plant Services Ltd

Regulators, Pressure
Bryan Donkin Co Ltd
Hunts Fluid Power Ltd
Martonair Ltd
Pneumatic Plant Services Ltd

Regulators Self-Relieving
Hunts Fluid Power Ltd

Regulators, not Self-Relieving
Hunts Fluid Power Ltd

Regulators, Vacuum
Pneumatic Plant Services Ltd

Riveters, Aircraft
Ingersoll-Rand Co Ltd

Riveters, Industrial
Hunts Fluid Power Ltd
Pneumatic Plant Services Ltd

Riveters, Squeeze
Hunts Fluid Power Ltd
Stewart Warner Ltd — Thor Tools Div

Riveters, Yoke
Stewart Warner Ltd — Thor Tools Div

Rock Drills, Deep hole
Pneumatic Plant Services Ltd
Stewart Warner Ltd — Thor Tools Div

Rock Drills
Pneumatic Plant Services Ltd

Rock Drills, Drifters
Pneumatic Plant Services Ltd

Rock Drills, Dustless (wet)
Ingersoll-Rand Co Ltd
Pneumatic Plant Services Ltd

Rock Drills, General Duty
Ingersoll-Rand Co Ltd
Pneumatic Plants Services Ltd
Stewart Warner Ltd — Thor Tools Div

Rock Drills, Hand-held
Ingersoll-Rand Co Ltd

Rock Drills, Leg mounted
Ingersoll-Rand Co Ltd

Rock Drills, Mobile
Ingersoll-Rand Co Ltd

Rock Drills, Pushers
Ingersoll-Rand Co Ltd

Rock Drills, Self-propelled
Ingersoll-Rand Co Ltd

Rock Drills, Sinkers
Ingersoll-Rand Co Ltd
Stewart Warner Ltd — Thor Tools Div

Rock Drills, Stopers
Ingersoll-Rand Co Ltd

Sanders, Orbital
Hunts Fluid Power Ltd
Ingersoll-Rand Co Ltd

Sanders, Rotary
Hunts Fluid Power Ltd
Ingersoll-Rand Co Ltd
Stewart Warner Ltd — Thor Tools Div

Saws, Chain
Pneumatic Plant Services Ltd

Saws, Circular
Ingersoll-Rand Co Ltd

Saws, Reciprocating
Ingersoll-Rand Co Ltd

Screwdrivers
Hunts Fluid Power Ltd
Ingersoll-Rand Co Ltd
Stewart Warner Ltd — Thor Tools Div

Sequence Controllers, Pneumatic
Systematic Design & Developments Ltd

Sequence Controllers — Programmable, Pneumatic
Systematic Design & Developments Ltd

Silencers, Compressor Delivery
Hunts Fluid Power Ltd
Pneumatic Plant Services Ltd

Silencers, Compressor Intake
Hunts Fluid Power Ltd
Pneumatic Plant Services Ltd

Silencers, Pneumatic Cylinders & Valve
Hunts Fluid Power Ltd
Martonair Ltd
Pneumatic Plant Services Ltd

Silencers, Pneumatic Tool
Pneumatic Plant Services Ltd
Stewart Warner Ltd — Thor Tools Div

Solenoids
Hunts Fluid Power Ltd

Spark Plug Cleaners
Hunts Fluid Power Ltd

Spraying Equipment, Agricultural
Pneumatic Plant Services Ltd

Switches, Pressure and Vacuum, Electric (see also Alarm Warnings)
Martonair Ltd
Pneumatic Plant Services Ltd

Tampers, Backfill
Pneumatic Plant Services Ltd
Stewart Warner Ltd — Thor Tools Div

Tampers, Railway Tie
Pneumatic Plant Services Ltd

Tappers, Thread
Stewart Warner Ltd — Thor Tools Div

Timers, Delay
Systematic Design & Developments Ltd

Timers, Pneumatic (Dial, sequence, delay, process, automatic and hand reset)
Martonair Ltd
Systematic Design & Developments Ltd

Tool Steels, Chisel Steels
Pneumatic Plant Services Ltd
Stewart Warner Ltd — Thor Tools Div

Tool Steels, Digger Steels
Pneumatic Plant Services Ltd
Stewart Warner Ltd — Thor Tools Div

Tool Steels, Dressing Tools
Pneumatic Plant Services Ltd

Tool Steels, Feathers
Pneumatic Plant Services Ltd

Tool Steels — Moil Points
Pneumatic Plant Services Ltd
Stewart Warner Ltd — Thor Tools Div

Tool Steels, Plug and Feathers
Pneumatic Plant Services Ltd

Tool Steels, Spades
Pneumatic Plant Services Ltd
Stewart Warner Ltd — Thor Tools Ltd

Tool Steels, Tamping Pads and Stems
Pneumatic Plant Services Ltd
Stewart Warner Ltd — Thor Tools Div

Tool Steels, Pick Steels
Pneumatic Plant Services Ltd

Tool Steels, Punches
Pneumatic Plant Services Ltd

Torque Actuators
Systematic Design & Developments Ltd

Traps, Water (see also Filters and Separators)
Hunts Fluid Power Ltd
Pneumatic Plant Services Ltd

Turbo Blowers
Bryan Donkin Co Ltd

Tyre Inflators
Hunts Fluid Power Ltd

Unloaders
Pneumatic Plant Services Ltd

Vacuum Pumps
Bryan Donkin Co Ltd
Hunts Fluid Power Ltd
Ingersoll-Rand Co Ltd
Pneumatic Plant Services Ltd

Valve Actuators
Hunts Fluid Power Ltd
Systematic Design & Developments Ltd

Valves, Compressor, Suction & Discharge
Pneumatic Plant Services Ltd

Valves, Directional Control for Controlling Pneumatic Cylinders, etc.
Fluid Automation Systems SA
Martonair Ltd

Valves, Manual
Hunts Fluid Power Ltd
Martonair Ltd

Valves, Mechanical
Fluid Automation Systems SA
Martonair Ltd
Systematic Design & Developments Ltd

Valves, Needle (see also Regulators, Flow)
Hunts Fluid Power Ltd
Martonair Ltd

Valves, Non-Return or Check
A.P.E.—Belliss Ltd
Bryan Donkin Co Ltd
Hunts Fluid Power Ltd
Martonair Ltd
Pneumatic Plant Services Ltd

Valves, Pressure Reducing (see also Regulators)
Bryan Donkin Co Ltd
Hunts Fluid Power Ltd
Martonair Ltd
Pneumatic Plants Services Ltd

Valves, Quick Exhaust
Hunts Fluid Power Ltd
Martonair Ltd

Valves, Relief, Pressure
Bryan Donkin Co Ltd
Hunts Fluid Power Ltd
Martonair Ltd
Pneumatic Plant Services Ltd

Valves, Relief, Vacuum
Hunts Fluid Power Ltd
Pneumatic Plant Services Ltd

Valves, Remote
Fluid Automation Systems SA
Systematic Design & Developments Ltd

Valves Remote Control
Hunts Fluid Power Ltd

Valves, Rotary
Hunts Fluid Power Ltd

Valves, Safety
Hunts Fluid Power

Valves, Sequence
Hunts Fluid Power Ltd
Systematic Design & Developments Ltd

Valves, Shuttle
Hunts Fluid Power Ltd
Martonair Ltd

Valves, Solenoid Operated
Hunts Fluid Power Ltd
Martonair Ltd

Valves, Solenoid Operated, flameproof
Fluid Automation Systems SA
Martonair Ltd

Valves, Speed Control (see also Regulators, flow)
Hunts Fluid Power Ltd
Martonair Ltd

Valves, Stop or Shut-off
Hunts Fluid Power Ltd
Martonair Ltd
Pneumatic Plant Services Ltd

Vibrators, Concrete Immersion
Ingersoll-Rand Co Ltd
Pneumatic Plant Services Ltd

Vibrators, External
Hunts Fluid Power Ltd
Pneumatic Plant Services Ltd

Vibrators, Internal
Pneumatic Plant Services Ltd

Wire Brush Machines
Stewart Warner Ltd — Thor Tools Div

Wrenches, Impact
Hunts Fluid Power Ltd
Ingersoll-Rand Co Ltd
Stewart Warner Ltd — Thor Tools Div

Wrenches, Torque
Ingersoll-Rand Co Ltd
Pneumatic Plant Services Ltd
Stewart Warner Ltd — Thor Tools Div

ALPHABETICAL LIST OF MANUFACTURERS WITH ADDRESSES, TELEPHONE NUMBERS, TELEGRAM ADDRESSES OF HEAD OFFICE, WORKS AND BRANCHES

A.P.E.-Belliss Ltd, Icknield Square, Birmingham B16 0QL, Tel: 021-454 3531 Grams APEBIR.G

Bryan Donkin Co Ltd, Derby Road, Chesterfield, Tel: Chesterfield 3153, Grams: Donkin Chesterfield, Telex: 54135
Branch: 46 Westminster Palace Gardens, Artillery Row, London S.W.1. Tel: Abbey 1096

Fluid Automation Systems SA, 124 route de l'Etraz, CH 1290 Versoix, Switzerland.
Tel: 55 26 91 & 55 14 93, Telex: 22 194

Hunts Fluid Power Limited, Nuffield Industrial Estate, Fleetsbridge, Poole, Dorset.
Tel: Poole (02013) 77423
Branches: Hunts Fluid Power Limited, Southmoor Lane, Havant, Hants.
Tel: Havant (07012) 76741
Avon Pneumatics Ltd, Emery Road, Brislington Trading Estate, Bristol.
Tel: Bristol (0272) 773671
Avon Pneumatics Ltd, Cannon Rd, Heathfield Industrial Estate, Newton Abbot, Devon.
Tel: (0626) 832866
Commair Engineering Ltd, 477 Malton Ave, Slough Trading Estate, Slough, Berks.
Tel: (0753) 70034

Ingersoll-Rand Co Ltd, 165 Queen Victoria Street, London E.C.4 Tel: 01-236 2300
Branches: Station Road, Coleshill, Birmingham. Tel: Coleshill 3737
Banks Road, off McMullen Rd, Darlington, Co. Durham. Tel: Darlington 69241
John F. Kennedy Drive, Bluebell, Dublin 12. Tel: Dublin 505934
26 Shawlands Arcade, Glasgow, S.1. Tel: Langside 9253
Richmond Rd, Trafford Park, Manchester 17. Tel: Trafford Park 3696

Martonair Limited, St Margarets Road, Twickenham, Middx. TW1 1RJ. Tel: 01-892 4411
Grams: Martonair Twickenham
Branches: 221 Bridge Street West, Birmingham B19 2YU. Tel: 021-359 6941/2/3
21 Ashton Vale Rd, Ashton Gate, Bristol. BS3 2HA. Tel: 0272 494551
Martonair House, Carlisle St, Gateshead, Tyne & Wear. NE10 0LD. Tel: 0632 692248
84 Kilbirnie St, Glasgow, G5 8JD. Tel: 041-429 7579 & 7570
9 Wallace Way, Hitchin. SG4 0SE. Tel: 0462 2243
Station Rd, Southwater, Horsham, West Sussex. Tel: 0403 731171 & 731451
Felnex Crescent, Pontefract Lane, Leeds LS9. Tel: 0532 450241
11 Cannock St, Leicester. LE4 7HR. Tel: 0533 769993
James Dunlop (N.I.) Ltd, 26 Roughfort Rd, Mallusk, Newtownabbey, Co. Antrim. BT36 8RE. Tel: 023 13 44721
39 Victoria Road, Romford, Essex. RM1 2LH. Tel: 0708 42256
Unit 7, Sherdley Rd Trading Estate, Sherdley Rd., St. Helens, Merseyside WA9 5AA Tel: 0744 55296/7
439/443 Penistone Road, Sheffield. S6 2GR. Tel: 0742 347666
Bankfield Trading Estate, South Reddish, Stockport. SK5 7PA. Tel: 061-480 1231/2
Hare St, Bilston, Wolverhampton, West Midlands. WV14 7JS. Tel: 0902 44877
Colchester Trading Estate, Colchester Ave, Cardiff. CF3 7AP. Tel: 0222 494551

Pneumatic Plant Services Ltd, Church Road, Mitcham, Surrey. Tel: 01-648 0275 Grams Pneumatic Croydon

Serck Heat Transfer, P.O. Box 598B Warwick Road, Birmingham B11 2QY. Tel: 021-772 4353
Grams. Serck Trans
Branches: H.O. Serck Div. Crossley St, Gorton, Manchester M18 8BH. Tel: 061-223 5166
60 Trafalgar Square, London WC2N 5DS. Tel: 01-839 2682
11 Sandyford Place, Glasgow G3 7NB. Tel: 041-248 6991
Serck GmbH, 2000 Hamburg 70, Tilsiter Strasse 90, West Germany. Tel: 040-6935011

Systematic Design & Developments Ltd, Incorporating Pneumethods of Cumbria, P.O. Box 3, Workington, Cumbria. Tel: Workington (0900) 5303

Stewart-Warner Limited – Thor Tools Division, Tynemouth, Tyne 9 Wear, England NE29 7UE
Tel: North Shields (08945) 73181. Grams Thortools Tynemouth Tyne & Wear Telex 53602

Index

P

Q

R

S

Index to Advertisers